ANALOG DESIGN ESSENTIALS

ANALOG DESIGN ESSENTIALS

by

Willy M. C. Sansen

Catholic University, Leuven, Belgium

 Springer

A C.I.P. Catalogue record is available from the Library of Congress.

Additional material to this book can be downloaded from http://extras.springer.com

ISBN 978-1-4899-7891-2 ISBN 978-0-387-25747-1 (eBook)
DOI 10.1007/978-0-387-25747-1

Published by Springer,
PO Box 17, 3300 AA Dordrecht, The Netherlands.

www.springer.com

Printed on acid-free paper

Dedication

This book is dedicated to my wife
Hadewych Hammenecker

Contents

Analog design is art and science at the same time.

It is art because it requires creativity to strike the right compromises between the specifications imposed and the ones forgotten.

It is also science because it requires a certain level of methodology to carry out a design, inevitably leading to more insight in the compromises taken.

This book is a guide through this wonderful world of art and science. It claims to provide the novice designers with all aspects of analog design, which are essential to this understanding.

As teaching is the best way to learn, all slides are added on a CD-ROM, with and without the comments added as notes in the pdf files. The reader is suggested to try to explain parts of this course to his fellow designers. This is the way to experience and to cultivate the circles of art and science embedded in this book.

All design is about circuits. All circuits contain transistors. Hand-models are required of these devices in order to be able to predict circuit performance. CAD tools such as SPICE, ELDO, SPECTRE, etc. are then used to verify the predicted performance. This feedback loop is essential to converge to a real design. This loop will be used continuously in this book.

Comparison of MOST and Bipolar transistor models

Willy Sansen
KULeuven, ESAT-MICAS
Leuven, Belgium

willy.sansen@esat.kuleuven.be

Willy Sansen 10-05 011

For the design of analog integrated circuits, we need to be able to predict the performance by means of simple expressions. As a result, simple models are required. This means that the small-signal operation of each transistor must be described by means of as few equations as possible. Clearly the performance of the circuit can then only be described in an approximate way. The main advantage however, is that transistor

Table of contents

Ref.: W. Sansen : Analog Design Essentials, Springer 2006

Willy Sansen 10-05 012

sizing and current levels can easily be derived from such simple expressions. They can then be used to simulate the circuit performance by means of a conventional circuit simulator such as SPICE or ELDO.

In these simulators, models are used which are much more accurate but also much more complicated. These simulations are required afterwards to verify the circuit performance. The initial design with simple models is the first step in the design procedure. They are aimed indeed at the determination of all transistor currents and sizes, according to the specifications imposed.

We start with MOST devices, although the bipolar transistor are historically first. Nowadays the number of MOS transistors integrated on chips, vastly outnumber the bipolar ones.

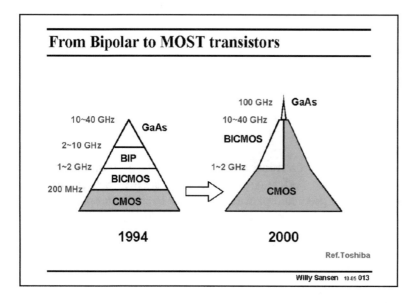

013
Indeed, previously CMOS devices were reserved for logic as they offer the highest density (in gates/mm^2). Most high-frequency circuitry was carried out in bipolar technology. As a result, a lot of analog functions were realized in bipolar technology. The highest-frequency circuits have been realized in exotic technologies such as GaAs and now InP technologies. They are quite expensive however and really reserved for the high frequency end.

The channel length of CMOS transistors shrinks continuously however. In 2004, a channel length of 0.13 micrometer is standard but several circuits using 90 nm have already been published (see ISSCC). This ever decreasing channel length gives rise to ever increasing speeds. As a result, CMOS devices are capable of gain at ever higher frequencies.

Today CMOS and bipolar technologies are in competition over a wide frequency region, extending all the way to 10 and even 40 GHz, as predicted in this slide. For these frequencies the question is indeed, which technology fulfills best the system and circuit requirements at a reasonable cost. BICMOS is always more expensive than standard CMOS technology. The question is, whether the increase in cost compensates the increase in performance?

The SIA roadmap

Year	Lmin μm	Bits/chip Gb/chip	Trans/chip millions/chip	Clock MHz	Wiring
1995	0.35	0.064	4	300	4 - 5
1998	0.25	0.256	7	450	5
2001	0.18	1	13	600	5 - 6
2004	0.13	4	25	800	6
2007	0.09	16	50	1000	6 - 7
2010	0.065	64	90	1100	7 - 8

2003

Semiconductor Industry Association

Willy Sansen 10.05 **014**

014

This ever decreasing channel length has been predicted by the SIA roadmap. It tries to predict what the channel length will be in a few years, by extrapolating the past evolution.

It is clear however, that the shrinking of the channel length has been carried out much faster than predicted. For example, the 90 nm technology was originally expected only in 2007, but was already offered in 2003. This technology was expected to allow 50 million transistors to be integrated on one single chip. Present day processors and memories offer double that amount. Moreover, this technology was expected to give rise to clock speeds around 1 GHz. High end PC's already clock speeds beyond 3 GHz!

The law of Moore

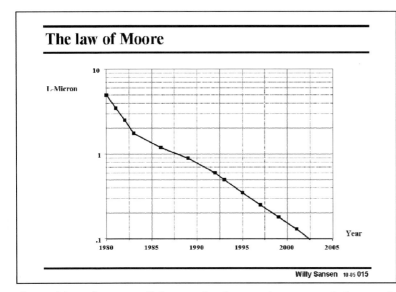

Willy Sansen 10.05 **015**

015

This ever decreasing channel length has also been predicted by the curve of Moore. This is simply a sketch of channel length versus time. It is a graphic representation of the numbers of the SIA road map. Indeed 90 nm is reached in 2003!

The slope of that curve has not always been the same. Indeed, the slope was higher in the early eighties, but has declined a bit as a result of economic recessions. Also, the cost of the production equipment and the mask making grows exponentially, delaying the introduction of ever newer technologies somewhat.

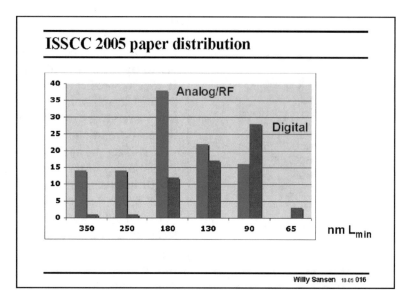

016

Which are the most used channel lengths today?

To explore this, the number of papers is shown of the last IEEE ISSCC conference (held at San Francisco in February) for two categories, the digital circuits and the analog or RF circuits.

It is clear that the digital circuits peak at 90 nm channel length, whereas the analog ones lag behind by two generations; they peak at about 180 nm.

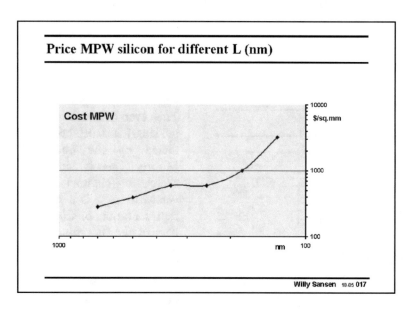

017

Indeed for small quantities, silicon foundries offer silicon at higher cost if the channel lengths are smaller. This is clearly illustrated by this cost of a Multi Project Wafer chip versus channel length. In such a MPW run, many designs are assembled and put together in one single mask and run. As a result the total cost is divided over all participants of this run. This has been the source of cheap silicon for many universities and fabless design centers.

The cost in $/mm^2 is reasonable up to about 0.18 μm. From 0.13 μm on the cost increases dramatically, depriving many universities from cheap silicon.

What the cost will be of 90 nm and 65 nm is easily found by extrapolation! This shows very clearly that a crisis is at hand!

018

Let us have a closer look now at a MOST device. What are the main parameters involved, and what are the simplest possible model equations that still describe the transistor models in an adequate way for hand analysis.

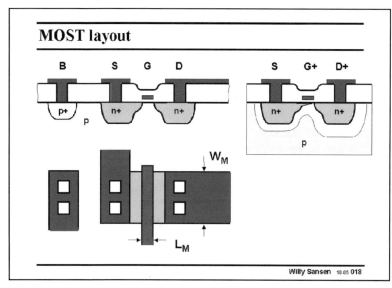

The cross-section of a MOS transistor is shown with its layout. On the left, the MOST is shown without biasing. On the right, voltages are applied to Gate and Drain.

The main dimensions of a MOST are the Length and Width. Both are drawn dimensions on the mask. In practice they are usually a bit smaller. This is a result of underdiffusion and some more technological steps. In this layout the W/L on the mask is about 5.

Application of a positive voltage at the Gate V_{GS} causes a negatively charged inversion layer, which connects the Source and Drain n+ islands. It is a conducting channel between Source and Drain and thus acts as a resistor between Source and Drain.

Application of a positive voltage at the Drain V_{DS}, with respect to the Source, allows some current to flow from Drain to Source (or electrons from Source to Drain). This current is I_{DS}. As a result, the channel becomes non-homogeneous. It conducts better on the Source side than on the Drain side. The channel may even disappear on the Drain side. Nevertheless, the electrons always manage to make it to the other side, because they have acquired sufficient speed along the channel.

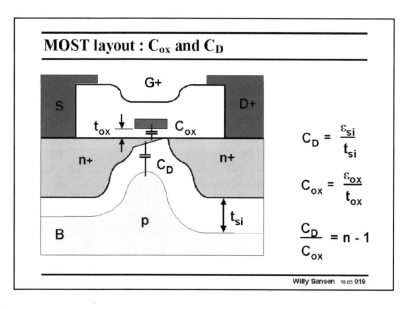

019

Zooming in on the channel region, disappears once V_{DS} is too high.

The channel region, together with the two n+ islands of Source and Drain, are enveloped by an isolation layer. Indeed, in a pn junction the p and n regions are always separated by an isolation region, which is called depletion region. The silicon is depleted of electrons or holes; it is non-conducting, it is an isolator, very much as oxide an isolator is.

The oxide has a thickness t_{ox}, whereas the depletion layer has a thickness of t_{si}. Both give rise to capacitances C_{ox} and C_D, respectively. Both have dimensions F/cm^{-2}. Normally C_D is about one third of C_{ox} as we will calculate in detail on the next slide. Their ratio is n-1 [Tsividis].

It is mportant, however, to note that the channel inversion layer is coupled to the Gate by means of C_{ox}, but as much coupled to the Bulk by C_D. Changing the Gate voltage will thus change the conductivity of the channel and hence the current I_{DS}. In a similar way, changing the Bulk voltage will thus also change the conductivity of the channel and will thus change the current I_{DS} as well. The top gate gives the MOST operation, whereas the bulk gives JFET operation. Indeed, a Junction FET is by definition a FET in which the current is controlled by a junction capacitance.

All MOST devices are thus parallel combinations of MOSTs and JFETs. We normally use only the MOST whereas the JFET is called the body effect, and is treated as a parasitic effect.

MOST layout : C_{ox} and C_D values

$$C_D = \frac{\varepsilon_{si}}{t_{si}} \qquad t_{si} = \sqrt{\frac{2\varepsilon_{si}(\phi - V_{BD})}{qN_B}}$$

$\varepsilon_{si} = 1 \text{ pF/cm}$

$\varepsilon_{ox} = 0.34 \text{ pF/cm}$

$\phi \approx 0.6 \text{ V}$

$q = 1.6 \ 10^{-19} \text{ C}$

$N_B \approx 4 \ 10^{17} \text{ cm}^{-3}$

Example : L = 0.35 μm; W/L = 8

$V_{BD} = -3.3 \text{ V}$: $t_{si} = 0.1 \ \mu m$

$C_D \approx 10^{-7} \text{ F/cm}^2$

$t_{ox} = \dfrac{L_{min}}{50}$ $t_{ox} = 7 \text{ nm}$

$C_{ox} \approx 5 \ 10^{-7} \text{ F/cm}^2$ $\dfrac{C_D}{C_{ox}} = n - 1 \approx 0.2$

Willy Sansen 10-05 0110

0110

The width of the depletion region depends to a large extent on doping levels and the voltage across it. The larger the doping levels on both sides of the junction, the narrower the depletion region is. On the other hand, the larger the voltage is across the depletion region, the wider this region becomes, as shown by the equation.

It includes the silicon dielectric constant ε_{si}, the junction built-in voltage φ, the charge of an electron q and the bulk doping level N_B. Values are given in this slide.

For example for a 0.35 mm technology, a drain-bulk voltage V_{BD} yields a depletion layer thickness of about 0.1 nm. It is about 14 times thicker than the gate oxide. This is offset somewhat by the fact that the silicon dielectric constant is three times higher than the oxide one. Silicon is three times more efficient to make capacitors with than oxide. Silicon capacitances are very nonlinear because they depend on the voltage, whereas oxides capacitances do not.

The ratio $n-1$ is then about 0.2. Most values of n are indeed between 1.2 and 1.5, depending on the value of t_{si}. Parameter n is thus never known accurately as it depends on biasing voltages.

Note that all capacitances are in F/cm². For a Gate area of WL of $5 \times 0.35 \times 0.35$ mm² the total Gate oxide capacitance is thus $C_{ox}WL \approx 5$ fF, which is quite a small value indeed!

0111

The bulk doping level N_B is not the same for a nMOST and a pMOST device. Indeed normally nMOST devices are implemented directly on the p-substrate. This substrate is thus common to all nMOS transistors on that chip.

The pMOS transistor has a p-channel however and has to be implemented in a n-tub or n-well, which is always higher doped than the common p-substrate. The disadvantage is that the bulk doping for a pMOST is always higher than the bulk doping of a nMOST. The pMOST

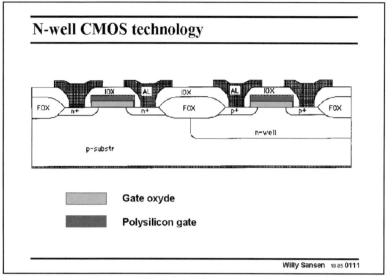

N-well CMOS technology

Gate oxyde

Polysilicon gate

Willy Sansen 10.05 0111

C_D will be higher and so is its n factor. The advantage of a pMOST however is that its bulk is isolated from the substrate and can be used to control the transistor current I_{DS}. Such pMOST devices have two gates, i.e. a top gate and a bottom gate. Both can be driven independently.

Most technologies are n-well CMOS technologies although some p-well ones are still around.

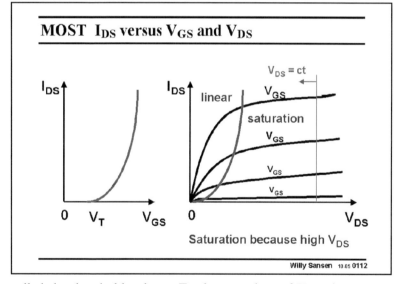

MOST I_{DS} versus V_{GS} and V_{DS}

Saturation because high V_{DS}

Willy Sansen 10.05 0112

0112

Application of a positive Gate voltage V_{GS} causes an inversion layer (or channel) which connects Source to Drain. Application of a positive voltage V_{DS} causes some current I_{DS} to flow from Drain to Source. Now we want to find simple expressions for this current, so that we can use them for design purposes.

The curve of I_{DS} versus V_{GS} is sketched on the left. The current starts flowing as soon as V_{GS} exceeds V_T, called the threshold voltage. For larger values of V_{GS}, the current increases in a nonlinear way. How much we actually exceed V_T is $V_{GS} - V_T$; this will be the most important design parameter later on!

The curve of I_{DS} versus V_{DS} is sketched on the left. For small values of V_{DS}, the current increases linearly. Indeed, the transistor behaves as a resistor. This is called the linear region.

For larger values of V_{DS} the current stops increasing but levels off towards nearly constant values: the current is said to saturate. This is called the saturation region. Curves are given for four different values of V_{GS}.

The linear and saturation regions are separated by a parabola, which is described by $V_{DS} = V_{GS} - V_T$. We will concentrate on the linear region first.

Table of contents

Willy Sansen 10-05 0113

0113

In many applications a MOST is simply used as a switch. Its voltage V_{DS} is then very small. The MOST is then operating in the linear region (sometimes called the ohmic region). In this region the MOST transistor is really a small resistor. It provides a linear voltage-current characteristic. The channel has the same conductivity at both sides – the Source side and the Drain side.

Let us investigate what the actual resistance then is.

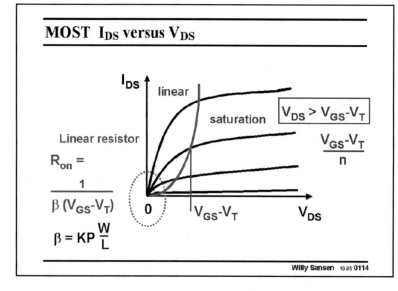

MOST I_{DS} versus V_{DS}

Linear resistor

$$R_{on} = \frac{1}{\beta\,(V_{GS}-V_T)}$$

$$\beta = KP\,\frac{W}{L}$$

$$V_{DS} > V_{GS}-V_T$$

$$\frac{V_{GS}-V_T}{n}$$

Willy Sansen 10-05 0114

0114

Zooming in on the corner, for very small values of V_{DS}, we find that indeed the $I_{DS}-V_{DS}$ curves are very linear. The MOST behaves as a pure resistor.

The resistance value R_{on} is given in this slide. In addition to the dimensions W and L, a technological parameter appears, called KP.

This parameter characterizes a certain CMOS technology as will be explained on the next slide. Its dimension is A/V^2.

It is clear that the transistor turns very nonlinear when we apply larger V_{DS} voltages. The crossover value towards the saturation is reached for $V_{DS} = V_{GS} - V_T$, or more accurately for $V_{DS} = (V_{GS} - V_T)/n$. We will drop this factor n however, as a kind of safety factor. We will, from now on, assume that a transistor is operating in the saturation region provided $V_{DS} > V_{GS} - V_T$.

0115

For sake of illustration let us a have a closer look at this resistor "in the corner". For this purpose we have to find an easy approximation for KP. It is given in this slide. Factor β (Greek beta) contains both the parameter KP and the dimensions of the resistor W and L.

Actually, KP contains the oxide capacitance C_{ox}, and the mobility μ (Greek mu). This factor

MOST parameters β , KP , C$_{ox}$, ...

$$\beta = KP \frac{W}{L}$$

$$KP = \mu \, C_{ox}$$

$$C_{ox} = \frac{\varepsilon_{ox}}{t_{ox}}$$

$$t_{ox} = \frac{L_{min}}{50}$$

$KP_n \approx 300 \ \mu A/V^2$

$C_{ox} \approx 5 \ 10^{-7} \ F/cm^2$

$\varepsilon_{ox} = 0.34 \ pF/cm$

$\varepsilon_{si} = 1 \ pF/cm$

$t_{ox} = 7 \ nm$

$L_{min} = 0.35 \ \mu m$

$\mu_p \approx 250 \ cm^2/Vs$

$\mu_n \approx 600 \ cm^2/Vs$

Willy Sansen 10-05 0115

shows what speed (cm/s) an electron can develop, subject to an electric field (V/cm). It is given in cm²/Vs. Electrons travel about twice as fast as holes.

Values for a standard 0.35 μm CMOS technology are given in this slide.

Note that the oxide thickness is about L/50, as has been checked on most standard CMOS processes of the last 20 years.

As a rule of thumb, the resistor of a square transistor (W/L=1) for a drive voltage $V_{GS} - V_T = 1$ V is about 3.4 kΩ in 0.35 μm CMOS.

For deeper submicron CMOS technologies, KP is higher because of C$_{ox}$. This square resistor now decreases!

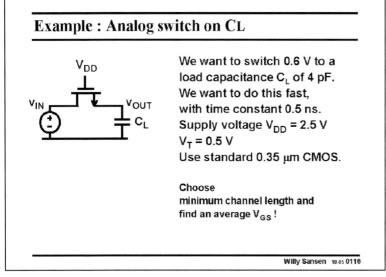

Example : Analog switch on CL

We want to switch 0.6 V to a load capacitance C$_L$ of 4 pF. We want to do this fast, with time constant 0.5 ns. Supply voltage V$_{DD}$ = 2.5 V
V$_T$ = 0.5 V
Use standard 0.35 μm CMOS.

Choose minimum channel length and find an average V$_{GS}$!

Willy Sansen 10-05 0116

0116

To have a time constant of 0.5 ns with 4 pF we need a switch resistance of 125 Ω.

This will, to a large extent, depend on the value of $V_{GS} - V_T$ used. Indeed, as soon as the switch turns on, the output voltage is still at zero Volt and $V_{GS} - V_T = 2$ V. At the end of the switching in, the output voltage has risen to 0.6 V: it has become the same as the input voltage. The $V_{GS} - V_T$ has decreased by 0.6 V towards $V_{GS} - V_T = 1.4$ V.

The average value is now $V_{GS} - V_T = 1.7$ V.

For a transistor size W/L=1, the on-resistance is thus 2 kΩ (using $KP = 300 \ \mu A/V^2$). This is 8 × larger than what we can allow. We thus have to take a W/L of 8.

Note that we will have great difficulties in switching large input voltages. Indeed, for $v_{OUT} = v_{IN} = 2$ V, the V_{GS} has become zero. As a result, the resistor has become infinity: the switch cannot be switched on any more!!

Note also that we have forgotten to take into account the bulk effect. Indeed, V_{BS} is not zero, it is 0.6 V. The parasitic JFET will play as well as we will see later.

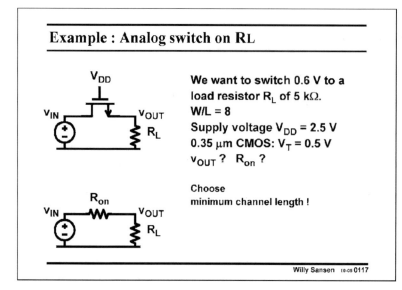

Example : Analog switch on RL

We want to switch 0.6 V to a load resistor R_L of 5 kΩ.
W/L = 8
Supply voltage V_{DD} = 2.5 V
0.35 µm CMOS: V_T = 0.5 V
v_{OUT} ? R_{on} ?

Choose
minimum channel length !

Willy Sansen   10-05 0117

0117

For a transistor size W/L = 8, the KP × W/L product is 2.4×10^{-3} S (using KP = 300 µA/V²). Taking the switch as a resistor with value R_{on}, as shown in this slide and substitution of R_{on} by its expression, requires an iteration, which yields a value of R_{on} of 216 Ω and an output voltage of 0.575 V.

Note that we have forgotten to take into account the bulk effect. Indeed, V_{BS} is not zero, it is about 0.575 V. The parasitic JFET will become active as well, as we will see next.

Body effect - Parasitic JFET

$$V_T = V_{T0} + \gamma \left[\sqrt{|2\Phi_F| + V_{BS}} - \sqrt{|2\Phi_F|} \right]$$

$$n = \frac{\gamma}{\sqrt{|2\Phi_F| + V_{BS}}} = 1 + \frac{C_D}{C_{ox}}$$

$|2\Phi_F| \approx 0.6$ V

$n \approx 1.2 ... 1.5$

$\gamma \approx 0.5 ...0.8$ V$^{1/2}$

Reverse v_{BS} increases $|V_T|$ and decreases $|i_{DS}|$!!!

n = 1/κ subthreshold gate coupling coeff. Tsividis

Willy Sansen   10-05 0118

0118

The drain-source current I_{DS} and the channel resistance R_{on} show the influence of V_{GS} in an explicit way, but not that of the bulk-source voltage V_{BS}. Indeed, the effect of V_{BS} is embedded in the threshold voltage V_T.

Increasing the V_{BS} will increase the depletion layer width under the channel and will increase V_T. More reverse biasing that junction will increase V_T in absolute value and decrease the current. For zero V_{BS}, V_T evidently equals V_{T0}.

Parameter γ (Greek gamma) has to do with the junction depletion region and is linked to parameter n. Actually, factor γ depends on the technology used (such as the bulk doping N_B) but is not voltage dependent. The denominator of n now shows explicitly the voltage dependence of n.

Some approximate parameter values are given as well for a 0.7 µm CMOS.

Ex. : Analog switch with nonzero VBS

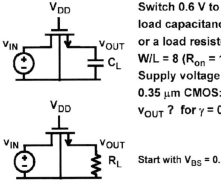

Switch 0.6 V to a
load capacitance C_L of 4 pF
or a load resistor R_L of 5 kΩ.
W/L = 8 (R_{on} = 125 Ω @ V_{BS} = 0)
Supply voltage V_{DD} = 2.5 V
0.35 μm CMOS: V_T = 0.5 V
v_{OUT} ? for γ = 0.5 V^{-1}

Start with V_{BS} = 0.

Willy Sansen 10-05 0119

0119

For a transistor size W/L = 8, the KP × W/L product is again 2.4×10^{-3} S (using KP = 300 μA/V^2).

Taking the switch as a resistor with value R_{on}, as shown in this slide, and substitution of R_{on} by its expression requires another iteration, as now V_T depends on the output voltage. This yields a value of R_{on} which is now larger. It is 291 Ω, instead of 216 Ω.

Also, the output voltage is a little bit lower. It is 0.567 V instead of 0.575 V. The time constant is now simply the product of 216 Ω and 4 pF.

Table of contents

- **Models of MOST transistors**
  - MOST as a resistor
  - MOST as an amplifier in strong inversion
  - Transition weak inversion-strong inversion
  - Transition strong inversion-velocity saturation
  - Capacitances and f_T
- **Models of Bipolar transistors**
- **Comparison of MOSTs & Bipolar transistors**

Willy Sansen 10-05 0120

0120

In most applications the MOST is used as an amplifier. This means that its V_{DS} is larger than $V_{GS} - V_T$. Its transconductance is then higher than at low V_{DS} values. It is used to generate gain.

The value of $V_{GS} - V_T$ itself, however, determines in which region the MOST is operating. For medium currents the MOST is operating in the strong inversion region. This is used most of the time.

At lower currents the MOST ends up in weak inversion. This is especially important for portable and low-power applications in general.

If we bias the MOST at the highest possible transconductance (for example in RF applications and very-low-noise applications), then the current densities are higher. The transconductance of the NMOST is then limited by velocity saturation. Again, another model is required.

All three regions are now discussed.

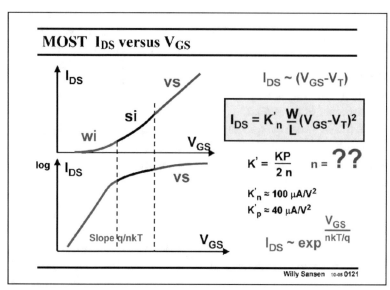

0121

In most amplifiers the MOST operates in the saturation region, i.e. we maintain $V_{DS} > V_{GS} - V_T$ at all times. We obtain the $I_{DS} - V_{GS}$ curve shown before. A closer look however, reveals that this curve has three distinctive regions. The one in the middle is called the strong-inversion region or square-law region as the current expression contains the factor $(V_{GS} - V_T)^2$.

At lower currents we find the weak-inversion region, or exponential region because the current expression now contains an exponential in V_{GS}. Indeed a $\log(I_{DS})$ curve is linear in that region.

At higher currents the $I_{DS} - V_{GS}$ curve becomes linear, because of several physical phenomena. The most important one is velocity saturation: all electrons reach their maximum speed v_{sat}.

Most transistors are biased in the strong-inversion region because this is a good compromise between current efficiency and speed, as explained later. In this region, the current expression is simply proportional to $(V_{GS} - V_T)^2$, but includes a technological parameter as well K'.

This parameter K' is linked to the one in the linear region KP by the ratio 2n. It is thus always smaller than KP. It is not very accurately known however, because of the mobility (in KP) and especially n. Remember that n depends on biasing voltages and so does K'.

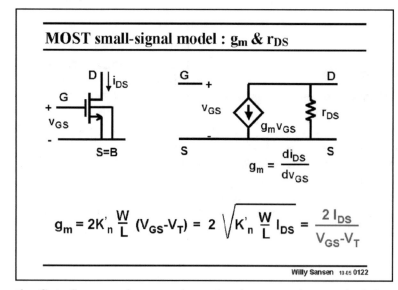

0122

Let us now make an amplifier using one single nMOST.

We assume that transistor is biased at some DC current I_{DS}. We want to know now what is the small-signal or AC current i_{DS} superimposed on it by application of a small-signal input voltage v_{GS}.

For this purpose we have to find the transistor transconductance g_m. This is nothing else than the derivative of the drain current to the Gate-Source voltage, as shown by the expression on the left.

However substitution of $V_{GS} - V_T$ from the current expression provides another expression for g_m.

Finally, substitution of W/L from the current expression provides a third expression for g_m.

The last one is the best known. It does not contain any technological parameter such as K' and is the most precise one. This is why it is highlighted.

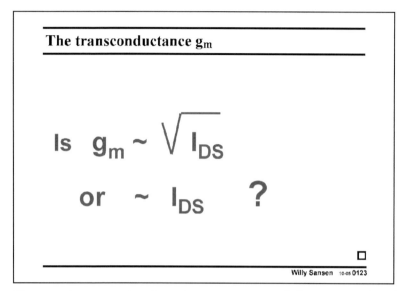

The transconductance g_m

$$\text{Is} \quad g_m \sim \sqrt{I_{DS}}$$

$$\text{or} \quad \sim I_{DS} \quad ?$$

Willy Sansen 10-05 0123

0123

Having three expression for only one single parameter g_m causes some ambiguity. The question asked is whether g_m is proportional to the square root of the current or to the current itself. The expressions show that both seem to be possible.

The answer becomes clear by checking the other parameters involved. During measurements, when the transistor size W/L is obviously constant, the middle expression prevails.

Then g_m is proportional to the square root of the current. Doubling the biasing current will increase the g_m only by 41%.

However, during the design procedure, the designer will fix the $V_{GS} - V_T$ at a certain value, for example at 0.2 V. Then g_m is proportional to the current itself. Doubling the biasing current will double the g_m as well.

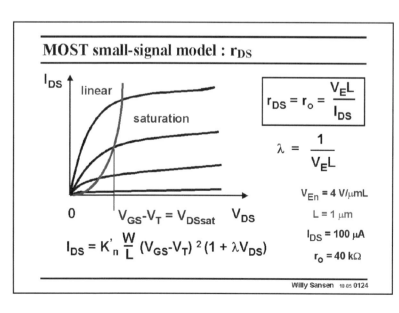

MOST small-signal model : r_{DS}

I_{DS}

linear

saturation

$$r_{DS} = r_o = \frac{V_E L}{I_{DS}}$$

$$\lambda = \frac{1}{V_E L}$$

$V_{En} = 4 \text{ V/}\mu\text{mL}$

$L = 1 \text{ }\mu\text{m}$

$I_{DS} = 100 \text{ }\mu\text{A}$

$r_o = 40 \text{ k}\Omega$

$0 \quad V_{GS} - V_T = V_{DSsat} \quad V_{DS}$

$$I_{DS} = K'_n \frac{W}{L} (V_{GS} - V_T)^2 (1 + \lambda V_{DS})$$

Willy Sansen 10-05 0124

0124

The small-signal model of a MOST also contains a finite output resistance r_{DS}. Indeed the $i_{DS} - v_{DS}$ curves in saturation are not quite flat. They exhibit thus a finite output resistance denoted by r_{DS} or r_o.

An additional parameter λ has to be included in the current expression to show that the current rises somewhat for increasing v_{DS}. Unfortunately, this parameter is not a constant. It depends on the channel length. Therefore, we prefer to use instead another parameter V_E. It is constant for a certain technology. It is different for a nMOST and a pMOST. Its dimension is V/μm.

The output resistance is then easily described. An example is given.

In models used for simulators (such as SPICE) several parameters are required to describe the output resistance. This model based on parameter V_E, is the simplest one and is only used for hand calculations. It only provides limited accuracy.

Parameter V_E is the fourth technological parameter that we find: we have had up till now n, V_T, KP and V_E.

Design parameters up till now are L and $V_{GS} - V_T$.

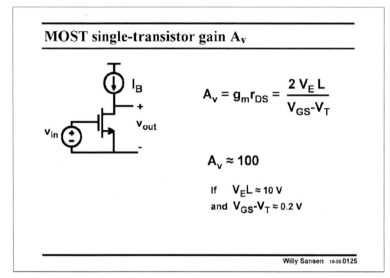

0125

Let us now investigate how much gain can be provided by a single-transistor amplifier, biased by a current source with value I_B.

The voltage gain is simply given by $g_m r_{DS}$ or the expression in this slide. Note that the current drops out as both parameters are current dependent.

It is clear that if we are interested in large gain, we will have to choose a large channel length, normally much larger than the minimum channel length of the technology used. Also we have to choose a value of $V_{GS} - V_T$ which is as small as possible. A reasonable value is 0.2 V. Reasons for this choice will be given later.

In order to obtain a voltage gain of 100, relatively large channel lengths are now necessary. If for some other reason we want to use the minimum channel length (for example for speed) then we will have to use circuit techniques to enhance the gain. Examples are cascodes, gain boosting, current starving, bootstrapping, etc.

Deep submicron CMOS technologies only provide very limited gain. All possible circuit techniques will have to be used to provide large gain.

Finally, note that such an amplifier (as most amplifiers are inverting), the output voltage increases when the input voltage decreases. This is why some authors add a minus sign to the expression of the gain.

0126

Indeed, for each transistor in the signal path, two independent choices will have to be made in the design procedure. They are the values of L and $V_{GS} - V_T$. A single-transistor amplifier can give a large amount of gain provided its L is large and its $V_{GS} - V_T$ is small. This will apply to all applications where high gain, low noise and low offset are most important, such as in operational amplifiers.

Design for high gain :

	High gain	High speed
V_{GS}-V_T	Low (0.2 V)	
L	High	

V_{GS}-V_T sets the ratio g_m/I_{DS} !

Willy Sansen 10-05 0126

Expressions cannot be used to establish these values. They have to be chosen at the very start of the design procedure.

Unfortunately, for high speed, we will see that exactly the opposite conclusions will have to be drawn. For high speed, a transistor in the signal path requires the L to be small and the $V_{GS}-V_T$ large. This will apply to all RF circuits such as Low Noise Amplifiers (LNA's), Voltage Controlled Oscillators (VCO's), Mixers, etc.

This compromise is one of the most basic compromises in analog CMOS design. After all it is gain versus speed!

Finally, note that the value of $V_{GS}-V_T$ sets the ratio g_m/I_{DS}. However, we need to have a look at weak inversion first. Choosing the value of $V_{GS}-V_T$ or choosing the value of g_m/I_{DS}, will ultimately be the same choice.

Example: single-transistor amplifier

We want to realize a three-stage amplifier
with a total gain of 10.000.
We use three single-transistor stages in series.
What minimum lengths do we have to use in
an advanced 65 nm CMOS technology
with V_E = 4 V/μm ?

Choose
V_{GS}-V_T = 0.2 V !

Willy Sansen 10-05 0127

0127

There is no reason to allocate more gain to one amplifier compared to the other. The gain per amplifier is thus 21.5 since 21.5 × 21.5 × 21.5 ≈ 10 000. As a result we need a $V_E L$ product of 2.15 V.

For this technology, we will need a channel length of L ≈ 0.5 μm. If we used the minimum channel length of 65 nm, the gain would only be about 2.6!!

Deep submicron CMOS only provides very low voltage gains!!

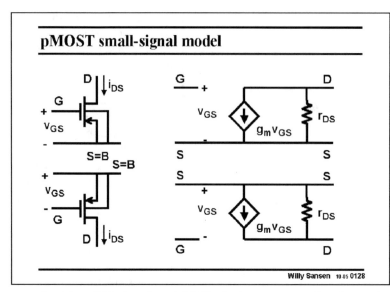

0128

The small-signal model of a pMOST is exactly the same as for a nMOST. For the same biasing current and the same $V_{GS} - V_T$, it also provides the same transconductance g_m. The output resistance may be different, depending somewhat on the values of parameter V_E.

Care must be taken however, on how to represent this small-signal model. Normally a nMOST device requires a positive V_{DS}, whereas a pMOST device a negative V_{DS}. This is why pMOST devices are usually shown inverted, i.e. with its Source on top. Indeed, only positive supply voltages are used nowadays. pMOS transistors are usually shown inverted.

In this case we have to be careful how to include the signs and the current direction. It is shown in this slide.

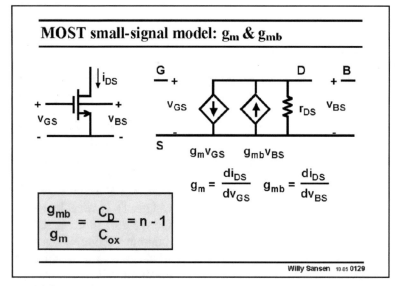

0129

pMOST devices in a n-well CMOS technology can also be driven at the Bulk. The Bulk is then used as an input instead of the Gate. This is much more dangerous as there is always a risk to forward that channel-bulk pn junction by incident. This must be avoided at all times and may probably require extra protection circuitry.

For a bulk input voltage, another transconductance must be added g_{mb}. Its value is proportional to the channel-bulk junction capacitance, in exactly the same way as the g_m is proportional to the gate oxide capacitance. In other words, the ratio of transconductances equals the ratio of the controlling capacitances, which equals $n - 1$. This is a very powerful relationship, but it never provides an accurate value, as n depends on some biasing voltages.

Table of contents

Willy Sansen 10-05 0130

0130

At low currents, the MOST operates in the weak-inversion region. This means that the channel conductivity has become very small. Actually the channel has ceased to exist. The channel has vanished.

The drift current, which flowed through the channel in strong inversion, is now replaced by a diffusion current (Ref. Tsividis). As a result the model is very different. It contains an exponential rather than a square-law characteristic.

Even more important is to know where exactly the strong-inversion region is substituted by the weak inversion region. Actually this region is quite wide. It is also called the moderate-inversion region. For the designer it is essential to know what is the $V_{GS} - V_T$ at which this transition occurs and especially what the current level is. This is why considerable attention is paid to this cross-over point.

I_{DS} & g_m versus V_{GS} : weak inversion

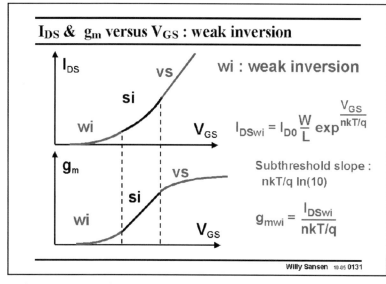

wi : weak inversion

$$I_{DSwi} = I_{D0} \frac{W}{L} \exp^{\frac{V_{GS}}{nkT/q}}$$

Subthreshold slope :
nkT/q ln(10)

$$g_{mwi} = \frac{I_{DSwi}}{nkT/q}$$

Willy Sansen 10-05 0131

0131

Now that we know how to describe the current of a MOST in the middle-current region (or strong inversion region), we have to focus on the low-current region (weak inversion) and then at the high-current region (velocity saturation). We are especially interested to figure out where the crossover values of V_{GS} are between these regions.

At low currents we have the weak-inversion region, also called the sub-threshold region as most of it occurs below V_T. It is also called the exponential region as we have an exponential current-voltage relationship.

The scaling factor is nkT/q, which is very close to the one of a bipolar transistor where it is kT/q. The factor is the Boltzmann factor and q the charge of an electron such that kT/q is about 26 mV at room temperature (actually at 300 K or 27°C). The difference is factor n however, the same n as before. We remember that this n depends on biasing voltages and is thus never accurately known. This is a considerable disadvantage of a MOST compared to a bipolar device.

In this region, the transconductance is again the derivative of the current with respect to V_{GS}, which is also exponential. Again the only difference with the g_m of a bipolar transistor is this factor n. It is now always lower.

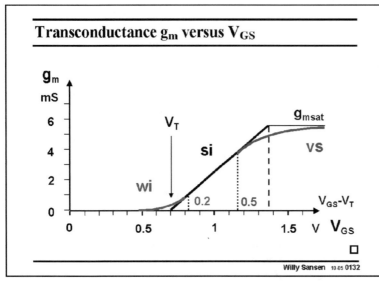

Transconductance g_m versus V_{GS}

Willy Sansen   10-05 0132

0132

What values of V_{GS} are commonly used?

At the higher end, we know that we never want to enter the high-current or velocity-saturation region. We stay away from the transition point to velocity saturation. The approximate value will be calculated further. It is about 0.5 V however, for present day technologies.

At the low-current end, we do not want to use the weak-inversion current region either. The absolute values of the currents and of the transconductances become so small that the noise becomes exceedingly large. Moreover, only low speeds can be obtained. There are some applications where low signal-to-noise ratios and low speed are no problem. Examples are biomedical applications and biotelemetry. In most other applications however, we need better noise performance and higher speeds. Therefore, we do want to work close to weak inversion but not in it. Typical values are $V_{GS} - V_T$ values between 0.15 and 0.2 V.

The reason why, is explained next.

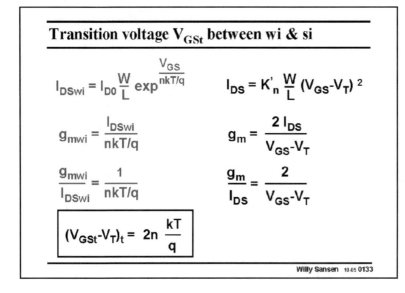

Transition voltage V_{GSt} between wi & si

$$I_{DSwi} = I_{D0} \frac{W}{L} \exp^{\frac{V_{GS}}{nkT/q}} \qquad I_{DS} = K'_n \frac{W}{L} (V_{GS}-V_T)^2$$

$$g_{mwi} = \frac{I_{DSwi}}{nkT/q} \qquad g_m = \frac{2 I_{DS}}{V_{GS}-V_T}$$

$$\frac{g_{mwi}}{I_{DSwi}} = \frac{1}{nkT/q} \qquad \frac{g_m}{I_{DS}} = \frac{2}{V_{GS}-V_T}$$

$$(V_{GSt}-V_T)_t = 2n \frac{kT}{q}$$

Willy Sansen   10-05 0133

0133

Let us now explore where the actual crossover or transition point is between the middle- and low-current regions. We will denote this value of V_{GS} by V_{GSt}.

The transition between strong and weak inversion is reached by simply equating the expressions of the current and their first derivatives, or transconductances.

Indeed, equating their g_m/I_{DS} ratios we obtain the transition value of $V_{GSt} - V_T$. It is simply $2nkT/q$.

Because of n, we cannot obtain an accurate value for this $V_{GSt} - V_T$. An approximate value of 70–80 mV is taken. This means that the transistor model changes from weak inversion to strong inversion at a value of V_{GS} of about $V_T + 70$ mV. For a V_T of 0.6 V, this would be a V_{GS} of about 0.67 V.

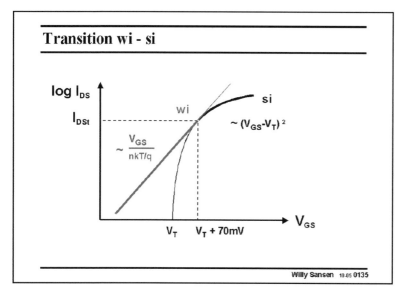

Transition Voltage V_{GSt} for different L

$$(V_{GSt}-V_T)_t = 2n \frac{kT}{q}$$

$$I_{DSt} \approx K'_n \frac{W}{L}\left(2n \frac{kT}{q}\right)^2$$

Is independent of channel length L
Is still true in ... years !

$$(V_{GSt}-V_T)_t = 2n \frac{kT}{q} \approx 70 \text{ mV}$$

$$I_{DSt} \approx 2 \text{ μA for } \frac{W}{L} = 10$$

for nMOST

Willy Sansen 10-05 0134

0134

Much more important than the absolute value of V_{GSt} is the fact that V_{GSt} does not depend on the channel length. As a result, this transition value $2nkT/q$ will not change for future CMOS technologies. Consequently, the value of $V_{GS} \approx 0.2$ V, chosen before, to stay out of weak inversion will remain the same for the years to come. This is a very comforting thought indeed!

The current obtained at the transition point is now easily calculated. Substitution of V_{GS} by V_{GSt} in the current expression in the square-law region, provides I_{DSt}. It is the transition current.

Obviously it depends on W/L. For a W/L = 10, a current is reached in the μA region. Clearly nA's can only be reached in the weak inversion region.

Transition wi - si

$\log I_{DS}$

I_{DSt}

wi

si

$\sim (V_{GS}-V_T)^2$

$\sim \frac{V_{GS}}{nkT/q}$

V_T $V_T + 70$mV

V_{GS}

Willy Sansen 10-05 0135

0135

How both models are connected is even better illustrated on a logarithmic scale for the current.

The exponential relationship in weak inversion (wi) is then a straight line. The square-law characteristic of the strong inversion region (si) on the other hand is very nonlinear. It goes to minus infinity where $V_{GS} = V_T$.

Both curves touch at the transition point of $V_{GSt} \approx V_T + 70$ mV. At this point the transistor jumps from the wi curve to the si curve.

Even if both curves are not quite touching each other, so that the transistor has to "jump" from one to the other, this is still a fairly accurate representation of the transition region.

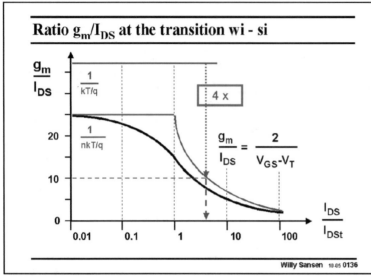

0136
This transition region is clearly visible when we plot g_m/I_{DS}. This ratio is nearly as important as the g_m itself, as it explains what the current efficiency is of a MOST.

This ratio is plotted versus current, normalized with respect to the transition current I_{DSt}. As a result for unity I_{DS}/I_{DSt} both models provide the same g_m/I_{DS} ratio, as expected.

At lower currents this ratio is constant and is given by $1/(nkT/q)$. This value is about 1/40mV or 25 V^{-1}. It is always smaller than for a Bipolar transistor where it is 1/26mV or 38 V^{-1} or about 40 V^{-1}.

At higher currents this g_m/I_{DS} ratio decreases as it is inversely proportional to $V_{GS}-V_T$. For example, at $V_{GS}=0.2$ V, this ratio is about 10. More accurate descriptions will follow.

It is already clear however, that a MOST provides much less transconducance than a bipolar transistor at the same current. The ratio is a factor of about four. In other words, for the same transconductance, a bipolar transistor only needs four times less current than the MOST. Portable applications will now consume less current if they are realized by means of bipolar transistors!

Finally, note that a real MOST does not follow the two models: it follows a smooth line from one to the other, as explained next.

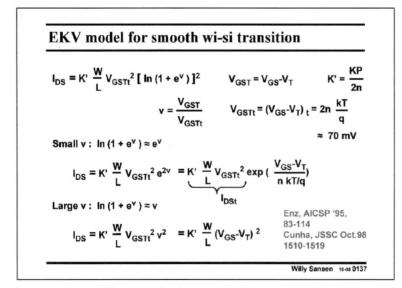

0137
This smooth transition between weak and strong inversion is best
   described by the EKV model [Enz], explained in this slide. It uses a function which contains the square of a natural log function of an exponential. The variable is v, which is $V_{GS}-V_T$, normalized to a quantity V_{GSTt} or simply $2nkT/q$. This includes factor n, which depends on some biasing voltages, which is usually between 70 and 80 mV.

In weak inversion, or for small v, the log function is approximated by a power series, and limited to its first term. The exponential function emerges, which is typical for the weak inversion region. The subthreshold slope is nkT/q. Also, the current for $V_{GS} = V_T$, or zero $V_{GS} - V_T$ is called I_{DSt}. It is called the transition current, as already found before.

In strong inversion, or large v, the log function cancels the exponential and v emerges by itself. The square-law expression is found, describing the current-voltage characteristic of a MOST in strong inversion.

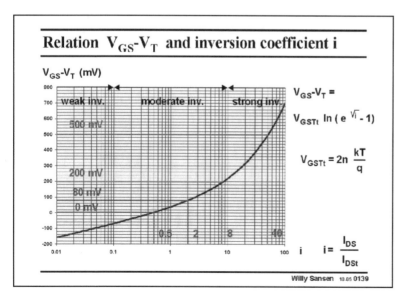

Transition current I_{DSt} between wi & si

$$I_{DSt} = I_{DS}\Big|_{\substack{V=1 \\ i=1}} = K' \frac{W}{L} V_{GSTt}^2 \qquad I_{DSt} = 2\,\mu A \text{ for } W/L = 10$$

$$i = \frac{I_{DS}}{I_{DSt}} = [\ln(1 + e^v)]^2 \qquad \text{inversion coefficient}$$

$$v = \ln(e^{\sqrt{i}} - 1)$$

$$V_{GS} - V_T = V_{GSTt} \ln(e^{\sqrt{i}} - 1)$$

$$V_{GSTt} = 2n \frac{kT}{q} \approx 70\,mV$$

Willy Sansen 10-05 0138

0138

This transition current I_{DSt} is the current at which both MOST models coincide, i.e. at which the currents are equal and also the transconductances. This current I_{DSt} can be used to normalize the current I_{DS}. This ratio is denoted by i and is called inversion coefficient. Remember that this current I_{DSt} is about 2 μA for $W/L = 10$ for a nMOST. It is now about 0.2 μA per unit W/L. At this current $i = 1$ and $v = 0.54$. For a V_{GSt} of 70 mV the value of $V_{GS} - V_T$ is about 38 mV.

The normalized voltage v, or the voltage $V_{GS} - V_T$ itself can now easily be described in terms of the inversion coefficient i. A plot of this relationship is given next.

Relation V_{GS}-V_T and inversion coefficient i

V_{GS}-V_T (mV)

weak inv. moderate inv. strong inv.

500 mV

200 mV

80 mV

0 mV

$$V_{GS} - V_T = V_{GSTt} \ln(e^{\sqrt{i}} - 1)$$

$$V_{GSTt} = 2n \frac{kT}{q}$$

$$i = \frac{I_{DS}}{I_{DSt}}$$

Willy Sansen 10-05 0139

0139

This plot shows the relationship between the voltage drive $V_{GS} - V_T$ of a MOST and its normalized current i. The strength about this relationship is that the transistor size is completely hidden by the transition current I_{DSt}. This curve is independent of transistor sizes.

Moreover, it is easy to find out what exactly happens at the weak-inversion transition. For large currents ($i > 10$), the transistor obviously operates in strong inversion. This is true for values of $V_{GS} - V_T$ larger than about 0.2 V, corresponding to inversion coefficients larger than 8. This corresponds to currents of 1.6 µA per unit W/L for a nMOST. This current is multiplied by five if a $V_{GS} - V_T$ is used of 0.5 V.

For very small currents (i < 0.1), the transistor is clearly in the weak-inversion region. Its characteristic is a straight line because of the semilog scales.

The region between (0.1 < i < 10) is called the moderate-inversion region. It provides a smooth transition between both regions.

An interesting point is at $V_{GS} - V_T = 0$ V. This is the point where the input voltage V_{GS} equals the threshold voltage V_T. At this point the current is half of the transition current I_{DSt}. It is about 0.1 µA per unit W/L. This is an often used way to measure the threshold voltage V_T. It is the voltage V_{GS} at a current of 0.1 µA per unit W/L.

Another interesting point is at $V_{GS} - V_T = 80$ V. This is the point where the actual crossover occurs from weak to strong inversion. At this point the current is about twice the current I_{DSt}.

Transconductance g_m between wi & si

$$i = \frac{I_{DS}}{I_{DSt}} = [\ln(1 + e^v)]^2 \qquad g_m \approx \dots.$$

$$GM = \frac{g_m}{I_{DS}} \frac{nkT}{q} = \frac{1 - e^{-\sqrt{i}}}{\sqrt{i}}$$

Large i : $GM = \frac{1}{\sqrt{i}}$

Small i : $GM = 1 - \frac{\sqrt{i}}{2}$

Alternative approximation :

$$GM = \frac{1}{\sqrt{1 + 0.5\sqrt{i} + i}}$$

Large i : $GM = \frac{1}{\sqrt{i}}$

Small i : $GM = 1 - \frac{\sqrt{i}}{4}$

Willy Sansen 10-05 0140

0140

The transconductance g_m is now easily calculated, as it is the derivative of the current. Instead the ratio g_m/I_{DS} is taken and normalized to nkT/q. It is denoted by GM. The approximations for strong inversion (i > 10) and weak inversion (i < 0.1) are easily obtained.

The full expression is plotted in the next slide.

Sometimes an alternative function is used to describe the transition region. It is a little closer to measured data. It involves square roots however, instead of exponentials. For strong inversion, it provides the same approximation. However, for weak inversion, it is slightly smaller.

Both functions are plotted in the next slide.

0141

Both models for the normalized transconductance to current ratio are plotted versus the inversion coefficient in this slide. Both go through the same point at i = 1. The difference is small. This is why the expression with the exponential will be used further.

Note that, at an inversion coefficient of 8, where the $V_{GS} - V_T$ is 0.2 V, the normalized GM is about 1/3 of the maximum value in weak inversion. The corresponding g_m/I_{DS} is about 4.2 V^{-1}. This is nearly ten times less than for a bipolar transistor. In order to boost this ratio, we must choose a smaller $V_{GS} - V_T$. For a value of 0.15 V, the ratio g_m/I_{DS} is already 5.4 V^{-1} and for 0.1 V close to 7 V^{-1}.

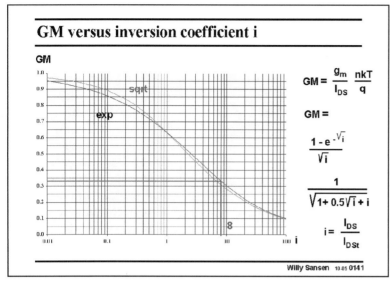

GM versus inversion coefficient i

$$GM = \frac{g_m}{I_{DS}} \frac{nkT}{q}$$

$$GM =$$

$$\frac{1 - e^{-\sqrt{i}}}{\sqrt{i}}$$

$$\frac{1}{\sqrt{1 + 0.5\sqrt{i} + i}}$$

$$i = \frac{I_{DS}}{I_{DSt}}$$

Willy Sansen 10-05 0141

Note that around a $V_{GS} - V_T$ of 0.2 V the g_m/I_{DS} is 4.2 V^{-1}, which is less than half of what is predicted by the expression of $g_m/I_{DS} = 2/(V_{GS} - V_T)$ in strong inversion, which gives 10 V^{-1}.

Values around 0.2 V are a good compromise between high transconductance and high current. They will be used in many design examples later on.

0142

At high currents however, the MOST transistor changes region again. This time velocity saturation emerges. Most electrons now move through the channel at maximum speed. The result is that the current increases linearly with the voltage drive and the transconductance levels off.

Again, a different model is required. We are especially interested in finding out where the transition point is between strong inversion and velocity saturation.

This is explained next.

0143

At the higher current end, several phenomena cause the current to become more linear versus V_{GS}. One of the most important ones is velocity saturation. This means that because of the high electric fields in the channel, all electrons move at maximum speed v_{sat}. This speed is a result of a collision process of the electrons in the channel. Its average value is about 10^7 cm/s.

The resulting expression of the current shows a linear dependence now of the current versus V_{GS}. The transconductance g_m, which is again the derivative, becomes very simple now. It does not include the channel length any more! It is the highest transconductance that this MOST can

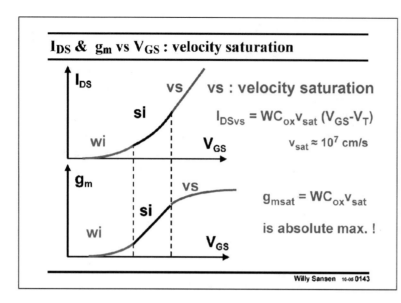

I_{DS} & g_m vs V_{GS} : velocity saturation

$I_{DSvs} = WC_{ox}v_{sat}(V_{GS}-V_T)$

$v_{sat} \approx 10^7$ cm/s

$g_{msat} = WC_{ox}v_{sat}$

is absolute max. !

Willy Sansen 10-05 **0143**

ever achieve. Its g_{msat}/W ratio only depends on technology (C_{ox}) and physics (v_{sat}).

Note however, that this g_{msat} does not depend any more on V_{GS} either. It has become a constant. This is why this region is never used by analog designers. The transconductance does not increase any more but the current consumption does!! This is why the highest values of V_{GS} used are the ones in the middle current region, close to the velocity saturation region.

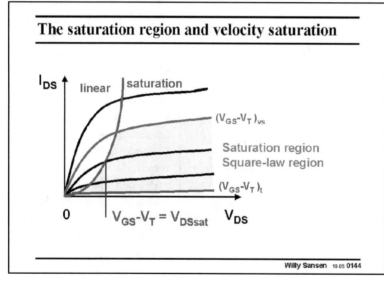

The saturation region and velocity saturation

$(V_{GS}-V_T)_{vs}$

Saturation region
Square-law region

$(V_{GS}-V_T)_t$

$V_{GS}-V_T = V_{DSsat}$

Willy Sansen 10-05 **0144**

0144

At this point, it is important to distinguish saturation from saturation velocity.

A MOST is in saturation when its V_{DS} is sufficiently large or when V_{DS} is larger than V_{DSsat}, which equals $V_{GS}-V_T$. Any operating point to the right of the parabola, separating the saturation region from the linear region is in the saturation region of that transistor.

In that region the lower currents correspond to the weak inversion region. The highest currents on the other hand correspond to the velocity saturation region. These transistors operate in saturation and are in the velocity saturation region.

The middle current region, which is shaded, corresponds to the region, which is characterized by the square-law model. These transistors operate in saturation and are in the square-law region.

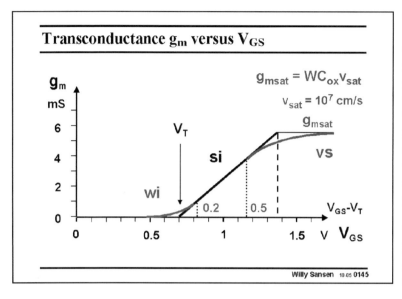

Transconductance g_m versus V_{GS}

$g_{msat} = WC_{ox}v_{sat}$

$v_{sat} = 10^7$ cm/s

Willy Sansen 10-05 0145

0145
Remember that we never want to enter the high-current or velocity-saturation region. As a designer we want to stay away from the transition point to velocity saturation. We want to calculate now the value of this transition point. It will come out to be about 0.5 V for present day technologies.

Typical values for $V_{GS} - V_T$ are now between 0.15–0.2 V at the low end and about 0.5 V at the high end.

The reason is explained next.

Velocity saturation : v_{sat} & θ

$$I_{DS} = \frac{K'_n \frac{W}{L} (V_{GS}-V_T)^2}{1 + \theta (V_{GS}-V_T)}$$

[large V_{GS}]

$$\approx \frac{K'_n}{\theta} \frac{W}{L} (V_{GS}-V_T)$$

$$g_{msat} \approx 2K'_n \frac{W}{L} (V_{GS}-V_T)^2 \frac{1 + \frac{\theta}{2}(V_{GS}-V_T)}{[1 + \theta (V_{GS}-V_T)]^2} \approx \frac{K'_n}{\theta} \frac{W}{L}$$

$$= WC_{ox}v_{sat}$$

$$\theta L = \frac{\mu}{2n} \frac{1}{v_{sat}} = \frac{1}{E_c} \quad E_c \text{ is the vertical critical field !}$$

$\theta L \approx 0.2$ μm/V : For L = 0.13 μm $\theta \approx 1.6$ V^{-1}

Willy Sansen 10-05 0146

0146
The linearization of the current and the reduction of the transconductance because of velocity saturation can also be described by parameter θ (Greek theta). It introduces a term in the denominator of the current expression in strong inversion, as shown in this slide. Indeed for high $V_{GS} - V_T$, the 1 drops out and the expression of the current becomes linear in V_{GS}, as shown at the right.

This parameter θ has the advantage that it lumps all physical phenomena which cause linearization of the current. One of these phenomena is the effect of the vertical electric field E_c across the oxide. It is used mainly by designers, not by technologists.

The transconductance g_{msat} in velocity saturation, can be obtained by taking the derivative of the current again. For large $V_{GS} - V_T$, a simple expression is found for g_{msat}. This must obviously equal the original expression of g_{msat} with v_{sat} in it.

Equation of both expressions of g_{msat} shows that this parameter θ depends on the channel length. The smaller L, the larger θ. It depends here on technology. It becomes quite large for nanometer CMOS.

Velocity saturation : θ & R_S & v_{sat}

$$I_{DS} = \frac{K'_n \frac{W}{L} (V_{GS}\text{-}V_T)^2}{1 + \theta (V_{GS}\text{-}V_T)}$$

[large V_{GS}]

$$g_{msat} \approx \frac{K'_n}{\theta} \frac{W}{L} \qquad g_{mRs} = \frac{g_m}{1 + g_m R_S} \approx \frac{1}{R_S}$$

$$\boxed{R_S = \frac{\theta}{K'_n W/L}} \qquad R_S \approx \frac{\mu}{2n} \frac{1}{W K'_n v_{sat}} \approx \frac{1}{W C_{ox} v_{sat}}$$

Willy Sansen 16-05 0147

0147

There is a third way to model the linearization of the current and the reduction of the transconductance, because of velocity saturation, by simply inserting a small resistor R_S is series with the Source.

Indeed this reduces the transconductance g_m by a factor $1 + g_m R_S$.

For large values of $V_{GS} - V_T$, or for large values of g_m, the resulting g_{mRS} becomes simply $1/R_S$. Equation of this g_m to the one with θ yield the relationship between θ and R_S. It is clear that the required value of R_S depends on the dimensions of the MOST and on K′.

It is also obviously easy to find the relationship between v_{sat} and R_S. Elimination of parameter θ between the two bottom expressions yields an expression in which only the transistor width W occurs, together with several technological parameters. It is a well known expression indeed.

Transition Voltage V_{GSTvs} between si and vs

$$I_{DS} = \frac{K'_n \frac{W}{L} (V_{GS}\text{-}V_T)^2}{1 + \theta (V_{GS}\text{-}V_T)} \qquad I_{DSsat} = W C_{ox} v_{sat} (V_{GS}\text{-}V_T)$$

$$g_{msat} = W C_{ox} v_{sat} \approx \frac{K'_n}{\theta} \frac{W}{L}$$

$$\boxed{(V_{GS}\text{-}V_T)_{vs} = \frac{1}{\theta} \approx 2nL \frac{v_{sat}}{\mu}} \quad \textbf{Is proportional to channel length L !!!}$$

$$\approx \textbf{5 L} \approx 0.62 \text{ V} \text{ if } L = 0.13 \text{ } \mu m$$

Willy Sansen 16-05 0148

0148

In order to find the transition voltage drive $(V_{GS} - V_T)$ V_{GSTvs}, we have to equate the expression of the transconductance g_{msat} in velocity saturation to the one in strong inversion. The resulting value V_{GSTvs} is simply $1/\theta$ which is also $2nLv_{sat}/\mu$, as shown on the previous slide.

For typical values, this V_{GSTvs} is just about 5L Volt with L in micrometer. It is clear that this transition voltage V_{GSTvs} is not constant, it decreases with decreasing channel length. For a 0.13 μm CMOS technology, it is only about 0.62 V. It is only a matter of time and it will also be 0.2 V!

Parameter θ or the inverse of this transition voltage $1/V_{GSTvs}$ can also be called the velocity saturation coefficient, although it includes several other phenomena associated with the high electric fields.

Transition Current I_{DSvs} between si and vs

$$I_{DSvs} \approx K' \, WL \left(\frac{2n \, v_{sat}}{\mu}\right)^2 \approx 100 \, n \, \varepsilon_{ox} \, W \, \frac{v_{sat}^2}{\mu}$$

$$\frac{I_{DSvs}}{W} \approx 10 \ A/cm \qquad K' = \frac{\mu C_{ox}}{2n}$$

$$C_{ox} = \frac{\varepsilon_{ox}}{t_{ox}} \qquad t_{ox} = \frac{L}{50}$$

W = 2.6 μm & L = 0.13 μm :

$I_{DSvs} \approx 2.6$ mA

$v_{sat} = 10^7$ cm/s
n = 1.4
μ = 500 cm²/Vs

Willy Sansen 16-05 0149

0149

The transition current I_{DSvs} is easily obtained by means of the transition voltage V_{GSTvs}. It seems to contain both W and L, in addition to some technological parameters.

Remember, however, that K′ contains C_{ox}, which depends on the oxide thickness t_{ox}, which is actually about L/50. The expression of I_{DSvs} can be simplified even further, provided the minimum channel length L is used. This clearly evidences the influence of the technology, but also contains transistor width W.

Substitution of the technological parameters for a nMOST ($n \approx 1.4$, $\mu = 500$ cm²/Vs) yields a very simple expression for the transition current density I_{DSvs}/W. Obviously, its value also depends on the dimensions used, but now through W the minimum value of which is actually linked to L.

Transconductance g_m between si and vs

$$g_{msat} = W \, C_{ox} \, v_{sat} \qquad g_{msat} \approx 17 \cdot 10^{-5} \ W/L \ S/cm$$

$$g_{mK'} = 2K' \, \frac{W}{L} \underbrace{(V_{GS}\text{-}V_T)}_{V_{GST}} \qquad g_{mK'} \approx 1.2 \cdot 10^{-9} \, V_{GST} \, W/L^2 \ S/cm$$

$$\frac{1}{g_m} = \frac{1}{g_{mK'}} + \frac{1}{g_{msat}} \qquad g_m \approx \frac{W}{L} \ \frac{17 \cdot 10^{-5}}{1 + 2.8 \cdot 10^4 \, L \, / \, V_{GST}} \bigg|_{\substack{L \\ in \ cm}}$$

If $V_{GST} = 0.2$ V, v_{sat} takes over for L < 65 nm (If 0.5 V for L < 0.15 μm)

Willy Sansen 16-05 0150

0150

The transconductance g_m is easily found in both regions; g_{msat} is the one in velocity saturation, whereas $g_{mK'}$ is the one in strong inversion. They are given in this slide.

Again, substitution of the technological parameters yields expressions for both transconductances in which only W and L are left explicitly. An expression of the transconductance g_m covering both regions can be obtained by simply putting both "in parallel". The total transconductance g_m is then always smaller than the smaller one of both.

In this way an expression is obtained with both W and L, and obviously $V_{GS} - V_T$. This expression will be used later on to optimize some high-speed circuits, but also operational amplifiers.

For a specific $V_{GS} - V_T$, this total transconductance is easily plotted versus channel length L, as shown next.

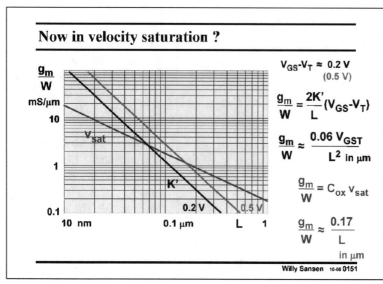

0151

For a specific $V_{GS} - V_T$, the total transconductance is easily plotted versus channel length L, as shown next. Actually, both transconductances are plotted separately.

The transconductance in saturation is in blue; the one in strong inversion for $V_{GST} = 0.2$ V in black and for $V_{GST} = 0.5$ V in red. For $V_{GST} = 0.2$ V, the crossover is around 65 nm channel length. This means that nowadays for analog designs in 0.13 μm and 0.18 μm, we can still use the model in strong inversion and that a small factor of θ is sufficient to provide accurate values of transconductance.

For high-frequency applications however, in which $V_{GST} = 0.5$ V or higher, the crossover value of channel length is about 0.15 mm. Transistors in a 90 nm technology operate mainly in velocity saturation. Parameter θ is then much larger, and certainly cannot be neglected.

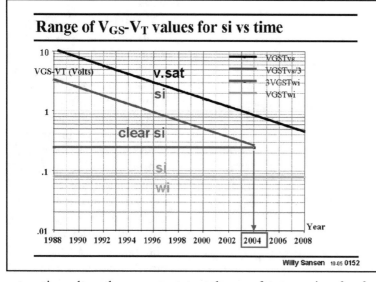

0152

As the channel lengths become smaller all the time, as predicted by the curve of Moore, given at the beginning of this Chapter, they can be predicted for the near future.

For each different channel length we can estimate the transition voltage V_{GSTvs} between strong inversion and velocity saturation. It decreases versus channel length L. It is given by the top black line.

To stay out of velocity saturation altogether we want to take a safety margin of a factor of three in V_{GST} or ten in current I_{DS}. This is the blue line. It obviously also decreases versus channel length or time.

The transition voltage V_{GSt} between weak and strong inversion on the other hand, is independent of the channel length. It is always about 70 to 80 mV. This is the green line. To stay out of the region (of moderate inversion) close to this transition point, again a safety margin is taken of three in V_{GST} or ten in current. This is the red line.

The region between the blue and red line is thus the range of V_{GST} values in which a square-law characteristic can easily be identified. As clearly shown, this region has just about disappeared. This seems to have happened in 2004!

In other words, in future, MOST models are required which are no more extrapolations of a square-law model but which connect an exponential to a linear curve, as bipolar models do!

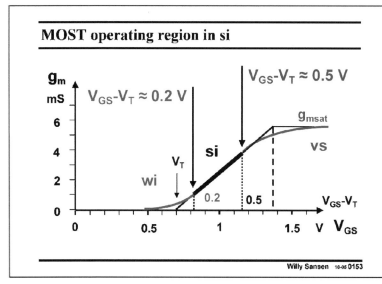

0153

As a result, the prediction of the current and the trans-conductance cannot be accurate any more.

Indeed, in the middle of the square-law region, around $V_{GS} - V_T = 0.2$ V, parameter K' contains the mobility and factor n, which are not accurate at all indeed.

Moreover, for lower values of $V_{GS} - V_T$, we end up too close to weak inversion, complicating the model considerably.

On the other side of the $V_{GS} - V_T$ range, velocity saturation is moving in closer and closer all the time, requiring another level of complexity.

It has now become very difficult to use simple models for hand calculations. The only good chance we have is around $V_{GS} - V_T \approx 0.2$ V for high gain and somewhat more (≈ 0.5 V) for high frequency applications. In all other regions elaborate models have to be used, which are only available from foundries.

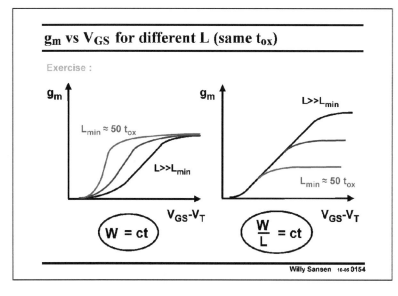

0154

As a conclusion let us try to understand how some of these important parameters change with the drive voltage $V_{GS} - V_T$. As they are meant as exercises, only a few comments are given.

In both cases we have to try to figure out how the transconductance g_m changes if we change the drive voltage $V_{GS} - V_T$. Both are for the same technology (constant oxide thickness and minimum channel length).

On the left however, we keep W constant, whereas on the right we keep W/L constant. Obviously the operating points are allowed to shift from weak inversion, over strong inversion to velocity operation.

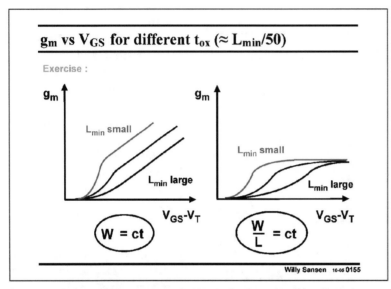

0155

Similar curves can be generated if we jump from technology to technology. This would answer the question, what happens to the performance of an amplifier which has been designed in one technology and which is then transferred to another one without changing the layout. How would the transconductances change the high-frequency performance, etc.?

Again, the operating points are allowed to shift from weak inversion, over strong inversion to velocity operation.

This kind of reasoning is excellent to develop insight in the relation between several operating regions and the fundamental design parameters such as the transconductance.

Table : MOST I_{DS} , g_m & g_m/I_{DS}

Summary :

TABLE 1-4   EXPRESSIONS OF I_{DS}, g_m AND g_m/I_{DS} FOR MOST

	I_{DS}	g_m	$\frac{g_m}{I_{DS}}=f(v_{GS}-V_T)$	$\frac{g_m}{I_{DS}}=f(I_{DS})$
wi	$I_{D0}\frac{W}{L}\exp\left(\frac{v_{GS}}{nkT/q}\right)$ (1-25a)	$\frac{I_{D0}}{nkT/q}\frac{W}{L}\exp\left(\frac{v_{GS}}{nkT/q}\right)$ (1-25b)	$\frac{1}{nkT/q}$ (1-26b)	$\frac{1}{nkT/q}$ (1-26b)
ws			$(v_{GS}-V_T)_{ws}=2n\frac{kT}{q}$	$I_{DSws}=\frac{KP}{2n}\frac{W}{L}\left(2n\frac{kT}{q}\right)^2$
si	$\frac{KP}{2n}\frac{W}{L}(v_{GS}-V_T)^2$ (1-18c)	$2\frac{KP}{2n}\frac{W}{L}(v_{GS}-V_T)$ (1-22a)	$\frac{2}{v_{GS}-V_T}$ (1-26a)	$2\sqrt{\frac{KP}{2n}\frac{W}{L}\frac{1}{I_{DS}}}$ (1-26a)
sv			$(v_{GS}-V_T)_{sv}=\frac{2nLC_{ox}v_{sat}}{KP}$	$I_{DSsv}=\frac{2WLC_{ox}^2v_{sat}^2}{KP/2n}$
vs	$WC_{ox}v_{sat}(v_{GS}-V_T)$ (1-38b)	$WC_{ox}v_{sat}$ (1-39)	$\frac{1}{v_{GS}-V_T}$	$\frac{WC_{ox}v_{sat}}{I_{DS}}$

Ref.: Laker, Sansen : Design of analog ..., MacGrawHill 1994; Table 1-4

Willy Sansen   10-05 **0156**

0156

As a summary, this overview Table is copied from the reference.

The expressions for the current, I_{DS}, the transconductance g_m and the g_m/I_{DS} ratio's are given horizontally for weak inversion, strong inversion and velocity saturation. Moreover, the transition or crossover values are denoted by ws for the transition between weak inversion and strong inversion, and by sv for the transition between strong inversion and velocity saturation.

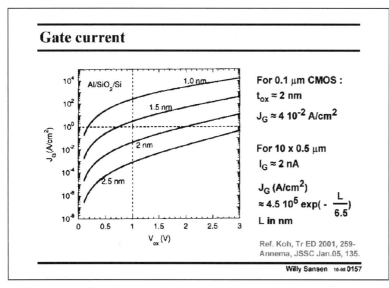

Gate current

For 0.1 μm CMOS :
$t_{ox} \approx 2$ nm
$J_G \approx 4 \cdot 10^{-2}$ A/cm^2

For 10 x 0.5 μm
$I_G \approx 2$ nA

J_G (A/cm^2)
$\approx 4.5 \cdot 10^5 \exp(-\dfrac{L}{6.5})$
L in nm

Ref. Koh, Tr ED 2001, 259-
Annema, JSSC Jan.05, 135.

Willy Sansen 10-05 0157

0157

As a final word of caution on these MOST models, before we tackle the capacitances, we have to have a look at the Gate current. For thin oxides, Gate leakage occurs. It is a result of tunneling through the oxide. It is thus proportional to the Gate area.

It increases exponentially with thinner and thinner oxides, as illustrated in this slide. A numerical approximation is given for an oxide voltage of 1 V.

It is clear that for CMOS technologies below 90 nm, this current cannot be neglected. This current behaves somewhat as a base current in a bipolar transistor. It still flows however, when the transistor is switched off, which is not the case in a bipolar transistor.

Table of contents

- **Models of MOST transistors**
  - MOST as a resistor
  - MOST as an amplifier in strong inversion
  - Transition weak inversion-strong inversion
  - Transition strong inversion-velocity saturation
  - Capacitances and f_T
- **Models of Bipolar transistors**
- **Comparison of MOSTs & Bipolar transistors**

Willy Sansen 10-05 0158

0158

At high frequencies, capacitances play a role.

First, we identify all possible capacitances present in MOST devices. We then take them together into terminal capacitances.

Finally, we have a look at the parameter f_T which is a kind of a high frequency limit beyond which a MOST cannot provide gain. Again we will try to find an expression of f_T in all three regions, weak inversion, strong inversion and velocity saturation.

0159

The most important capacitances are shown on this cross-section.

The Gate oxide capacitance C_{oxt} is the most important one as it controls the transistor current.

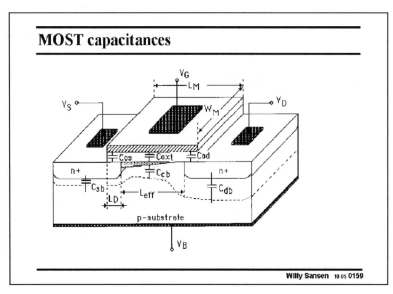

MOST capacitances

Willy Sansen 10-05 0159

The total oxide capacitance C_{oxt} equals about WL C_{ox}. It overlaps both the Source and the Drain and causes overlap capacitances Cos and Cod respectively.

The channel has a depletion layer (or junction) capacitance to the substrate C_{cb}. It now depends on the voltage across it.

Both Source and Drain regions have depletion capacitances C_{sb} and C_{db} to the substrate as well.

All capacitances are now described in detail.

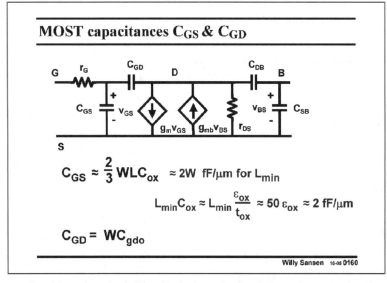

MOST capacitances C_{GS} & C_{GD}

$$C_{GS} \approx \frac{2}{3} WLC_{ox} \approx 2W \text{ fF/}\mu\text{m for } L_{min}$$

$$L_{min}C_{ox} \approx L_{min}\frac{\varepsilon_{ox}}{t_{ox}} \approx 50\,\varepsilon_{ox} \approx 2 \text{ fF/}\mu\text{m}$$

$$C_{GD} = WC_{gdo}$$

Willy Sansen 10-05 0160

0160

The Gate-Source capacitance C_{GS} includes the oxide capacitance C_{oxt} and the Gate-Source overlap capacitance C_{gso}. It is usually taken to be only the C_{oxt} itself, which is a good average value as it is somewhat overestimated but also somewhat underestimated.

It is overestimated because the Gate-Source capacitance C_{GS} is actually only about 2/3 of C_{oxt} [Ref. Tsividis]. Indeed the channel has vanished at the Drain side. Electric-field calculations have shown that a reduction of the C_{oxt} by a factor 2/3 is about right.

The Gate-Source capacitance C_{GS} is also underestimated because it must include the Gate-Source overlap capacitance C_{gso}. This is normally only a fraction (20–25%) of C_{oxt}.

Taking just $C_{oxt} = $ WL C_{ox} is thus a good estimate. It is clear that CAD models provide more accurate values for these capacitances. These simple ones are good enough for hand calculations.

If a minimum channel length L_{min} is used, then the C_{GS} can easily be estimated to be only 2W, in fF, in which W is in micrometer. Indeed the oxide thickness t_{ox} of standard CMOS processes is very close to L/50. The expression of C_{GS} can thus be simplified to 2W.

This shows that the input capacitance C_{GS} of a MOST only depends on its width, at least if the minimum channel length is always used.

This rule of thumb will be used quite often to calculate parasitic capacitances in opamps (see Chapter 6).

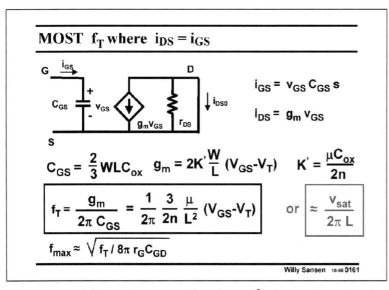

MOST f_T where $i_{DS} = i_{GS}$

$i_{GS} = v_{GS}\, C_{GS}\, s$

$i_{DS} = g_m\, v_{GS}$

$C_{GS} = \frac{2}{3} W L C_{ox}$ $g_m = 2K' \frac{W}{L} (V_{GS} - V_T)$ $K' = \frac{\mu C_{ox}}{2n}$

$f_T = \frac{g_m}{2\pi\, C_{GS}} = \frac{1}{2\pi} \frac{3}{2n} \frac{\mu}{L^2} (V_{GS} - V_T)$ or $\approx \frac{v_{sat}}{2\pi\, L}$

$f_{max} \approx \sqrt{f_T / 8\pi\, r_G C_{GD}}$

Willy Sansen   10-05 0161

0161
Parameter f_T is the frequency for which the output current i_{DS}, though a short, equals the input current i_{GS}. It is determined by the time constant of C_{GS} and the transconductance g_m.

This parameter is a kind of Figure-of-merit for high frequency performance, taking into account only the intrinsic transistor parameters.

Substitution of C_{GS} and g_m in f_T shows that this frequency is proportional to $V_{GS} - V_T$ and inversely proportional to L^2. Decreasing the channel length allows an higher frequency performance indeed.

In velocity saturation however, the time that the electrons need to cross the channel length is L/v_{sat}. The frequency f_T in velocity saturation is, to put it simply $v_{sat}/2\pi L$. This is the highest frequency that can be obtained with a MOST. It is obviously obtained for high values of $V_{GS} - V_T$. It does not increase with smaller L^2 however, but only with L.

Another figure-of-merit exists which includes some more extrinsic parameters. It is the maximum frequency of oscillation f_{max}. It is obviously related to f_T but includes parameters such as the Gate series resistance r_G and the Gate-Drain capacitance C_{GD}. It can be higher or lower than f_T.

Design for high speed :

	High gain	High speed
$V_{GS}\text{-}V_T$	Low (0.2 V)	High (0.5 V)
L	High	Low

$V_{GS}\text{-}V_T$ sets the ratio g_m/I_{DS} !

Willy Sansen   10-05 0162

0162
It is striking that for high-frequency design, a large $V_{GS} - V_T$ is required and the lowest possible value of channel length L. This is exactly the opposite of what is required for high gain (and later low noise and offset).

This is probably one of the most basic compromises in analog CMOS design. High-gain devices such as at the input of operational amplifiers, have to be designed for small $V_{GS} - V_T$

and for large channel length L. For high frequency designs, such as VCO's and LNA's, it is exactly the opposite.

Remember that setting the values of $V_{GS} - V_T$ is the same as setting the ratio of g_m/I_{DS} or the same as setting the inversion coefficient $i = I_{DS}/I_{DSt}$.

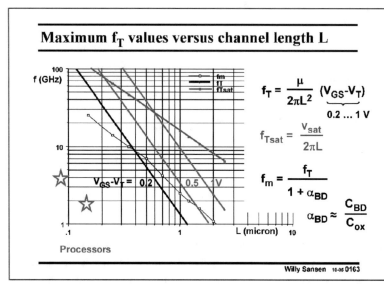

0163

For a specific $V_{GS} - V_T$, the frequencies f_T are easily plotted versus channel length L, as shown in this slide. The one for velocity saturation is independent of $V_{GS} - V_T$ but has a lower slope.

The frequencies f_T for saturation is in blue; the ones in strong inversion are for $V_{GST} = 0.2$ V in black, for $V_{GST} = 0.5$ V in red, and for $V_{GST} = 1$ V in magenta.

It is striking to see that f_T values in this slide 100 GHz are readily available, at least for CMOS processes with channel lengths below 0.13 μm. Note also that from this technology, velocity saturation dominates even for $V_{GS} - V_T$ values of only 0.2 V.

For $V_{GST} = 0.2$ V, the crossover is somewhere below 90 nm channel length. For high-frequency applications however, in which $V_{GST} = 0.5$ V or higher, the crossover value in L is about 0.25 μm.

For hand calculations, we need a model for f_T that encompasses both the strong-inversion region and the velocity-saturation region. This is given next.

Note, however, that some experimental upper frequencies are added of VCO's and LNA's. They are on a curve labeled f_{max}, which are about 1/5 of f_T. Also some clock frequencies are added for some present-day microprocessors. They are at about 1/100 of f_T!

0164

A simple model for f_T which spans both regions of operation, the strong-inversion and the velocity-saturation region, is simply obtained by substitution of the transconductance g_m by its expression covering both regions. Also C_{GS} can be substituted by 2W.

The resulting expression is a useful tool to predict high-frequency performance. It will be used in many later Chapters to establish upper limits of frequency performance.

Obviously, the transition values of channel length L, for specific values of $V_{GS} - V_T$ are the same as for the transconductance g_m. For $V_{GS} - V_T = 0.2$ V, channel lengths below 65 nm provide operation fully in velocity saturation.

f_T model in si and velocity saturation

$$f_T = \frac{g_m}{2\pi\, C_{GS}} \qquad C_{GS} = kW \qquad k = 2\ fF/\mu m = 2\ 10^{-11}\ F/cm$$

$$g_m = \frac{W}{L}\ \frac{17\ 10^{-5}}{1 + 2.8\ 10^4\ L/V_{GST}} \qquad L\ in\ cm$$

$$f_T = \frac{1}{L}\ \frac{13.5}{1 + 2.8\ L/V_{GST}}\ GHz \qquad L\ in\ \mu m$$

If $V_{GST} = 0.2\ V$, v_{sat} takes over for $L < 65\ nm$
If $V_{GST} = 0.5\ V$ for $L < 0.15\ \mu m$

Willy Sansen 10-05 0164

---

f_T model in si and weak inversion

$$f_T = \frac{g_m}{2\pi\, C_{GS}}$$

$$GM = \frac{g_m}{I_{DS}}\ \frac{nkT}{q} = \frac{1 - e^{-\sqrt{i}}}{\sqrt{i}}$$

$$g_m = \frac{I_{DS}}{nkT/q}\ \frac{1 - e^{-\sqrt{i}}}{\sqrt{i}} \qquad but\ I_{DS} = i\ I_{DSt}$$

$$g_m = \frac{I_{DSt}}{nkT/q}\ \sqrt{i}\ (1 - e^{-\sqrt{i}})$$

$$\frac{f_T}{f_{TH}} = \sqrt{i}\ (1 - e^{-\sqrt{i}})$$

$$f_{TH} = \frac{I_{DSt}}{2\pi\, C_{GS}\, nkT/q} = \frac{K'\, V_{GSTt}^2\, W/L}{2\pi\, WL\, C_{ox}\, nkT/q}$$

$$\approx i\ for\ small\ i\ !$$

$$= \frac{4\, K'\, nkT/q}{2\pi\, C_{ox}\, L^2} = \frac{2\, \mu\, kT/q}{2\pi\, L^2}$$

Willy Sansen 10-05 0165

0165

At low currents, the absolute values of the transconductance g_m are low, yielding low values for frequency f_T as well. The question is, what absolute values of f_T can actually be obtained for MOSTs in weak inversion?

A simple model for f_T which spans the weak-inversion and the strong-inversion region, is again obtained by substitution of the transconductance g_m by its expression covering both regions.

At high currents (for which this model is actually not valid as the effect of velocity saturation is not included), the highest value of f_T is reached, which is $\sqrt{i}\ f_{TH}$. The lower the current (or the inversion coefficient), the lower f_T is, as plotted versus inversion coefficient.

Indeed for very small values of i, f_T equals about i f_{TH}.

Note however, that reference frequency f_{TH} increases with L^{-2} and can now reach very attractive values.

0166

Both parameters GM (which is g_m/I_{DS} normalized to nkT/q) and f_T (normalized to f_{TH}) are shown versus inversion coefficient i. For weak inversion the g_m/I_{DS} ratio is high but the f_T is low.

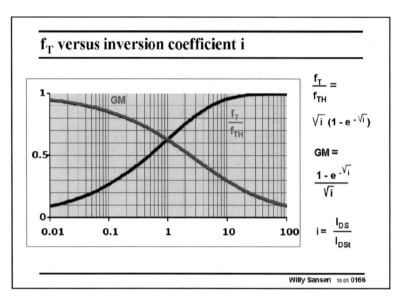

f_T versus inversion coefficient i

$$\frac{f_T}{f_{TH}} = \sqrt{i}\,(1 - e^{-\sqrt{i}})$$

$$GM = \frac{1 - e^{-\sqrt{i}}}{\sqrt{i}}$$

$$i = \frac{I_{DS}}{I_{DSt}}$$

Willy Sansen 10-05 0166

The gain-speed compromise is very well illustrated by this curve!

However, for deep weak-inversion ($i = 0.01$) the f_T value has dropped to only about 10%. If a transistor is taken with 130 nm channel length, with an f_T of about 100 GHz, its resulting f_T is still about 10 GHz!

In other words, very reasonable high-frequency performance is possible in weak inversion, provided nanometer CMOS is available.

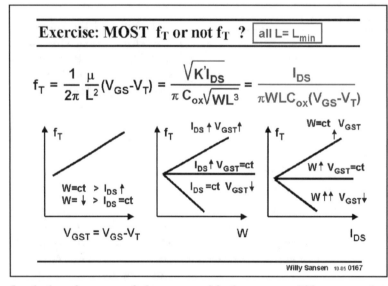

Exercise: MOST f_T or not f_T ? $\boxed{\text{all } L = L_{min}}$

$$f_T = \frac{1}{2\pi}\frac{\mu}{L^2}(V_{GS} - V_T) = \frac{\sqrt{K'I_{DS}}}{\pi\, C_{ox}\sqrt{WL^3}} = \frac{I_{DS}}{\pi W L C_{ox}(V_{GS} - V_T)}$$

$W = ct > I_{DS}\uparrow$
$W = \downarrow > I_{DS} = ct$
$V_{GST} = V_{GS} - V_T$

$I_{DS}\uparrow V_{GST}\uparrow$
$I_{DS}\uparrow V_{GST} = ct$
$I_{DS} = ct \; V_{GST}\downarrow$
W

$W = ct \; V_{GST}$
$W\uparrow V_{GST} = ct$
$W\uparrow\uparrow V_{GST}\downarrow$
I_{DS}

Willy Sansen 10-05 0167

0167

As an exercise, let us try to find out whether the frequency f_T increases or decreases with current.

Substitution of the $V_{GS} - V_T$ brings in the width W and length L of the transistor. Various different expressions can now be obtained. The most important one is probably the last one, which shows that an answer to this question can only be obtained, provided one more parameter is fixed.

Indeed, if $V_{GS} - V_T$ is fixed, then frequency f_T increases with the current. If however, the $V_{GS} - V_T$ is decreased accordingly, then the f_T does not change with the current. It is even possible to decrease the $V_{GS} - V_T$ in such a way that frequency f_T decreases with increasing current.

This conclusion has been reached before. In a MOST transistor it is never possible to find out how one parameter changes with respect to one other one. Another parameter always has to be fixed. Indeed, a MOST has two degrees of freedom. Normally $V_{GS} - V_T$ (or inversion coefficient) and L are chosen by the designer, setting all other parameters.

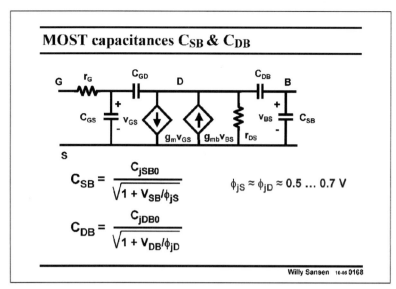

MOST capacitances C_SB & C_DB

$$C_{SB} = \frac{C_{jSB0}}{\sqrt{1 + V_{SB}/\phi_{jS}}}$$

$$C_{DB} = \frac{C_{jDB0}}{\sqrt{1 + V_{DB}/\phi_{jD}}}$$

$$\phi_{jS} \approx \phi_{jD} \approx 0.5 \dots 0.7 \text{ V}$$

Willy Sansen 10-05 0168

0168

Both islands, the Source and the Drain, have a junction capacitance with respect to the bulk (or substrate). They can easily be added to the small-signal diagram as shown in this slide.

Junction capacitances normally show a voltage dependence described by a square root, when they are reverse biased.

The most important one is doubtless the Drain-Bulk capacitance C_{DB} as it appears at the output of the MOST amplifier. The Source-Bulk capacitance C_{SB} is normally shorted out.

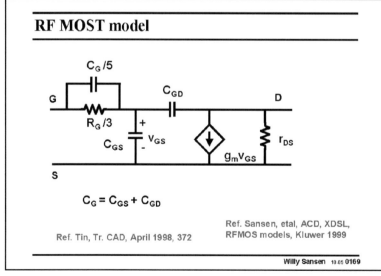

RF MOST model

$$C_G = C_{GS} + C_{GD}$$

Ref. Tin, Tr. CAD, April 1998, 372

Ref. Sansen, etal, ACD, XDSL, RFMOS models, Kluwer 1999

Willy Sansen 10-05 0169

0169

At very high frequencies, for example beyond $f_T/3$, the input impedance cannot be modeled by a simple Gate resistor R_G and input capacitances C_{GS} and C_{GD}. It can be corrected by a small capacitance shunting resistor R_G. This capacitance is normally obtained by fitting the measured input impedance, obtained from s-parameters, to the model. This usually leads to a reduction of the Gate resistor R_G (here to 1/3) and a shunt capacitor, which is a fraction (here 1/5) of the input capacitance.

These are obviously fitting measurements. They can vary from designer to designer depending on the actual frequency range used. Remember that this additional shunt capacitance is only required for the highest frequency range. Only designs of high-frequency VCO's and LNA's may require this.

Single-page MOST model

$$I_{DS} = K'_n \frac{W}{L} (V_{GS} - V_T)^2 \qquad V_{GS} - V_T \approx 0.2\ V \qquad \begin{array}{l} K'_n \approx 100\ \mu A/V^2 \\ K'_p \approx 40\ \mu A/V^2 \end{array}$$

$$g_m = 2K'_n \frac{W}{L} (V_{GS} - V_T) = 2 \sqrt{K'_n \frac{W}{L} I_{DS}} = \frac{2\ I_{DS}}{V_{GS} - V_T}$$

$$r_{DS} = r_o = \frac{V_E L}{I_{DS}} \qquad V_{En} \approx 5\ V/\mu mL \quad V_{Ep} \approx 8\ V/\mu mL$$

$$v_{sat} = 10^7\ cm/s$$

$$f_T = \frac{1}{2\pi} \frac{3}{2n} \frac{\mu}{L^2} (V_{GS} - V_T) \quad \text{or now} \approx \frac{v_{sat}}{2\pi\ L}$$

Willy Sansen 10-05 0170

0170

As a conclusion, a summary is given of the model parameters which will be used throughout this text. As they fit on one page, they are called the single-page MOST model.

It is suggested to know these simple expressions by heart. This will not be very difficult as they will be used many times.

Growing number of parameters !

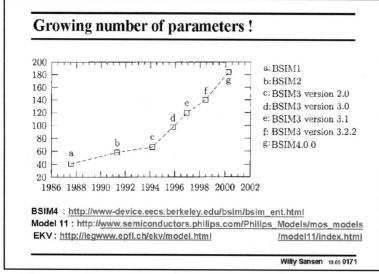

a: BSIM1
b: BSIM2
c: BSIM3 version 2.0
d: BSIM3 version 3.0
e: BSIM3 version 3.1
f: BSIM3 version 3.2.2
g: BSIM4.0.0

BSIM4 : http://www-device.eecs.berkeley.edu/bsim/bsim_ent.html
Model 11 : http://www.semiconductors.philips.com/Philips_Models/mos_models
EKV : http://legwww.epfl.ch/ekv/model.html   /model11/index.html

Willy Sansen 10-05 0171

0171

Several more elaborate and precise models are being used. The three most used ones are probably the BSIM models, the Philips models and the EKV models. References are given on all of them.

It is striking however, that the number of model parameters seems to increase all the time. A curve is given for the BSIM model. This is a result of the addition of corrections on a corrected parameter, starting with a simple model in strong inversion. As a result, the number of parameters explodes.

A new approach has been taken in the EKV model, where a model in strong and weak inversion is introduced with less model parameters. The total number of parameters is then less as well.

Ultimately, it is the standardization of the model that is of primordial importance. The model that is most used is the most important one, whatever the quality is of that model.

0172

If a new model is being used, it is wise to check a few weaknesses which used to be around in older models.

A short list is given in the slide.

Benchmark tests

1. Weak inversion transition for I_{DS} and g_m/I_{DS} ratio
2. Velocity saturation transition for I_{DS} and g_m/I_{DS} ratio
3. Output conductance around V_{DSsat}
4. Continuity of currents and caps around zero V_{DS}
5. Thermal and 1/f noise
6. High frequency input impedance (s_{11}) and

   transimpedance (s_{21})

Willy Sansen 10-05 0172

The two transition points on either side of the strong-inversion region must be checked. This means that we have to have a look at the continuity of the current itself and its first derivative. Also the second and third derivative must be continuous if we are interested in distortion (see Chapter 18).

Continuity must also be ensured of the current and its first derivative at all transition points for the voltage applied. They are

– transition between linear and saturation region (for increasing v_{DS})
– transition around zero v_{DS}

The models must also be verified if noise models are used.

Finally, for high-frequency design, the input impedance (derived from parameter s_{11}) and the gain (from s_{21}) had better be verified versus frequency. Of course the other two s parameters can be added as well.

Table of contents

- Models of MOST **transistors**
- • Models of Bipolar transistors
- Comparison of MOSTs & Bipolar transistors

Willy Sansen 10-05 0173

0173

Considerable attention has been paid up till now to models of MOST devices, despite the fact that the bipolar transistors are historically first. Bipolar transistors are easily found nowadays in BiCMOS processes in which MOST devices and bipolar transistors are integrated together.

We will have a look at models for bipolar transistors first, and carry out a comparison in performance afterwards.

0174

A bipolar transistor consists of two diodes back to back, very close together. The first one is forward biased, whereas the other one reverse biased. It is a device which is very different from

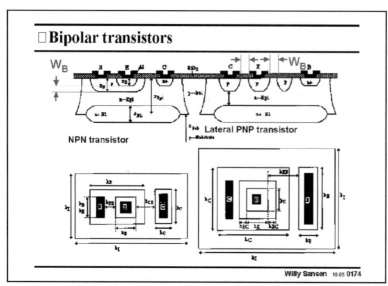

a MOST as it is actually two diodes, rather than a resistor.

The two diodes are easily recognized in the vertical npn transistor on the left. In this structure, a p-region is diffused in the n-epitaxial layer (called Collector), and forms the Base, with depth x_B. A n-region is then diffused in the Base to form the Emitter, with depth x_E. The difference in depth is then the actual base width W_B.

Another n-region is diffused in the Collector n-epitaxial region to make good contact. Moreover, a buried layer is added underneath the transistor to lower the series collector resistance. Quite often more diffusions are added to lower, for example, the Base resistance.

The transistor on the left is a vertical npn transistor. The one on the right however is a vertical pnp transistor. The Base width is now the horizontal spacing between the two p-islands being the Emitter and the Collector, in which the latter one normally surrounds the Emitter in the middle. This is clearly visible in the layout. It is not used very often because of its low current gain. Vertical pnp's are much better.

For both transistors, the substrate is always reverse biased. If only one single positive supply voltage is available, the substrate is connected to ground.

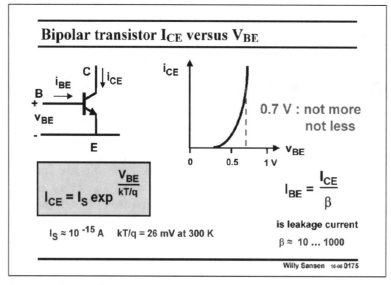

0175

When the Base-Emitter diode is forward biased, a large number of electrons flow from the n-side (Emitter) to the p-Base. Because the Base width W_B is so small, most electrons are collected by the Collector. Actually they diffuse from Emitter to Collector.

The resulting Collector current is nearly the same as the Emitter current. It is denoted by I_{CE} and it is given as a function of the Base-Emitter voltage V_{BE} by a similar exponential as for a forward-biased diode. Indeed the V_{BE} is also scaled by kT/q in which k is the Boltzmann constant and q the charge of an electron

$(1.6 \times 10^{-19}$ C). Parameter T is the absolute temperature (in Kelvin or in degrees Celsius plus 273). At 20°C (or 86°F) kT/q is 25.86 mV. We will usually take 26 mV as the actual operating temperature may be higher. In power application this temperature may be a lot higher!

The beauty of this scaling constant kT/q is that it is not changing from one technology to the next. It only contains fundamental physical parameters and the absolute temperature. This is one of the major advantages of the bipolar transistor. The second advantage of a bipolar transistor is that its characteristic is an exponential, which is mathematically the steepest curve around. Its derivative or the transconductance will be higher than for any other semiconductor device.

Its main disadvantage however is that a Base current is flowing, which is a fraction $1/\beta$ of the collector current, and which is never well known.

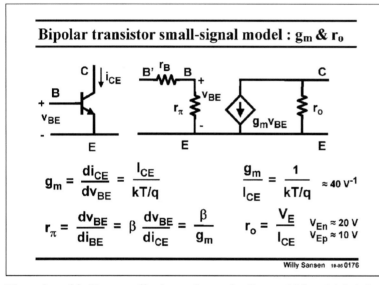

Bipolar transistor small-signal model : g_m & r_o

$$g_m = \frac{di_{CE}}{dv_{BE}} = \frac{I_{CE}}{kT/q}$$

$$\frac{g_m}{I_{CE}} = \frac{1}{kT/q} \approx 40 \text{ V}^{-1}$$

$$r_\pi = \frac{dv_{BE}}{di_{BE}} = \beta \frac{dv_{BE}}{di_{CE}} = \frac{\beta}{g_m}$$

$$r_o = \frac{V_E}{I_{CE}} \quad V_{En} \approx 20 \text{ V} \quad V_{Ep} \approx 10 \text{ V}$$

Willy Sansen 10-05 0176

0176

The small-signal model of a bipolar transistor is now easily derived.

The transconductance g_m is the derivative of the $I_{CE} - V_{BE}$ curve. The g_m/I_{CE} ratio is simply one over kT/q. This is close to 40 per V, which is a lot larger than for a MOST. This is clearly one of the advantages of a bipolar transistor.

The bipolar transistor also has a finite output resistance r_o. It can be modeled by an Early Voltage V_E.

Note that this V_E actually depends on the Base width, which is however, not a degree of freedom in a vertical npn, only in a lateral pnp as in MOST design. This value of V_E is given by the technology description. It decreases for narrower Base widths.

In order to accommodate the small-signal Base current, a resistor is added at the input. It is called r_π. It is the derivative of the $I_{BE} - V_{BE}$ curve. As a result it contains both the transconductance g_m and current gain factor β. As this factor β is never accurately known, this resistor r_π is not well known either. It certainly depends on the current. For a current of 0.1 mA and a β of about 100, this resistor is about 26 kΩ. This is a rather low value and imprecise. It cannot be allowed to play an important role in high-precision circuitry.

Finally, a Base resistor r_B has to be added in series with the input. It is an ohmic resistance which is mainly the resistance of p-Base region between the Emitter side and the Base contact.

0177

As a bipolar transistor is built up with many junctions, it contains many junction capacitances. The most important ones are the Base-Emitter junction capacitance C_{jBE} and the Collector-Substrate junction C_{CS}.

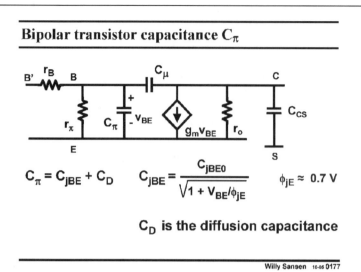

Bipolar transistor capacitance C_π

$$C_\pi = C_{jBE} + C_D \qquad C_{jBE} = \frac{C_{jBE0}}{\sqrt{1 + V_{BE}/\phi_{jE}}} \qquad \phi_{jE} \approx 0.7 \text{ V}$$

C_D is the diffusion capacitance

Willy Sansen   10-05 0177

The Collector-Base junction capacitance is called C_μ.

However, the total input capacitance C_π, also includes the diffusion capacitance C_D. Its origin is explained next.

Note however, that the Base-Emitter junction is forward biased. The value of the Base-Emitter junction capacitance C_{jBE} is thus larger than its value at zero Volt, by a factor of 2 to 3.

0178

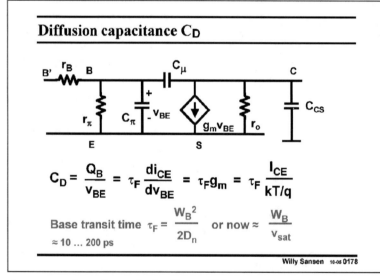

Diffusion capacitance C_D

$$C_D = \frac{Q_B}{v_{BE}} = \tau_F \frac{di_{CE}}{dv_{BE}} = \tau_F g_m = \tau_F \frac{I_{CE}}{kT/q}$$

Base transit time $\tau_F = \dfrac{W_B^2}{2D_n}$ or now $\approx \dfrac{W_B}{v_{sat}}$

$\approx 10 \ldots 200$ ps

Willy Sansen   10-05 0178

The diffusion capacitance C_D models the transition of the electrons flowing through the Base region, as a result of the Base-Emitter voltage V_{BE}. Indeed these electrons are minority carriers in the p-Base region. Their passing through the Base, represents a charge Q_B in the base, which constantly disappears toward the Collector but which is also constantly replenished from the Emitter. These electrons take an average time τ_F to flow through the Base. This time is called the Base transit time. This time constant τ_F allows this diffusion capacitance C_D to be described by $\tau_F g_m$, which obviously depends on the current I_{CE}.

The larger the current, the more C_D is dominant in C_π. On the other hand, for small currents, C_π consists mainly of the junction capacitance Q_{jBE}.

This Base transit time τ_F strongly depends on the Base width W_B, as shown in this slide. Parameter D is the diffusion constant of the electrons in the p-Base region. In high-speed devices however, all electrons move at maximum speed, which is the speed of velocity saturation v_{sat} (10^7 cm/s), very much as in a high-speed MOST. The time needed by the electrons to flow through the Base is then easily calculated. It is in the order of ps. For example for a W_B of 0.2 μm, τ_F is a mere 2 ps!!!

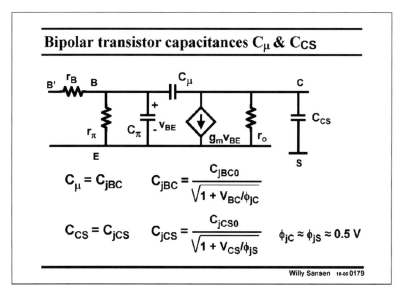

The other capacitances are all junction capacitances indeed, all reverse biased.

The Collector-Base junction capacitance is called C_μ. It is usually quite small.

The Collector-Substrate junction C_{CS} on the other hand, is usually quite large as the whole transistor structure is embedded in its collector tub. Moreover, it is connected at the Collector, which usually functions at the output of the transistor. It plays an important role once high-frequency performance is envisaged.

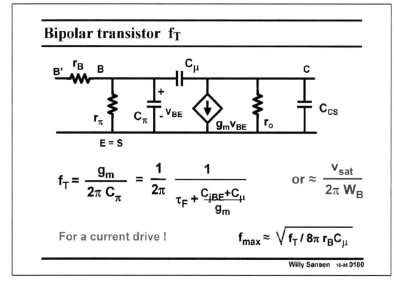

In the same way as for a MOST, parameter f_T represents the high-frequency capability of a bipolar transistor. It is the frequency where the current gain is unity, in an output short. It is also the frequency of the time constant consisting of $1/g_m$ and input capacitance C_π.

Substitution of these two parameters yields the expression given in this slide.

It clearly shows that f_T heavily depends on the current. The maximum value of f_T is reached for high currents. This is hardly a surprising result indeed!

This maximum value is simply linked to the Base transit time τ_F. For example, for a τ_F of 2 ps, f_T becomes 160 GHz, which is quite high indeed!

At lower currents, the two junction capacitances take over. The values of f_T are then smaller. They are plotted next.

Note that f_{max} includes again the parameters r_B and C_μ, as for a MOST.

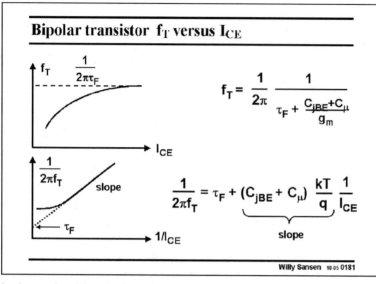

0181

The values of f_T are plotted versus current I_{CE} twice.

In the top diagram, both scales are linear. It is clear that the maximum value of f_T is reached for high currents. Its value is then determined by the Base transit time τ_F.

It is more instructive however to plot the inverse of f_T versus the inverse of the current. The extrapolated value at zero $1/I_{CE}$ is then again τ_F, but the slope of the curve, which is now linear, is determined by the junction capacitances C_{jBE} and C_μ.

Remember that capacitance C_{jBE} is taken in forward bias. This is one of the few ways to measure this capacitance in forward bias!

Single-page Bipolar transistor model

$$I_{CE} = I_S \exp^{\dfrac{V_{BE}}{kT/q}}$$

$I_S \approx 10^{-15}\,A \quad kT/q = 26\ mV\ at\ 300\ K$

$$g_m = \dfrac{I_{CE}}{kT/q} \qquad r_o = \dfrac{V_E}{I_{CE}}$$

$V_{En} \approx 20\ V \quad V_{Ep} \approx 10\ V$

$$f_T = \dfrac{1}{2\pi} \dfrac{1}{\tau_F + \dfrac{C_{je}+C_{jc}}{g_m}} \qquad or \approx \dfrac{v_{sat}}{2\pi\ W_B}$$

0182

In the same way as for a MOST, all important expressions which are required to be known by heart, are collected under the name "Single-page bipolar model".

They will be used throughout this text. It will now be quite easy to memorize them.

0183

Now that the models of both the MOST and the bipolar transistor have been explained in simple terms, the question arises how do they compare for circuit performance.

Indeed, both of them are available in BiCMOS processes. For each function, a choice can be made between both types.

Table of contents

- **Models of MOST transistors**
- **Models of Bipolar transistors**
- **Comparison of MOSTs and Bipolars**

Willy Sansen 10-05 0183

0184

In order to organize the comparison, a Table is used listing most important parameters, under "Specification". The relevant arguments are then collected in two columns, one for MOSTs and one for bipolar transistors.

It is clear that the zero input current is a major advantage for a MOST. As a result, its input impedance is infinity. It is now possible to store a charge on a capacitance and read it with

---

Comparison MOST - Bipolar

TABLE 2-8 COMPARISON OF MOSTS AND BIPOLAR TRANSISTORS

	Specification		MOST	Bipolar transistor	
1.	I_{IN}		0	I_C/β	β ?
	R_{IN}		∞	$r_\pi + r_B$	
2.	V_{DSsat}		$V_{GS} - V_T = \sqrt{\dfrac{I_{DS}}{K'W/L}}$	few kT/q	
3.	$\dfrac{g_m}{I}$	wi	$\dfrac{1}{nkT/q}$	$\dfrac{1}{kT/q}$	$n = 1 + \dfrac{C_D}{C_{ox}}$
		si	$\dfrac{2}{V_{GS} - V_T}$	$\dfrac{1}{kT/q}$	$4\dots 6$ x
		vs	$\dfrac{1}{V_{GS} - V_T}$	$\dfrac{1}{kT/q}$	

Ref. Laker Sansen Table 2-8

Willy Sansen 10-05 0184

---

a nMOST. This is used in switched-capacitor filters (see Chapter 17) but also in Sample-and-hold circuits in front of ADC's.

In future nanometer technologies however, Gate current may flow in a similar way as for bipolar transistors. This is an enormous disadvantage for many circuits!

The second consideration is on the minimum V_{DSsat}. It is the minimum output voltage for which the MOST operates in saturation, exhibiting a large output resistance r_o, and thus large gain.

0185

The minimum V_{DSsat}, for which the MOST operates in saturation, is little less than $V_{GS} - V_T$, which also sets the current depending on the transistor dimensions W and L.

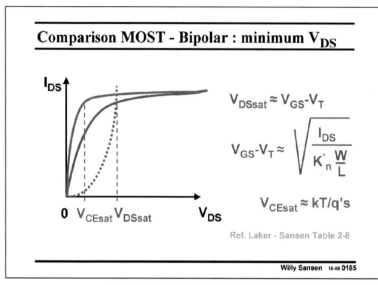

Comparison MOST - Bipolar : minimum V_{DS}

$$V_{DSsat} \approx V_{GS} - V_T$$

$$V_{GS} - V_T \approx \sqrt{\frac{I_{DS}}{K'_n \frac{W}{L}}}$$

$$V_{CEsat} \approx kT/q's$$

Ref. Laker - Sansen Table 2-8

Willy Sansen   10-05 **0185**

For high-gain stages the $V_{GS} - V_T$ is chosen to be small, such as 0.15–0.2 V. The voltage loss V_{DSsat} is also small.

For high-speed stages on the other hand, this $V_{GS} - V_T$ may be as high as 0.5 V. The Drain-Source voltage can never be less than 0.5 V without losing gain. This creates a severe limitation in output swing.

Bipolar transistors always exhibit a minimum output voltage of a few times kT/q. This is more of the order of

magnitude of 0.1 V, whatever purpose the transistor is used for, high gain of high speed. This is clearly an advantage for the bipolar transistor. It is still the favorite device for very low supply voltages.

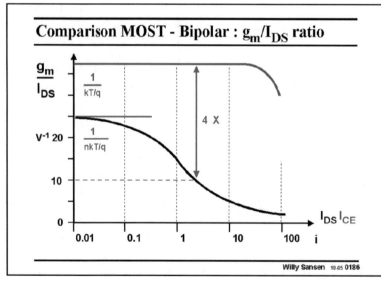

Comparison MOST - Bipolar : g_m/I_{DS} ratio

Willy Sansen   10-05 **0186**

0186

The g_m/I_{DS} ratio is also better with a bipolar transistor. Indeed, this ratio continues to decrease for a MOST, at increasing current (or inversion factor i). For a bipolar transistor this ratio is larger to start with, even at very low currents, but also remains large for intermediate currents. For a $V_{GS} - V_T$ of 0.15–0.2 V, the g_m/I_{DS} ratio is around a factor of 4 better for a bipolar transistor than for a MOST. For the same trans-

conductance, a bipolar transistor only requires 25% of the current of a MOST. This is a considerable advantage for portable applications, where battery power is expensive.

0187

The difference in design plan is a fourth consideration.

A MOST has two expressions for the current I_{DS} and the transconductance g_m. However, they contain five variables (or design parameters). They are I_{DS}, g_m and $V_{GS} - V_T$, W and L.

Most often the transconductance g_m is imposed as a result of circuit specifications. This leaves

Design plan for g_m :

$$I_{DS} = K'_n \frac{W}{L} (V_{GS} - V_T)^2$$

$$g_m = 2K'_n \frac{W}{L} (V_{GS} - V_T) = 2\sqrt{K'_n \frac{W}{L} I_{DS}} = \frac{2 I_{DS}}{V_{GS} - V_T}$$

4 variables with 2 equations >> 2 free variables

Choose $V_{GS} - V_T$ and L !

□

Willy Sansen 10-05 0187

us with 4 variables. The ones with the smallest ranges are better chosen up front. They are $V_{GS} - V_T$ and L.

The first choice to be made is the $V_{GS} - V_T$ as it sets the operating point of the transistor. It fixes the g_m/I_{DS} ratio and the inversion and velocity coefficients, all at the same time. It is clearly the first choice to be made. In general, for high-gain stages the $V_{GS} - V_T$ is chosen between 0.15 and 0.2 V, for high-speed stages the $V_{GS} - V_T$ is chosen around 0.5 V.

Once g_m and $V_{GS} - V_T$ are known, the I_{DS} is known as well, from which W/L is readily calculated. From the choice of L, W is calculated. In general, for high-gain stages the L is chosen 4–8 times the minimum L, for high-speed stages the L is chosen minimal.

Comparison MOST - Bipolar

4.	Design planning		$L, V_{GS} - V_T$	kT/q
5.	*I*-range		1 decade	7 decades
6.	Max f_T	low *I*	C_{GS}, C_{GD}	C_{jEt}, C_μ
		high *I*	v_{sat}/L_{eff}	v_{sat}/W_B
7.	Noise $\overline{dv_i{}^2}$	Therm.	$4kT\left(\frac{2/3}{g_m} + R_G\right)$	$4kT\left(\frac{1/2}{g_m} + R_B\right)$
		1/*f*	10×	
	Offset		10×	

$v_{sat} \approx 10^7$ cm/s

Ref. Laker Sansen Table 2-8

Willy Sansen 10-05 0188

0188

After the difference in design plan (number 4), we have a look at the accuracy of the models (number 5).

A MOST needs three models to cover the whole current range, and many transistor parameters, which is expanding continuously. A bipolar transistor only needs one single model, which is quite precise. Moreover, it has had the same model for years. This single model is valid over a large number (more than 7), of decades in current, contrary to a MOST, where one model is valid over only 1 or 2 decades in current.

Speed is another point of comparison. At high currents all electrons move at the speed of v_{sat}. As a result the highest speed device is the one with the smaller dimension the channel length L or the Base width W_B. The channel length is being decreased continuously, which is not the case for the Base width. Nanometer CMOS technologies are capable of a very-high-frequency performance indeed!

Noise is next. The expressions of the thermal noise are nearly the same for both transistor types. The factor 2/3 is close to 1/2 indeed. However, for the same current the g_m of a bipolar transistor is larger and hence its thermal noise is lower.

This is very different for 1/f noise. A bipolar transistor is a bulk device. Its 1/f equivalent input noise voltage is smaller by an order of magnitude and so is its offset voltage.

Table of contents

- **Models of MOST transistors**

- **Models of Bipolar transistors**

- **Comparison of MOSTs and Bipolars**

Ref.: W. Sansen : Analog Design Essentials, Springer 2006

Willy Sansen 10-05 **0189**

0189

This concludes the comparison between MOSTs and bipolar transistors. It is clear that a bipolar transistor is better for high-precision circuitry, but a MOST is more compatible with CMOS logic. A BiCMOS process seems to be the ideal compromise, but is more expensive unfortunately.

Now that models for hand calculations have been derived, they will be used in elementary circuits.

Reference books on Transistor models

T. Fjeldly, T. Ytterdal, M. Shur, "Introduction to Device Modeling and Circuit Simulation", Wiley 1998.

D. Foty, "MOSFET Modeling with SPICE, Prentice Hall

K. Laker, W.Sansen, "Design of Analog Integrated Circuits andSystems", MacGrawHill. NY., Febr.1994.

A. Sedra, K.Smith, "Microelectronic Circuits", CBS College Publishing, 2004.

Y. Taur, T. Ning, "Fundamentals of Modern VLSI Devices" Cambridge Univ. Press, 1998.

Y. Tsividis, "Operation and modeling of the MOS transistor", McGraw-Hill, 2004.

A. Vladimirescu "The SPICE book", Wiley, 1994

Willy Sansen 10-05 **0190**

0190

Several books are available which provide many more details on MOST models. The best known one is undoubtedly the one of Y. Tsividis.

Some other books concentrate on the models as they are implemented in SPICE.

References on Analog Design

P.Allen, D.Holberg, "CMOS Analog Circuit Design", Holt, Rinehart and Winston. 1987, Oxford Press 2002

P.Gray, P.Hurst, S. Lewis, R.Meyer, "Analysis and Design of Analog Integrated Circuits", Wiley, 1977/84/93/01

R.Gregorian, G.Temes, "Analog MOS Int. Circuits for Signal Processing", Wiley, 1986.

Huijsing, Van de Plassche, Sansen, "Analog Circuit Design", Kluwer Ac.Publ. 1993/4/5....

D.Johns, K.Martin, "Analog integrated circuit design", Wiley 1997.

K.Laker, W.Sansen, "Design of Analog Integrated Circuits and Systems", McGraw Hill. NY., Febr.1994.

H.W.Ott, "Noise reduction techniques in Electronic Systems", Wiley, 1988.

B. Razavi, "Design of analog CMOS integrated circuits", McGraw Hill. NY., 2000.

A.Sedra, K.Smith, "Microelectronic Circuits", CBS College Publishing, 1987.

Willy Sansen 10-05 0191

0191

Other books are more general. They give an introduction on models but are mainly focused on analog circuit design.

One of the main "classics" is Gray and Meyer, now Gray, Hurst, Lewis and Meyer.

The ones under Huijsing, Van de Plassche and Sansen are actually Proceedings of the annual European workshop "Advances in Analog Circuit Design", which was organized the first time in 1993. All the other ones are textbooks.

021

Amplifiers, Source followers & Cascodes

Willy Sansen

KULeuven, ESAT-MICAS

Leuven, Belgium

willy.sansen@esat.kuleuven.be

Willy Sansen 10.05 021

All analog circuits are built up by means of a limited number of elementary building blocks. Thorough knowledge of these blocks is therefore essential to obtain insight into more complicated circuit schematics. This is why they are considered separately and analyzed in great detail. Several overview slides are added so as not to lose the context out of sight.

022

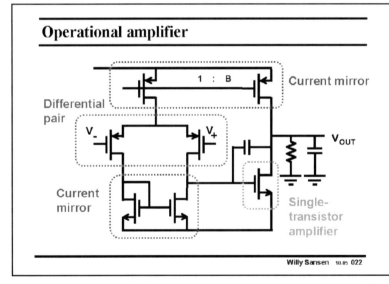

Operational amplifier

1 : B — Current mirror

Differential pair

V_- V_+ V_{OUT}

Current mirror

Single-transistor amplifier

Willy Sansen 10.05 022

In all analog electronics, an operational amplifier is the most versatile circuit block. It consists of a differential input and a single output. Its gain is very large. It is normally used in a feedback loop. A simple realization is shown in this slide.

Such an operational amplifier, usually shortened to opamp, contains several elementary building blocks. In the first stage we can identify a differential pair loaded by a current mirror. The second stage is little more than a single-transistor amplifier, with a DC current source as load, which is also part of a current mirror.

These elementary building blocks will now be studied in great detail.

023

After all, only a few elementary circuit blocks can be identified. One single transistor can be used as an amplifier, a source follower or a cascode. Moreover, gain boosting can be applied to a cascode, which is little more than a cascode combined with an amplifier. Obviously a MOST can also be used as a switch.

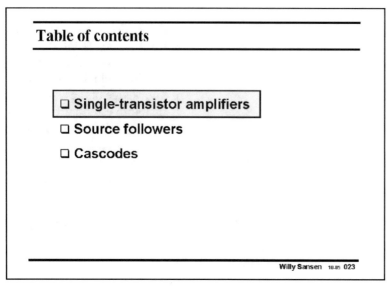

Table of contents

Willy Sansen 10.05 023

With two transistors two more configurations can be identified. They are the differential pair and the current mirror. Combined, they give rise to a full four-transistor differential voltage and current amplifier. One version of this differential current amplifier has as many as four inputs, which makes it a very versatile building block indeed.

We start with an amplifier, using only one single transistor.

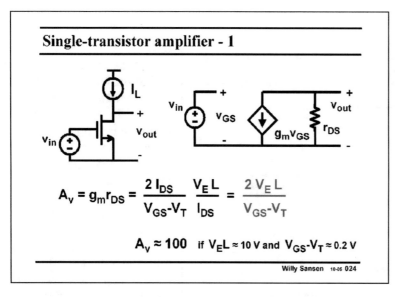

Single-transistor amplifier - 1

$$A_v = g_m r_{DS} = \frac{2 I_{DS}}{V_{GS}-V_T} \frac{V_E L}{I_{DS}} = \frac{2 V_E L}{V_{GS}-V_T}$$

$$A_v \approx 100 \quad \text{if } V_E L \approx 10 \text{ V and } V_{GS}-V_T \approx 0.2 \text{ V}$$

Willy Sansen 10-05 024

024

A single-transistor amplifier is biased by a voltage source V_{IN}, on which a small-signal input voltage v_{in} is superimposed. Biasing will be discussed later.

However, it is important to note that an amplifier is normally loaded by a DC current source. In this way the maximum gain can be obtained. Indeed, an ideal current source has an infinite output resistance. It is therefore not visible in the small-signal equivalent circuit, shown on the right.

The voltage gain A_v is easily found by this equivalent circuit. It is $g_m r_{DS}$. Both parameters depend on the current, however. As a result, the gain A_v becomes independent of the current. It depends not only on a technological parameter V_E but also on two parameters, which can be chosen by the designer. They are $V_{GS}-V_T$ and channel length L.

It is obvious that for large gain A_v, we must choose $V_{GS}-V_T$ as small as possible and L as large as possible.

025

These are really important conclusions. Large gain A_v can only be achieved by choosing large channel length L and by making $V_{GS}-V_T$ as small as possible.

As a result, the minimum channel length is never used in an analog amplifier. Usually we limit the value of L to at least 4–5 times the minimum value.

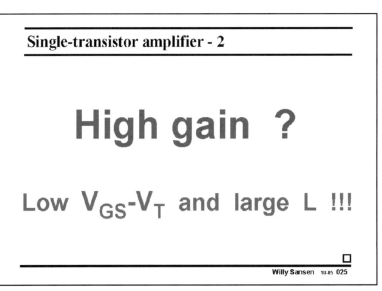

Single-transistor amplifier - 2

High gain ?

Low V_{GS}-V_T and large L !!!

Willy Sansen 10.05 025

Also, we take the take the value of $V_{GS}-V_T$ as small as possible. A typical value is 0.15–0.2 V. We cannot go much lower as we then end up in the weak inversion region. The absolute value of the current and the transconductance then become so small that the noise becomes too large. Noise will be explained in detail in Chapter 3.

Small values of the current inevitably lead to larger noise and smaller signal-to-noise ratios (SNR). For SNR values below 40 dB, weak inversion can be used. Sensor interface and biomedical preamplifiers are comfortable with this. Communication amplifiers usually require more than 70 dB SNR. As a result we have to stay in the region between weak and strong inversion. A value of $V_{GS}-V_T$ between 0.15 V and 0.2 V is therefore a good compromise.

MOST or bipolar amplifier ?

MOST $\quad A_v = \dfrac{V_E L}{(V_{GS}-V_T)/2}$

$A_v \approx 100$ if $V_E L \approx 10$ V and V_{GS}-$V_T \approx 0.2$ V

Bipolar $\quad A_v = \dfrac{V_E}{kT/q}$

3 vs 2 stages for 10^6

$A_v \approx 1000$ if $V_E \approx 26$ V since $kT/q = 26$ mV

Willy Sansen 10.05 026

026

Deep submicron CMOS devices provide less and less voltage gain however. Indeed for a relatively large channel length of 2.5 µm (and a $V_E = 4$ V/µm) the $V_E L$ factor is about 10 V, which yields a gain of 100, using a $V_{GS}-V_T = 0.2$ V.

For a minimum channel length of 90 nm however, parameter V_E does not change all that much, and the voltage gain is now only 3.6!! In practice it is a little bit more, but not much, i.e. about 6. As a result we will have to devise all possible tricks to enhance the gain. Cascodes offer such capability, and gain boosting, as explained later in this chapter.

Even with a voltage gain of 100, three stages would be needed to obtain a voltage gain of 10^6, as is common in operational amplifiers. A bipolar transistor can provide the same gain in only two stages.

Indeed in a bipolar transistor, the input voltage is scaled to kT/q rather than to $(V_{GS}-V_T)/2$. We now gain a factor of about 4! The other reason for larger gain is the slightly higher value of the V_E parameter.

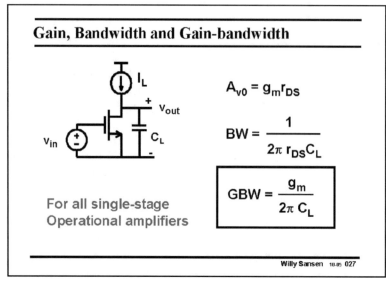

Gain, Bandwidth and Gain-bandwidth

$$A_{v0} = g_m r_{DS}$$

$$BW = \frac{1}{2\pi\, r_{DS} C_L}$$

$$GBW = \frac{g_m}{2\pi\, C_L}$$

For all single-stage
Operational amplifiers

Willy Sansen 10.05 027

027

At higher frequencies, the voltage gain will fall off as a result of all capacitances. There are three positions where a capacitance can occur. Normally the load capacitance is the largest one, as it consists of all interconnect capacitance to the next stage and the feedback capacitances applied (as in switched-capacitance filters).

Here only a load capacitance C_L is present. The low frequency gain A_{v0} is as before. The pole frequency, at which the gain starts decreasing is called the bandwidth BW or the −3dB frequency. It is simply determined by the output RC time constant.

The product of the low-frequency gain and the bandwidth is called the Gain-Bandwidth product GBW. It is by far the most important quality factor of an amplifier. Actually it is the most important specification of an amplifier. Later on we will compare amplifiers by means of a Figure of Merit (FOM), which quotes how much GBW can be obtained for a certain load capacitance and power consumption.

The GBW itself is readily obtained. It only depends on the transistor transconductance and the load capacitance, not on the output resistance. We will see that this expression is valid for all possible single-stage amplifiers. It is therefore a very important expression!

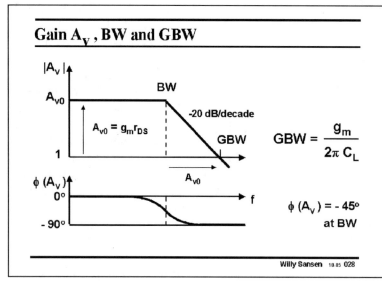

Gain A_v, BW and GBW

$$GBW = \frac{g_m}{2\pi\, C_L}$$

$$\phi (A_v) = -45°$$
at BW

Willy Sansen 10.05 028

028

To have a better view on how the frequencies BW and GBW are related to the low-frequency gain, a Bode diagram is included.

Such a Bode diagram consists of two diagrams, both versus frequency on a logarithmic scale. The top one shows the logarithm of the amplitude of the gain, denoted by $|A_v|$. The bottom one plots the phase of the gain A_v.

It is clear that the GBW is indeed the product of A_{v0} and the BW. At the BW itself, the phase shift is −45°. At higher frequencies however, the phase shift increases to −90°.

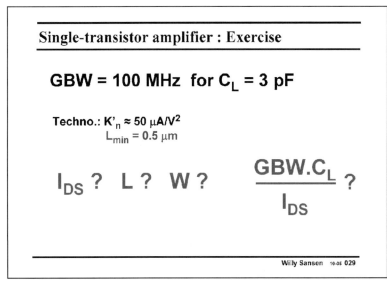

Single-transistor amplifier : Exercise

GBW = 100 MHz for C_L = 3 pF

Techno.: $K'_n \approx 50 \ \mu A/V^2$
L_{min} = 0.5 μm

I_{DS} ? L ? W ? $\dfrac{GBW \cdot C_L}{I_{DS}}$?

Willy Sansen   10-05 029

029

As an exercise, let us design a single-transistor amplifier for a GBW = 100 MHz and a $C_L = 3$ pF.

The technology is shown in this slide. The minimum channel length is 0.5 μm. The required g_m is readily calculated to be about 2 mS (or 2 mMhos).

We choose $V_{GS} - V_T = 0.2$ V. This transconductance of 2 mS thus requires 0.2 mA current. Using the current expression in strong inversion yields W/L = 100.

We choose $L \approx 4 \times L_{min} = 2$ μm, to obtain some gain.

As a result W = 200 μm.

The FOM is 1500 MHzpF/mA, which is not bad! After all only one transistor is involved!! Most opamps later on will have FOMs between 100 and 200, at least the better ones!

Gain, Bandwidth and Gain-bandwidth

$A_{v0} = g_m r_{DS}$

$BW = \dfrac{1}{2\pi R_S C_{GS}}$

$GBW = \dfrac{g_m}{2\pi C_{GS}} \dfrac{r_{DS}}{R_S} = f_T \dfrac{r_{DS}}{R_S} \sim \dfrac{1}{W C_{ox}} \dfrac{1}{V_{GS} - V_T}$

W ? L ? V_{GS}-V_T ?

Willy Sansen   10-05 0210

0210

If no load capacitance is present, but a large input capacitance C_{GS}, then the bandwidth is determined at the input. This is also true for many sensor and biomedical preamplifiers where the source resistance can be quite high (>1 MΩ).

In this case, the BW is simply given by the RC product at the input.

The GBW on the other hand, is not as simple as before. Many transistor parameters now play a role.

Some of them can be bundled in the high-frequency parameter f_T.

For high-frequency performance, it is not sufficient however to make f_T large. Rather the product $f_T r_{DS}$ must be optimized, which is a technological challenge indeed!

The channel length does not seem to play a role; both W and $V_{GS} - V_T$ must be made small!!! All surprising results indeed!

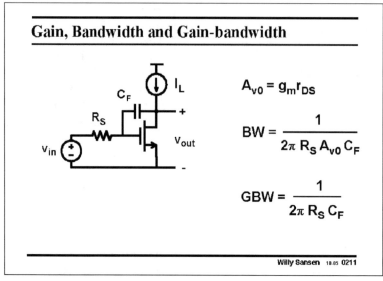

Gain, Bandwidth and Gain-bandwidth

$$A_{v0} = g_m r_{DS}$$

$$BW = \frac{1}{2\pi R_S A_{v0} C_F}$$

$$GBW = \frac{1}{2\pi R_S C_F}$$

Willy Sansen 10.05 0211

0211

Finally, the third and final possible addition of a single capacitance to this circuit, is shown in this slide. It is a feedback capacitance C_F from output to input. It is also called a Miller capacitance.

Since this capacitance is connected from output to input, it gives the same time constant with the source resistor at the input, as a capacitance C_{GS}, but multiplied by the gain A_{v0}. Indeed the output has a signal amplitude which is A_{v0} times larger than the input. Seen at the input, the capacitance C_F also seems to be A_{v0} times larger.

The GBW is now completely independent of any transistor parameters. This is expected! Indeed with feedback, the gain becomes independent of the amplifier parameters, and only dependent on the external feedback elements.

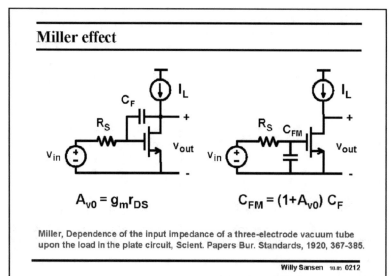

Miller effect

$$A_{v0} = g_m r_{DS}$$

$$C_{FM} = (1+A_{v0}) C_F$$

Miller, Dependence of the input impedance of a three-electrode vacuum tube upon the load in the plate circuit, Scient. Papers Bur. Standards, 1920, 367-385.

Willy Sansen 10.05 0212

0212

The same conclusions can be better visualized, in the diagram in this slide. The bandwidth is determined by the same time constant with the source resistor at the input, as a capacitance C_{GS}, but multiplied by the gain A_{v0}.

This Miller effect only applies to the impedance seen at the input. It is not valid for the impedance at the output.

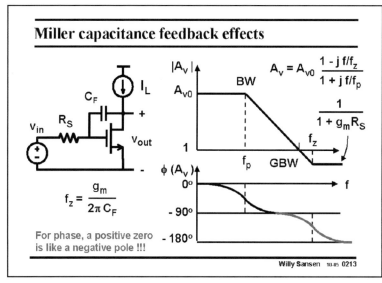

Miller capacitance feedback effects

$$A_v = A_{v0} \frac{1 - j\, f/f_z}{1 + j\, f/f_p}$$

$$\frac{1}{1 + g_m R_S}$$

$$f_z = \frac{g_m}{2\pi\, C_F}$$

For phase, a positive zero is like a negative pole !!!

Willy Sansen 10-05 0213

0213

Actually, a Miller capacitance also causes a zero in the transfer characteristic.

A full small-signal analysis reveals that the pole (the BW) is followed by a zero at high frequencies. It is a positive zero. It occurs at the right hand side of the polar diagram. It causes a $-180°$ phase shift at higher frequencies.

A $-180°$ phase shift would also be caused by a second pole. We come to the remarkable result that a single capacitance can cause the same phase shift as two poles. Normally only one single pole is attributed per capacitance. This is a truly exceptional situation indeed.

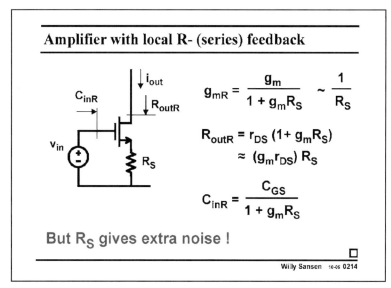

Amplifier with local R- (series) feedback

$$g_{mR} = \frac{g_m}{1 + g_m R_S} \sim \frac{1}{R_S}$$

$$R_{outR} = r_{DS}(1 + g_m R_S)$$
$$\approx (g_m r_{DS})\, R_S$$

$$C_{inR} = \frac{C_{GS}}{1 + g_m R_S}$$

But R_S gives extra noise !

Willy Sansen 10-05 0214

0214

Quite often, local series feedback is applied to a single-transistor amplifier, by means of a resistor R_S. In principle, the effect of this resistor can be calculated by common feedback theory (see Chapter 13). The so called loop gain is then $(1 + g_m R_S)$. It affects all other circuit specifications. It is a very simple case however, which allows us to calculate all these effects directly.

The transconductance is reduced by this loop gain. For large resistors, the transconductance is reduced to $1/R_S$. It is independent of the current in contrast with g_m.

A major effect is that the output resistance increases drastically. Indeed, it goes up by the same loop gain. An easy way to remember this expression is to take the series resistor R_S, multiplied by the intrinsic transistor gain itself $g_m R_{DS}$.

This increased output resistance will be used to increase the gain!

The input capacitance is decreased by the feedback. The larger the resistor R_S, the smaller the input capacitance. If R_S were replaced by a DC current source, the input capacitance would be negligible. Actually, the gain would be negligible as well. This is a source follower, as discussed later.

The main problem of R_S is its noise. This is why in low-noise RF circuits, an inductor is used instead.

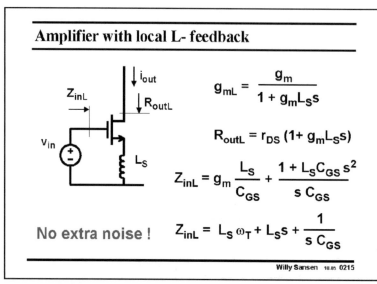

0215

Inductors and capacitances do not give noise, at least not as long as their series loss resistance is zero. Only resistances give noise among the passives!

Inserting an inductor makes both the transconductance and the output resistance frequency dependent.

The input impedance Z_{inL} however, which was capacitive without series R or L, becomes now purely resistive with value $g_m L_S/C_{GS}$ or $L_S \omega_T$. Indeed, the input capacitance C_{GS} is tuned out by the inductor. Its input resistance can be easily designed to be 50 Ω to match an incoming 50 Ω transmission line (cable, antenna, etc.). In this way, a very-high-frequency low-noise amplifier can be designed (see Chapter 23).

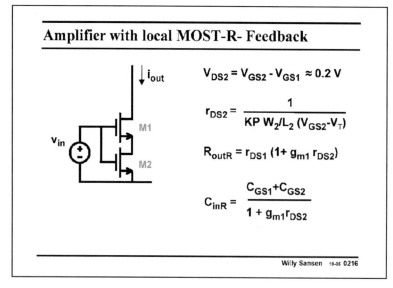

0216

An easy way to realize such a series feedback resistor is to use a nMOST in the linear region. This is only possible however, if its V_{DS2} is quite small, between 100 and 200 mV. This is also the difference between the V_{GS}'s of the two transistors.

It is not that easy to track the parameters of both transistors. Indeed, MOST M1 works in saturation, involving parameter K', whereas transistor M2 operates as a resistor, with parameter KP. They differ by parameter n, which is always uncertain.

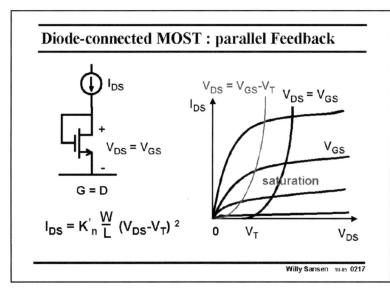

Diode-connected MOST : parallel Feedback

$$I_{DS} = K'_n \frac{W}{L} (V_{DS}-V_T)^2$$

Willy Sansen 10.05 0217

0217

Parallel feedback can be applied to a single transistor as well, which results in a diode-connected transistor.

Connecting collector to base in a bipolar transistor, gives us a real base-emitter diode. In a MOST however, there is no gate-source diode. And yet, connecting the drain to the gate gives us something similar.

Indeed the current voltage characteristic is obtained by shifting the curve, separating the linear and the saturation region, which is at $V_{DS} = V_{GS} - V_T$, to the right by V_T.

As a result, we can indeed use the current-voltage characteristic of a MOST in saturation. The resulting curve is very nonlinear however. It resembles somewhat, a diode characteristic.

We will use this simple circuit to convert current to voltage.

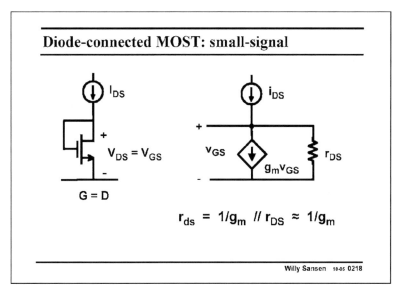

Diode-connected MOST: small-signal

$$r_{ds} = 1/g_m \mathbin{/\!/} r_{DS} \approx 1/g_m$$

Willy Sansen 10-05 0218

0218

If we add a small-signal current to the DC current, then we find the small-signal equivalent circuit, as shown.

The small-signal resistance r_{ds} is then $1/g_m$ in parallel with the transistor output resistance r_{DS}. Since the latter one is always significantly larger than $1/g_m$, it can be simplified to $1/g_m$ itself.

The small-signal resistance of a diode connected MOST is thus always $1/g_m$, very much as for a bipolar transistor.

0219

At high frequencies, this voltage-to-current converter performs quite well.

Indeed adding the C_{GS} and C_{DS}, which are the two most important capacitances of a MOST,

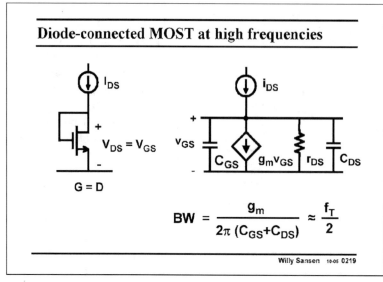

Diode-connected MOST at high frequencies

$$BW = \frac{g_m}{2\pi(C_{GS}+C_{DS})} \approx \frac{f_T}{2}$$

Willy Sansen 10-05 0219

yields a bandwidth BW, which is quite high. This BW is determined by $1/g_m$ and the sum of the two capacitances. These capacitances are very similar in size however. They have been taken to be equal in size.

The bandwidth is thus well approximated by $f_T/2$.

For a high bandwidth, we must therefore design a transistor with high f_T, which requires a large $V_{GS}-V_T$ and minimum channel length L.

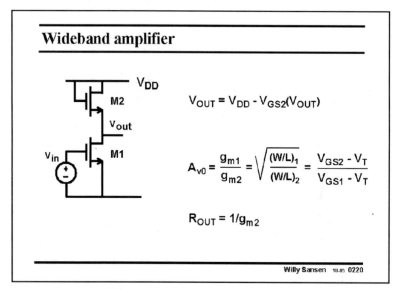

Wideband amplifier

$$V_{OUT} = V_{DD} - V_{GS2}(V_{OUT})$$

$$A_{v0} = \frac{g_{m1}}{g_{m2}} = \sqrt{\frac{(W/L)_1}{(W/L)_2}} = \frac{V_{GS2}-V_T}{V_{GS1}-V_T}$$

$$R_{OUT} = 1/g_{m2}$$

Willy Sansen 10-05 0220

0220

As a load for this amplifier, a resistor can be used. In this case however, the gain is very low, somewhere between 3 and 5, depending on the supply voltage.

Suitable resistors are not always available. In a digital CMOS process, none of them can provide reasonably large resistance values. This is why many circuit schematics have been proposed, using MOSTs as loads. One of them is shown in this slide.

It uses a nMOST connected as a diode. Its small-signal resistance is thus $1/g_{m2}$. The gain is the ratio of the transconductances. It is small but fairly accurate as it is mainly set by the ratio of the transistor sizes.

Its main advantage is that no pMOSTs are used. This amplifier can achieve high bandwidths, also because the output impedance is quite small!

Its main disadvantage is that the DC output voltage is connected to the supply line over the V_{GS2}. Because of the body effect of transistor M2, this DC output voltage is not well defined. The biasing of the next stage may suffer from this.

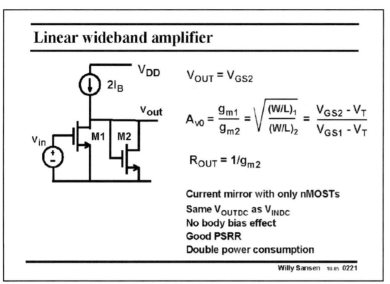

0221
A better solution for the biasing is shown in this slide. There is a DC input voltage which gives rise to an equal DC output voltage. In this case the current is divided over both transistors and the gain is accurately given by the transistor size ratio or the $V_{GS} - V_T$ ratio's.

An additional advantage is that the circuit can easily be put in series with many more similar stages. This may be needed because the gain is small. A transistor ratio of 25 gives only a gain of 5. For larger gains, several more stages like this have to be cascaded.

The body effect does not apply a role any more as all bulk contacts are grounded.

Again only nMOSTs are used for higher frequency performance.

The main drawback of this amplifier solution is that the current consumption is twice that of the previous circuit. However, it is quite often used as a wideband amplifier, in optical receivers, etc.

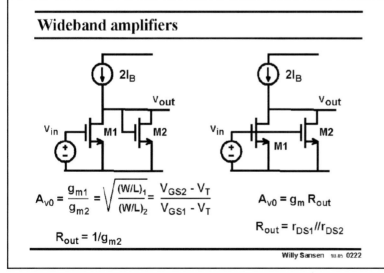

0222
The question arises which amplifier would be better, the first one or the second, where the same current is used but now with transistor M2 in parallel with M1?

Clearly in the second amplifier, the output resistance is higher and so is the gain. Accordingly, the bandwidth will be smaller.

Let us have a closer look at this amplifier. Normally the current source is realized by means of another transistor, a pMOST device, the gate of which is connected to a voltage reference, which sets all DC currents.

We still have two possibilities as shown next.

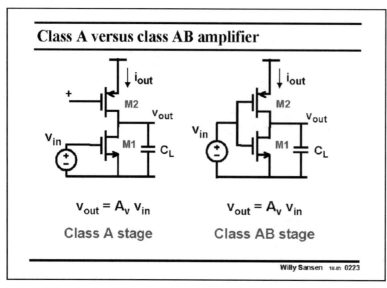

Class A versus class AB amplifier

$v_{out} = A_v v_{in}$

Class A stage

$v_{out} = A_v v_{in}$

Class AB stage

Willy Sansen 10.05 0223

0223

The first amplifier has a constant DC current as the gate of current source transistor M2 is connected to a DC reference. At low frequencies, the load capacitance C_L does not apply. In this case, the DC current through transistors M1 and M2 does not change with the signal level. It is by definition a class A amplifier.

The result is very different if we connect and drive both gates together. Depending on the input signal level, the currents in both transistors varies greatly. It is a class AB amplifier.

Actually, this amplifier is used for both digital and analog input signals.

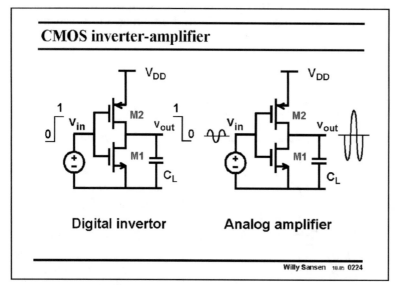

CMOS inverter-amplifier

Digital invertor

Analog amplifier

Willy Sansen 10.05 0224

0224

Indeed, this is the well known digital inverter. When the input goes high (from a digital 0 to 1), the output goes low, and vice-versa. In both cases, no current can flow. This is the main advantage of this digital inverter. It only consumes power during switching. Many millions can now be integrated on a single silicon chip, without excessive heat dissipation.

As an analog amplifier, the input biasing voltage is such that the output voltage sits at an acceptable value between the supply voltage V_{DD} and ground. A small-signal input voltage is then amplified (and inverted) to the output.

This is examined in more detail next.

0225

In order to establish the actual transfer curve and the transistor currents, we have to realize that both transistors always carry the same DC current. Indeed, no DC current can escape through the capacitance. This is also true for AC currents at low frequencies.

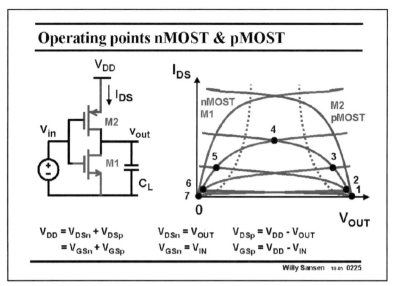

Operating points nMOST & pMOST

$V_{DD} = V_{DSn} + V_{DSp}$
$\quad = V_{GSn} + V_{GSp}$

$V_{DSn} = V_{OUT}$
$V_{GSn} = V_{IN}$

$V_{DSp} = V_{DD} - V_{OUT}$
$V_{GSp} = V_{DD} - V_{IN}$

Willy Sansen 10.05 0225

Also, note that the sum of the V_{DS}'s is V_{DD}, and that the sum of the V_{GS}'s is also V_{DD}. For a low input voltage, the V_{GSn} is also low and the V_{GSp} is high. In this case, the nMOST is off and the pMOST is on. The crossing point of their $I_{DS} - V_{DS}$ curves is thus at point 1. Indeed the pMOST is biased as a small resistor, and the output voltage is the same as the supply voltage.

Increasing the input voltage will cause the crossing point to shift from point 1, to 2, and so on, all the way to point 7. At this latter point the pMOST is off and the nMOST has a large V_{GS}, but small V_{DS}. It is in the linear region and so behaves as a small resistor. The output voltage is now zero. The transistor currents are zero as well.

When the input voltage is about halfway, the transistor current flows, and the output voltage is about halfway as well. This is point 4. This is the normal biasing for this circuit as an analog amplifier.

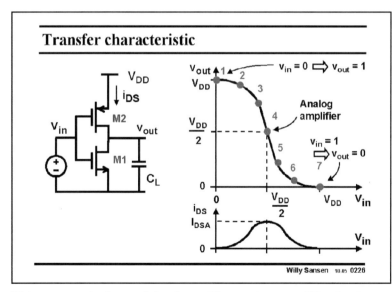

Transfer characteristic

$v_{in} = 0 \Rightarrow v_{out} = 1$

Analog amplifier

$v_{in} = 1 \Rightarrow v_{out} = 0$

Willy Sansen 10.05 0226

0226

The transfer characteristic is now easily reconstructed. It is a plot of the output voltage versus input voltage. For each input voltage, the output voltage is found by the points 1 to 7 and added to the plot. The corresponding transistor current is added underneath.

It is clear that no current flows in the points 1 and 7, when the output is at a digital "1" or "0".

The current reaches its maximum in the middle. This current is denoted by I_{DSA}. The analog amplifier is always biased in this point. It is its quiescent current.

The exact calculation of the whole transfer characteristic is not that easy. It is only easy in the extreme points 1, 4 and 7. This exact calculation is better left to a circuit simulator such as SPICE.

We now concentrate on this amplifier, biased in the middle point 4.

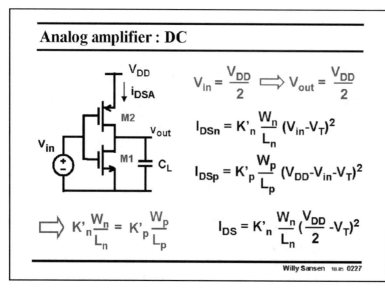

Analog amplifier : DC

$$V_{in} = \frac{V_{DD}}{2} \implies V_{out} = \frac{V_{DD}}{2}$$

$$I_{DSn} = K'_n \frac{W_n}{L_n} (V_{in}-V_T)^2$$

$$I_{DSp} = K'_p \frac{W_p}{L_p} (V_{DD}-V_{in}-V_T)^2$$

$$\implies K'_n \frac{W_n}{L_n} = K'_p \frac{W_p}{L_p} \qquad I_{DS} = K'_n \frac{W_n}{L_n} (\frac{V_{DD}}{2} -V_T)^2$$

Willy Sansen 10.05 0227

0227

We firstly have to determine the exact biasing. We normally require the output voltage to be half the supply voltage, when the input voltage is half the supply voltage. It does not have to be like that, but this is a better way, for example, if several of these stages have to cascaded.

In this case, both transistors have the same V_{GS}, which is $V_{DD}/2$. They also have the same current. This is only possible if their W/L values are inversely proportional to their K' values. Since K'_n is usually twice as large as K'_p, the pMOST transistor has usually twice the W/L value compared to the nMOST.

The expression of the transistor current is then easily obtained, with V_{GS} substituted by $V_{DD}/2$.

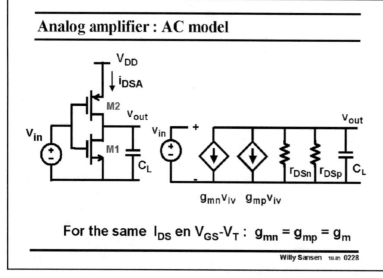

Analog amplifier : AC model

$g_{mn}v_{iv}$ $g_{mp}v_{iv}$

For the same I_{DS} en $V_{GS}-V_T$: $g_{mn} = g_{mp} = g_m$

Willy Sansen 10.05 0228

0228

In order to obtain the expression of the voltage gain, we need to draw the small-signal equivalent circuit of this amplifier. Note that the supply voltage is always AC ground. Also note that both transistors have equal small-signal models and even equal transconductances.

It is clear from this small-signal circuit, that both transistors are actually in parallel, for small-signal operation (for DC operation or biasing they are in series). They provide equal contributions to the small-signal output current and to the gain. The total transconductance is thus twice the transconductance of a single transistor.

0229

In order to calculate the gain at low frequencies, we need to know the output resistance. It is the parallel combination of both output resistances. This is highest when both resistances are

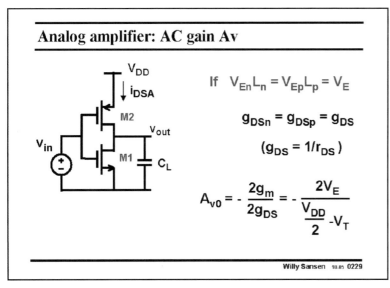

Analog amplifier: AC gain Av

If $\quad V_{En}L_n = V_{Ep}L_p = V_E$

$$g_{DSn} = g_{DSp} = g_{DS}$$

$$(g_{DS} = 1/r_{DS})$$

$$A_{v0} = -\frac{2g_m}{2g_{DS}} = -\frac{2V_E}{\dfrac{V_{DD}}{2} - V_T}$$

Willy Sansen 10.05 0229

equal. This is why normally both $V_E L$ products are also made the same. The total output resistance is then $r_{DS}/2$, also written as $2/g_{DS}$.

The voltage gain is then simply given by the product of the total transconductance with the total output resistance, which is g_m/g_{DS}. This can easily be rewritten as shown in this slide.

Note that the current is not included, a result which we had found already for a single-transistor amplifier. This is not surprising as here we have two transistors in parallel, for small-signal operation.

Also note, that the voltage gain goes up when the supply voltage goes down. An optimum supply voltage is thus found where the V_{GS} values are about 0.2 V. This supply voltage is thus $2(V_T + 0.2)$, which is 1.1 V for a V_T of 0.35 V. In deep submicron CMOS this is quite a reasonable value indeed!

If the supply voltage is larger, only small values of the gain are possible. More complicated circuits are then needed to enhance the voltage gain. Cascodes can be used for example, and gain-boosting, but also bootstrapping and current-cancellation and -starving techniques.

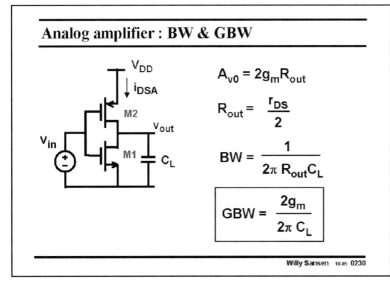

Analog amplifier : BW & GBW

$$A_{v0} = 2g_m R_{out}$$

$$R_{out} = \frac{r_{DS}}{2}$$

$$BW = \frac{1}{2\pi R_{out}C_L}$$

$$GBW = \frac{2g_m}{2\pi C_L}$$

Willy Sansen 10.05 0230

0230

We can also leave the output resistance in the expression of the gain. This makes it a bit easier to see how this same output resistance determines the output pole or the bandwidth.

Moreover, this output resistance drops out in the calculation of the GBW. This was also the case for a single-transistor amplifier. However the GBW is now twice as high as we have double the transconductance of one single transistor.

This circuit is therefore a simple case of current reuse.

The GBW is again the most important specification. It indicates how much voltage gain can be expected at any frequency. It depends on the current, though transconductance g_m.

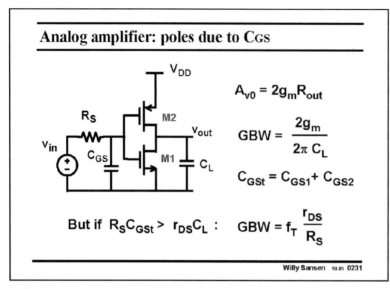

0231

The BW and GBW are easily determined since there is only one large capacitor at the output. If this load capacitor is smaller, then the transistor capacitances start to apply.

For example, if the source resistor R_S is large, then clearly the input capacitance $2C_{GS}$ will cause an additional time constant $2R_SC_{GS}$, which will generate another pole. This is called the non-dominant pole.

This latter non-dominant pole can even become dominant if R_S is very large, or more importantly, if the R_SC_{GSt} product is larger than $r_{DS}C_L$. This is easily calculated from the small-signal equivalent circuit.

In this case, the GBW is determined by the R_SC_{GS} product. As was already the case for a single-transistor amplifier, the GBW now depends on the f_T and a resistor ratio.

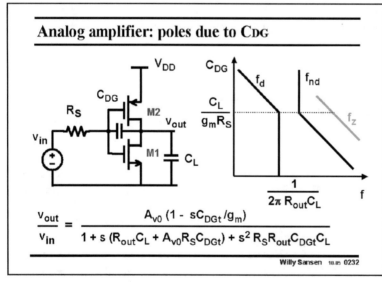

0232

It is also possible that the Miller effect is present. For example if R_S is large, and the gain is large, then $2C_{DGt}A_v$ may be larger than $2C_{GS}$. The non-dominant pole is then determined by the time constant $2R_SC_{DGt}A_v$.

This latter non-dominant pole can even become dominant if R_S or C_{DGt} is very large. In order to see this, we have to derive the full expression of the gain, from the small-signal equivalent

circuit. It is given in this slide. It has two poles and one zero.

The dominant pole is at the output, at least whilst C_{DGt} is small. The best way to show this, is to verify a pole-zero position diagram with C_{DGt} as a variable. This is nothing more than bilogarithmic diagram of the poles and zeros versus parameter C_{DGt}. It is asymptotic to show clearly the positions of all break points. More details on such diagram are given in slide N0536 of Chapter 5.

The dominant pole frequency f_d is clearly determined by the output time constant. For higher values of C_{DGt} than $C_L/(g_mR_S)$, the Miller effect dominates.

The non-dominant pole f_{nd} is at higher frequencies.
The positive zero f_z is at very high frequencies.

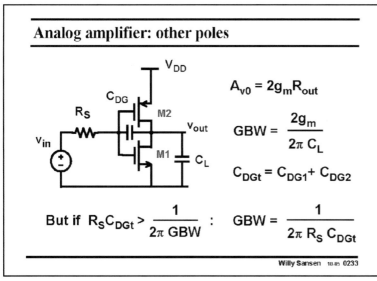

0233
The Miller effect is dominant if R_S is very large, or more importantly, if the $R_S C_{DGt}$ product is larger than $1(2\pi GBW)$. This is also easily calculated from the small-signal equivalent circuit, or from the expression of the gain.

In this case, the GBW is determined by the $R_S C_{DGt}$ product, as was already the case for a single-transistor amplifier with only a Miller capacitance. Indeed, only the feedback elements will

then determine the GBW, not the transistor parameters.

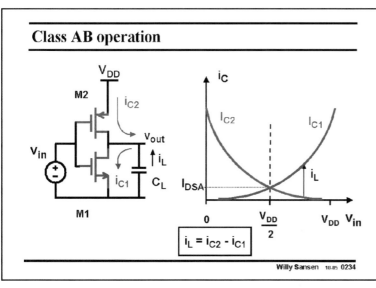

0234
This same small-signal amplifier also behaves as a class-AB amplifier for large input signals. This is especially needed to drive larger load capacitances.

Switching the input voltage from low to high, will cause the nMOST current i_{C1} to be much larger than the quiescent current I_{DSA}. At this point the pMOST current i_{C2} has become very small. The current i_L in the load capacitance will now be nearly equal to the

nMOST current i_{C1} and will be large, discharging that load capacitance quite fast.

This is only one of the simplest class-AB amplifiers however, with lot of disadvantages. One of the most notable disadvantages is that the currents strongly depend on the supply voltage. Better ones are discussed in Chapter 12.

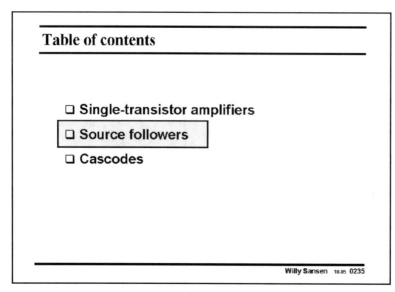

0235

Up till now, only amplifiers have been discussed. There are two more functions that can be carried out by means of a single-transistor. They are a source-follower and a cascode.

Of course we can still use a transistor as a switch, as we will do in Chapter 14 on Switched-capacitor filters.

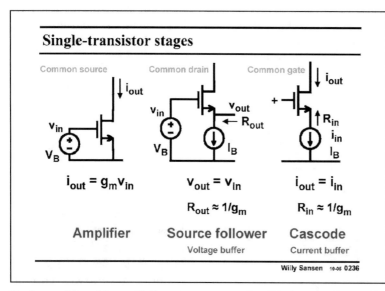

0236

A good overview of all the single-transistor stages is given in this slide.

The first one, the amplifier, is also called common-source stage. It converts an input voltage into an output current, by means of transconductance g_m. The biasing is done by means of a voltage source V_B at the input, which is the V_{GS}.

The other two stages are called respectively a Source follower and Cascode. They have the same biasing setup. Both are biased by a DC current source I_B. Their V_{GS} then adjusts itself such that the current can flow effectively.

In a Source follower the small-signal input voltage is applied at the Gate. The output is at the Source. Since the current is kept constant by the current source, the V_{GS} is also constant. As a result any small-signal variation at the input will give rise to an equal small-signal at the output. The voltage gain is therefore unity. This is why a Source follower is also called a Voltage buffer.

It is a buffer because its input resistance is infinity but its output resistance is only $1/g_m$. This stage will now be used to transfer a voltage in an accurate way from a high impedance to a lower one. This is required for microphone amplifiers, biopotential preamplifiers, etc. Their internal impedances can be many hundreds of $M\Omega$'s!

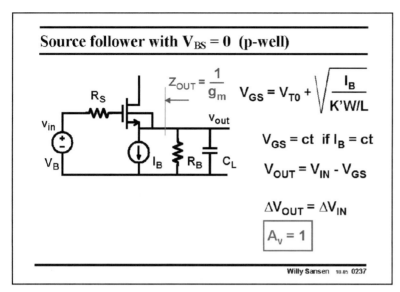

Source follower with $V_{BS} = 0$ (p-well)

$$Z_{OUT} = \frac{1}{g_m}$$

$$V_{GS} = V_{T0} + \sqrt{\frac{I_B}{K'W/L}}$$

$$V_{GS} = ct \text{ if } I_B = ct$$

$$V_{OUT} = V_{IN} - V_{GS}$$

$$\Delta V_{OUT} = \Delta V_{IN}$$

$$A_v = 1$$

Willy Sansen 10-05 0237

0237

An nMOST source follower is sketched in this slide. Note that firstly we connect the Bulk contact to the Source. This is only possible in a p-well CMOS technology. In a n-well CMOS technology, which is a lot more prevalent, we would have to take a pMOS source follower. Otherwise we cannot connect the Bulk to the Source.

For a DC current source I_B, the resultant V_{GS} is easily found from the current expression. The DC output voltage will now be the DC gate voltage V_B minus this V_{GS} value. We can try to optimize this value of V_B for the largest possible output swing. Quite often, a Source follower has to handle only small signals. When used in the output stage of a power amplifier however, the optimization of the output swing is a definite requirement, as will be explained in Chapter 12 on class AB amplifiers.

It is clear that as long as V_{GS} is constant, the output follows the input for small signals. The gain is therefore unity.

Moreover, substitution of the MOST by its small-signal model, readily shows that the output resistance is $1/g_m$. We can usually neglect r_{DS} with respect to $1/g_m$, obviously depending on the values of $V_{GS} - V_T$ and channel length L.

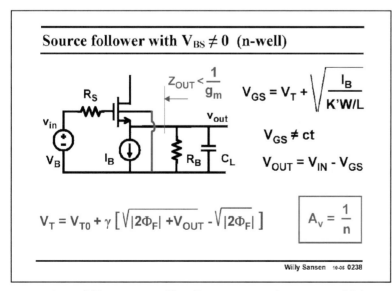

Source follower with $V_{BS} \neq 0$ (n-well)

$$Z_{OUT} < \frac{1}{g_m}$$

$$V_{GS} = V_T + \sqrt{\frac{I_B}{K'W/L}}$$

$$V_{GS} \neq ct$$

$$V_{OUT} = V_{IN} - V_{GS}$$

$$V_T = V_{T0} + \gamma \left[\sqrt{|2\Phi_F| + V_{OUT}} - \sqrt{|2\Phi_F|} \right]$$

$$A_v = \frac{1}{n}$$

Willy Sansen 10-05 0238

0238

In a nMOST source follower in a n-well CMOS technology, the Bulk can never be connected to the Source. As a result, the MOST has a V_{BS} voltage, which is actually the output voltage V_{OUT}. Consequently, the V_{GS} is no longer constant. It receives a small-signal contribution, which is not zero.

If we now use the expression of the V_T as a function of V_{BS} (shown in this slide), we obtain a nonlinear expression of V_{OUT} versus V_{IN}. Parameter γ shows up, which represents the body effect or the parasitic JFET.

Taking the derivative of V_{OUT} versus V_{IN} yields the small-signal gain. Surprisingly this gain is very simple, just $1/n$. Since n is a rather unpredictable value, so is the gain. It is certainly smaller than unity, somewhere between 0.6 and 0.8.

The output impedance is then somewhat smaller as it is now the parallel combination of $1/g_m$ and $1/g_{mB}$. This shows clearly that at the output we see the MOST and the parasitic JFET in parallel.

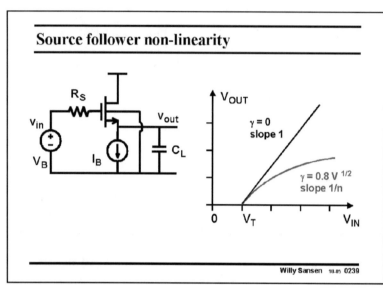

0239

The nonlinearity of the expression of V_{OUT} versus V_{IN} is clearly visible in the transfer characteristic.

If the Bulk can be connected to the Source ($\gamma = 0$) then the slope of the characteristic is unity: the Source follower has a unity gain.

If however, the Bulk is connected to ground ($\gamma > 0$) then clearly the curve is very nonlinear. A lot of distortion is then generated. In addition, it is not clear what the gain is, as the slope depends on the DC biasing voltages.

It is preferable that this stage is never used. In a n-well CMOS process, only pMOST source followers can be used, provided their Bulk is connected to the Source.

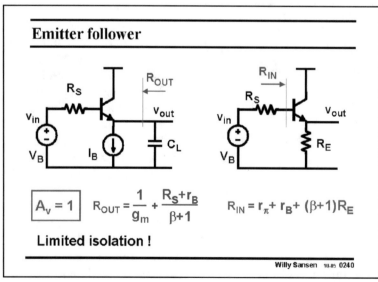

0240

A Source follower is an ideal buffer in the sense that it translates an input voltage unattenuated (if the Bulk is connected to the Source) from an infinite impedance at the input to a mere $1/g_m$ to the output.

This is not true for its bipolar equivalent, an emitter follower.

A bipolar transistor has Base current. As a result it has a finite input resistance (r_π). This causes an additional term in the expression of the output resistance. This term is the sum of all the resistances, seen at the input, divided by

$\beta + 1$ or simply β. A bipolar transistor only provides a limited amount of isolation between input and output, which is about β. The higher the beta the better the isolation. A MOST, which has a beta of infinity also provides infinite isolation.

The same is true for the input resistance. It can never be infinite, as for a MOST source follower. It will be the resistance seen at the emitter, multiplied by the beta. This may be insufficient for some preamplifiers (microphone, etc). This is why, quite often, a few emitter followers are cascaded.

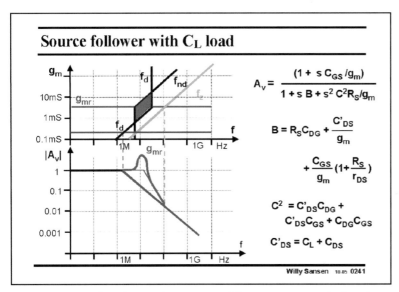

0241

At higher frequencies, a source follower looses its buffer capabilities. Moreover, it can show peaking.

Indeed, including all three transistor capacitances in the small-signal model, yields an expression for the gain which is quite complicated. The best way to explore this, is to draw a pole-zero position diagram, with one particular parameter as a variable. Here the transconductance is chosen, as it depends directly on the current. The current stands for the power consumption.

Two poles can be extracted from the roots of the denominator. One zero is also present.

In the middle, a hatched region emerges, where the lines of f_d and f_{nd} cross. In this region the poles are complex. They cause peaking in the Bode diagram.

It is clear that there is an optimum current for such a Source-follower that is actually quite small. For very low currents, the f_d (or bandwidth) is too small, and increases with the current. For higher currents, complex poles develop and for even higher currents the bandwidth ceases to increase. The optimum is thus at the bottom point of the complex pole region. Let us find the corresponding current.

0242

The value of the transconductance, which goes through the middle of the complex-pole region is denoted by g_{mr}. It is given in this slide. It depends on the source resistor and the capacitance C_{DG} from gate to ground. Adding capacitance to the gate allows us to reduce the transconductance or current where the peak is the worst.

The width of that region is given by Δg_{mr}. It can be made small by increasing indeed the capacitance C_{DG} at the gate. This is called "compensating the source follower". What actually happens is that a low-pass filter is added before the source follower, avoiding complex poles to appear.

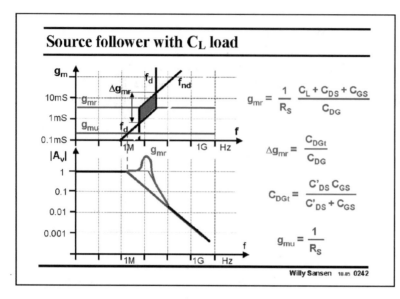

Source follower with C_L load

$$g_{mr} = \frac{1}{R_S} \frac{C_L + C_{DS} + C_{GS}}{C_{DG}}$$

$$\Delta g_{mr} = \frac{C_{DGt}}{C_{DG}}$$

$$C_{DGt} = \frac{C'_{DS}\, C_{GS}}{C'_{DS} + C_{GS}}$$

$$g_{mu} = \frac{1}{R_S}$$

Willy Sansen 10.05 0242

The actual value of g_{mr} at the bottom of the hatched region is then simply g_{mr} divided by the square root of Δg_{mr}.

Another point of interest is the crossing point of the nondominant pole with the zero (blue). At this point, we obtain a pure first-order roll-off, albeit with a lower dominant pole. The corresponding transconductance is denoted by g_{mu}. It only depends on the source resistor.

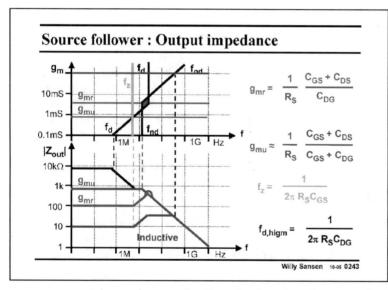

Source follower : Output impedance

$$g_{mr} = \frac{1}{R_S} \frac{C_{GS} + C_{DS}}{C_{DG}}$$

$$g_{mu} \approx \frac{1}{R_S} \frac{C_{GS} + C_{DS}}{C_{GS} + C_{DG}}$$

$$f_z = \frac{1}{2\pi\, R_S C_{GS}}$$

$$f_{d,higm} = \frac{1}{2\pi\, R_S C_{DG}}$$

Willy Sansen 10-05 0243

0243

The output impedance of a source follower, also shows this region of complex poles. for a particular current. Indeed the pole-zero position diagram of the output impedance, with the transconductance as a parameter, shows that around g_{mr}, complex poles occur, leading to peaking in the Bode diagram. The current corresponding to this g_{mr}, is therefore to be avoided.

However, another kind of peaking occurs. In fact, the zero occurs at lower frequencies than both poles. As a result, for large currents, the output impedance first rises, before it decreases. It is inductive!

A perfect first-order characteristic can be obtained by tuning the current to realize g_{mu}. This is the transconductance where the zero cancels out the first pole. A wide band source follower at low current results. The output impedance is then resistive up to high frequencies, and equals $1/g_{mu}$.

This value of g_{mu} depends mainly on the source resistance. For small resistances, the current can be become excessive. In this case, the Source follower can probably be omitted altogether.

0244

Both kinds of peaking are even more pronounced in an emitter follower. Indeed, the pole-zero position diagram of the output impedance, with the transconductance as a parameter, shows

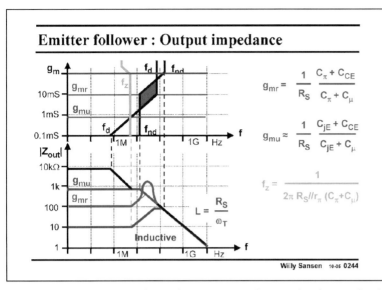

that around g_{mr} complex poles occur again, leading to severe peaking in the Bode diagram. The current corresponding to this g_{mr} should be avoided.

Again the zero occurs at lower frequencies than both poles. As a result, for large currents, the output impedance first rises before it levels off and then decreases. It is severely inductive!

Again a perfect first-order characteristic can be obtained by tuning the current to realize g_{mu}. This is the transconductance where the zero cancels out the first pole. A wide band source follower at low current results with an output impedance which is resistive up to high frequencies, and with value $1/g_{mu}$.

Especially for bipolar transistor emitter followers, a buffer with purely resistive output impedance up to high frequencies is of added value in RF circuits, for example to be connected to a $50\,\Omega$ transmission line.

This value of g_{mu} depends mainly on the source resistance. For small resistances, the current may become excessive.

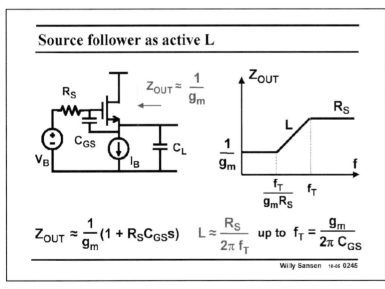

0245

The inductive output impedance of both source and emitter followers can be used wherever an inductor is required. They have been used in oscillators but also to add a bit of peaking to an amplifier, to compensate for an early roll-off because of all the transistor capacitances.

In its simplest form, such output inductor can be obtained by a source follower with high source resistor R_S and biased at a large DC current I_B. The inductor has a value given in this slide.

The output impedance is only inductive for frequencies between $f_T/(g_m R_S)$ and f_T itself. The quality factor of this inductor is low however, as its series resistor is R_S itself.

At low frequencies the output resistance is obviously $1/g_m$. For higher frequencies, capacitance

C_{GS} acts as a short, As a result the output resistance gradually becomes R_S. When R_S is larger than $1/g_m$, the output impedance must therefore rise, which causes the output impedance to be inductive. Obviously, if R_S were equal to $1/g_m$, the output impedance would be resistive up to very high frequencies.

The output impedance is thus inductive because R_S is much larger than $1/g_m$.

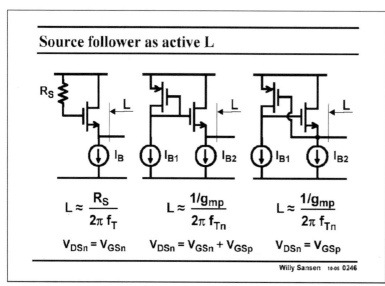

0246

Many other circuits are possible to realize this high value of source resistor R_S. A diode connected MOST can be taken, as in the circuit in the middle. The source resistor then has the value $R_S = 1/g_{mp}$.

The preferred circuit is on the right. This is a feedback circuit, in which the drain of the pMOST is still shorted to its gate, over a nMOST source follower. As a result, the source resistor R_S still equals $1/g_{mp}$.

The advantage of the latter circuit is that the voltage drop across is lower than for the middle circuit. It is only V_{DSn}. This is obviously important for deep submicron CMOS circuits, where the supply voltage is little over 1 V.

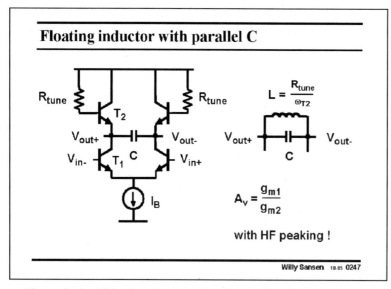

0247

A good example of a wideband (differential) amplifier is shown in this slide. The capacitances from the output terminals to ground cause a reduction in bandwidth. Adding resistors in the bases of the top bipolar transistors T_2, causes their output impedances to be inductive, with value L. As a result, peaking occurs, increasing the bandwidth. This bandwidth increase can thus be tuned by means of these base resistors.

The gain itself is simply given by the ratio of the two transconductances. Its value is low, which is typical for wideband amplifiers, e.g. for the transimpedance input amplifier of an optical receiver.

Table of contents

Willy Sansen 10.05 0248

0248

With a single transistor, we can make amplifiers, source followers and also cascodes. If we do not consider transistors used as switches, we now have the third and last single-transistor circuit possible.

Later, we will combine these three elementary single-transistor transistors to two or three elementary two-transistor circuits.

Most analog circuits are built up by means of these single-and two transistor circuits. A thorough understanding of these elementary blocks is therefore of vital importance.

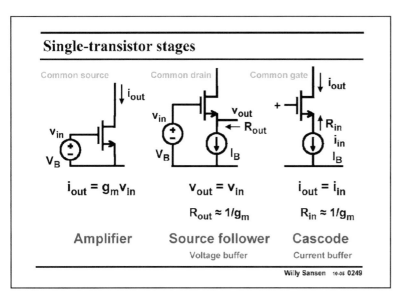

0249

Once more, an overview of the three single-transistor stages is given in this slide.

The first one, the amplifier, converts an input voltage into an output current, by means of transconductance g_m.

The second stage is a source follower. It is biased by a DC current source I_B. The voltage gain is unity. This is why a source follower is called a voltage buffer.

The third stage is a cascode. It is also biased by a DC current source I_B. A small-signal input current is superimposed on the source current. The output is at the drain. Since no current can escape, the current gain is not unity. This is why a cascode is also called a current buffer.

It is a buffer because its input resistance is small but its output resistance is high. This stage will therefore be used to transfer a current accurately from a low impedance to a higher one. This is required for current output sensors, such a photodiodes and potentiostatic sensors. Their internal impedances can be hundreds of MΩ's, and we need an impedance converter to reduce values!

Note that the input impedance of a cascode is the same as the output impedance of a source follower!

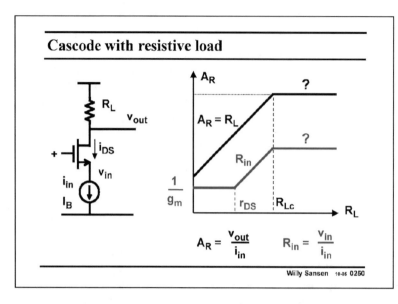

0250

Most often, the output current of a cascode is converted into a voltage by means of a resistance R_L. We thus obtain a current-to-voltage converter or transresistance amplifier. The higher the resistance R_L the higher the gain, since the transresistance $v_{out}/i_{in} = A_R$ is simply R_L itself. This is true whilst R_L is not too high.

For not too high values of R_L, the input resistance $v_{in}/i_{in} = R_{in}$ is then low, or simply $1/g_m$.

We will try to make the value of R_L as high as possible, in order to achieve a high gain. This can be done by cascoding a few transistors as a load. The question then is, what will the gain be?

Indeed, for very high values of R_L, it is not clear what the gain will be. For example, for infinite R_L, the current i_{in} cannot flow any more. What can the ratio v_{out}/i_{in} be?

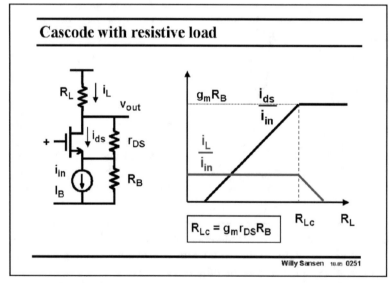

0251

The only possible way to respect the laws of Kirchoff, is to add either an output resistance to the input current source or to limit the load resistor to finite values. Let us see what happens when we make the load resistor infinity and we add an output resistance R_B in parallel with the input current source. Moreover, we separate the output resistor r_{DS} from the transistor itself, to determine where the currents are actually flowing.

Two currents are now calculated, the current i_{ds} in the transistor itself and the current through the load resistor i_L.

This latter current i_L is constant as imposed by the input current source. However, if R_L is larger than a specific value R_{Lc}, then this current has to go to zero!

The transistor current i_{ds} on the other hand, increases with the load resistor, to become constant once R_L is larger than R_{Lc}. The transistor thus amplifies the input current to fairly large values! Normally, $g_m R_B$ is much larger than unity!

Note that R_{Lc} contains all the parameters involved, the ones of the transistor such as g_m and r_{DS} but also R_B. It contains the actual gain $g_m r_{DS}$ of the transistor itself.

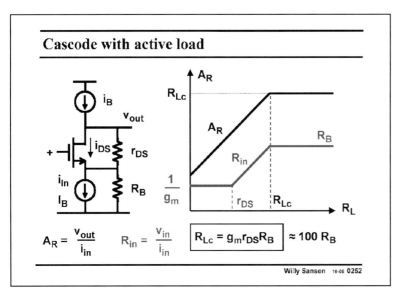

Cascode with active load

$$A_R = \frac{v_{out}}{i_{in}} \qquad R_{in} = \frac{v_{in}}{i_{in}} \qquad \boxed{R_{Lc} = g_m r_{DS} R_B} \approx 100\, R_B$$

Willy Sansen 10-05 0252

0252

Let us now look again at the actual transresistance A_R and at the input resistance R_{in}.

For very large values of the load resistor, A_R becomes R_{Lc} itself. The gain A_R includes the gain of the transistor $g_m r_{DS}$, and is therefore quite high. It is the highest transresistance gain that can be achieved by a single-transistor cascode. For that purpose, the load resistor must be larger than that very same transresistance R_{Lc}.

For these large load resistances, the input resistance of the transistor itself becomes infinity. This is expected as the load resistor is also infinity – no current can pass through the transistor.

The resistance seen by the input current source thus becomes resistor R_B, which is actually its output resistance. It is thus by no means small, as suggested by the $1/g_m$ low load resistances.

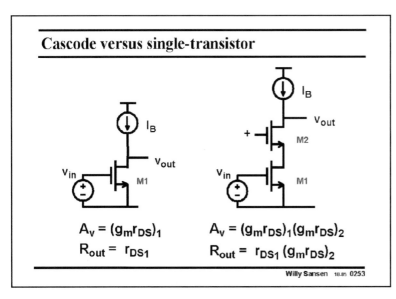

Cascode versus single-transistor

$$A_v = (g_m r_{DS})_1 \qquad\qquad A_v = (g_m r_{DS})_1 (g_m r_{DS})_2$$
$$R_{out} = r_{DS1} \qquad\qquad R_{out} = r_{DS1}(g_m r_{DS})_2$$

Willy Sansen 10-05 0253

0253

Substituting the input current source by another transistor, yields the two-transistor cascode, shown at the right. Since a transistor acts as a current source, with output resistor r_{DS}, it can take the place of the input current source with output resistor R_B. The maximum gain (for very large resistive load) is this given in this slide,

This shows that the total voltage gain now equals the product of the gains of both transistors.

Obviously this is a result of the increase of the output resistance. For one single transistor (left), this output impedance is only r_{DS1} but for a transistor with another r_{DS} in its Source (right), it is $r_{DS1} g_{m2} r_{DS2}$, as shown in this slide.

The larger the output impedance, the larger the gain.

Putting a cascode on top of a single-transistor amplifier is one of the most common design techniques to increase the gain. It will be used whenever we require more gain!

Other gain techniques will be gain boosting, bootstrapping and current cancellation and starving schemes.

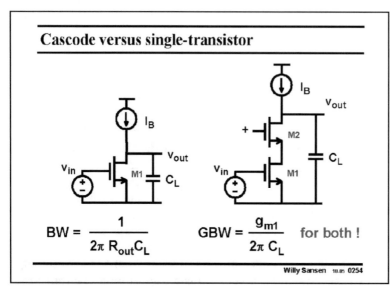

0254

Now that we know that a cascode enhances the gain considerably, we would like to know up tol what frequencies, a cascode manages to do this.

When the output capacitance is the main capacitance, the GBW is as given in this slide.

The Bandwidth for both cases must be different, as the output resistances differ widely.

The GBW is the same for both, however.

Indeed, in the case with a cascode, the gain is much larger but the bandwidth is equally lower. The GBW is thus the same. This is shown next.

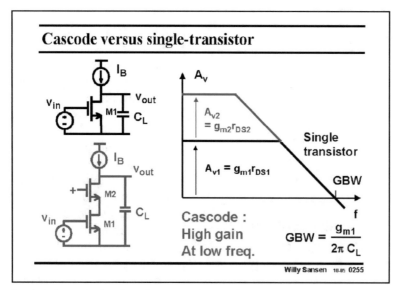

0255

It is clear that a cascode enhances the gain at low frequencies.

Compared to a single-stage amplifier, the gain at low frequencies is as large as the bandwidth is smaller. As a result the GBW is the same.

This is only true if the load resistance is very high. In some wideband amplifiers, the load resistor is small, e.g. 50 Ω. In this case, the output resistance and the bandwidth are the same for both. The gain is also the same, but quite small, as is the GBW.

In analog integrated circuits we normally use DC current sources as loads, in order to increase the gain. Therefore, we can conclude that such cascode stages only enhance the gain at low frequencies!

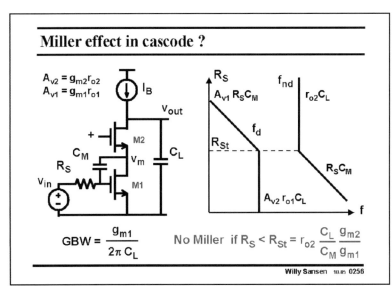

0256

For cascode amplifiers with a DC current source as load, the gain is quite large. The gain across the input transistor is also large. The impedance at the point between both transistors is the output resistance r_{DS1} (or r_{o1}). The gain from the input to the middle point is thus $g_{m1}r_{DS1}$, which is the gain of transistor M1. The Miller effect of M1 can thus also play a role, especially when source resistor R_S is higher. The question is, which pole is now dominant, the one caused by C_L or the one caused by C_M?

Substitution of the transistors by their small-signal models (with g_m and r_{DS}), and calculation of the total gain A_v, yields a second-order equation, the roots of which are easily found. They are the two poles. They are plotted in a pole-zero position diagram with R_S as a variable as shown in this slide.

For low values of R_S, the load capacitance C_L is obviously dominant. This dominant-pole frequency f_d is simply determined by the load capacitance and the output resistance $r_{o2}g_{m2}r_{o1}$ or $A_{v2}r_{o1}$. There is a nondominant pole however at f_{nd}, caused by time constant R_SC_M!

For high values of R_S however, the Miller effect of this capacitance C_M dominates. The dominant pole frequency f_d now decreases with increasing resistor R_S.

The cross-over value of R_S is R_{St}. It is in the order of tens of MΩ's as C_{SL} is normally much larger than C_M, which is marginally more than C_{DG}.

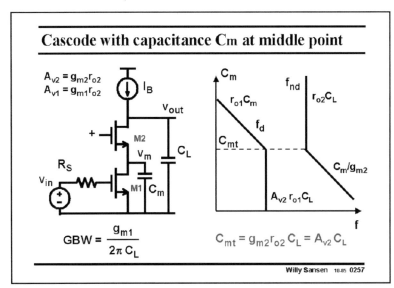

0257

Another small capacitance may also play a role. Capacitance C_m from the middle point to ground consists of C_{DS1} and C_{GS2}. They are not that small. Do they create a pole of importance?

Again, we have only two capacitances, that yields a second-order equation, the roots of which give two poles. They are plotted in a pole-zero position diagram with C_m as a variable as shown in this slide.

For low values of C_m, the

load capacitance C_L is obviously dominant again. Again, there is a non-dominant pole at f_{nd}, caused by time constant C_m/g_{m2}. If g_{m2} is similar to g_{m1}, then this non-dominant pole is a lot higher than the GBW, as C_L is definitely a lot larger than C_m.

For high values of C_m however, the time constant $r_{o1}C_m$ takes over. This is never likely to occur, as for this region C_m must be larger than $A_{v2}C_L$, which is a very high capacitance indeed.

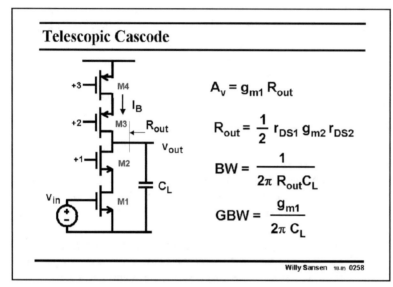

Telescopic Cascode

$$A_v = g_{m1} R_{out}$$

$$R_{out} = \frac{1}{2} r_{DS1} g_{m2} r_{DS2}$$

$$BW = \frac{1}{2\pi R_{out}C_L}$$

$$GBW = \frac{g_{m1}}{2\pi C_L}$$

Willy Sansen 10.05 0258

0258

Ideal DC current sources do not exist, however. They have to be realized by means of transistors. Such a realization is shown in this slide. MOSTs M3 and M4 are used in series to give a DC current source with an high output resistance.

If the output resistance of transistors M1/M2 is about the same as the output resistance of transistors M3/M4, then the total output resistance R_{out} is about half of that.

The bandwidth is obviously determined by the load capacitance. The GBW is then the same as for a single-transistor amplifier.

This cascode configuration is called the Telescopic cascode, as all transistors are in series between supply line and ground.

The main disadvantage of this cascode is that all transistors must be kept in the saturation region. This means that the minimum voltage v_{DS} across each transistor is about $V_{GS} - V_T$. If we take $V_{GS} - V_T \approx 0.2$ V, we find that the maximum output voltage cannot be higher than the supply voltage minus 0.4 V, and cannot be lower than 0.4 V. The maximum swing is thus 0.8 V lower than the supply voltage. This is a major loss for low supply voltages!

0259

A folded cascode is used more often, because it has some advantages when used as a differential circuit (see Chapter 8). It is called "folded" because the cascode transistor is now a pMOST rather than a nMOST. The small-signal current i_{ds1}, which is determined by the input transistor M1, now flows upwards rather than downwards through the cascode transistor M2.

Normally the biasing current I_{B1} is equally split up over the two transistors, so that the DC currents in both transistors are the same. The two current swings are also the same.

All other specifications such as the gain, the bandwidth and the GBW, are the same as for a telescopic cascode. Note, however, that the current consumption is twice that of a telescopic cascode.

Moreover, the maximum output voltage swing is about the same. If we realize the current

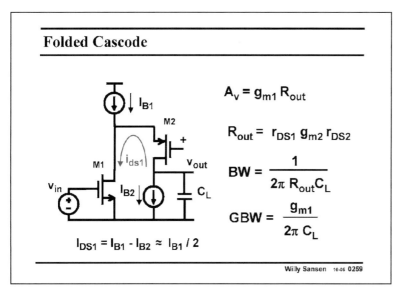

Folded Cascode

$$A_v = g_{m1} R_{out}$$

$$R_{out} = r_{DS1} g_{m2} r_{DS2}$$

$$BW = \frac{1}{2\pi R_{out} C_L}$$

$$GBW = \frac{g_{m1}}{2\pi C_L}$$

$$I_{DS1} = I_{B1} - I_{B2} \approx I_{B1} / 2$$

Willy Sansen 10-05 **0259**

source I_{B1} with one single MOST and the current source I_{B2} with two MOSTs in series, then the maximum output swing is again 0.8 V smaller than the supply voltage.

We now find a good argument to use a folded cascode rather than a telescopic one. Twice the current consumption for a folded cascode is a real drawback!!

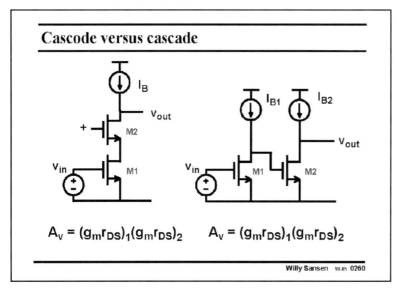

Cascode versus cascade

$$A_v = (g_m r_{DS})_1 (g_m r_{DS})_2 \qquad A_v = (g_m r_{DS})_1 (g_m r_{DS})_2$$

Willy Sansen 10-05 **0260**

0260

Note also that we are dealing here with a cascode, not a cascade amplifier. Such a cascade amplifier simply consists of two consecutive single-stage amplifiers.

The voltage gains are equal but the bandwidths and GBW are very different.

Moreover, the power dissipation of a cascade or two-stage amplifier is much higher.

0261

Indeed a cascode is actually a single-stage amplifier with gain enhancement (at low frequencies). This means that one single node is at high impedance. This is the node where the gain is realized and where the signal swing is large. This is also evident from the node where the load capacitance determines the dominant pole.

In such a single-stage amplifier the GBW is always determined by that load capacitance and the input transconductance.

A cascade amplifier is actually a two stage amplifier. This means that two nodes in that circuit are at high impedance. This also means that there are two nodes where the capacitance to ground gives a pole. Two poles can create stability problems. This problem can be eliminated

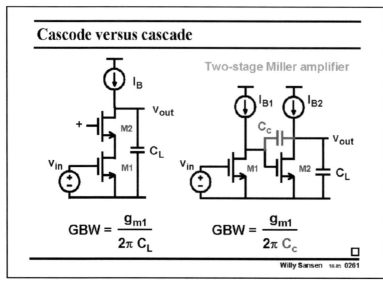

Cascode versus cascade

Two-stage Miller amplifier

$$\text{GBW} = \frac{g_{m1}}{2\pi\,C_L}$$

$$\text{GBW} = \frac{g_{m1}}{2\pi\,C_c}$$

Willy Sansen 10-05 0261

by adding a Miller capacitance across the second stage, as explained in Chapter 5. It is called a compensation capacitance C_c.

In such a two-stage amplifier the GBW is always determined by that compensation capacitance and the input transconductance.

Obviously, any addition of capacitance usually increases the power consumption. From this point of view, a single-stage amplifier is usually better. One exception may be the output swing. If only one transistor is used for the current source I_{B2}, then the output swing is only 0.4 V less than the supply voltage.

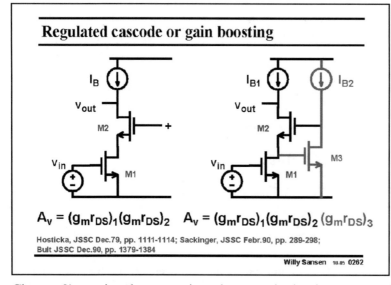

Regulated cascode or gain boosting

$$A_v = (g_m r_{DS})_1 (g_m r_{DS})_2$$

$$A_v = (g_m r_{DS})_1 (g_m r_{DS})_2 (g_m r_{DS})_3$$

Hosticka, JSSC Dec.79, pp. 1111-1114; Sackinger, JSSC Febr.90, pp. 289-298;
Bult JSSC Dec.90, pp. 1379-1384

Willy Sansen 10-05 0262

0262

For deep submicron CMOS the gain contributed by two transistors may not be sufficient. For MOSTs with gate lengths of 90 nm and less the gain of one single transistor is less than 10!

To obtain more gain, feedback can be applied around the cascode transistor. It is then called a regulated cascode (Hosticka, Sackinger). This is also called gain boosting (Bult).

This feedback is actually parallel-series feedback (see Chapter 8), causing the output impedance to rise by the amount of feedback gain. The gain goes up by the same amount.

If we realize that feedback amplifier by just one transistor M3, the gain of that transistor is added to the total gain.

An additional advantage of that feedback amplifier is that the impedance in the middle (at the gate of M3) is reduced by the same feedback gain.

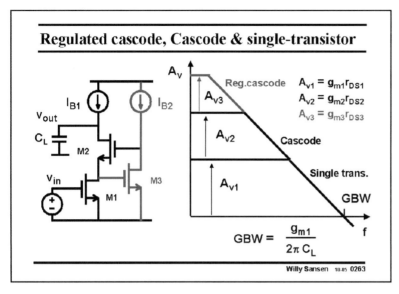

0263

Note, however, that gain boosting adds another gain enhancement at low frequencies. It does not alter the GBW.

It is therefore good to compare the performance of a regulated cascode, as cascode and a single-stage amplifier. All of them have the same GBW but widely different gains and bandwidths.

Clearly, if this amplifier is to be only used at high frequencies, then there is no need for cascodes!

Regulated cascodes will always be used whenever we have to combine high GBW with high gain. For high-frequency performance we need to use high $V_{GS} - V_T$ and minimum channel length, which inevitably leads to lower gains. By means of gain boosting we have a circuit trick, by which we can increase the gain.

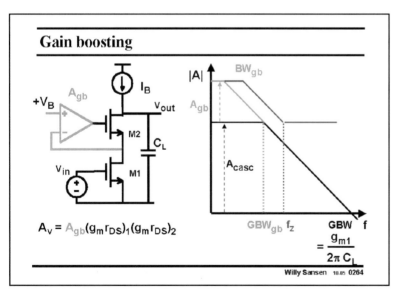

0264

It is obvious that the gain boosting amplifier, which was just one transistor before, can be replaced by a full operational amplifier with large gain A_{gb}. Its non-inverting input is connected to a biasing voltage V_B. The feedback loop will ensure that the voltage at the source of transistor M2 is kept as constant as possible.

In this case, full gain is added on top of the gain characteristic of the cascode. Obviously, this is only required if real high gains are required at real low frequencies. This is the case, for example, in low-distortion amplifiers at audio frequencies and below.

However, the design of such a gain boosting stage is not trivial. It has its own gain A_{gb}, bandwidth BW_{gb}, and hence GBW_{gb}. We have to make sure that the GBW coincides exactly with the BW of the original cascode amplifier, if not a pole-zero doublet occurs. Such a doublet is lethal for the settling time of that amplifier, as explained next.

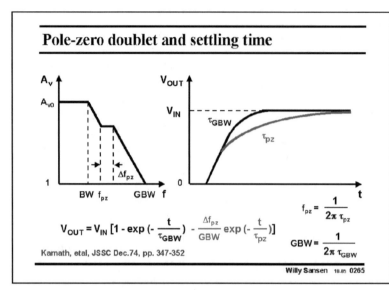

Pole-zero doublet and settling time

$$V_{OUT} = V_{IN} \left[1 - exp\left(- \frac{t}{\tau_{GBW}}\right) - \frac{\Delta f_{pz}}{GBW} exp\left(- \frac{t}{\tau_{pz}}\right) \right]$$

Kamath, etal, JSSC Dec.74, pp. 347-352

$$f_{pz} = \frac{1}{2\pi \tau_{pz}}$$

$$GBW = \frac{1}{2\pi \tau_{GBW}}$$

Willy Sansen 10.05 0265

0265

If a pole-zero doublet occurs in the middle of the Bode diagram, then the settling time is ruined.

The settling is the time required to reach the final value of the output voltage, within a certain error. For example, when we take an opamp with unity-gain feedback, then the bandwidth BW coincides with the GBW. When we apply a step waveform to that amplifier, we expect the output to follow with a time constant $1/(2\pi GBW)$, as described by an exponential.

To reach a settling time of 0.1%, we need to wait $\ln(1000)$ or 6.9 times this time constant.

When we have a pole-zero doublet at fairly low frequencies f_{pz}, with a large spreading Δf_{pz}, then an additional exponential shows up in the time response, with a much larger time constant $1/(2\pi f_{pz})$.

It will take much more time to reach the final output voltage within 0.1%. The 0.1% settling time is much longer.

In all switching applications, such as switched-capacitor filters, the settling time determines what is the minimum width of the clock pulses. It determines the maximum frequency of such a system (see Chapter 14). In such systems, pole-zero doublets must be avoided at all cost!

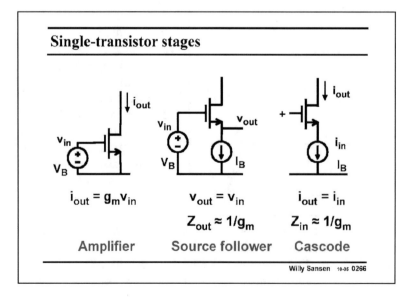

Single-transistor stages

$i_{out} = g_m v_{in}$

Amplifier

$v_{out} = v_{in}$

$Z_{out} \approx 1/g_m$

Source follower

$i_{out} = i_{in}$

$Z_{in} \approx 1/g_m$

Cascode

Willy Sansen 10-05 0266

0266

Finally, an overview is given of all three single-transistor stages. Their gains, input- and output impedances are added under all different circumstances of source and load impedance.

Remember that there are three of them. The transconductance controls the voltage-to-current gain in an amplifier.

A source follower has unity voltage gain but low output resistance.

The cascode has unity current gain but high output resistance.

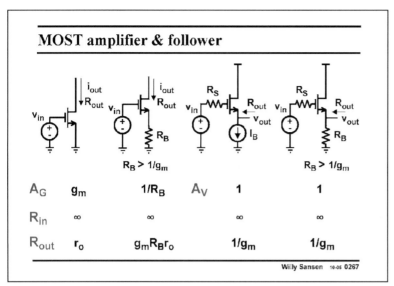

0267
The gains of a single-transistor amplifier and a cascode are given in this slide. The input and output impedances are also added.

It is clear that the input impedance is always infinity. The output impedance is resistive as no capacitances are included. It differs widely depending on the actual circuit configuration.

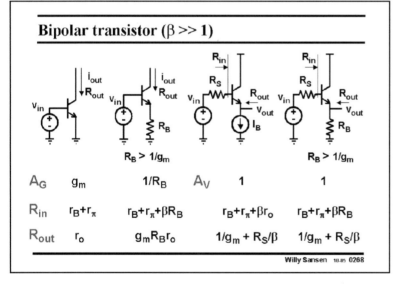

0268
The same can be done for a bipolar transistor. The beta of the bipolar transistor is always assumed to be much larger than unity.

However, the input impedances are no more at infinity. They involve r_π and base resistance r_B.

The output impedances are also very different.

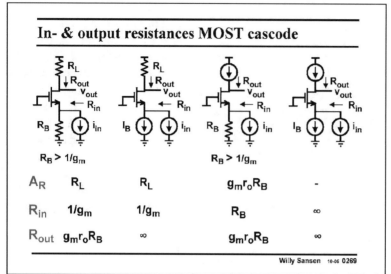

0269

For a MOST cascode, the gains, the input and output impedances are calculated as well, again without capacitances.

As a gain, the output voltage is taken versus input current. The results are very different depending on whether the input current source has an infinite output resistance or an output resistance R_B.

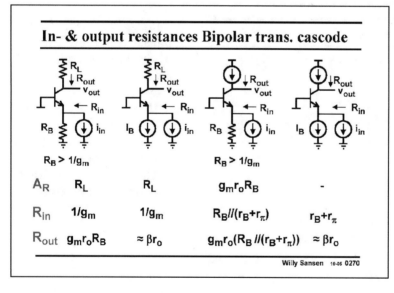

0270

Finally, the gains, input and output impedances are given for a bipolar transistor cascode.

They are clearly even more important than with a MOST.

Calculation of A_R for a MOST cascode

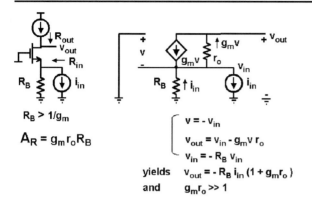

$R_B > 1/g_m$

$A_R = g_m r_o R_B$

$$v = -v_{in}$$
$$v_{out} = v_{in} - g_m v \, r_o$$
$$v_{in} = -R_B \, v_{in}$$

yields $\quad v_{out} = -R_B \, i_{in} (1 + g_m r_o)$

and $\quad g_m r_o \gg 1$

Willy Sansen 10.05 0271

0271

All calculations have been carried out by substituting the transistor by its small-signal equivalent circuit (with only g_m and r_o), and solving the Kirchoff equations. An example is given in this slide. It is the derivation of the transresistance (output voltage to input current ratio) of a MOST cascode.

Note that a DC current source can simply be omitted in the small-signal equivalent circuit whereas a DC voltage source is simply a short.

Table of contents

☐ **Single-transistor amplifiers**

☐ **Source followers**

☐ **Cascodes**

Willy Sansen 10.05 0272

0272

Now that we know nearly everything about the three single-transistor configurations, it is time to have a look at the elementary two-transistor configurations.

The first one is a current mirror, this is followed by a differential pair.

Most of the analog electronics can be built up by means of these two elementary circuits.

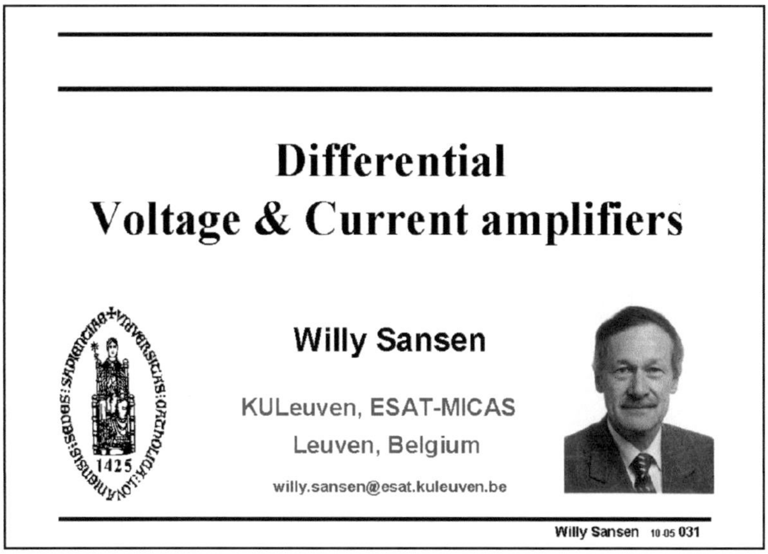

031

Differential
Voltage & Current amplifiers

Willy Sansen

KULeuven, ESAT-MICAS
Leuven, Belgium

willy.sansen@esat.kuleuven.be

Willy Sansen 10 05 **031**

All analog circuits are built up by use of a very limited number of elementary building blocks. Thorough knowledge of these simple blocks is therefore essential in obtaining insight into more complicated circuit schematics. This is why they are considered separately and analyzed in great detail.

We now have a complete knowledge of single-transistor circuits. Now we will concentrate on the study of current mirrors and differential pair. These form the cornerstone of all analog design.

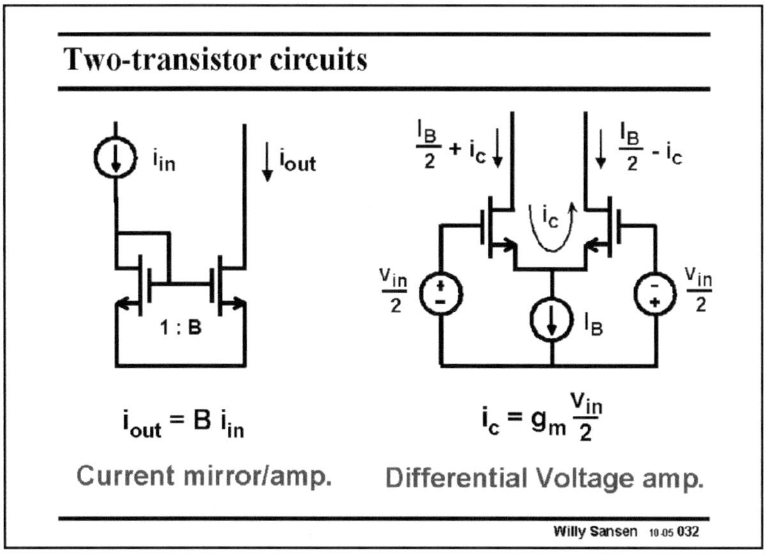

032

Two-transistor circuits

$i_{out} = B \, i_{in}$

Current mirror/amp.

$i_c = g_m \dfrac{v_{in}}{2}$

Differential Voltage amp.

Willy Sansen 10 05 **032**

A current mirror, in its simplest form, is illustrated on the left. A differential pair is shown on the right.

A current mirror is actually a combination of a diode-connected transistor followed by a single-transistor amplifier. The first one converts the input current into a voltage whereas the latter one converts the voltage into a current.

The current ratio will be quite accurate, simply because the non-linearity of the diode-connected MOST is compensated by the non-linearity of the amplifying MOST.

If the ratio of their W/L's is B, then the current ratio is also B. Indeed, these MOSTs have the same V_{GS} and hence $V_{GS} - V_T$. This ratio is the current gain.

A differential pair consists of two equal transistors, which are both operating as single-transistor amplifiers. The input voltage is differential now, as is the output voltage.

We consider the current mirror first.

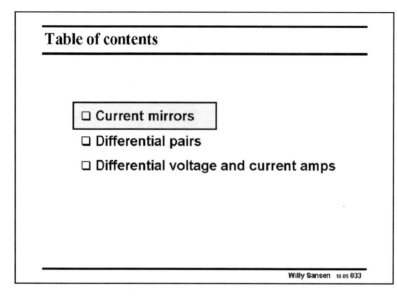

033

We begin with the current mirror first because it is a simpler circuit to understand.

Later on we will discuss the differential pair as a first full-differential amplifier. It will be followed by many more differential amplifiers. We need them because they are insensitive to substrate noise and ground bounces. They are therefore necessary in all mixed-signal design, where analog circuits are added on the substrate as digital systems.

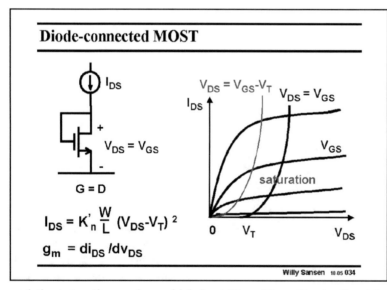

034

The input of a current mirror consists of a diode-connected transistor.

Connecting Collector to Base in a bipolar transistor, gives us a real base-emitter diode. In a MOST however, there is no Gate-Source diode. However, connecting the Drain to the Gate provides us with something similar.

Indeed, the current voltage characteristic is obtained by shifting the curve, separating the linear and the saturation region, which is at $V_{DS} = V_{GS} - V_T$, to the right by V_T.

As a result, we can use the current-voltage characteristic of a MOST in saturation. The resulting curve is very nonlinear, however. It resembles somewhat, a diode characteristic.

We will use this simple circuit to convert current to voltage.

035

Addition of a single-transistor amplifier to the diode connected MOST, yields a current mirror. The amplifier compensates the nonlinearity of the diode to provide a perfectly linear current ratio B.

This circuit will be utilized for both biasing and for wideband current amplification.

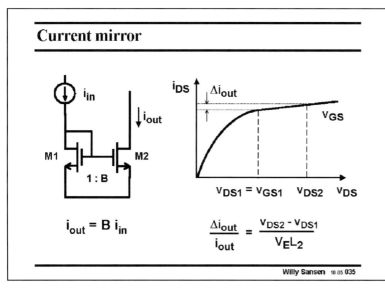

Current mirror

$i_{out} = B \, i_{in}$

$$\frac{\Delta i_{out}}{i_{out}} = \frac{v_{DS2} - v_{DS1}}{V_E L_2}$$

Willy Sansen 10-05 035

In practice, the current ratio is not that accurate. Indeed, both transistor may not operate at the same v_{DS} voltage. A difference in v_{DS} will give a difference in current, as the $i_{DS} - v_{DS}$ characteristic is not that flat.

This difference is easily calculated. It is related to the Early voltage and hence to the channel length L_2. The larger the channel length, the flatter the curve and the smaller the current difference will be.

It is difficult to make that v_{DS} voltage difference zero. This is why we resort to circuit techniques.

036

Now let us focus on how to make the current difference zero. For this purpose we have to add two transistors M3 and M4, which have as main goal to make the voltages v_{DS} across the current mirror devices M1 and M2 as equal as possible. We have two realizations.

The left one consists of a voltage divider by means of two diode connected transistors M1 and M3, connected to a cascoded amplifier M2 and M4.

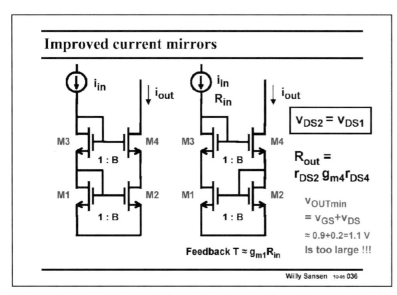

Improved current mirrors

$v_{DS2} = v_{DS1}$

$R_{out} = r_{DS2} \, g_{m4} \, r_{DS4}$

$v_{OUTmin} = v_{GS} + v_{DS} \approx 0.9 + 0.2 = 1.1 \text{ V}$ Is too large !!!

Feedback $T \approx g_{m1} R_{in}$

Willy Sansen 10-05 036

Transistors M4 and M2 have the same W/L ratio as M2 and M1, which is B. Since M1 and M2 have the same V_{GS}, transistors M3 and M4 must have the same V_{GS} as well. The currents through M3 and M4 have the same ratio B as their W/L ratios. As a result the voltages v_{DS1} and v_{DS2} must also be the same. The current ratio will therefore be quite accurate.

A similar reasoning is valid for the current mirror on the right. There is one major difference however. The circuit on the right is a feedback amplifier with loop gain T. Since all time constants in that loop are of the same order of magnitude, they create a system with several poles. As a result, peaking can occur in the current transfer characteristic.

An important disadvantage of both current mirrors, is that their minimum output voltage (compliance voltage) is quite high! As a result they cannot be used in low-voltage applications!

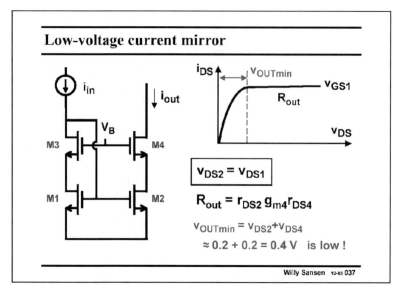

037

For low supply voltages, this current mirror is an ideal choice!

This is actually a conventional two-transistor current mirror, in which two cascodes M3 and M4 have been added. We find the same advantages as in both previous current mirrors:

1. Both voltages v_{DS1} and v_{DS2} are equal resulting in an accurate current ratio
2. The output resistance is very high as for any cascode configuration

We now have a considerable advantage in that we can set the biasing voltage such that only 0.2 Volt is left across M2 and M4, keeping both of them in saturation! For a V_T of 0.7 V, the V_{GS} is about 0.9 V. Hence V_B must be about 1.1 V.

The compliance voltage V_{OUTmin} is thus now reduced from 1.1 V to 0.4 V, which is a significant difference.

This is also a disadvantage, we now need an external biasing voltage V_B.

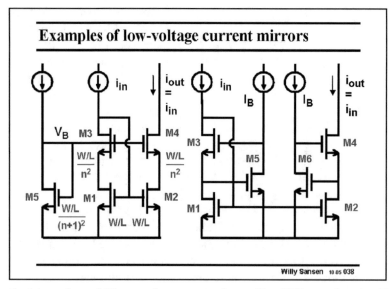

038

Two examples are given of the low-voltage current mirror described previously. Both of them use cascodes. Both of them provide feedback from the Drain of transistor M3 to the Gates of current-mirror devices M1 and M2. Both of them provide a large output impedance and allow a large output voltage swing.

On the left, an example is given of how to provide the biasing voltage V_B of the cascode devices. It is clear that transistor M5 must have a much smaller W/L so that its V_{GS} is about 0.2 V larger than that of cascode transistors M3 and M4. We need about 0.2 V as a V_{DS} for both current mirror devices M1 and M2. A good value for the parameter is approximately 5.

The current mirror on the right is simply the low-voltage current mirror given previously, in which gain-boosting is applied to the cascode transistors. It allows the same output swing as the circuit on the left but the output impedance is even higher.

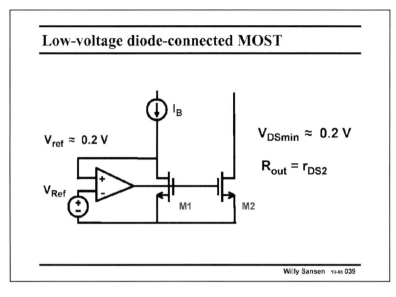

Low-voltage diode-connected MOST

$V_{ref} \approx 0.2 \, V$

V_{Ref}

I_B

$M1$ $M2$

$V_{DSmin} \approx 0.2 \, V$

$R_{out} = r_{DS2}$

Willy Sansen 10-05 039

039

If the compliance voltage of 0.4 V is still too high, one single output transistor can be used.

An opamp is then used to provide feedback around the input device M1. Its drain voltage is then only 0.2 V, which is a lot less than the 0.9 V in a conventional current mirror (for a V_T of 0.7 V).

The output impedance is not very high. It is simply the output resistance of the output transistor M2.

There are better uses for an opamp as shown next.

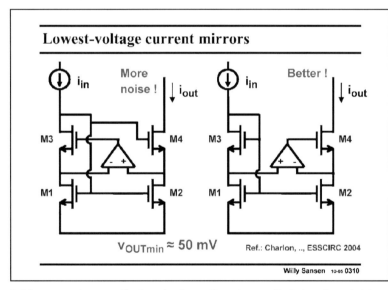

Lowest-voltage current mirrors

i_{in}

More noise !

i_{out}

i_{in}

Better !

i_{out}

M3 M4 M3 M4

M1 M2 M1 M2

$V_{OUTmin} \approx 50 \, mV$ Ref.: Charlon, .., ESSCIRC 2004

Willy Sansen 10-05 0310

0310

Two current mirrors are shown which exploit the opamp for both purposes, i.e. to reduce the compliance voltage as much as possible and also to increase the output resistance as much as possible. Both of them are derived from the 4-transistor current source described previously.

The difference is that an opamp is either added to the left cascode or to the right cascode.

On the right one, it is fairly easy to see that gain boosting is applied to the right cascode, increasing the output resistance by the gain of that opamp.

It is not so obvious that the compliance voltage can be as low as a few tens of mV's. Indeed the gain of the feedback loop is so high, due to the opamp, that the transistors M1 and M2 can enter the linear region. Moreover cascode transistor M4 can also enter the linear region. The output resistance will be reduced but the opamp provides sufficient gain to compensate for this loss in output resistance.

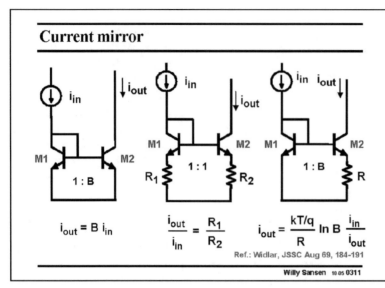

Ref.: Widlar, JSSC Aug 69, 184-191

0311

All MOST current sources can be duplicated with bipolar transistors. Historically, the bipolar ones were developed first and the MOST circuits were copies of the bipolar ones.

The first one is very conventional current mirror.

The second one uses series resistors to set the current ratio's to the resistor ratio, instead of the transistor ratio. It is easier to make a current ratio more accurate (see Chapter 12).

In MOST current mirrors there is no need to include in series resistors, as increasing the $V_{GS}-V_T$ values has the same effect (see next Chapter).

Moreover, the output resistance is increased by use of the series resistors.

The third circuit is not a current mirror but a current reference. Indeed, leaving out one of the two resistors, and taking M2 a factor B larger than M1, provides an output current that is independent of the input current, at least if we use another current mirror on top to ensure that $i_{out}=i_{in}$. This will all be repeated in Chapter 13.

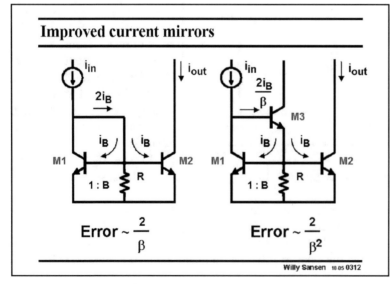

0312

One additional problem is encountered with bipolar transistors: they carry a base current i_B.

These base currents are both subtracted from the input current source. The error between output and input current is thus about $2/\beta$. We assume resistor R to be really large.

Addition of another transistor M3 reduces this error by the another beta. Even for small beta values, this error can be made small.

The role of this resistor is then more clear. It increases the current in transistor M3. This will increase its beta somewhat. Indeed, in some older bipolar technologies, the beta drops rapidly for small collector currents. In BiCMOS technologies however, the processing is a lot cleaner. The beta hardly drops at low currents and this resistor can then be omitted.

In a BiCMOS technology, a MOST could be used for transistor M3. Its Gate current is zero and so would be the current error. We have to be careful however with the M1, M3 feedback loop. If we reduce the capacitance C_{BE3} too much, by replacing it with a small MOST, we may find some peaking in the current transfer function, as a result of too many poles too close together! A pole-zero position diagram of the current gain with C_{BE3} as a variable is the best way to study this phenomenon.

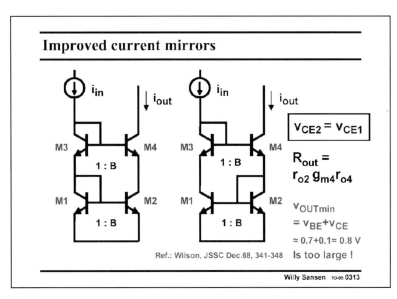

0313

The same techniques can be used as before to make the current difference zero. Two transistors M3 and M4 are added, which have as a first objective to make the voltages v_{CE} across the current mirror devices M1 and M2 as equal as possible. We have again two realizations.

The left one is a passive circuit as it consists of a voltage divider utilizing two diode connected transistors M1 and M3, connected to a cascoded amplifier M2 and M4.

Transistors M4 and M2 have the same W/L ratio as M2 and M1, which is B. Transistors M3 and M4 must have the same V_{BE} because their currents have the same ratio B. As a result, the voltages v_{CE1} and v_{CE2} must also be the same. The current ratio will therefore be quite accurate.

The same is true for the current mirror on the right. It is an active circuit however. It is a feedback amplifier with loop gain T. Multiple poles occur, leading to peaking in the current transfer characteristic.

An important disadvantage of both current mirrors, is again that their minimum output voltage (compliance voltage) is quite high! As a result they cannot be used in low-voltage applications! One of the low-voltage current mirrors should be used!

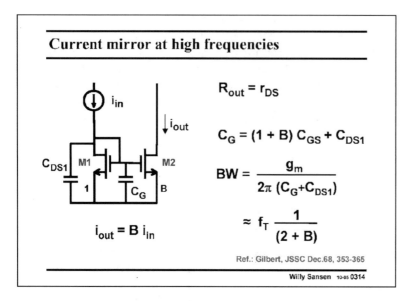

0314

Let us now have a closer look at the high-frequency performance.

The most important capacitances are shown explicitly. The C_{GS} of transistor M2 is obviously B times larger than the one of transistor M1. This is why the total capacitance at the inner node is C_G.

Again we assume C_{DS1} to be very similar in size to C_{GS}. As a result, the bandwidth of this current amplifier, with current gain B, is again a fraction of f_T.

If we design the two MOSTs with large $V_{GS} - V_T$ and smallest possible channel length L, then a very high bandwidth can be expected.

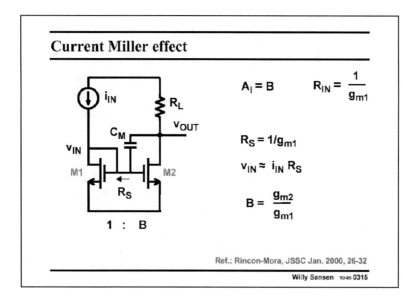

0315

Such a current mirror or amplifier can also cause capacitance multiplication by the Miller effect, very much as a voltage amplifier causes capacitance multiplication.

The Miller capacitance is C_M, and is connected from the voltage output, realized by means of a load resistor, and the middle point.

The current gain is again B. The input resistance is again $1/g_{m1}$. It serves as a source resistor $_{RS}$ for the amplifying transistor.

For the calculation of the capacitance multiplication, we simplify the circuit, as shown next.

0316

The bandwidth denoted by f_{-3dB} is now readily calculated.

It clearly shows that capacitance C_M is indeed multiplied by the voltage gain A_{v2}. This latter gain is also determined by current gain factor B.

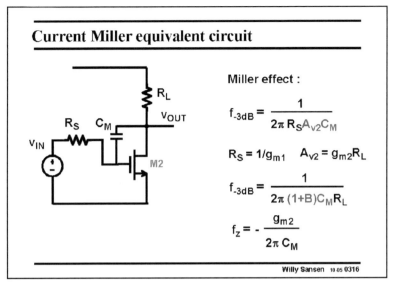

Current Miller equivalent circuit

Miller effect :

$$f_{-3dB} = \frac{1}{2\pi R_S A_{v2} C_M}$$

$$R_S = 1/g_{m1} \quad A_{v2} = g_{m2}R_L$$

$$f_{-3dB} = \frac{1}{2\pi (1+B)C_M R_L}$$

$$f_z = -\frac{g_{m2}}{2\pi C_M}$$

Willy Sansen 10 05 0316

There is also a zero as for the voltage Miller effect (see Chapter 2). This zero occurs at real high frequencies however and is usually negligible.

This phenomenon is called the current Miller effect. It will be used to realize large capacitances on a chip. A well known application is the compensation of operational amplifiers (see Chapter 5).

Table of contents

- ☐ **Current mirrors**
- ☐ **Differential pairs**
- ☐ **Differential voltage and current amps**

Willy Sansen 10 05 0317

0317

The second and more important two-transistor elementary circuit is a differential pair. Actually, it is two single-transistor amplifiers in parallel, in order to reject disturbing common-mode signals. It will thus be the basis for all fully-differential circuits.

In mixed-signal circuitry, only fully-differential circuits can be allowed, in order to reject ground noise, supply line spikes, etc.

First of all, wee start with a simple differential pair. Addition of two load resistors yields a voltage differential amplifier

0318

In such a voltage differential amplifier, two equal transistors are used and two equal load resistors. We will never be able to make these transistors exactly identical. Small differences will always exist. They will give rise to mismatch, offset, etc. This will be discussed in Chapter 12. In this chapter we will always assume that the transistors are exactly the same and so are the resistors.

Such an amplifier is biased by a DC current source I_B.

There are two input voltages and two output voltages. The input voltages are referred to ground. This is only possible if two supply lines are used, for example V_{SS} at -5 V and V_{DD} at

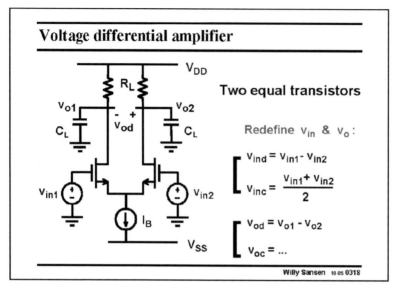

Voltage differential amplifier

Two equal transistors

Redefine v_{in} & v_o :

$$v_{ind} = v_{in1} - v_{in2}$$
$$v_{inc} = \frac{v_{in1} + v_{in2}}{2}$$

$$v_{od} = v_{o1} - v_{o2}$$
$$v_{oc} = ...$$

Willy Sansen 10-05 0318

5 V. This total supply voltage of 10 V can also be used as a single supply voltage with respect to ground. In this case however, the inputs voltages must be referred to a DC reference voltage, somewhere between 10 V and ground. Actually, this has now become common. Supply voltages are used such as 1.8 and 2.5 V, depending on the technology. An internal reference must then be derived, for example 1 V, to make sure that the input devices are properly biased.

Whatever the input voltages are, we will redefine them, to gain insight into the operation of this circuit. We define the differential input voltage and the common-mode or average input voltage, as shown.

The same is true for the two output voltages.

We will be interested mainly in the differential output/input voltages!

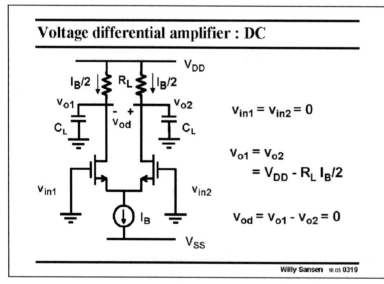

Voltage differential amplifier : DC

$$v_{in1} = v_{in2} = 0$$

$$v_{o1} = v_{o2}$$
$$= V_{DD} - R_L I_B/2$$

$$v_{od} = v_{o1} - v_{o2} = 0$$

Willy Sansen 10-05 0319

0319

Let us first have a closer look at the DC operation.

When both input voltages are zero, then both transistors have exactly the same v_{GS}. They must now have the same current. Since the sum of the currents is always I_B, each transistor must carry a DC current $I_B/2$.

Since both load resistors are the same, the voltages across them must be the same as well. The differential voltage is therefore zero.

As could be expected, zero differential input voltage gives zero differential output voltage.

0320

When we now apply a differential input voltage, divided equally over both inputs, both input voltages are the same in amplitude, but opposite in sign.

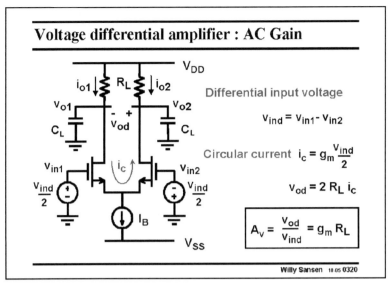

Voltage differential amplifier : AC Gain

Differential input voltage

$$v_{ind} = v_{in1} - v_{in2}$$

Circular current $i_c = g_m \dfrac{v_{ind}}{2}$

$$v_{od} = 2 R_L i_c$$

$$A_v = \frac{v_{od}}{v_{ind}} = g_m R_L$$

Willy Sansen 10 05 0320

If we assume that the left gate voltage increases, then the left transistor current also increases. The right transistor current must decrease by the same amount, as the sum of the currents is still I_B.

This increase in current is the AC current or rather the circular current. It is added to the current $I_B/2$ on the left and subtracted on the right. It flows as indicated by the arrow.

This circular current will now be converted into a differential output voltage v_{od}. Indeed, the circular current flows through both load resistors and develops v_{od}.

The gain is now easily calculated. The circular current depends on the input voltage by the transconductance. The output voltage v_{od} depends simply on this current.

Note that there are several "factors of two" involved.

The resulting expression of the voltage gain is exactly the same as for a single-transistor amplifier.

Note also that the AC current circles around through both transistors and load resistors. It does not flow through the supply lines!

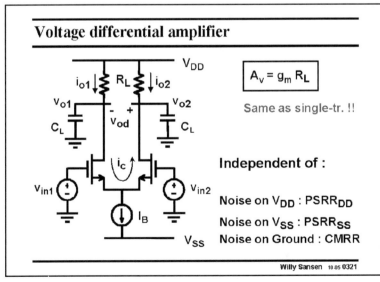

Voltage differential amplifier

$$A_v = g_m R_L$$

Same as single-tr. !!

Independent of :

Noise on V_{DD} : PSRR$_{DD}$

Noise on V_{SS} : PSRR$_{SS}$

Noise on Ground : CMRR

Willy Sansen 10 05 0321

0321

The resulting voltage gain has indeed exactly the same expression as for a single-transistor amplifier. The transconductance is now only half however, as the current per transistor is only half (for equal $V_{GS} - V_T$).

The main advantage of this stage is that this voltage gain is completely decoupled from the supply line and ground disturbances. Indeed, noise on the positive supply is also present on both output voltages to ground. It is cancelled out from the differential output voltage. The power-supply rejection ratio will thus be very high. This is in effect, the ratio of two gains, the gain from power supply to the output and the gain from input to output.

The same is true for the negative supply V_{SS}. The same is also true for the ground. Ground noise, caused by digital circuits on the same substrate, will be added to both inputs and will thus be rejected by the differential input. As a result the CMRR (Common-mode Rejection Ratio) will be high. All these factors, PSRR and CMRR will be discussed in more detail in Chapter 12.

A differential pair is therefore nothing more than a single-transistor amplifier, carried out in a differential way, in order to be able to cope with common-mode disturbances.

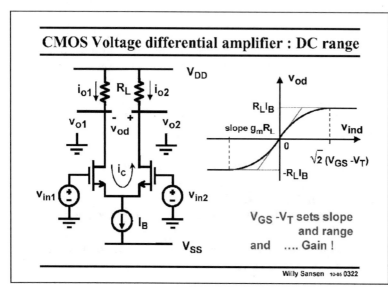

0322

The output voltage versus input voltage is illustrated in this slide.

For small input voltages, the gain is simply $g_m R_L$. It is the slope of the characteristic around zero.

For larger input voltages, the output level saturates to a voltage $R_L I_B$. For even larger input voltages the output voltage is constant. The curve in between is actually a parabola, described by the equation on the previous slide.

For small input voltages, the factor under the square root is simply unity. What remains in that expression is simply g_m.

For an input voltage equal to $\sqrt{2}(V_{GS} - V_T)$ the top of the parabola is reached. The tangent at this point provides a very smooth transition to a constant output voltage.

It is clear that $V_{GS} - V_T$ is the parameter that controls both the g_m and the range. For a small $V_{GS} - V_T$, the gain is high (with a steep slope) but the range is small. For RF receivers a large $V_{GS} - V_T$ is preferred; the range is wide but the gain is small. Moreover, a large $V_{GS} - V_T$, provides a highfrequency response.

0323

Bipolar transistors have an even lower $V_{GS} - V_T$, or specifically kT/q. The gain is thus very high but the range very small. For small input voltages, the gain is again $g_m R_L$. It is the slope of the characteristic around zero.

For larger input voltages, the output level saturates again to a voltage $R_L I_B$. For even larger input voltages the output voltage is constant. The curve in between consists of exponentials such that a smooth transition is guaranteed. The actual curve will be calculated on the next slide.

It is clear that there is no way to control both the g_m and the range. They are always set by kT/q. The only way to change this ratio is to insert emitter resistors. The larger the resistor, the wider the range but the smaller the small-signal gain becomes.

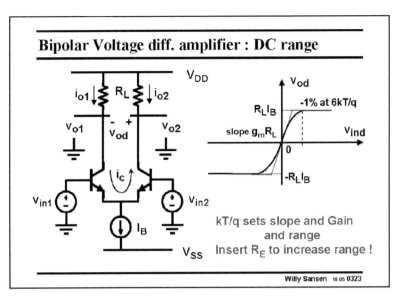

Bipolar Voltage diff. amplifier : DC range

kT/q sets slope and Gain
and range
Insert R_E to increase range !

Willy Sansen 10 05 0323

MOST Voltage diff. amplifier : large input signals

$$\frac{i_{Od}}{I_B} = \frac{v_{Id}}{(V_{GS}-V_T)} \sqrt{1 - \frac{1}{4}\left(\frac{v_{Id}}{V_{GS}-V_T}\right)^2}$$

v_{Id} is the differential input voltage

i_{Od} is the differential output current ($g_m v_{Id}$) or

twice the circular current $g_m v_{Id}/2$

I_B is the total DC current in the pair

Note that $g_m = \dfrac{I_B}{V_{GS}-V_T} = K' W/L (V_{GS} - V_T)$

Willy Sansen 10 05 0324

0324
When larger input signals are applied, an increasing amount of current flows on one side. For a very large input signal (as for a digital input drive) all current flows in one transistor and the other ones is off. In this case the maximum output voltage is reached, i.e. $R_L I_B$. The differential pair therefore also behaves as a limiter, when overdriven.

In this case we can no longer use the transconductance, we have to use the full current expressions. Taking into account that

– the input voltage v_{Id} is applied between both v_{GS}'s and that
– the sum of the currents is still I_B,

we can fairly easily find the differential output current i_{Od}, which is twice the Drain current in one transistor. The differential output voltage is then this current multiplied by R_L.

It is clear that the output voltage is fairly nonlinear, as shown next. We firstly want to derive the transfer characteristic of a differential pair, for comparison purposes.

Bipolar Voltage diff. amp. : large input signals

$$\frac{i_{Od}}{I_B} = \tanh\frac{V_{Id}}{2\,kT/q} \qquad \tanh x = \frac{e^x - e^{-x}}{e^x + e^{-x}} = \frac{2e^x - 1}{2e^x + 1}$$

v_{Id} is the differential input voltage
i_{Od} is the differential output current ($g_m v_{Id}$) or
      twice the circular current $g_m v_{Id}/2$
I_B is the total DC current in the pair

Note that $g_m = \dfrac{I_B}{2\,kT/q}$

Willy Sansen 10 05 0325

0325

For a differential pair with bipolar transistors, a similar reasoning can be applied. For a large input signal (as for a digital input drive) all current flows in one transistor and the other ones is off. The maximum output voltage is reached, i.e. $R_L I_B$. The differential pair therefore behaves as a limiter, when overdriven.

In this case we have to use the full current expression. Taking into account that

– the input voltage is applied between both v_{BE}'s and that
– the sum of the currents is still I_B,
  we can fairly easily find the differential output current i_{Od}, which is twice the Collector current in one transistor. This expression consists of exponentials. It is clear that this differential output current i_{Od} is very nonlinear, as shown next.

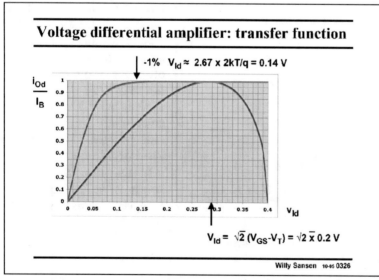

Voltage differential amplifier: transfer function

-1% $V_{Id} \approx 2.67 \times 2kT/q = 0.14$ V

$V_{Id} = \sqrt{2}\,(V_{GS}\text{-}V_T) = \sqrt{2} \times 0.2$ V

Willy Sansen 10-05 0326

0326

The differential output current i_{Od} is plotted for both a MOST differential pair (with $V_{GS} - V_T = 0.2$ V) and for a bipolar differential pair.

For a MOST differential pair, it is clear that this curve is only valid up until $v_{Id} = \sqrt{2}(V_{GS} - V_T)$ which is about 0.283 V in this example. For larger values of v_{Id}, $i_{Od} = I_B$. For small values of the input voltage, the slope is the transconductance g_m.

For a bipolar differential pair, the effect of the exponentials is clearly seen. The differential output current i_{Od} reaches its final value within 1% after about 6 times kT/q or 0.14 V.

If we consider a MOST in moderate inversion with a $V_{GS} - V_T = 52$ mV, as for a bipolar transistor, then the transconductances would be the same. The differential output current in the MOST pair then reaches its maximum value I_B already at $\sqrt{2} \times 52 \approx 70$ mV, which is 70 mV earlier than the -1% point of a bipolar differential pair!

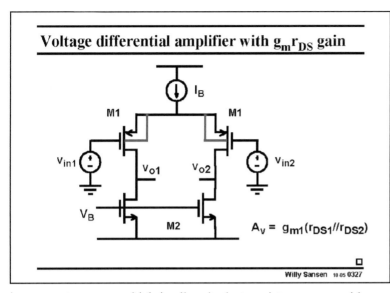

0327
The problem with the previous differential amplifier is that the gain $g_m R_L$ is quite small. We need larger values of load resistor to increase this gain. For small supply voltages this is not possible.

Sometimes resistors are used, when they need to be trimmed to reduce the effect of mismatch. See Chapter 12 for more details.

A first solution may be to use the output resistance of a MOST as a load, as shown in this slide. The gain increases to $g_m r_{DS}$, which is all a single transistor can provide.

We could also put in cascodes, as shown previously.

We will now see two more techniques to enhance the gain. They are current cancellation and bootstrapping.

Before we go into detail, we have noticed that the circuit in this slide is not possible from the view of biasing. Both voltage V_B and current I_B intend to provide biasing currents to the amplifying devices M1. This is not possible. We will have to apply common-mode feedback as explained in Chapter 9.

Finally, note that the input devices are pMOSTs, and that their Bulks are connected to their n-well. This is the most common input stage configuration in a n-well CMOS process. Connecting Bulk to Source improves the matching of the input devices, as will be explained in Chapter 12. Note that the Bulk connections are not always shown!

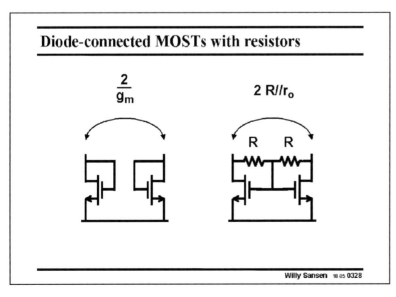

0328
If large-valued resistors are available, another self-biasing load can be found. It is shown in this slide.

Actually, a virtual ground is realized between both transistors. This point does not carry any AC signal.

The output resistance will thus be twice what is seen on either side. It is thus twice resistor R in parallel with the output resistance r_{DS} or r_o of the transistor.

It is not as high as previous. Moreover, we need to

be able to identify large resistors in a MOST process. This may be the case in an analog CMOS process but certainly not in a digital one.

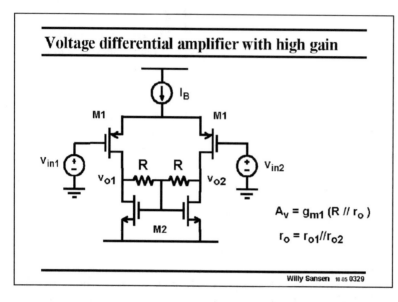

0329

Application of this load to a differential pair, gives the high-gain amplifier shown in this slide.

It is clear that the differential output signals are cancelled out at the common-Gate point of the transistors M2. The self-biasing is actually a case of common-mode feedback, as explained in Chapter 8.

The gain is moderately high.

There is no biasing problem. Current source I_B determines all currents.

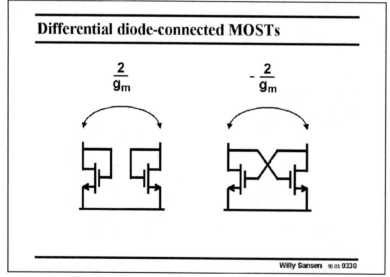

0330

Even more gain can be obtained by current cancellation

For this purpose, we need to take differential diode-connected MOSTs. Two of them provide a differential small-signal resistance of $2/g_m$.

Cross coupling them creates a positive feedback amplifier with two stages. As a result the differential resistance becomes negative or $-2/g_m$.

This negative resistance is easily controlled through the current. It is often used in oscillators, RF voltage-controlled oscillators and in wideband amplifiers.

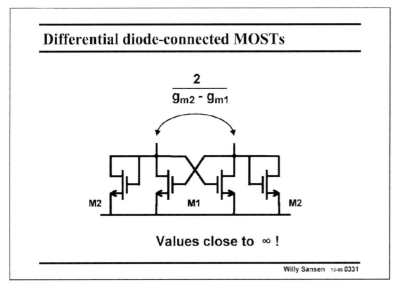

0331

Connection of a set of positive diode resistances with negative ones, allows the realization of any resistance between $1/g_m$ and infinity.

Making all devices the same cancels the AC current in this circuit, causing the differential to be very high (infinity if matching is perfect).

It is therefore an ideal load for a differential amplifier.

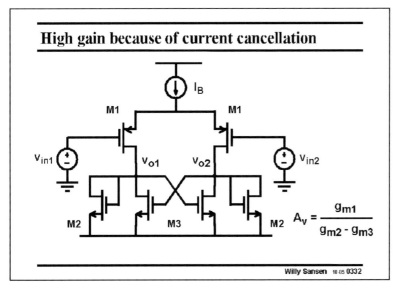

0332

Such high-gain amplifier is shown in this slide. All load MOSTs M2 and M3 have equal size. Their transconductances are now also equal. As a result they exhibit quasi-infinite resistance to the input devices, resulting in large voltage gains.

The biasing problem is also solved. Only I_B sets the currents. This load is said to be self-biasing.

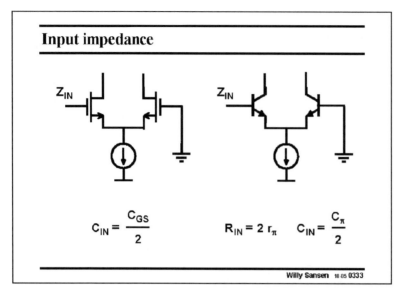

0333
The input impedance of a differential pair is two times the input impedance of a single transistor. This means that the input resistance must be doubled, as in a bipolar transistor pair on the right. The input capacitance must be halved, however.

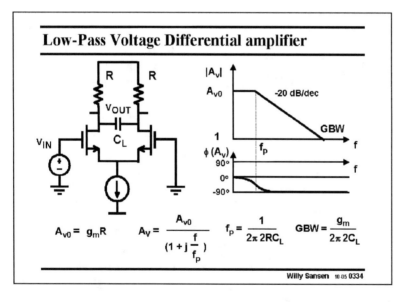

0334
In order to conclude this section on differential voltage amplifiers, a few simple circuits are added playing with capacitances.

Obviously the load capacitance C_L generates a low-pass filter characteristic. The pole time constant is two times RC_L.

Since this is a first-order filter, the slope is 20 dB/decade.

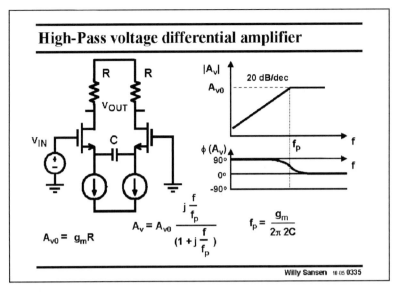

High-Pass voltage differential amplifier

$A_{vo} = g_m R$

$A_v = A_{vo} \dfrac{j\dfrac{f}{f_p}}{(1 + j\dfrac{f}{f_p})}$

$f_p = \dfrac{g_m}{2\pi \, 2C}$

Willy Sansen 10 05 0335

0335

Connecting that capacitance C between both sources, yields a high-pass characteristic.

The pole time-constant is at $g_m/2C$ because the capacitance sees two resistances $1/g_m$ in series in order to reach ground.

Again the slope is 20 dB/ decade.

The actual calculation of this gain characteristic is described in the next slide.

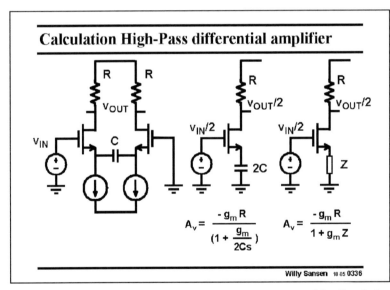

Calculation High-Pass differential amplifier

$A_v = \dfrac{-g_m R}{(1 + \dfrac{g_m}{2Cs})}$

$A_v = \dfrac{-g_m R}{1 + g_m Z}$

Willy Sansen 10 05 0336

0336

The calculation of the gain characteristics of a differential amplifier can always performed on the "differential half circuit".

For this purpose, all bridge elements have to be multiplied or divided by two. For example capacitance C must be multiplied by 2. A resistor would have to be divided by 2!

Also, both the input and output voltage are halved, by halving the circuit itself.

The gain is now easily calculated. Impedance Z provides feedback by means of feedback factor $1 + g_m Z$. Substitution of Z by $1/Cs$ or $1/Cj\omega$ then provides the final gain expression.

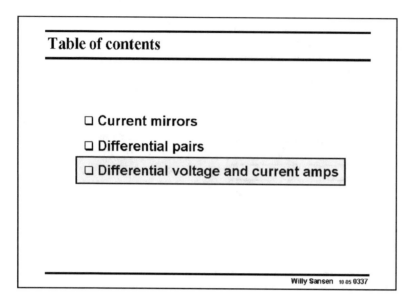

Willy Sansen 10 05 0337

0337

Now that we have a good understanding of current mirrors and voltage pairs, let us combine them in differential voltage and current amplifiers.

All subsequent circuits are actually combinations of current mirrors and differential pairs, with as the primary requirement, to increase the gain but also to convert the differential output in a single-ended one. This is necessary to avoid differential pairs in a second stage of an amplifier.

Most opamps will indeed use just a single-transistor amplifier as a second stage. The simplest amplifier usually provides the widest bandwidth. A single transistor is hard to beat!

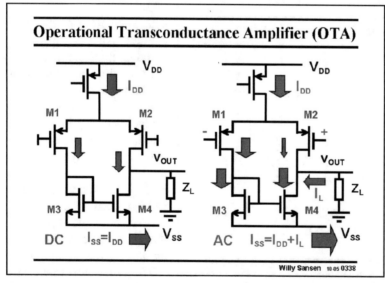

Willy Sansen 10 05 0338

0338

The amplifier in this slide is called an OTA. It consists of one differential pair loaded by a simple current mirror. It has thus one single output indeed.

On the left, the DC current flows are shown. The width of the arrow corresponds to the size of the current. The current source I_{DD} is divided by two over both input devices, and is taken up by the negative supply. Obviously, both supply currents are the same, as no current can escape. Indeed, no current can escape at the output!

On the right, the currents are shown when an input voltage is applied. This voltage is such that the current in M1 increases, causing a decrease by the same amount in the other transistor M2. The current source receives this larger current in M3 and forces the same current through M4.

At the output there is a big difference between what is offered by M2 and what is required by M4. This difference is the output current. It flows through the load impedance Z_L, and causes an output voltage.

Again the supply currents and the current through load Z_L must add up correctly, to keep Kirchoff happy!

Single-stage OTA: operation

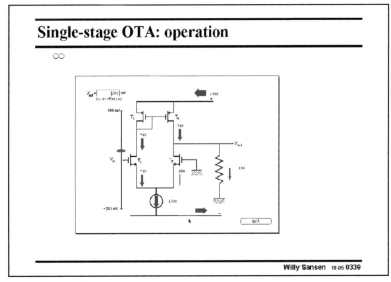

Willy Sansen 10.05 0339

0339
The same current flows are given for this inverted OTA.

An input voltage of several mV's will cause a difference in current in the input devices. The current in T1 is 74% of the total current in the current source, leaving 26% for the other transistor T2. This difference in current has previously been called circular current. It is thus 24% of the current of the DC current source or 48% of the DC transistor current!

This current of 74% flows through three devices, whereas the fourth transistor T2 only carries 26%. This is a very asymmetrical gain stage indeed. This is a result of the single-ended nature of this circuit.

The current through the load is then the difference between 74% of T4 and 26% of T2, which is 48%. The load resistance converts this current into an output voltage.

Normally there is no load resistor. It actually consists of the two output resistances r_{DS} of the transistors T2 and T4 in parallel.

Single-stage OTA

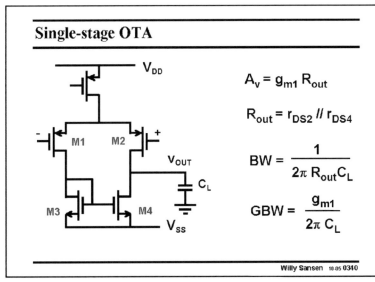

$$A_v = g_{m1} R_{out}$$

$$R_{out} = r_{DS2} // r_{DS4}$$

$$BW = \frac{1}{2\pi R_{out}C_L}$$

$$GBW = \frac{g_{m1}}{2\pi C_L}$$

Willy Sansen 10.05 0340

0340
The gain is now easily calculated.

If we denote the output resistance by R_{out}, which is again the parallel combination of the two output resistances r_{DS} of the transistors T2 and T4, then the gain is simply $g_{m1} R_{out}$.

The bandwidth is cause by the output capacitance and the same resistance R_{out}.

As a result, the GBW is again the same as for a single-transistor amplifier.

Indeed, this is a single-stage amplifier because only one single high-impedance point can be distinguished, i.e. at the output. All other nodes are on the $1/g_m$ impedance level.

The gain is not very high however. We could use cascodes to increase the gain. Gain boosting and current cancellation have also been discussed. We will now introduce a fourth technique to enhance gain, bootstrapping.

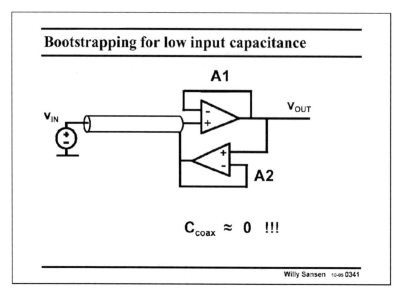

Bootstrapping for low input capacitance

A1

V_{IN}

V_{OUT}

A2

$C_{coax} \approx 0$!!!

Willy Sansen 10-05 0341

0341
Bootstrapping has actually been introduced to nullify the effect of the input capacitance of a coax cable in front of a preamplifier.

Indeed the second buffer A2 will ensure that the outer screen of the coax cable is at the same potential as the inner conductor. As a result there is no voltage difference between inner and outer wire of that cable. Its capacitance (of up to 1pF/cm) is thus nullified. The capacitance of the cable is "bootstrapped out".

The input capacitance is either quite small or the input impedance is exceedingly high!

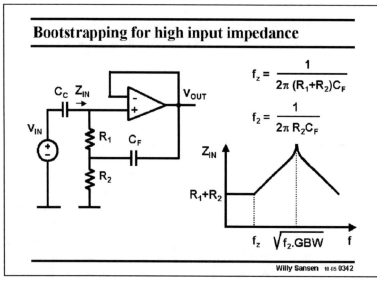

Bootstrapping for high input impedance

C_C Z_{IN}

V_{OUT}

V_{IN}

R_1 C_F

R_2

$$f_z = \frac{1}{2\pi (R_1+R_2)C_F}$$

$$f_2 = \frac{1}{2\pi R_2 C_F}$$

Z_{IN}

R_1+R_2

f_z $\sqrt{f_2.GBW}$ f

Willy Sansen 10-05 0342

0342
This buffer has a very high input impedance, despite the biasing resistors R_1 and R_2. It is often used to measure bio-impedances. For safety, a coupling capacitor has to be used. Biasing resistors are thus needed to define the DC conditions of the opamp. The +input is now at ground and so is the output.

These resistors would also draw AC current, which is not allowed. For this purpose resistor R_1 is bootstrapped out by use of feedback capacitor C_F.

For sufficient high gain in the opamp, the output follows the input voltage. The voltage across resistor R_1 is approximately zero. It is thus bootstrapped out. It presents a resistance of infinity. This phenomenon starts where capacitance C_F starts to take effect, which is at the zero frequency f_z.

In practice however the gain of the opamp is limited. Its gain decreases towards the GBW. As a consequence, the input impedance will not continue to rise. It settles around the frequency halfway between f_2 and the GBW (on a logarithmic scale).

For example, if two resistors are taken of 1 MΩ and a capacitor C_F of 0.1 μF, the zero frequency f_z is 0.8 Hz and the peak frequency occurs at 1.3 kHz for a GBW of 1 MHz. At this point Z_{IN} is about 1.6 GΩ.

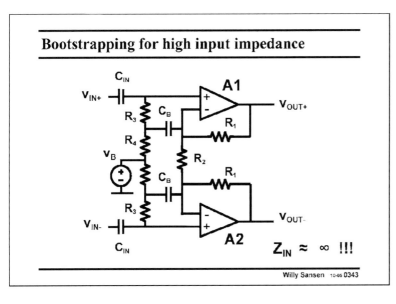

Bootstrapping for high input impedance

$Z_{IN} \approx \infty$!!!

Willy Sansen 10-05 0343

0343

Bootstrapping has been introduced to increase the input impedance of biopotential preamplifiers.

Both opamps A1 and A2 form an instrumentation amplifier with the three resistors R_1 (twice) and R_2. Its gain is set to a precise value of $2R_1/R_2$ by these resistors.

Moreover, the input impedances, at the + terminals of the opamps are very high. This is necessary not to draw any current from the electrodes (sensors).

The input sensors are always decoupled from the opamp inputs by means of capacitors C_{IN}. However, the + inputs of the opamps must be biased at a particular voltage. This is done by means of the resistors R_3/R_4. As a result the average output voltage is set by the biasing voltage V_B.

The input impedance is reduced considerably by these resistors R_3/R_4. This why the bootstrap capacitances C_B are added. They bootstrap the resistors R_3. Because of the feedback action of the opamp, the voltage across resistor R_3 is nearly the same. It looks like a resistor with value infinity. It is bootstrapped out.

The input impedance is then exceedingly high!

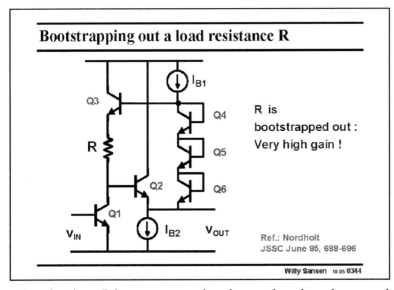

Bootstrapping out a load resistance R

R is
bootstrapped out :
Very high gain !

Ref.: Nordholt
JSSC June 85, 688-696

Willy Sansen 10-05 0344

0344

In a similar way, a load resistor R of an amplifier can be bootstrapped out, in order to make its effective value much higher, also rendering the voltage gain much higher. An example is given in this slide.

The amplifier simply consists of transistor Q1, followed by an emitter follower with transistor Q2. Its voltage gain would normally be $g_{m1}R$. Depending on the actual DC current, this gain is not all that high.

Load resistor R is not connected to the supply voltage however, but to another emitter follower Q3, which is connected to the output voltage over three diode connected transistors, each carrying about 0.6 V. The DC voltage across resistor R is thus also about 0.6 V.

However, the AC voltage at the output of emitter follower Q3 is about the same as the actual output voltage v_{OUT}, which is the same as the AC voltage at the collector of the input transistor Q1. There is now no AC voltage across resistor R. It is bootstrapped out. It appears to be infinitely high. As a result the voltage gain is quite high.

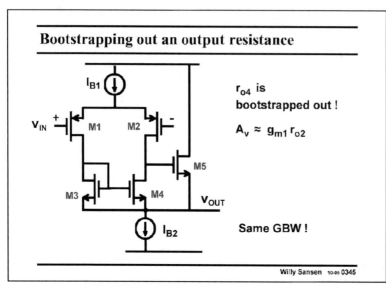

Bootstrapping out an output resistance

r_{o4} is bootstrapped out !

$A_v \approx g_{m1} r_{o2}$

Same GBW !

Willy Sansen   10-95 0345

0345

In a similar way, transistor M5 bootstraps out the output resistance of transistor M4 such that only r_{o2} is left in the gain expression.

Transistor M5 actually functions as a source follower. Its gain approximates to unity, as for the buffer A2 in the previous slide. Transistor M5 sees thus the same AC voltage at Drain and Source. As a result, its output resistance r_{o4} is bootstrapped out.

In order to really increase the gain, we would have to add cascodes in series with transistors M1 and M2, or design them with large channel lengths.

Note that this gain enhancement technique does not affect the GBW, as we have seen before with all gain enhancement techniques.

Bootstrapping is a fourth technique for gain enhancement. In practice we will use combinations of all these four techniques. We will need them as we go deeper and deeper into deep submicron CMOS!

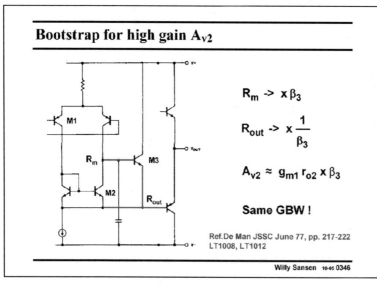

Bootstrap for high gain A_{v2}

$R_m \rightarrow x\,\beta_3$

$R_{out} \rightarrow x\,\dfrac{1}{\beta_3}$

$A_{v2} \approx g_{m1} r_{o2} \times \beta_3$

Same GBW !

Ref.De Man JSSC June 77, pp. 217-222
LT1008, LT1012

Willy Sansen   10-95 0346

0346

A practical example of this bootstrapping technique is shown in this slide. It uses bipolar transistors. This example was published a long time ago. Since the beta is an additional parameter in a bipolar transistor, it also appears in the gain expressions.

Note that the output impedance, which was already low because we are dealing with an emitter follower, is reduced further. It is now quite low. This is an

ideal situation in view of the need to develop the next stage.

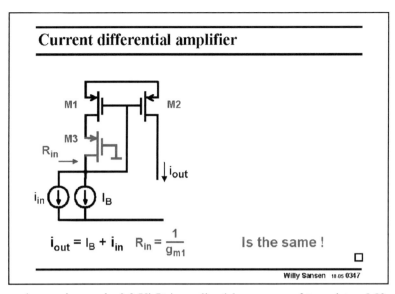

0347

After we have discussed the OTA, which is in effect a single-ended differential voltage amplifier, we focus on the most used single-ended differential current amplifier.

It is derived from a conventional current mirror, to which a cascode is added.

This quite easy as the V_{GS1} of the input MOST M1 is fairly large, for example 0.7 V if $V_T = 0.5$ V. This gives us plenty of room for cascodes. One single cascode requires only 0.2 V! It is realized by means of transistor M3.

This additional cascode does not change anything in the current mirror. The diode-connected transistor will be more accurate as more feedback loop gain is now available.

The two input current sources, one for DC denoted by I_B and one for AC i_{in}, will be mirrored just as in the situation without cascode.

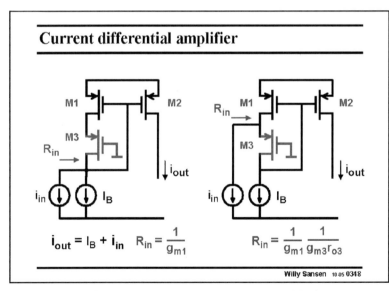

0348

The addition of this cascode creates an additional node however. This node is more attractive as an input than the original one.

The reason is that the impedance at this point is $g_{m3}r_{o3}$ times lower than at the original input. This is a result of the feedback loop. This factor is the additional loop gain as a result of the addition of the cascode.

Therefore, the input signal current source is led to the Source of the cascode, rather than to its Drain. It is easier to realize an ideal current source when the input impedance is smaller. Moreover, the input capacitance at this point will be smaller.

Note that the input signal current flows through M1, as there is no outlet through M3. This current is then mirrored by the current mirror M1-M2.

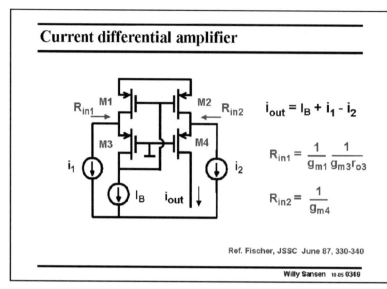

Current differential amplifier

$$i_{out} = I_B + i_1 - i_2$$

$$R_{in1} = \frac{1}{g_{m1}} \frac{1}{g_{m3}r_{o3}}$$

$$R_{in2} = \frac{1}{g_{m4}}$$

Ref. Fischer, JSSC June 87, 330-340

Willy Sansen 10-05 0349

0349

Addition of another cascode by means of M4, leads to the current differential amplifier in this slide.

This allows for the application of another input current i_2, which flows to the output through M4. The output current contains the differential input current, superimposed on the DC biasing current I_B.

The input impedances for the input currents, are not the same. For current i_1, the input impedance is quite small whereas for input current i_2, it will depend on the load impedance. For a small load impedance, the input resistance will be $1/g_{m4}$.

This is probably the most used current differential amplifier. It will be used in many operational amplifiers to convert the differential output signal to a single-ended one.

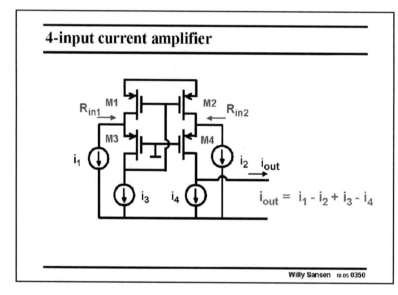

4-input current amplifier

$$i_{out} = i_1 - i_2 + i_3 - i_4$$

Willy Sansen 10-05 0350

0350

This same current differential amplifier can easily be expanded to a four input current amplifier.

It is sufficient to add an AC signal to the current sources i_3 and i_4. In this way a current difference-difference amplifier can be built or any other multiple-input analog processing block up to high frequencies.

Note that only the output impedance is high. All other nodes are at $1/g_m$ level or below.

0351

An important additional advantage of this current differential amplifier is that it can operate at very low supply voltages. This is shown in this slide for a 1 V supply voltage.

For a V_T of 0.7 V, the $V_{GS} - V_T$ has to be deceased to 0.15 V rather than 0.2 V, to be able to cope with a supply voltage of 1 V. Indeed all V_{GS} values are then 0.85 V. This gives plenty of headroom for the input current sources.

Low voltage operation

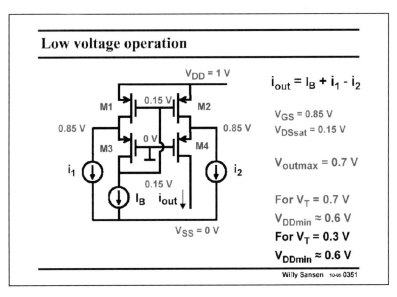

$i_{out} = I_B + i_1 - i_2$

$V_{GS} = 0.85$ V
$V_{DSsat} = 0.15$ V

$V_{outmax} = 0.7$ V

For $V_T = 0.7$ V
$V_{DDmin} \approx 0.6$ V
For $V_T = 0.3$ V
$V_{DDmin} \approx 0.6$ V

Willy Sansen 10-05 0351

Note that the Gates of the cascodes M3 and M4 are now at ground, which is the lowest voltage available.

The maximum output voltage V_{outmax} is 0.7 V as we need at least 0.15 V V_{DS} per transistor.

It is clear that for deep submicron CMOS, where the V_T decreases to as little as 0.3 V, the supply voltage can be as low as 0.6 V. Quite a low value indeed!!

Table of contents

Willy Sansen 10-05 0352

0352

This concludes the Chapter on two-transistor circuits such as current mirrors and differential pairs. They have been used to build up the most common voltage and current differential amplifiers with single-ended output. We will use them routinely in the operational amplifiers.

041

Before we go into more detail in opamp design, we want to know more about the limits of operation. At the low end this is noise, whereas at the high end it is distortion. This latter topic is discussed in Chapter 15. We now want to learn more about noise.

SNR and SNDR

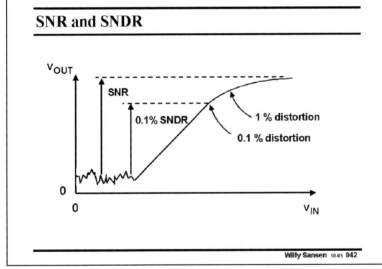

Willy Sansen 10 05 042

042

Every amplifier has noise. When we apply a small-signal input to an amplifier, then the output signal is an accurate replica of the input, but amplified. The output is proportional to the input, at least in the middle range.

When we increase the input signal amplitude, the output amplitude levels off. We generate distortion. The higher we increase the input the flatter the response becomes and the more distortion is generated. In most systems we can allow something like 0.1% distortion. In some audio amplifiers and high-performance analog-to-digital converters we would want rather less than 0.001% distortion!!

When we decrease the signal amplitude, it gets lost in noise. The Signal-to-Noise Ratio (SNR) is the highest possible range of input signals we can obtain regardless of distortion.

The Signal-to-Noise-and-Distortion Ratio (SNDR) on the other hand, limits this SNR to a certain amount of distortion. Above the 0.1% SNDR is shown, for which the distortion is limited to 0.1%. Clearly the SNDR is always smaller than the SNR.

117

Table of contents

Willy Sansen 10 05 043

043

We will firstly examine how we can describe noise. After all, this is a quantity that comes as a power, not a voltage or current!

Then we will discuss all circuits that we have seen hitherto from the point of view of noise performance.

The main points will be how to optimize noise performance. For a resistive input source, we will optimize the current flowing. For a capacitive input source, we will perform what is called "capacitive noise matching".

Noise versus time

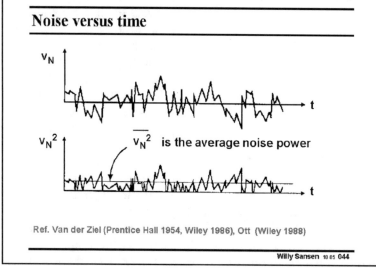

$\overline{v_N^2}$ is the average noise power

Ref. Van der Ziel (Prentice Hall 1954, Wiley 1986), Ott (Wiley 1988)

Willy Sansen 10 05 044

044

Noise is an arbitrary signal. We never know what to expect next. The signal amplitude is unpredictable. As a result there is a Gaussian spreading around zero and the average value is zero. Also, all possible pulses seem to be present, i.e. sharp ones and wide ones.

In order to be able to describe noise we have to take the power of the noise. For this purpose, we square the voltage or we rectify the voltage as shown in this slide. The average versus time is now the average noise power. It is clearly not zero but it averages out the peaks over a specific amount of time.

045

Fourier has explained how to convert any signal versus time into a signal versus frequency. For noise, we obtain a curve which consists of two regions:

– The region of white noise is flat and extends to very high frequencies (10^{13} Hz).

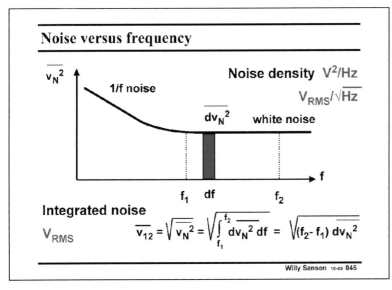

Noise versus frequency

$\overline{v_N^2}$

1/f noise

Noise density V^2/Hz

$V_{RMS}/\sqrt{Hz}$

$\overline{dv_N^2}$ white noise

f_1 df f_2 f

Integrated noise

V_{RMS}

$$\overline{v_{12}} = \sqrt{\overline{v_N^2}} = \sqrt{\int_{f_1}^{f_2} \overline{dv_N^2}\, df} = \sqrt{(f_2 - f_1)\, \overline{dv_N^2}}$$

Willy Sansen 10-05 045

– The region at low frequencies, which is called pink or 1/f noise as the noise power is inversely proportional to the frequency.

The **noise density** is now the noise power in an elementary small frequency band df. Its dimension is V^2/Hz. Taking the square root yields $VRMS/\sqrt{Hz}$.

The **integrated noise** is the total noise power between two frequencies. This integral is easily taken when the noise is white or flat. It is proportional to the difference in frequency. Noise is therefore integrated on a linear frequency axis. Bode diagrams are always presented with a logarithmic axis. They now give the wrong picture. They overemphasize the low-frequency noise.

The integrated noise (or also total noise) has V^2 as a dimension or V_{RMS} when the square root is taken.

The calculation of the integrated noise in the 1/f noise region involves another integral. The main point is not to omit the lower bound frequency. The 1/f noise would go to infinity and so does the integrated noise. It would take an infinite time to measure that however!

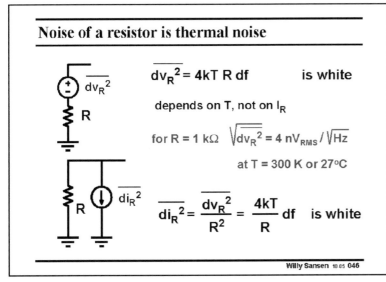

Noise of a resistor is thermal noise

$\overline{dv_R^2}$

R

$\overline{di_R^2}$

R

$$\overline{dv_R^2} = 4kT\,R\,df \qquad \text{is white}$$

depends on T, not on I_R

for R = 1 kΩ $\sqrt{\overline{dv_R^2}} = 4\,nV_{RMS}/\sqrt{Hz}$

at T = 300 K or 27°C

$$\overline{di_R^2} = \frac{\overline{dv_R^2}}{R^2} = \frac{4kT}{R}\,df \quad \text{is white}$$

Willy Sansen 10-05 046

046

Let us now determine which noise sources we can find in an electronic circuit. Resistors and junctions give noise. We will take resistors first.

A resistor gives thermal noise. It is modeled by a voltage source in series with the resistor or a current source in parallel.

The noise voltage is proportional to that resistor, and to the absolute temperature (in Kelvin). It does not depend on the actual current flowing through the resistor. Cooling down will therefore reduce the noise (as in many space applications!).

For a resistor of 1 kΩ, the thermal noise density is about $4\,nV_{RMS}/\sqrt{Hz}$ at room temperature.

This is proportional to the square root of the resistor value. A resistor of 100 kΩ would give $40\,nV_{RMS}/\sqrt{Hz}$ but a resistor of 10 kΩ only $4 \times \sqrt{10}$ or $12\,nV_{RMS}/\sqrt{Hz}$.

Over a bandwidth from 20 Hz to 20 kHz, the total noise of a 100 kΩ resistor would be $40 \times \sqrt{20\,000}\,nV_{RMS}$ or $5.6\,\mu V_{RMS}$. For a maximum signal amplitude of $100\,mV_{RMS}$, this would be SNR of 17 700 or 85 dB.

Note that the lower bound frequency is usually negligible. Indeed, either subtracting 20 Hz from 20 kHz or not, will not make any difference.

Finally, note that a noise current can be used in parallel. The larger this parallel resistor, the lower the noise.

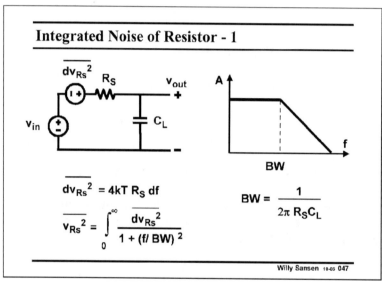

047

The upper frequency limit of a resistor R_S is normally determined by a capacitance to ground, denoted by C_L. Their time constant determines the bandwidth BW.

If we now want to calculate the integrated noise of this resistor-capacitance combination, we have to take the integral over all frequencies. This integral obviously contains the first-order filter transfer function, related to this bandwidth BW.

We now have to perform this integration. Luckily this is a well known integral.

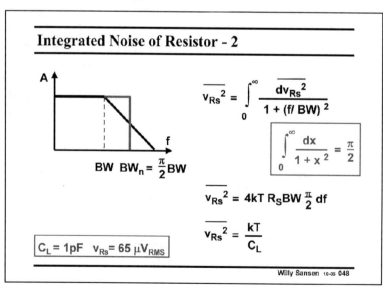

048

Indeed the integral gives $\pi/2$ as a value, provided the variables are correct. The integrated noise is now simply $4kTR_SBW\ \pi/2$, which is the noise of the resistor itself, multiplied by BW $\pi/2$. This latter bandwidth is called the **noise bandwidth** BW_n.

The ratio of the BW_n to the BW is 1.57. This extra 57% is required to take into account the integral of the region with the first-order slope of 20 dB/decade. If we took a steeper filter (3rd order or higher), then the BW_n and the BW nearly coincide.

However, the bandwidth BW also contains R_S. As a result R_S is cancelled out in the expression of the integrated noise. A very simple result emerges. The integrated noise is simply kT/C_L.

For 1 pF the integrated noise is 65 μV_{RMS}. For less noise, we have to increase the size of the capacitance. For 10 pF that would be $65/\sqrt{10}$ or about 21 μV_{RMS}.

This can be understood by realizing that for larger resistors, the noise increases but the bandwidth is reduced by exactly the same amount.

Filters all use the integrated noise as a specification or the SNR. They all use large capacitances to increase their SNR, and an amplitude which is as high as possible.

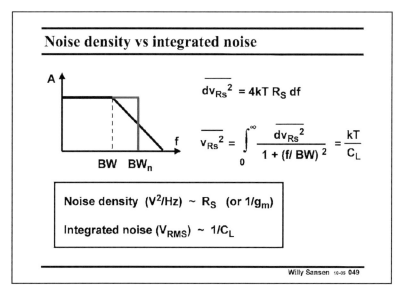

Noise density vs integrated noise

$$\overline{dv_{Rs}^2} = 4kT\, R_S\, df$$

$$\overline{v_{Rs}^2} = \int_0^\infty \frac{\overline{dv_{Rs}^2}}{1 + (f/\,BW)^2} = \frac{kT}{C_L}$$

Noise density (V^2/Hz) ~ R_S (or $1/g_m$)

Integrated noise (V_{RMS}) ~ $1/C_L$

Willy Sansen 10-05 049

049

In this way we reach the important conclusion that noise density depends on the resistor and integrated noise depends on the capacitance.

For low noise density we need a small series resistance (or large parallel resistance).

For low integrated or total noise, we need a large capacitance to ground.

Both obviously depend on absolute temperature.

A designer will therefore have to determine whether he deals with a narrow-band system (receivers, bandpass filters, ...) or a wide-band system (switched-capacitor filters, low-pass filters, ...).

In a narrow-band system, the narrow band noise (spot noise) is the noise density, multiplied by the bandwidth. The resistances determine the noise performance.

In a wide-band system, the kT/C noise is dominant. For low noise, we have to increase the capacitances. Doing this will obviously increase the power consumption.

This is a rule of thumb. Low noise requires smaller resistances and or larger capacitances. Both increase the power consumption. Increasing the noise performance inevitably leads to larger power consumption. A low-noise low-power circuit requirement certainly leads to a severe compromise!

0410

A resistor also exhibits 1/f noise.

In general, 1/f noise is dependent on size and the quality of conduction (or homogeneity).

This is why the expression for the 1/f noise is a result of fitting lots of data. It contains the DC voltage across the resistor, the size (A_R or WL) and a factor KF_R, that indicates what material is to be used.

For example a n-well resistor is single-crystalline silicon. This material is very homogeneous. As a result the KF_R factor is small.

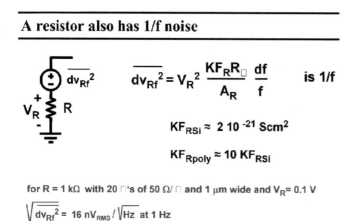

A resistor also has 1/f noise

$$\overline{dv_{Rf}^2} = V_R^2 \frac{KF_R R_\square}{A_R} \frac{df}{f} \quad \text{is 1/f}$$

$$KF_{RSi} \approx 2\ 10^{-21}\ Scm^2$$

$$KF_{Rpoly} \approx 10\ KF_{RSi}$$

for R = 1 kΩ with 20 ⬚'s of 50 Ω/⬚ and 1 μm wide and V_R= 0.1 V

$$\sqrt{\overline{dv_{Rf}^2}} = 16\ nV_{RMS}/\sqrt{Hz}\ \text{at 1 Hz}$$

Ref. Vandamme, ESSDERC '04

Willy Sansen 10-05 0410

The same applies to all diffused resistances. Poly-silicon resistors on the other hand are much worse. Discrete carbon resistors are probably the worst for 1/f noise.

If we use the same resistor with much larger dimensions (same W/L but larger WL), then the 1/f noise is significantly reduced.

Finally reducing the DC voltage to zero by putting a capacitance in series, also kills the 1/f noise. This is probably the most widely used technique in low-noise preamplifiers!

Note that the 1/f noise is normally specified at 1 Hz. Since the noise voltage is inversely proportional to the square root of the frequency, it decreases only slowly.

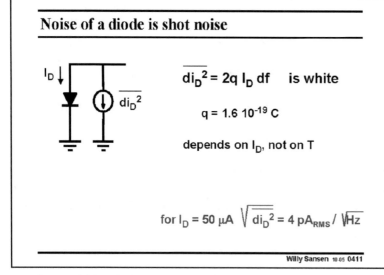

Noise of a diode is shot noise

$$\overline{di_D^2} = 2q\ I_D\ df \quad \text{is white}$$

$$q = 1.6\ 10^{-19}\ C$$

depends on I_D, not on T

for I_D = 50 μA $\sqrt{\overline{di_D^2}} = 4\ pA_{RMS}/\sqrt{Hz}$

Willy Sansen 10-05 0411

0411

The other source of white noise is the shot noise, generated by a junction, or diode.

This is white noise but not thermal noise.

Indeed this noise current is proportional to the current but is independent of the temperature. Cooling therefore, does not help!

This applies to any junction (or diode) which carries current. It is irrelevant whether the current flows in the forward direction (as in any forward biased diode) or in the reverse direction (as in a photodiode). The same expression is still valid.

0412

Obviously, a diode also exhibits 1/f noise.

Again the amount of noise depends on the current through it, the size and the material used.

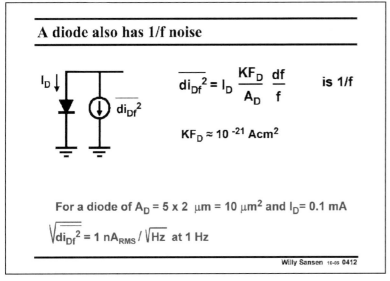

A diode also has 1/f noise

$$\overline{di_{Df}^2} = I_D \frac{KF_D}{A_D} \frac{df}{f} \quad \text{is 1/f}$$

$$KF_D \approx 10^{-21} \, Acm^2$$

For a diode of $A_D = 5 \times 2 \; \mu m = 10 \; \mu m^2$ and $I_D = 0.1$ mA

$$\sqrt{\overline{di_{Df}^2}} = 1 \, nA_{RMS} / \sqrt{Hz} \text{ at 1 Hz}$$

Willy Sansen 10-05 0412

For example, a small thin-film diode gives a significant amount of 1/f noise.

A large (power) diode in single-crystalline silicon is not too bad. This corresponds to the value given in this slide.

Obviously, the spreading on this value is quite large! A measurement on one single device is not relevant. Averages have to be taken over hundreds of devices!

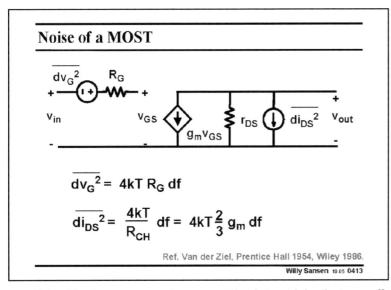

Noise of a MOST

$$\overline{dv_G^2} = 4kT \, R_G \, df$$

$$\overline{di_{DS}^2} = \frac{4kT}{R_{CH}} df = 4kT \frac{2}{3} g_m \, df$$

Ref. Van der Ziel, Prentice Hall 1954, Wiley 1986.

Willy Sansen 10 05 0413

0413

A MOST has a resistive channel. As a result, it exhibits thermal noise, just like any other resistor.

The channel is not very homogeneous, as it conducts well at the Source side, but is pinched off at the Drain side. This is why some integrals will be required to calculate the channel resistance and the noise generated by it.

Nevertheless, the channel noise can be represented by a noise current source in parallel with the g_m current source. The integral leads to a effective channel resistor R_{CH} of $3/2g_m$. The 4kT factor clearly shows that we are dealing with thermal noise.

In deep submicron or nanometer CMOS devices, velocity saturation appears. As a result the coefficient 2/3 increases (Ref. Han, JSSC, March '05, 726–735). It is about 50% larger for 0.18 μm CMOS and doubles for 0.13 μm CMOS.

The poly Gate resistor R_G cannot be discounted, however. Even if the Gate material is highly doped, it can make a large contribution, depending on the actual dimensions, and on how long the Gate line is prolonged outside the active region.

0414

Normally we refer both white noise sources to the input, in order to be able to calculate the SNR at the input.

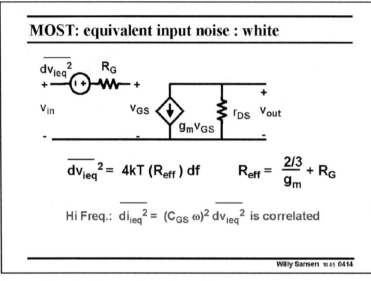

MOST: equivalent input noise : white

$$\overline{dv_{ieq}}^2 = 4kT\,(R_{eff})\,df \qquad R_{eff} = \frac{2/3}{g_m} + R_G$$

Hi Freq.: $\overline{di_{ieq}}^2 = (C_{GS}\,\omega)^2\,\overline{dv_{ieq}}^2$ is correlated

Willy Sansen 10.05 0414

The channel noise current can easily be shifted to the input by dividing it by g_m (and the power by g_m^2).

The two noise powers are then added at the input. In this way we obtain a thermal noise resistance R_{eff}, which is the sum of both sources. The channel noise gives the first contribution. The Gate resistor R_G is the other one. The input noise voltage is called the **equivalent input noise voltage**. It is inversely proportional to transconductance g_m, at least whilst the Gate resistor is small.

At very high frequencies, a capacitance C_{GS} appears across the Gate-Source terminals. As a result, for a low source resistance (typically 50 Ω) a noise current can flow through capacitance C_{GS} as shown in this slide.

This current is obviously correlated to the equivalent input voltage. Their powers cannot be added up. However, this noise current is only relevant at very high frequencies, beyond $f_T/5$. It is only relevant in the noise optimization of LNA's, VCO's and RF mixers.

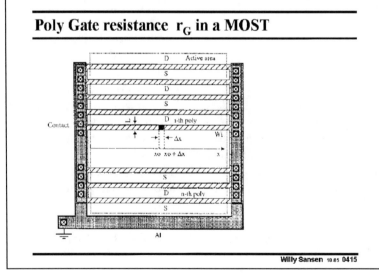

Poly Gate resistance r_G in a MOST

Willy Sansen 10.05 0415

0415

The equivalent input noise voltage of a MOST can be made small by increasing the transconductance. This will require a small $V_{GS} - V_T$ but a large W/L and current.

This is only true if the Gate resistance is small. In this layout a MOST is shown with very large W/L. The poly Gate lines between the Sources and Drains have become very long, generating a fairly large resistance.

In order to avoid this resistance, it is better to divide the layout in several smaller blocks, such that the Gate contacts can be much closer to the middle of the Gate lines.

The Gate resistance therefore depends very much on the layout style.

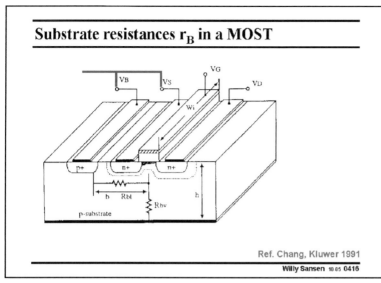

Again, the value of this resistance R_B depends on the layout. For low-noise performance, a large substrate contact will have to be designed.

0416

This is also very much the case for the Substrate resistance.

Even if the Substrate contact is shorted to the Source contact, it is impossible to avoid the presence of a Substrate resistance. Its value evidently depends on whether the substrate has a back contact or not. It is difficult to calculate its value because it is a distributed resistance. It will always be present.

This resistance gives noise, which has to be included in the small-signal diagram.

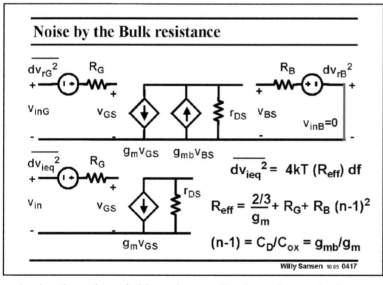

0417

The noise of the Bulk or Substrate resistance R_B is included. It is multiplied by the Bulk transconductance g_{mb}, to be transferred to the output. Then it has to be divided again by the transconductance g_m itself, to be transferred to the Gate. The ratio g_{mb}/g_m applies in the parameter $n-1$.

The total equivalent input noise voltage includes the contribution of the substrate resistance R_B, multiplied by $(n-1)^2$.

0418

The last contribution to the equivalent input noise voltage is given by the series resistance in the Source itself. Normally, this is quite small, but depending on the effective channel length.

It is easy to show (see next slide) that the noise of the Source resistor can simply be added to the noise of the Gate resistor. It is as if the Source resistor itself can be added to the Gate resistor.

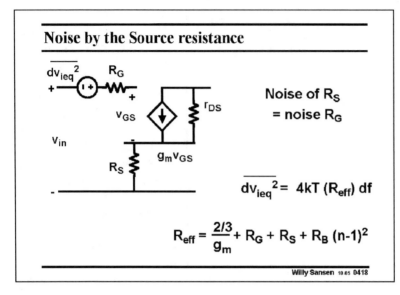

Noise by the Source resistance

Noise of R_S
= noise R_G

$$\overline{dv_{ieq}^2} = 4kT\,(R_{eff})\,df$$

$$R_{eff} = \frac{2/3}{g_m} + R_G + R_S + R_B\,(n-1)^2$$

Willy Sansen 10-05 0418

The realization of a large low-noise MOST requires the simultaneous minimization of all four contributions. Quite often the bulk resistor R_B is forgotten in the design plan!!

Nowadays, for deep submicron CMOS, the channel noise contribution seems to have a coefficient larger than 2/3. Values of up to 2 have been measured. It is not yet clear, whether this is because of the upcoming effect of velocity saturation or with breakdown or punch-through.

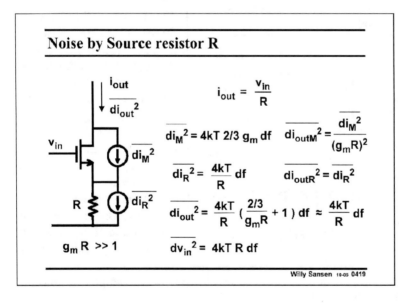

Noise by Source resistor R

$$i_{out} = \frac{v_{in}}{R}$$

$$\overline{di_M^2} = 4kT\,2/3\,g_m\,df \qquad \overline{di_{outM}^2} = \frac{\overline{di_M^2}}{(g_m R)^2}$$

$$\overline{di_R^2} = \frac{4kT}{R}\,df \qquad \overline{di_{outR}^2} = \overline{di_R^2}$$

$$\overline{di_{out}^2} = \frac{4kT}{R}\,\left(\frac{2/3}{g_m R} + 1\right)\,df \approx \frac{4kT}{R}\,df$$

$g_m R \gg 1$

$$\overline{dv_{in}^2} = 4kT\,R\,df$$

Willy Sansen 10-05 0419

0419
The noise of the Source resistor R can simply be added to the noise of the Gate resistor. The calculation is given in this slide. We firstly assume that this resistor is larger than $1/g_m$. Otherwise it has no effect!

We now calculate the contribution of the channel noise to the output. It is the channel noise itself divided by $(g_m R)^2$.

We then calculate the contribution of the resistor noise to the output. It is simply $4kT/R$.

When we take the sum of both, we see that the channel noise of the transistor has become negligible, compared to the resistor noise. Indeed, $g_m R$ is much larger than unity.

As a result, the equivalent input noise voltage is dominated by the resistor noise. Moreover, the expression is the same as for the Gate resistance noise.

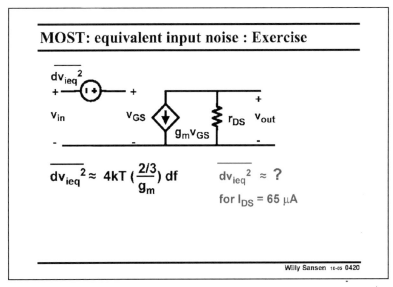

MOST: equivalent input noise : Exercise

$$\overline{dv_{ieq}^2} \approx 4kT \left(\frac{2/3}{g_m}\right) df \qquad \overline{dv_{ieq}^2} \approx ?$$

$$\text{for } I_{DS} = 65 \ \mu A$$

Willy Sansen 10-05 0420

0420

As an exercise, let us try to calculate the equivalent input noise voltage of a MOST biased at 65 μA. Only thermal noise is taken into account and the layout is such that all Gate and Bulk resistances can be ignored. The expression is repeated in this slide. A transistor with 65 μA current, generates a transconductance of $g_m = 0.65$ mS (if $V_{GS} - V_T = 0.2$ V is chosen). The noise resistance $2/3/g_m$ is then about 1 kΩ. The equivalent input noise voltage is therefore 4 $nV_{RMS}/\sqrt{Hz}$.

MOST: equivalent input noise : 1/f noise

$$\overline{dv_{ieqf}^2} = \frac{KF_F}{WL \ C_{ox}^2} \frac{df}{f}$$

pMOST $KF_F \approx 10^{-32}$ C^2/cm^2

nMOST $KF_F \approx 4 \ 10^{-31}$ C^2/cm^2

pJFET $KF_F \approx 10^{-33}$ C^2/cm^2

W & L in cm; C_{ox} in F/cm^2

Willy Sansen 10-05 0421

0421

A MOST device also exhibits a lot of 1/f noise. This is due to the surface states. The silicon has a crystal structure, which is cut off at the surface, where Gate oxide is grown on top. This causes surface states which contribute to the 1/ noise.

Several expressions are in use. The one with C_{ox}^2 in the denominator has the advantage, that coefficient KF_F is nearly independent of the technology. Actually, all technology effects are represented by the C_{ox}^2. If we use a KF with C_{ox} only, then we lose this advantage.

Obviously, the transistor size WL (not W/L) is also included. A MOST with a thin oxide or a small channel length, and a large WL product shows little 1/f noise.

We also note that a p-JFET is the transistor with lowest 1/f noise. A pMOST is about ten times worse. A nMOST is by far the worst transistor for 1/f noise. It is 30–60 times larger than for a pMOST of the same size. The expression gives a factor of 40, but there is a large spreading on it!

It is for this reason that some audio preamplifiers still want JFETs at their inputs. This also applies to some radiation detection circuitry.

Finally note that the equivalent input 1/f noise voltage does not depend on the DC biasing current. The output current depends on the DC current, but not the equivalent input voltage. A small current dependency may sometimes be detected. This is usually negligible.

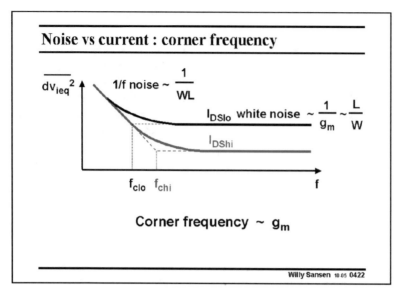

Noise vs current : corner frequency

Corner frequency ~ g_m

Willy Sansen 10.05 0422

0422

When we examine the white and 1/f noise together, we notice a very slow crossover from one to the other. Their asymptotes meet at frequency f_c, called the corner frequency.

Obviously, this frequency depends on the DC biasing current. The higher the current, the larger the transconductance and the lower the white noise.

Since the 1/f noise does not change, the corner frequency shifts to the right, for higher DC currents.

Actually, the corner frequency is a very strange measure for 1/f noise. The lower the white noise, the higher the corner frequency.

Also, note that the thermal noise depends on the ratio W/L, whereas the 1/f noise depends on the WL product.

Finally, note that there are circuit techniques to reduce the 1/f noise such as chopping and correlated-double-sampling (Ref. Enz, Temes, Proc. IEEE, Nov.96, 1584–1614). Also switching the biasing of the transistors can reduce the 1/f noise somewhat, as much as 10 dB (Ref. Gierkink, JSSC July 1999, 1022–1025).

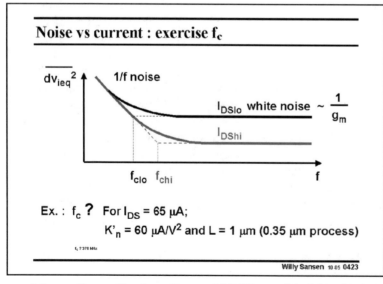

Noise vs current : exercise f_c

Ex. : f_c ? For I_{DS} = 65 μA;
K'$_n$ = 60 μA/V^2 and L = 1 μm (0.35 μm process)

f_c 7.370 MHz

Willy Sansen 10.05 0423

0423

As an exercise, we want to calculate what the corner frequency f_c is for the nMOST of the previous exercise, with a DC current of 65 μA.

We use a technology of 0.35 μm CMOS with a K' as given; the channel length has been chosen to be three times larger (for gain).

Simply plugging all the values in the 1/f noise expression yields an f_c of 370 kHz. It is clear that small-sized MOST are no

good for audio applications. Power MOSTs are ideal for that!

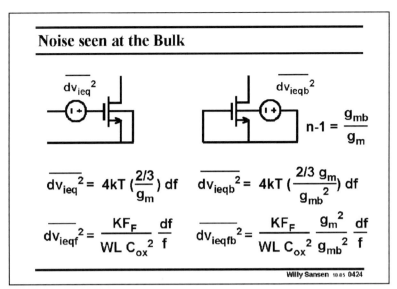

Noise seen at the Bulk

$$\overline{dv_{ieq}}^2 = 4kT \left(\frac{2/3}{g_m}\right) df$$

$$\overline{dv_{ieqb}}^2 = 4kT \left(\frac{2/3\ g_m}{g_{mb}^2}\right) df$$

$$\overline{dv_{ieqf}}^2 = \frac{KF_F}{WL\ C_{ox}^2} \frac{df}{f}$$

$$\overline{dv_{ieqfb}}^2 = \frac{KF_F}{WL\ C_{ox}^2} \frac{g_m^2}{g_{mb}^2} \frac{df}{f}$$

$$n-1 = \frac{g_{mb}}{g_m}$$

Willy Sansen 10-05 0424

0424

When a MOST is driven at its Bulk rather than at its Gate, then the Bulk transconductance g_{mb} has to be used rather than the transconductance g_m. Remember that their ratio is $n-1$ (see Chapter 1), which is not known accurately, but lies somewhere between 0.3 and 0.5.

The equivalent input noise voltages that are present at the Bulk, are then different. They are repeated at the Gate on the left. They are given by the expressions on the right. Both expressions contain the Bulk transconductance g_{mb} squared. Since g_{mb} is always smaller than g_m, the Bulk equivalent input noise is always larger than the Gate equivalent input noise. This evidently applies to both the white noise and the 1/f noise.

A Bulk drive is therefore not a good solution for noise.

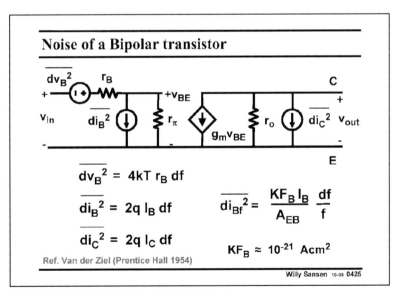

Noise of a Bipolar transistor

$$\overline{dv_B}^2 = 4kT\ r_B\ df$$

$$\overline{di_B}^2 = 2q\ I_B\ df$$

$$\overline{di_C}^2 = 2q\ I_C\ df$$

$$\overline{di_{Bf}}^2 = \frac{KF_B\ I_B}{A_{EB}} \frac{df}{f}$$

$$KF_B \approx 10^{-21}\ Acm^2$$

Ref. Van der Ziel (Prentice Hall 1954)

Willy Sansen 10-05 0425

0425

A bipolar transistor has two pn-junctions, through which current flows. As a result two sources of shot noise will have to be present. They are added in the small-signal model in this slide. They are white noise sources.

One collector shot noise current source is added between collector and emitter. It is proportional to the collector current.

The other one is between base and emitter and is proportional to the base current.

Finally, a resistive base resistance noise voltage has to be added in series with the base input.

Normally, the 1/f noise is added to the base shot noise current source. For a silicon transistor, an average value of the KF factor is given in this slide. Normally, the 1/f noise of a bipolar transistor is much lower than of a MOST because the current flows in the bulk, not at the surface.

The 1/f noise is again inversely proportional to the emitter size A_{EB}.

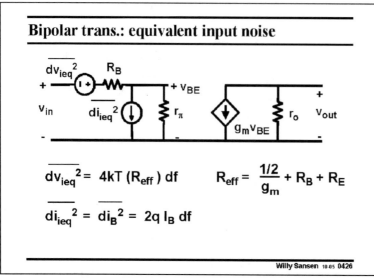

0426
Again the noise sources can be combined at the input, in order to be able to compare them to the input signal.

The collector shot noise has to be divided by g_m^2 in order to be translated into an input voltage.

The base shot noise remains where it is.

As a result, two equivalent noise sources are found, a voltage noise source and a current noise source, which is actually the base shot noise. Which one is dominant will depend on the source impedance, as we will see later.

The equivalent input noise voltage obviously also includes the base and emitter resistances.

Note that the expression of the equivalent input voltage is very similar to the one for MOST. The only difference is that now the coefficient of $1/g_m$ is $1/2$ instead of $2/3$. This is small difference indeed. We cannot forget however that for the same DC current the transconductance of a bipolar transistor is about 4 times larger than for a MOST. Its equivalent input noise voltage will therefore decrease.

Table of contents

+ **Definitions of noise**
+ Noise of an amplifier
+ Noise of a follower
+ Noise of a cascode
+ Noise of a current mirror
+ Noise of a differential pair
+ Capacitive noise matching

Willy Sansen 10.05 0427

0427
Now that we know what the noise models are for a MOST and a bipolar transistor, we want to use them to calculate the SNR of an amplifier, a cascode, etc.

Moreover we have used ideal current sources for the biasing and as active loads. All the current sources have to be realized by means of transistors, which also exhibit noise.

0428
Addition of an active load M2 to a single-transistor amplifier M1, gives the circuit in this slide.

The equivalent input noise source of the load transistor M2 is shown explicitly. It is in series

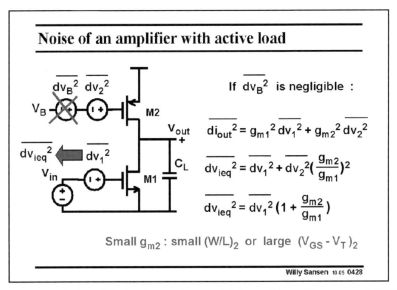

Noise of an amplifier with active load

If $\overline{dv_B^2}$ is negligible :

$$\overline{di_{out}^2} = g_{m1}^2 \overline{dv_1^2} + g_{m2}^2 \overline{dv_2^2}$$

$$\overline{dv_{ieq}^2} = \overline{dv_1^2} + \overline{dv_2^2}\left(\frac{g_{m2}}{g_{m1}}\right)^2$$

$$\overline{dv_{ieq}^2} = \overline{dv_1^2}\left(1 + \frac{g_{m2}}{g_{m1}}\right)$$

Small g_{m2} : small $(W/L)_2$ or large $(V_{GS} - V_T)_2$

Willy Sansen 10 05 0428

with the noise coming from the biasing voltage V_B. Normally, this biasing voltage is followed by a large decoupling capacitance to ground, such that the noise from it can be ignored.

The noise of the load transistor M2 is now amplified by g_{m2} towards the output. It has to be divided by g_{m1} to be referred to the input. The noise of transistor M2 is therefore multiplied by a factor g_{m2}/g_{m1}.

To make the noise contribution of M2 negligible, we must design this load transistor with large $V_{GS} - V_T$ or small W/L, which is actually the same. Both transistors now carry the same DC current. Transconductance g_{m2} can only be made smaller if it is designed for a larger $V_{GS} - V_T$, such as 0.5 V. The input transistor then keeps 0.2 V as a $V_{GS} - V_T$.

This is an important conclusion, which will be repeated many times. Current source and current mirror devices must be designed for small size W/L and hence for large $V_{GS} - V_T$!

Note that only the white noise sources have been considered here.

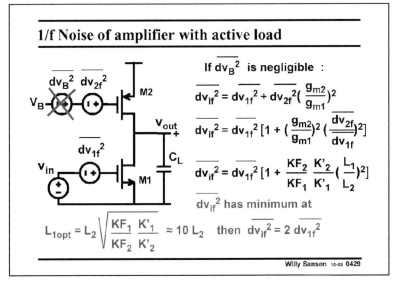

1/f Noise of amplifier with active load

If $\overline{dv_B^2}$ is negligible :

$$\overline{dv_{if}^2} = \overline{dv_{1f}^2} + \overline{dv_{2f}^2}\left(\frac{g_{m2}}{g_{m1}}\right)^2$$

$$\overline{dv_{if}^2} = \overline{dv_{1f}^2}\left[1 + \left(\frac{g_{m2}}{g_{m1}}\right)^2\left(\frac{\overline{dv_{2f}}}{\overline{dv_{1f}}}\right)^2\right]$$

$$\overline{dv_{if}^2} = \overline{dv_{1f}^2}\left[1 + \frac{KF_2}{KF_1}\frac{K'_2}{K'_1}\left(\frac{L_1}{L_2}\right)^2\right]$$

$\overline{dv_{if}^2}$ has minimum at

$$L_{1opt} = L_2\sqrt{\frac{KF_1}{KF_2}\frac{K'_1}{K'_2}} \approx 10 \, L_2 \quad \text{then} \quad \overline{dv_{if}^2} = 2\,\overline{dv_{1f}^2}$$

Willy Sansen 10-05 0429

0429
The same analysis can be repeated for 1/f noise.

However, all 1/f noise sources contain the area WL of the transistors. The resulting expression are therefore more cumbersome.

Moreover the equivalent input noise voltage shows a minimum, if the input channel length is taken as a variable. It shows that the input transistor channel length L_1 must be about 10 times larger than load transistor channel length L_2. This is not a problem as the load transistor has normally a small W/L. It is normally a small square device.

The drawback could then be that the gain is reduced as a result of the small channel length L_2. Cascodes will therefore be needed to alleviate this problem.

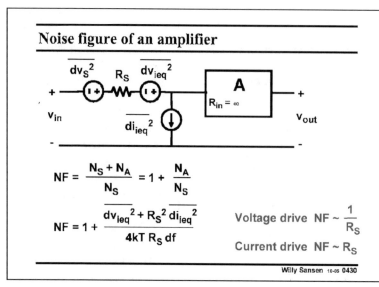

Willy Sansen 10-05 0430

0430
An amplifier which is used in a system with a characteristic impedance, such as $600\,\Omega$ in audio and $50\,\Omega$ in RF circuits, is characterized by a Noise Figure NF.

The Noise Figure is defined as the ratio of the total input noise to the noise of the source impedance R_S. It actually indicates how much noise is added by the amplifier to the noise already present by the characteristic source impedance R_S.

If we now take an amplifier with both an input noise voltage and current (understanding that a MOST amplifier has no input noise current), then we can easily determine the expression of the Noise Figure.

The noise source that is now dominant, depends on the value of the source resistor R_S.

For a small R_S, the amplifier is voltage-driven. In this case $R_S di_{ieq}$ in the numerator is negligible. The noise voltage is then dominant.

This is not at all surprising, a noise performance of a voltage-driven amplifier is governed by the equivalent input noise voltage. The Noise Figure then decreases for increasing R_S. For a large R_S, the amplifier is current-driven. In this case, dv_{ieq} in the numerator is negligible. The noise current is then dominant and the Noise Figure increases for increasing R_S.

There must therefore be a minimum, versus R_S.

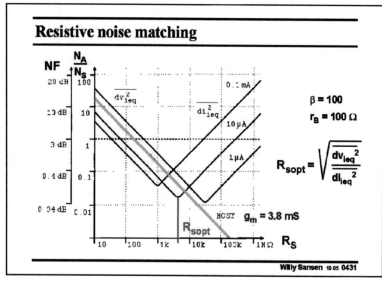

Willy Sansen 10-05 0431

0431
This minimum is clearly visible when the NF, together with the N_A/N_S ratio, is plotted versus R_S.

For small values of R_S, the NF decreases, but increases again for large values of R_S. The minimum in the middle is simply obtained for the ratio of the two equivalent input noise sources.

Operating a bipolar amplifier in this minimum is called Resistive Noise Matching.

Note that for a voltage drive, the larger current gives the larger g_m and the lower NF. On the other hand, for a current drive, the larger current gives a larger base shot noise and hence a larger NF. This is exactly the opposite result.

A MOST does not have an input noise current. Its NF is a straight line, decreasing continuously.

It is clear that for large source impedances, a bipolar transistor will never be used. The base noise current would flow in this large source resistance and kill the noise performance. Large source resistances are very common. Examples are preamplifiers for photodiodes, biopotentials, but also low-frequency capacitive pressure sensors, etc.

Table of contents

- ◆ **Definitions of noise**
- ◆ **Noise of an amplifier**
- ◆ **Noise of a follower**
- ◆ **Noise of a cascode**
- ◆ **Noise of a current mirror**
- ◆ **Noise of a differential pair**
- ◆ **Capacitive noise matching**

Willy Sansen 10 05 0432

0432
Now that we know how an amplifier performs with respect to low noise, let us observe the two other single-transistor stages. They are the source follower and the cascode.

We will find that the noise performance of a source follower is really poor whereas a cascode performs beautifully!

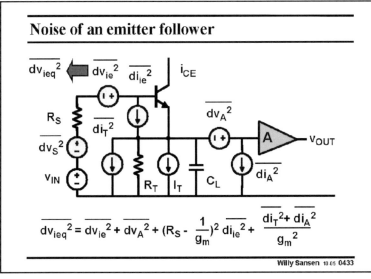

Noise of an emitter follower

$$dv_{ieq}^2 = dv_{ie}^2 + dv_A^2 + (R_S - \frac{1}{g_m})^2 \, di_{ie}^2 + \frac{di_T^2 + di_A^2}{g_m^2}$$

Willy Sansen 10 05 0433

0433
A source follower, or an emitter follower in this case, has a gain of at most unity. As a result, the noise of the next amplifier is not attenuated when we refer its noise to the input.

When we identify all noise sources for a bipolar realization, not forgetting the output noise of the DC biasing current source I_T, we then easily find their contributions to the output. Dividing all output components by the gain gives us the total

equivalent input noise voltage.

It is clear that the input noise voltages of both the emitter follower and the amplifier, appear together at the input. Moreover, the input noise current of the emitter follower flows through the large resistor R_S. If this resistor weren't large, why would we need an emitter follower?

Finally, the current of the emitter follower cannot be too small or the noise current of the current source comes into effect.

It can be concluded that the noise performance of a source follower is really poor. It should never be used at the input of a low-noise amplifier!

0434

The noise performance of a cascode is excellent. It does not impair the noise performance at all, as shown next.

Willy Sansen 10 05 0434

0435

Indeed, we like using cascodes because they add a lot of gain without increasing the current consumption. The question now is whether cascode transistor M2, which heavily contributes to the gain, will also contribute to the noise. The answer is negative!

For both transistors the equivalent input noise voltage is in series with the Gate.

The noise voltage of the cascode transistor M2, will be visible at its Source terminal but cannot influence the current through it. Actually, M2 acts as a source follower for its noise source dv^2. As a result the output current is insensitive to the noise of the cascode.

Noise of a cascode amplifier

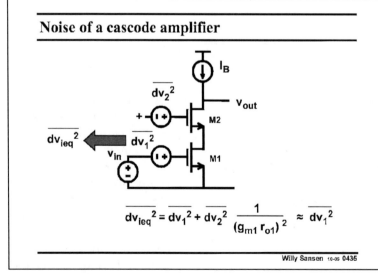

$$\overline{dv_{ieq}^2} = \overline{dv_1^2} + \overline{dv_2^2}\,\frac{1}{(g_{m1}\,r_{o1})^2} \approx \overline{dv_1^2}$$

Willy Sansen 10-05 0435

A calculation of the gain, using small-signal models for the transistors (with g_m and r_o), shows that the equivalent input noise voltage of the cascode has to be divided by the gain of the input transistor, squared. Even if this gain is very low, as in deep submicron CMOS, the squaring will make sure that the noise of M2 is neglected.

As a result, the noise of a two-transistor cascode is the same as for the input transistor. And yet the cascode does contribute to the gain! This is why cascodes are so frequently used!

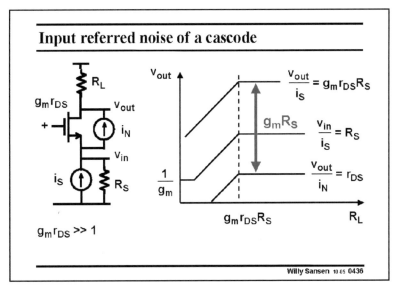

0436

Another way to look at the noise performance is to concentrate on the cascode only.

Current source i_S is the input current source, coming eventually from another transistor. Current source i_N is the cascode noise current.

For large values of load resistor R_L, we find that the cascode noise current i_N is simply multiplied by R_S toward the output. The input signal current i_S is multiplied by a much larger factor, i.e. $g_m r_{DS} R_S$, or $g_m R_S$ times greater than i_N.

This factor $g_m R_S$ is therefore the value by which the input current is more amplified than the cascode noise current. This is why the noise of the cascode is negligible.

This factor $g_m R_S$ depends on R_S however. If the cascode is not driven by a real current source, with large output resistance, then this noise reduction decreases, as shown next.

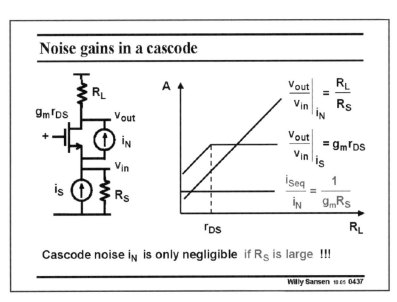

0437

Indeed, the voltage gains across the cascode are shown, versus load resistance R_L.

For a high load resistance, the gain reduction $g_m R_S$ is clearly visible.

This means that the cascode cannot be driven by a low resistance, or by a voltage source, as suggested in some wideband amplifiers. The noise performance then deteriorates really badly.

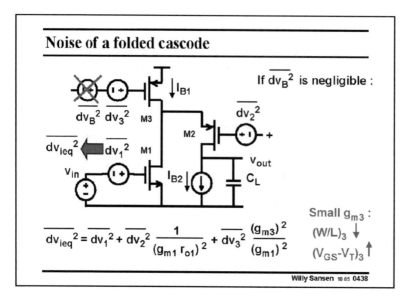

0438

A folded cascode contains two DC current sources. The top one, with transistor M3, distributes the DC currents over both stages. The other one acts as an active load.

The noise of the top current source is very similar as for an active load. Transistor M3 must be designed with large $V_{GS} - V_T$ or small W/L. Indeed, the full expression shows the same g_m ratio as for a single-transistor amplifier with active load.

The noise of the cascode can be neglected altogether, as shown previously. Its noise is reduced by the gain of the input transistor.

A folded cascode is a low-noise amplifier, provided the noise of transistor M3 can be reduced by proper sizing or other techniques.

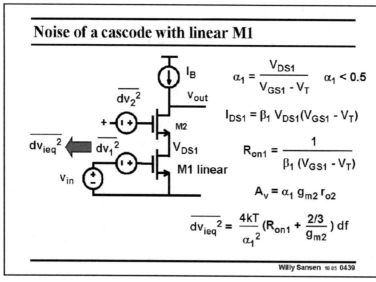

0439

Cascodes are sometimes used with an input transistor M1 in the linear region, to avoid distortion at the input.

In this case, it is no longer clear whether the noise of the cascode transistor is still negligible.

For this purpose we have to introduce a parameter to indicate how deeply the input transistor M1 works in the linear region. This is parameter α_1. For $\alpha_1 = 1$, the transistor is on the onset of saturation. Normally α_1 is about 0.5 or less.

The expression describing a MOST in the linear region is given, followed by its on resistance R_{on1}.

The voltage gain is smaller than before. Actually it is only a fraction of the gain of the cascode by itself!

The total equivalent input noise voltage is due to the on resistance R_{on1} and the thermal noise of the cascode. The latter noise source is no longer negligible! Moreover the input noise voltage is even larger because of the α_1 factor in the denominator.

The actual small-signal model and calculations are given next.

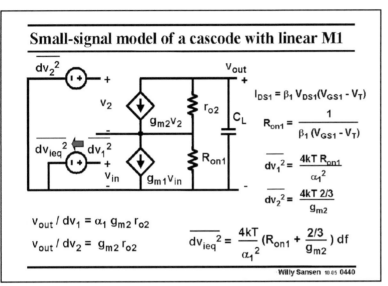

Small-signal model of a cascode with linear M1

$I_{DS1} = \beta_1 V_{DS1}(V_{GS1} - V_T)$

$R_{on1} = \dfrac{1}{\beta_1 (V_{GS1} - V_T)}$

$\overline{dv_1}^2 = \dfrac{4kT\,R_{on1}}{\alpha_1^2}$

$\overline{dv_2}^2 = \dfrac{4kT\,2/3}{g_{m2}}$

$v_{out} / dv_1 = \alpha_1\, g_{m2}\, r_{o2}$

$v_{out} / dv_2 = g_{m2}\, r_{o2}$

$\overline{dv_{ieq}}^2 = \dfrac{4kT}{\alpha_1^2}\left(R_{on1} + \dfrac{2/3}{g_{m2}}\right) df$

Willy Sansen 10 05 0440

0440

The small-signal model of the cascode with an input transistor M1 in the linear region is shown in this slide.

Note that the small-signal model of transistor M1, in the linear region, is a current source $g_{m1} v_{in}$ with a small on resistance R_{on1} in parallel. The output noise current is simply due to the on resistance R_{on1}. It can be referred to the input, as a input noise voltage, by dividing this noise current by g_{m1}^2.

The total equivalent input noise voltage is now the sum of the input noise of transistor M1 and the noise of the cascode.

For this calculation, we need to find the gains to the output, of the input noise voltage of the input transistor dv_1, and of the input noise voltage of the cascode transistor dv_2. They are both given at low frequencies.

Table of contents

Willy Sansen 10 05 0441

0441

After the noise analysis of the single-transistor configurations, we now consider the two-transistor configurations.

A current mirror is taken first now, followed by differential pairs.

We already know that a current source transistor should be designed with large $V_{GS} - V_T$ or small W/L. Let us see whether this applies to all current sources and mirrors.

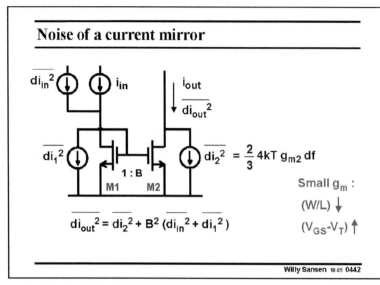

0442

A simple current mirror is shown in this slide. Its current gain factor is B.

All possible noise sources are shown by mean of current sources. The input signal has an input noise source in parallel and both transistors have noise current sources, which are proportional to g_m.

The total output noise is given below. Clearly all input noise is multiplied by B^2.

The only way to make the output noise small is to design all devices with large $V_{GS} - V_T$ or small W/L, as expected. For later matching (Chapter 12), we will also need to design the current mirrors with small W/L. This is therefore a very attractive result.

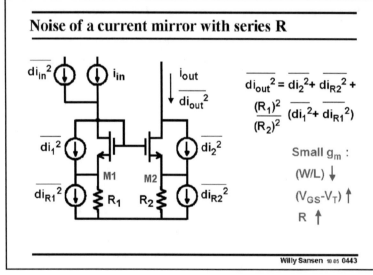

0443

The noise of a current source can be further reduced by inserting series resistances, as shown in this slide.

At first sight, it is a strange concept to add noisy resistors to reduce the noise, but it works.

Indeed let us first add all relevant noise sources. The resistor noise sources are modeled by currents, inversely proportional to the resistor values.

The total output noise current is given. Instead of trying to analyze this result, let us plot the total output noise versus series resistor.

It is shown next.

0444

This log-log plot shows the contributions of the transistors themselves, marked by M, and of the series resistors, marked by R. The current factor B is taken as unity for simplicity. Also both resistors are identical.

It is clear that for low resistors the noise of the transistors is dominant. The output noise is

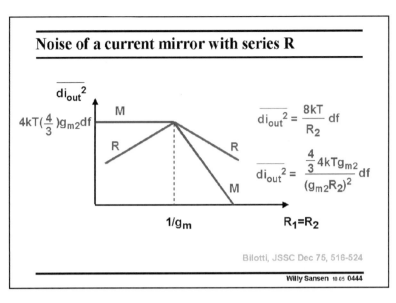

Noise of a current mirror with series R

$$\overline{di_{out}^2} = \frac{8kT}{R_2}\,df$$

$$\overline{di_{out}^2} = \frac{\frac{4}{3}4kTg_{m2}}{(g_{m2}R_2)^2}\,df$$

Bilotti, JSSC Dec 75, 516-524

Willy Sansen 10 05 0444

simply the sum of the noise power of the two transistors.

However, if we increase the resistors to beyond $1/g_m$, we find two important results:

– The transistor noise is decreased as a result of the feedback factor $g_{m2}R_2$. The power is decreased by the square of this factor.
– Also the total output noise is smaller than before; it has decreased to smaller values.

Actually, the resistor noise has taken over the transistor noise. As a result the output noise power decreases inversely proportional to that resistor value. Also their $1/f$ noise may be much lower.

This is a remarkable result. Series resistors will always be used to realize current sources with ultra-low output noise.

This is a result that has been known for bipolar transistor current sources since 1975. Why then, is this technique not used routinely for MOST's?

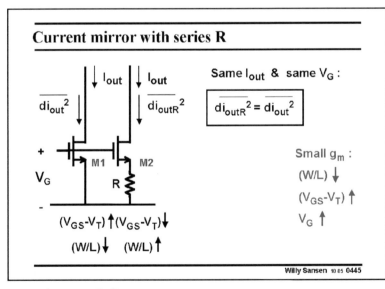

Current mirror with series R

Same I_{out} & same V_G :

$$\boxed{\overline{di_{outR}^2} = \overline{di_{out}^2}}$$

Small g_m :

$(W/L)\downarrow$

$(V_{GS}-V_T)\uparrow$

$V_G\uparrow$

$(V_{GS}-V_T)\uparrow(V_{GS}-V_T)\downarrow$

$(W/L)\downarrow$ $(W/L)\uparrow$

Willy Sansen 10 05 0445

0445

MOST devices have already a technique to reduce the output noise current. They don't need series resistors.

This technique consists of taking large values of $V_{GS}-V_T$ or small values of W/L, has already been explained. This gives the same effect as adding series resistors.

Indeed let us compare two transistor amplifiers with different $V_{GS}-V_T$ and W/L. They have the same Gate voltage V_G. They also carry the same DC current.

The first one with transistor M1 has a large $V_{GS}-V_T$ and hence small W/L. The second one with transistor M2 has a smaller $V_{GS}-V_T$ and much larger W/L. The difference in V_{GS} is taken up by a series resistor R.

The question is whether they have the same gain or same output noise.

The second one has a larger g_m because its $V_{GS}-V_T$ is smaller. The feedback of R, reduces the gain. As a result they have the same gain.

The same is also true for the output noise. The second one generates more output noise current because of the larger g_m, but it is reduced because of the feedback of R. As a result both output noise values are the same.

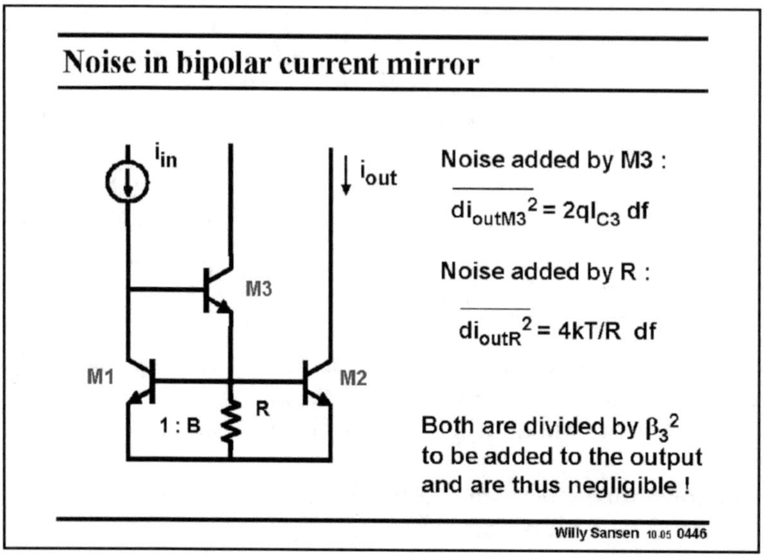

Noise in bipolar current mirror
i_in
i_out
M3
M1
M2
R
1 : B
Noise added by M3 :
di_outM3^2 = 2qI_C3 df
Noise added by R :
di_outR^2 = 4kT/R df
Both are divided by β3^2
to be added to the output
and are thus negligible !
Willy Sansen 10-05 0446

0446

The output noise current of the popular bipolar current mirror is now examined.

Clearly we already know what output noise current is produced by the two current mirror devices M1 and M2. The question is how much noise is contributed by transistor M3 and by the resistor R.

The expressions are given in this slide. Transistor M3 gives shot noise whereas the resistor R gives thermal noise.

Both inject noise at the common point. As a result they are subject to the feedback loop formed by M1 and M3. This is why their contributions to the output noise power have to divided by β_3^2.

As a result they are negligible.

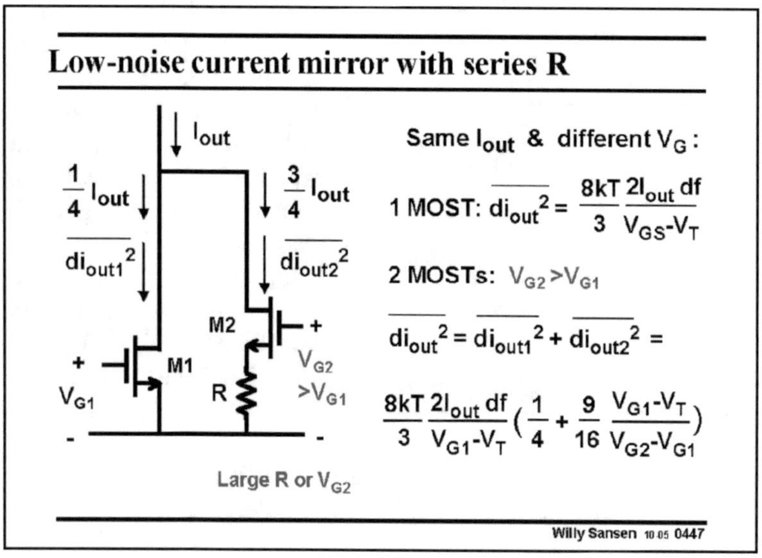

Low-noise current mirror with series R
I_out
1/4 I_out
3/4 I_out
di_out1^2
di_out2^2
M1
M2
R
+
V_G1
-
+
V_G2
>V_G1
-
Large R or V_G2
Same I_out & different V_G :
1 MOST: di_out^2 = 8kT/3 · 2I_out df / (V_GS-V_T)
2 MOSTs: V_G2 > V_G1
di_out^2 = di_out1^2 + di_out2^2 =
8kT/3 · 2I_out df / (V_G1-V_T) · (1/4 + 9/16 · (V_G1-V_T)/(V_G2-V_G1))
Willy Sansen 10-05 0447

0447

A circuit technique that is sometimes used to reduce the output noise of a current source is given in this slide.

The transistor is split up into parts. One of them, usually the one with the larger current, then receives a series resistor to reduce its noise output. In this example, this latter one takes 3/4 of the total current.

How well does this work? Obviously in order to be able to accommodate this series resistor, we need to take a V_{G2} biasing voltage that is quite large. It is much larger than V_{G1}. We compare two cases. In the first case we have only one single MOST, which takes all the DC current. Its output noise current is then known.

In the other case, we take two MOSTs. One has a much larger V_{G2} than the other. In this case, we have two contributions to the output noise current. The first one is due to transistor M1, which runs at 1/4 of the DC current. The other one is due to the resistor R. This resistor however, must have a value $(V_{G2} - V_{G1})/(0.75 I_{out})$.

Two terms appear in the total noise of the second case. The first one is the transistor noise whereas the second one is due to the resistor. It is clear that we need a really large V_{G2} to make this work. If we applied that same V_{G2} to the single MOST of the first case, we would obtain similar noise performance. This technique therefore does not work!

Table of contents

- Definitions of noise
- Noise of an amplifier
- Noise of a follower
- Noise of a cascode
- Noise of a current mirror
- Noise of a differential pair
- Capacitive noise matching

Willy Sansen 10 05 0448

0448

After the noise analysis of the current mirrors, we will discuss the noise performance of a differential pair.

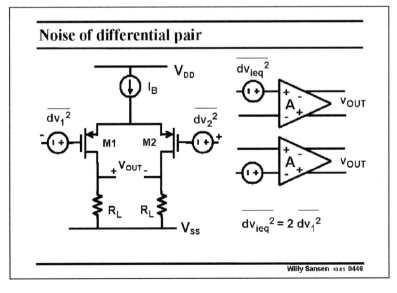

Willy Sansen 10 05 0449

0449

A simple differential amplifier is illustrated in this slide.

Both transistors exhibit the same amount of noise, as they carry equal currents. These noise powers are not correlated. They have to be summed up. The question is on which side?

The symbol of such a differential amplifier is shown on the right. Do we have to insert the equivalent noise voltage at the positive or at the negative input?

The answer is obvious.

Both are equivalent! Indeed the noise voltage is squared at the end of the calculations. As a result, it does not make any difference whether the noise is applied to the positive or negative side. We prefer the side where it is easier to do the calculations!

As a result, the total equivalent input noise source is simply twice the noise voltage power of one single transistor.

A differential amplifier always gives $\sqrt{2}$ or 41% more input noise voltage than a single amplifier. The lowest-noise amplifiers are single-input. On the other hand, these single-input amplifiers are much more sensitive to substrate noise. The debate on whether to use differential input amplifiers in RF receivers for example, is therefore still ongoing.

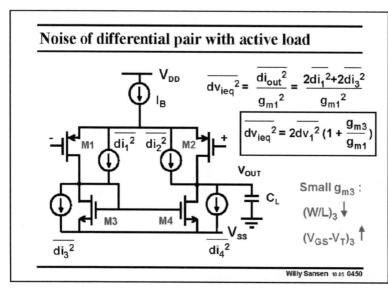

0450

A voltage differential amplifier with single-ended output is shown in this slide. The noise sources of all four transistors are added by their current sources.

This is a circuit with two equal halves. If we know the input noise power for one halve, we simply multiply by two.

Moreover each half consists of an amplifying transistor loaded by a current source. We already know how to reduce the noise contribution of the current source. We simply design it with a larger $V_{GS} - V_T$.

The resulting equivalent input voltage is now what we expected. It contains a factor of two for the two halves. Also it contains the g_m ratio, which is typical for an active load.

If we succeed in making the load $V_{GS} - V_T$ small, then we can limit the input noise to the two input transistors only. However, if we choose the same $V_{GS} - V_T$ for all transistors, or if we have bipolar transistors, then the noise of all 4 transistors is equally important!

0451

Differential pairs with large $V_{GS} - V_T$ values are actually called transconductors. If we cannot make the $V_{GS} - V_T$ sufficiently large, because of velocity saturation, we can still add resistors. In both cases we assume equal resistors and we also take $g_m R > 1$.

For small signals both realizations are equivalent. Indeed their gains are the same.

How about noise?

First of all, we notice that in the first case, DC current $I_B/2$ flows through the resistors R, which is not true in the second case. We therefore need a larger DC supply voltage.

Moreover, the noise performance is quite different. In the first case the equivalent input noise voltage is the noise contributed by both resistors. This is because for $g_m R > 1$, the transistor noise is negligible.

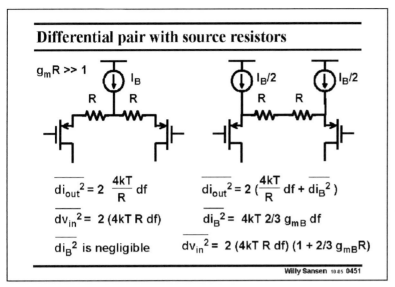

Differential pair with source resistors

$g_m R \gg 1$

$$\overline{di_{out}^2} = 2\,\frac{4kT}{R}\,df$$

$$\overline{dv_{in}^2} = 2\,(4kT\,R\,df)$$

$$\overline{di_B^2} \text{ is negligible}$$

$$\overline{di_{out}^2} = 2\,(\frac{4kT}{R}\,df + \overline{di_B^2}\,)$$

$$\overline{di_B^2} = 4kT\,2/3\,g_{mB}\,df$$

$$\overline{dv_{in}^2} = 2\,(4kT\,R\,df)\,(1 + 2/3\,g_{mB}R)$$

Willy Sansen 10 05 **0451**

Note that the noise contributed by the DC current source I_B is altogether negligible, as it is a common-mode signal, cancelled by the differential output.

In the second case, the noise from the DC current sources $I_B/2$ (with transconductance g_{mB}) is not negligible. On the contrary it is the dominant noise source. This is the main disadvantage of the second case!

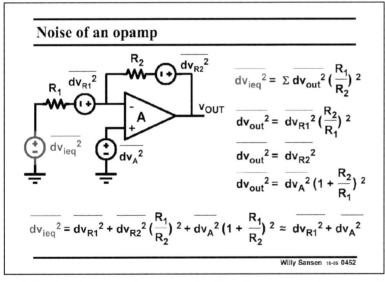

Noise of an opamp

$$\overline{dv_{ieq}^2} = \Sigma\,\overline{dv_{out}^2}\,(\frac{R_1}{R_2})^{\,2}$$

$$\overline{dv_{out}^2} = \overline{dv_{R1}^2}\,(\frac{R_2}{R_1})^{\,2}$$

$$\overline{dv_{out}^2} = \overline{dv_{R2}^2}$$

$$\overline{dv_{out}^2} = \overline{dv_A^2}\,(1 + \frac{R_2}{R_1})^{\,2}$$

$$\overline{dv_{ieq}^2} = \overline{dv_{R1}^2} + \overline{dv_{R2}^2}\,(\frac{R_1}{R_2})^{\,2} + \overline{dv_A^2}\,(1 + \frac{R_1}{R_2})^{\,2} \approx \overline{dv_{R1}^2} + \overline{dv_A^2}$$

Willy Sansen 10-05 **0452**

0452

Finally, let us have a look at the noise performance of an opamp with resistive feedback.

We assume that the overall voltage gain is large, i.e. that R_2 is much larger than R_1.

We can distinguish three sources of noise, i.e. the two resistors and the opamp itself, which has an input noise voltage source v_A.

Calculation of the contributions of these three voltage sources to the output and division by the voltage gain R_2/R_1, allows us to determine the total equivalent input noise voltage power.

For large gain, the noise voltage of the input resistor R_1 and of the opamp are the dominant noise sources.

This is to be expected. The input resistor R_1 is in series with the input signal. Also the opamps noise voltage is its equivalent input noise voltage. At the input, it appears unaltered.

0453

For most of the amplifiers, a resistive source impedance was assumed. Indeed, many sensors are resistive such as Wheatstone bridge pressure sensors.

Table of contents

Ref.: Z.Y.Chang, W.Sansen, Low-noise wide-band amplifiers, Kluwer AP, 1991

Willy Sansen 10.05 0453

Many sensors are capacitive, however. Photodiodes and radiation detectors are capacitive, but so are capacitive accelerometers, microphones, etc. The question is then, what transistor biasing provides the best noise performance? This is called capacitive noise matching.

Quite often this analysis leads to complicated expressions. They have been simplified to the bare minimum. Also the design plan has been simplified to a single equation.

Capacitive-source amplifier

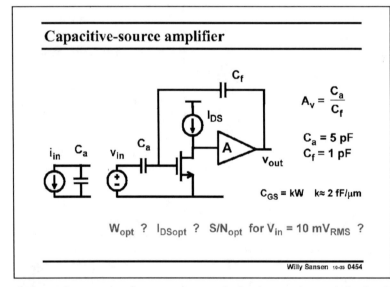

$$A_v = \frac{C_a}{C_f}$$

$$C_a = 5 \text{ pF}$$
$$C_f = 1 \text{ pF}$$

$$C_{GS} = kW \quad k \approx 2 \text{ fF}/\mu m$$

$$W_{opt} \; ? \quad I_{DSopt} \; ? \quad S/N_{opt} \text{ for } V_{in} = 10 \text{ mV}_{RMS} \; ?$$

Willy Sansen 10-05 0454

0454

A capacitive sensor can be represented by a current source in parallel or a voltage in series. The latter model is chosen as the calculations are a bit simpler. The optimization provides the same results.

The first preamplifier consists of a single transistor loaded by an ideal (or very-low-noise) current source. It is usually followed by another amplifier to have more gain.

The feedback loop is carried out by means of a capacitance. Indeed capacitances don't give any noise. Since the source is capacitive, the feedback element should also be capacitive!

The gain A_v is then simply given by the ratio of the two capacitances.

In this case, the input transistor is the only noisy component. The question then is, what must be its channel width W_{opt} (for minimum channel length L) and its current I_{DSopt}, for minimum noise?

What would be its SNR for an input signal of 10 mV_{RMS}?

We cannot forget that the input impedance of the MOST is also capacitive. Its C_{GS} is proportional to the width W, for minimum channel length L. We will use minimum channel length L, because we will end up with very large W/L ratio's. It is preferable to use minimum channel length then!

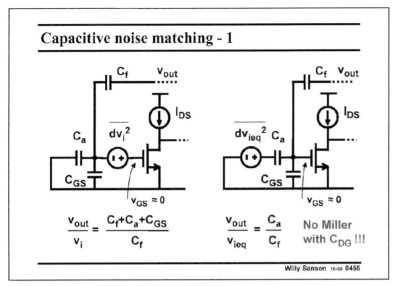

Capacitive noise matching - 1

$$\frac{V_{out}}{V_i} = \frac{C_f + C_a + C_{GS}}{C_f}$$

$$\frac{V_{out}}{V_{ieq}} = \frac{C_a}{C_f} \quad \text{No Miller with } C_{DG} \text{!!!}$$

Willy Sansen 10-05 0455

0455

Now we have to transfer the noise source from the Gate of the MOST to the input of the amplifier. Once the noise source is at the input, the SNR is readily calculated.

In order to do so, we calculate the gain from the equivalent input noise voltage to the output, we calculate the gain from the input to the output, and we equate both.

The gain from the noise source to the output is the most difficult. Since the feedback loop is still closed by capacitance C_c, the capacitance ratio determines the gain. Indeed, the Gate itself acts as a virtual ground.

The gain from input to output is simply C_a/C_f.

Equation of both provides the noise contribution of the MOST, transferred to the input.

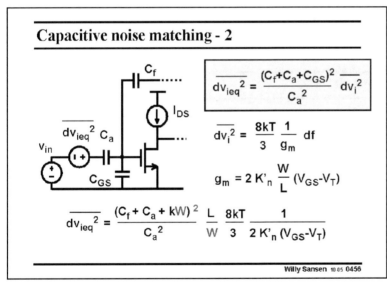

Capacitive noise matching - 2

$$\overline{dv_{ieq}^2} = \frac{(C_f + C_a + C_{GS})^2}{C_a^2} \overline{dv_i^2}$$

$$\overline{dv_i^2} = \frac{8kT}{3} \frac{1}{g_m} df$$

$$g_m = 2 K'_n \frac{W}{L} (V_{GS} - V_T)$$

$$\overline{dv_{ieq}^2} = \frac{(C_f + C_a + kW)^2}{C_a^2} \frac{L}{W} \frac{8kT}{3} \frac{1}{2 K'_n (V_{GS} - V_T)}$$

Willy Sansen 10 05 0456

0456

The noise of the MOST is actually transferred to the input by means of a capacitive transformer. It is clearly amplified, by that capacitance ratio. Note however, that this capacitive ratio depends on the transistor size or transistor width, as C_{GS} is part of it.

Rewriting this expression, in terms of transistor width, shows that there is minimum of the input noise versus width W. For small W, it drops out of the numerator and the noise goes down with W. For large W, the kW term overcompensates the W factor in the denominator and the noise goes up with W.

It is clear that we will try to make that up-transformation factor as small as possible. The numerator contains all possible capacitances connected to that node. A long coaxial wire for example between the photodiode and the amplifier would add a lot of capacitance, heavily deteriorating the noise performance. This is why all low-noise capacitive sensors have to be integrated together with their preamplifier.

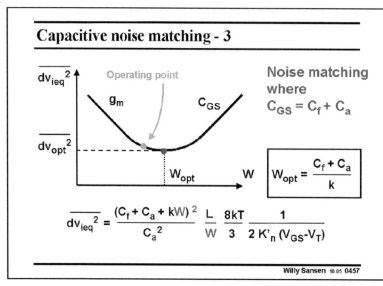

0457
This expression is given again, together with a plot of the input noise versus width W.

There is clearly a minimum. It is obtained at the point where the transistor input capacitance C_{GS}, equals the sum of capacitances, seen by the transistor.

For example if the sensor capacitance $C_a = 5$ pF, a feedback capacitance of $C_f = 1$ pF provides a voltage gain of 5. The optimum

width is the $W_{opt} = 6/0.002 = 3000$ μm or 3 mm.

In practice we prefer an operating point on the left of this optimum. The noise is not much worse but the size can be as much as half!

Now that we know the transistor width, we have to know in which technology (L and K′) we will realize this amplifier. Choosing a $V_{GS} - V_T = 0.2$ V then gives us the current I_{DSopt} and the transconductance g_{mopt}.

For example for $L = 0.13$ μm, for which $K'_n = 150$ μA/V^2, $I_{DSopt} = 138$ mA and $g_{mopt} = 1.38$ S. The noise resistance $2/3g_{mopt}$ is then 0.48 Ω. This is a very low value indeed!

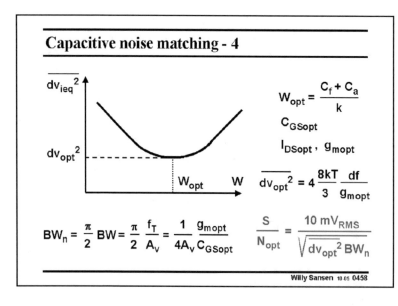

0458
In order to obtain the SNR, we must find the total or integrated noise. The bandwidth BW is approximated by the f_T of that input device, divided by the gain. Parameter f_T is determined by the input capacitance C_{GSopt} and the transconductance g_{mopt}. In this example, it is 36 GHz.

The noise bandwidth is now 57% larger or 57 GHz.

Finally the SNR is then the ratio of the input signal to the total noise. The result is easily calculated!

At the optimum the noise power is twice as high, corresponding to $0.48\sqrt{2}$ or 0.68 Ω. The noise density is then about 0.1 nV$_{RMS}$/$\sqrt{Hz}$. The integrated noise is then 24 μV$_{RMS}$.

The SNR is finally 417 or 52 dB. This is not band for a 36 GHz amplifier!

Table of contents

- ◆ **Definitions of noise**
- ◆ **Noise of an amplifier**
- ◆ **Noise of a follower**
- ◆ **Noise of a cascode**
- ◆ **Noise of a current mirror**
- ◆ **Noise of a differential pair**
- ◆ **Capacitive noise matching**

Willy Sansen 10.05 0459

0459

In this Chapter, we have carried out simplified analyses on the noise performance of all elementary circuits. This will allow us to obtain the equivalent input noise on most of the circuitry that follows.

051

Stability of Operational amplifiers

Willy Sansen

KULeuven, ESAT-MICAS
Leuven, Belgium

willy.sansen@esat.kuleuven.be

Willy Sansen 10-05 051

The operational amplifier (opamp) is certainly the main building block in all of the analog electronics. It is usually implemented in a feedback loop to provide stable and predictable gain and low noise.

In this chapter we will review what is required to make sure that this amplifier is stable under all conditions of feedback. An opamp is required not only to be stable but also to provide a well-behaved response. For example, peaking in the frequency domain is normally to be avoided. Also, when a square waveform is applied to an opamp, we don't want any ringing.

All these requirements will impose some fairly precise settings on the positions of the poles and zeros of this amplifier. Adding the requirement that the power consumption be minimized, we will find that it is quite easy to find an optimum in performance, in view of a certain GBW and capacitive load.

However, there are many more specifications in an opamp. They will be postponed to the Chapter following this one.

052

Table of contents

Willy Sansen 10-05 052

First of all, we have to review some of the terminology of an operational amplifier, such as open-and closed-loop gain. Also, we want to review some basics of second-order systems. These are the first two Sections. We put emphasis on two-stage amplifiers. We have dealt with single-transistor stages already in Chapter 2.

We will focus especially on that positive zero which shows up in any two-stage opamp. They can be avoided by means of additional current consumption. It is much more elegant however, to use some circuit tricks which allow the reduction of the total biasing current.

Finally, we will extended the compensation techniques, adopted for a two-stage amplifier, to three-stage amplifiers. Many class-AB amplifiers have three stages. Moreover, three stages become a necessity once the gain per stages has gone down to real low values, as in nanometer CMOS.

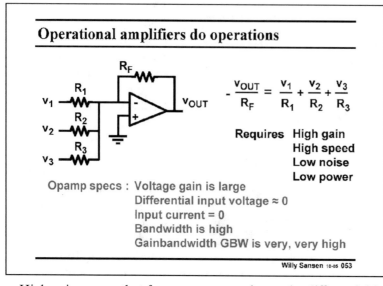

053

Operational amplifiers have been used to carry out operations on analog signals with great precision. They allow the addition, subtraction, multiplication, etc. of analog voltages. This is shown in this slide for three input voltages. The output voltage is a precise sum of the input voltages, scaled by the corresponding resistors.

This only works well provided the opamp itself has high gain up to high frequencies, with low noise, etc.

High gain means that for any output voltage, the differential input voltage is about zero.

The input currents are always zero if we use MOSTs and no bipolar transistors. In nanometer CMOS some Gate current may show up, giving rise to problems with the input currents!

This means that the most important specification of an opamp is its gain and bandwidth or its gain bandwidth product GBW.

We will optimize the GBW of an opamp for a certain capacitive load, towards minimum power consumption.

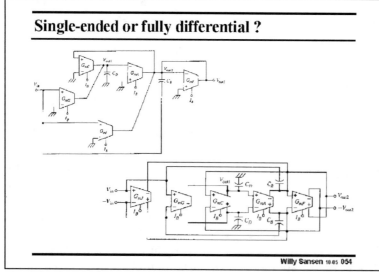

054

Most simple opamps only have one single-output. This has been the case in most discrete electronics. All voltages are referred to ground, which is easy to reach on a printed circuit board, for example.

In integrated circuits, more and more analog functions are integrated together with digital blocks. As a result, the substrate is polluted with clock spikes and logic noise. In this mixed-signal environment, all circuits must be fully-differential. This doubles the power consumption but rejects the common-mode noise.

We will discuss single-ended amplifiers first and then construct fully-differential amplifiers in Chapter 9.

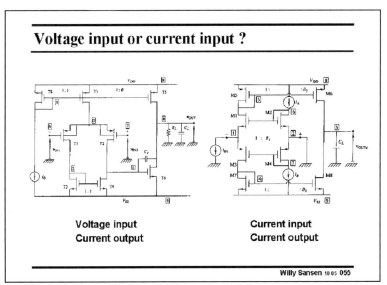

Voltage input or current input ?

Voltage input
Current output

Current input
Current output

Willy Sansen 10-05 055

055

An opamp can have a voltage input or a current input.

A voltage opamp has a differential voltage amplifier as an input (see left). It senses an input voltage.

It usually contains a single-transistor as a second stage. The output itself is at a high impedance level, at least at low frequencies. It therefore acts as a current source for the load.

As a result, we have a voltage-current amplifier or transconductance amplifier.

It is often called Operational Transconductance Amplifier.

The other amplifier (see right) has a current input, because the first transistor at the input is a cascode. It also has a current output. It is thus a current-current amplifier.

Needless to say that the characteristics are very different. They are difficult to compare as they involve external resistors.

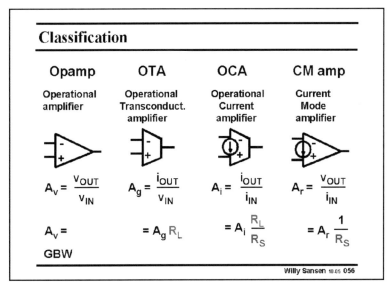

Classification

Opamp	OTA	OCA	CM amp
Operational amplifier	Operational Transconduct. amplifier	Operational Current amplifier	Current Mode amplifier
$A_v = \dfrac{v_{OUT}}{v_{IN}}$	$A_g = \dfrac{i_{OUT}}{v_{IN}}$	$A_i = \dfrac{i_{OUT}}{i_{IN}}$	$A_r = \dfrac{v_{OUT}}{i_{IN}}$
$A_v =$ GBW	$= A_g R_L$	$= A_i \dfrac{R_L}{R_S}$	$= A_r \dfrac{1}{R_S}$

Willy Sansen 10-05 056

056

All possible combinations for voltage/current input and output are depicted in this slide. The operational transconductance amplifier is the second one. If we add a (class AB) output to this OTA, we obtain a voltage output. This is now a conventional operational amplifier. Its voltage gain is normally very high.

It is not that easy to compare an OTA to an opamp, as the voltage gain of an OTA depends on the load R_L.

Both amplifiers can be realized with a current input. They have different names depending on who is talking.

In order to compare a current input amplifier to a voltage-input amplifier is more difficult.

Indeed, sometimes a current-input amplifier is driven from an input voltage source, having a source resistance R_S. Clearly, this resistance shows up in the comparison with a voltage amplifier. Obviously the smaller we can make R_S, the better.

We will start with an OTA.

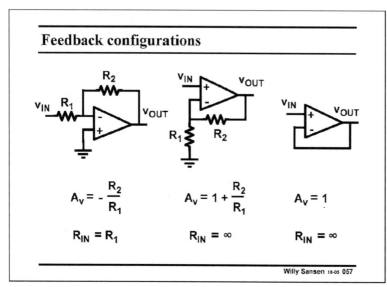

057

Opamps and OTA's are used with feedback. Normally, resistors are used, but also (switched) capacitors and sometimes even inductors.

Some easy configurations are sketched in this slide. The first one is the inverting amplifier, the second one the non-inverting amplifier.

However, they have all gains which are easy to set precisely. They all have different input resistances.

The last configuration is a buffer. The gain is unity but the input impedance is very high and the output impedance low.

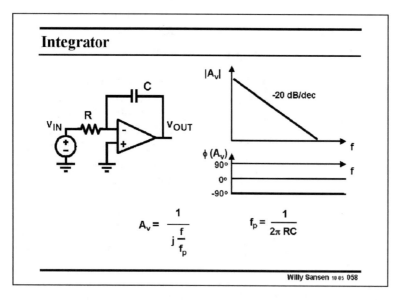

058

Opamps are used to make all kinds of filters as well.

The simplest one is probably the integrator.

At ever lower frequencies, the gain increases continuously until the value is reached of the open-loop gain of the opamp itself.

At all other frequencies, a constant slope is obtained of -20 dB/decade, and a constant phase shift of $90°$.

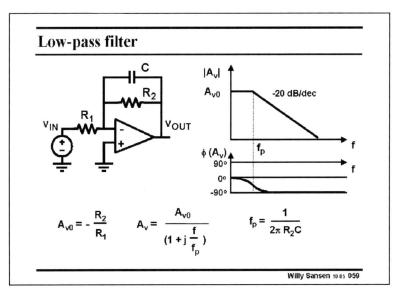

Low-pass filter

$$A_{v0} = -\frac{R_2}{R_1} \qquad A_v = \frac{A_{v0}}{(1 + j\frac{f}{f_p})} \qquad f_p = \frac{1}{2\pi R_2 C}$$

Willy Sansen 10-05 059

059

A first-order low-pass filter is an integrator with a resistor across the capacitor. It is also called a lossy integrator.

At lower and lower frequencies, the gain is now set by the ratio of the two resistors.

At the pole frequency, or the bandwidth, the phase shift is exactly halfway between zero and 90°, which is 45°.

Note that a first-order decreasing characteristic, which is called a pole, always shows a slope of -20 dB/decade and $-90°$ phase shift.

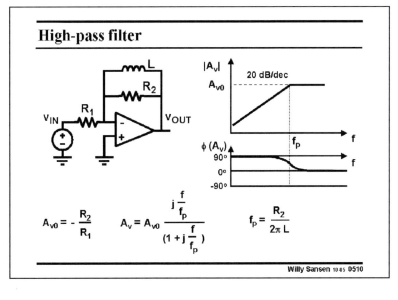

High-pass filter

$$A_{v0} = -\frac{R_2}{R_1} \qquad A_v = A_{v0} \frac{j\frac{f}{f_p}}{(1 + j\frac{f}{f_p})} \qquad f_p = \frac{R_2}{2\pi L}$$

Willy Sansen 10-05 0510

0510

Inductors can obviously be used as well.

Substitution of the capacitor by an inductor provides a high-pass characteristic. At high frequencies, the gain will be reduced by the internal poles of the opamp itself, as we will see later in this Chapter.

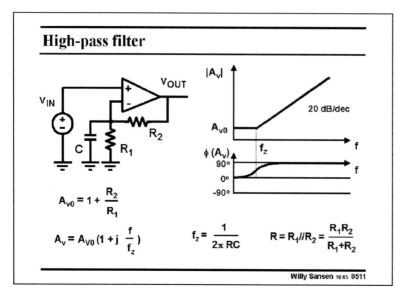

0511

Another high-pass filter is shown in this slide. It uses a capacitor. The transfer characteristic is very different.

At very low frequencies, the gain is constant. It starts increasing at the zero frequency.

Note that a first-order increasing characteristic, which is caused by a zero, always shows a slope of 20 dB/decade and a 90° phase shift.

Obviously, at high frequencies, the gain will decrease because of the internal poles of the opamp itself.

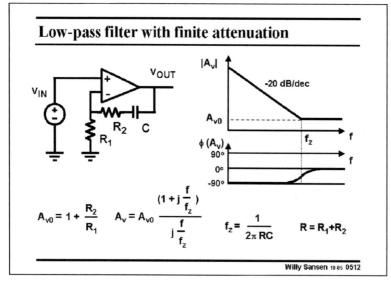

0512

Another filter with a low-pass characteristic is shown in this slide. At high frequencies it now has a constant gain.

Since we are dealing with a pole at zero frequency and a zero here, the phase shift is different.

Many more filters can be realized by means of opamps. These few examples have been added to illustrate this point. We will now focus on the poles and zeros within the opamp itself.

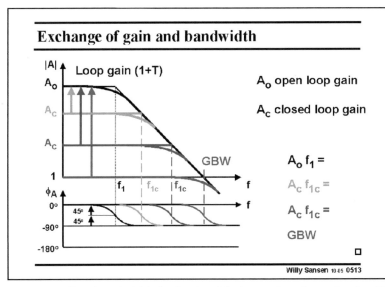

Exchange of gain and bandwidth

A_o open loop gain

A_c closed loop gain

$$A_o f_1 =$$
$$A_c f_{1c} =$$
$$A_c f_{1c} =$$
GBW

Willy Sansen 10-05 0513

0513

An opamp always has one internal dominant pole. It occurs at frequency f_1. It is normally caused by one of the bigger capacitances inside the amplifier.

The product of the open-loop gain A_o and this pole frequency f_1 is the GBW. The GBW is the product of the gain and the bandwidth, for each setting of the gain.

Indeed, the ratio of the two resistors, in an inverting amplifier for example, sets the closed-loop gain A_c. The corresponding bandwidth is the f_{1c}. Their product is again the GBW.

In the case of a unity-gain buffer, the bandwidth coincides with the GBW, which is the maximum frequency at which this opamp can be used.

An opamp allows therefore, an exchange gain for bandwidth. The lower the closed-loop gain, the higher the bandwidth. The product is always the GBW.

An opamp is a very versatile building block.

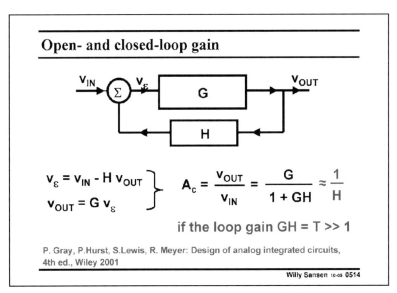

Open- and closed-loop gain

$$v_\varepsilon = v_{IN} - H\, v_{OUT}$$
$$v_{OUT} = G\, v_\varepsilon$$

$$A_c = \frac{v_{OUT}}{v_{IN}} = \frac{G}{1 + GH} \approx \frac{1}{H}$$

if the loop gain GH = T >> 1

P. Gray, P.Hurst, S.Lewis, R. Meyer: Design of analog integrated circuits, 4th ed., Wiley 2001

Willy Sansen 10-05 0514

0514

Using an opamp with feedback is only a special case of a feedback system. However, in such a system the gain block G has a lot of gain, which is not very precise.

The feedback elements, which are resistors and capacitors, determine the closed-loop characteristic. They are very precise.

The loop gain is the ratio of the open-loop gain and the closed-loop gain. It is the gain, going around in the loop. It is the quantity that determines all the properties of a feedback system. It is the quantity which indicates how the input and output impedances change. This will be explained in a more rigorous way in Chapter 8.

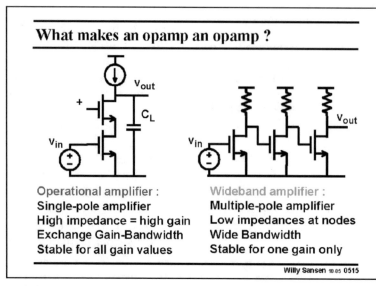

What makes an opamp an opamp ?

Operational amplifier :
**Single-pole amplifier
High impedance = high gain
Exchange Gain-Bandwidth
Stable for all gain values**

Wideband amplifier :
**Multiple-pole amplifier
Low impedances at nodes
Wide Bandwidth
Stable for one gain only**

Willy Sansen 10.05 0515

0515

An opamp is really a single-pole system. As a result, it allows exchange of gain with bandwidth, within a specific GBW.

Because there is only one dominant pole, there can be only one internal node at high impedance. If there are more nodes at high impedance, then we have more poles. In this case we have to add capacitance or increase the currents, such that this second pole, the nondominant pole, is at sufficiently high frequencies, beyond the GBW.

All two-stage amplifiers have two high-impedance nodes and hence two poles. Therefore, all amplifiers with two high-impedance nodes are called two-stage amplifiers, irrespective of the number of transistors.

We will have to compensate these two-stage amplifiers, i.e. we will have to add capacitance or increase currents to shift the non-dominant pole out to sufficiently high frequencies. As a result, the amplifier resembles again a single-pole system.

Wideband amplifiers are very different. They consist of more stages, each of them having a pole. They are normally compensated at one particular setting of the gain. They are not meant to exchange gain for bandwidth. On the contrary, at that gain setting, they are optimized for maximum bandwidth. More about such amplifiers is given in Chapter 8.

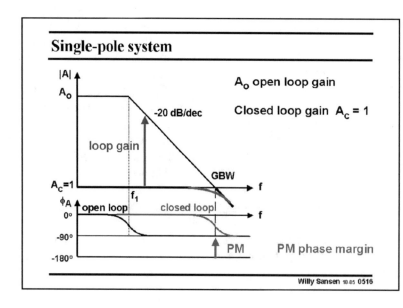

Single-pole system

A_o open loop gain

Closed loop gain $A_c = 1$

Willy Sansen 10.05 0516

0516

If the operational amplifier is truly a single-pole amplifier, then it can never show peaking or any other form of instability. Indeed the slope of -20 dB/decade is then maintained to frequencies beyond its GBW. Also, its phase of $-90°$ is constant for all frequencies beyond the bandwidth f_1.

Application of unity-gain feedback results in an amplifier, the bandwidth of which coincides with the GBW.

There is no trace of peaking.

Peaking or onset of oscillation would only be possible if the phase characteristic approached the $-180°$ line. In this case the negative feedback would be converted into positive feedback and oscillation would be possible.

We will have to verify how far away the phase is from that critical $-180°$. This is why this phase distance has received a name. It is called **phase margin**. It is taken at the frequency where the loop gain is unity. In this case this frequency is the GBW.

Clearly, a phase margin of 90° is large enough not to find peaking, or any form of oscillation.

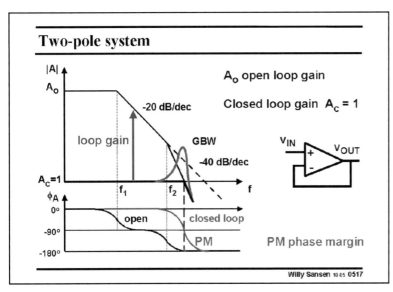

0517

This is very different for an opamp with two poles at frequencies f_1 and f_2.

Each pole causes a phase shift of $-90°$. As a result, we find a phase shift at high frequencies. This means that at high frequencies, the signal is inverted. There is still a loop gain slightly larger than unity. Negative feedback turns into positive feedback with a little bit of gain. We obtain therefore, an oscillator rather than an amplifier!

At the frequency where the loop gain becomes unity (which here is the GBW), the phase margin PM is not quite zero. If it were zero we would have a real oscillator. It is not quite zero but very small. This is why this amplifier shows a tendency for oscillation. It shows a large amount of peaking at that frequency.

We do not want peaking because such a peak is very irreproducible. Moreover the noise is deteriorated by that peak. Remember, noise has to be looked at on a linear frequency axis. Such a peak then extends over most of the frequency range.

The question is, how far do we have to stay with our phase characteristic from this critical $-180°$; how large can the phase margin PM be allowed to increase to avoid this peaking?

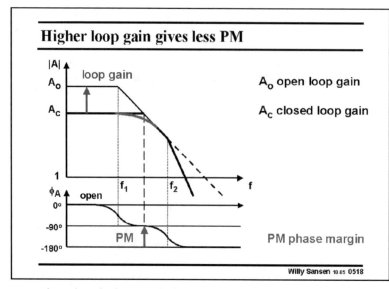

0518

Nevertheless, the same amplifier, with two poles, can be used without peaking. It is sufficient to use it at a higher closed-loop gain A_c.

The closed-loop gain is quite high now. As a result, the loop gain is much smaller. Moreover, when we check where the loop gain becomes unity, we find a frequency at which the phase margin PM is quite high.

This is why there is no peaking. The amplitude curve is quite nicely rounded.

It is clear that the same amplifier can show peaking or not, depending on the actual closed-loop gain A_c. It is also clear that for unity-gain feedback, the frequency where we have to check the phase margin is the highest. The phase margin is the smallest at the highest frequency. Unity-gain amplifiers give the highest amount of peaking!

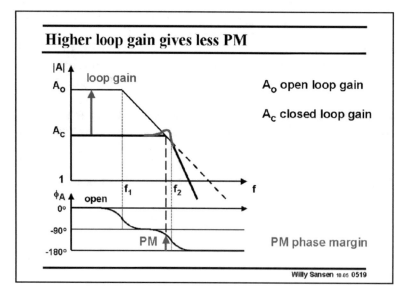

0519

Let us now gradually lower the closed loop gain A_c.

The loop gain increases, and the frequency at which we have to read the phase margin increases. The phase margin thus decreases and the peaking is gradually showing up!

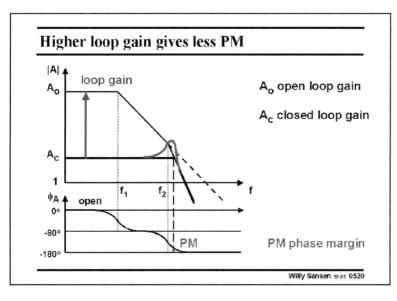

0520

For even lower closed-loop gain A_c, the loop gain increases further, and the frequency at which we have to read the phase margin increases even more. The phase margin thus decreases further and the peaking therefore becomes more severe!

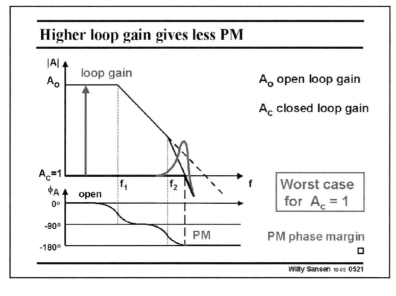

0521

Finally, we have again reached the point of unity-gain closed-loop gain A_c. The loop gain is the same as the open-loop gain. The frequency at which we have to read the phase margin is now the GBW. The phase margin has become very small and the peaking is most severe!

It is clear that the worst peaking has been obtained for the largest loop gain, i.e. for the lost closed-loop gain or for unity gain. This is clearly the worst case.

We will try to avoid this peaking by adding a compensation capacitance or by increasing the currents. It is clear that an amplifier which has been compensated at unity gain, is overcompensated at most other settings of closed-loop gain.

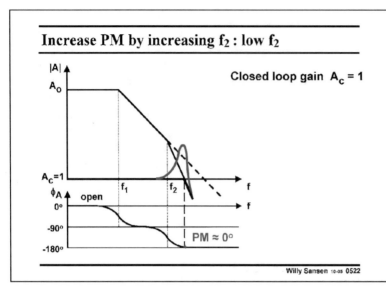

0522

How can we compensate a two-pole amplifier?

The objective is quite straightforward. We have to find a means to shift the second or non-dominant pole f_2 to higher frequencies. The extra $-90°$ attributed to this pole has now disappeared. The phase margin is now $90°$.

To illustrate the effect of shifting the non-dominant pole to higher frequencies, we repeat the Bode diagrams for the same two-pole amplifier and the same unity-gain, but with three different positions of the nondominant pole f_2.

In the Bode diagrams shown in this slide, the second pole f_2 is clearly too close to the first one f_1. Large peaking results.

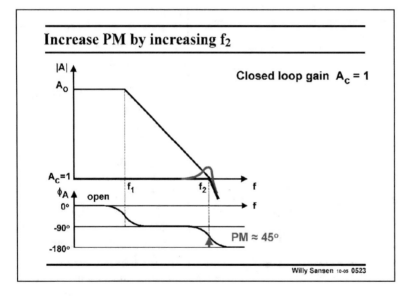

0523

Shifting the second pole to higher frequencies increases the phase margin and decreases the peaking.

In this case the non-dominant pole f_2 coincides with the GBW. As a result the phase margin is $45°$. The peaking is smaller indeed.

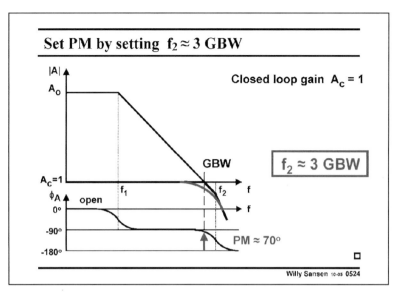

Set PM by setting $f_2 \approx 3$ GBW

Closed loop gain $A_c = 1$

$\boxed{f_2 \approx 3 \text{ GBW}}$

PM ≈ 70°

Willy Sansen 10-05 0524

0524

Finally, we have shifted the non-dominant pole f_2 to a value which is about three times the GBW.

In this case the phase margin is close to 70°. The peaking is gone altogether. The amplitude curve is nicely rounded.

This is what we want to design all our opamps for, a phase margin of around 70°, such that no peaking occurs.

Where is this factor of three coming from?

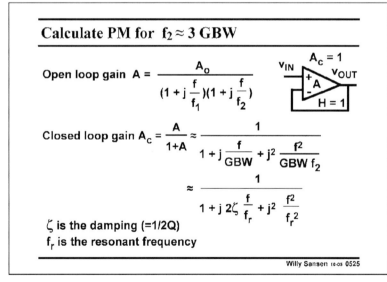

Calculate PM for $f_2 \approx 3$ GBW

$A_c = 1$

Open loop gain $A = \dfrac{A_o}{(1 + j\dfrac{f}{f_1})(1 + j\dfrac{f}{f_2})}$

Closed loop gain $A_c = \dfrac{A}{1+A} \approx \dfrac{1}{1 + j\dfrac{f}{GBW} + j^2\dfrac{f^2}{GBW\,f_2}}$

$\approx \dfrac{1}{1 + j\,2\zeta\dfrac{f}{f_r} + j^2\dfrac{f^2}{f_r^2}}$

ζ is the damping (=1/2Q)
f_r is the resonant frequency

Willy Sansen 10-05 0525

0525

This factor of three is actually a result of a calculation of the peaking and phase margin, of a two-pole system to which unity-gain feedback is applied. This is explained in all textbooks on feedback or control theory!

When we take the expression of amplifier A with low-frequency gain A_0 and two poles, we have to plug it in the feedback expression for unity gain. Actually, this feedback expression was

$G/(1+GH)$ but here $H=1$ and $G=A$. The closed-loop gain A_c is unity at low frequencies.

In this feedback expression, we can rewrite the coefficients in terms of resonant frequency f_r and damping ζ (Greek letter d, zeta).

Often parameter Q is used instead of ζ, then $Q = 1/2\,\zeta$.

It is clear that f_r gives the frequency at which peaking or resonance occurs. Parameter ζ determines how high the peaking is. For zero ζ, the term in s vanishes and we obtain a zero denominator at frequency f_r. Therefore, we obtain an oscillator at frequency f_r.

We will need values of ζ between 0.5 and 1 to avoid peaking. Actually when $\zeta = 1$, we have one double pole.

0526
The actual values of the phase margin and peaking are now easily calculated. They are given in this slide. Also, the amount of peaking in the frequency domain P_f is given, followed by the amount of peaking in the time domain P_t.

Relation PM, damping and f_2/GBW

$$f_r = \sqrt{GBW\,f_2} \qquad PM\,(°) = 90° - \arctan\frac{GBW}{f_2} = \arctan\frac{f_2}{GBW}$$

$\dfrac{f_2}{GBW}$	PM (°)	$\zeta = \dfrac{1}{2}\sqrt{\dfrac{f_2}{GBW}}$	P_f (dB)	P_t (dB)
0.5	27	0.35	3.6	2.3
1	45	0.5	1.25	1.3
1.5	56	0.61	0.28	0.73
2	63	0.71	0	0.37
3	72	0.87	0	0.04

Willy Sansen 10.05 **0526**

For a ratio of three between the non-dominant pole and the GBW, the phase margin is 72°. The corresponding ζ is 0.87 (or Q=0.57). No peaking occurs in the Bode diagram.

We could be allowed to reduce the non-dominant pole a bit, to two times the GBW. The phase margin decreases to 63°, decreasing the ζ as well to 0.71 (and Q=0.71) and we still do not have peaking.

We cannot forget, however, that we are using hand calculations here. After this part of the design procedure, we want to verify the circuit performance by means of a numerical simulator such as SPICE, Then all parasitic capacitances come in, pushing the non-dominant pole to lower values and

decreasing the phase margin. A value of three is thus a good safety position to start with.

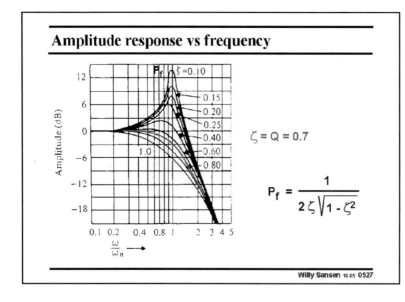

Willy Sansen 10.05 **0527**

0527
A better sketch of the peaking in the frequency domain (or in the Bode diagram) is shown in this slide.

The value of the maximum peaking P_f is given as well. It is clear that for a damping ζ of 0.7, we obtain a maximally-flat response. Going to larger ζ reduces the bandwidth too much.

Also, going to smaller values of ζ causes peaking indeed.

For zero ζ, we would have a peak to infinity, which is typical for an oscillator.

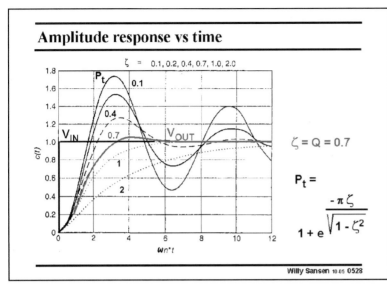

Amplitude response vs time

ζ = 0.1, 0.2, 0.4, 0.7, 1.0, 2.0

$\zeta = Q = 0.7$

$$P_t = \frac{}{1 + e^{\dfrac{-\pi\zeta}{\sqrt{1-\zeta^2}}}}$$

Willy Sansen 10 05 0528

0528

Peaking in the frequency domain corresponds to ringing in the time domain.

To the same amplifier, we apply now an input voltage v_{IN} which has a square waveform. The output voltage v_{OUT} follows with some delay.

For small values of ζ however, the output voltage overshoots, followed by ringing. The system is underdamped. The peak of the first overshoot P_t is given.

By taking a damping ζ of 0.7, the overshoot is very light and there is no ringing. A value of ζ of 0.87 would not give any overshoot at all.

These values of ζ for no ringing are clearly similar to the values for no peaking. Ringing in the time domain and peaking in the frequency domain are clearly equivalent.

The settling time is the time required to obtain the final value with a certain error. For example, a square waveform applied to a first-order system, gives an exponential with a certain time constant. For settling within 0.1%, we need to wait ln(1000) or 6.9 time constants.

For a slightly underdamped two-pole system, it is not so obvious to find the 0.1% settling time. Certainly a ζ between 0.7 and 0.8 gives the best compromise between rise time and settling time.

Table of contents

- **Use of operational amplifiers**
- **Stability of 2-stage opamp**
- **Pole splitting**
- **Compensation of positive zero**
- **Stability of 3-stage opamp**

Willy Sansen 10 05 0529

0529

Now that we know what it means to have a stable amplifier, or an amplifier without peaking or ringing, let us apply this theory now to a conventional two-stage amplifier.

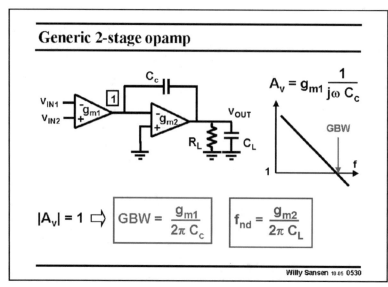

0530
A generic 2-stage amplifier is shown in this slide.

It consists of a differential input stage, which converts the differential input voltage into a current, by transconductance g_{m1}. A second-stage follows, which is usually little more than a single-transistor amplifier, and which has a transconductance g_{m2}. The output load consists of both a resistor and a capacitor.

The second stage has a feedback capacitor C_c. It will be used to compensate this opamp. This is why it is called **compensation capacitance**. We will now try to find the gain, bandwidth and the gain-bandwidth product GBW.

The gain is readily found by realizing that the second stage is actually a transresistance amplifier which converts the input current into the output voltage, by means of the impedance of capacitor C_c.

The gain A_v is then simply the product of the input g_{m1} with the impedance of C_c. Obviously, this gain A_v decreases with frequency, and crosses the unity-gain line at the frequency GBW.

The gain does not go to infinity at very low frequencies. It stops somewhere depending on whether cascodes are used, etc. The low-frequency gain is not that important after all. The higher frequency region is much more important, since feedback is always applied.

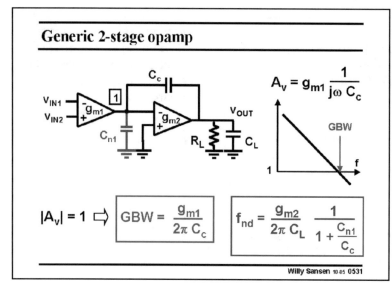

0531
The GBW is thus given by the frequency where the voltage gain is unity. Its expression is valid for all two-stage amplifiers!

Remember that a single-transistor amplifier has a similar expression for the GBW. However, note that it contains the load capacitance. This two-stage opamp contains the compensation capacitance C_c instead.

For stability, we have to know the position of the non-dominant pole.

This pole f_{nd} is determined by the other capacitance, i.e. the load capacitance C_L. The time constant is given by the product of this load capacitance C_L and the resistance seen by it. This is resistor R_L but especially resistance $1/g_{m2}$, offered by the second stage, across which C_c acts as a short-circuit at these high frequencies, where f_{nd} is expected to occur.

Indeed the second stage is usually a single transistor. Its Drain is then connected to its Gate. Its resistance is simply $1/g_{m2}$.

Therefore, the non-dominant pole is mainly determined by time constant C_L/g_{m2}. An exact calculation reveals that we have to take into account that a small capacitance C_{n1} is present at node 1. The capacitive division C_{n1}/C_c is a kind of correction factor. We normally choose capacitor C_c to be at least three times larger than C_{n1}.

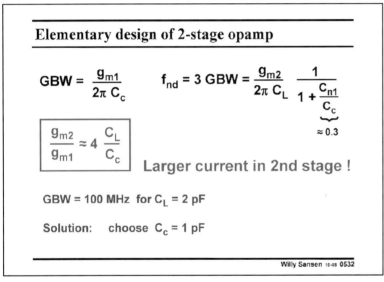

0532

Both the GBW and the non-dominant pole are linked by the stability requirement. The f_{nd}/GBW must be about three!

Rewriting this provides an important relationship between the trans-conductances on one hand, and the capacitances on the other.

The correction factor C_{n1}/C_c is simply taken to be 0.3. Combining this with the ratio f_{nd}/GBW of three, we obtain a factor of 4.

This relationship shows why the current in the second stage of a 2-stage opamp always consumes much more power than the first stage. Indeed for a specific $V_{GS} - V_T$ (such as 0.2 V), the transconductances represent the currents. Normally, we choose the compensation capacitance to be smaller than the load capacitance. It is usually 2–3 times smaller.

As a result, the current in the second stage is 8–12 times larger than the current in an input transistor.

This relationship also shows that a redesign for larger C_L, requires either a larger C_c or a larger g_{m2}. These are the two techniques, that we will use to compensate a two-stage opamp.

As an example, for a specific GBW and C_L, the equations can easily be solved to find the two g_m's, provided we firstly choose C_c!

0533

The stability requirement forces us to position the non-dominant pole at sufficiently high frequencies (around 3 GBW).

Table of contents

Willy Sansen 10-05 0533

The question now is, how to carry out the design? Which parameters are we going to use in the design plan to shift the non-dominant pole.

We will find that there are two possible design plans, both with advantages and disadvantages. Both of them lead to pole splitting, which allows us to move this non-dominant pole to even higher values.

Generic 2-stage opamp : Miller OTA

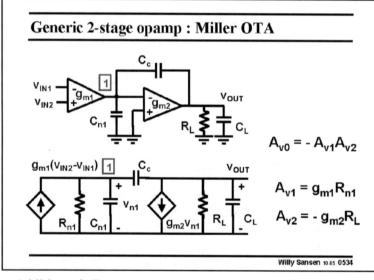

$$A_{v0} = - A_{v1}A_{v2}$$

$$A_{v1} = g_{m1}R_{n1}$$

$$A_{v2} = - g_{m2}R_L$$

Willy Sansen 10-05 0534

0534

Let us now take the two-stage operational amplifier. We substitute the g_m-blocks by their voltage controlled current sources. Also, node resistances are added on each node, representing the output impedances. Therefore, the small-signal equivalent circuit is obtained.

The low-frequency gains A_{v1} and A_{v2} are easily derived as they are merely products of g_m's and output resistances. The total gain A_v is then their product.

Addition of all capacitances provides an expression of the gain versus frequency, which is of second order. It is only of second order, despite the fact that we distinguish three capacitances, because the three capacitances form a capacitive loop. Disrupting this loop, for example by putting in a series resistor somewhere in this loop, would raise the order of the gain expression to three. The analysis would then become much more cumbersome.

0535

The full expression of the gain A_v is given in this slide. Only two approximations have been taken. We have assumed that the gain is larger than unity and that node resistance resistor R_{n1} is larger than load resistance R_L.

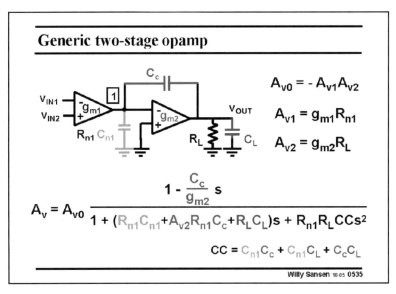

Generic two-stage opamp

$$A_{v0} = -A_{v1}A_{v2}$$

$$A_{v1} = g_{m1}R_{n1}$$

$$A_{v2} = g_{m2}R_L$$

$$A_v = A_{v0} \frac{1 - \frac{C_c}{g_{m2}}s}{1 + (R_{n1}C_{n1} + A_{v2}R_{n1}C_c + R_LC_L)s + R_{n1}R_LCCs^2}$$

$$CC = C_{n1}C_c + C_{n1}C_L + C_cC_L$$

Willy Sansen 10 05 0535

The denominator is of second order in s or jω. It now has two roots, which are the two poles. They can be real or complex.

The numerator has one root only, which is the zero. It is a positive zero (in the polar diagram).

The question is again how we have to dimension C_c and/or g_{m2} to shift the non-dominant pole out to higher frequencies?

For this purpose, we have to take a look at the positions of the poles when C_c is varied.

This is not so obvious as C_c occurs at several places in the denominator. In order to figure out how the two poles (and the zero) change as we vary C_c, we draw the pole-zero position diagram.

Approximate poles and zeros

$$A = A_0 \frac{1 - cs}{1 + as + bs^2}$$

$$\text{Zero } s = \frac{1}{c}$$

$$\text{Pole } s_1 = -\frac{1}{a}$$

$$s_2 = -\frac{a}{b} \quad \text{if } s_2 \gg s_1$$

Willy Sansen 10 05 0536

0536

It is not so difficult to obtain the two poles. After all, they are the roots of the denominator, which is just a second-order expression.

In most cases however, there is an easy way to obtain the poles. We assume that the poles have values which are widely different. Indeed, we expect to find a dominant pole and a non-dominant pole which are very different.

In this case, the dominant pole can be obtained by dropping the term in s^2 in the denominator. The dominant pole is simply $-1/a$.

The non-dominant pole is also easy to find. It is obtained by dropping the term 1 in the denominator, and by taking out one s. The non-dominant pole is simply $-a/b$.

Clearly, as coefficient a changes as a result of a parameter change somewhere, both poles are then affected, but in opposite ways. If the dominant pole decreases, the non-dominant pole must increase.

This will be the basis for pole splitting.

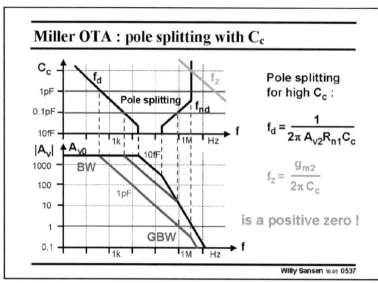

Miller OTA : pole splitting with C_c

Pole splitting for high C_c :

$$f_d = \frac{1}{2\pi A_{v2}R_{n1}C_c}$$

$$f_z = \frac{g_{m2}}{2\pi C_c}$$

is a positive zero !

Willy Sansen 10-05 0537

0537

The pole-zero position diagram is a plot of the poles and zero versus frequency, for one of the design parameters as a variable.

The frequency axis is the same as in the Bode diagram.

In this example, compensation capacitance C_c is taken as a design variable. Indeed, we want to see how C_c can be used to shift out the second pole to higher frequencies than GBW.

Clearly for a C_c smaller than 10 fF, two poles occur. They are fairly close to each other. The Bode diagram is now easily sketched.

However, for larger C_c (larger than about 20 fF in this example) the poles split. The dominant pole f_d becomes ever more dominant.

The non-dominant pole f_{nd} now shifts out, as intended.

The Bode diagrams are added for values of C_c of 0.1 and 1 pF. It is clear that for a C_c of 1 pF, sufficient pole splitting has been achieved. Indeed, the f_{nd} is about three times larger than the GBW, which is about 1 MHz.

The expression of the dominant pole is easily extracted from the expression of the gain. This is given in this slide. It is clearly due to the Miller effect of this (fairly large) capacitor C_c.

However, we also have a positive zero!

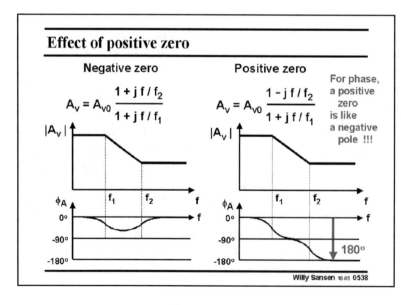

Effect of positive zero

Negative zero

$$A_v = A_{v0} \frac{1 + jf/f_2}{1 + jf/f_1}$$

Positive zero

$$A_v = A_{v0} \frac{1 - jf/f_2}{1 + jf/f_1}$$

For phase, a positive zero is like a negative pole !!!

Willy Sansen 10-05 0538

0538

In order to better understand what a positive zero means, we have to compare the effect on the phase of a positive zero, with that of a negative zero.

There is no need to have a second-order system for this purpose. A first-order system can be taken as well. The Bode diagrams are sketched for a first-order system with one single pole and one single zero. In the second case the zero is positive.

Both have obviously the same amplitude. Consequently, amplitudes are not affected by signs.

They have a very different phase characteristic, however. In the first Bode diagram the phase returns to zero for high frequencies. For the second diagram however, for a positive zero, the phase goes to $-180°$.

This is like having a second pole, rather than a zero!

This completely ruins our phase margin!!!

We have tried to limit the phase contribution of the non-dominant pole to about $20°$ by carefully locating this non-dominant pole beyond the GBW. A positive zero now shows up, which brings in another $-90°$. This will ruin the phase margin.

Moreover, the larger we make the compensation capacitance C_c, the more this zero shifts to lower frequencies. Large values of C_c are therefore not allowed!

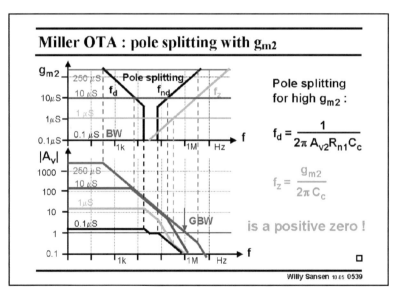

0539

Another way to realize the same pole splitting is to use g_{m2}, set by the current in the second stage. It works even better than with C_c!

Increasing the current from low values to higher one causes the dominant pole to be exactly the same as before. In the Miller effect, the g_{m2} is as much present as C_c itself.

However, the non-dominant has a better behavior. It keeps on increasing to high values, for high values of g_{m2}.

The main advantage is that the positive zero moves out to higher values, when g_{m2} increases. Therefore, this zero disappears!

As a result, it is a lot easier to realize pole splitting by increasing g_{m2} than it is with C_c.

The major drawback of compensating an opamp with g_{m2} is that the current consumption increases drastically.

For low-power designs we therefore prefer to compensate an opamp by increasing the compensation capacitance C_c. We now have to find other ways of dealing with the positive zero!

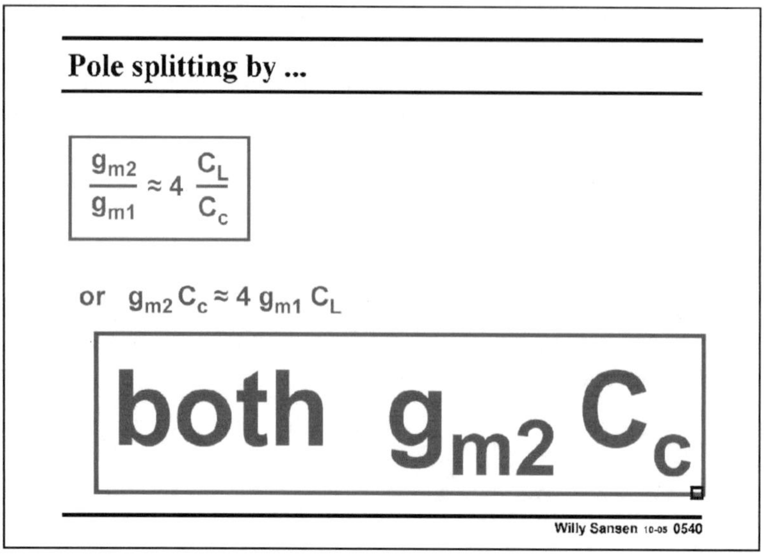

Pole splitting by ...
$\frac{g_{m2}}{g_{m1}} \approx 4 \frac{C_L}{C_c}$
or $g_{m2} C_c \approx 4 g_{m1} C_L$
both g_{m2} C_c
Willy Sansen 10-05 0540

0540

This conclusion is what we have already reached previously.

Increasing the load capacitance will force us to either increase the compensation capacitance or to increase the current in the second stage, as imposed by the stability requirement. Both help!

Therefore, if during some design effort, the capacitance C_c does not provide sufficient phase margin, then we have to increase the current in the second stage somewhat. This always helps!

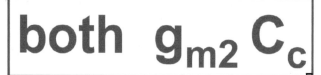

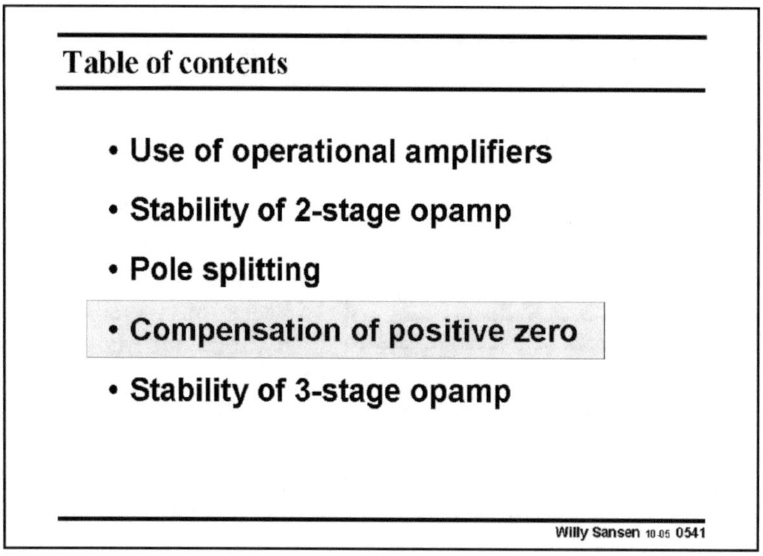

Table of contents
• Use of operational amplifiers
• Stability of 2-stage opamp
• Pole splitting
• Compensation of positive zero
• Stability of 3-stage opamp
Willy Sansen 10-05 0541

0541

Nevertheless, we do not want to increase the current in the second stage. All analog design is towards low-power design.

This is why we have to find all possible techniques to remove this positive zero.

They are listed next.

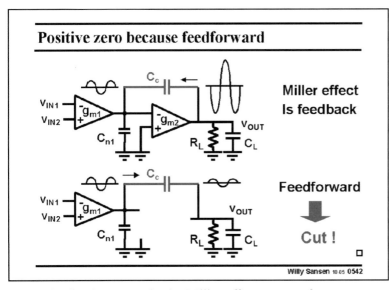

Positive zero because feedforward

Miller effect
Is feedback

Feedforward

Cut !

Willy Sansen 10.05 0542

0542

In order to understand how we can abolish the positive zero, we have to try to understand what its origin is, which adds another $-90°$ phase shift.

This is a result of the feedforward through the compensation capacitance.

Indeed, the compensation capacitance is bidirectional after all, as most capacitances. This means that feedback current and feedforward current flow at the same time.

The feedback current is the Miller effect current from output to input. It flows between two nodes which are opposite in phase.

The feedforward current is only easy to see when we leave out the amplifier itself. We now notice a feedforward current through C_c which causes a small output signal which is in phase with the input. This is the current which causes the zero. It is a positive zero because it provides an output signal which is of opposite phase compared with the amplified output signal.

To abolish this positive zero, we have to make that compensation capacitance unidirectional. In other words, we have to put a transistor in series, which cuts the feedforward path.

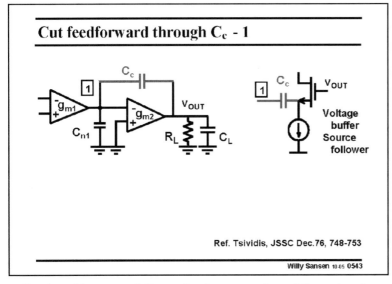

Cut feedforward through C_c - 1

Voltage buffer
Source follower

Ref. Tsividis, JSSC Dec.76, 748-753

Willy Sansen 10.05 0543

0543

There are three ways to abolish the positive zero. For the first two, it is easy to see that the feedforward current is blocked. The third technique is more difficult to understand.

The first technique consists of putting a source follower in series with the compensation capacitance. The feedback is still present but the feedforward current flows through the source follower to the positive supply, without affecting the output.

Putting this source follower in the expression of the gain, gives as a result that the numerator has gone. There is no more zero. It is now an easy way to solve this problem. However, we need some biasing current through the follower. This may not be the real solution for low-power design.

The second technique is using a cascode instead. Its DC current is pulled out of the Source. A pMOST can be used as well, provided DC current is injected into the Source.

Again, the AC feedback current can flow but not the feedforward current.

The zero simply vanishes, but again at the cost of some additional biasing current. This problem can be solved as shown on the next slide.

The third technique does not require any biasing current. This is why it is often preferred in low-power amplifiers.

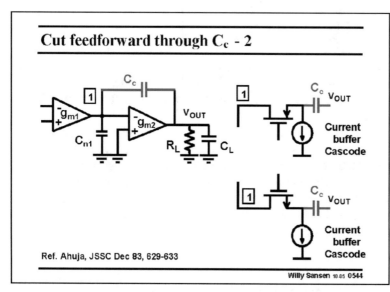

0544

There are three ways to abolish the positive zero. For the first two, it is easy to see that the feedforward current is blocked. The third technique is more difficult to understand.

The first technique consists of putting a source follower in series with the compensation capacitance. The feedback is still present but the feedforward current flows through the source follower to the positive supply, without affecting the output.

Putting this source follower in the expression of the gain, gives as a result that the numerator has gone. There is no more zero. It is now an easy way to solve this problem. However, we need some biasing current through the follower. This may not be the real solution for low-power design.

The second technique is using a cascode. Its DC current is pulled out of the Source. A pMOST can also be used, provided DC current is injected into the Source.

Again, the AC feedback current can flow but not the feedforward current.

The zero simply vanishes, but again at the cost of some additional biasing current. This problem can be solved as shown on the next slide.

The third technique does not require any biasing current. This is why it is often preferred in low-power amplifiers.

0545

A good example of the use of the second technique is shown in this slide.

It is a telescopic cascode followed by a single-transistor amplifier as a second stage.

The compensation capacitance C_c is no longer connected between the output and input of the second stage. It takes a path through one of the cascodes, avoiding the positive zero.

The second technique is used quite often as it does not require any additional biasing current!

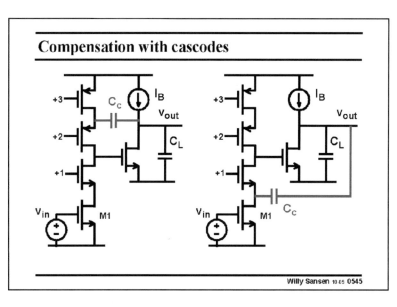

Compensation with cascodes

Willy Sansen 10 05 0545

It is not so easy to work out which cascode solution of the two to use. In principle, the second one is a little bit better as it takes the side of the input transistor. There are less higher-order poles and zeros in this case!

Some designers take half the compensation capacitance to both sides, which is very similar indeed!

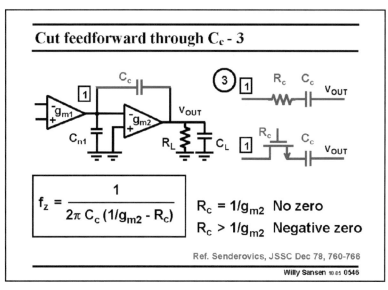

Cut feedforward through C$_c$ - 3

$$f_z = \frac{1}{2\pi C_c (1/g_{m2} - R_c)}$$

$R_c = 1/g_{m2}$ No zero

$R_c > 1/g_{m2}$ Negative zero

Ref. Senderovics, JSSC Dec 78, 760-766

Willy Sansen 10 05 0546

0546

The third technique to abolish the positive zero, is to insert a small resistor R$_c$, in series with C$_c$.

This resistor causes some cancellation of the effect of the feedforward by the feedback.

The expression of the zero is now modified as shown. It is clear that for a resistance R$_c$ equal to $1/g_{m2}$, the zero is at infinity. It has vanished.

It is not so easy, however, to match a resistor to a g$_m$ value. Especially if the resistor is realized by means of a MOST in the linear region, then the matching is more difficult.

There is a simple solution to this problem, however. We increase the size of the resistor.

This zero now turns into a negative zero. In other words the minus sign in the expression of the zero compensates the minus sign in the gain expression.

This negative zero is positioned between the negative poles and can therefore be used to compensate one of them.

0547

For a much larger resistor, the expression of the zero can be simplified to the one given in this slide. In order to be effective, we have to position this zero close to the GBW, for example at 2–3

Negative zero compensation

$$R_c \gg 1/g_{m2} \quad \Longrightarrow \quad f_z = -\frac{1}{2\pi\, C_c\, R_c}$$

$$f_z = 3\ GBW \quad \Longrightarrow \quad R_c = \frac{1}{3\, g_{m1}}$$

Final choice :

$$\frac{1}{g_{m2}} < R_c < \frac{1}{3 g_{m1}}$$

Willy Sansen 10-05 **0547**

times the GBW where the nondominant poles are.

This yields a new expression for R_c, which is now related to g_{m1} instead of gm2.

We simply position R_c between both the values obtained, with preference to be closer to $1/3g_{m1}$.

A numerical example will show that it is quite easy to position the resistor R_c as indicated and that the tolerance on this resistor can be allowed to be quite large!

Exercise of 2-stage opamp

GBW = 50 MHz for C_L = 2 pF
Find I_{DS1}; I_{DS2} ; C_c and R_c !

Choose C_c = 1 pF > g_{m1} = $2\pi\, C_c GBW$ = 315 μS
$\qquad\qquad\qquad\quad I_{DS1}$ = 31.5 μA & $1/g_{m1} \approx$ 3.2 kΩ

f_{nd} = 150 MHz > g_{m2} = $2\pi\, C_L 4GBW$ = $8g_{m1}$ = 2520 μS
$\qquad\qquad\qquad\quad I_{DS2}$ = 252 μA & $1/g_{m2} \approx$ 400 Ω

400 Ω < R_c < 1 kΩ : R_c = $1/\sqrt{2.5} \approx 400\sqrt{2.5} \approx$ 640 $\Omega \pm$ 60%

Willy Sansen 10-05 **0548**

0548

As an example, let us take a two-stage opamp with a GBW of 50 MHz for a C_L of 2 pF.

We resort to the same design plan as before. We choose C_c to be 1 pF, half the value of C_L. We will refine this choice in the next Chapter.

For a chosen C_c, the g_{m1} is easily calculated. For a $V_{GS} - V_T$ of 0.2 V, its current is ten times larger, or 31 μA. Note that $1/g_{m1}$ is 3.2 kΩ and $1/3g_{m1}$ is about 1 kΩ.

For the second stage we recall that f_{nd} must be three times higher than the GBW. Its g_{m2} is also readily calculated. The current is now eight times higher than the current in the input transistor or 252 μA and $1/g_{m2}$ is 400 Ω.

We now have to position R_c between 400 Ω and 1 kΩ. Doing this on a logarithmic scale means that we have to take a harmonic average. This gives about 640 Ω.

The advantage of doing this is that the tolerance is larger and the same in both directions. This resistor can have an absolute tolerance of 60%, which makes it quite easy to achieve!

Table of contents

Willy Sansen 10-05 0549

0549

As we have become familiar with the design of a two-stage amplifier, we can easily extend this design method to three-stage amplifiers. Indeed, we will use the Miller effect twice.

The resulting power consumption will be relatively high as we have to deal with the two non-dominant poles. In Chapter 10 on multistage amplifiers, we will present some more multistage amplifiers with much lower power consumption.

1-stage CMOS OTA

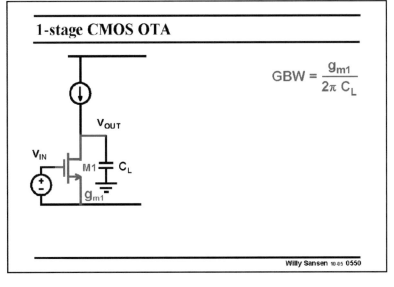

$$GBW = \frac{g_{m1}}{2\pi \, C_L}$$

Willy Sansen 10-05 0550

0550

Before we expand towards a three-stage amplifier, let us review the principles of a single- and two-stage amplifier.

A single-stage amplifier has only one high-impedance point, usually at the output. The load capacitance C_L therefore determines the GBW.

It is also the input g_{m1} which determines this GBW.

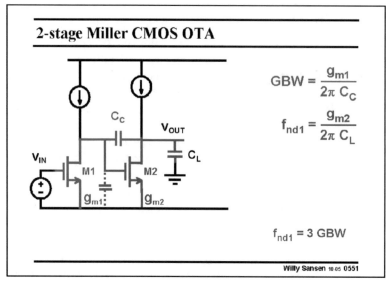

0551

A two-stage amplifier has two high-impedance points. These two points must be connected by a compensation capacitance C_C to carry out polesplitting. This compensation capacitance therefore determines the GBW.

Again it is the input g_{m1} which determines this GBW.

The non-dominant pole is now determined by the load capacitance C_L at the output.

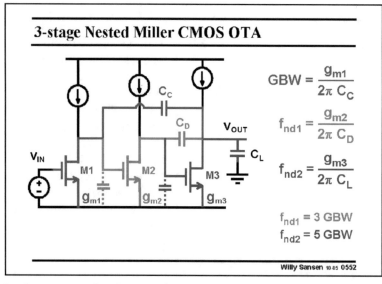

0552

A three-stage amplifier has three high-impedance points. These three points must be connected by two compensation capacitances C_C and C_D to carry out pole splitting.

The most efficient way to achieve this is to nest these two Miller capacitances, resulting in the **Nested-Miller configuration**. Capacitance C_C is the overall compensation capacitance. This is why it determines the GBW. The other one merely leads to a non-dominant pole.

Again, it is the input g_{m1} which determines this GBW. There are now two non-dominant poles. One of them is again determined by the load capacitance C_L, as for a two-stage amplifier. The other is determined by the additional compensation capacitance C_D. The question now is, how to deal with two non-dominant poles rather than one?

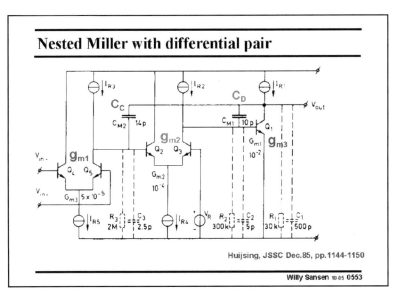

Nested Miller with differential pair

Huijsing, JSSC Dec.85, pp.1144-1150

Willy Sansen 10.05 0553

0553

We have to be careful spelling out the nested-Miller compensation of a three-stage amplifier. The first two stages are normally realized by means of differential pairs.

The first stage is a differential pair because we need a differential input.

The second stage is realized by means of a differential pair because we need a non-inverting amplifier. Otherwise, capacitance C_C would provide positive feedback! An alternative for this second stage would be a current mirror as explained in Chapter 10 on multistage amplifiers.

One of the first bipolar realizations is shown in this slide. The transconductances and compensation capacitances are easily identified.

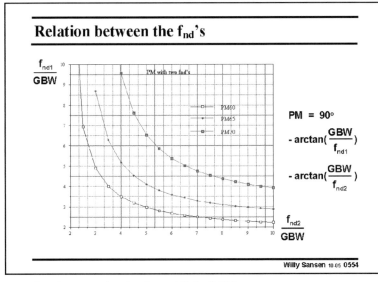

Relation between the f_{nd}'s

$$PM = 90° - \arctan\left(\frac{GBW}{f_{nd1}}\right) - \arctan\left(\frac{GBW}{f_{nd2}}\right)$$

Willy Sansen 10.05 0554

0554

If two non-dominant poles contribute to the phase margin, then we have to extend the expression of the Phase Margin PM with another pole, as given in this slide.

The curves for equal PM of respectively 60°, 65° and 70° are now easily calculated. Taking the tangent of the expression of PM allows the simplification of the calculations!

Let us focus on the relationship between the two non-dominant poles f_{nd} for a PM of 60°.

It is clear that if we have a f_{nd} fairly close to the GBW. The other one will then be far away. They can also be at an equal distance from the GBW, at about 3.5–4 times the GBW.

Normally, one f_{nd} is put at about 3 times the GBW and the other one at about 5 times. We obviously prefer to have the $5 \times GBW$ at the output because the higher frequency requires the higher current. It is more advantageous to have the higher current at the output. More current can be taken out by the load.

A good choice for a PM of 60° is therefore a ratio f_{nd}/GBW of three for the middle pole and five for the output pole.

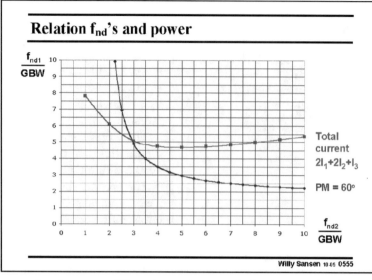

Relation f$_{nd}$'s and power

Total current $2I_1 + 2I_2 + I_3$

PM = 60°

Willy Sansen 10-05 0555

0555

If we select non-dominant poles which are widely different, then the current required to position the higher-frequency pole may become excessive.

To illustrate this point, the relationship between the two non-dominant poles is plotted in this slide, followed by a plot of the total current consumption.

This shows that once we select a higher-frequency pole beyond 6–7 times the GBW, the total power consumption increases. There is a minimum indeed around values of $5 \times \text{GBW}$, requiring the other pole to be around $3 \times \text{GBW}$, exactly as suggested in the previous slide.

The minimum is not very pronounced, however. The designer is still relatively free in his choice of non-dominant pole positions.

Elementary design of 3-stage opamp

$$\text{GBW} = \frac{g_{m1}}{2\pi \, C_C} \qquad f_{nd1} = 3\,\text{GBW} = \frac{g_{m2}}{2\pi \, C_D}$$

$$f_{nd2} = 5\,\text{GBW} = \frac{g_{m3}}{2\pi \, C_L}$$

Choose $C_D \approx C_C$!

$$\frac{g_{m2}}{g_{m1}} \approx 3 \qquad \frac{g_{m3}}{g_{m1}} \approx 5\,\frac{C_L}{C_C}$$

Even larger current in output stage !

Willy Sansen 10-05 0556

0556

The design plan itself is then fairly straightforward.

Remember that for the two-stage amplifier, we have chosen the value of the compensation capacitance. It was taken to be about 2–3 times smaller than the load capacitance.

For this three-stage amplifier we adopt a similar strategy. Both compensation capacitances are chosen to be equal, and to be about 2–3 times smaller than the load capacitance.

Of course, they can be taken differently. It is clear that according to the equations describing the stability, we would obtain the same Phase Margin. It is not clear yet however, how this would affect some other specifications.

Having chosen the two compensation capacitances, we simply have to solve three equations with three variables g_{m1}, g_{m2} and g_{m3}.

Exercise of 3-stage opamp

GBW = 50 MHz for C_L = 2 pF
Find I_{DS1}; I_{DS2}; I_{DS3}; C_C and C_D !

Choose $C_C = C_D$ = 1 pF > $g_{m1} = 2\pi C_C GBW = 315\ \mu S$
$$I_{DS1} = 31\ \mu A$$

f_{nd1} = 150 MHz > $g_{m2} = 2\pi C_D 3GBW = 3g_{m1} = 945\ \mu S$
$$I_{DS2} = 95\ \mu A$$

f_{nd2} = 250 MHz > $g_{m3} = 2\pi C_L 5GBW = 10g_{m1} = 3150\ \mu S$
$$I_{DS3} = 315\ \mu A$$

Willy Sansen 10.05 0557

0557

As an example, let us design an opamp with the GBW and C_L given in this slide.

The compensation capacitances have been set at half the load capacitance or 1 pF. The g_m's are now easily calculated. It is obvious that because of the higher-frequency pole f_{nd2} at the output,

the output transistor M3 also consumes most of the current.

Comparison 1, 2 & 3 stage designs

GBW = 50 MHz for C_L = 2 pF

Single stage : I_{DS1} = 31 µA $I_{TOT} = 2I_{DS1}$ = 62 µA

Two stages :   Choose C_C = 1 pF
I_{DS1} = 31 µA I_{DS2} = 252 µA $I_{TOT} = 2I_{DS1} + I_{DS2}$ = 314 µA

Three stages :   Choose $C_C = C_D$ = 1 pF
I_{DS1} = 31 µA I_{DS2} = 95 µA I_{DS3} = 315 µA
$$I_{TOT} = 2I_{DS1} + 2I_{DS2} + I_{DS3} = 567\ \mu A$$

Willy Sansen 10.05 0558

0558

For sake of comparison, we have designed a single-stage, two-stage and a three-stage opamp for the same specifications.

It is obvious that the single-stage opamp is the champion in power consumption. The addition of compensation capacitances invariably leads to excessive power consumption.

However, a single-stage opamp does not provide a lot of gain. The addition of cascodes and gain boosting reduces the output swing. This is where a two-stage Miller opamp offers considerable advantages.

A three-stage amplifier will be used whenever we need a class-AB output stage. Also when the supply voltage is so low that there is no room for cascodes, we may have to go for cascading rather than cascoding!

Table of contents

Willy Sansen 10-05 **0559**

0559

We have now acquired sufficient knowledge about the stability of an operational transconductance amplifier, to be able to implement it by means of MOSTs or bipolar transistors.

This is discussed in the next Chapter.

Systematic Design of Operational Amplifiers

Willy Sansen

KULeuven, ESAT-MICAS
Leuven, Belgium

willy.sansen@esat.kuleuven.be

Willy Sansen 10 05 **061**

061

In the previous Chapter we learned all we need about the stability of an operational transconductance amplifier, to be able to implement it by means of MOST devices.

Now we would like to refine the design plan. Up till now we have arbitrarily chosen the compensation capacitances. The question is, how can we improve on that?

Finally, we have now focused on a few specifications only, which are the GBW and the Phase Margin. Surely there are many more specifications. They are elaborated on in this Chapter.

Table of contents

- **Design of Single-stage OTA**

- **Design of Miller CMOS OTA**

- **Design for GBW and Phase Margin**

- **Other specs:** Input range, output range, SR, ...

Ref.: Sansen : Analog design essentials, Springer 2006

Willy Sansen 10 05 **062**

062

The transistor implementation of a single-transistor amplifier is given first.

It is followed by a two-transistor Miller OTA. A detailed design plan is discussed, to give rise to the lowest possible power consumption.

Moreover, the maximum GBW is estimated for a specific CMOS technology. It is shown that GHz values are easily reached.

Finally, all other specifications are listed, a few of which are discussed for the two-stage Miller OTA.

The same CMOS Miller OTA is used throughout. All numerical values relate to the same amplifier. As a result, the reader can give himself a good idea about the appropriate orders of magnitudes.

The design of a single-stage OTA is approached first.

063

A differential voltage amplifier is given in this slide.

The expressions describing the gain, bandwidth and GBW of a single-stage OTA, have been previously discussed.

181

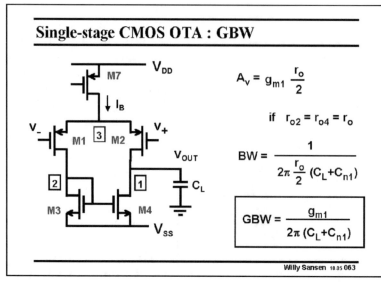

Single-stage CMOS OTA : GBW

$$A_v = g_{m1} \frac{r_o}{2}$$

if $r_{o2} = r_{o4} = r_o$

$$BW = \frac{1}{2\pi \frac{r_o}{2}(C_L + C_{n1})}$$

$$GBW = \frac{g_{m1}}{2\pi(C_L + C_{n1})}$$

Willy Sansen 10 05 063

The GBW is obviously what could have been expected. Note that the load capacitance also contains some parasitic capacitances, which are due to the transistor capacitances. They are summarized as C_{n1}, the sum of the transistor capacitances at node 1.

Nevertheless, this circuit also contains a second node, and even a third one. Do we have to take the capacitances of these nodes to ground into account?

Indeed, a capacitance to ground gives a pole. Do we therefore have two additional non-dominant poles?

The answer is negative. First of all, at node three, no AC signal is present when the stage is driven differentially. Node 3 does not come in.

At node 2 we do have a non-dominant pole. There are two reasons however, why it can be neglected.

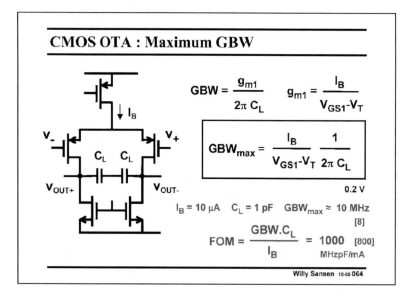

CMOS OTA : Maximum GBW

$$GBW = \frac{g_{m1}}{2\pi C_L} \qquad g_{m1} = \frac{I_B}{V_{GS1} - V_T}$$

$$GBW_{max} = \frac{I_B}{V_{GS1} - V_T} \frac{1}{2\pi C_L}$$

0.2 V

$I_B = 10\ \mu A \quad C_L = 1\ pF \quad GBW_{max} \approx 10\ MHz$

[8]

$$FOM = \frac{GBW \cdot C_L}{I_B} = 1000 \quad [800]$$

MHzpF/mA

Willy Sansen 10-05 064

064

Before we focus on the non-dominant poles, let us find out how much GBW can be expected from such a simple amplifier. We know already that it is the best for low-power consumption. What does this mean in actual numbers?

The simplest Figure of Merit (FOM) for opamps is the one with the GBW, the C_L and the power consumption. Later on, some other specifications could be added, such as noise, or swing, etc.

Instead of the power consumption, the current consumption can also be taken.

For this purpose, a differential configuration has been taken as sketched in this slide. The effective load capacitance is now only half of what we would have for a single-ended output. On the other hand, there is also no current mirror to double the output current. This OTA is representative for a single-stage amplifier!

We find that for a 10 μA total current, a load capacitance of 1 pF can still yield a GBW of

10 MHz. Even easier to remember is that with 1 μA, a 1 MHz GBW can be achieved for a 1pF load capacitance.

This gives a FOM of 1000 in MHzpF/mA. Actually, it is slightly less, i.e. 800. We will see however, that this is an excellent result, when we compare with other OTA's in the next Chapter.

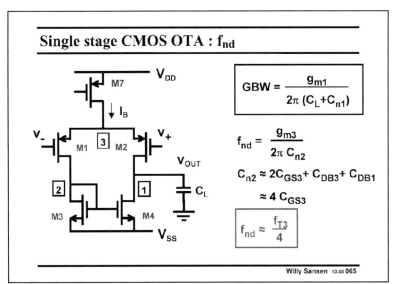

065

Let us now focus again on non-dominant poles.

The position of this non-dominant pole is easily established. The resistance at node 2 is simply $1/g_{m3}$. The node capacitance C_{n2} is the sum of all transistor capacitances, connected to node 2.

For a MOST, the capacitance C_{DB} is about equal to its C_{GS}. This is why all four capacitances in C_{n2} are taken equal. This is a gross approximation but good enough to deal with this pole. Capacitance C_{n2} is about $4 \times CGS3$.

The non-dominant pole f_{nd} is about $f_{T3}/4$. This is the first reason why node 2 does not affect us too much. This non-dominant pole is simply located at too high frequencies, compared to the GBW.

The second reason is on the next slide.

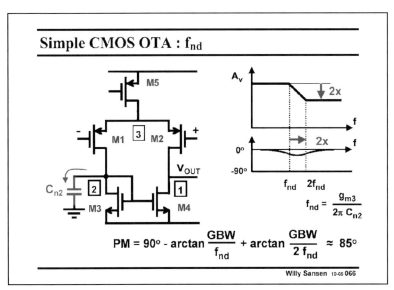

066

The capacitance C_{n2} on node 2 does create a pole indeed, but also a zero.

Indeed, a capacitance to ground on the other side of a differential amplifier with a single output, creates a pole f_{nd} and a zero at twice the frequency $2 \times f_{nd}$.

At higher frequencies, the output current is divided by two, since the current mirror does not receive any more current. A division by two can only be represented by a pole-zero doublet, the zero

of which is a factor of two higher than the pole.

This is illustrated for the voltage amplifier in which all other capacitances are omitted.

The advantage of this zero is that it greatly compensates the phase shift of the pole. The net result is a small change in phase shift.

The influence on the Phase Margin of this pole-zero pair is therefore negligible.

As a result, the capacitance C_{n2} at node 2 can be ignored. We have now found two reasons for that.

067

Single stage CMOS OTA : Design 1

GBW = 100 MHz for C_L = 2 pF

Techno: L_{min} = 0.35 μm; K'_n = 60 μA/V² & K'_p = 30 μA/V²

I_{DS} ? W ? L ?

g_m = GBW 2π C_L = 1.2 mS

V_{GS}-V_T = 0.2 V $I_{DS} = g_m \dfrac{V_{GS}\text{-}V_T}{2} = \dfrac{g_m}{10} = 0.12$ mA

$\dfrac{W}{L} = \dfrac{I_{DS}}{K'(V_{GS}\text{-}V_T)^2} = 100$ $L_p = L_n = 1$ μm **GAIN !**

$W_p = 100$ μm; $W_n = 50$ μm

Willy Sansen 10-05 067

As a design example, let us take a GBW and C_L as indicated.

The g_m is readily calculated, and so is the current, provided the $V_{GS} - V_T$ is chosen appropriately (to be 0.2 V).

By means of the K′ factor we can calculate the required W/L.

Now the L must be selected. In order to achieve some gain, we take about 3 times the minimum channel length or 1 μm.

The widths are then easily added. The nMOS has a smaller width as its K′ is larger.

We could try to verify whether the pole at node 2 is indeed negligible. For this purpose we must find f_T or rather the input capacitance C_{GS2}.

A MOST has a $C_{GS} = kW$ with $k = 2$ fF/μm if the minimum length is used. A MOST with a W/L = 100 for a L = 0.35 μm would have a W = 35 μm and hence a C_{GS} of 70 fF. Now both the L and W are three times larger. The C_{GS2} is $70 \times 3 \times 3 = 630$ fF. Its f_{T2} is about 300 MHz and $f_{T2}/4 \approx 76$ MHz. Luckily, this pole is followed by zero!!!

If we really do not want this pole-zero doublet below the GBW, we have to make the transistors smaller, deteriorating the gain. Another possibility is to make the transistors smaller and to add cascodes to increase the gain!

Table of contents

Willy Sansen 10 05 068

068

Now we focus on the design plan of a CMOS Miller OTA, realized with transistors. Again we will verify the gain, bandwidth and the GBW. Moreover, we have to make sure that the compensation capacitance generates enough polesplitting to position the f_{nd} beyond the GBW.

Miller CMOS OTA

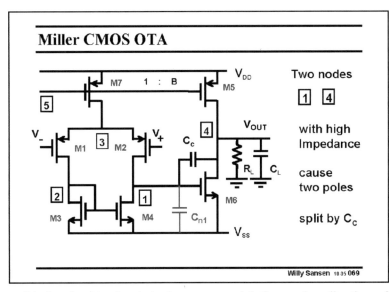

Two nodes

1 4

with high Impedance

cause two poles

split by C_c

Willy Sansen 10 05 069

069

A CMOS OTA is shown in this slide. The input devices are normally pMOST devices as they give better matching (see Chapter 12).

The first block converts the differential input voltage into a current by means of transconductance g_{m1}.

The second stage is a transimpedance amplifier, converting this current into a voltage. Actually, only one transistor M6 takes care of that, together with capacitance C_c.

Obviously this circuit is the most straightforward realization of the two-stage OTA discussed in the previous Chapter.

Indeed, nodes 1 and 4 cause two poles, which are split by C_c. Parasitic capacitance C_{n1} is also shown. It consists mainly of the input capacitance C_{GS6} of transistor M6. This latter transistor is a big transistor as it carries a much larger current than the input transistors.

0610

In order to be able to calculate the gains, etc, we draw on the small signal equivalent circuit. It is shown below.

The 4-transistor input stage is represented by the g_{m1} generator, and the second stage by the g_{m6} generator.

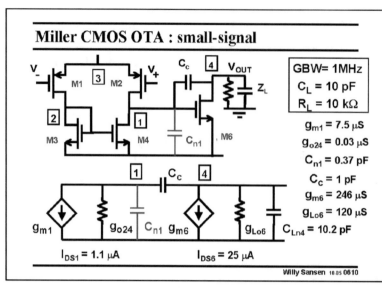

The input stage has an output resistance which is the inverse of g_{o24}, which is a short way of stating $g_{o2} + g_{o4}$.

This circuit can be simplified to a two-node circuit with two transconductances and a RC circuit from each node to ground. Obviously, there are two nodes, but also a compensation capacitance C_c to split them up.

All values given are for 1 MHz/10 pF realization, which will be used for all numerical examples from now on. These values have also been used for the pole-zero position diagrams in the previous Chapter.

How exactly these values have been obtained, will be explained when we discuss the design plans next.

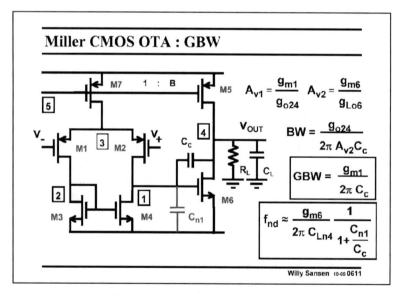

0611

The gains are now easily calculated. We have two stages so we have two gains A_{v1} and A_{v2}, and a total gain A_v.

The bandwidth is obviously due to the Miller effect of capacitance C_c, as expected.

The GBW is now the product of both. It is exactly what we have anticipated.

Also, the non-dominant pole is again what we have derived in the previous Chapter. Again, normally C_c is about 3 times C_{n1}.

0612

For this particular amplifier, the pole-zero position and Bode diagrams are sketched in this slide.

For zero C_c, two poles are found, which are clearly too close together. Peaking would occur if feedback is applied.

This capacitance has been increased to about 1 pF. In this case the dominant pole has decreased a lot but what is more important is that the non-dominant pole has moved out until

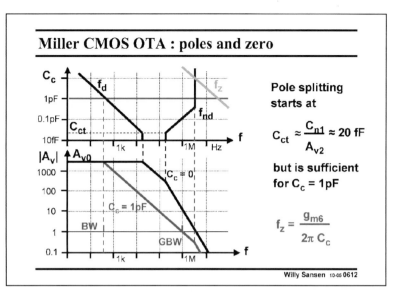

Miller CMOS OTA : poles and zero

Pole splitting starts at

$$C_{ct} \approx \frac{C_{n1}}{A_{v2}} \approx 20 \text{ fF}$$

but is sufficient for $C_c = 1pF$

$$f_z = \frac{g_{m6}}{2\pi C_c}$$

Willy Sansen 10-05 0612

it is almost three times the GBW.

The zero is still too far to bother us!

The result is a CMOS Miller OTA with a gain of about 3000 or 70 dB, a bandwidth of about 300 Hz and a GBW of 1 MHz.

The total power consumption is 27 µA. Its FOM is therefore 370 MHzpF/mA which is an excellent value for a two-stage amplifier. Actually, anything that is better than 100 is good!

Table of contents

- **Design of Single-stage OTA**
- **Design of Miller CMOS OTA**
- **Design for GBW and Phase Margin**
- **Other specs:** Input range, output range, SR, ...

Willy Sansen 10-05 0613

0613

The question remains. How has this Miller OTA been designed. Can we do better? Can we obtain a better FOM?

We will see that there are three ways to obtain the same optimum in terms of minimum power consumption. Any design plan can now be adopted.

0614

Since we only have two specifications up till now, we only have two equations, one for the GBW and one for the stability.

As a result, when we require a specific GBW for a specific C_L, we simply have to solve these two equations.

The problem is that we have three variables. They are the current in the first stage (or g_{m1}), the current in the second stage (or g_{m6}) and the compensation capacitance C_c.

Up till now, we have chosen the compensation capacitance, which allows us to solve the two equations, since they only have two variables g_{m1} and g_{m6}.

This is indeed the first possible design plan.

Miller CMOS OTA: Design plan

$$GBW = \frac{g_{m1}}{2\pi\,C_c}$$

GBW = 100 MHz and C_L = 2 pF

$$f_{nd} \approx \frac{g_{m6}}{2\pi\,C_{Ln4}}\;\frac{1}{1+\dfrac{C_{n1}}{C_c}}$$

Two **equations for**

Three **variables** g_{m1}, g_{m6} and C_c ?!?

Solution : **choose** g_{m1} **or** g_{m6} **or** C_c !!!

Willy Sansen 10-05 **0614**

There are two more design plans, i.e. choosing the g_{m1} first and then solve the two equations, or choosing the g_{m6} first and then solve the two equations.

All three design plans lead to the same optimum.

The same two equations can obviously be used in any direction. One could wonder, for example, how much GBW can be obtained for a C_L of 5 pF with only 0.2 mA? We also ask, how much load capacitance can we drive with 1 mA for a GBW of 200 MHz?

What is wrong with

choosing $\boxed{C_c = 1\ pF}$ **?**

◻

Willy Sansen 10-05 **0615**

0615

All three design plans lead to the same optimum indeed.

There is a slight preference for choosing the compensation capacitance C_c first as this capacitance can only have a small range of values. Indeed, it cannot be smaller than about $3 \times C_{n1}$.

On the other hand, it cannot be larger than the load capacitance C_L divided by 2–3. It is now fairly easy to choose this compensation capacitance about right.

This is why so many designers simply choose C_c as a starting point for their design procedure. Of course, even better is to try a few different capacitance values!

Let us work out an example first!

0616

As an example, let us find the currents in both stages for a given GBW and

C_L. An appropriate compensation capacitance is selected of 1 pF, which immediately yields the two g_m's. The currents and W/L's are now easily calculated.

Miller CMOS OTA: Design vs C_c

Choose $C_c \approx 3\, C_{n1}$ $\quad GBW = \dfrac{g_{m1}}{2\pi\, C_c}$

$$3\,GBW \approx \dfrac{g_{m6}}{2\pi\, C_{Ln4}} \dfrac{1}{1.3}$$

$$\dfrac{g_{m6}}{g_{m1}} \approx 4\, \dfrac{C_L}{C_c}$$

GBW = 100 MHz and C_L = 2 pF

Choose $C_{n1} < C_c < C_L$

Choice C_c = 1 pF gives g_{m1} = 0.6 mS and g_{m6} = 4.8 mS

Willy Sansen 10-05 0616

A problem may be that it is not so obvious how to find the value of capacitance C_{n1}. It is mainly C_{GS6}. As long as we do not know g_{m6}, we do not know C_{GS6} either.

Moreover, it would be interesting to try out a few different values of C_c.

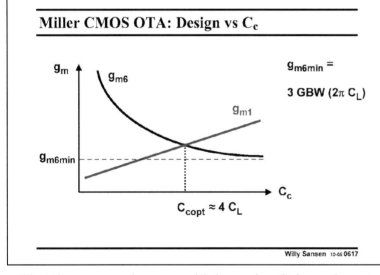

Miller CMOS OTA: Design vs C_c

g_{m6min} =

3 GBW ($2\pi\, C_L$)

$C_{copt} \approx 4\, C_L$

Willy Sansen 10-05 0617

0617

These are the reasons why it is better to try a few different values of C_c and solve the two equations for g_{m1} and g_{m6}. The currents of both stages can now be plotted versus capacitance C_c, together with the total current consumption. This is sketched in this slide.

What is obvious, is that we obtain a minimum in power consumption!

The g_{m1} increases with C_c is easy to see from the expression of the GBW.

That the g_{m6} now decreases with increasing C_c is maybe not so obvious, and yet it is clearly given by the expression of the non-dominant pole f_{nd}.

Indeed for a constant GBW and hence constant f_{nd}, g_{m6} decreases for increasing C_c. Actually this is not all unexpected. After all g_{m6} and C_c ensure stability together. If one is larger, the other one can be smaller!

Both curves cross at a value of C_c which is fairly large. Remember however that we will select a value of C_c, which is 2–3 times smaller than C_L. This is very close to the optimum but just left of it!

For very large values of C_c, g_{m6} reaches a minimum. The current in the output stage reaches its minimum value. This value of g_{m6} is given in this slide. It is obviously proportional to both the GBW and the C_L.

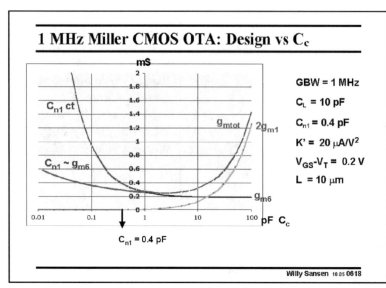

0618

As an example, we take the same CMOS Miller OTA of 1 MHz for 10 pF, which has been used before.

The currents in both stages, i.e. $2g_{m1}$, g_{m6} and g_{mtot} are plotted versus compensation capacitance C_c. Remember that they stand for $2I_{DS1}$, I_{DS6} and I_{tot} but are 10 times larger (for $V_{GS}-V_T=0.2$ V).

It is clear that the minimum has a different shape from the previous illustration.

This is an ideal plot to select a value of C_c. It shows that the value of 1 pF is indeed a good choice, at least if no other specifications have to be taken into account.

We could have also taken 2 or even 3 pF as well. Indeed, the additional current consumption is still less than the current through the output stage. A larger capacitance would reduce the noise, as we will see later in this Chapter.

Finally, note that for these plots a constant value of C_{n1} has been assumed. This is not quite true, however. For larger g_m, the current is also larger and so is the size W/L and the input capacitance. As a result C_{n1} increases with g_{m6}. If we introduce this relationship, then g_{m6} is much flatter versus C_c for small values of C_c. This does not change our choice of compensation capacitance however, since we have to select a value of at least 2–3 times C_{n1}.

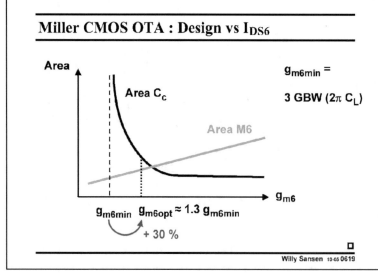

0619

The second design plan is to choose g_{m6} or the current in the output stage.

This is also very easy. Indeed, we already know what the minimum value of g_{m6} is. It is repeated in this slide. It is the value of g_{m6} obtained for infinite C_c.

We now simply take a g_{m6} value which is 30% larger. This means that C_c will end up at a value which is about 3 times C_{n1}, as explained by the expression of the non-dominant pole.

This value of g_{m6} should be close to the minimum in area, which is illustrated in this slide.

The advantage of taking g_{m6} as an independent variable, is that we can now easily calculate C_{n1}, which is little more than C_{GS6}.

Moreover, it is more evident to start with g_{m6} in this design plan. After all, the current in the output stage is by far the larger one. We therefore want to focus on the minimization of this current first.

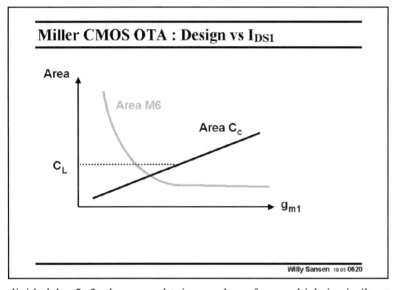

Miller CMOS OTA : Design vs I_{DS1}

Willy Sansen 10 05 0620

0620

Finally, the third design plan is to start with a selection of the current in the input stage.

A possible reason to go for this design plan, could be the noise performance. After all the equivalent input noise voltage will depend on the transconductances of the MOSTs in the input stage.

Again, a minimum is reached when we plot the total area versus g_{m1}.

Once C_c becomes C_L divided by 2–3, then we obtain a value of g_{m1} which is similar to the values obtained by the previous design plans.

However, it is not worked out any further.

Optimum design for high speed Miller OTA - 1

$$GBW = \frac{g_{m1}}{2\pi \, C_c}$$

$$f_{nd} = \frac{g_{m6}}{2\pi \, C_L} \cdot \frac{1}{1 + C_{n1}/C_c}$$

$$C_L = \alpha \, C_c \qquad \alpha \approx 2$$

$$C_c = \beta \, C_{n1} = \beta \, C_{GS6} \qquad \beta \approx 3$$

$$f_{nd} = \gamma \, GBW \qquad \gamma \approx 2$$

$$C_{GS} = kW \qquad k = 2 \; 10^{-11} \; F/cm$$

$$GBW = \frac{f_{nd}}{\gamma} = \frac{g_{m6}}{2\pi \, C_L} \cdot \frac{1}{\gamma \, (1 + 1/\beta)}$$

$$C_L = \alpha \, C_c = \alpha \, \beta \, C_{n1} = \alpha \, \beta \, C_{GS6} = \alpha \, \beta \, kW_6 \qquad W_6 \Uparrow \text{ if } C_L \Uparrow$$

Willy Sansen 10-05 0621

0621

This design procedure can now easily be formalized, for such a Miller CMOS OTA.

The goal is to find out what maximum GBW can be reached within a certain CMOS technology. Also, we want to find the shortest way to design this OTA.

First of all, a number of design choices have to be made. They have all been previously used. We list them again and introduce design parameters α, β and γ.

Parameter α sets the ratio between C_L and C_c. Take 2 as an example.

Parameter β sets the ratio between C_c and C_{n1} or C_{GS6}. Take 3 as an example. Also, remember that C_{GS} can easily be described as a function of the transistor width; the k parameter is about $2\,fF/\mu m$.

Parameter γ sets the ratio between f_{nd} and C_c. We have taken 3 many times, let us take 2 for this example.

The maximum GBW can now easily be described as a fraction γ of the non-dominant pole.

Also, note that C_L can be described in terms of the width of the output transistor. Obviously, the larger the C_L, the more current we will need to drive it, and the larger the transistor width becomes!

Optimum design Miller for high speed OTA - 2

Elimination of C_L yields

$$GBW = \underbrace{\frac{g_{m6}}{2\pi\, kW_6}}_{f_{T6}} \frac{1}{\alpha\,\beta\,\gamma\,(1 + 1/\beta)} \qquad g_m = \frac{W}{L}\,\frac{17\ 10^{-5}}{1 + 2.8\ 10^4\ L\,/V_{GST}}$$

$$W, L \text{ in cm}$$

$$GBW = \frac{1}{2\pi\, L_6}\,\frac{1}{\alpha\,\beta\,\gamma\,(1 + 1/\beta)}\,\frac{8.5\ 10^6}{1 + 2.8\ 10^4\,L_6\,/\,V_{GST6}} \qquad L \text{ in cm}$$

GBW is not determined by C_L, only by L (and V_{GST}) !!

f_T is also determined by L !!!

Willy Sansen 10.05 0622

0622

The last expression of C_L can now be substituted in the expression of the GBW. We now take the general expression of g_m from the first Chapter on models. Remember that this is the expression which spans both the strong inversion and velocity saturation regions. These are the regions which we will use for high speed.

Substitution of this g_m expression in the one of the GBW, yields an expression in which only $V_{GS} - V_T$ and L are left as parameters.

This is not surprising at all. We have known all along that these are the two choices that we have to make for any transistor in the signal path.

It is surprising, however, to find that the maximum GBW does not depend on the load capacitance.

Actually, increasing the load capacitance increases the width of the output transistor and its current. The speed of the output transistor mainly depends on its length!

The speed of a MOST is better represented by parameter f_T. This is why we now try to substitute the transistor parameters $V_{GS} - V_T$ and L by parameter f_T.

0623

The expression of parameter f_T for both the weak inversion and velocity saturation regions are taken from Chapter 1. Clearly, it depends on $V_{GS} - V_T$ and L, very much as the GBW does.

Substitution now yields a final expression of the GBW with only f_T as a parameter.

For the values previously chosen, we find that the maximum GBW is about $1/16$ of the f_T of the output device.

A two-stage Miller CMOS OTA can have a GBW of 5 GHz, provided we select a CMOS technology where an f_T of 80 GHz can be obtained. Checking the f_T curves of Chapter 1, we find

Optimum design Miller for high speed OTA - 3

Substitution for f_T yields

$$f_T = \frac{g_m}{2\pi C_{GS}}$$

$$\boxed{GBW = \frac{f_{T6}}{\alpha\,\beta\,\gamma\,(1 + 1/\beta)}}$$

$$f_T = \frac{1}{L}\,\frac{1.35}{1 + 2.8\ 10^4\ L / V_{GST}}$$

L in cm
f_T in MHz

GBW is not determined by C_L, only by f_T

f_T is determined by L (and V_{GST}) !!!

If $V_{GST} = 0.2\,V$, v_{sat} takes over for L < 65 nm (If 0.5 V for L < 0.15 μm)

Willy Sansen 10 05 0623

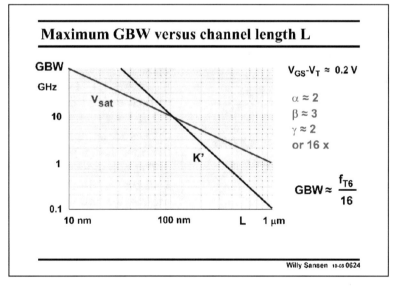

Maximum GBW versus channel length L

$V_{GS}\text{-}V_T \approx 0.2\,V$

$\alpha \approx 2$
$\beta \approx 3$
$\gamma \approx 2$
or 16 x

$$GBW \approx \frac{f_{T6}}{16}$$

Willy Sansen 10-05 0624

that a 80 nm CMOS is required for that (for $V_{GS} - V_T = 0.2\,V$), but only 0.1 μm technology if we make $V_{GST} = 0.5\,V$.

The actual power consumption will depend on the capacitive load. The larger the load, the higher the power consumption!

The optimum design plan has now become fairly simple, as shown next.

0624
However, before the design plan is spelled out, let us see what values of GBW are available for different values of channel length.

For this purpose, the values of f_T have to be obtained from Chapter 1. They refer to transistor M6. A value is taken of $V_{GS} - V_T = 0.2\,V$.

The parameters α, β and γ have to be selected. They are taken the same as before, giving rise to a ratio of 16 between f_{T6} and the maximum GBW.

The plot shows that for large channel lengths, where the mobility (K') model is still valid, a GBW is obtained if the channel length L is not larger than 0.35 μm.

For smaller channel lengths, the velocity saturation model evidently prevails. A 10 GHz GBW can be realized provided the minimum channel length is now 90 nm or less. This is also the cross-over value of the channel length from one model to the other!

0625
Let us now draw up a design plan.
The three design choices have to be selected first.

Design optimization for high speed Miller OTA

- **Choose** $\alpha \, \beta \, \gamma$
- **Find minimum f_{T6} for specified GBW**
- **Choose maximum channel length L_6 (max. gain)**
  for a chosen V_{GS6}-V_T
- **W_6 is calculated from C_L ,**
  and determines I_{DS6}
- **C_c is calculated from C_L through α**
- **g_{m1} and I_{DS1} are calculated from C_c**
- **Noise is determined by g_{m1} or C_c**

Willy Sansen 10.05 **0625**

We must find the minimum f_T which can handle the process. The higher f_T, the smaller the channel length will be and the more expensive the CMOS technology required. A minimum f_T leads to a minimum channel length L.

We now choose the actual channel length. It can be the minimum channel length or a somewhat larger value, depending on the gain required. Also, the value of $V_{GS} - V_T$ must be selected.

The capacitive load now determines the output transistor width, and its current.

All other values are now easily derived.

Design Ex. for GBW = 0.4 GHz & CL = 5 pF

• **Choose $\alpha \, \beta \, \gamma$**	**2 3 2**
• **Minimum f_{T6} for GBW = 0.4 GHz**	f_{T6} = **6.4 GHz**
• **Maximum channel length L_6**	L_6 = **0.5 μm**
for a chosen V_{GS6}-V_T = 0.2 V	
• **L_6 is taken to be the minimum L**	
• **W_6 is calculated from C_L ,**	W_6 = **417 μm**
and determines I_{DS6} (K'_n = 70 μA/V^2)	I_{DS6} = **2.3 mA**
and determines C_{n1} (k = 2 fF/μm)	C_{n1} = **0.83 pF**
• **C_c is calculated from C_L through α**	C_c = **2.5 pF**
• **g_{m1} and I_{DS1} are calculated from C_c**	I_{DS1} = **0.63 mA**

Willy Sansen 10.05 **0626**

0626

A numerical example is worked out in this slide.

The very first design choices to be made are for the three design parameters.

The minimum f_T value is a direct result from these choices. We now have to discover which channel length can deliver such high f_T values. Note that this is the actual channel length used, which may be 2–3 times higher than the minimum channel length of a particular CMOS technology. Here they are taken the same, because we do not have a lot of room left: 80 GHz f_T is quite high indeed.

The transistor width is a direct result of the load capacitance. It determines both the current and the value of C_{n1}.

The compensation capacitance C_c is a fraction α of C_{L1}.

Clearly, C_{n1} comes out to be 1/3 of C_c since β was 3.

From the GBW we finally obtain g_{m1} and I_{DS1}.

The total current consumption is 3.56 mA, which is quite high because of the large GBW and load capacitance. Its FOM however, is 561 MHzpF/mA, which is not bad at all!

Optimum design Miller for low speed OTA

$$GBW = \frac{f_{T6}}{\alpha \beta \gamma (1 + 1/\beta)}$$

$$\frac{f_T}{f_{TH}} = \sqrt{i}\ (1 - e^{-\sqrt{i}}) \approx i \text{ for small i}$$

$$f_{TH} = \frac{2\ \mu\ kT/q}{2\pi\ L^2}$$

GBW is not determined by C_L, only by f_T

f_T is determined by L and i !!!

Willy Sansen 10-05 0627

0627

The expression of parameter f_T for both the weak inversion and strong inversion regions is taken from Chapter 1. Clearly, it depends on inversion coefficient i and L.

The expression of the GBW with only f_T as a parameter, is the same as previously mentioned.

For the values chosen before, we find that the maximum GBW is about 1/16 of the f_T of the output device.

A two-stage Miller CMOS OTA can have a small GBW, provided we select proper values of L and i.

The actual power consumption will depend on the capacitive load. The larger the load, the higher the power consumption!

The optimum design plan has become fairly simple now, as shown next.

Design optimization for low speed Miller OTA

- **Choose $\alpha \beta \gamma$**
- **Find minimum f_{T6} for specified GBW**
- **Choose channel length L_6 (max. gain), which gives f_{TH6}**
- **Calculate i_6**
- **W_6 is calculated from C_L ,**

    and determines I_{DST6} and I_{DS6}

- **C_c is calculated from C_L through α**
- **g_{m1} and I_{DS1} are calculated from C_c**
- **Noise is determined by g_{m1} or C_c**

Willy Sansen 10-05 0628

0628

The three design choices have to be selected first.

We choose the actual channel length L_6. It can be the minimum channel length or a somewhat larger value, depending on the gain required. However, this value of L_6 sets the frequency f_{TH6}.

The value of i is now easily calculated as f_{TH}/f_{TH6}.

The capacitive load now determines the output transistor width, and its current.

All other values are now easily derived.

0629

A numerical example is worked out below.

The values speak for themselves.

Design Ex. for GBW = 1 MHz & CL = 5 pF

- Choose $\alpha \beta \gamma$ 2 3 2
- Minimum f_{T6} for GBW = 1 MHz f_{T6} = 16 MHz
- Maximum channel length L_6 L_6 = 0.5 μm
  gives f_{TH6} f_{TH6} = 2 GHz
- Inversion coefficient i is   i = 0.008
- W_6 is calculated from C_L, W_6 = 417 μm
  and determines I_{DST6} (K'_n = 70 μA/V^2) I_{DST6} = 0.33 mA
  and determines I_{DS6} I_{DS6} = 2.7 μA
  and determines C_{n1} (k = 2 fF/μm) C_{n1} = 0.83 pF
- C_c is calculated from C_L through α C_c = 2.5 pF
- g_{m1} and I_{DS1} are calculated from C_c I_{DS1} = 1.6 μA

Willy Sansen 10 05 **0629**

The FOM of this last opamp is 575 MHzpF/mA, which is quite impressive indeed.

Table of contents

- **Design of Single-stage OTA**

- **Design of Miller CMOS OTA**

- **Design for GBW and Phase Margin**

- **Other :** SR, Output Impedance, Noise, ...

Willy Sansen 10 05 **0630**

However, the value of L_6 could have been taken larger than 0.5 μm. This would have decreased the value of f_{TH6} and increased the inversion coefficient i_6.

For example, doubling the channel length to 1 μm, decreases f_{TH6} to 480 MHz, and increases i to 0.033. This halves I_{DST6} to 0.16 μA. Current I_{DS6} doubles to 5.5 μA, leaving the input stage current I_{DS1} unchanged at 1.6 μA. Also, the compensation capacitance is the same at 2.5 pF.

0630
Up till now, the GBW and stability conditions have been studied in great detail. However, an OTA has many more specifications. Actually, the shortlist of the specifications, for which we have to carry out a design, is one of the major design decisions. The other specifications, which are not shortlisted, are to be verified later, hoping that none of them have to be upgraded to that shortlist, so that we do not have to start all over again.

First of all, we will try to make a systematic list of all possible specifications. We are not sure that this is possible. For an analog circuit, some more specifications can always be added. There is always something extra that can be achieved with an analog circuit.

Nevertheless, this list of specifications is fairly complete. Of course none of them will be used. It is just an attempt to list them all.

Specifications of commercial amplifiers do not follow this list. They are actually a summary of measurement results. Some of them are missing. Some of them are contradictory.

Afterwards, a few of the most important specifications are analyzed in detail.

Miller CMOS OTA: Specifications 1

1. Introductory analysis
1.1 DC currents and voltages on all nodes
1.2 Small-signal parameters of all transistors

2. DC analysis
2.1 Common-mode input voltage range vs supply Voltage
2.2 Output voltage range vs supply Voltage
2.3 Maximum output current (sink and source)

Willy Sansen 10 05 0631

0631

As an introduction, it is good to carry out a DC analysis of an amplifier somewhere in the middle of the design space. This will not give accurate data, but is a good starting point to have some idea about possible DC currents and voltages.

This is followed by a small-signal analysis, so that some elementary knowledge is obtained about orders of magnitude of transconductances, output resistances, capacitances, etc.

Experienced designers can leave out these two introductory steps.

DC analysis comes next.

Perhaps one of the most important specifications is the common-mode input range over which an amplifier can operate. This is actually the average input voltage range. For smaller and smaller supply voltages, this specification has become one of the most important.

The maximum output voltage range is a lot easier to achieve. A rail-to-rail output range is quite feasible provided the output loads are purely resistive and provided only two transistors are used in the output stage with no cascodes.

The maximum output current is normally the DC current of the output stage. In Chapter 11 we will add class-AB output stages to be able to deliver more current.

Miller CMOS OTA: Specifications 2

3. AC and transient analysis
3.1 AC resistance and capacitance on all nodes
3.2 Gain versus frequency : GBW, ...
3.3 Gainbandwidth versus biasing current
3.4 Slew rate versus load capacitance
3.5 Output voltage range versus frequency
3.6 Settling time
3.7 Input impedance vs frequency (open & closed loop)
3.8 Output impedance vs frequency (open & closed loop)

Willy Sansen 10 05 0632

0632

The AC analysis has been partially carried out.

It is good practice to verify the impedance on all the nodes. It gives an idea on where to expect additional poles, etc.

The gain versus frequency is actually the only specification that we have fully studied.

The GBW versus biasing is a misleading specification. When we have minimized the current consumption for a certain GBW, we can then

no longer modify it. Changing the current would render that amplifier unstable or generated overconsumption in current. As a result, spec 3.3 does not actually exist. For an overdesigned amplifier it is possible though, to tune the GBW by means of the biasing current.

The Slew Rate and output voltage at high frequencies will be discussed in more detail. They are just too important.

The settling time is really important for Analog-to-digital converters and all switching applications. It will be discussed at a later stage.

The input impedance of a CMOS OTA is purely capacitive as the two input C_{GS} capacitances appear in series. For a bipolar OTA, resistances have to be added. No more attention will be paid to them.

The output impedance will be examined, as there is no class-AB output stage.

Miller CMOS OTA: Specifications 3

4. Specifications related to offset and noise

4.1 Offset voltage versus common-mode input Voltage

4.2 CMRR versus frequency

4.3 Input bias current and offset

4.4 Equivalent input noise voltage versus frequency

4.5 Equivalent input noise current versus frequency

4.6 Noise optimization for capacitive/inductive sources

4.7 PSRR versus frequency

4.8 Distortion

Willy Sansen 10 05 0633

0633

Many specifications are dealing with offset and noise.

Offset has to do with mismatch between transistors, capacitors, etc. It can be reduced by increasing the size. The CMRR is related to it. In a sense we have two specifications, originating from the same phenomena. Actually, the PSRR is also related to it. They are all discussed in Chapter 15.

The input bias current is certainly of importance for a bipolar opamps and maybe in the future for CMOS opamps, if Gate current is present.

The equivalent input noise voltage and current will be discussed in detail.

Capacitive noise matching has been briefly mentioned in Chapter 4. Inductive source impedances are too cumbersome and are left out here.

Finally, distortion is gaining in importance. This is why a whole Chapter is devoted to it (Chapter 18).

0634

Finally, we have a number of more exotic aspects of opamps.

Connection of an inductor instead of a capacitance as a load may impair the stability. A good example is a loudspeaker.

Switching a whole opamp in and out is even more exotic. And yet, this is a technique to realize switched-capacitor filters down to really low supply voltages. Indeed, for these voltages it is no longer possible to implement switches. Switching the opamp itself is a possible solution.

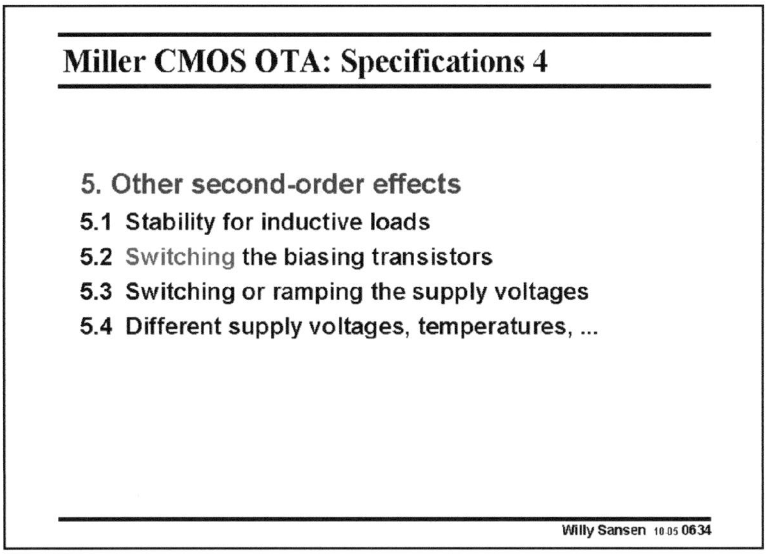

Miller CMOS OTA: Specifications 4

5. Other second-order effects

5.1 Stability for inductive loads

5.2 Switching the biasing transistors

5.3 Switching or ramping the supply voltages

5.4 Different supply voltages, temperatures, ...

Willy Sansen 10 05 0634

This can be done either by switching in and out the biasing transistors or by switching the supply voltage itself. The stability and recovery are different in both cases. This will be discussed in Chapter 21.

Repeating all specifications at different supply voltages and at different temperatures are obvious specifications to be added, depending on the application.

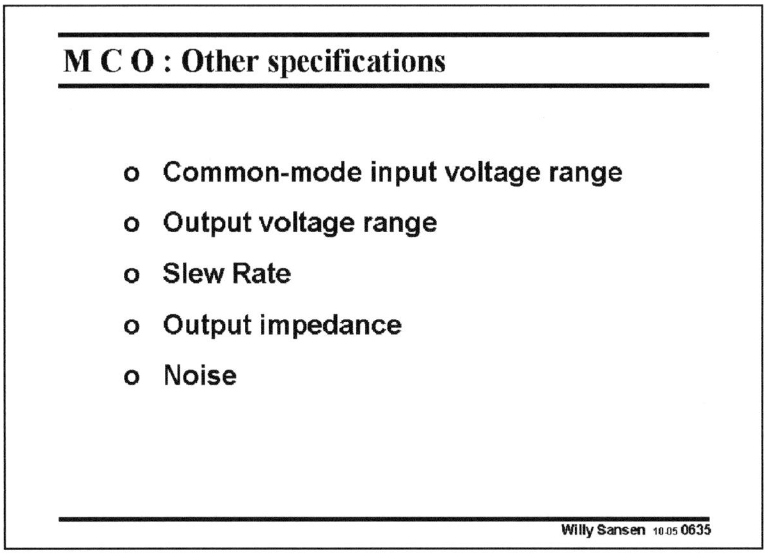

M C O : Other specifications

o **Common-mode input voltage range**

o **Output voltage range**

o **Slew Rate**

o **Output impedance**

o **Noise**

Willy Sansen 10 05 0635

0635

A few of these specifications will now be discussed.

The numerical results will be given for the same Miller CMOS OTA. We start with the common-mode input voltage range.

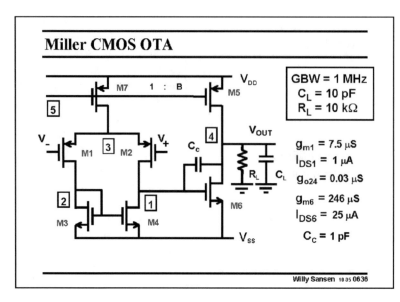

0636
For all specifications, the same amplifier will be used throughout. It is the Miller CMOS OTA of 1 MHz GBW for a 10 pF load.

In this way, the reader can give himself a fairly good picture on how all these numerical specifications fit together.

The circuit is shown again in this slide.

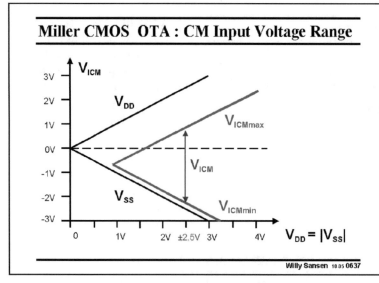

0637
The Common-mode input voltage is the average input voltage. Its range is limited by the supply voltages.

For the CM input voltage going up, the V_{GS1} of the input devices, added to the V_{DS7} of the DC current source, do not allow operation up to the positive supply voltage. The maximum CM input voltage is therefore $V_{DD} - V_{GS1} - V_{DS7}$.

The values obviously depend on which $V_{GS} - V_T$ values have been used. For the input devices the $V_{GS} - V_T$ value is small (0.2 V) but it is a lot larger (0.5 V) for the DC current source.

For a supply voltage of ± 2.5 V, the maximum CM input voltage is illustrated in this slide.

The lowest possible CM input voltage can come much closer to the negative supply voltage. Indeed, it is given by $V_{SS} + V_{GS3} + V_{DS1} - V_{GS1}$. This value can be quite close to the V_{SS} supply but can never actually reach it, whatever $V_{GS} - V_T$ values are chosen.

The total CM input voltage range is only a fraction of the rail-to-rail span. Some other amplifiers (in Chapter 11) will be capable of input rail-to-rail performance.

Also, note that this amplifier could still operate at ± 1 V provided the input transistors are biased at about -0.7 V.

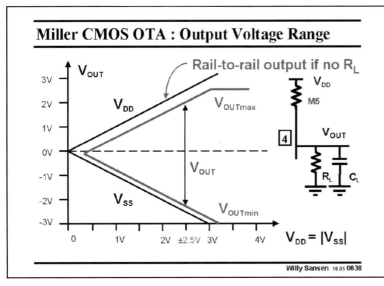

0638

The output voltage range is a lot better!

It depends on whether a resistive load is added to the capacitive one or not. Normally, there is no resistive load on-chip. Blocks are put in series such that they only see Gates as loads. This depends on the application however.

If no resistive loads are present, then the output can go rail-to-rail. Indeed, even for output voltages close to the positive rail, when the output transistor ends up in the linear region, the capacitor is still charged further until the output voltage reaches the positive supply voltage. Of course, in this region, the gain will have decreased. Some distortion will now show up. Nevertheless, the supply rail can be reached! The same applies to the negative rail.

If there is a resistive load, a resistive divider is created as soon as the output transistor M5 enters the linear region, as illustrated in this slide. In this case, the supply rail can never be reached. The output can get close though, depending on the size of the output transistor.

Note that this is the ideal output structure for a wide-swing opamp. No cascodes are used. Moreover, the output devices are connected drain-to-drain. This is why class AB output drivers use this output configuration. Only the Gate drive circuits differ (see Chapter 12).

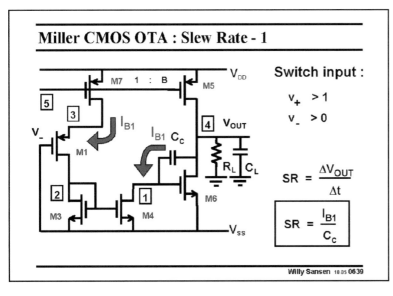

0639

Whenever an opamp is driven with a large input voltage, slewing occurs at the output.

Large input voltages are used to try to make the opamp go faster.

In this case, the input transistors are overdriven, i.e. one is on and the other is off, as illustrated in this slide.

The input transistor which is on, now operates as a cascode, driven by the total input stage current I_{B1}. The current mirror then draws the same current from the compensation capacitance C_c.

We have a situation where a capacitance C_c is driven by a constant current. As a result, the voltage slope across it, is constant and called the Slew Rate SR.

This SR appears at the output as the V_{GS6} is still about constant. Indeed, transistor M6 still conducts as if nothing has happened!

This SR limits the steepest possible slope at the output of the opamp.

Clearly, this phenomenon works in both directions.

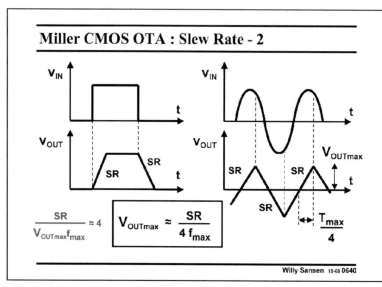

0640

This steepest possible slope at the output is clearly visible when the opamp is driven by a square waveform with large amplitude.

A trapezoidal output waveform results, clearly showing the limited slopes.

This slewing is also visible when a sinusoidal waveform is applied with large amplitude. In this case, triagonal distortion results.

This latter waveform shows that there is a direct link between the maximum

amplitude V_{OUTmax}, the frequency f_{max} and the Slew Rate SR. During a period $T_{max}/4$, the SR allows a maximum amplitude V_{OUTmax}. The maximum output voltage V_{OUTmax} is therefore heavily limited by the Slew Rate, as given by this expression.

This is plotted next.

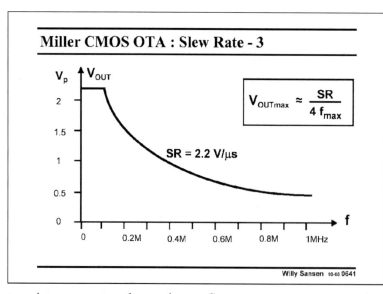

0641

This plot shows that for higher frequencies, close to the GBW, only small output voltage amplitudes can be expected.

This curve is also called the large-signal bandwidth of the OTA.

In this design example, a SR of 2.2 V/μs only provides 0.4 V_{peak} at the GBW of 1 MHz, rather than 2.5 V_{peak} at low frequencies.

It is not clear how to remedy this, as both GBW and SR depend on the input

transistor current and capacitance C_c.

Actually, we want to achieve a larger SR for the same GBW. Which transistor parameters have to be adjusted?

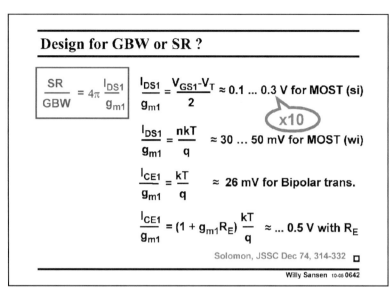

Design for GBW or SR ?

$$\frac{SR}{GBW} = 4\pi \left| \frac{I_{DS1}}{g_{m1}} \right|$$

$$\frac{I_{DS1}}{g_{m1}} = \frac{V_{GS1} - V_T}{2} \approx 0.1 \ldots 0.3 \text{ V for MOST (si)}$$

x10

$$\frac{I_{DS1}}{g_{m1}} = \frac{nkT}{q} \approx 30 \ldots 50 \text{ mV for MOST (wi)}$$

$$\frac{I_{CE1}}{g_{m1}} = \frac{kT}{q} \approx 26 \text{ mV for Bipolar trans.}$$

$$\frac{I_{CE1}}{g_{m1}} = (1 + g_{m1}R_E) \frac{kT}{q} \approx \ldots 0.5 \text{ V with } R_E$$

Solomon, JSSC Dec 74, 314-332 □

Willy Sansen 10-05 0642

0642

In order to achieve a larger SR for the same GBW, we take the ratio and rewrite it in terms of transistor parameters.

We find that for a MOST, this ratio is simply proportional to its $V_{GS} - V_T$. The larger we make the input transistor $V_{GS1} - V_T$, the larger the SR will be for the same GBW. Clearly, SR has to do with high speed. This result is not unexpected. We have known all along that for high speed we need to take large values of $V_{GS1} - V_T$. The noise performance will suffer from that, but again we have known this all along.

Clearly, the use of a bipolar transistor at the input reduces the SR by a factor of 10 for the same GBW. This also applies to a MOST in weak inversion.

In order to improve the SR for a bipolar amplifier, we have to insert series resistors. The SR increases accordingly. Clearly, the noise performance suffers from that as well.

To realize an amplifier with high speed or SR, and low noise at the same time is a real compromise!

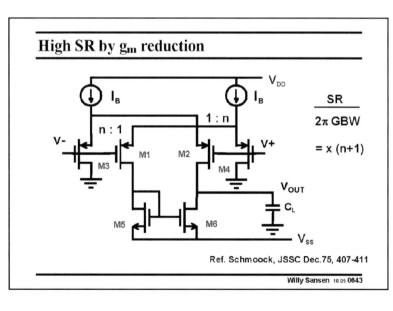

High SR by g_m reduction

$$\frac{SR}{2\pi \text{ GBW}} = x \ (n+1)$$

Ref. Schmoock, JSSC Dec.75, 407-411

Willy Sansen 10-05 0643

0643

There is a possibility to decouple SR from GBW by using cross-coupling as shown in this slide (actually realized for bipolar transistors). However, the cost is increased power consumption.

For small signals, transistors M1 and M2 provide gain and GBW as usual. They run at a fairly small DC current however, such that their g_m is also small. Their DC current is only $1/(n+1)$ of the biasing current I_B.

For large input signals the input transistors switch on or off. As a result all current I_B can flow to the output, to slew the output voltage. The current is now $n+1$ times larger. The ratio SR/GBW is also $n+1$ times higher.

The main drawback is obviously that for small signals, a large current I_B is flowing, only a small fraction of which is used to generate transconductance. Most of the DC current is now actually wasted.

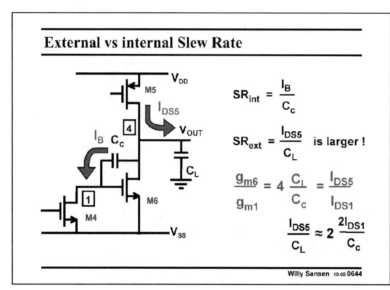

0644

Normally, it is the internal Slew Rate which establishes the limit. However, it can also be the external Slew Rate.

The load capacitance needs to be charged as well. All current of current source M5 is now used to slew the output voltage.

In the design procedures presented in this Chapter, the output stage current is a lot larger than the input stage current, whereas the compensation capacitance is only a factor 2–3 times smaller than the load capacitance, the internal SR is the limiting factor, by at least a factor of 2.

This is not always the case however and should be verified. In the example of the 1MHz Miller CMOS OTA, the internal SR is 2.2 V/μs but the external SR is only 2.5 V/μs, which is barely larger!

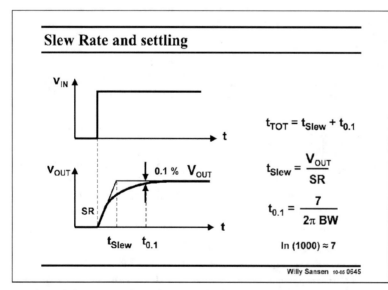

0645

The Slew Rate is also part of the total settling time.

When we apply a square waveform with large amplitude, the output will slew first until it reaches the final output voltage within a large percentage, let us say 10%. From then on the small-signal operation is taking over and the bandwidth (or GBW) is determining the settling time. The cross-over from SR to BW limiting behavior is not clear.

The minimum settling time would be based on a calculation in which we forget about the SR altogether. In this case, 0.1% settling takes 6.9 (or 7) time constants.

The maximum settling time is obtained by addition of the time required to slew from zero to the final output voltage, to the 7 time constants required to reach 0.1% accuracy.

For example, if we take the same 1 MHz Miller CMOS OTA, set at a closed-loop gain of 10. The BW is 100 kHz and the corresponding time constant is 1.6 µs. Settling to 0.1% would take 7 time constants or 11.2 µs.

To reach an output voltage of 1 V with a SR of 2.2 V/µs takes 0.45 µs. The total settling time is now between 11.2 and 11.6 µs. The actual settling time takes up most of the time!

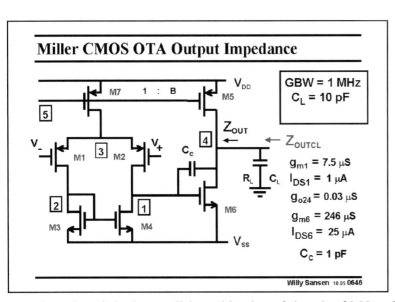

0646

The output impedance is examined next.

For this purpose, we have left out the external resistive load. They are normally absent anyway. We can distinguish two output impedances, i.e. the one of the amplifier itself Z_{OUT} and the output impedance including the load capacitance Z_{OUTCL}. The latter one is the impedance at the interconnect to the next stage.

The output impedance Z_{OUT} itself is high at low frequencies, where it is the parallel combination of the r_o's of M5 and M6.

At high frequencies however, the compensation capacitance C_c acts as a short. The output impedance Z_{OUT} becomes resistive with value $1/g_{m6}$. This is shown next.

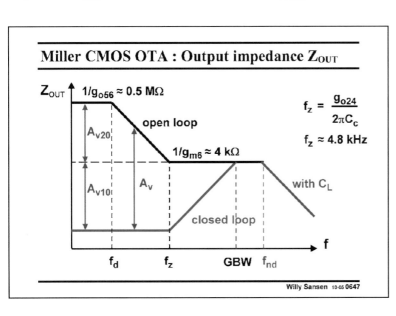

0647

Without feedback, the open-loop output impedance Z_{OUT} is high indeed at low frequencies. It decreases until it reaches the level $1/g_{m6}$.

The pole f_d of this Z_{OUT} is about the same as for the open-loop gain characteristic. The zero f_z however, is a new characteristic frequency. It is nowhere visible in the gain characteristic. It only appears in the output impedance and in the noise characteristic, which comes next.

Actually, this zero is at the frequency where the output resistance of the first stage is taken over by the impedance of the compensation capacitance C_c. It is the frequency where the gain of the first stage starts decreasing.

Anyway, for most of the frequency region, the output impedance is quite low. As a result, we do not need a class AB stage. These latter stages are only necessary to drive off-chip loads. Also, a line connection at an impedance of a few kΩ's does not pick up a lot of noise. It is a good value for interconnect.

Application of unity-gain feedback causes the output impedance to be divided by the total gain. Since the pole f_d is the same for both, it disappears. The zero remains, indicating a wide region which is inductive.

With the load capacitance C_L included, the impedance Z_{OUTCL} starts decreasing at the non-dominant pole f_{nd}.

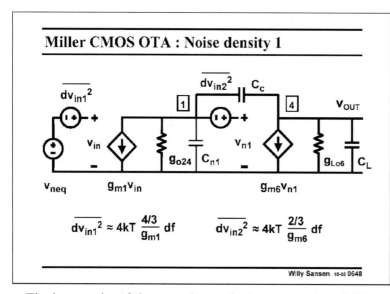

Miller CMOS OTA : Noise density 1

$$\overline{dv_{in1}}^2 \approx 4kT \frac{4/3}{g_{m1}} df \qquad \overline{dv_{in2}}^2 \approx 4kT \frac{2/3}{g_{m6}} df$$

Willy Sansen 10-05 0648

0648

In order to find the equivalent input noise voltage, we have to introduce the input noise voltages for the first and second stages.

They are included in the small-signal equivalent circuit shown in this slide.

The input noise of the input stage is given as well. Only the two input transistors themselves are included. We have assumed that the current mirror devices have been designed for low noise.

The input noise of the second stage is due to transistor M6 only. It is embedded in the circuit in this slide. Note that it has taken a fairly strange position, in series with the Gate. Therefore, its effect is not found that easily.

We now have to shift the input noise voltage of the second stage to the input. For this purpose, we calculate its contribution to the output, and then divide it by the total gain.

The result is shown next.

0649

Both contributions of the noise sources to the output noise are shown in this slide for unity-gain feedback. The contribution of the input stage drops off at frequencies beyond the GBW.

The second-stage noise however, becomes dominant at high frequencies. The behavior at these frequencies is most important as noise has to be looked at on a linear frequency scale.

At low frequencies the noise density of the second stage is clearly negligible. It can be divided by the gain of the first stage squared.

Beyond the zero frequency f_z, which is the same as for the output impedance, the noise contribution of the second stage starts rising, until it becomes dominant at the highest frequencies.

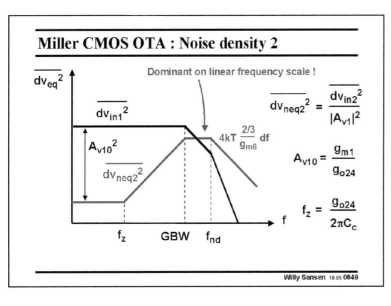

The noise starts rising because from the zero frequency on, the gain of the first stage goes down. As a result, the noise goes up.

The result is not so bad in the sense that the noise density of the second stage never really takes over below the GBW. The maximum noise contribution of the second stage is actually the input noise voltage, as given in the previous slide. Since g_{m6} is much larger than g_{m1}, the noise of the second stage is always negligible with respect to the noise of the first stage.

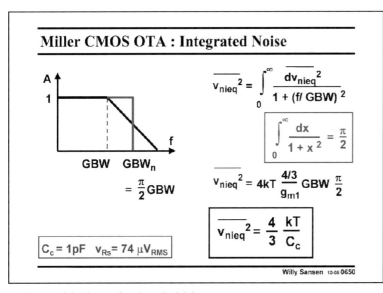

0650

In order to find the integrated noise of this OTA, we have to integrate the input noise density over all frequencies. Since we are dealing with a first-order roll-off, we will find again this increase of the bandwidth to the noise bandwidth by a factor of $\pi/2$.

A unity-gain situation is used. The bandwidth equals the GBW.

The total integrated noise is now simply the product of the equivalent input noise voltage with the noise bandwidth.

Since transconductance g_{m1} determines both, the input noise voltage and the GBW, it cancels out. As a result, the total integrated noise only contains the compensation capacitance C_c.

The noise of the second stage has been neglected although its contribution is close to 20%.

Increasing the total integrated noise performance, will require larger capacitances, and hence larger currents.

Low noise always leads to higher power consumption!

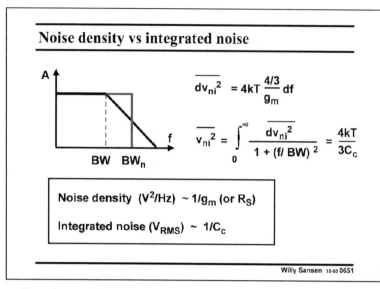

Noise density vs integrated noise

$$\overline{dv_{ni}^2} = 4kT\frac{4/3}{g_m}df$$

$$\overline{v_{ni}^2} = \int_0^\infty \frac{\overline{dv_{ni}^2}}{1 + (f/BW)^2} = \frac{4kT}{3C_c}$$

Noise density $(V^2/Hz) \sim 1/g_m$ (or R_S)

Integrated noise $(V_{RMS}) \sim 1/C_c$

Willy Sansen 10-05 0651

0651

As for resistive noise, OTA noise leads to similar conclusions.

Noise density always depends on resistors or transconductances, whereas integrated noise depends on the main capacitance. This is C_L for a single-stage amplifier but C_c for a two- or three-stage amplifier.

They are linked, however. Larger capacitances will lead to larger currents, which will yield larger transconductances.

Care has to be exerted however, to make sure that the noise of the output stage does not become dominant at the highest frequencies of interest. For this purpose the output g_m must always be larger than the input g_m. This is a requirement which has originated from stability considerations as well!

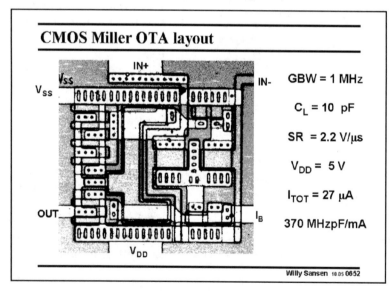

CMOS Miller OTA layout

GBW = 1 MHz
$C_L = 10$ pF
SR = 2.2 V/μs
$V_{DD} = 5$ V
$I_{TOT} = 27$ μA
370 MHzpF/mA

Willy Sansen 10.05 0652

0652

As a conclusion on this Miller CMOS OTA of 1 MHz GBW for a 10 pF load capacitance, we need to look at the layout.

The input devices at the right in the middle, are fairly big to suppress the 1/f noise somewhat.

The output devices at the left. They are obviously larger. The compensation capacitance of 1 pF is clearly visible, although quite small.

0653

As a conclusion to this Chapter, a design exercise is launched of a Miller CMOS OTA.

The specifications are fairly conventional.

It is suggested to adopt the second design plan, in which the minimum current in the output stage is to be calculated first.

Miller CMOS OTA : Exercise

GBW = 50 MHz for C_L = 2 pF : use min. I_{DS6} !

Techno: L_{min} = 0.5 μm; K'_n = 50 μA/V² & K'_p = 25 μA/V²

C_{GS} = kW (= C_{ox}WL) and k = 2 fF/ μm

$V_{GS} - V_T = 0.2$ V

Find
g_{m6} I_{DS6} W_6 $C_{n1} = C_{GS6}$ C_c g_{m1} I_{DS1} $\overline{dv_{ineq}^2}$ v_{inRMS}

Willy Sansen 10 05 **0653**

The sequence suggested to design this amplifier is given. After g_{m6} and I_{DS6} we have to extract W_6 as this provides C_{GS6} which is taken to be equal to C_{n1}.

The compensation capacitance C_c is now easily obtained from the expression for the non-dominant pole.

Finally, we design the input stage as we know C_c.

Noise density and integrated noise are easy extensions in this design exercise.

Conclusion : Table of contents

- **Design of Single-stage OTA**
- **Design of Miller CMOS OTA**
- **Design for GBW and Phase Margin**
- **Other specs:** Input range, output range, SR, ...

Willy Sansen 10 05 **0654**

0654

In this Chapter, a considerable amount of design detail has been given on a Miller CMOS operational transconductance amplifier.

It has been shown that an optimum can be reached in terms of power consumption and that this optimum can be reached along various different design procedures.

In practice, too many specifications have to be fulfilled however, for too few design variables. Many compromises are thus to be taken. Increasing the complexity of the circuit configuration is only one way to deal with these compromises.

This is why we want to examine more complicated circuits and different realizations of operational amplifiers.

Important opamp configurations

Willy Sansen

KULeuven, ESAT-MICAS

Leuven, Belgium

willy.sansen@esat.kuleuven.be

Willy Sansen   10 05 071

071

The list of different operational amplifiers is endless. And yet it is possible to classify them in a limited number of important categories. Examples are symmetrical opamps and folded cascodes. They are being reused and redesigned continuously. They are the kings of the list of important amplifiers.

Many other opamps can be included in this list because they highlight some cleverness in design or because they excel in performance.

In this Chapter, a review is given on many important opamp circuits. In many cases the design compromises are discussed, together with their limits in terms of speed or noise or some other specifications.

Table of contents

- **Simple CMOS OTA**
- **CMOS Miller OTA**
- **Symmetrical CMOS OTA**
- **Folded cascode OTA**
- **Other opamps**

Willy Sansen   10 05 072

072

In this Chapter the trade offs between standard CMOS and BiCMOS are also discussed. This is why some known schematics are also included.

Most of the discussion is on the symmetrical OTA and the folded cascode OTA as they are so often used.

Finally, a list of published opamps are discussed to show that the design principles are applicable to most of them.

We recall the simplest differential voltage amplifier that we have observed.

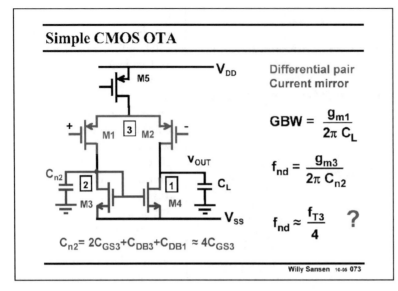

073
This single-stage CMOS OTA is well known.

Because it has a simple configuration means that it can be used up to really high frequencies. Its only possible second pole is negligible because of two reasons. it occurs at values related to f_T and secondly because this node is on the other side of the output. For a single-ended amplifier this means that this second pole is followed by a zero at twice the frequency. As a result, it is negligible.

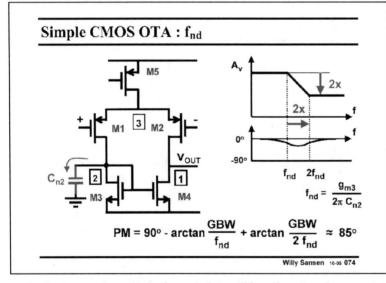

074
Even if we connected a large external capacitance at node 2, we would still find half of the circular current, generated by M1 and M2, through the output load.

Whatever the size of the capacitor is, we always retain half of the current through the output load. This factor of two can only be explained by a pole-zero doublet with spreading two. Its effect on the phase margin is therefore marginally small.

Each time we have a single-ended amplifier, the capacitances on the other side than the output, will therefore be negligible for the Phase Margin.

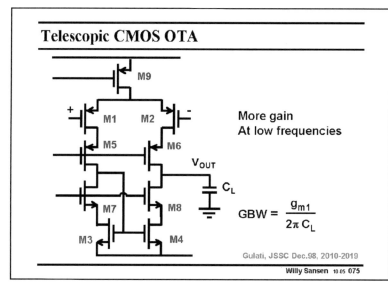

Telescopic CMOS OTA

More gain
At low frequencies

V_{OUT}

$GBW = \dfrac{g_{m1}}{2\pi C_L}$

Gulati, JSSC Dec.98, 2010-2019

Willy Sansen 10 05 075

075

The gain of such a voltage amplifier is rather limited as the gain per transistor can be quite small for nanometer MOST devices.

This is why cascodes are better added. Four cascode MOSTs are added M5–8 in series with the input devices and current mirror, as shown in this slide. Note that cascode M7 is included in the feedback loop around transistor M3, which allows a larger output swing.

This is called the telescopic CMOS OTA. The impedance at the output node increases considerably, but not the GBW, as shown next.

Obviously the power consumption does not increase.

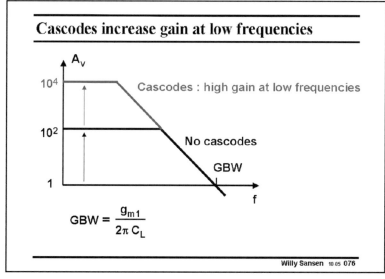

Cascodes increase gain at low frequencies

A_v

10^4

Cascodes : high gain at low frequencies

10^2

No cascodes

GBW

1

f

$GBW = \dfrac{g_{m1}}{2\pi C_L}$

Willy Sansen 10 05 076

076

Without cascodes, the gain is moderate. With cascodes however, the gain is increased, but only at low frequencies.

Cascode transistors are now mainly used for more gain at low frequencies, for example for lower distortion at low frequencies.

Another "hat" can be put on top of this characteristic by application of gain boosting to the cascode transistors M6 and M8.

For deep submicron or nanometer CMOS this has always become a necessity, as the gain per transistor has become less than 10.

Table of contents

Willy Sansen 10 05 077

077

A two-stage amplifier such as the CMOS Miller OTA needs more power to reach similar values of GBW, compared to a single-stage amplifier.

BiCMOS can now be considered to save power.

The design plans have been previously discussed. They will be applied.

Miller CMOS OTA

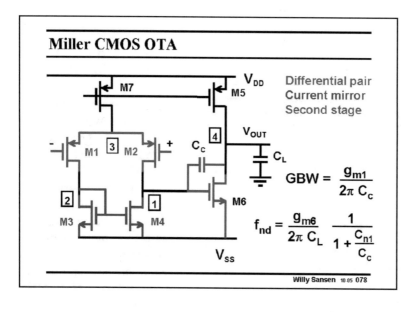

Differential pair
Current mirror
Second stage

$$GBW = \frac{g_{m1}}{2\pi\, C_c}$$

$$f_{nd} = \frac{g_{m6}}{2\pi\, C_L}\; \frac{1}{1 + \dfrac{C_{n1}}{C_c}}$$

Willy Sansen 10 05 078

078

A Miller OTA in CMOS technologies has been discussed in great detail. The two governing expressions are listed for the last time. They are dealing with the GBW and with the non-dominant pole.

In each design plan, it is always better to start with the highest frequencies first, which here is the non-dominant pole.

Decisions about g_{m6} and C_c are always first, as explained before.

079

Can BiCMOS provide additional advantages?

A typical BiCMOS realization of a Miller OTA is shown in this slide.

The second stage uses a bipolar transistor as its g_m is the same as for a MOST but with 4 times less DC current. Since the current in the second stage is by far the larger one, big savings in power consumption are achieved.

The input resistance of a bipolar transistor is too small, however. It reduces the resistance at node 1 considerably, such that there is little gain left (if any) in the first stage.

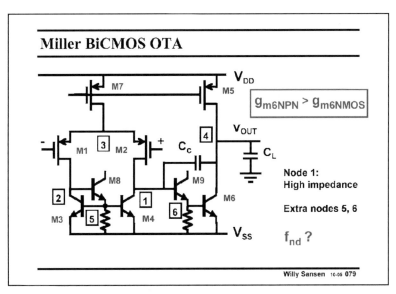

Miller BiCMOS OTA

$g_{m6NPN} > g_{m6NMOS}$

Node 1:
High impedance

Extra nodes 5, 6

f_{nd} ?

Willy Sansen 10-05 079

This is why we need an Emitter follower between the first stage and the input transistor of the second stage. It is realized with transistor M9. The input resistance is now beta times higher and hopefully comparable to the output resistance of the first stage.

This Emitter follower raises the DC voltage at node 1 by one more V_{BE}. As a result, node 1 is about 2 V_{BE}'s or 1.3 V higher than V_{SS}.

In order to establish the same DC voltage at the other node 2, we use the three-transistor bipolar current mirror, explained in Chapter 2. Now both nodes 1 and 2 present similar DC voltages to the input pair, improving matching.

Table of contents

Willy Sansen 10 05 0710

0710

One of the most used OTA's is the symmetrical OTA. It is more symmetrical than the Miller OTA. As a result, matching is improved which provides better offset and CMRR specifications (see Chapter 15).

0711

A symmetrical OTA consists of one differential pair and three current mirrors. The input differential pair is loaded with two equal current mirrors, which provide a current gain B. It is sometimes called a load-compensated OTA as both loads are now the same.

In the case of a single-ended output we need another current mirror with gain 1 to reach this

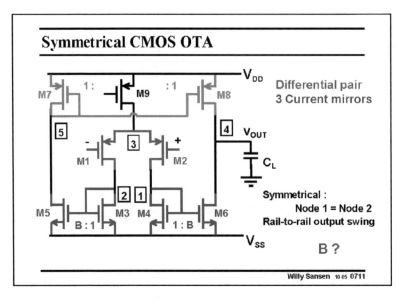

Symmetrical CMOS OTA

Differential pair
3 Current mirrors

Symmetrical :
Node 1 = Node 2
Rail-to-rail output swing

B ?

Willy Sansen 10 05 0711

output. In the case of two outputs (in the next Chapter), we do not need this current mirror any more. This analysis is carried out for a single-ended output.

It is clear that this OTA is symmetrical. The input devices see exactly the same DC voltage and load impedance. This is about the best that can be achieved with respect to matching.

Moreover, there is some extra gain, by current factor B. How far can we go with B?

0712

The gain at low frequencies is easily calculated. Indeed, the circular current, generated by the input devices, is amplified by B and flows in the output load.

The output resistance R_{n4} at node 4 is quite high. Actually it is the only high resistance in the circuit. All other nodes are at the $1/g_m$ level. As a result, this is a single-stage amplifier: there is only one high-resistance node, one single node where the gain is large, where the

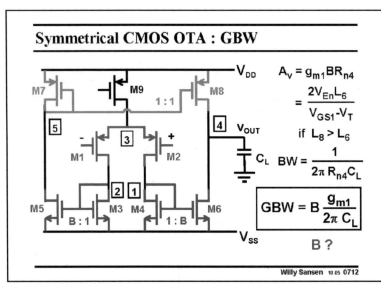

Symmetrical CMOS OTA : GBW

$A_v = g_{m1} B R_{n4}$

$= \dfrac{2 V_{En} L_6}{V_{GS1} - V_T}$

if $L_8 > L_6$

$BW = \dfrac{1}{2\pi R_{n4} C_L}$

$$GBW = B\, \dfrac{g_{m1}}{2\pi C_L}$$

B ?

Willy Sansen 10 05 0712

swing is large, and ultimately where the dominant pole is formed.

The voltage gain at low frequencies is now easily obtained.

The bandwidth is created at the same output node. The GBW is the product. It is the same as for a single-transistor amplifier but multiplied by current factor B.

Increasing B increases the GBW.

How far can we go with B?

0713

However, all other nodes create non-dominant poles. Since we find three other nodes 1, 2 and 5, do we have three non-dominant poles?

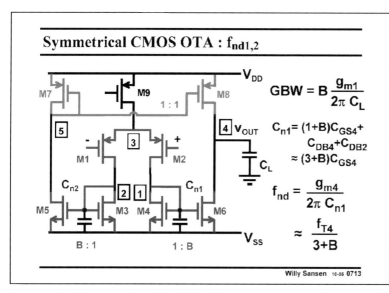

Symmetrical CMOS OTA : $f_{nd1,2}$

$$GBW = B \frac{g_{m1}}{2\pi C_L}$$

$$C_{n1} = (1+B)C_{GS4} + C_{DB4} + C_{DB2}$$
$$\approx (3+B)C_{GS4}$$

$$f_{nd} = \frac{g_{m4}}{2\pi C_{n1}}$$

$$\approx \frac{f_{T4}}{3+B}$$

Willy Sansen 10-05 **0713**

that node. They are listed in this slide. As a very crude approximation, we take them all to be the same, except for the current mirror. At node 1, transistor M6 offers an input capacitance which is B times larger than for transistor M4.

Finally, the non-dominant pole frequency can be rewritten in terms of f_T and current factor B. The larger B, the lower the non-dominant pole. This expression therefore provides the limit on B.

The answer is negative. We will see that only one non-dominant pole is playing a role. It is the one at nodes 1 and 2.

How can the non-dominant pole at node 1 be the same as at node 2? Actually, for a differential output voltage, it is fairly easy to show that these nodes together form just one pole (see next slide).

As a result, the non-dominant pole is determined by the resistance $1/g_{m4}$ and all capacitances connected to

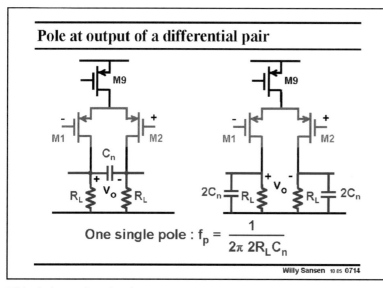

Pole at output of a differential pair

One single pole : $f_p = \dfrac{1}{2\pi \, 2R_L C_n}$

Willy Sansen 10 05 **0714**

0714

For a differential output, two transistors to ground provide only a first-order characteristic – there is only one single pole.

This is obvious for the circuit on the left. Since there is only one capacitance, only one pole can emerge.

However, this circuit can easily be converted to the circuit on the right. We take two capacitances in series with double the value and then ground the node between both capacitances.

This is how the circuit on the right is derived from the first one. For AC they are exactly the same. They have the same pole!

To make it slightly more intriguing, we could wonder what happens if there is some asymmetry. For example, if one capacitance is slightly larger than the other one, how can it create a pole with the same value?

In this case, we find two poles, but we also find a zero inbetween, to ensure a first-order roll-off.

The net result is that for a differential output, these two nodes establish one single pole only!

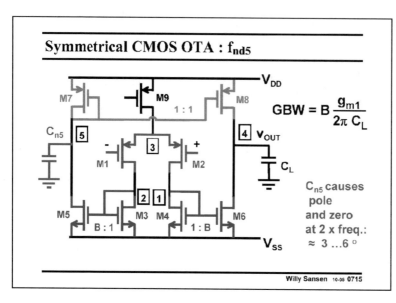

0715

How about the pole at node 5?

Remember that this is a node at the other side of a single-ended amplifier. Each time we have a single-ended amplifier, the capacitances on the other side than the output, will be negligible for the Phase Margin.

Indeed, a pole-zero doublet is created by the capacitances at this node 5. Its spreading is only 2. As a result, the effect on the Phase Margin is now negligible.

Despite the fact that we have three nodes at the $1/g_m$ level, we only have one single non-dominant pole!

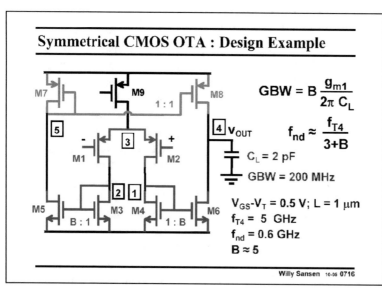

0716

As an example, let us design a CMOS symmetrical OTA for a GBW of 200 MHz and 2 pF load capacitance.

The expressions of the GBW and f_{nd} are repeated.

Obviously, for wide-band performance, we have to take a high-speed transistor for M4 and M6. This means that this current amplifier (or mirror) devices have to be designed for large $V_{GS} - V_T$ and small L.

Some values have been selected, depending on the CMOS process available. The resulting f_T is about 5 GHz.

The maximum value of B is found by equating f_{nd} to $3 \times$ GBW. The value of B is therefore 5. Many designers use between 3 and 5.

The input transconductance is now easily obtained from the GBW. It is $g_{m1} = 0.5$ mS, which requires about 50 µA. The total current consumption is now 0.6 mA.

The FOM of this amplifier is 670 MHzpF/mA, which is quite good, indeed!

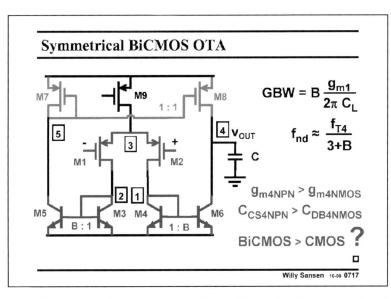

Symmetrical BiCMOS OTA

$$GBW = B \frac{g_{m1}}{2\pi C_L}$$

$$f_{nd} \approx \frac{f_{T4}}{3+B}$$

$$g_{m4NPN} > g_{m4NMOS}$$
$$C_{CS4NPN} > C_{DB4NMOS}$$

BiCMOS > CMOS **?**

Willy Sansen 10-05 0717

0717

Can BiCMOS provide similar power savings as for a Miller OTA?

The answer is negative.

The current sources are the only candidates to be implemented in bipolar. The input devices are better MOSTs. They provide less input biasing current and higher Slew Rate.

There are two considerations:

1. The npn transistors certainly have a higher g_m but this advantage is not really exploited in a current mirror. They also have a higher f_T, at least within a particular BiCMOS process. They may not have a higher f_T however, than the nMOSTs in a more advanced standard CMOS process, offered at the same time.

2. Bipolar transistors have a relatively large collector-substrate capacitance C_{CS}. As a result, the parasitic capacitance at nodes 1 and 2 are probably a lot larger than those given by f_T.

As a conclusion, a BiCMOS symmetrical OTA is probably not faster than a CMOS equivalent. For the same GBW it probably does not draw less current than the CMOS.

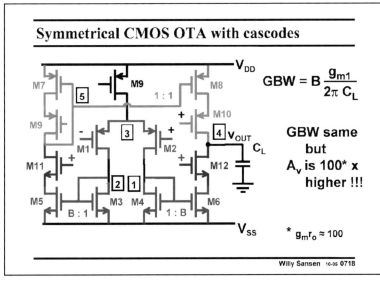

Symmetrical CMOS OTA with cascodes

$$GBW = B \frac{g_{m1}}{2\pi C_L}$$

GBW same
but
A_v is 100* x
higher !!!

$$* \ g_m r_o \approx 100$$

Willy Sansen 10-05 0718

0718

The previous symmetrical OTA's all had too little gain. Cascodes are now added to increase the gain. Gain boosting could even be applied to the output cascodes M10 and M12 to boost the gain even more.

Note that cascodes are added on both sides, to preserve symmetry. Also note that a current mirror is taken (with M7–M10) which allows a large output swing. Indeed the output voltage can swing to within 0.4 V of the supply voltage, without transistors M8/M10 or M6/M12 entering the linear region.

The insertion of cascodes increases the gain but not the GBW. The cascodes only increase the gain at low frequencies, as previously shown. Moreover, the gain at low frequencies can be increased even more by application of gain boosting to the cascodes M10 and M12. This is a general practice for nanometer CMOS where the gain per transistor has become quite small, i.e. less than 10.

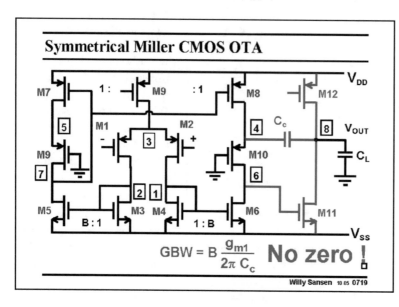

0719

A two-stage Miller operational amplifier is easily built by means of a symmetrical OTA as a first stage, as shown in this slide.

Its GBW also includes C_c and B.

The compensation capacitance C_c is obviously not connected directly from Drain to Gate, but takes a path through cascode transistor M10 to avoid the positive zero.

As a result, the current through the output stage M11/M12 can be taken smaller, saving power.

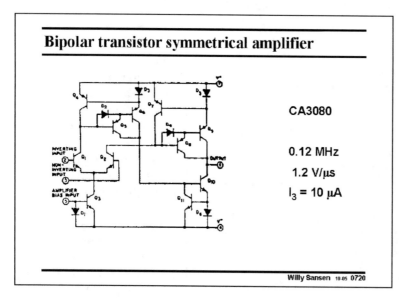

0720

An earlier realization of such a symmetrical OTA is shown in this slide. It is a bipolar realization.

The current mirrors are quite elaborate, to obtain precise current mirroring and good matching.

The specifications obviously depend on the actual DC currents flowing. Actually, this circuit block can be tuned to any value of GBW by modifying the current through Q3. The non-dominant pole tracks the GBW.

0721

Another way to increase the gain is by current starving.

This is actually a fully-differential symmetrical OTA. As no cascodes are used, the voltage gain is very modest.

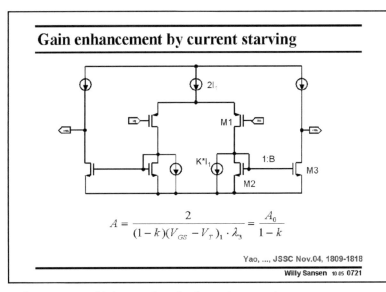

Gain enhancement by current starving

$$A = \frac{2}{(1-k)(V_{GS}-V_T)_1 \cdot \lambda_3} = \frac{A_0}{1-k}$$

Yao, ..., JSSC Nov.04, 1809-1818

Willy Sansen 10 05 0721

However, the addition of two DC current sources with values KI_1 increases the gain considerably. A typical value for k is 0.8.

In this case, 80% of the DC current provided by the input transistors M1 is taken away by the DC current sources. Only 20% of the DC current, together with the signal current is injected into the transistors M2 of the current mirrors.

Because the DC currents in the output transistors M3 are also lower, the output resistances are higher and so is the voltage gain.

This technique cannot be pushed too far as mismatch will occur. Moreover, the resistance at the inner node of the current mirrors determines the non-dominant pole. It cannot be increased too much.

Table of contents

- **Simple CMOS OTA**
- **CMOS Miller OTA**
- **Symmetrical CMOS OTA**
- **Folded cascode OTA**
- **Other opamps**

Willy Sansen 10 05 0722

0722

The other "most used" operational transconductance amplifier is the folded cascode OTA.

Many designers limit themselves to folded cascode OTA's only. It is therefore important to figure out what exactly the advantages and disadvantages are. Also, which design plan delivers the best performance in terms of power consumption, noise, etc.

0723

A folded cascode OTA consists of an input differential pair, two cascodes and one current mirror. The latter current mirror will not be necessary when we have two outputs, as explained in the next Chapter.

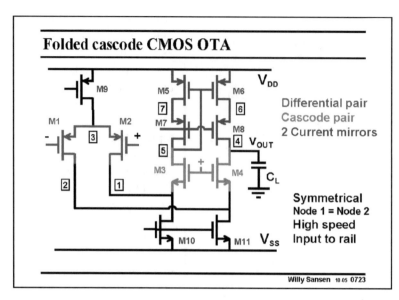

It is a high-swing current mirror again.

This circuit is as symmetrical as the symmetrical OTA as both input devices see exactly the same DC voltage and impedance, at nodes 1 and 2.

The output is again the only point at high resistance. Indeed all other nodes are at $1/g_m$ level. It is again a single-stage amplifier, despite its complexity.

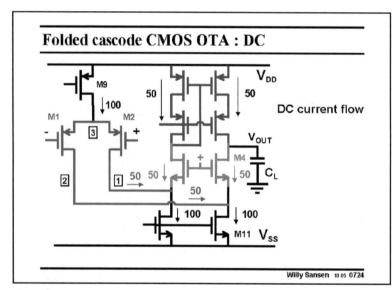

0724

Let us first examine how the DC operation actually works.

The input devices are biased by a current source (with M9) at for example 100 μA. Both input devices carry 50 μA.

At node 2, transistor M11 draws 100 μA. The difference between this current and what is coming from M1, is then pulled from cascode transistor M4.

This current flows through both cascode transistors. The current source on top mirrors this current. There is no way that DC current could flow out, even if the output node would be connected to ground.

Normally, all currents in the input and cascode devices are the same, i.e. 50 μA. This is not a necessity but is certainly the best way to avoid all kinds of artifacts, such as asymmetrical swing, Slew-Rate, etc.

0725

The small-signal operation is easily understood.

The input transistors create a circular current, which flows through the cascode transistors to the high-impedance node. The output resistance at node 4 is again R_{n4}.

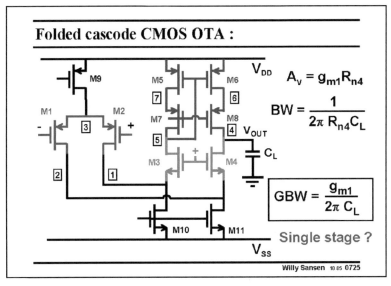

Folded cascode CMOS OTA :

$A_v = g_{m1}R_{n4}$

$BW = \dfrac{1}{2\pi\,R_{n4}C_L}$

$GBW = \dfrac{g_{m1}}{2\pi\,C_L}$

Single stage ?

Willy Sansen 10 05 **0725**

The voltage gain at low frequencies is now easily obtained. Note that this gain is high because cascodes are used. Gain boosting could be applied to the cascodes M4 and M8 to increase the gain even further.

The bandwidth is created at the same output node. The GBW is the product. It is exactly the same as for a single-transistor amplifier. Of course, the input transconcuctance is smaller here, as only half of the current flows in the input stage.

What is the advantage of this folded OTA? It consumes twice the current of a telescopic cascode stage!

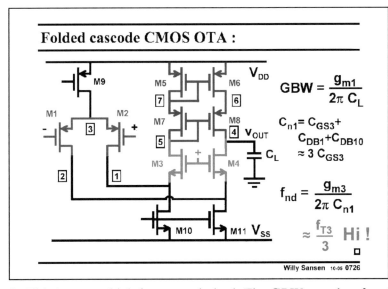

Folded cascode CMOS OTA :

$GBW = \dfrac{g_{m1}}{2\pi\,C_L}$

$C_{n1} = C_{GS3} + C_{DB1} + C_{DB10}$
$\approx 3\,C_{GS3}$

$f_{nd} = \dfrac{g_{m3}}{2\pi\,C_{n1}}$

$\approx \dfrac{f_{T3}}{3}$ **Hi !**

Willy Sansen 10-05 **0726**

0726

In order to find the advantages of a folded cascode OTA, we have to verify the high-frequency performance.

The non-dominant pole is created at the nodes 1 and 2. They form together one single non-dominant pole. The resistance at node 1 is $1/g_{m3}$ and the capacitance at this node is C_{n1}. It is a sum of three small capacitances, which are all similar in size.

The non-dominant pole occurs at about one third of

f_T. This is a very high frequency indeed. The GBW can therefore be quite high.

This is the first advantage of a folded OTA.

Finally, note that the current mirror with transistors M5-M8 can also be used. Remember, however, that this current source requires more than 1 V to keep all transistors in saturation. This is the loss in output swing at each supply line. The previous current mirror is a lot better for low-voltage applications.

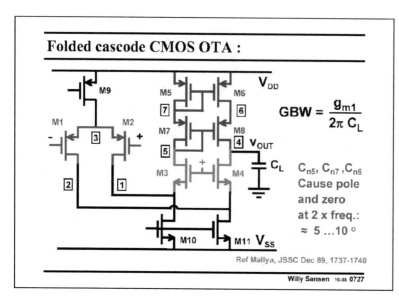

Folded cascode CMOS OTA :

$$GBW = \frac{g_{m1}}{2\pi \, C_L}$$

C_{n5}, C_{n7}, C_{n6}
Cause pole
and zero
at 2 x freq.:
$\approx 5 \ldots 10°$

Ref Mallya, JSSC Dec 89, 1737-1740

Willy Sansen 10-95 0727

0727
The capacitances at the top nodes 5, 6 and 7 also cause non-dominant poles. They are followed by zeros however, at double the frequency. Indeed, each time we have a single-ended amplifier, the capacitances on the other side than the output, will be negligible for the Phase Margin.

As a conclusion, this OTA has only one single non-dominant pole. It is fairly easy to design. It also has the advantage that it is very fast.

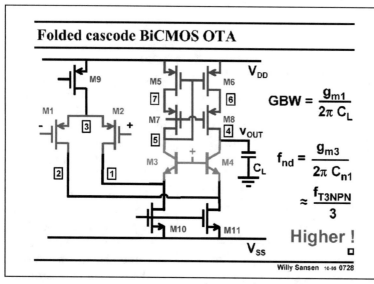

Folded cascode BiCMOS OTA

$$GBW = \frac{g_{m1}}{2\pi \, C_L}$$

$$f_{nd} = \frac{g_{m3}}{2\pi \, C_{n1}}$$

$$\approx \frac{f_{T3NPN}}{3}$$

Higher !

Willy Sansen 10-95 0728

0728
Would a BiCMOS folded OTA be even faster?

The only good position for bipolar transistors to be plugged in, is as cascode devices. Indeed, this is where we need the highest-speed devices. The non-dominant pole at nodes 1 and 2 are linked to the f_T of the cascodes. This is higher than the f_T of the nMOST f_T within a particular BiCMOS technology. Remember that this is not necessarily higher than for nMOSTs in a more recent standard CMOS technology.

Never use bipolar devices for the transistors M10 and M11 in the DC current sources. Their collector-substrate capacitances would reduce the non-dominant pole at nodes 1 and 2 too much!

0729
The second important advantage of a folded cascode OTA is that the input transistors can operate with their Gates beyond the supply lines. The common-mode input voltage range can include one of the supply rails!

In the circuit in this slide, the pMOST devices at the input still operate when the Gates are connected to ground, or even below ground. The V_{GS} values of the input transistors are easily

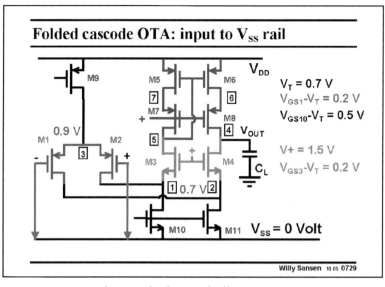

0.9 V (for $V_T = 0.7$ V), which is more than sufficient to accommodate the V_{DS1} and V_{DS10}.

If V_{DS1} is about 0.2 V and $V_{DS10} = 0.5$ V, then the input transistors can still operate with their gates at 0.2 V below ground!

A folded cascode opamp include the ground rail. This is why they have been often used for single-supply systems such as automotive applications before, but now also for all mixed-signal applications, in which the processors use only one single supply line.

Moreover, connecting two folded cascodes in parallel, one with pMOSTs at the input and another one with nMOSTs at the input, allows coverage of the full rail-to-rail range. This is how rail-to-rail input opamps are put together! These are discussed in Chapter 11.

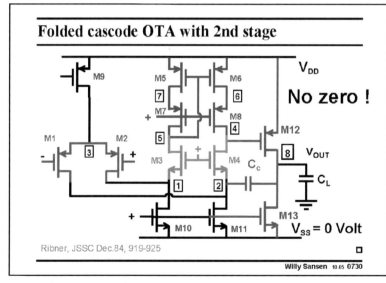

0730

The folded cascode OTA is also an excellent first stage for a two-stage Miller CMOS OTA. As usual the second stage is just one single transistor with active load.

As a result, the GBW is set by g_{m1} and C_c. Now there are two nondominant poles, however. The lower-frequency one is normally at the output. The other, at nodes 1 and 2 are usually at the highest frequency.

Because of the second stage, the output swing can be rail-to-rail. Indeed, even when the output voltage is very close to the positive supply voltage, and the output transistor M12 enters the linear region, and loses its gain, there is still sufficient gain remaining in the first stage to suppress the distortion.

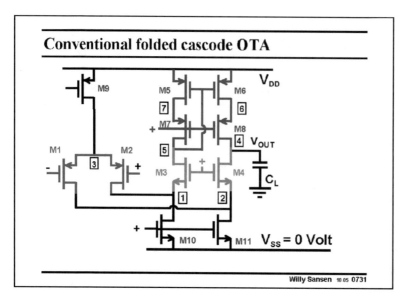

0731

For sake of comparison, the conventional folded cascode OTA is repeated.

The top current mirror M5-M8 can also be realized in a different way, as shown next.

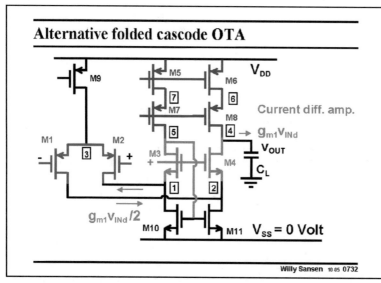

0732

In this alternative folded cascode OTA, the current mirroring is carried out around the cascodes themselves.

Indeed, transistors M3/M4 are also the cascodes in the differential current amplifier formed by M3/M4 and M10/M11. Such an amplifier provides the difference between the input currents, as explained in Chapter 2. These input currents are the same in amplitude but opposite in phase, as they come directly from the input pair.

The output current in the load capacitance is simply $g_{m1}v_{ind}$.

Let us try to discover what the differences are with the previous conventional folded cascode OTA.

Clearly, the gain and output impedance are the same. The number of biasing lines is one less as M5–6 and M9 can share the same Gate line.

The main difference however, is in the impedance seen by the input transistors. In the conventional folded cascode, the input devices see exactly the same impedance. In the alternative configuration, transistor M1 sees $1/g_{m4}$ but transistor M2 sees $1/g_{m3}$ divided by the gain of transistor M3 or $g_{m3}r_{o3}$. This is much smaller!

The alternative folded cascode OTA is therefore a bit less symmetrical. This will be visible in the higher-order poles and zeros, which are of less concern to us!

Comparison amplifiers

	I_{TOT} mA	$\overline{dv_{in,eq}}^2$ $\dfrac{8/3\ kT\ df}{g_{m1}}$	Swing
Volt. OTA (4 Ts)	0.25	4	avg.
Symmetrical (B= 3)	0.33	16	max.
Telescopic	0.25	4	small
Folded casc.	0.5	4	avg.
Miller 2-stage (C_L/C_c= 2.5)	1.1	4	max.

GBW = 100 MHz C_L = 2 pF V_{GS}-V_T = 0.2 V   Fully differential

Willy Sansen 10 05 0733

0733

For sake of comparison, a short table is given listing the main advantages and disadvantages.

The four-transistor single-stage voltage differential amplifier is the first on the list. It is followed by a symmetrical CMOS OTA. Then we have two cascode CMOS OTA's. Finally, a two-stage Miller CMOS OTA is added.

It is clear that the Miller CMOS OTA takes the highest power consumption. The best one is a telescopic cascode.

For high output swing, the telescopic cascode OTA is the worst one. The best are the Miller CMOS OTA and the symmetrical one, at least if no cascodes are used!

The symmetrical OTA is the worst for noise, however.

This shows that even for as few as three specifications, not one single amplifier can be called the best. Many designers prefer a folded cascode OTA, which is certainly a good compromise.

Table of contents

Willy Sansen 10 05 0734

0734

Many more opamp configurations can be added. They are all included in

this list because they have some peculiarity which is worth investigating. We will always try to recognize the design principles which we have studied before. We will try to find out whether one single or two stages are used. We want to know which trick has been used to abolish the positive zero. Also, we want to recognize the structure of a symmetrical or folded cascode, etc.

Some of these are fully-differential. This means that they have two outputs and require common-mode feedback. This will be explained in the next Chapter, however. We focus here on their circuit configuration, without being affected by their differential nature.

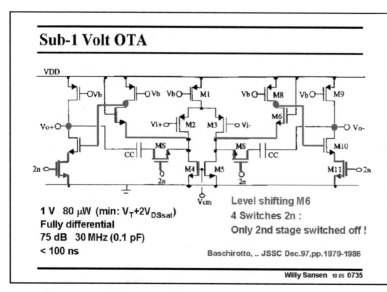

The first one in the list is an OTA which works on a mere 1 V supply voltage.

Moreover, this OTA can be switched in and out. To verify the operation, we must close all four (blue) switches.

Clearly, a two-stage Miller CMOS OTA emerges, with a folded-cascode as a first stage. Because of the low supply voltage, transistor M8 does not have a cascode. The gain will therefore not be that high. A second stage provides a good remedy, however.

Obviously, the compensation capacitance CC does not connect Drain to Gate directly around output transistor M10. It takes a path through cascode device M6, in order to avoid a positive zero.

Finally, note that the common-mode input voltage range is just about zero. Indeed the sum of V_{DS1} and V_{GS3} is about 1 V. The Gates of the input devices can only operate around zero.

On the other hand, the average output voltage will be 0.5 V to maximize the output swing. As a consequence, the output can never be directly connected to the input, to make a buffer for example. A level shifter over 0.5 V will have to be inserted between output and input.

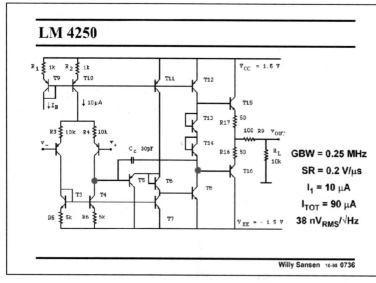

A very conventional two-stage Miller opamp with bipolar transistors is shown in this slide. It can be used for supply voltages down to ±1.5 V.

Each high-impedance point is indicated by means of a red dot. It is clearly a two-stage amplifier with a class AB output stage.

The GBW is obviously determined by the input transconductance and the 30 pF compensation capacitance. Bipolar transistors have sufficient transconductance not to have problems with positive zeros.

With bipolar transistors, an emitter follower is required between input and second stage. This transistor is T5. A level shifter T6 then follows to reduce the voltage to about 0.7 V, the V_{BE} of transistor T8. This level shifter is also required to reach this low supply voltage.

In the input stage, series resistors of 10 kΩ are used to increase the Slew-Rate.

The output stage consists of two emitter followers. As a consequence twice a V_{BE} of about 0.7 V is lost in the output swing. For such large supply voltages, we do not mind so much. For smaller supply voltages or larger output swings we must use two Collector-to-collector output devices, as we have previously seen in most opamps.

Diodes T13/T14 are used to set the quiescent current in the output devices.

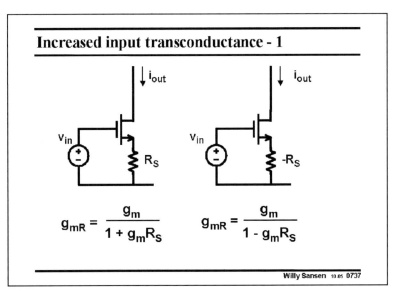

0737
An attempt to raise the transconductance of MOSTs to higher values is shown here.

It starts with inserting a series resistor. This is obviously going to decrease the transconductance.

It can increase the transconsuctance, provided we can make the resistor negative. For example, if we can make $g_m R_S = 0.8$, then g_{mR} will be 5 times g_m.

It is too difficult, however, to match a g_m to a resistor R_S to obtain an accurate value of, for example, 0.8.

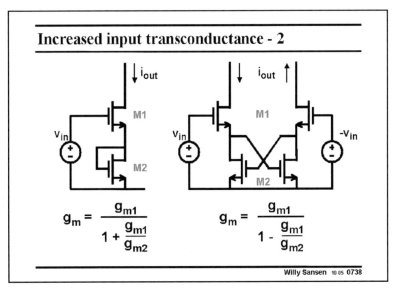

0738
This matching is quite feasible provided we take two nMOST devices, the bottom of which is connected as a diode. Since both carry equal DC currents, the g_m ratio is now the square root of the W/L ratio, or the $V_{GS} - V_T$ ratio, and this can be made quite accurate.

Moreover, a negative resistance is easily realized in differential form, as shown in this slide.

If we can make a layout with a W/L ratio of 0.5, for

example, then the g_m ratio is 0.71 and the transconductance goes up by a factor of about 3. We could push this a bit more of course!

A full OTA realization is shown next.

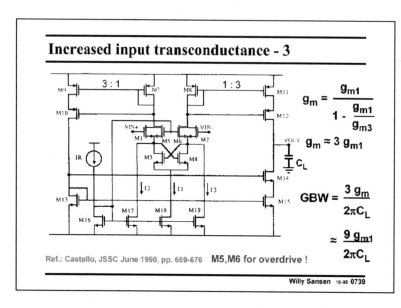

Increased input transconductance - 3

$$g_m = \frac{g_{m1}}{1 - \frac{g_{m1}}{g_{m3}}}$$

$$g_m \approx 3\, g_{m1}$$

$$GBW = \frac{3\, g_m}{2\pi C_L}$$

$$\approx \frac{9\, g_{m1}}{2\pi C_L}$$

Ref.: Castello, JSSC June 1990, pp. 669-676   **M5,M6 for overdrive !**

Willy Sansen 10-95 0739

0739

This is clearly a symmetrical CMOS OTA, with a B factor of three. The input devices have negative series resistances to increase the input transconductance. Two more transistors M5 and M6 are added to avoid latch-up of the input stage. After all, negative resistors are used in oscillators, in comparators, in flip-flops, etc., because they are regenerative. They may cause latch-up when overdriven. Transistors M5 and M6 have to prevent this.

Such an Operational Transconductor Amplifier is also called Transconductor, because it allows larger input voltages with low distortion.

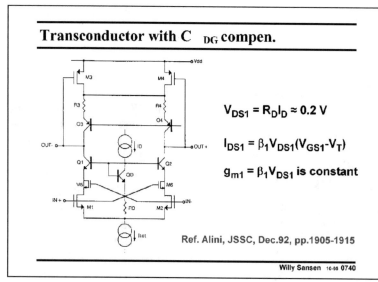

Transconductor with C $_{DG}$ compen.

$$V_{DS1} = R_D I_D \approx 0.2\ V$$

$$I_{DS1} = \beta_1 V_{DS1}(V_{GS1} - V_T)$$

$$g_{m1} = \beta_1 V_{DS1}\ \text{is constant}$$

Ref. Alini, JSSC, Dec.92, pp.1905-1915

Willy Sansen 10-95 0740

0740

Another transconductor with high-speed capability is shown in this slide.

It is little more than a differential single-stage voltage amplifier with cascodes.

However, the input transistors operate in the linear region. This is achieved on purpose to avoid distortion, when driven with large input signal levels.

Indeed, in the linear region the current is proportional to V_{GS}, not to V_{GS}^2. As a result, the transconductance is constant provided V_{DS} can be kept constant.

This is achieved by fixing the voltage across resistor R_D by means of a constant current I_D.

Obviously, the transconductance in the linear region is smaller than in saturation. Lower distortion always goes together with lower gain though! Feedback does this too, exchanging gain for low distortion.

In order to boost the high-frequency performance, two small capacitances are added by means of transistors M5 and M6. They are added to compensate the input capacitances C_{GS} of the input transistors. They are connected to nodes at opposite polarity. For a size of about one third, compensation can be achieved. This is why they are drawn smaller!

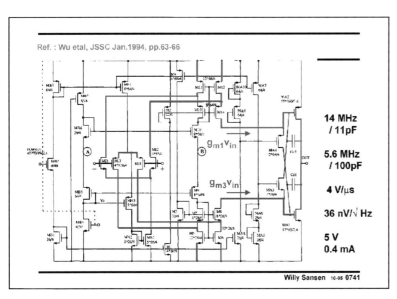

0741

This is the first fully rail-to-rail amplifier that we will discuss, is shown in this slide. It provides rail-to-rail capability at both the input and the output. It can be connected in unity gain as a buffer. It has a class AB output stage to be able to provide large output currents indeed.

At the input two folded cascode stages are connected in parallel. The common-mode input range thus includes both supply lines.

Their outputs are applied to two differential current amplifiers, ending up at the Gates of the two large output devices. These devices are the output stage.

We have a two-stage Miller opamp. The compensation capacitances are clearly distinguished. However, they connect directly Drain to Gate. Perhaps it is better to find a path through one of the cascodes. For example C_{c2} may be better connected to the source of M14!

The Gates of the output transistors are at very high impedance. It may not appear like that because these nodes are also connected to two Sources of transistors MA3 and MA4. Sources suggest impedance levels of $1/g_m$. This is not the case here however, as these two transistors are bootstrapped out. This is explained later.

The rail-to-rail input stage can cause large variations in GBW, however. This is examined first.

0742

Indeed, for common-mode or average input signals in the middle, both input stages are in operation. The total transconductance is now the sum of the transconductances of the nMOSTs and pMOSTs at the input.

For higher common-mode input signals however, the pMOSTs are shut off. The transconductance is now only half. The same is true for low common-mode input signals. The total transconductance has a bell shape versus the input common-mode input voltage, as shown in this slide, and so does the GBW.

This gives a lot of distortion.

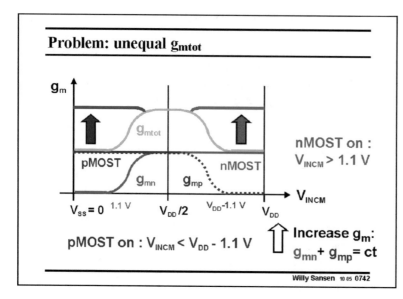

Some circuitry has to be added to equalize the transconductance over the full common-mode input range. In other words, for low common-mode input voltages, we need to double the transconductance of the pMOSTs; for high voltages, we need to double the transconductances of the nMOSTs.

Various circuits can be devised to carry out such a task. One of these is explained next. The others are discussed in Chapter 11.

0743

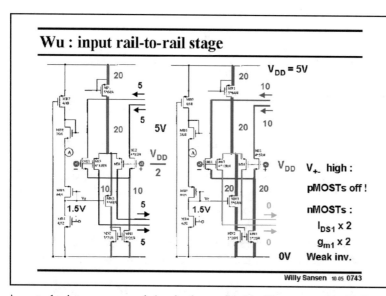

The input stage of the rail-to-rail opamp is repeated in this slide.

The current which flows in a branch is indicated by the thickness of line. The input circuit is repeated twice, once for a common-mode input voltage halfway the supply voltages. The other is the situation when the inputs are connected to the positive supply.

When the inputs are halfway the supply voltage, the DC currents through all input devices are equal (and about 5 μA). However the DC current source for the pMOST differential pair carries a current of 20 μA. Indeed, half of this current flows through a cascode MN3 to a current mirror which serves the nMOST differential pair.

When the input voltage goes up, towards the positive supply voltage, then the V_{GS} of this cascode transistor MN3 is increased, such that it takes the full 20 μA. As a result, the nMOST differential pair receives the full 20 μA. On the other hand, the pMOST differential pair is left without DC current.

As a consequence, the current in the nMOST pair is multiplied by 2. This is not sufficient if the input devices work in saturation. This is sufficient if the input devices work in the weak inversion region. Doubling the current then doubles the transconductance.

The sizes of the input devices are so large that the input devices are more likely going to work in weak inversion. After all, the GBW is only 14 MHz, which is quite feasible in weak inversion.

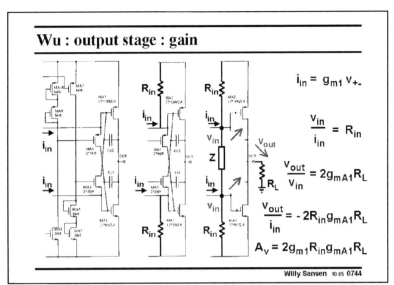

Wu : output stage : gain

$$i_{in} = g_{m1} v_{+-}$$

$$\frac{v_{in}}{i_{in}} = R_{in}$$

$$\frac{v_{out}}{v_{in}} = 2g_{mA1}R_L$$

$$\frac{v_{out}}{i_{in}} = -2R_{in}g_{mA1}R_L$$

$$A_v = 2g_{m1}R_{in}g_{mA1}R_L$$

Willy Sansen 10 05 0744

0744

For high gain, the output impedance of the first stage or the Gates of the output transistors have to be at very high impedance. It may not appear like that because these nodes are connected to two Sources, of transistors MA3 and MA4. Sources suggest impedance levels of $1/g_m$. This is not the case here however, as shown next.

The output stage is repeated three times. The first one is simply copied from the overall circuit diagram. In the second one, the output resistance of the first stage is represented by R_{in}. In the third one, the two transistors in parallel are substituted by an impedance called Z.

It is now easy to calculate the gain of this amplifier. The input stage provides a conversion of g_{m1}. The total gain also includes the transconductance of the output devices g_{mA1}. Impedance Z is not part of it. The reason is that the impedance Z is bootstrapped out.

We see on the third diagram, that the currents coming from the input stage have the same phase and therefore drive the output transistors with the same phase as well. This is typical for a class-AB stage. Both transistors have to be driven in phase to turn one output transistor and the other one off.

As a result, the voltages at the Gates of the output devices are nearly the same in amplitude. No AC voltage appears across the impedance Z. It does not carry any AC current. It looks like an infinite impedance and it is bootstrapped out.

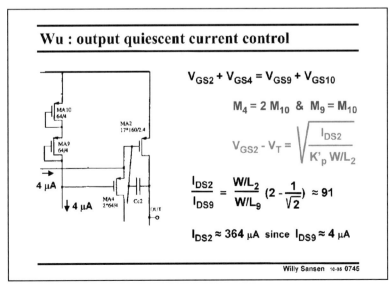

Wu : output quiescent current control

$$V_{GS2} + V_{GS4} = V_{GS9} + V_{GS10}$$

$$M_4 = 2 M_{10} \quad \& \quad M_9 = M_{10}$$

$$V_{GS2} - V_T = \sqrt{\frac{I_{DS2}}{K'_p \, W/L_2}}$$

$$\frac{I_{DS2}}{I_{DS9}} = \frac{W/L_2}{W/L_9} \left(2 - \frac{1}{\sqrt{2}}\right) \approx 91$$

$$I_{DS2} \approx 364 \, \mu A \text{ since } I_{DS9} \approx 4 \, \mu A$$

Willy Sansen 10-05 0745

0745

What is the purpose then of these two transistors MA3 and MA4 when they do not play a role for the gain?

They are there to set the quiescent current in the output transistors.

The output transistor MA2 forms with transistor MA4 a translinear loop with transistors MA9 and MA10. The sum of their V_{GS}'s are the same as spelled out in this slide.

The DC currents in three of these four devices are constant and set by DC current sources to be 4 –5 µA. As a result, the DC current through the forth one (MA2) is also set to be constant.

All four transistors have the same V_T and K'_p. The ratio of the DC current through output transistor MA2 is now about 100 times the current in MA4.

This is an easy way of controling the DC current in a class-AB stage, as will be explained in more detail in Chapter 12.

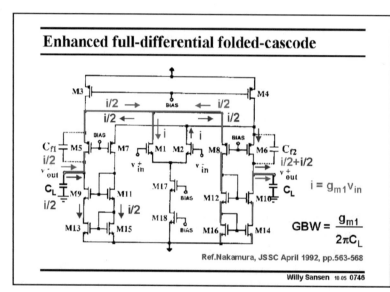

0746

An even more symmetrical folded cascode OTA than the folded cascode OTA is shown in this slide.

The Drain of each input transistor sees exactly the same impedance, even at the highest frequencies.

At each Drain the circular current i is split up in two equal parts i/2. One part goes directly to the output, whereas the other is mirrored first.

Because of this perfect matching at high frequencies, this amplifier has a perfect cancellation of higher order poles and zeros. As a result, it has a much higher CMRR. Moreover, the Slew Rate is perfectly symmetrical.

It is therefore an ideal building block for higher frequencies.

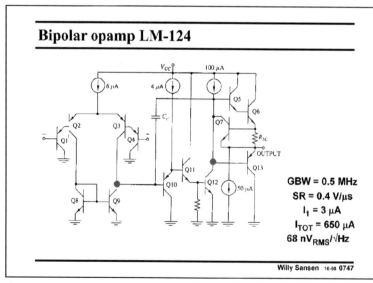

0747

Another bipolar opamp is also capable of taking input voltages below ground. Remember that this is an advantage of a folded cascode.

This input differential pair is preceded by two Emitter followers. As a result the inputs can go below ground. For V_{BE}'s of 0.6 V, and V_{CE}'s of 0.1 V, we see that the inputs can go 0.5 V below ground.

This is an opamp which is suited very well for single-supply applications such as most of the automotive applications, etc.

Moreover, this opamp takes very little power. As a consequence, the GBW is fairly low.

The noise is very high, however. The reasons are that the currents are small but especially that emitter followers are used at the input. They do not give voltage gain. As a result, all six transistors of the input stage contribute to the equivalent input noise.

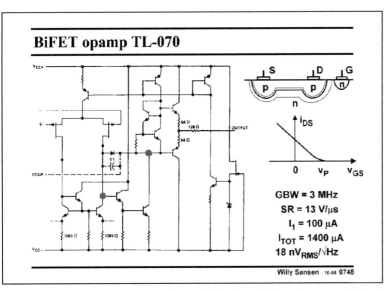

0748

A two-stage bipolar opamp with JFETs at the input is shown in this slide.

JFETs behave as MOSTs but with larger input currents. Actually, their input currents are leakage currents, because of the reverse biased input pn junctions. They are much smaller than for bipolar transistors though. Also, their threshold voltages are negative. They are depletion devices rather than enhancement devices such as MOSTs.

They conduct at zero V_{GS}. Also, their threshold voltage, called pinch-off voltage V_P, is usually several Volts.

These p-channel JFETs substitute the pnp transistors which were originally in this circuit. After all, this is just a two-stage operational amplifier with Miller compensation. With bipolar transistors at the input however, the Slew Rate is too small. JFETs have been used instead to increase the Slew Rate.

They also give very little 1/f noise, which is an additional advantage for low-frequency circuits such as high-performance audio amplifiers.

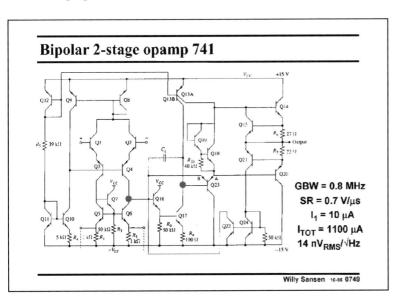

0749

This is a two-stage opamp which has been the workhorse for all discrete analog electronics over decades of years.

The only difference with any two-stage Miller compensated opamp is the input stage.

Lateral pnp transistors have a low beta and cannot be used as input transistors. On the other hand, we definitely want to use high-speed npn transistors in the second stage to shift the non-dominant pole to high frequencies. As a result, the current mirror in the input stage must be realized by means of npn devices as well.

This is why non transistors are used at the input. They give small input base currents. These input npn's are now put in series with lateral pnp's, to be able to drive the npn current mirror.

Since all input transistors carry the same current, they all have the same transconductance. The input transconductance is now reduced by two to $g_{m1}/2$. This is only a small loss.

The pnp transistors in the input stage are biased by a common-mode feedback loop. Indeed, this loop is closed over the input devices and the current mirror Q8/Q9. This loop desensitizes the DC currents of the input devices from the pnp beta's.

However, the performance of this bipolar opamp is rather moderate.

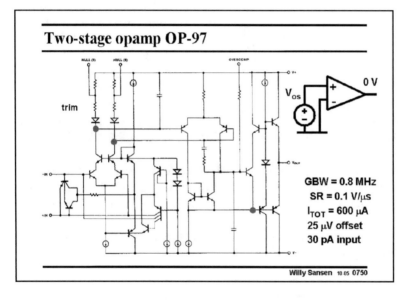

0750

A high performance bipolar opamp is shown in this slide.

It is again a two-stage opamp as suggested by the red dots. Its GBW is moderate but its gain is very high. Its offset is trimmed to very low values.

This is achieved by using resistive loads in the first stage. These resistors can be trimmed by laser or other techniques, to very small values, improving the Common-mode Rejection Ratio considerably (see Chapter 15).

The resistors in the input stage are not as good as active loads, however. They lead to lower gain. This is why a second stage is used with very high gain.

This second stage consists of a differential voltage amplifier, to which an emitter follower has been added, as explained next.

0751

The second stage of this amplifier is taken separately. It consists indeed of a differential voltage amplifier, to which an Emitter follower has been added, as shown in this slide.

This Emitter follower M3 bootstraps out the output resistor r_{o2} of transistor M2. As a result, only the output resistance of the input pnp plays a role for the gain. They are lateral devices in which the output resistance can be made as large as needed.

Moreover, the output impedance R_{out} will also be smaller.

Bootstrap for high gain A_{v2}

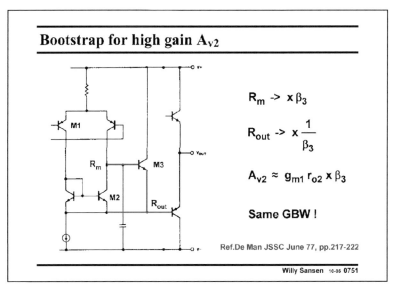

$R_m \rightarrow \times \beta_3$

$R_{out} \rightarrow \times \dfrac{1}{\beta_3}$

$A_{v2} \approx g_{m1}\, r_{o2} \times \beta_3$

Same GBW !

Ref.De Man JSSC June 77, pp.217-222

Willy Sansen 10-05 0751

An accurate analysis shows that the gain is actually multiplied by the beta β_3 of transistor M3.

This is an attractive technique to boost the gain. Since the gain becomes smaller for smaller channel lengths, all possible gain-boosting techniques will become necessary. Bootstrapping resistances to high values is certainly among them.

Table of contents

Willy Sansen 10.05 0752

0752

In this Chapter, a wide variety of possible operational amplifiers have been discussed. Most design effort has gone to the symmetrical amplifier and the folded cascode.

However, most of them have a single-ended output. They cannot be used in a mixed-signal environment. For this purpose, they must be fully differential. They must have two outputs, as introduced in the next Chapter.

081

Fully-differential amplifiers

Willy Sansen

KULeuven, ESAT-MICAS

Leuven, Belgium

willy.sansen@esat.kuleuven.be

Willy Sansen 10-05 081

Fully-differential circuits have two differential outputs. They need them to be able to reject the common-mode disturbances generated by the digital circuits, the class-AB drivers, the clock drivers, etc.

As a consequence, all mixed-signal circuitry requires the amplifiers to be fully-differential. However, his will cost a lot of additional power consumption. Consequently, an extra amplifier will be required to stabilize the average or common-mode output level. It is called the common-mode feedback (CMFB) amplifier. It obviously takes additional current.

One of the most important specifications will therefore be at which extra current an amplifier can be made fully-differential.

082

Table of contents

- **Requirements**
- **Fully-diff. amps with linear MOSTs**
- **FDA's with error amp.& source followers**
- **Folded cascode OTA without SF's**
- **Other fully-differential amps**
- **Exercise**

Willy Sansen 10-05 082

Besides the power consumption, some other specifications may come in, which are typical for CMFB amplifiers. For example, the input range is an important characteristic.

All requirements of CMFB amplifiers will be reviewed first.

They are followed by a discussion of the three most important types of CMFB amplifiers. All of them have advantages and disadvantages. None of them provides an ideal solution.

After this discussion, a number of practical realizations are examined. Their compromises are discussed.

Finally, an exercise is launched to generate better insight. It includes the comparison of a CMOS with a BiCMOS solution.

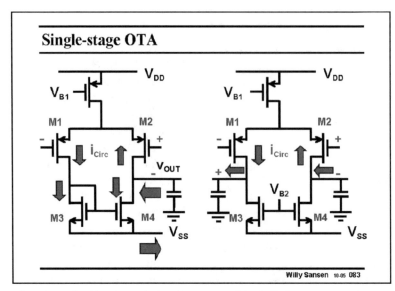

083
The simplest fully-differential amplifier is certainly this single-stage OTA, shown on the right. It is very similar to the differential voltage amplifier discussed in Chapter 2, shown on the left. However, the current mirror is substituted by two DC current sources.

The circular current, generated by the differential input voltage is indicated with arrows.

It is clear that this fully-differential OTA is even simpler than the single-ended voltage amplifier. Only two transistors participate in the small-signal operation. It can therefore reach higher frequencies!

On the other hand, it is also clear that this amplifier has a biasing problem. Both biasing voltages V_{B1} and V_{B2} try to set the DC currents, which is one too many!

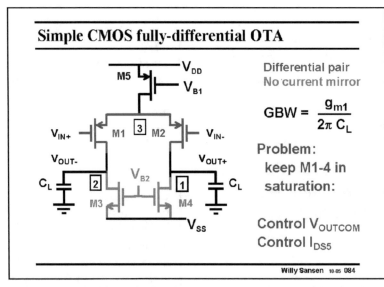

084
The biasing voltages V_{B1} and V_{B2} have to be such that all transistors are in the saturation region. Otherwise they would exhibit a small output resistance which would deteriorate the gain.

The problem of the two biasing voltages is that they have to be matched to such a degree that the average output voltages are somewhere halfway between the supply voltages, to keep all transistors in saturation, even for a large output swing.

For example, if V_{B1} is fixed, then a value of V_{B2}, which is 20 mV too high would reduce both output voltages by 1 V (if the gain of the nMOSTs is 50). Even worse, when the V_{B2} is larger, the average output voltages are so low that the nMOSTs M3/M4 end up in the linear region, killing the gain!

We have the same problem when biasing voltage V_{B2} is too low. Now the average output voltages are too high and the pMOSTs M1/M2 end up in the linear region, killing the gain as well!

This kind of matching is impossible to realize. This is why we need an additional amplifier to tune V_{B2} to the required average or common-mode output voltages. This amplifier only works on common-mode signals. It is called the common-mode feedback amplifier.

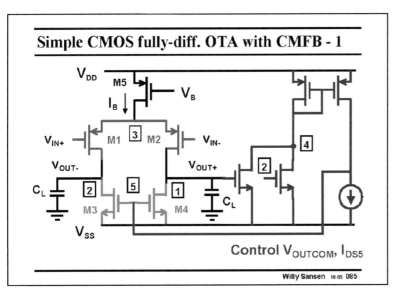

085

One example of such a CMFB amplifier is shown in this slide.

Both output voltages are measured. Since we only want feedback on the common-mode signals, we have to cancel out the differential signals. This is done at node 4.

Now we have to close the loop with an amplifier, and feed it to a common-mode point. Any biasing point in the circuit can be used for that. For this amplifier it is node 5.

Clearly, part of the circuit belongs to both the common-mode and the differential amplifier. For example, transistors M3 and M4 are DC current sources for the differential signals, but single-transistor amplifiers for the common-mode signals.

Also, the CMFB amplifier is always connected in unity-gain feedback. Nodes 1 and 2 are at the same time the input and output of the CMFB amplifier. It may thus require more power to ensure stability.

The differential amplifier is evidently shown without feedback.

Biasing voltage V_B is the independent biasing voltage. This could well have been the Gates of the NMOSTs M3/M4, as shown next.

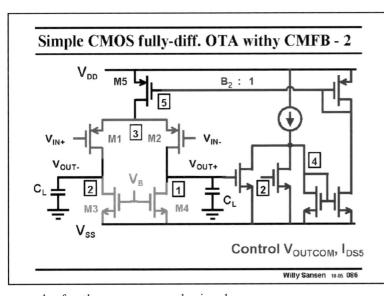

086

Another CMFB amplifier is shown here. It now closes the feedback loop to the top current source. This is indeed good!

Again, the output voltages are measured. The differential signals are cancelled out and the CMFB loop is closed by means of an amplifier.

Now transistors M1 and M2 are common to both amplifiers. They function as a (differential) amplifier for differential signals, but as cascodes for the common-mode signals.

In order to gain somewhat more insight in this CMFB amplifier it is sketched separately in the next slide.

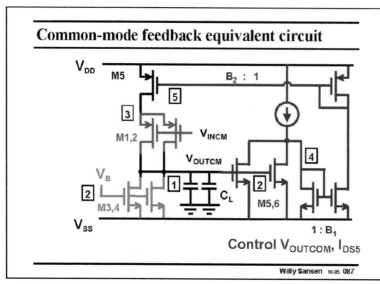

Common-mode feedback equivalent circuit

Willy Sansen 10-05 087

087

The common-mode equivalent circuit is easily found by putting all differential devices in parallel and connecting them to the common-mode input signals.

It is clear that node 1 is at the same time the input and the output of the

CMFB amplifier. This is also the circuit that will be used, to derive the common-mode gain, bandwidth and GBW_{CM}.

Actually, the open-loop gain is $B_1 B_2 g_{m5} R_{n1}$, in which B_1 and B_2 are the current gain factors of the two current mirrors. This gain is not so high but only a small amount of gain is needed. The stabilization of the common-mode output voltage does not need to be so accurate. The outputs will both be at $V_{GS5,6}$ above the negative supply. For large swing, we increase the size of these V_{GS}'s.

The GBW_{CM} will evidently be given by the $B_1 B_2 g_{m5}/(2\pi C_L)$. We have two input transistors M5 and M6 but also two load capacitors. The GBW_{CM} can therefore be made quite high, at the cost of a lot of power consumption though!

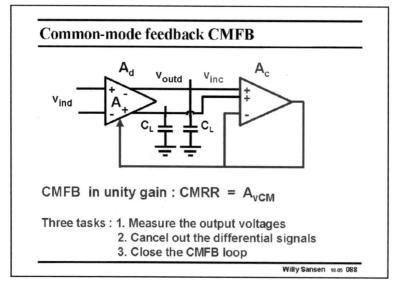

Common-mode feedback CMFB

CMFB in unity gain : CMRR = A_{vCM}

Three tasks : 1. Measure the output voltages
2. Cancel out the differential signals
3. Close the CMFB loop

Willy Sansen 10-05 088

088

As a summary, let us repeat what the three tasks are of a CMFB amplifier.

They have to measure the output voltages, cancel the differential signals and close the feedback loop.

Also, the CMFB amplifier always operates in unity-gain.

Finally, it is important to note that the gain of the CMFB amplifier is used to increase the Common-mode rejection ratio (as shown in Chapter 15).

Requirements fully-differential amplifiers

- High speed : $GBW_{CM} > GBW_{DM}$

- Matching

- Output swing limited by :

  - Output swing of differential-mode amp

  - Input range of common-mode amp

- Low power $P_{CM} < P_{DM}$

Willy Sansen 10-05 089

089

Let us now look at the main requirements of CMFB amplifiers..

The first one requires the common-mode GBW_{CM} to be higher than the differential GBW_{DM}. However, this depends on the application.

Indeed, if the common-mode amplifier were slow, only providing DC biasing, then a high-speed spike on the supply line or the substrate would throw the input devices or the active loads in the linear region. Slow common-mode feedback would then take too much time to restore the biasing in the input stage. During all this time, the high-speed differential amplifier would be out of operation. This why this specification comes first.

In some specific circuits such as some sigma-delta converters, high speed amplifiers are only used in the low-frequency region. In this case this specification can be relaxed considerably!

Requiring the GBW_{CM} to be as large as the GBW_{DM} will require a lot of power, directly conflicting with the last specification. We will see that there is no easy way of avoiding this compromise. In principle, a fully-differential amplifier simply doubles the power consumption.

Finally, the output swing is also a problem. It is limited by both the output swing of the differential amplifier and by the common-mode input range of the CMFB amplifier (whichever is smaller).

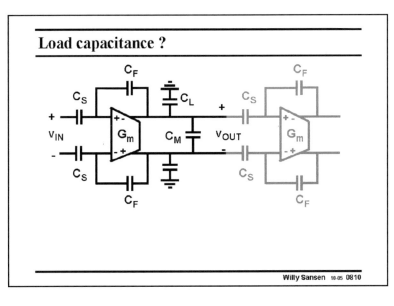

Load capacitance ?

Willy Sansen 10-05 0810

0810

The differential and common-mode amplifiers don't necessarily have the same load capacitances.

A situation is sketched in this slide where we have two consecutive fully-differential amplifiers. Both are using differential feedback to set the gain and the bandwidth.

The capacitances C_L are parasitic capacitances to ground. They obviously depend on the length and nature of the interconnect. Capacitance C_M is a mutual capacitance.

Now both the differential and common-mode load capacitances are derived.

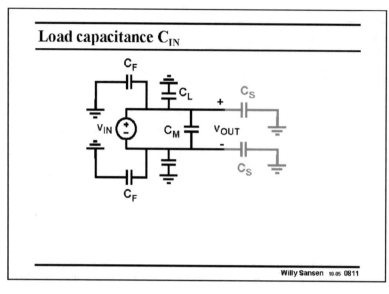

0811

All the capacitances which are present at the output terminals of one amplifier are sketched in this slide. The input capacitances of the next amplifier are included as well.

The virtual grounds are taken as real grounds.

An input voltage is added to measure the total capacitance at the outputs. It will be a differential input voltage to find the differential load capacitance, but a common-mode input voltage for the load capacitance of the CMFB amplifier.

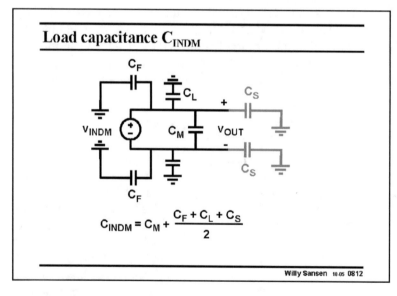

0812

For differential operation, the differential input voltage sees a load capacitance C_{INDM}, as indicated in this slide.

The differential load capacitance is quite small. It only contains the mutual capacitance and half of all the others.

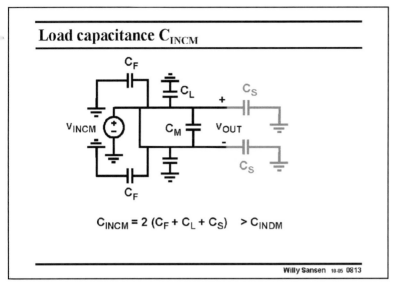

Load capacitance C$_{INCM}$

$$C_{INCM} = 2 (C_F + C_L + C_S) \quad > C_{INDM}$$

Willy Sansen 10-05 0813

0813

For common-mode operation, the common-mode input voltage sees a load capacitance C$_{INCM}$, which is a lot larger.

The feedback capacitors C$_F$ and also the sampling capacitors C$_S$ are all doubled.

The CMFB amplifier has to drive larger load capacitances than the differential one. Moreover, it is always connected in unity gain, for which stability is harder to achieve. These are two reasons for trying to reduce the power consumption as much as possible.

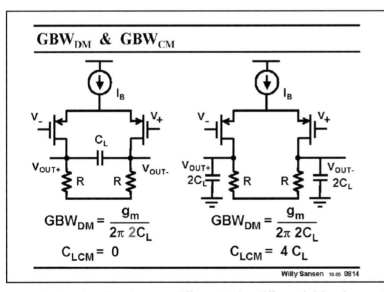

GBW$_{DM}$ & GBW$_{CM}$

$$GBW_{DM} = \frac{g_m}{2\pi\, 2C_L}$$

$$C_{LCM} = 0$$

$$GBW_{DM} = \frac{g_m}{2\pi\, 2C_L}$$

$$C_{LCM} = 4\, C_L$$

Willy Sansen 10-05 0814

0814

Even with the simplest, single-stage, fully-differential amplifier, we have to be careful with the definition of the load capacitance. After all, this load capacitance determines the GBW!

With a floating capacitance (at the left) we have to include this capacitance twice in the differential GBW$_{DM}$. Moreover there is no common-mode load capacitance! Its GBW$_{CM}$ would therefore be infinity.

With two capacitors to ground, the situation is very different. The differential load capacitance is smaller; it is C$_L$ itself. The common-mode load capacitance is not at zero. It is twice C$_L$!

Table of contents

Willy Sansen   10-05 0815

0815

Now that we know how to derive the specifications of a fully-differential amplifier, let us have a look at the three most used ones. The first one uses MOSTs in the linear region for the common-mode feedback.

CMFB amplifier with linear MOSTs

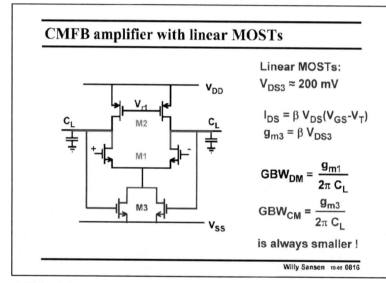

Linear MOSTs:
$V_{DS3} \approx 200$ mV

$I_{DS} = \beta\, V_{DS}(V_{GS}-V_T)$
$g_{m3} = \beta\, V_{DS3}$

$$GBW_{DM} = \frac{g_{m1}}{2\pi\, C_L}$$

$$GBW_{CM} = \frac{g_{m3}}{2\pi\, C_L}$$

is always smaller !

Willy Sansen   10-05 0816

0816

This is probably the simplest possible CMFB amplifier.

It consists of a differential pair, the current source of which consists of two transistors M3 in the linear region, with a $V_{DS3} < V_{GS3} - V_T$.

The three functions of the CMFB amplifier are clearly distinguished.

The output voltages are measured by the two transistors M3. Their Drains are connected to cancel the differential signal and the feedback loop is closed.

Transistors M3 are thus the input devices of the CMFB amplifier. This is why their g_{m3} is in the expression of the GBW_{CM}.

The input transistors M1 of the differential pair function as cascodes for the CMFB.

It is clear that in the linear region, the transconductances are much smaller than in the saturation region. The common-mode GBW_{CM} is therefore smaller than the GBW_{DM}. This is a disadvantage!

Why do the transistors M3 operate in the linear region?

There are two reasons. First of all, to have an output voltage in the middle, we need a large V_{GS3}. So as not to loose a large voltage drop, we need a small V_{DS3}. Clearly M3 must be in the linear region.

The other reason is linearity. We need a linear cancellation of the differential signal to avoid the reduction of the differential gain because of the feedback. Transistors in the linear region are very linear indeed!

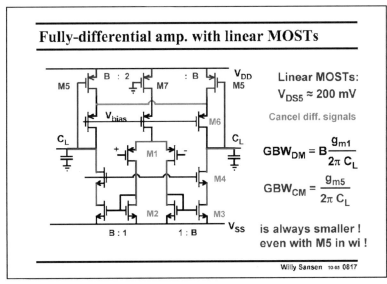

0817

Another fully-differential amplifier with CMFB is shown in this slide.

The differential amplifier is a symmetrical amplifier, whereas the CMFB amplifier is the same as before. It uses transistors M5 in the linear region.

Again, the outputs are measured. The differential signal is cancelled out with the (green) line and the loop is closed.

Transistors M6 are cascodes in both amplifiers.

Note that the independent biasing is taken care of by transistor M7 in the middle. It has a large V_{GS7} (half the total supply voltage) and a small V_{DS7}.

By matching the transistors M5 to M7, the output voltages will be around zero. Assume then, that we have a B factor of three. Transistor M5 is then 50% larger than transistor M7. Its current is also 50% larger than in transistor M7. Their V_{DS} voltages are the same because of cascodes M6. Their V_{GS} values must also be the same. Since the Gate of M7 is connected to ground, the output voltages must also be around ground.

Moreover, the output voltages are better defined (by matching) than in the previous circuit, where they depend on transistor sizes.

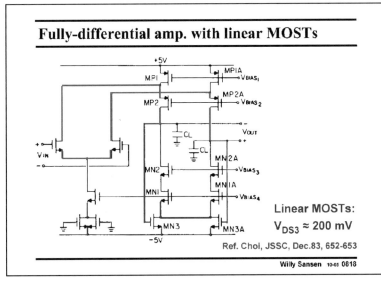

0818

The same CMFB amplifier can be applied to a folded cascode stage as well.

Now we have a folded cascode OTA for the differential operation. The same CMFB amplifier is applied as before.

Transistors MN3 operate in the linear region. They carry out the three functions again.

The output voltages will be around 0 V because the transistors MN3 are matched to the two transistors in the current source at the input. Their Gates are connected to ground. The currents in all branches are the same. The V_{GS} values are also the same. The output voltages are therefore around ground.

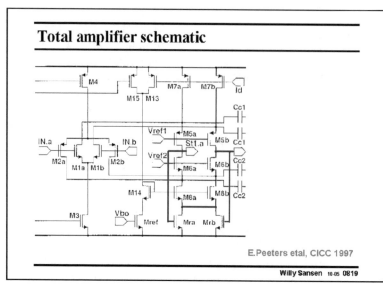

Total amplifier schematic

E.Peeters etal, CICC 1997

Willy Sansen 10-05 0819

0819

A fully-differential rail-to-rail amplifier is shown in this slide. It consists of two folded cascodes in parallel.

The CMFB amplifier is as before. Transistors Mra and Mrb are in the linear region and define the GBW_{CM}. This is smaller than the GBW_{DM}, which may be a disadvantage.

This CMFB amplifier has the same advantage as before. The average output voltage is set by matching. However, the transistors Mra/Mrb are matched to transistors Mref. Its Gate voltage Vbo sets the average voltage of the outputs.

The independent biasing is carried out by the transistors M3/M4 and M15. All currents are set by these sources.

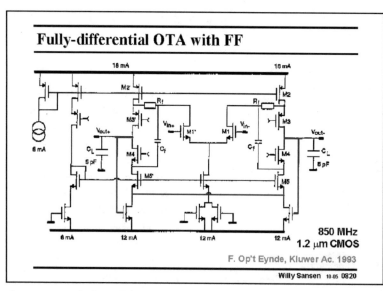

Fully-differential OTA with FF

850 MHz
1.2 μm CMOS

F. Op't Eynde, Kluwer Ac. 1993

Willy Sansen 10-05 0820

0820

Another example of CMFB with MOSTs in the linear region is shown in this slide.

It is a high-speed amplifier. A GBW of 850 MHz in two 5 pF capacitors can be reached thanks to the large currents, despite the old 1.2 μm CMOS technology.

It is a folded cascode for a differential operation. The only additional feature is the feedforward around the slower pMOST cascodes through capacitors C_f.

The CMFB amplifier is the same as before. The outputs are around 0 V, because the Gates of the nMOSTs providing DC current to the input pair are at zero ground.

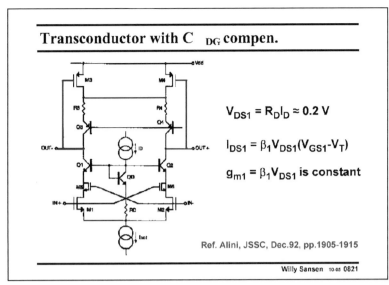

Transconductor with C$_{DG}$ compen.

$$V_{DS1} = R_D I_D \approx 0.2\ V$$

$$I_{DS1} = \beta_1 V_{DS1}(V_{GS1}\text{-}V_T)$$

$$g_{m1} = \beta_1 V_{DS1}\ \text{is constant}$$

Ref. Alini, JSSC, Dec.92, pp.1905-1915

Willy Sansen 10-05 0821

0821

As a last example of this type of common-mode feedback, a simple single-stage voltage amplifier is repeated in this slide.

It is however, a transconductor for differential operation. The input devices operate in the linear region to avoid distortion.

The CMFB amplifier also consists of devices in the linear region, in order to provide accurate cancellation of the differential signal. The loop is closed over pnp transistors with some Emitter degeneration.

This CMFB amplifier has one disadvantage however, which is also present for the very first single-stage voltage amplifier discussed on slide 16 of this Chapter. The average output voltage is not so well defined.

The average or common-mode output voltage is set by the V_{GS} values of the top transistors M3 and M4. These values will depend on the currents imposed by current source I_{tot}, the sizes of transistors M3/M4 and their KP values. They are therefore not that accurate.

Whether this is a problem depends on the next circuit, which inherits its DC biasing from this stage.

Table of contents

Willy Sansen 10-05 0822

0822

The main advantage of this first principle of CMFB is that no additional power is required. The main disadvantage is that this CMFB amplifier may not be sufficiently fast, depending on the application.

This second principle has exactly the opposite characteristics. It takes two more stages, resulting in much more power consumption. The speed however, can be set independently of the differential amplifier, and can be made as large as the designer likes.

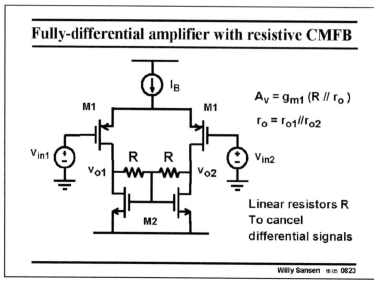

0823

This fully-differential amplifier uses two equal resistors R to cancel out the differential signals. In this way a common-mode biasing voltage is obtained for the Gates of transistors M2.

Consequently, there is no biasing problem. Current source I_B determines all currents.

The main disadvantage is that the resistors R must be increased in size to ensure a lot of gain. An easy solution is to insert source followers between the Drains of the input transistors M1 and the two resistors R, as shown next. Smaller values of R are then possible.

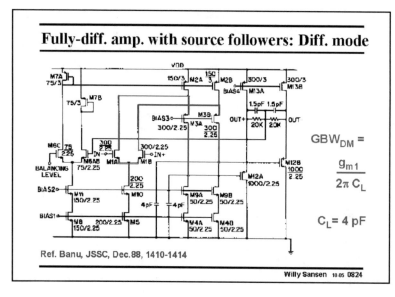

0824

The circuit schematic is shown in this slide.

The differential amplifier is just a folded cascode OTA, loaded with 4 pF capacitors.

Its GBW_{DM} is then readily obtained.

Two more source followers are added at the outputs, which are required for the CMFB, but which can be used as outputs for differential operation as well.

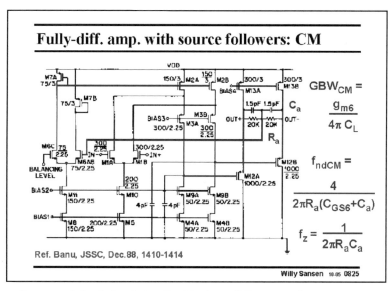

Fully-diff. amp. with source followers: CM

Ref. Banu, JSSC, Dec.88, 1410-1414

Willy Sansen 10-05 0825

$$GBW_{CM} = \frac{g_{m6}}{4\pi C_L}$$

$$f_{ndCM} = \frac{4}{2\pi R_a(C_{GS6}+C_a)}$$

$$f_z = \frac{1}{2\pi R_a C_a}$$

0825

The CMFB amplifier is highlighted on this slide.

It takes two more stages to realize this feedback loop. The first one consists of two source followers to provide low-impedance outputs. They are needed to be able to connect two resistors R_a to the outputs, to provide accurate cancellation of the differential signals.

The other stage on the left, is the error amplifier. It compares the average output voltage to a reference voltage, and feeds back to the biasing of the cascode stage.

It is clear that a lot of power is required to bias these two extra stages. Minimization of the power consumption is of the utmost importance.

The GBW_{CM} is easily obtained, as transistors M6 are the input devices. The source followers only have a gain of unity. A factor of two is lost because only one output is taken of the error amplifier. Since the GBW_{CM} is set independently of the differential input, it can be set at any value larger than the GBW_{DM}.

Obviously, we are very worried about the stability of this multistage CMFB amplifier. The nodes with non-dominant poles are the outputs, the Gate of M6AB and the Gates of current mirror M2/M7.The most important one is probably at the Gate of M6AB, depending on the choice of R_a. At this node the resistors R_a are in parallel, and the input capacitances C_{GS6} in series, which gives twice a factor of two in the f_{ndCM}. A zero f_z is introduced with C_a to compensate this f_{ndCM}.

Table of contents

- Requirements
- Fully-diff. amps with linear MOSTs
- FDA's with error amp.& source followers
- Folded cascode OTA without SF's
- Other fully-differential amps
- Exercise

Willy Sansen 10-05 0826

0826

The source followers in the previous realization take a lot of power. This is why they are left out in this third type of CMFB.

This one will take less power. However, some other disadvantage will show up. The output swing will be limited by the CMFB amplifier, not by the differential one!

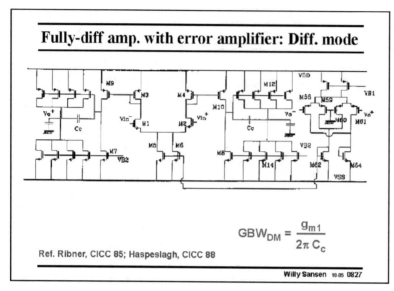

Fully-diff amp. with error amplifier: Diff. mode

$$GBW_{DM} = \frac{g_{m1}}{2\pi\,C_c}$$

Ref. Ribner, CICC 85; Haspeslagh, CICC 88

Willy Sansen 10-05 0827

0827

The error amplifier for the CMFB is now on the right. The differential amplifier is just a conventional two-stage Miller amplifier with a symmetrical amplifier as an input stage.

The GBW_{DM} is therefore easily obtained.

The biasing of the input pair is provided by the CMFB as shown next.

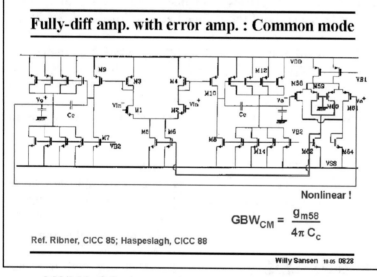

Fully-diff amp. with error amp. : Common mode

Nonlinear !

$$GBW_{CM} = \frac{g_{m58}}{4\pi\,C_c}$$

Ref. Ribner, CICC 85; Haspeslagh, CICC 88

Willy Sansen 10-05 0828

0828

The error amplifier consists of two differential pairs M58–61, each of them being connected to an output of the differential amplifier. They compare the outputs directly to ground, which is obviously halfway the supply voltages.

The average output voltage is well defined.

The cancellation of the differential signals is now carried out at the outputs of these two differential pairs and fed back to the current source M5/M6 of the input pair.

The GBW_{CM} is thus determined by the common-mode pairs M58–61, as given in this slide. Again a factor of two is lost because only one output is taken of these pairs.

It can be set at any value though, higher than the GBW_{CM} if required.

The main advantage of this CMFB configuration is that it takes less power and yet provides a wide-band CMFB amplifier. However, the only non-dominant pole added is at the gates of M5/M6.

The main disadvantage of this solution is that the output swing is limited by the common-mode input range of the CMFB amplifier. The differential pair M58/M59 limits the range to about $2.8(V_{GS}-V_T)$. The output swing of the differential amplifier is rail-to-rail.

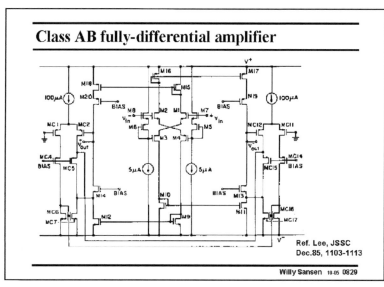

Class AB fully-differential amplifier

Ref. Lee, JSSC
Dec.85, 1103-1113

Willy Sansen 10-05 0829

0829

An example of such common-mode feedback is the amplifier shown in this slide.

The input stage consists of class AB differential pairs and will be discussed in Chapter 12. Their outputs are current mirrored (M15/M18, M10/M11, M16/M17, M9/M12) to the outputs over cascode transistors. It is thus a symmetrical OTA for a differential operation.

The common-feedback amplifier consists of two differential pairs on either side. Two connections are used to cancel the differential signal. The outputs are then fed back to the sources of cascodes M13 and M14 to the outputs.

It is clear that a lot of current is consumed in the CMFB amplifiers. Moreover, the output swings are limited by input ranges of the CMFB differential pairs, as explained in this slide.

Comparison

Criterion	Linear MOST	Error amp Source foll.	Error amp. Quad amp.
GBW$_{CM}$/GBW$_{DM}$	< 0.1	> 1	> 1
Required tol.	< 1 %	< 6 %	< 6 %
Diff.output swing Is limited by	0.8 V$_{DDSS}$ cascodes	0.4 V$_{DDSS}$ source foll.	0.4 V$_{DDSS}$ cm input
Power dissipation	1 amp	3 amps	2 amps

Willy Sansen 10-05 0830

0830

As a comparison, the three types of CMFB amplifiers are listed in this slide.

It is clear that the first type, with the MOSTs in the linear region, has the lowest power consumption. It does not offer the same wide-band performance of the other two. It only uses one single amplifier.

Its output swing is limited by the use of cascodes. It can still be about 80% of the total supply voltage, which is not too bad at all.

The third type of CMFB amplifier needs one amplifier more, i.e. the error amplifier. Its speed is good but its output swing is limited by the input common-mode range.

In addition, the middle CMFB amplifier needs source followers. It now requires three amplifier stages which takes a lot of power. Its output swing is now mainly limited by the source followers. Indeed their V$_{GS}$ values have to be subtracted directly from the output swings.

Adding more differential pairs in the CMFB pairs makes the requirements on the matching more severe. The first solution with transistors in the linear region require stiffer tolerances than the other ones. It would take an elaborate analysis to show this, which, however, is left out.

Table of contents

- Requirements
- Fully-diff. amps with linear MOSTs
- FDA's with error amp.& source followers
- Folded cascode OTA without SF's
- Other fully-differential amps
- Exercise

Willy Sansen 10-05 0831

0831
Most fully-differential amplifiers use one of the three types of CMFB shown in this slide. They may also use some variations on these configurations, as shown next.

Also some switching solutions are available, which are used in sampled-data systems.

Fully-differential amplifier with gain boosting

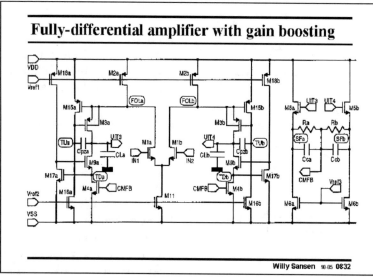

Willy Sansen 10-05 0832

0832
The first example is shown in this slide.

It consists of a folded-cascode OTA for differential operation. Gain boosting is added to all cascodes. Its GBW_{DM} is of the order of hundreds of MHz, and yet its gain can be as high as 100 dB.

The CMFB amplifier starts with source followers. It is therefore of type two. However, no error amplifier is used to reduce the power consumption. The extra gain of the error amplifier is not required as the common-mode feedback is applied to the Gates of transistors M4, which have gain-boosted cascodes to the outputs as well. The common-mode loop gain is sufficiently high indeed.

0833

Another good example of a fully-differential amplifier of type 2 is shown in this slide.

Two-stage Miller opamps are used now.

This gives the additional advantage that the output impedance is not so high at middle frequencies. The source followers can now be left out. Only an error amplifier is required.

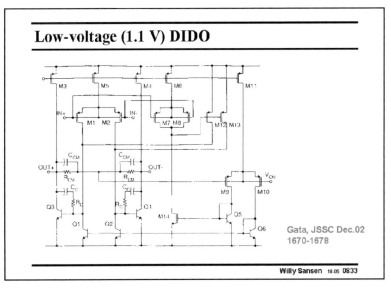

Low-voltage (1.1 V) DIDO

Gata, JSSC Dec.02
1670-1678

Willy Sansen 10-05 0833

0834

An example of the CMFB amplifier of type 3 is shown in this slide.

The main disadvantage of such a CMFB amplifier is that its input range is too limited. This input range can be enhanced by addition of Source resistors R as shown in this slide. The linearity is therefore greatly improved.

Note, the average output voltage is the sum of the reference voltage VDC and V_{GS7}.

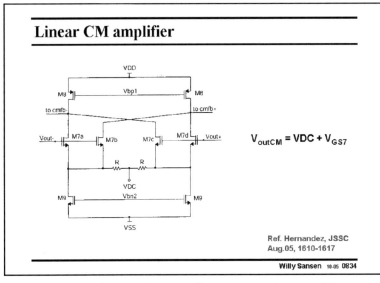

Linear CM amplifier

$$V_{outCM} = VDC + V_{GS7}$$

Ref. Hernandez, JSSC
Aug.05, 1610-1617

Willy Sansen 10-05 0834

Also, the transistors M7 are split up in equal parts. All four DC currents through transistors M7 are equal. The cancellation of the differential signal is now carried out by cross-coupling.

An additional advantage of the insertion of the resistors R is that the input capacitance loading offered to the differential amplifier is minimal. In this way, high-frequency, fully-differential opamps can be obtained.

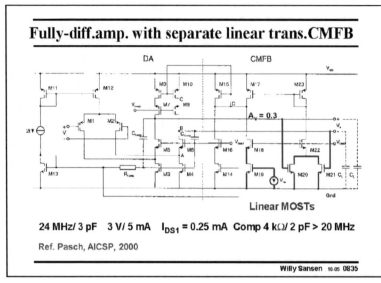

Fully-diff.amp. with separate linear trans.CMFB

DA   CMFB

$A_v = 0.3$

Linear MOSTs

24 MHz/ 3 pF   3 V/ 5 mA I_{DS1} = 0.25 mA   Comp 4 kΩ/ 2 pF > 20 MHz

Ref. Pasch, AICSP, 2000

Willy Sansen 10-05 0835

0835

This fully-differential amplifier uses a CMFB of type one, i.e. with transistors in the linear region (on the right).

The differential amplifier is just a folded cascode OTA. Additional capacitances C_{comp} with R_{comp} are used to boost its high-frequency performance (above 20 MHz).

The CMFB amplifier consists of MOSTs M20/M21 in the linear region and current mirrors M23/M17 and M15/M9,10. This is a wide-band amplifier, but at the cost of additional power consumption.

The outputs will end up around reference voltage V_{ref}, as the transistors M19–21 have sizes which are matched to their currents.

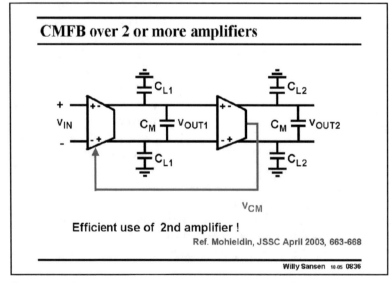

CMFB over 2 or more amplifiers

C_{L1} C_{L2}

$+$
V_{IN} C_M V_{OUT1} C_M V_{OUT2}
$-$

C_{L1} C_{L2}

V_{CM}

Efficient use of 2nd amplifier !

Ref. Mohieldin, JSSC April 2003, 663-668

Willy Sansen 10-05 0836

0836

When several fully-differential amplifiers are used in series, for high-order filters for example, there is no need to have CMFB around each stage separately. In order to save power, CMFB can be applied over two differential amplifiers in series, as shown in this slide.

Now, the average output voltage is measured at the output of the second amplifier, and applied to the common-mode input of the first amplifier. Since the common-mode output of the first amplifier is used to set the biasing of the second one, all common-mode levels are well defined.

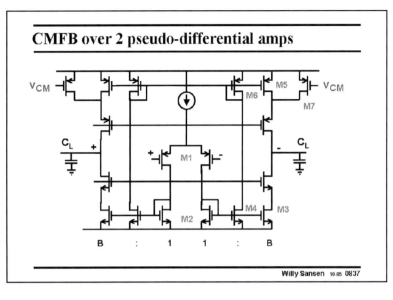

0837

Common-mode feedback can also be embedded in the differential amplifier as shown in this slide.

After all, this is a symmetrical OTA with current factor B.

However, the current mirrors M2/M3 have an additional output current with transistors M4, to cancel the differential signals at the diode-connected MOSTs M6. These latter devices are the inputs of current mirrors M5/M6 which close the feedback loop.

The actual output voltages can be modified by injecting currents with transistors M7 driven by some other CMFB loop.

The additional power consumption in this solution, is quite modest. Moreover, the output swing is not hampered by the CMFB amplifier.

As a consequence, this is an attractive solution indeed!

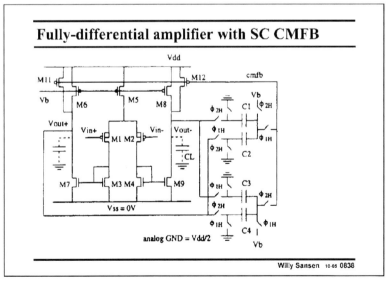

0838

When a clock is available, as in all sampled-data circuitry, it can be used towards a low-power CMFB loop.

This amplifier is just a symmetrical OTA.

The outputs are measured by a number of switches and capacitors, which provide common-mode feedback to the gates of transistors M11/M12.

This feedback is not applied directly to the Gates of M8/M6 to make sure that not all current can be switched in the output devices. For common-mode disturbance in the supply lines or substrate, large transients at the outputs can be avoided in this way.

In order to see how this circuit works, it is copied when all transistors are on, driven by clock Φ_1, and then when all transistors are on, driven by clock Φ_2.

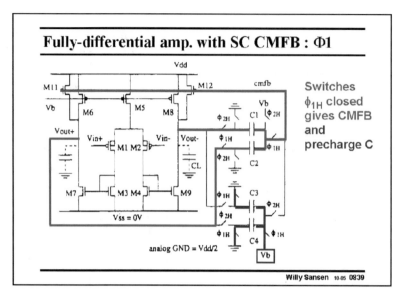

0839
This is exactly the same circuit as before.

In this slide, clock Φ_1 is high, all the transistors, which are driven by this clock, are on. The other ones are off. Thick lines indicate which paths the signals can take.

Clearly, the outputs are AC coupled by capacitors C1/C2. Their differential content is then cancelled out. Finally, this signal is then applied to the Gate of transistor M12 to close the feedback loop.

The GBW_{CM} is set by g_{m12} and the output load capacitor.

The DC level at the Gate of M12 is not defined because of the coupling capacitances C1/C2.

This is why the other two capacitances C3/C4 are precharged to the proper DC voltages. Their left side is set to analog ground (Vdd/2), whereas their right side is set at a biasing voltage, which is the same as at the Gates of current mirror transistors M5/M6/M8.

On the next phase, the capacitors are swapped around, as shown in the next slide.

Continuous CMFB is thus ensured.

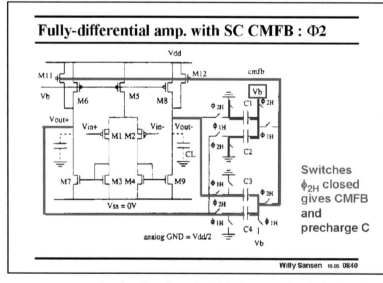

0840
In the phase Φ_2, the other transistors are on. Capacitors C3/C4 now provide common-mode feedback, whereas the other ones C1/C2 are reset or precharged.

It is clear that this solution does not take any power at all except for the switching power consumption of all switches and capacitors.

However, there are a few disadvantages.

First of all, the clock frequency appears in the signal path. This is a result of clock injection and charge redistribution, which are typical for all sampled-data circuits such as switched-capacitor filters. This is explained in detail in Chapter 17.

As a result, we can only use this solution at frequencies well below the clock frequency.

Moreover, these clock injection and charge redistribution signals severely limit the dynamic range. Intermodulation (and folding) of these signals provide an upper limit to the signal-to-noise ratio.

Finally, the switched capacitors increase the capacitive load of the CMFB amplifier. As a result, the GBW_{CM} will be reduced and the common-mode settling time increased.

Table of contents

- **Requirements**
- **Fully-diff. amps with linear MOSTs**
- **FDA's with error amp.& source followers**
- **Folded cascode OTA without SF's**
- **Other fully-differential amps**
- **Exercise**

Willy Sansen 10-05 0841

0841

Now that most types of CMFB amplifiers have been presented, this is a good opportunity to launch a design exercise.

Only the design task is described. The solution can be obtained from the editor, in the exercise solution book.

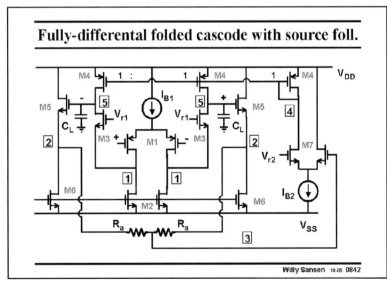

Fully-differential folded cascode with source foll.

Willy Sansen 10-05 0842

0842

This fully-differential amplifier is of type 2. It is a folded cascode. It has both source followers and an error amplifier for the common-mode feedback.

The outputs are evidently at nodes 5 or 2.

The circuit has been simplified somewhat. For example, the current mirror transistors M4 have all the same size. Moreover, the CMFB resistors R_a are used without the parallel capacitors C_a. They can be added afterwards.

Fully-diff. amp. : Specifications

Techn: CMOS $L_{min} = 0.8 \ \mu m$; $V_T = 0.7$ V

$K'_n = 60 \ \mu A/V^2$ & $K'_p = 30 \ \mu A/V^2$

$V_{En} = 4 \ V/\mu m$ & $V_{Ep} = 6 \ V/\mu m$

Specs: $GBW_{DM} = 10$ MHz $C_L = 3$ pF

$GBW_{CM} = 20$ MHz

all PM > 70°

$V_{DD}/V_{SS} = \pm 1.5$ V

Maximum $V_{swingptp} = V_{outmax} - V_{outmin}$

Minimum I_{tot}

Verify: Slew Rate,   Noise, ...

Willy Sansen 10-05 0843

0843

The specifications to which this amplifier has to be designed, are given in this slide. The GBW's are quite moderate, but so is the technology.

Note that the GBW_{CM} is a factor of two larger than the GBW_{DM}.

Note also that the total supply voltage is only 3 V. This is why we want to maximize the output swing. Obviously we always take minimum currents.

For these specifications, only one possible design results. Only one set of currents and transistor sizes emerges.

All other specifications such as Slew Rate, noise density, etc., can be verified afterwards.

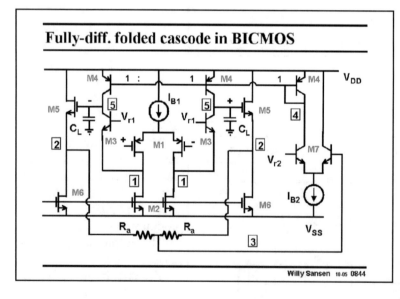

Willy Sansen 10-05 0844

0844

A BiCMOS equivalent is shown in this slide.

The CMFB amplifier is the highest-speed circuit block. This is why it is realized by means of npn transistors. Also the current mirrors M4 are realized with pnp transistors. They only need a V_{CE} of 0.1 V compared with a V_{DS} of 0.2 V for MOSTs. We gain 0.1 V in output swing.

Fully-diff. amp. : Specifications

Techn: BICMOS
$L_{min} = 0.8 \ \mu m$; $V_T = 0.7$ V

$K'_n = 60 \ \mu A/V^2$ & $K'_p = 30 \ \mu A/V^2$

$V_{En} = 4$ V/μm & $V_{Ep} = 6$ V/μm

$f_{Tn} = 12$ GHz & $f_{Tp} = 4$ GHz

Specs:
$GBW_{DM} = 10$ MHz $\quad C_L = 3$ pF

$GBW_{CM} = 20$ MHz

all PM > 70°

$V_{DD}/V_{SS} = \pm 1.5$ V

Maximum $V_{swingptp} = V_{outmax} - V_{outmin}$

Minimum I_{tot}

Verify: Slew Rate, Noise, ...

Willy Sansen 10-05 0845

0845

The specifications to which this amplifier has to be designed are the same.

Again, the total supply voltage is 3 V. We want to maximize again the output swing. All currents have to be minimized.

Only one possible design emerges.

The other specifications such as Slew Rate, noise density, etc., are then verified for comparison.

As a hint for this design, the following steps have to be taken.

The maximum output swing is the difference between the largest voltage at the output and the smallest one. This determines the output swing, but also the average output voltage. This leads to the upper value of the reference voltage V_{r1} and the value of V_{r2}.

From the values of the GBW, the g_m can easily be derived. Choices for $V_{GS} - V_T$ and L lead to values of the current and the transistor size.

In this way, all currents are known, except for the currents in the source followers M5.

These currents, through transistors M5 determine the swing across resistors R_a, which has been derived before. The resistors R_a determine on their turn, the non-dominant poles.

Table of contents

- **Requirements**
- **Fully-diff. amps with linear MOSTs**
- **FDA's with error amp.& source followers**
- **Folded cascode OTA without SF's**
- **Other fully-differential amps**
- **Exercise**

Willy Sansen 10-05 0846

0846

In this Chapter fully-differential amplifiers have been discussed in great detail. Design procedures have been discussed for all CMFB amplifiers.

Finally an exercise is launched to test the comprehension of the reader.

Design of Multistage Operational amplifiers

Willy Sansen

KULeuven, ESAT-MICAS

Leuven, Belgium

willy.sansen@esat.kuleuven.be

Willy Sansen 10-05 091

091

In many applications, operational amplifiers are required which have three stages. For example, class AB amplifiers have an output stage which provide little gain but a lot of current. The first two stages are now needed to provide gain.

Also, when the supply voltage is less than 1 V, more stages are required as cascoding is no longer possible.

In this Chapter, the principles are discussed to stabilize three-stage amplifiers. Moreover, circuit principles are devised to reduce the power consumption.

Table of contents

- **Design procedure**
- **Nested-Miller designs**
- **Low-power designs**
- **Comparison**

Ref.: W. Sansen : Analog Design Essentials, Springer 2006

Willy Sansen 10-05 092

092

With three stages, stability is less obvious than two stages. This is why the first topic in this Chapter is devoted to stability requirements. A specific design procedure has to be followed to ensure stability with minimum power consumption.

Nested-Miller designs are discussed next. They have been used for some time now. However, they suffer from excessive power consumption in the output stage.

This is why several low-power designs are introduced and compared. They allow power savings of up to 40 times. They are therefore useful for low-power and portable systems.

Why three-stage amplifiers ?

1. Each MOST only gives $g_m r_o \approx 15$ or 24 dB :
   High gain requires three stages !

2. For drivers (small R_L) : $g_m R_L$ is very low :
   High gain requires three stages !

3. For low V_{DD}, no cascoding but cascading !
   High gain requires three stages !

Willy Sansen 10-05 093

093
The main reason for more than two stages is gain. Despite the availability of gain boosting, bootstrapping, etc. to increase the gain, quite often the supply voltage is just too low. In this case, three stages or more may be required.

For small channel lengths, the gain per transistor, which is $g_m r_o$, has become quite small. For example, for 130 nm CMOS, less than about 15 (or 24 dB) per transistor can be expected.

For large gains, three or more stages are therefore required.

This is certainly true if a small resistor or large capacitance must be driven. In this case, the output stage provides little gain. Two more stages are then required to drive the output stage with a lot of gain.

For very low supply voltages less than 1 V, cascoding is no longer possible because of the reduced output swings. Cascading is then required, as shown next.

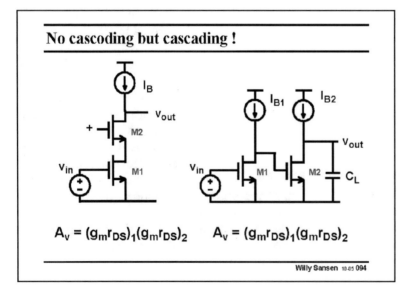

No cascoding but cascading !

$A_v = (g_m r_{DS})_1 (g_m r_{DS})_2$ $A_v = (g_m r_{DS})_1 (g_m r_{DS})_2$

Willy Sansen 10-05 094

094
Both a cascode stage and a cascaded stage provide the same gain because both circuits have the same number of transistors in the signal path. However, the latter one takes more current.

Only current sources are used here as ideal loads. Despite the larger current, the cascade stage can provide a rail-to-rail output swing, which may be very desirable for low supply voltages.

The speed is not higher however, as shown next.

095

Addition of the same load capacitance C_L, shows that the GBW of the cascode stage simply depends directly on this capacitance.

A cascaded amplifier however, has two stages. A compensation capacitance is thus required to provide pole splitting for stability. As a consequence the output pole g_{m2}/C_L will have to be two to three times larger than the GBW. More current will therefore be consumed. In general, addition of capacitances leads to additional current consumption.

As said before, the cascaded stage allows a rail-to-rail output swing, which is a considerable advantage.

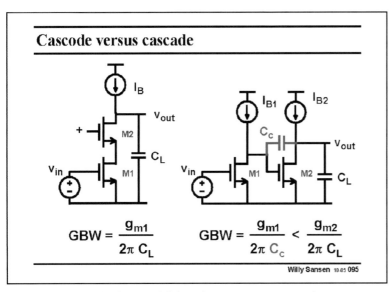

Cascode versus cascade

$$GBW = \frac{g_{m1}}{2\pi \, C_L}$$

$$GBW = \frac{g_{m1}}{2\pi \, C_c} < \frac{g_{m2}}{2\pi \, C_L}$$

Willy Sansen 10-05 095

096

In order to create a feeling on how a three-stage amplifier is stabilized, the principles are reviewed for a single-and two-stage amplifier.

A single-stage amplifier has only one single high-impedance point at the output. The output capacitance C_L determines the GBW.

In the case of variable load capacitances, as in switched-capacitor filters, we do not want the GBW to depend on the load. A two-stage opamp is then better used.

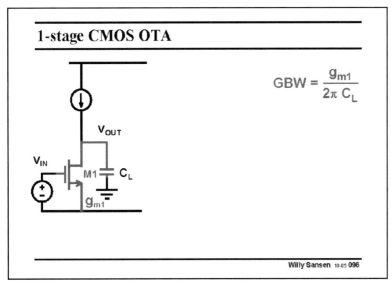

1-stage CMOS OTA

$$GBW = \frac{g_{m1}}{2\pi \, C_L}$$

Willy Sansen 10-05 096

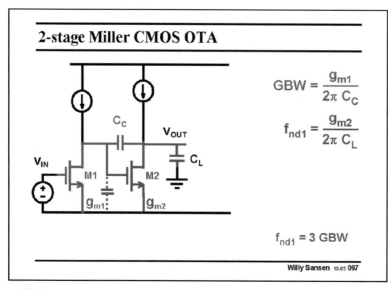

2-stage Miller CMOS OTA

$$GBW = \frac{g_{m1}}{2\pi \, C_C}$$

$$f_{nd1} = \frac{g_{m2}}{2\pi \, C_L}$$

$$f_{nd1} = 3 \, GBW$$

Willy Sansen 10-05 **097**

097

A two-stage amplifier has two high-impedance points. They need to be connected by a compensation capacitance C_c to provide pole splitting and to generate a dominant pole. The GBW is therefore determined by this compensation capacitance C_c.

For stability, the non-dominant pole is now determined by the load capacitance C_L. It must be sufficiently large compared to the GBW, to provide a sufficient phase margin. A ratio of three is taken for a phase margin of about 70°.

Since we have two stages, we have two time constants. They are the ones for the GBW and the one for the non-dominant pole. The latter one is the output time constant. It is the output g_{m3} divided by the load capacitance C_L. It is normally set at two to three times the GBW depending on the phase margin required.

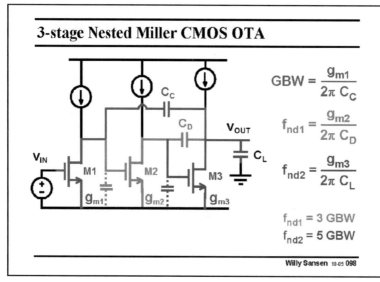

3-stage Nested Miller CMOS OTA

$$GBW = \frac{g_{m1}}{2\pi \, C_C}$$

$$f_{nd1} = \frac{g_{m2}}{2\pi \, C_D}$$

$$f_{nd2} = \frac{g_{m3}}{2\pi \, C_L}$$

$$f_{nd1} = 3 \, GBW$$
$$f_{nd2} = 5 \, GBW$$

Willy Sansen 10-05 **098**

098

We now have three stages, we can expect three time constants. They will be the ones for the GBW and now for two non-dominant poles.

Since three high-impedance points are present, two compensation capacitances are required for stability. Both are connected to the output. This is called nested-Miller compensation.

Indeed the GBW has the same expression as before. The reason is that the compensation capacitance connects the output to the output of the input transistor. It shunts both transistors M2 and M3. Transistor M2 is a kind of driver for transistor M3. Together they form the output stage.

The output time constant is the same as well. It is again the output g_{m3} divided by the load capacitance C_L, exactly as for a two-stage amplifier.

The middle stage now brings in another time constant, also given by its g_{m2} divided by its own output capacitance C_D.

As a result, we find two non-dominant poles. Both have to be positioned sufficiently far beyond the GBW, such that together they provide a reasonable phase margin. Ratio's of 3 and 5 have been taken in this slide. Why these values is shown next.

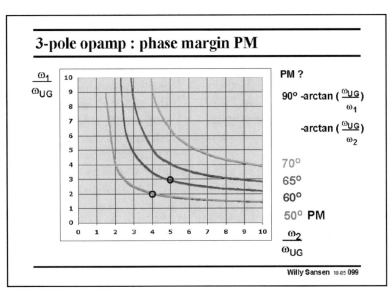

099

The curves for equal phase margin PM for different combinations of the two non-dominant poles are shown in this slide. The (circle) frequencies of the non-dominant poles are scaled by one of the GBW.

It is clear that for a PM of 60°, one non-dominant pole has to be positioned at 3 times the GBW and the other one at 5 times the GBW (see dot). Obviously, all other combinations of the same (blue) curve pro-

vide the same PM of 60°. This means that all these combinations give about the same amount of peaking in the transient response. Ratio's of 3.5 and 4 are therefore perfectly acceptable. Ratio's of 2.5 and 7 would also be acceptable but positioning a non-dominant pole at 7 times the GBW would probably require too much power. This combination is better avoided.

It is also clear that a PM of about 60° is sufficiently high, even if a bit of peaking occurs. A PM of 70° would require non-dominant poles at too high frequencies, and would consume too much power.

Many designers take an even higher risk by requiring a PM of only 50°. In this case, the transient response is really on the edge of peaking. The non-dominant poles can then be positioned at 2 and 4 times the GBW only (see dot)! This is called the Butterworth response. It provides a maximally flat response once the feedback loop is closed towards unity gain. It is often used for the design of three-stage amplifiers although it may not yield the lowest power consumption.

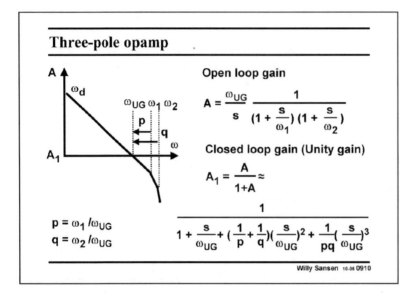

0910

The positions of the two non-dominant poles determine the open-loop response as sketched in this slide. How is the closed-loop response for unity gain?

The expression of the open-loop gain A is approximated in this slide. The dominant pole at ω_d is situated somewhere at low frequencies. We concentrate on the frequency region around the GBW. Therefore, ω_d is not visible. Moreover it is left out of the expressions for simplification.

The non-dominant poles are at (circle) frequencies ω_1 and ω_2. The GBW occurs at unity gain or at (circle) frequency ω_{UG}. The ratios of the non-dominant poles to the GBW are parameters p and q. Values of 3 and 5 have been used before for a phase margin of 60°.

When the loop is closed towards unity gain, the expression of the gain A_1 is now of third order. Three poles occur quite close together, one at the GBW and two more at slightly higher frequencies

What are the best values of p and q for a smooth response in closed-loop?

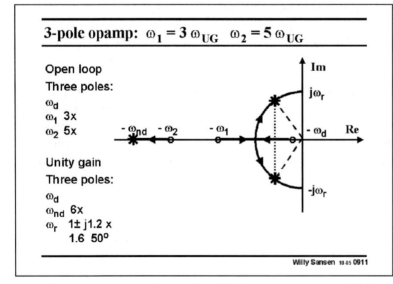

0911

For values of p and q of 3 and 5, the situation is sketched in this slide in the complex plane (or polar diagram).

In open loop, the dominant pole and two non-dominant poles are all on the real axis. All poles are negative. A stable system thus results.

When the loop is closed, the root locus shows where the poles end up. The two first poles ω_d and ω_1 form complex poles, with resonant frequency at ω_r, whereas the third one ω_2 moves to higher frequencies, to ω_{nd} at about 6 times the open-loop GBW.

The complex poles have a resonant frequency ω_r at about 1.6 times the open-loop GBW. They are complex with a real part about equal to the GBW and a real part of about 1.2 times the GBW.

This shows clearly that even with real non-dominant poles in an open loop, complex poles usually result in a closed loop.

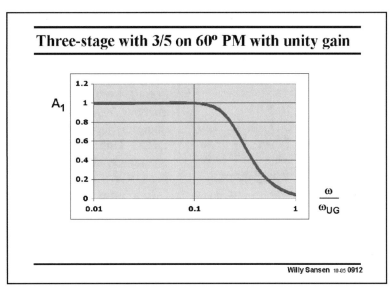

0912

A plot of the gain A1 in unity-gain is shown in this slide. It is merely the amplitude of the expression given in this slide.

It is clear that a very flat response is obtained indeed. This is why this combination of non-dominant poles is chosen by many designers. They can go a bit further however, as shown next.

Note also that only a fraction of the open-loop GBW is available in closed loop. The -3 dB point now occurs at about 0.3 times the GBW only.

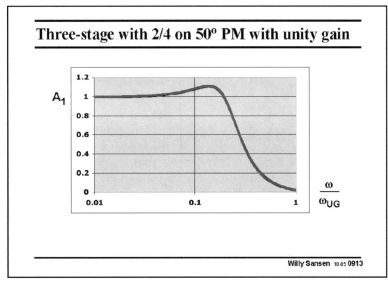

0913

Taking a lower phase margin of only 50°, with two non-dominant poles at 2 and 4 times the GBW yields a little bit of peaking. It is about 10% or 0.8 dB. However, the curve drops even faster. The -3 dB point is reached at about 0.24 times the open-loop GBW.

The main advantage of this response is that the non-dominant poles occur at lower frequencies, such that the power consumption may be lower. However, a little bit of peaking is the price to pay!

We can still go a little bit further. We could allow complex poles already in open loop. The power consumption could then be less, as these poles are closer to the GBW than before. Obviously, we will find complex poles in the closed-loop response as well!

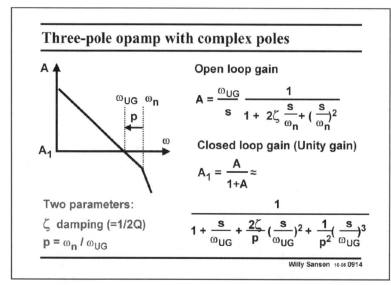

0914
The expression is given for the open-loop gain of a three-pole opamp. Again the dominant pole is left out because it occurs at very low frequencies.

However, the two non-dominant poles are now complex. They are characterized by a resonant frequency ω_n and by a damping factor ζ (or $Q = 1/2\zeta$). This resonant frequency is at a ratio p of the GBW.

It is clear that we again have two parameters, this time not two real non-dominant poles but one pair of complex poles. The parameters are now not p and q but p and ζ.

In a unity-gain closed loop, the expression of the gain A_1 is easily obtained. It is obviously of third order.

The question now is, what is the best choice of the two parameters p and ζ to ensure a maximally flat response?

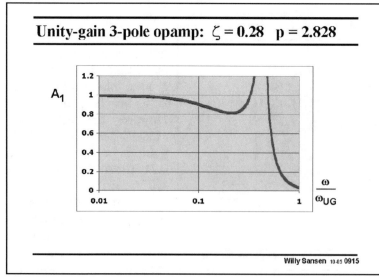

0915
In order to illustrate how these two parameters play towards the frequency response, several values of damping factor ζ are tried out for equal distance p of the resonant frequency to the GBW.

Perhaps the chosen values of p and ζ seem to be a bit awkward. A few slides later, it will become clear why these have been selected.

One thing is obvious. The damping is too low. A huge peak occurs, even beyond the vertical scale. As we have kept the same scale for all examples, the peak exceeds the scale.

The damping must therefore be increased in order to flatten the response.

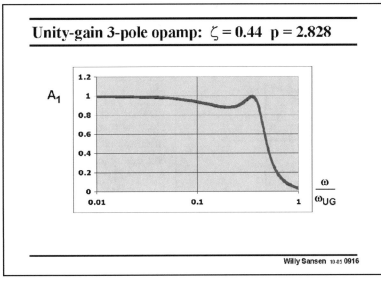

Unity-gain 3-pole opamp: $\zeta = 0.44$ p = 2.828

Willy Sansen 10-05 0916

0916

Increasing the damping factor ζ reduces the peaking. At this value of ζ the peaking is about the same as the low-frequency value of the gain. There is a dip of about 10%, which could be acceptable depending on the application.

Because of the peaking, the -3dB frequency (at a value of 0.707) is slightly larger. It occurs at about 0.43 times the open-loop GBW.

Let us now increase the damping even more. Several values of ζ are now tried, until a very flat response is obtained. Thy are collected on the next slide.

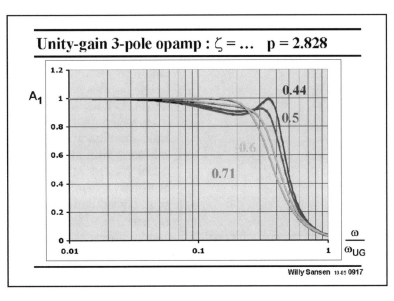

Unity-gain 3-pole opamp : $\zeta = ...$ p = 2.828

Willy Sansen 10-05 0917

0917

The responses are sketched for different values of damping factors ζ. The one for $\zeta = 0.44$ is repeated for reference. Values are taken of 0.5, 0.6 and finally of 0.71.

It is clear that the curve for $\zeta = 0.71$ (actually $1/\sqrt{2}$) a maximally flat response is obtained. This is the third-order Butterworth response. It occurs for a p factor of $2\sqrt{2}$ and a ζ of $1/\sqrt{2}$. In an open loop, the non-dominant poles are already complex. They are certainly complex in a closed loop.

This positioning of non-dominant is quite popular with designers of three-stage amplifiers. The non-dominant poles are at relatively low frequencies. The power consumption is relatively small.

It is now clear why a p factor of 2.828 has been chosen. It is exactly the p factor required for a third-order maximally flat Butterworth response.

Finally, note that the maximally flat response leads to a -3 dB frequency, which is somewhat lower than before. It is now about 0.3 times the open-loop GBW.

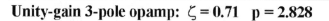

Unity-gain 3-pole opamp: $\zeta = 0.71$ $p = 2.828$

Closed loop
Butterworth response :
maximally flat

Poles :

$\omega_{-3dB} = \dfrac{\omega_c}{2}$

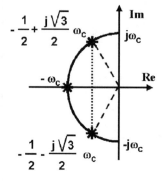

$-\dfrac{1}{2} + \dfrac{j\sqrt{3}}{2}\,\omega_c$

$-\omega_c$

$-\dfrac{1}{2} - \dfrac{j\sqrt{3}}{2}\,\omega_c$

Willy Sansen 10-05 **0918**

0918
The complex plane corresponding to this situation, is sketched in this slide. The poles are only shown in closed loop.

All poles lie on a semicircle. The resulting -3 dB frequency is half this pole frequency ω_c, which is obviously related to the open-loop GBW.

This response will now be used to compare several three-stage opamp configurations.

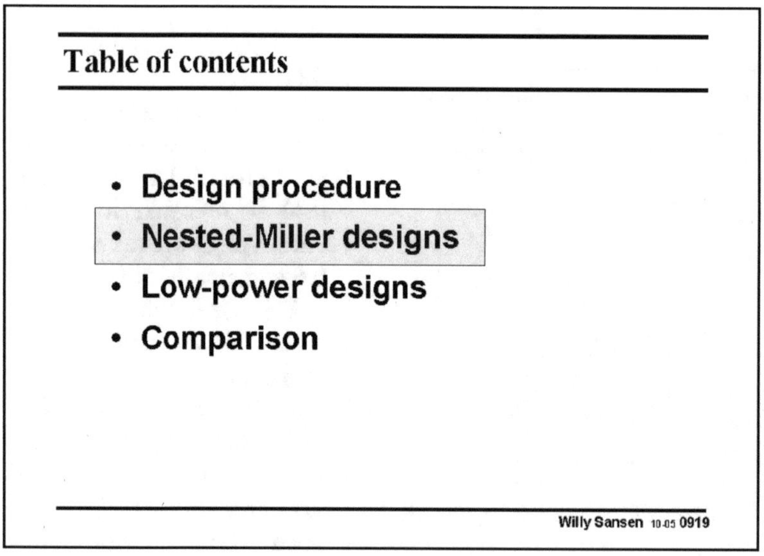

Table of contents

- **Design procedure**
- **Nested-Miller designs**
- **Low-power designs**
- **Comparison**

Willy Sansen 10-05 **0919**

0919
Now that we know how to stabilize a three-stage amplifier, that we know how to choose the two variables p and q, or even better p and ζ – we use this procedure to design such an amplifier.

However, there are several configurations. We will start with the conventional Nested Miller Compensation (NMC) and then gradually add other branches such as feedforward, multi-path branches, etc.

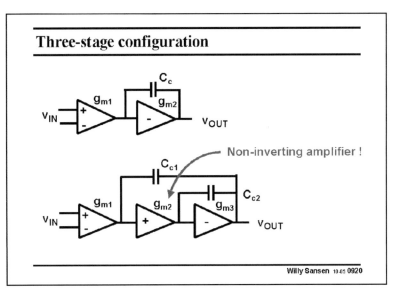

0920

When we focus on configurations, we have to make a decision first on which non-inverting amplifier to use.

Indeed, the second stage of a three-stage amplifier is normally single-ended, as the output stage is single-ended as well.

However, a single-transistor amplifier is always inverting. It cannot be used as a second-stage of a three-stage amplifier. Otherwise compensation capacitance C_{c1} would provide positive feedback!

It is clear that such a non-inverting amplifier is not required in a two-stage amplifier, only in a three-stage amplifier.

There are two possible solutions. The original one is shown next.

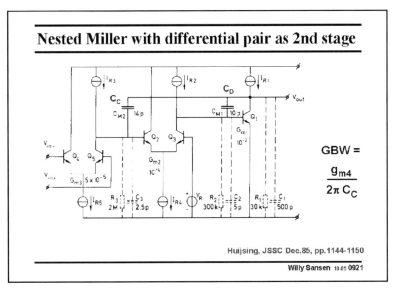

Huijsing, JSSC Dec. 85, pp. 1144-1150

0921

In this three stage amplifier, which is fully realized in bipolar technology, a differential pair is used as a second stage. The output "on the other side" is non-inverting indeed.

The other functions of the three stages are easily recognized. The nested compensation capacitances are 14 pF and 10 pF respectively. The first one determines the GBW.

An alternative is shown next.

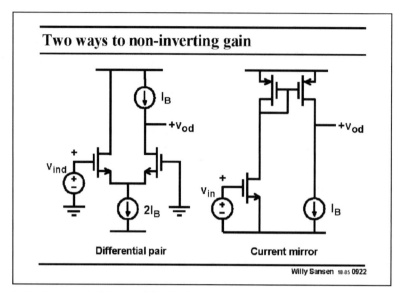

Two ways to non-inverting gain

I_B

$+v_{od}$

v_{ind}

$2I_B$

$+v_{od}$

v_{in}

I_B

Differential pair   Current mirror

Willy Sansen 10-05 0922

0922

After all, a current mirror is capable of providing a non-inverting gain. Both a differential pair and a current mirror are elementary circuit blocks.

If all transistors carry equal DC currents, then the power consumption is the same as well.

Both are simple and can provide gain up to high frequencies. It is not clear which one to prefer!

Perhaps the current mirror has a slight edge as it is the circuit block with the highest bandwidth.

A comparative study will be carried out to be able to find out what the difference really is!

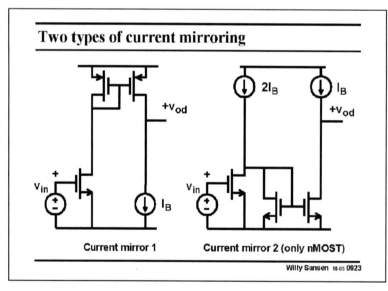

Two types of current mirroring

$+v_{od}$

v_{in}

I_B

$2I_B$ I_B

$+v_{od}$

v_{in}

Current mirror 1   Current mirror 2 (only nMOST)

Willy Sansen 10-05 0923

0923

More sophisticated current mirrors can be used as well. In the second one, only nMOST devices are used. It most probably has a larger bandwidth than the first one. However, it consumes 50% more current!

Would the increase in bandwidth be offset by the increase in power consumption?

Probably, the difference between both is minor.

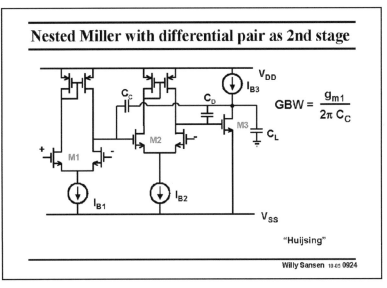

Nested Miller with differential pair as 2nd stage

$$GBW = \frac{g_{m1}}{2\pi C_C}$$

"Huijsing"

Willy Sansen 10-05 0924

0924

In this original realization, a differential pair is used as a second stage. Only MOSTs are now used, rather than bipolar transistors.

The two compensation capacitances are easily found.

As this circuit originates from a bipolar version of Huijsing, it is denoted by "Huijsing".

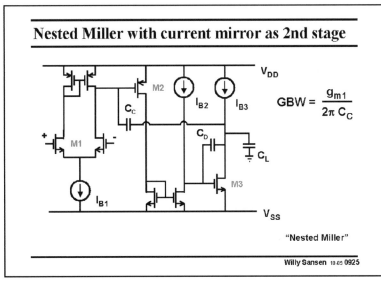

Nested Miller with current mirror as 2nd stage

$$GBW = \frac{g_{m1}}{2\pi C_C}$$

"Nested Miller"

Willy Sansen 10-05 0925

0925

The only difference between the previous amplifier and this one is the use of a current mirror as a second stage. The input and output stage are the same. Also the load capacitance and the currents are the same.

The only question is therefore, which one of these amplifiers can achieve a higher GBW. This will obviously depend on the non-dominant poles associated with the second stage.

This amplifier is denoted by a "nested Miller" although both of them are nested Miller amplifiers.

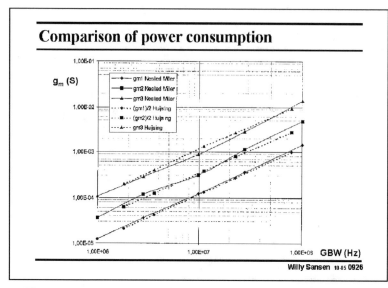

Comparison of power consumption

0926
Both amplifiers have been designed for the same load capacitance. They are designed for different values of the GBW, using a Butterworth 3rd order response. Afterwards they are optimized somewhat by means of a few additional simulations.

The curves give the values of the transconductances of the three stages. The currents are proportional to them, depending on the chosen values of VGS-VT.

These curves show that there is only a minor difference between both amplifiers. The main difference is found at higher frequencies. For values of the GBW around 100 MHz, the opamp with a differential pair as a second stage has more difficulty in reaching this 100 MHz than the one with a current mirror. This shows that the latter approach is more advantageous for higher frequencies.

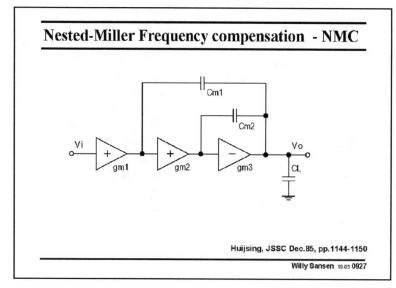

Nested-Miller Frequency compensation - NMC

Huijsing, JSSC Dec.85, pp.1144-1150

Willy Sansen 10-05 0927

0927
Now several more NMC opamps will be discussed.

For sake of comparison, the conventional three-stage NMC amplifier is repeated.

Its main disadvantage is the current consumption of the output stage. Indeed, compensation capacitance C_{m2} shunts the output stage. Its gain is now reduced at higher frequencies. Therefore, transconductance g_{m3} must be enlarged to ensure pole splitting.

The design procedure is repeated next.

NMC equations in open loop

$$A_v(s) = \frac{A_{dc}\left(1 + \frac{s}{\omega_3} + \frac{s^2}{\omega_3\omega_4}\right)}{\left(1 + \frac{s}{\omega_d}\right)\left(1 + \frac{s}{\omega_1} + \frac{s^2}{\omega_1\omega_2}\right)}$$

$$A_{dc} = g_{m1}g_{m2}g_{m3}R_1R_2R_3$$

$$\omega_d = -\frac{1}{C_{m1}g_{m2}g_{m3}R_1R_2R_3}$$

$$\omega_{UG} = \frac{g_{m1}}{C_{m1}}$$

$$\omega_1 = \frac{g_{m2}}{C_{m2}}$$

$$\omega_2 = \frac{g_{m3}}{C_L}$$

$$\omega_3 = -\frac{g_{m3}}{C_{m2}}$$

$$\omega_4 = \frac{g_{m2}}{C_{m1}}$$

Willy Sansen 10-05 0928

0928

In an open loop, three poles occur and two zeros. They are given on the right.

The open-loop gain A_{dc} is large because three stages are present. The resistance of each node I to ground is denoted by R_i.

The capacitance of each node to ground is small. Actually, the minimum the compensation capacitances are taken to be at least three times larger than those parasitic node capacitances. This is why the node capacitances do not show up in the approximate expressions in this slide.

The dominant pole is caused by the Miller effect of the overall compensation capacitance C_{m1}. The GBW is determined by this capacitance C_{m1} and the input transconductance.

The design procedure is given next.

NMC stability

$$A_v(s) = \frac{A_{dc}\left(1 + \frac{s}{\omega_3} + \frac{s^2}{\omega_3\omega_4}\right)}{\left(1 + \frac{s}{\omega_d}\right)\left(1 + \frac{s}{\omega_1} + \frac{s^2}{\omega_1\omega_2}\right)}$$

$g_{m1} < g_{m2} < g_{m3}$

Butterworth 3rd order : $\zeta = 0.7$; $p = 2.8$

$\omega_2 = 2\,\omega_1 = 4\,\omega_{UG}$

Zero's negligible

$$\omega_1 = \frac{g_{m2}}{C_{m2}}$$

$$\omega_2 = \frac{g_{m3}}{C_L}$$

$$\omega_3 = -\frac{g_{m3}}{C_{m2}}$$

$$\omega_4 = \frac{g_{m2}}{C_{m1}}$$

Willy Sansen 10-05 0929

0929

We can normally assume that the transconductances increase towards the output. For a 3rd-order Butterworth response, the two non-dominant poles ω_1 and ω_2 have to be put at 2 and 4 times the GBW.

Both zeros are usually negligible. Note that one of them is in the right half of the complex plane.

Note that we still have to solve a system of three equations (for the GBW, ω_1 and ω_2) and five variables.

Normally, the two compensation capacitances are chosen. The first one C_{m1} can be chosen as small as possible, i.e. at least three times the node capacitance at the output of the first stage but not so small that the input noise is too high.

The other capacitance C_{m2} can also be chosen as small as possible, i.e. at least three times the node capacitance at the output of the second stage.

Minimum values of these compensation capacitances are always chosen to reduce the power consumption as much as possible. Of course, if noise is an important specification, then these capacitances may have to be increased, increasing the power consumption. As expected, low noise always leads to larger power consumption.

Designers often take C_{m1} and C_{m2} to be the same and give them a low value. This is clearly never an optimum design!

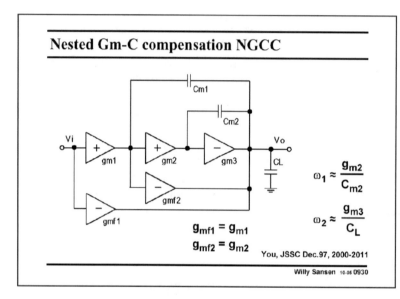

0930
This three-stage configuration has nested Miller compensation with additional Gm blocks for realizing two feedforward paths. If their transconductances are chosen to be the same as their corresponding stage transconductances then both zeros' are cancelled out.

The two resulting non-dominant poles are then the same as for a conventional NMC amplifier.

Little is gained in terms of power consumption.

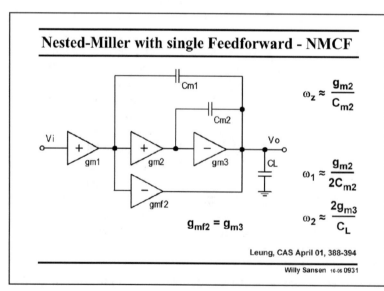

0931
One single feedforward path can be used as well. Its transconductance g_{mf2} however, equals the transconductance g_{m3} of the output stage. It is used to generate a zero ω_z in the left half of the complex plane, to cancel out a non-dominant pole. It therefore increases the phase margin.

If $g_{m3} > g_{m2}$, then the non-dominant poles can be approximated as shown in this slide.

On the other hand, the power consumption is also increased, as the feedforward stage and the output stage take the largest currents.

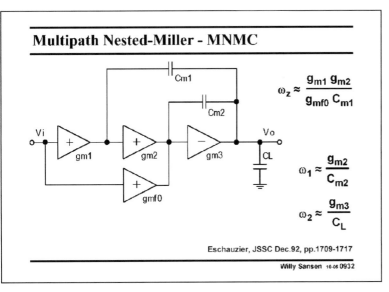

Multipath Nested-Miller - MNMC

$$\omega_z \approx \frac{g_{m1}\,g_{m2}}{g_{mf0}\,C_{m1}}$$

$$\omega_1 \approx \frac{g_{m2}}{C_{m2}}$$

$$\omega_2 \approx \frac{g_{m3}}{C_L}$$

Eschauzier, JSSC Dec.92, pp.1709-1717

Willy Sansen 10-05 0932

0932

Another compensation technique is the multi-path nested Miller, shown in this slide.

The feedforward stage now bridges the input and the second stage. It is not connected to the output.

This feedforward stage is used to generate a zero ω_z in the left half of the complex plane, to cancel out a non-dominant pole. It therefore increases the phase margin.

If $g_{m3} > g_{m2}$, then the non-dominant poles can be approximated as shown in this slide. They are the same as for a conventional NMC amplifier.

Comparison Nested-Miller solutions

Topology	Stages	PM	GB=2πGBW	T_{eL}	$T_{eL}/T_{eL\,(NMC)}$
Single	One	$<90°$	(g_m/C_L)	1.0	4.0
SMC	Two	$<63°$	$0.5(g_{m2}/C_L)$	0.5	2.0
NMC	Three	$\approx60°$	$0.25(g_{m3}/C_L)$	0.25	1.0
NGCC	Three	$\approx60°$	$0.25(g_{m3}/C_L)$	0.25	1.0
NMCF	Three	$>60°$	$<0.5(g_{m3}/C_L)$	<0.5	<2.0
MNMC	Three	$\approx63°$	$\approx0.5(g_{m3}/C_L)$	≈0.5	≈2.0

Willy Sansen 10-05 0933

0933

For sake of comparison all previous amplifiers are listed in a table. They are preceded by a single-stage and a two-stage Miller amplifier.

All of them have a comparable phase margin except for the single-stage amplifier. The resulting GBW for a 3rd-order Butterworth design is also listed as a fraction of the output time constant g_{m3}/C_L.

The next column shows what GBW can be expected, compared to a single-stage amplifier. The conventional NMC amplifier only achieves about 1/4 of what a single stage can achieve.

When compared to a NMC amplifier (in the last column), it is clear that both last amplifiers which use a single feedforward stage do better in terms of power consumption.

However, the improvement is at best a factor of 2.

This is why we will now concentrate on three-stage amplifiers, which provide much better current savings.

Table of contents

- **Design procedure**
- **Nested-Miller designs**
- **Low-power designs**
- **Comparison**

Willy Sansen 10-05 0934

0934

For really low power consumption, other configurations must be found. They will all try to generate extra zeros in the left half of the complex plane, in order to cancel one of the non-dominant poles, without however, taking too much extra power consumption.

Three configurations will be discussed. The last two achieve considerable savings in power consumption, beyond that which can be achieved with a single stage.

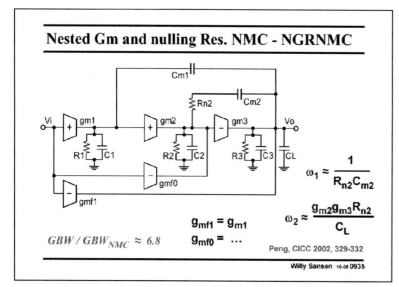

Nested Gm and nulling Res. NMC - NGRNMC

$GBW / GBW_{NMC} \approx 6.8$

$g_{mf1} = g_{m1}$

$g_{mf0} = \cdots$

$\omega_1 \approx \dfrac{1}{R_{n2}C_{m2}}$

$\omega_2 \approx \dfrac{g_{m2}g_{m3}R_{n2}}{C_L}$

Peng, CICC 2002, 329-332

Willy Sansen 10-06 0935

0935

The first configuration is close to the NGCC shown previously. It also uses two feedforward stages. The only difference with the NGCC amplifier is the nulling resistor R_{n2} in series with the second compensation capacitance. This is a well known technique of generating a zero in the left-half-plane, at least when this resistor R_{n2} is made larger than $1/g_{m3}$ (see Chapter 4).

Again, the feedforward stages are used to cancel out the zero's. This time, however, transconductance g_{mf0} is made larger in order to generate left-hand zeros. As a result, the complex non-dominant poles are made to cancel with a pair of complex left-hand zero's.

For a 3rd-order Butterworth realization, this gives a GBW which is 6.8 times larger than a NMC realization with the same load capacitance and power consumption.

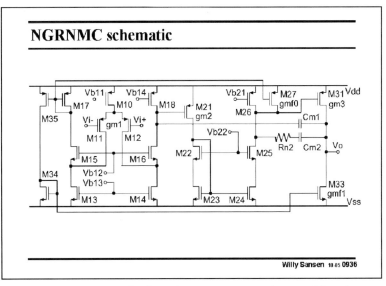

0936

A circuit realization of the NGRNMC amplifier is shown in this slide.

The input stage is a folded cascode. Current mirrors are used as noninverting amplifiers.

Current mirror M27/M17 is used to set accurately the value of g_{mf0}. Also, current mirror M33/M34 is used to set g_{mf1}.

This latter feedforward stage is also used for better large-signal performance. It increases the Slew Rate of the output stage drastically. The output stage is now biased as a class-AB stage. The quiescent current is set accurately by transistor M34. However, he maximum output currents are a lot larger.

The Slew Rate will now be limited by the DC current of the first stage into compensation capacitance C_{m1}.

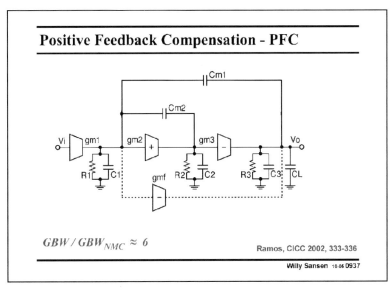

0937

A second low-power nested Miller compensation technique uses positive feedback with capacitance C_{m2} around the second-stage. Its purpose is again to introduce a left-hand zero in order to cancel out one of the nondominant poles.

A single feedforward stage is used to turn the output stage into a class-AB stage. In this way the Slew Rate will be limited by the DC current in the first stage.

PFC Equations

$$A_v(s) = \frac{A_{dc}\left(1 + \dfrac{s}{\omega_1} + \dfrac{s^2}{\omega_1 \omega_3}\right)}{\left(1 + \dfrac{s}{\omega_d}\right)\left(1 + \dfrac{s}{\omega_1} + \dfrac{s^2}{\omega_1 \omega_2}\right)}$$

$$A_{dc} = g_{m1} g_{m2} g_{m3} R_1 R_2 R_3$$

$$\omega_d = -\frac{1}{C_{m1} g_{m2} g_{m3} R_1 R_2 R_3}$$

$$\omega_{UG} = \frac{g_{m1}}{C_{m1}}$$

$$\omega_1 = \frac{g_{m2}}{2C_{m2}}$$

$$\omega_2 = \frac{2g_{m3}}{C_L}$$

$$\omega_3 = -\frac{2g_{m3}}{C_{m1}}$$

Stability : $\dfrac{2g_{m3}}{C_L} > \dfrac{g_{m2}}{C_{m1}}$

Willy Sansen 10-05 0938

0938

The approximate expression of the open-loop gain is given in this slide. As expected, three poles show up and two zero's. The GBW is the same as for any three-stage amplifier with an overall Miller capacitance C_{m1}.

Note, however, that the first non-dominant pole coincides with the first zero. They now cancel each other.

For stability, transconductance g_{m3} must be sufficiently large or compensation capacitance C_{m1}.

Relatively large values are used for C_{m1} to avoid excessive power consumption.

For a 3rd-order Butterworth realization, this PFC configuration gives a GBW which is about 6 times larger than a NMC realization with the same load capacitance and power consumption.

PFC schematic

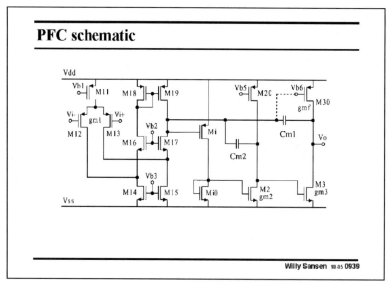

Willy Sansen 10-05 0939

0939

A circuit realization is shown in this slide. Again, a folded cascode is used at the input. A current mirror is used as a second stage. The compensation capacitances are easily identified.

The Gate of output transistor is connected to the output of the first stage, turning the output stage into a class-AB amplifier. This greatly improves the Slew Rate as it is now limited by the total DC current in the input stage into compensation capacitance C_{m1}.

0940

Another low-power three-stage amplifier is shown in this slide. Again the second compensation capacitance C_{m2} is taken away from shunting the output stage. It is used as capacitance C_a, which is driven by its own driver with transconductance g_{ma}. This driver is an AC boosting amplifier towards the second node of the amplifier, whence its name.

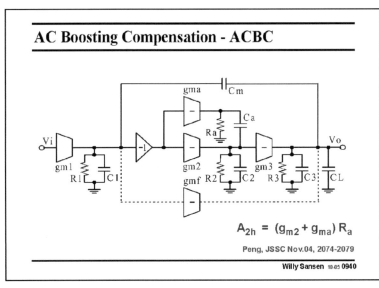

The main purpose is again not to shunt the output stage and to generate a left-hand zero to compensate one of the non-dominant poles. In this way the power consumption is expected to be less than a single-stage amplifier with the same GBW and C_L.

An important additional design parameter is A_{2h}. It is the gain of the additional amplifier when C_a acts as a short, or at high frequencies.

0941

The expression of the open-loop gain A_v is given in this slide. Now we have 4 poles and 3 zero's. The first non-dominant pole ω_1 coincides with the first zero however, and will be cancelled out. Moreover, the next non-dominant pole ω_2 increases with gain A_{2h}. Increasing this gain shifts this non-dominant pole to higher frequencies, allowing a higher GBW as well.

In the slide labeled "AC Boosting Compensation - ACBC":

$$A_{2h} = (g_{m2} + g_{ma})\, R_a$$

Peng, JSSC Nov.04, 2074-2079

Willy Sansen 10-05 0940

In the slide labeled "ACBC equations":

$$A_v(s) = \frac{A_{dc}\left(1 + \dfrac{s}{\omega_1} + \dfrac{s^2}{\omega_1\omega_4} + \dfrac{s^3}{\omega_1\omega_3\omega_4}\right)}{\left(1 + \dfrac{s}{\omega_d}\right)\left(1 + \dfrac{s}{\omega_1} + \dfrac{s^2}{\omega_1\omega_2} + \dfrac{s^3}{\omega_1\omega_2\omega_3}\right)}$$

$$A_{dc} = g_{m1}g_{m2}g_{m3}R_1R_2R_3$$

$$\omega_d = -\frac{1}{C_{m1}g_{m2}g_{m3}R_1R_2R_3}$$

$$\omega_{UG} = \frac{g_{m1}}{C_{m1}}$$

$$\omega_1 = \frac{1}{A_{2h}}\frac{g_{m2}}{C_a}$$

$$\omega_2 = A_{2h}\frac{g_{m3}}{C_L}$$

$$\omega_3 = \frac{1}{R_aC_2}$$

$$\omega_4 = -A_{2h}\frac{g_{m3}}{C_m}$$

$$A_{2h} = (g_{m2} + g_{ma})\, R_a$$

Willy Sansen 10-05 0941

ACBC stability

$$A_v(s) = \frac{A_{dc}\left(1 + \dfrac{s}{\omega_1} + \dfrac{s^2}{\omega_1\omega_4} + \dfrac{s^3}{\omega_1\omega_4\omega_3}\right)}{\left(1 + \dfrac{s}{\omega_d}\right)\left(1 + \dfrac{s}{\omega_1}\right)\left(1 + \dfrac{s}{\omega_2}\right)\left(1 + \dfrac{s}{\omega_3}\right)}$$

$$\omega_1 = \frac{1}{A_{2h}}\frac{g_{m2}}{C_a}$$

$$\omega_2 = A_{2h}\frac{g_{m3}}{C_L}$$

Stability : $\omega_3 > \omega_2 > \omega_1$

Pole and zero at ω_1 cancel

$$\omega_3 = \frac{1}{R_a C_2}$$

Design : $\omega_2 \approx 2\,\omega_{UG}$ **for 60° PM**

$$\omega_4 = -A_{2h}\frac{g_{m3}}{C_m}$$

$GBW \, / \, GBW_{NMC} \approx 17$

$$A_{2h} = (g_{m2} + g_{ma})\,R_a$$

Willy Sansen 10-06 0942

0942

First, for stability, we position the three non-dominant poles in an increasing order of magnitude. The non-dominant pole ω_3 is thus the highest one. This is possible as capacitance C_2 is a small parasitic node capacitance. Moreover, resistance R_a cannot be made too large. It will be realized by means of a diode-connected MOST.

In this case, it is clear that the first non-dominant pole ω_1 cancels the first zero. The other zero's are negligible.

The design itself starts by positioning the first non-dominant pole ω_2 at two times the GBW, as before for a maximally-flat 3rd-order Butterworth characteristic.

The result is a considerable savings in power. Indeed the GBW is a factor 17 times larger than for a NMC amplifier with the same load capacitance and power consumption.

ACBC schematic

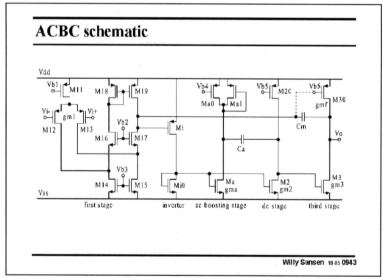

Willy Sansen 10-05 0943

0943

A possible circuit realization of a ACBC amplifier is shown in this slide. As usual, it starts with a folded cascode, with a current mirror as a second stage. The output stage is made class AB by connecting the Gate of transistor M30 to the output of the first stage. This greatly increases the output Slew Rate as well.

The gain boosting stage consists of transistors Ma and Ma1. The gain is fairly precise. For a GBW of 2 MHz and a C_L of 500 pF, this gain A_{2h} is about 9 with a C_a of 3 pF. Compensation capacitance C_m itself is 10 pF and the total current consumption 160 µA. The current through M11 is 18 µA and through M30 about 100 µA. The second stage has only 5 µA in each branch!

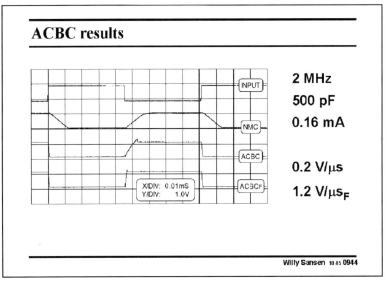

2 MHz
500 pF
0.16 mA

0.2 V/µs
1.2 V/µs$_F$

0944

These transient measurement results show how important it is to have the feedforward with class-AB stage combination at the output.

For a square input waveform, the output waveforms are displayed for the Gate of M30 connected to a DC biasing point (ACBC) and then for the Gate connected to the output of the first stage (ACBC$_F$), labeled with F of Feedforward.

In the latter case, the Slew Rate is much higher and the rise time is much shorter.

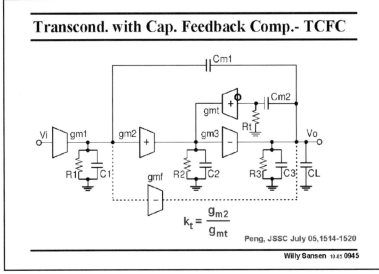

0945

A final example of a low-power three stage amplifier is the TCFC amplifier. It stands for Transconductance with Capacitance Feedback Compensation. The internal compensation capacitance does not short the second stage. It is fed back through a transconductance gmt. This is a cascode or current buffer, it has a low input resistance Rt = 1/gmt and a high output resistance.

The characteristic variable is ratio k_t, which can be set quite accurately as it is a ratio of two transconductances of pMOST transistors. Typical values are 2–3.

Again, a feedforward block is used with transconductance gmf to cancel out a zero and to provide class AB biasing in the output stage. This transconductance gmf is close to gm3 of the output stage.

TCFC equations

$$A_V(s) = \frac{A_{dc}\left(1 + \dfrac{s}{\omega_2} + \dfrac{s^2}{\omega_2{}^2} + \dfrac{s^3}{\omega_2{}^2\omega_4}\right)}{\left(1 + \dfrac{s}{\omega_d}\right)\left(1 + \dfrac{s}{\omega_1} + \dfrac{s^2}{\omega_1\omega_2} + \dfrac{s^3}{\omega_1\omega_2\omega_3}\right)}$$

$$A_{dc} = g_{m1}g_{m2}g_{m3}R_1R_2R_3$$

$$\omega_d = -\frac{1}{C_{m1}g_{m2}g_{m3}R_1R_2R_3}$$

$$\omega_{UG} = \frac{g_{m1}}{C_{m1}} \qquad k_t = \frac{g_{m2}}{g_{mt}}$$

$$\omega_1 = \frac{1}{1+k_t}\frac{g_{m2}}{C_{m2}}$$

$$\omega_2 = \frac{1}{k_t}\frac{g_{m2}}{C_{m2}}$$

$$\omega_3 = (1+k_t)\frac{C_{m2}}{C_2}\frac{g_{m3}}{C_L}$$

$$\omega_4 = -k_t\frac{C_{m2}}{C_2}\frac{g_{m3}}{g_{m1}}\omega_{UG}$$

Willy Sansen 10-05 0946

0946

The expression of the open-loop gain is given in this slide. Again, we find 4 poles and 3 zero's. The GBW is the same as for any three-stage amplifier. The second non-dominant pole ω_2 coincides with the first zero which occurs twice. One of them is now cancelled out. The other zero ω_2 will be used to improve the phase margin.

The right-hand zero ω_4 contains the GBW. This zero is always a lot higher than the GBW, by factor k_t and by the ratio's C_{m2}/C_2 and g_{m3}/g_{m1}, which are both quite high. Indeed, compensation capacitance C_{m2} is always chosen larger than the parasitic node capacitance C_2.

This right-hand zero ω_4 is negligible.

The stability conditions are discussed next.

TCFC stability

Stability $(k_t = 2)$:

$$\frac{C_{m2}}{C_2}\frac{g_{m3}}{C_L} > \omega_{UG} \text{ since } C_{m2} > C_2$$

Design :

$\omega_3 > \omega_1$ since $C_{m2} > C_2$; then $p_{nd} = -\omega_1$

set $\omega_1 \approx 2\,\omega_{UG}$ for 60° PM

$\omega_4 > \omega_{UG}$; then $z_{nd} = -\omega_2$

is $(1+k_t)/k_t$ larger than ω_1

GBW / GBW$_{NMC}$ ≈ 41

$$\omega_1 = \frac{1}{1+k_t}\frac{g_{m2}}{C_{m2}}$$

$$\omega_2 = \frac{1}{k_t}\frac{g_{m2}}{C_{m2}}$$

$$\omega_3 = (1+k_t)\frac{C_{m2}}{C_2}\frac{g_{m3}}{C_L}$$

$$\omega_4 = -k_t\frac{C_{m2}}{C_2}\frac{g_{m3}}{g_{m1}}\omega_{UG}$$

Willy Sansen 10-05 0947

0947

For stability, we need to have the third non-dominant pole ω_2 larger than the first one ω_1. This is always the case since k_t is larger than unity. In the design example later on, a value of k_t is taken of two.

The main stability condition specifies that the second non-dominant pole ω_3 must be larger than the GBW. This is easily satisfied as the ratio's C_{m2}/C_2 and g_{m3}/g_{m1} are large indeed. This circuit is therefore easy to stabilize!

The most important non-dominant pole is ω_1. For a 3rd-order Butterworth characteristic this pole is positioned at 2 times the ω_{UG} (or the GBW).

As discussed on the previous slide, the right-half-plane zero ω_4 is a lot larger than the GBW and can now be neglected. The only zero left is a zero ω_2 in the left half of the complex plane. It is only a little (actually $(1 + k_t)/k_t$ or 1.5 times in this design) larger than the non-dominant pole. It is therefore ideally positioned to improve the phase margin.

The resulting ratio in GBW compared to a conventional NMC amplifier is quite impressive.

TCFC schematic

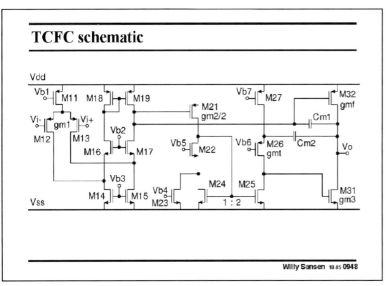

Willy Sansen 10-05 **0948**

0948

A circuit realization is shown in this slide.

As usual, it starts with a folded cascode. A current mirror is used as a second stage. Output transistor M32 is driven by the output of the first stage. The output stage operates in class AB. Hence, it does not limit the Slew Rate. The Slew Rate will now be determined by the DC current of the first stage and compensation capacitance Cm1.

The cascode is series with compensation capacitance Cm2 is realized with transistor M26. Its transconductance is twice the transconductance of transistor M21, which is the amplifying transistor of the second stage. Indeed, current mirror M24/M25 ensures a current ratio of two. Factor k_t is fairly precisely two as well.

Table of contents

Willy Sansen 10-05 **0949**

0949

As many different configurations have been discussed, they have to be compared with some kind of Figure of Merit.

Such a Figure of Merit normally includes the power dissipation for a certain GBW and load capacitance. However, two different FOMs exist. One of them takes into account the actual power dissipation in mW, whereas the other one takes only the current into account. The difference is the supply voltage used.

TCFC and other 3-stage opamps

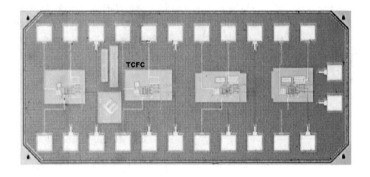

TCFC

Willy Sansen 10-05 **0950**

0950

Such a comparison has been made based on experimental results only.

In this layout, several opamps can be distinguished, among which is the TCFC one. They all have similar sizes, as the load capacitances were 150 pF for all of them. The technology was 0.35 μm CMOS.

The comparison is better carried out in a Table as shown next.

Comparison

No Capacitance

Nested Miller comp.

NMC with nulling R

Miller C Substitutions

AC Boosting comp.

Transconductance Capacitance FC

		IFOMs $\frac{MHz \cdot pF}{mA}$	IFOVL $\frac{Links \cdot pF}{mA}$	FOMs $\frac{MHz \cdot pF}{mW}$	FOML $\frac{Links \cdot pF}{mW}$	Tech.
NCs						
NCFF	[Tha03]	536	-	214	-	0.5μm CMOS
NMs						
NMC	[Esc92]	632	211	79	26	2GHz f_t BJT
MNMC	[Esc92]	1053	368	132	46	2GHz f_t BJT
NMCF	[Leu01a]	600	246	300	123	0.8μm CMOS
NGCC	[You97]	36	148	18	74	2μm CMOS
HNMC	[Esc94]	134	134	89	89	0.8μm CMOS
MHNMC	[Esc94]	401	467	267	311	0.8μm CMOS
DNMC	[Per93]	250	188	50	38	0.8μm CMOS
MRs						
NMCNR	[Leu01a]	410	168	205	84	0.8μm CMOS
IRNMC	[Ho03]	626	444	209	148	0.6μm CMOS
EFC	[Ng09]	817	1200	272	400	0.5μm CMOS
NGRNMC	[Peng03]	700	490	280	196	0.35μm CMOS
MSs						
PFC	[Ram03b]	1915	709	1276	473	0.5μm CMOS
DFCFC	[Leu01a]	1238	628	619	314	0.8μm CMOS
ACBs						
ACBC	[Peng04]	5981	2215	2991	1108	0.35μm CMOS
ACBC$_F$	[Peng04]	5864	3086	2932	1543	0.35μm CMOS
TCFs						
AFC	[Ahu83]	326	-	33	-	4μm CMOS
AFFC	[Lee03b]	2700	894	1350	447	0.8μm CMOS
DLPC	[Lee03a]	3818	1800	2545	1200	0.35μm CMOS
TCFC	[Peng05]	14250	5175	9500	3450	0.35μm CMOS

Willy Sansen 10-06 **0951**

0951

For this purpose the several three-stage opamps have been put in categories. All possible Figures-of-Merit have been added. Not only a choice is made between the current (mA) or the power (mW) consumption. Another alternative is to use either the GBW, which is a quality factor for small signals, or the Slew Rate, which is more important for large-signal operation, such as in switched-capacitor circuits.

The first category uses no compensation capacitance. It is clear that this is not the way to go!

The second category is the most used one. It lists all the NMC varieties. Several of them have been discussed in this Chapter. The FOMs are reasonably good. The best one is the multi-path opamp MNMC. However, its Slew Rate is not so good, because no feedforward stage is used to turn the output stage into a class-AB stage.

The addition of nulling resistances in series with the compensation capacitance does not increase the FOM a lot.

Real improvements are only achieved once left-hand-plane zero's can be generated, which compensate the main non-dominant pole. This is the case when positive feedback is used (PFC) or damping-factor control (DFCFC). This is even more the case when a AC boosting amplifier is used (ACBC) or additional transconductance blocks, as in Active feedback compensation (AFC) and especially transconductance and capacitances compensation (TCFC). The results are obvious.

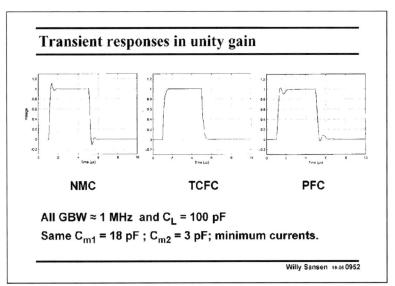

0952

The transient responses of these amplifiers are related to the pole-zero positions but not always in a straight-forward way. The responses of three amplifiers are given in this slide for the same GBW and load capacitance. They also use the same 3rd-order Butterworth frequency response and the same compensation capacitances. As a result, the current in the NMC is a lot larger than in the other two, as discussed before.

The time responses show that the NMC and PFC amplifier show some overshoot, whereas the TCFC amplifier does not. This is a result of small differences in the pole-zero positions.

0953

All important references are collected here. Most of them are from the *IEEE Journal of Solid-State Circuits*. A few are from the Transactions on Circuits and Systems and also from the Custom Integrated Circuits Conference.

References

J. Huijsing, D. Linebarger, "Low-voltage operational amplifier with rail-to-rail input and output ranges" *IEEE J. Solid-State Circuits*, vol. 20, pp. 1144-1150, Dec. 1985.

R. Eschauzier, L. Kerklaan, and J. Huijsing, "A 100-MHz 100-dB operational amplifier with multipath nested Miller compensation structure," *IEEE J. Solid-State Circuits*, vol. 27, pp. 1709-1717, Dec. 1992.

S,Pernici, G. Nicollini, R. Castello, "A CMOS low-distortion fully differential power amplifier with double nested Miller compensation", *IEEE J. Solid-State Circuits*, vol. 28, pp. 758-763, July 1993.

F. You, S. H. K. Embabi and E. Sánchez-Sinencio, "Multistage amplifier topologies with nested Gm-C compensation," *IEEE J. Solid-State Circuits*, vol. 32 pp. 2000-2011, Dec. 1997.

K. N. Leung and P. K. T. Mok, "Nested Miller compensation in low-power CMOS design," *IEEE Trans. Circuits Syst. II*, vol. 48, pp. 388-394, Apr. 2001.

K. N. Leung and P. K. T. Mok, W. H. Ki, and J. K. O. Sin, "Three-stage large capacitive load amplifier with damping-factor-control frequency compensation," *IEEE J. Solid-State Circuits*, vol. 35, pp. 221-230, Feb. 2000.

H. Lee and P. K. T. Mok, "Active-feedback frequency-compensation technique for low-power multistage amplifiers," *IEEE J. Solid-State Circuits*, vol. 38, pp. 511-520, Mar. 2003.

X. Peng and W. Sansen, "Nested feedforward gm-stage and nulling resistor plus nested Miller compensation for multistage amplifiers," CICC, May 2002, pp. 329-332.

J. Ramos and M. Steyaert, "Three stage amplifier with positive feedback compensation scheme," in *Proc. IEEE Custom Integrated Circuits Conf.*, Orlando, FL, May 2002, pp. 333-336.

X. Peng and W. Sansen, "AC boosting compensation schema for low-power multistage amplifiers", *IEEE J. Solid-State Circuits*, vol. 39, pp. 2074-2079, Nov. 2004.

X. Peng and W. Sansen, "Transconductance with capacitances feedback compensation for multi-stage amplifiers", *IEEE J. Solid-State Circuits*, vol. 40, pp.1514-1520, July 2005.

Willy Sansen 10-05 0953

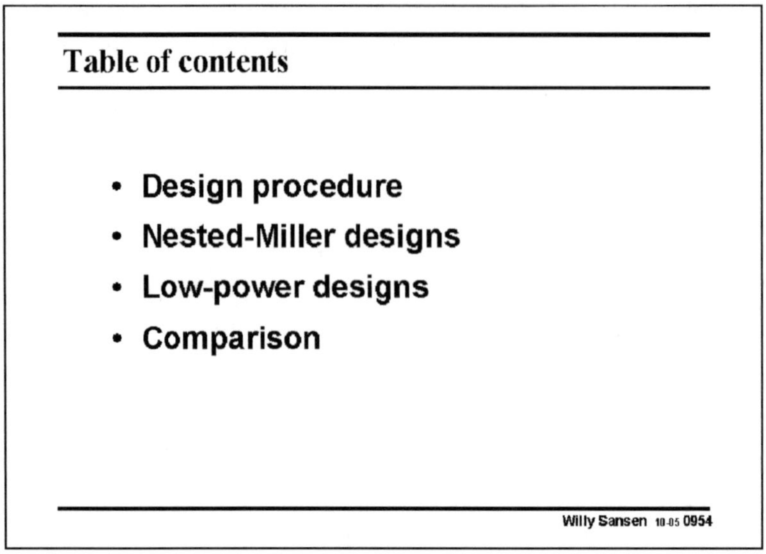

Table of contents

- **Design procedure**
- **Nested-Miller designs**
- **Low-power designs**
- **Comparison**

Willy Sansen 10-05 0954

0954

As a conclusion, this Chapter has been devoted to three-stage operational amplifiers. Considerable attention has been paid to the stability of such amplifiers. At least two zero's are always generated, in addition to three poles.

Conventional Miller compensation is discussed next, followed by several derivatives. Their main disadvantage is the power consumption in the third stage.

To reduce power consumption, several more configurations are discussed. They all have 4 poles and 3 zero's. They all exploit the cancellation effect of one non-dominant pole by a zero. Moreover, they all position a left-hand-plane zero close to the next non-dominant pole. In this way, the power consumption can be reduced considerably. In the best case (in a TCFC amplifier) the power consumption can be reduced by up to a factor of 40, compared to a conventional Nested Miller Compensation amplifier.

Current-input Operational Amplifiers

Willy Sansen

KULeuven, ESAT-MICAS
Leuven, Belgium

willy.sansen@esat.kuleuven.be

Willy Sansen   10-05 101

101

Instead of voltages, currents can be used at the inputs as well. They lead to current-input amplifiers.

They are discussed in this Chapter.

Table of contents

- **Operational current amplifier**
- **Configurations current amplifiers**

Willy Sansen   10-05 102

102

First of all, full operational amplifiers are discussed, after which some more configurations are following.

In this class, transresistance amplifiers can be added as well. They are discussed in Chapter 14 however, on shunt-series feedback at the input.

As a result this is a fairly short Chapter.

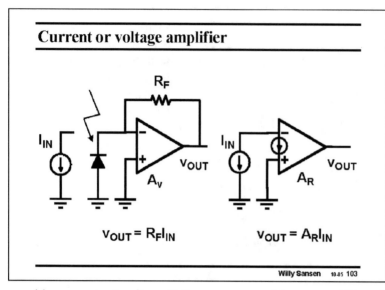

103

Current amplifiers are used for current sensors and for some high-frequency applications.

For example a photo-diode (pixel detector, radiation detector, ...) which receives light, provides a current proportional to this amount of light, provided it is reverse biased. It behaves as a current source with value I_{IN}.

In this case a voltage amplifier can be used with resistive feedback, which provides an output voltage $R_F I_{IN}$, or a current-input amplifier with transresistance A_R.

The question is, which one is better, from the point of view of gain (or sensitivity), speed and noise performance?

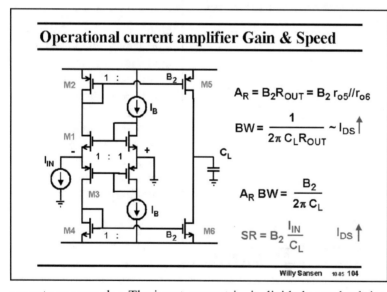

104

A current input assumes the use of cascodes, as shown in this slide.

This amplifier actually consists of two current mirrors, connected by their reference voltages. The outputs are then current mirrored to the output, with current factor B_2.

These two current mirrors provide the biasing. Both transistors M1 and M3 are biased at current I_B.

However, for small-signals, transistors M1 and M3 operate as cascodes. The input current i_{IN} is divided over both input cascodes, multiplied by B_2, and generates an output voltage in the output resistance R_{OUT} at the Drains of output transistors M5 and M6.

The current gain is very modest but the transresistance can be very large, especially if cascodes are used on transistors M5/M6 and gain boosting on these cascodes.

The bandwidth BW is determined by R_{OUT} as well. As a result, the transresistance-bandwidth product does not contain R_{OUT} any more.

It is not possible to compare the product A_RBW to the GBW of a voltage amplifier. They have totally different dimensions!

The main advantage of this amplifier is that the Slew-Rate is unlimited. Indeed, for a large input current, the SR is determined by this input current itself, multiplied by B2.

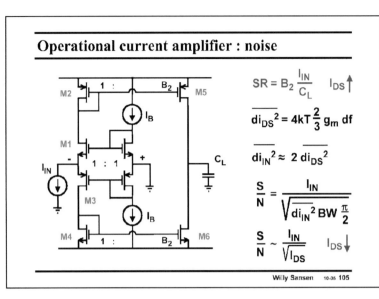

Operational current amplifier : noise

$$SR = B_2 \frac{I_{IN}}{C_L} \quad I_{DS}\uparrow$$

$$\overline{di_{DS}^2} = 4kT\frac{2}{3}g_m\,df$$

$$\overline{di_{IN}^2} \approx 2\,\overline{di_{DS}^2}$$

$$\frac{S}{N} = \frac{I_{IN}}{\sqrt{\overline{di_{IN}^2}\,BW\,\frac{\pi}{2}}}$$

$$\frac{S}{N} \sim \frac{I_{IN}}{\sqrt{I_{DS}}} \quad I_{DS}\downarrow$$

Willy Sansen   10-05 105

105
This large input current can be larger than the biasing current I_B. In this case, transistor M1 carries a large current and transistor M3 goes off. Transistor M1 operates in class-AB! The Slew-Rate is then very high, at the price of some distortion!

If we do not want this distortion, we have to choose a biasing current I_B, which is always larger than the peak signal current i_{IN}.

High-speed generally leads to bad noise performance!

This also applies to this current-input amplifier. A detailed analysis (on the next slide) shows that the noise of a cascode is always negligible, provided it is not driven by a voltage source and it does not have a low resistive load. This is exactly what we have here.

Cascode transistor M1 sees $1/g_{m3}$ as a source resistance and has $1/g_{m2}$ as a load. The output noise current of M1 flows unattenuated through both M2 and M3. Fortunately, they cancel out at the output. However, the noise current powers of M2 and M4 themselves add up. Together they make up the total equivalent input noise current.

The SNR is then easily calculated. The larger we take I_B, the worse the SNR becomes! Indeed, higher speed leads to more noise!

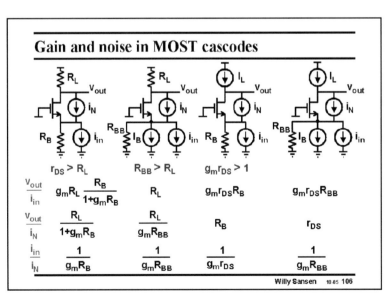

Gain and noise in MOST cascodes

	$r_{DS} > R_L$	$R_{BB} > R_L$	$g_m r_{DS} > 1$	
$\dfrac{V_{out}}{i_{in}}$	$g_m R_L \dfrac{R_B}{1+g_m R_B}$	R_L	$g_m r_{DS} R_B$	$g_m r_{DS} R_{BB}$
$\dfrac{V_{out}}{i_N}$	$\dfrac{R_L}{1+g_m R_B}$	$\dfrac{R_L}{g_m R_{BB}}$	R_B	r_{DS}
$\dfrac{i_{in}}{i_N}$	$\dfrac{1}{g_m R_B}$	$\dfrac{1}{g_m R_{BB}}$	$\dfrac{1}{g_m r_{DS}}$	$\dfrac{1}{g_m R_{BB}}$

Willy Sansen   10-05 106

106
For a MOST cascode, the contributions of the input current i_{in} and of the transistor current noise source i_N are compared. Both are calculated to the output. Finally, the ratio of input noise to the transistor noise current i_{in}/i_N is calculated. This shows what the equivalent input noise current is as a result of the transistor noise, which is the same in all cases, and which is proportional to g_m.

Four cases can be distinguished depending on whether the cascode has a real current drive (cases 2 and 4). Resistance R_{BB} is then the output resistance of the input current source i_{in}. It is larger than R_L unless R_L is substituted by a current source I_L.

When the cascode is driven by a low resistance R_B, it acts rather as a voltage source. Again, the load can be a small resistor R_L or large, as a current source I_L.

It is clear that the contribution of the transistor noise to the input current is always small, except when the cascode is voltage driven (or driven with another $1/g_m$) and at the same time, a small load resistor is used, or the input $1/g_m$ of a current mirror.

There is one single condition where the noise of a cascode is important. In all other cases it is negligible.

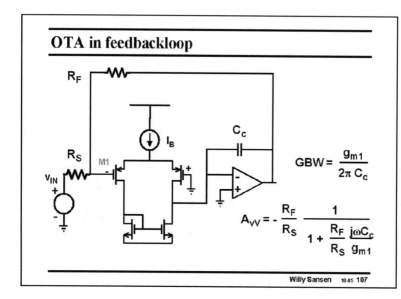

107
Another way to explore the high-speed capabilities of a current opamp is to close the feedback loop.

A two-stage voltage opamp is sketched in this slide, in which the loop is closed by means of two resistors R_S and R_F. They define the gain R_F/R_S, and also the bandwidth.

Indeed, a voltage opamp exchanges bandwidth for gain. It has a constant GBW, as is clearly shown by the expressions.

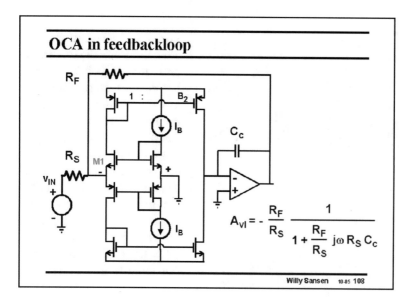

108
A similar two-stage opamp is sketched in this slide, but with a current-input stage as a first stage rather than with a voltage-input stage.

Again, the loop is closed by means of two resistors R_S and R_F, which again define the gain R_F/R_S.

However, the bandwidth is now different. It seems to be determined by the series resistor R_S. This resistor converts the input voltage into a current, as a result it appears in the expressions separately.

If we now keep this resistor R_S constant, and vary the feedback resistor R_F to set the gain, then we have exactly the same set of curves as for a voltage opamp.

This is shown next.

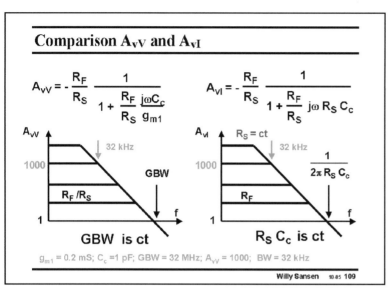

109

On the left, a voltage opamp is shown with a gain of 1000. Its GBW is 32 MHz.

On the right, a current opamp is shown, with the same gain. The input series resistor R_S is kept constant, at a value of 5 kΩ. As a result, this current opamp has a GBW of 32 MHz. The exchange of gain and bandwidth is the same as for a voltage opamp.

On the other hand, we can vary this resistor R_S to set the gain. This is shown next.

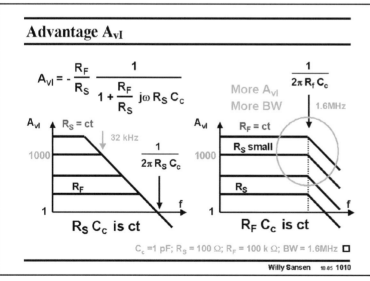

1010

On the left, the same voltage opamp is taken from the previous slide. On the right, we use the same current opamp but we now vary R_S to set the gain. The feedback resistor R_F is kept constant, at a value of 100 kΩ.

We now we have an amplifier with a constant bandwidth, rather than with constant gain-bandwidth. The bandwidth is determined by the $R_F C_c$ time constant.

It is clear that with this current opamp, we can reach combinations of large gain and high speed, which are not available with a voltage opamp. This is a clear advantage.

For a gain of 1000, the bandwidth now extends to 1.6 MHz, rather than to 32 kHz, which is an enormous increase.

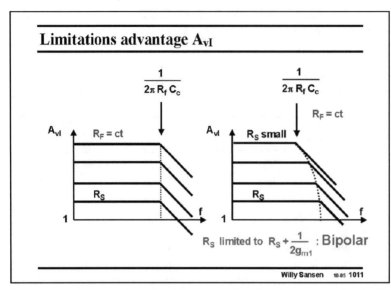

1011

For this gain of 1000, The value of R_S is now $100\,\Omega$. For higher gains, this resistor has to be even smaller. This is not so easy as the input resistance $1/2g_{m1}$ of the circuit starts playing a role.

As a consequence, these high speeds are not easily realized at high gain values. The curves bend to the left. The gain is still large in practice, but not as large as expected.

Also, bipolar transistors have lower input resistances. They can therefore be used towards higher gain-bandwidth combinations. This is why most commercial current opamps are in bipolar technology.

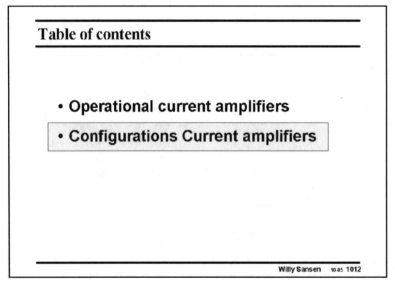

1012

Now that the advantages are known for current opamps, let us have a look at some realizations.

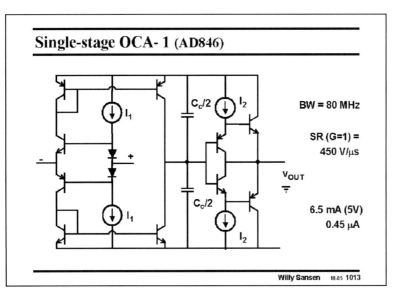

1013

This is a single-stage current operational amplifier in bipolar technology. The bandwidth is set by the two capacitors at the output. It is 80 MHz.

Obviously, the Slew Rate is large as well. For a closed-loop gain of 1, it is 450 V/μs!

The output stage consists of a double emitter follower.

As usual, with high-speed amplifiers, no noise specification is given.

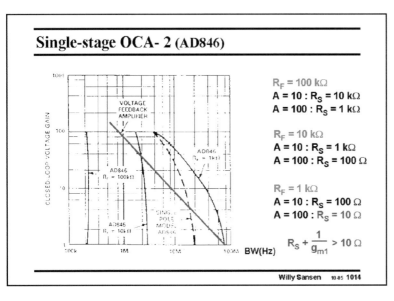

1014

The specifications of this amplifier clearly show the constant-bandwidth behavior.

For example, for a $R_F = 100 \, k\Omega$. The bandwidth is just over 200 kHz. It does not depend on the actual gain.

Curves are also given for $R_F = 10 \, k\Omega$ and 1 kΩ. It is clear that for the latter, a gain of 100 is not easily obtained. Indeed, for this value of R_F, a gain of 100 would necessitate a R_S of a mere 10 Ω!

In this case, the input resistance of the amplifier takes over. The curve bends to the left.

At a gain of 100, a bandwidth of 4 MHz results, which would require a GBW of 400 MHz if it were a voltage amplifier.

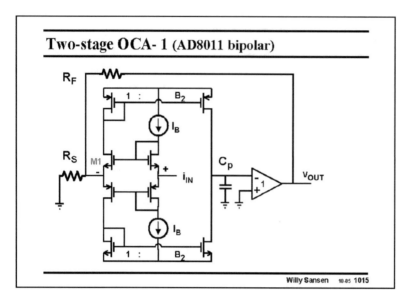

1015

A single-stage amplifier is easily extended to a two-stage amplifier as shown next.

This amplifier is copied from an earlier slide for sake of comparison.

Note that the emitter followers at the output are represented by a voltage amplifier with unity gain.

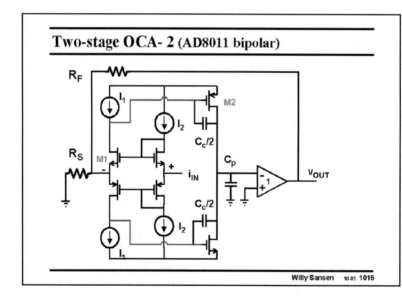

1016

The two-stage current amplifier is shown in this slide.

Obviously, a compensation capacitor is now required, which will cause more current to flow in the second stage.

The main advantage is however that the output swing can be larger. Also, more gain can be achieved without the use of cascodes.

Again, bipolar transistors are used at the input to have a smaller input resistance. Is this why smaller values can be used for the series input resistor R_S, yielding higher gain-bandwidth combinations?

The input cascode transistors determine the noise performance in single-stage current amplifiers. Active loads are now used for these cascodes. Is their noise still dominant?

The answer is positive. Indeed, at higher frequencies, the compensation capacitance acts as a short-circuit. The second stage is little more than an impedance $1/g_{m2}$. This is a low impedance. The noise of the input cascodes is therefore still dominant.

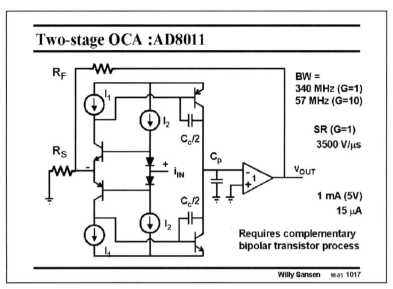

1017

In bipolar technology, this current opamp has excellent performance.

For a gain of 10, the bandwidth is 57 MHz. For a gain of 1, the BW is 340 MHz and the SR is an impressive 3500 V/μs.

The quiescent current is only 1 mA on a single 5 V supply.

The input current for these values is 15 μA.

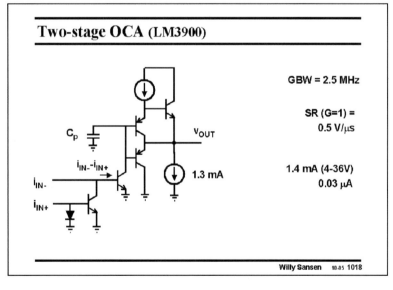

1018

Current amplifiers have existed before.

An earlier current amplifier is shown in this slide.

The current difference at the input is realized by addition of a current mirror on the non-inverting input terminal.

It is clearly a single-stage amplifier with modest performance.

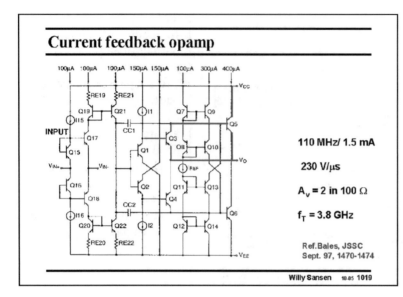

Current feedback opamp

110 MHz/ 1.5 mA

230 V/µs

A_v = 2 in 100 Ω

f_T = 3.8 GHz

Ref. Bales, JSSC
Sept. 97, 1470-1474

Willy Sansen   10.05 1019

1019

An integrated current amplifier is depicted in this slide.

It is a two-stage amplifier with current input. The output of the first stage is at the Drains of transistors Q21/Q22. Transistors Q1 and Q2 now serve as emitter followers to drive the second stage.

In this stage, pnp-npn composites are used to drive the output load. For example, the Q3/Q5 composite behaves as a super-pnp transistor. On the other side, Q4/Q6 are a super npn.

The compensation capacitances CC1 and CC2 are connected from the middle point of such a composite, which is a strange point indeed.

The performance is great: 110 MHz bandwidth and 230 V/µs, using bipolar transistors with an f_T of 3.8 GHz only!

Table of contents

• **Operational current amplifiers**

• **Other current amplifiers**

Willy Sansen   10.05 1020

1020

In this Chapter, current amplifiers are shortly discussed and compared to voltage amplifiers.

Current amplifiers provide higher-speed performance at the cost of higher equivalent input noise.

Rail-to-rail input and output amplifiers

Willy Sansen

KULeuven, ESAT-MICAS

Leuven, Belgium

willy.sansen@esat.kuleuven.be

111

Rail-to-rail amplifiers are a special category of amplifiers. They can take input signal swings from the positive to the negative supply rail. Their common-mode input range extends from rail to rail.

However, this is fairly easy to do at the output. This is quite complicated however at the input. In principle, only a folded cascode is capable of including a supply rail at the input. This will be the basis for all-rail-to-rail input amplifiers.

An alternative principle is to use depletion-mode nMOST devices. Leaving out an ion implant allows nMOST devices to obtain a negative threshold voltage. This allows rail-to-rail input stages with supply voltages down to 1 V. It is not considered any further however, as no standard CMOS can be used.

Let us first decide when we need such an input. What are the circuit configurations which allow a rail-to-rail input range?

Table of contents

- **Why rail-to-rail ?**
- **3 x Current mirror rtr amplifiers**
- **Zener diode rtr amplifiers**
- **Current regulator rtr amplifier on 1.5 V**
- **Supply regulating rtr amplifier on 1.3 V**
- **Other rtr amplifiers and comparison**

112

In addition, such amplifiers will have to operate at low supply voltages. We are thus looking for rail-to-rail input amplifiers at low supply voltages where we have the highest need indeed for large signal swings.

Why rail-to-rail amplifiers ?

- **For low supply voltages : use full range for maximum dynamic range**

- **Fully differential signal processing**

- **Rail-to-rail output is always required**

- **But not necessarily rail-to-rail-input !**

Willy Sansen 10-05 113

113

The first reason to require a rail-to-rail input range is to be able to maintain the same Signal-to-Noise ratio for smaller supply voltages. Indeed signal swings decrease with the supply voltage. A rail-to-rail swing at the input is the highest that can be obtained.

For maximum output swing, a fully-differential operation is a must as well.

At the output, it is easy to provide a rail-to-rail swing. We simply take two output transistors Drain to Drain. For a capacitive load, rail-to-rail is guaranteed.

At the input, however, we do not need rail-to-rail operation.

Symmetrical CMOS OTA

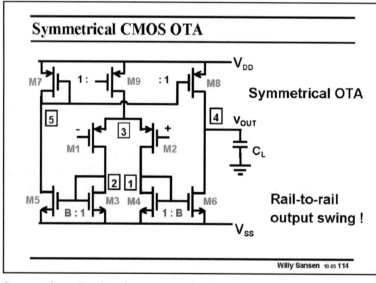

Symmetrical OTA

Rail-to-rail output swing !

Willy Sansen 10-05 114

114

To illustrate this last point, let us take this well known symmetrical opamp.

For a capacitive load, it is obvious that the output can reach rail-to-rail output voltages. At the extremities, the output transistor goes into the linear region and the gain is reduced leading to distortion. It is still rail-to-rail, however.

At the input, it is clear that a voltage is lost of $V_{GS1} + V_{DSsat9}$ for the input range, which is rather large.

Some other circuitry is now required.

Is this always required?

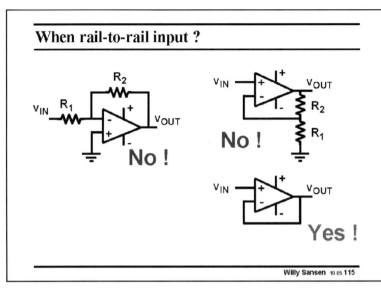

115

The answer is negative. Three amplifier realizations are shown in this slide. Only one of them requires a rail-to-rail input range.

An inverting amplifier has almost no swing at all at the inputs. The output signal is divided by the open-loop gain so that the minus input hardly sees any signal. At higher frequencies, the minus input sees a larger voltage but never rail-to-rail!

This also applies to the non-inverting amplifier. Both inputs have about the same swing. If some gain is required, then the inputs can never reach rail-to-rail swings.

The only configuration which needs rail-to-rail input swing is the buffer. Since the gain is unity, a rail-to-rail output requires the input to be able to follow.

Buffers are more often class-AB amplifiers. Most class-AB amplifiers have a rail-to-rail input. Many of the examples given, have a class-AB output.

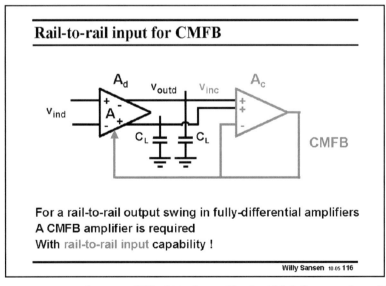

116

Another good application of a rail-to-rail input is the CMFB amplifier required in fully-differential amplifiers.

However, the output swing of the differential amplifier is followed by a CMFB amplifier. In order to be able to have rail-to-rail differential output swing we must have a rail-to-rail input swing for the CMFB amplifier.

This is rarely the case as rail-to-rail input amplifiers are more complicated. As a consequence, they are difficult to be realized at high frequencies with limited power consumption.

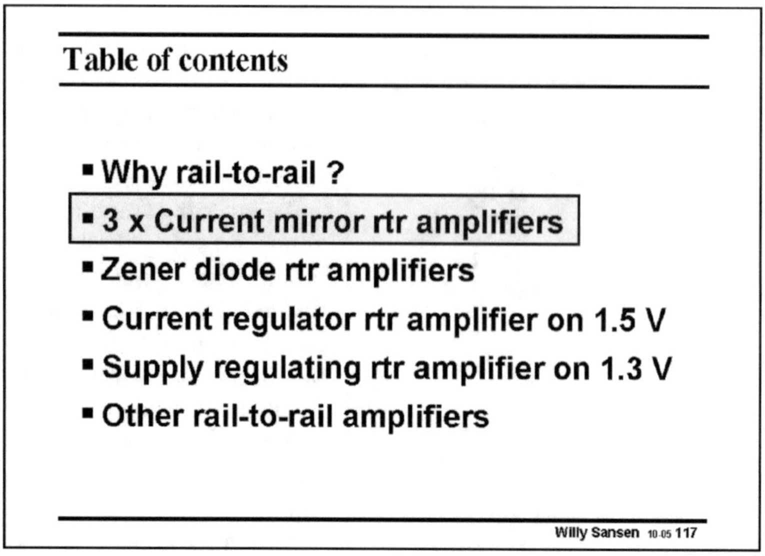

Table of contents

Willy Sansen 10-05 **117**

117
The first type of rail-to-rail input amplifier is the one that uses $3\times$ current mirrors, as shown next.

Problem ?

$V_{GS} \approx 0.9\,V$ & $V_{DSsat} \approx 0.2\,V$ >>> $V_{GSDS} = 1.1\,V$

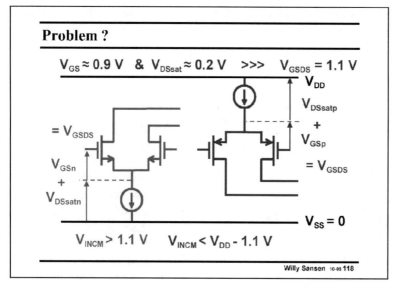

$V_{INCM} > 1.1\,V$ $V_{INCM} < V_{DD} - 1.1\,V$

Willy Sansen 10-05 **118**

118
The only way to reach a rail-to-rail input swing is to use two folded cascodes in parallel. Only the input transistors are shown.

Each of the input pairs needs a minimum input voltage of one $V_{GS}+V_{DSsat}$. For a V_T of 0.7 V, this is about 1.1 V (when $V_{GS}-V_T=0.2\,V$).

When both inputs are now connected in parallel, 2.2 V is required as a minimum value for the supply voltage!

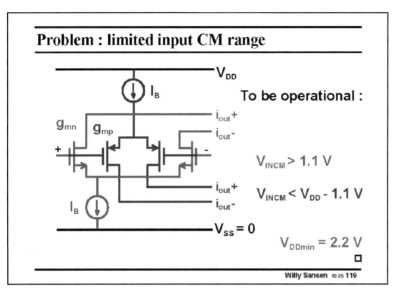

Willy Sansen 10-05 119

119

Connecting both inputs in parallel reduces the input common-mode range by 2.2 V. The minimum supply voltage must therefore be a minimum of 2.2 V. In practice the supply voltage will be 2.5 V.

We are a long way from a 1 V supply voltage! Even if V_T were only 0.3 V, the minimum supply voltage would still be 1.4 V.

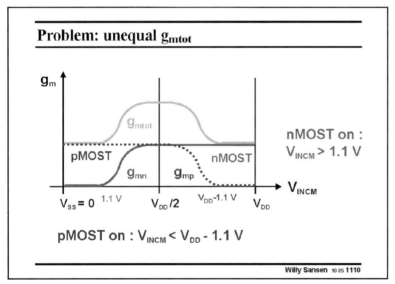

Willy Sansen 10-05 1110

1110

With a supply voltage of 2.5 V, there is a problem with the total transconductance and thus with the total GBW.

For common-mode input voltages with V_{INCM} in the middle (at half the supply voltage), both input pairs are operational. Their g_m's are now added. Actually, the g_m is doubled, as normally the nMOST transconductance equals the pMOST one.

On both extremities however, only one pair is operational and the g_m is not doubled. It is therefore half of what is obtained in the middle.

The transconductances are sketched in this slide for each pair separately and also summed. It is clear that the pMOST g_m goes to zero when the common-mode input V_{INCM} reaches the positive supply voltage within 1.1

V. Also, at zero V_{INCM}, the nMOSTs are off and only start working once V_{INCM} is larger than approximately 1.1 V.

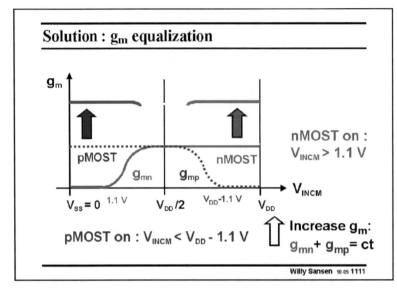

Solution : g_m equalization

nMOST on : $V_{INCM} > 1.1$ V

pMOST on : $V_{INCM} < V_{DD} - 1.1$ V

Increase g_m:
$g_{mn} + g_{mp} = ct$

Willy Sansen 10.05 **1111**

1111

We cannot allow the total transconductance g_{mtot} to change over a factor of two over the input common-mode range. This would give a large amount of distortion.

In order to equalize the g_{mtot}, we must increase the g_m of each pair by a factor of two at both ends. In general, we must find a way to make g_{mtot} or the sum of the g_m's constant over the whole common-mode range.

Equalize g_{mtot} in strong inversion

$$g_{mn} + g_{mp} = ct1$$

$$\sqrt{2 K'_n \frac{W_n}{L_n} I_{Bn}} + \sqrt{2 K'_p \frac{W_p}{L_p} I_{Bp}} = ct1$$

$$\sqrt{K'_n I_{Bn}} + \sqrt{K'_p I_{Bp}} = ct2$$

$$\sqrt{I_{Bn}} + \sqrt{I_{Bp}} = ct3$$

3 x Current mirror : $\sqrt{1} + \sqrt{1} = \sqrt{0} + \sqrt{4}$ $> 4 - 1 = 3$

Willy Sansen 10.05 **1112**

1112

How can we make the sum of the g_m's constant over the whole common mode range?

Substitution of the g_m's by their expression with the current and W/L shows that we have to adjust the currents. The MOSTs are assumed to work in strong inversion.

To double the g_m, we must multiply the current by four. In other words, we must add three times more current to the existing current than we already had.

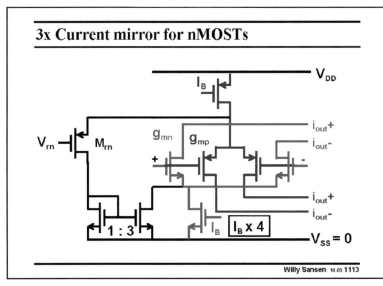

1113

A circuit that is able to add three times more current to the existing current than already present, is shown in this slide.

For a common-mode (or average) input voltage in the middle, both pairs are operational. The transistor M_{rn} is then off. Its gate voltage V_{rn} is too high compared to the common-mode input voltage.

When the input voltages increase however, the pMOSTs turn off, and all current I_B flows through M_{rn} to the $3\times$ current mirror, which is then added to the current I_B of the nMOST current source. This current is thus multiplied by four. The nMOST transconductances are now doubled.

The actual voltage at which the current is taken from the input pMOSTs towards M_{rn} is reference voltage V_{rn}. The transistor M_{rn} now forms a differential pair with both input pMOSTs. When the common-mode input voltage is exactly V_{rn}, half the current I_B flows through the pMOSTs and the other half through the transistor M_{rn}. When the average input voltage is higher than V_{rn}, all the current flows through M_{rn}. The input pMOSTs are now completely shut off.

Note that reference voltage V_{rn} must be about 1.1 V lower than the supply voltage if V_T is 0.7 V.

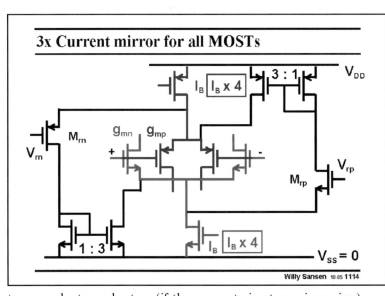

1114

The same arrangement can be made for the nMOSTs. Another reference voltage V_{rp} is now added, which is about 1.1 V.

When the average input voltage is less than 1.1 V, all current I_B of the nMOST current source is pulled through transistor M_{rp}. It is then multiplied by three and added to the current already flowing through the pMOSTs. The current is multiplied by four and the

transconductance by two (if they operate in strong inversion).

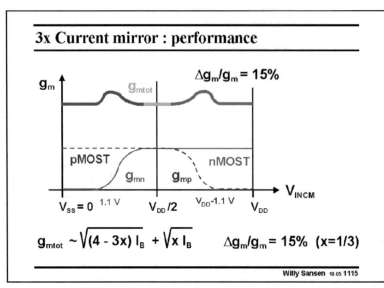

1115

The transconductance is then effectively equalized over the whole common-mode input range. During the transitions however, the sum of the g_m's is not constant. They show an error of about 15%.

An approximate expression of the actual g_{mtot} is given. It is derived from the simple square-law models of a MOST.

An error of 15% may not sound too bad. It all depends on the application.

However, We will see later in a comparative table (last slide), that this is the worst of all.

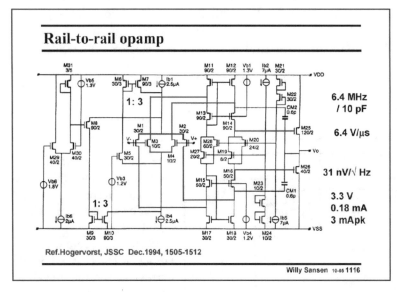

1116

A realization which has used this g_m equalization is shown in this slide.

The two input cascodes in parallel are easily found. The output transistors act as a second stage. Miller capacitances are therefore necessary. They go through cascodes to avoid positive zero's. Some specifications are also listed.

The $3 \times$ current mirrors are easily identified as well.

Table of contents

Willy Sansen 10-05 **1117**

1117
Another way to provide g_m equalization is to use Zener diodes. Although Zener diodes are not available, MOSTs are used instead.

Zener diodes

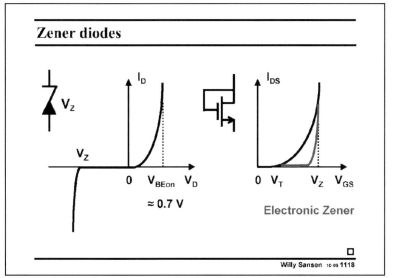

Electronic Zener

Willy Sansen 10-05 **1118**

1118
A Zener diode actually provides a more sharp breakdown at V_Z. A MOST does not provide the same sharpness. The addition of a few MOSTs can sharpen the corner somewhat. It is then called an electronic Zener diode.

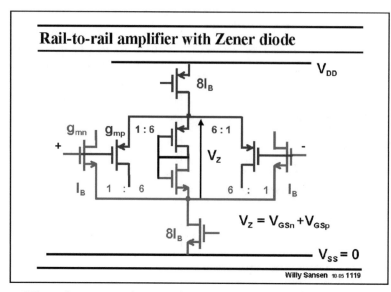

1119

The principle is shown in this slide.

For input common-mode voltages in the middle, all transistors are on. The two diode connected MOSTs in the middle conduct as well. Because they are $6 \times$ larger, they take six times more current. The total current in either current source is now $8I_B$, where I_B is the current in each input transistor.

Let us call the voltage across the two diodes V_Z. It is two times their V_{GS}.

When the average input voltage goes up, the pMOSTs drop out, as shown next.

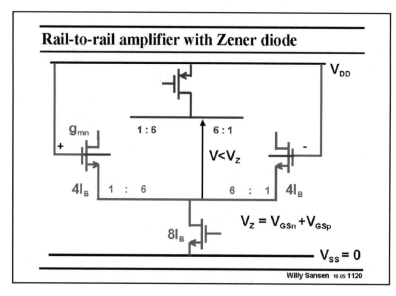

1120

When the average input voltage goes up, for example, up to the supply voltage, the pMOSTs drop out. Indeed, the voltage across the two diodes has dropped to below the V_Z voltage and drops out. They are omitted from the figure in this slide.

As a consequence, all the current $8I_B$ has to flow in the two nMOST transistors. They both carry a current of $4I_B$, which is four times larger than before. Their transconductance now doubles.

1121

A much better behavior is achieved by taking an electronic Zener. For this purpose, an amplifier is added in parallel with the nMOST Zener diode followed by a source follower.

The transitions are now much better defined. This means that the error in transconductance over the common-mode input range is smaller.

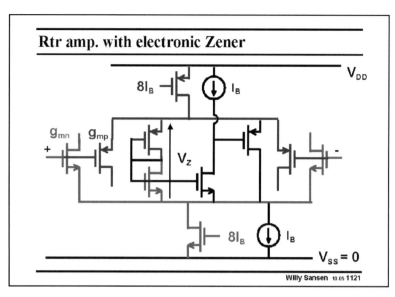

1122

However, the error in transconductance is only 6% in contrast with 25% without the amplifier.

The larger current in the input devices leads to a larger Slew Rate, which is a considerable additional advantage.

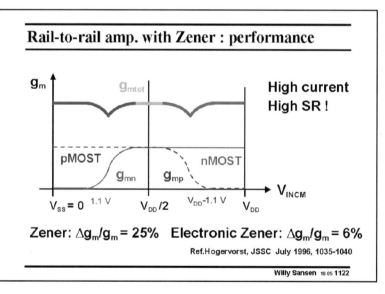

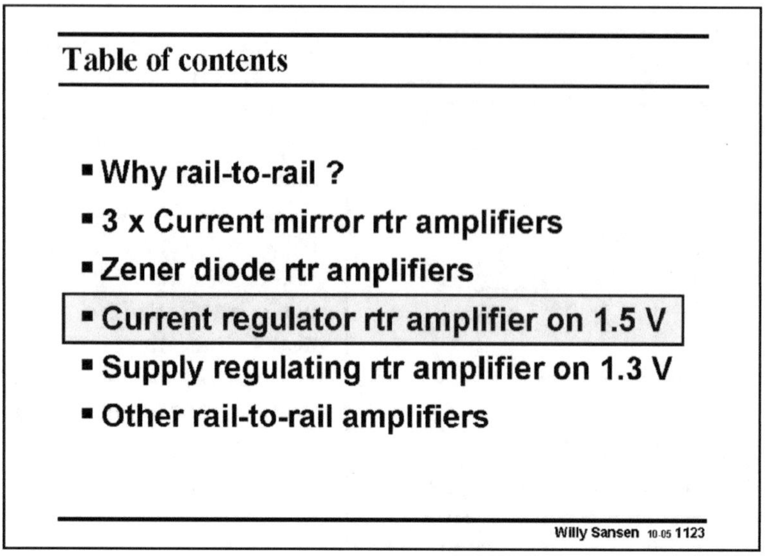

Table of contents

- **Why rail-to-rail ?**
- **3 x Current mirror rtr amplifiers**
- **Zener diode rtr amplifiers**
- **Current regulator rtr amplifier on 1.5 V**
- **Supply regulating rtr amplifier on 1.3 V**
- **Other rail-to-rail amplifiers**

Willy Sansen 10.05 1123

1123

The configurations in this slide exploit the square-law relationship of a MOST in strong inversion, to extract a doubling of the transconductance as a result of a multiplication of the current by four.

At lower currents, and at lower frequencies, weak inversion can be used. In this case, the transconductance is proportional to the current and current feedback loops can be used to equalize the transconductance.

Equalize g_{mtot} in weak inversion

$$g_{mn} + g_{mp} = ct$$

$$\frac{I_{Bn}}{2\,n_n\,kT/q} + \frac{I_{Bp}}{2\,n_p\,kT/q} = ct$$

$$I_{Bn} + \frac{n_n}{n_p}\,I_{Bp} = ct$$

$$n = 1 + \frac{C_D\,(V_{BS})}{C_{ox}}$$

Willy Sansen 10.05 1124

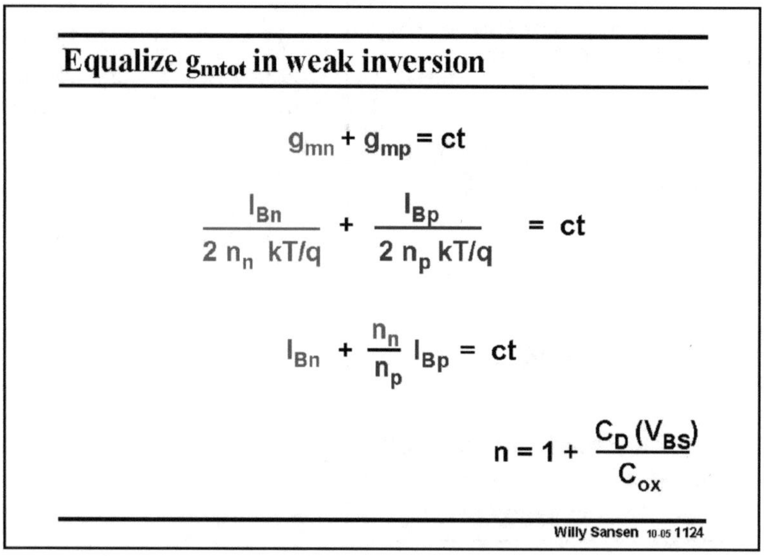

1124

In weak inversion, the transconductance is proportional to the current. Maintaining a constant current over the whole common-mode input range will provide a constant total transconductance.

There is one problem however. The factor n is different for nMOST and pMOST devices. Remember that this n factor is not very precise because it contains the depletion layer capacitance, which depends on biasing voltages.

Nevertheless, it is vital to try to compensate for this difference in the n factor. An error in transconductance will otherwise result.

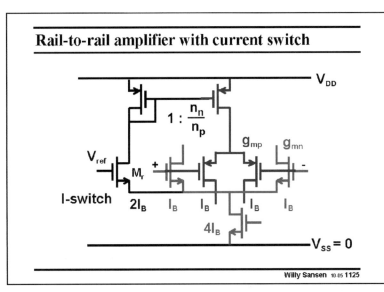

1125

A possible realization is shown in this slide. It only works for low average input voltages.

For input voltages close to V_{ref}, all four input MOSTs carry about the same current I_B. Transistor M_r takes about half the current $2I_B$ of the current source with current $4I_B$. Transistor M_r therefore acts as a kind of current switch.

For lower input voltages, the input nMOSTs are off and all the current $4I_B$ flows through transistor M_r. As a result, the pMOST input devices carry a current $2I_B$. Their current is doubled and so is their transconductance.

For high input voltages, the pMOSTs are off. The input nMOSTs then draw all the current $4I_B$. The current and the transconductance of the input nMOSTs is doubled.

The top pMOST current mirror is used to provide some compensation for the difference in n factor.

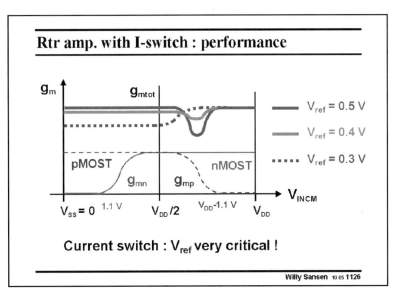

1126

However, the design of this rail-to-rail input stage is not obvious. The choice of the reference voltage V_{ref} is rather critical. This is shown in this slide.

Exact equalization of g_m is a real compromise indeed.

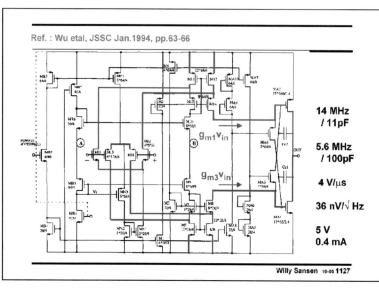

Ref. : Wu etal, JSSC Jan.1994, pp.63-66

14 MHz / 11pF

$g_{m1}v_{in}$

5.6 MHz / 100pF

$g_{m3}v_{in}$

4 V/μs

36 nV/√ Hz

5 V
0.4 mA

Willy Sansen 10-05 1127

1127

An example of this kind of g_m equalization is shown in this slide. This opamp has rail-to-rail input and output. So it can be used as a buffer.

The opamp itself has a double folded cascode as a first stage and two output transistors as a second stage. Miller capacitances are now required. They do not go through cascodes however, possibly leading to problems with positive zero's.

Some specifications are added.

The g_m equalization circuitry is highlighted on the next slide.

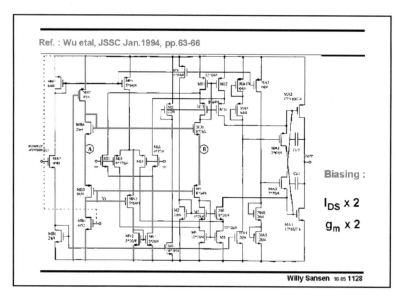

Ref. : Wu etal, JSSC Jan.1994, pp.63-66

Biasing :

I_{DS} x 2

g_m x 2

Willy Sansen 10-05 1128

1128

The reference voltage V_{ref} is generated at the Drain/Gate of transistor MB5. The current-switch transistor is MN3. Its current determines the current in the nMOST transistor pair through current mirror MN1/MN2.

In order to see how this circuit is working, it is repeated on the next slide. It is already clear that the current in one pair will double when the other one is off. As a result, the transconductance will double as well.

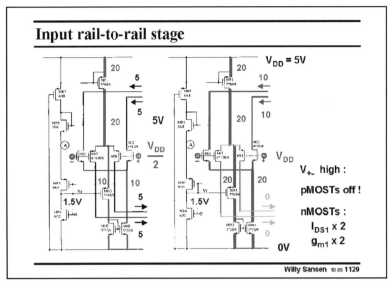

Willy Sansen 10.05 1129

1129

The input stage is shown twice, once with the inputs halfway at the supply voltage and once with the inputs at the supply voltage. In the first case, all four input transistors carry approximately equal currents (about 5 μA). The thickness of the line corresponds to the size of the current.

Note that the top current source has a fixed current. All other currents change with the average input voltage.

For a high input voltage, the pMOSTs go off. The full current is now available to the nMOST devices. Their current doubles and so does their transconductance, providing that they operate in weak inversion.

Large devices are used for small currents. The transistors operate in weak inversion indeed.

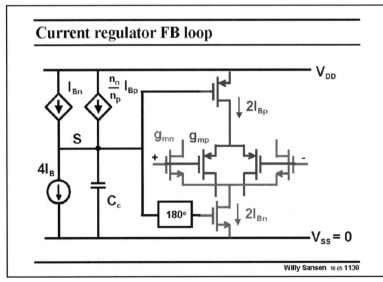

Willy Sansen 10.05 1130

1130

A very different approach is to use a current feedback loop, provided the transistors operate in weak inversion

In the example in this slide, the total current is measured in the pMOST input pair and mirrored to current generator I_{Bp}. Also, the total current is measured in the nMOST input pair and mirrored to current generator I_{Bn}. Both are summed after correction of the I_{Bp} current for the n factor in summation point S and compared to the reference current $4I_B$.

If the sum of the currents does not correspond to $4I_B$, the gates of the current sources have to be adjusted. If the Gate of the pMOST current source must go up, then the Gate of the nMOST current source must go down. This is why there is an inverter before its Gate.

Clearly this is a common-mode feedback loop. It must therefore be made stable. However, compensation capacitance cannot be too large. The common-mode feedback loop must be as fast as the differential circuit! This is explained in Chapter 8.

How can we measure the total current in each input pair?

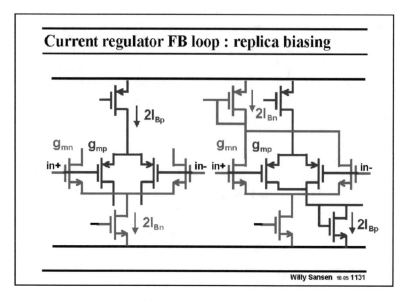

Current regulator FB loop : replica biasing

Willy Sansen 10-05 1131

1131
The feedback loop acts on a replica of the input stage.

The input stage is copied, with the same transistor sizes and currents. The corresponding inputs are connected. The outputs of the replicas are shorted to cancel out the differential signals. In this way, the total current is measured in such a pair, regardless of whether it is on, off or halfway.

The averaged outputs now lead to current mirrors to be summed towards point S.

Replica is an ideal way to bias transistors without actually contacting the sensitive nodes, and without feedback. A few more examples are given.

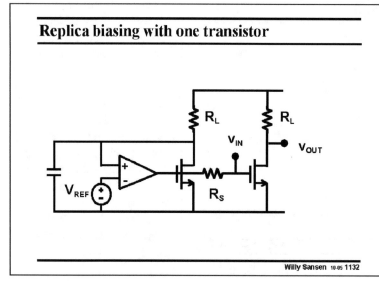

Replica biasing with one transistor

Willy Sansen 10-05 1132

1132
This circuit contains mainly one single-transistor amplifier.

How can we set the average output voltage without feedback on this transistor?

The answer is replica biasing. A transistor with the same characteristics is put in parallel. This one is now part of a feedback loop. Its DC output voltage will be V_{REF}. The average output voltage of the amplifier itself will also be V_{REF}.

The DC output voltage will not be exactly V_{REF} of course, because some mismatch is always present.

The capacitance is required to reduce the influence of the AC voltage on the biasing. The resistor R_S is to raise the AC input impedance of the amplifier.

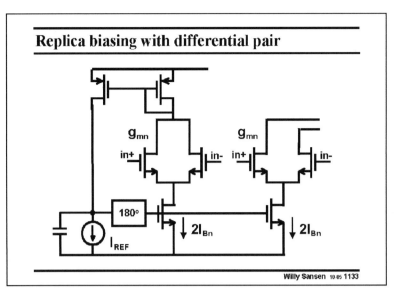

1133

Another example of replica biasing is given in this slide. It is much closer to the rail-to-rail amplifier discussed earlier.

The input pair is doubled. They share the voltage at the Gates of their DC current sources. The left pair however is included in a feedback loop, which ensures that the current equals I_{REF}.

Again, a capacitor is required to stabilize the common-mode feedback loop.

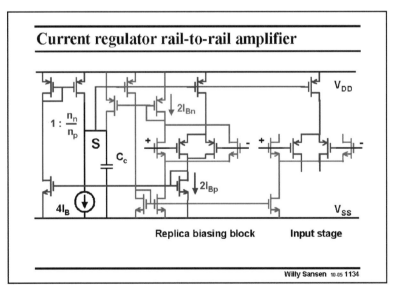

1134

A similar replica biasing is used in this rail-to-rail input stage.

Both input pairs have a replica stage, the outputs of which are shorted and fed to the Summation point by means of current mirrors.

At this point, the summed currents are compared to the reference current $4I_B$. This point directly drives the Gates of the pMOST current sources. An inverter is required however to drive the Gates of the nMOST current sources.

The correction of the n factor is applied to the pMOST current mirror on the left.

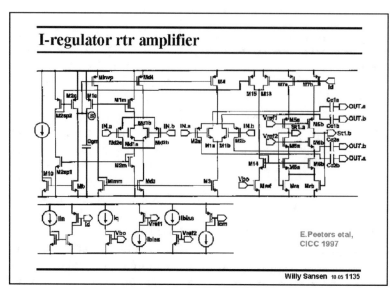

1135

The total first stage of this amplifier is given in this slide.

The cascodes transistors of the two folded cascodes are easily recognized. Transistors Mra and Mrb provide common-mode feedback to the differential outputs.

The four capacitances are Miller compensation capacitances, as shown next.

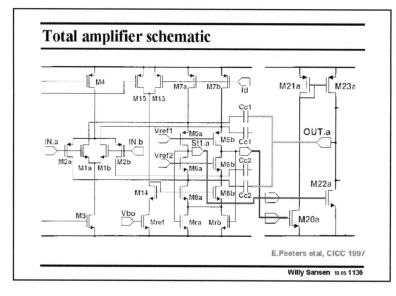

1136

One of the two output stages is shown, together with the first-stage folded cascodes, without the replica biasing.

The two second-stage transistors lead to the output. From the output two compensation capacitances lead to the output of the first stage through both cascodes. Two capacitors are needed because only one of the cascodes may be operational at one of the extremities.

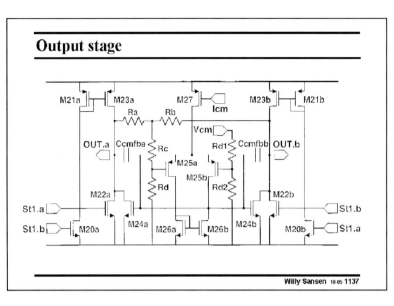

configurations (see Chapter 8). This is only one of them.

1137

The full second stage or output stage is given in this slide.

The output stage of the previous slide is now repeated twice, once with transistors M20a-M23a and once with transistors M20b-M23b.

The output stage is also differential. As a result, another common-mode feedback (CMFB) amplifier is required. The output is averaged with resistors Ra and Rb. The CMFB amplifier can take many different

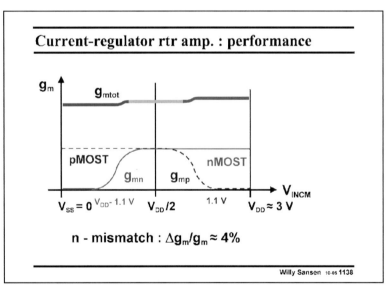

1138

The g_m equalization works quite well. The error of 4% is due to the mismatch in n factor.

The supply voltage used here is quite large. It can be reduced to about 2.2 V if we want to have both pairs on at the same time.

Is this required? The answer is negative as shown next.

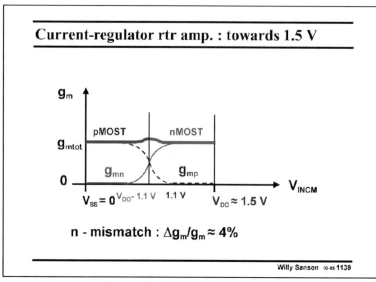

1139

There is no need to have an input range where both input pairs are on. The supply voltage could be reduced to the point where a perfect crossover is achieved. This means that the pMOST pairs goes off exactly at the point where the nMOST pair comes up.

This cross-over point is exactly at halfway the supply voltage for the average input. Also, this point is exactly where the g_m's are halved. As a result, there is no more sum to be taken, except around the middle of the input range. The total transconductance equals the transconductance of one pair taken only at the extremities.

The resultant supply voltage is 1.5 V in this design example. This is set experimentally of course. This voltage is now twice $V_{GS} + V_{DSsat}$. This is $V_T + 2(V_{GS} - V_T)$. For operation in weak inversion $V_{GS} - V_T = 50$ mV has been taken. The values of V_T must therefore have been 0.65 V.

For V_T values of 0.3 V, this rail-to-rail opamp would operate on a supply voltage of 0.8 V. This is below 1 V!

However, it is impossible to set this supply voltage beforehand, as it depends on the absolute value of V_T. Some tolerance must be added to this value. For V_T values of 0.3 V, this rail-to-rail opamp can certainly operate on a supply voltage below 1 V!

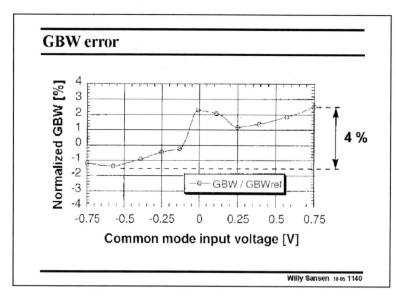

1140

The error in transconductance is obviously reflected in the error on the GBW. This has been measured as shown in this slide.

From left to right, the total deviation is 4% or maybe better $\pm 2\%$. The supply voltage is only 1.5 V.

Input offset voltage

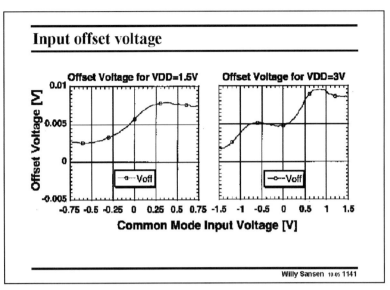

Offset Voltage for VDD=1.5V

Offset Voltage for VDD=3V

Willy Sansen 10-05 1141

1141

However, the variation in g_m or GBW is not the biggest problem. The biggest problem is the change in offset voltage from left to right.

Going from low to high common-mode input voltages, a different input pair is operational, pMOST on the left but nMOST on the right.

These pairs usually have different offset voltages. This change in offset voltage gives a lot of distortion. For a supply voltage of 1.5 V, the difference is about 5 mV. This can be regarded as an error signal (for example 1 V). The distortion can be as high as 0.5% or -50 dB. This is too much for most applications!

In bipolar technologies, this offset can be ten times smaller. The distortion is also ten times smaller or -70 dB, which is much more acceptable.

For a supply voltage of 3 V, the offset is averaged out in the middle. The distortion remains however.

Rail-to-rail Opamp with Current regulator

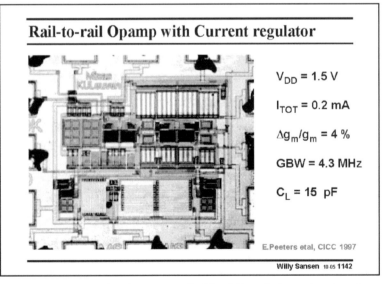

$V_{DD} = 1.5$ V

$I_{TOT} = 0.2$ mA

$\Delta g_m/g_m = 4$ %

GBW = 4.3 MHz

$C_L = 15$ pF

E.Peeters etal, CICC 1997

Willy Sansen 10-05 1142

1142

This layout is shown that most input stage devices are quite big. Indeed, they are used in weak inversion, which leads to large W/L ratios.

The specifications are added as well. They show a FOM of about 320 MHzpF/mA, which is very reasonable indeed. This shows that a replica input stage does not add all that much to the power consumption. After all, the current consumption in the first stage of a two-stage opamp is always small. Doubling it is not going to make a big difference in total power consumption.

Table of contents

Willy Sansen 10.05 **1143**

1143

The previous rail-to-rail amplifier had the advantage that the supply voltage can be as low as 1.5 V but had at the same time the disadvantage that this supply voltage is hard to predict.

In the next rail-to-rail amplifier an internal supply voltage is derived from the external one, which is always set at the minimum possible supply voltage. An exact cross-over situation is always maintained. In this way, a minimum supply voltage is reached which is barely 1.3 V.

Moreover, the PSRR is greatly improved as well.

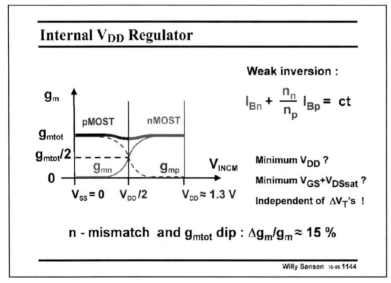

Internal V_{DD} Regulator

Weak inversion :

$$I_{Bn} + \frac{n_n}{n_p} I_{Bp} = ct$$

Minimum V_{DD} ?

Minimum $V_{GS} + V_{DSsat}$?

Independent of ΔV_T's !

n - mismatch and g_{mtot} dip : $\Delta g_m / g_m \approx 15\,\%$

Willy Sansen 10.05 **1144**

1144

The internal supply voltage V_{DD} will be such that the exact cross-over condition is always maintained automatically. This means that at half this supply voltage $V_{DD}/2$, the g_m's of both pairs are reduced to half. In this way the sum is unity over the whole common-mode input range.

This internal supply voltage will now be a result of two feedback loops.

The first one has a task to maintain the same current in both pairs. It will therefore be a current feedback which ensures equality of the currents and hence of the transconductances.

The second feedback loop will be a low dropout voltage regulator which maintains the minimum possible internal supply voltage V_{DD}. It will always equal the sum of the $V_{GS} + V_{DSsat}$ voltages of both pairs at half the current, or half the transconductance.

Whatever the V_T's are, the minimum internal supply voltage is always guaranteed, providing a constant g_m.

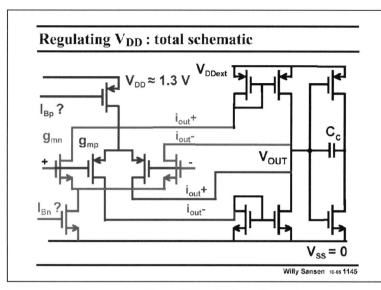

Regulating V$_{DD}$: total schematic

V$_{DD}$ ≈ 1.3 V

V$_{DDext}$

I$_{Bp}$?

g$_{mn}$ g$_{mp}$

i$_{out}$+

i$_{out}$-

C$_c$

V$_{OUT}$

+

-

i$_{out}$+

I$_{Bn}$?

i$_{out}$-

V$_{SS}$ = 0

Willy Sansen 10-05 1145

1145

The rail-to-rail opamp is a simple double input stage. Not even cascodes are used. The four outputs of the two input pairs are combined towards one single output node in the simplest possible way. This can be improved greatly. However, the focus is on the rail-to-rail performance of the input stage.

A second stage is added to drive the measurement system.

The two supply voltage are clearly separated. The internal supply voltage V$_{DD}$ will be derived from the external one V$_{DDext}$.

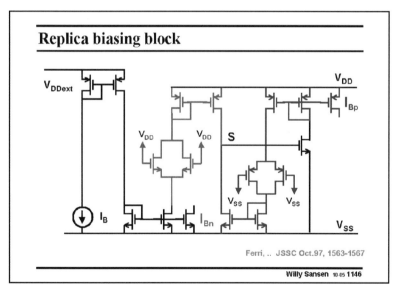

Replica biasing block

V$_{DDext}$

V$_{DD}$

I$_{Bp}$

V$_{DD}$ V$_{DD}$ S

V$_{SS}$ V$_{SS}$

I$_B$

I$_{Bn}$

V$_{SS}$

Ferri, .. JSSC Oct.97, 1563-1567

Willy Sansen 10-05 1146

1146

The first feedback has a task of maintaining the same current in both pairs. It is the current feedback which ensures equality of the currents, whatever happens to the transistors.

The independent biasing is provided by current source I$_B$. The same current also flows in the nMOST differential pair. This pair is a replica of the nMOST pair used in the amplifier itself. However, the input Gates are connected to the positive supply V$_{DD}$.

Its average current is measured and compared to the average current of the pMOST pair, the Gates of which are connected to the negative supply V$_{SS}$ (or ground).

The point of comparison, point S, is fed back to a current mirror which closes the feedback loop.

This circuit already provides the Gate drives for the nMOST and the pMOST current sources in the actual amplifier. They are labeled by I$_{Bn}$ and I$_{Bp}$.

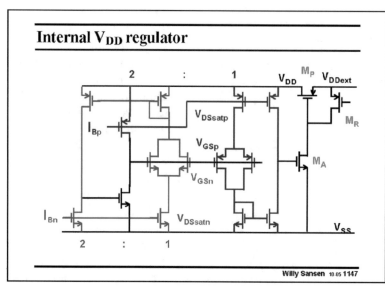

Internal V$_{DD}$ regulator

Willy Sansen 10-05 **1147**

1147

The other feedback loop is a low-dropout voltage regulator which maintains the minimum possible internal supply voltage V$_{DD}$. It must equal the sum of the V$_{GS}$ + V$_{DSsat}$ voltages of both pairs at half the current. The pass transistor M$_P$ is driven by amplifier M$_A$/M$_R$. This latter transistor acts as a resistive load.

The half current sources are now generated first. The factors of 2 are clearly visible. Moreover, the input pairs are duplicated once more, with their four Gates connected together.

A voltage regulator feedback circuit is added on the right, such that voltage V$_{DD}$ always equals the sum of the voltages V$_{DSsatp}$ + V$_{GSp}$ and the V$_{GSn}$ + V$_{DSsatn}$.

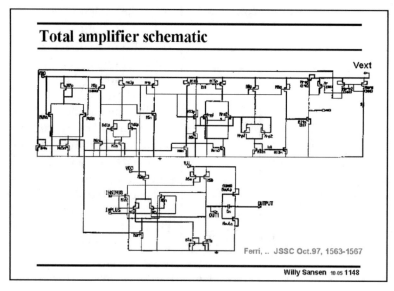

Total amplifier schematic

Ferri, .. JSSC Oct.97, 1563-1567

Willy Sansen 10-05 **1148**

1148

The whole amplifier is shown in this slide.

The actual amplifier is at the lower end. The top part is the biasing circuitry.

The input pairs have been duplicated twice. This adds a small amount to the power consumption, Since this is a two-stage amplifier however, the power consumption is not that bad, as will be shown in the comparative table at the end of this Chapter.

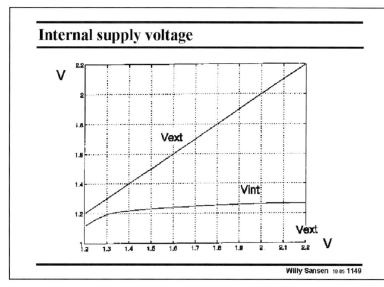

1149

The internal supply voltage V_{int}, is derived from the external V_{ext}, as shown by this measurement result.

The value of about 1.2–1.3 V depends obviously on the actual values of V_T. They must be about 100 mV lower than for the previous rail-to-rail amplifier.

The minimum external supply voltage is little more than the internal one, about 100 mV higher. This means that this amplifier can be used with an external supply voltage from 1.3 V.

The maximum external supply voltage is limited however. Folded cascodes can only go about 0.5 V above the positive supply voltage. The rail-to-rail performance is therefore limited to approximately 1.8 V.

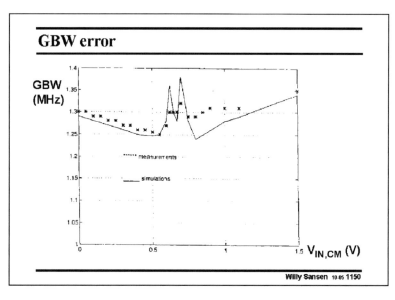

1150

The actual change in input transconductance or GBW over the common mode input range is about 6%. The actual crossover region shows the most irregularities.

In this design example, the change in transconductance may even be more important for the distortion than the change in offset.

Let us have a look at the offset indeed.

Rail-to-rail amp. with V_{DD} regulator : Specs

V_{DDmin} = 1.3 V

GBW = 1.3 MHz in C_L = 15 pF

g_{m1} = 200 μS

I_{DSn1} = 10 μA

W/L_{in} = 830

I_{TOT} = 354 μA

$V_{in,eq}$ = 25 nV$_{RMS}$/√Hz

$V_{in,offset}$ = 0.8 mV (3σ = 0.2 mV)

Ferri, .. JSSC Oct.97, 1563-1567

Willy Sansen 10-05 1151

1151

Some of the specifications are collected here.

Its FOM is not poor, as it is 56 MHzpF/mA. This shows again that duplicating the input stage of a two-stage opamp is more acceptable from the point of view of power consumption.

The input g_{m1}/I_{DS1} is about 20 V^{-1}, which shows that the transistor is in weak inversion. Because of the low currents used, the equivalent input noise is rather high.

In weak inversion, the transistors have large sizes. The offset is fairly small. The change of offset is fairly small as well!

Rtr Opamp with V_{DD}-regulator

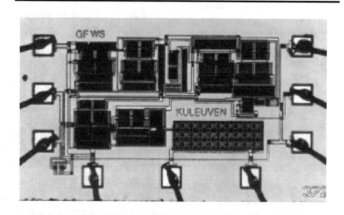

Willy Sansen 10-05 1152

1152

The area of this opamp is about 1.5×0.8 mm. This is quite large as many two replica input stages have been added. Moreover, all input transistors work in weak inversion such that they have large sizes. This is clearly visible.

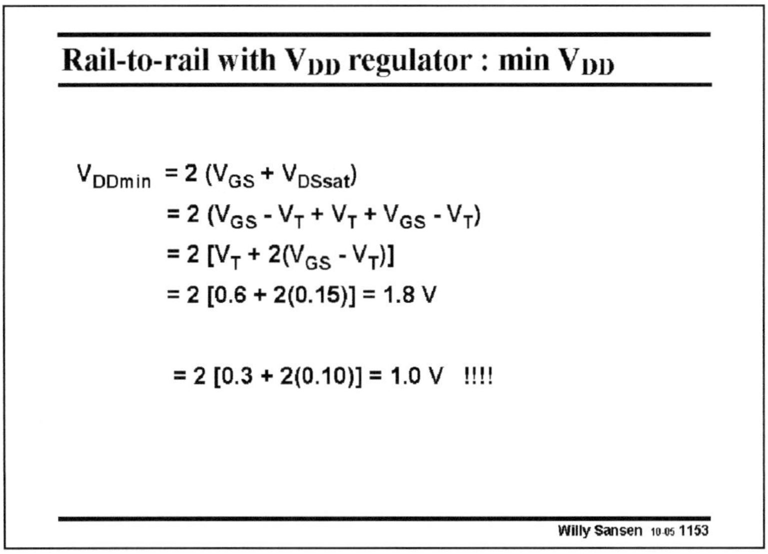

Rail-to-rail with V$_{DD}$ regulator : min V$_{DD}$

$$V_{DDmin} = 2\ (V_{GS} + V_{DSsat})$$
$$= 2\ (V_{GS} - V_T + V_T + V_{GS} - V_T)$$
$$= 2\ [V_T + 2(V_{GS} - V_T)]$$
$$= 2\ [0.6 + 2(0.15)] = 1.8\ V$$

$$= 2\ [0.3 + 2(0.10)] = 1.0\ V\ \ !!!!$$

Willy Sansen 10.05 1153

1153

All these rail-to-rail amplifiers need a double V$_{GSn}$ + V$_{DSsatn}$ as a supply voltage to be able to operate.

For example, for a V$_T$ of 0.6 V and a V$_{GS}$ − V$_T$ of 0.15 V, the supply voltage becomes 1.8 V. Reducing the V$_T$ to 0.3 V as is the case for CMOS technologies with 90 nm channel length and below, the supply voltage could reach the 1 V level indeed. The condition is however, that the transistor works more in weak inversion. Its V$_{GS}$ − V$_T$ of 0.10 V is very close to the 70–80 mV crossover value between strong and weak inversion (see Chapter 1).

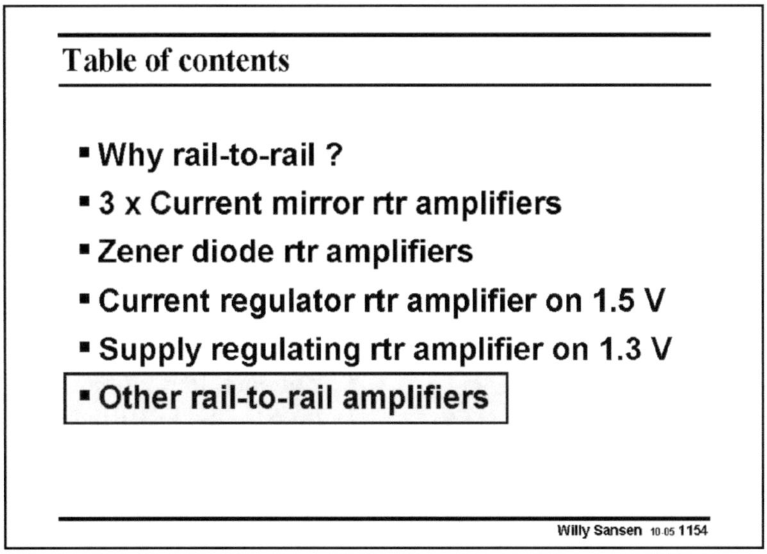

Table of contents

- **Why rail-to-rail ?**
- **3 x Current mirror rtr amplifiers**
- **Zener diode rtr amplifiers**
- **Current regulator rtr amplifier on 1.5 V**
- **Supply regulating rtr amplifier on 1.3 V**
- **Other rail-to-rail amplifiers**

Willy Sansen 10.05 1154

1154

Many more rail-to-rail input amplifiers exist. Many of them are found in combination with class-AB stages.

Some of them are discussed here. They have been selected on the basis that they offer some exciting design aspects.

At the end of this Chapter a comparative table is added.

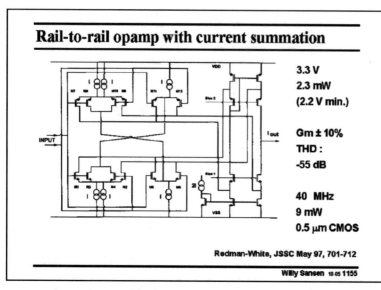

Rail-to-rail opamp with current summation

3.3 V
2.3 mW
(2.2 V min.)

Gm ± 10%
THD :
-55 dB

40 MHz
9 mW
0.5 μm CMOS

Redman-White, JSSC May 97, 701-712

Willy Sansen 10-05 1155

1155

Another way to keep the sum of the currents constant over the whole common-mode input range is shown in this slide. This will allow the sum of the transconductances to stay constant, provided the transistors are biased in weak inversion.

The input stage is shown in this slide. A second stage has to be added to make a full opamp. This input stage consumes 2.3 mW, whereas the full opamp is 9 mW (all on 3.3 V). The supply voltage can be as low as 2.2 V however, as demonstrated previously.

When the input voltages are halfway the supply voltage, then all input nMOSTs M1-M4 are carrying a current I/2 (and so do the input pMOSTs M7-M10). Transistors M3 and M4 pull all current I away from the pMOSTs M11-M12, such that these latter devices are off. In the same way, transistors M5 and M6 are off.

When the common-mode input voltage is high, pMOSTs M7-M10 go off. Transistors M9-M10 do not pull current away any more from M5 and M6, which now carry current I/2 as well. The nMOSTs M1-M2 and M5-M6 now contribute to the total transconductance.

For a high CM input voltage, transistors M5-M6 take over the role of M7-M8. This is in the same way as for low CM input voltages, M11-M12 takes over the role of M1-M2. The total current and transconductance is therefore constant over the whole common-mode input range.

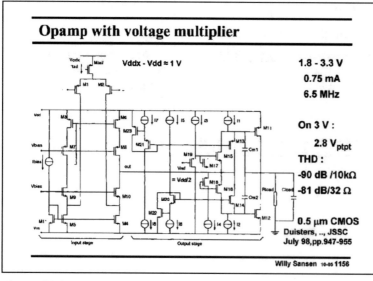

Opamp with voltage multiplier

Vddx - Vdd ≈ 1 V

1.8 - 3.3 V
0.75 mA
6.5 MHz

On 3 V :

2.8 V_ptpt
THD :
-90 dB /10kΩ
-81 dB/32 Ω

0.5 μm CMOS
Duisters, .., JSSC
July 98,pp.947-955

Willy Sansen 10-05 1156

1156

Distortion is the main problem of CMOS rail-to-rail input amplifiers. Without trimming, they cannot offer more than 40–50 dB signal-to-distortion ratio.

This amplifier provides a solution. Its signal-to-distortion ratio can be as high as 90 dB!

This is accomplished by using only one differential pair at the input. An internal voltage regulator is used, which provides an internal supply voltage, this is always higher than the external one by about 1 V. This Volt is sufficient to allow the input Gates to cover the full supply voltage. It is thus a rail-to-rail input amplifier indeed, but with low distortion.

The second stage is a class-AB amplifier, which will be discussed in Chapter 12.

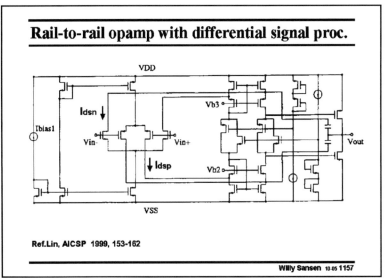

Rail-to-rail opamp with differential signal proc.

Ref.Lin, AICSP 1999, 153-162

Willy Sansen 10-05 1157

1157

This rail-to-rail opamp is well known. It consists of a double folded cascode followed by a class-AB second stage with Miller compensation through the cascodes.

The g_m – equalization still has to be added though. The principle is discussed next.

Note that for increasing Vin-, both currents I_{dsn} and I_{dsp} increase. One of them will disappear however, when the common-mode voltage V_{INCM} is high or low. When V_{INCM} is high, the input pMOSTs are off and I_{dsp} disappears but I_{dsn} survives.

If we now apply a maximum-current selecting circuit to I_{dsn} and I_{dsp}, the larger one will survive. This is the current that is passed on to the next stage.

We now have to insert a maximum-current selector between the Drains which carry I_{dsn} and I_{dsp} and the second stage.

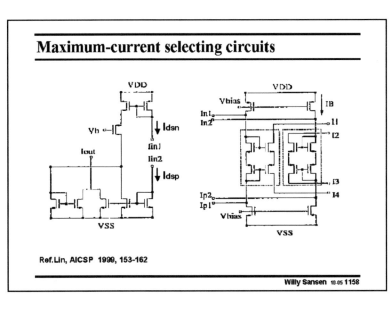

Maximum-current selecting circuits

Ref.Lin, AICSP 1999, 153-162

Willy Sansen 10-05 1158

1158

The g_m-equalization makes use of maximum-current selector circuits. The main difference with all previous rail-to-rail amplifiers is that all g_m – equalizers made use of common-mode circuitry. All added noise is therefore common-mode noise, which is cancelled by the differential output.

The g_m – equalizers discussed here, act on the differential circuits. All added noise now ends up in the signal path. It cannot be cancelled any more.

Two maximum-current selecting circuits are shown in this slide. The left one is a single-ended one, whereas the right one is floating.

The nMOST Drain with current I_{dsn} is connected to the Iin1 input, and the pMOST Drain with current I_{dsp} to the Iin2 input. Both currents are summed by the current mirrors towards Iout. As a consequence, the larger current wins. It is then fed to the second stage.

The maximum-current selecting circuits on the right is explained next.

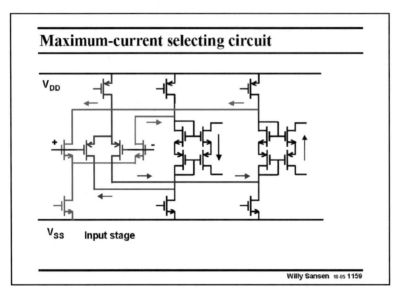

1159

The polarities of the circular currents are indicated. The +input is assumed to rise.

It is clear that both rising currents are led to a floating current mirror (most right), the output of which goes to the second stage.

Both currents of opposite polarity are flowing through the left floating current mirror. Its output goes to the other output transistor.

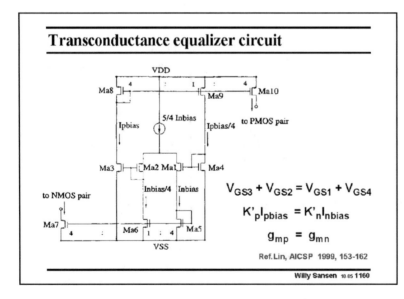

1160

Another circuit of interest is the biasing circuit which makes sure that the nMOSTs and pMOSTs have equal transconductances.

This circuit is shown in this slide. Transistors Ma1–4 form a translinear loop, as indicated by the expression with the V_{GS}'s. The V_T's drop out. What is left is an expression with the currents and the transistor sizes. All currents are indicated, and so are the W/K ratio's.

The result is that the $K'I_{DS}$ products for both a nMOST and a pMOST are the same. Their transconductances are the same as well.

This is therefore a transconductance equalizer circuit.

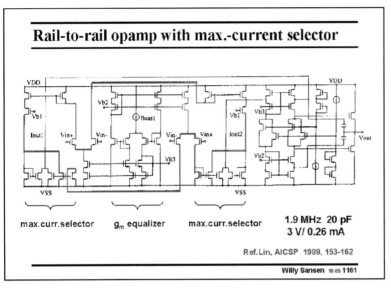

Rail-to-rail opamp with max.-current selector

max.curr.selector g_m equalizer   max.curr.selector

1.9 MHz 20 pF
3 V/ 0.26 mA

Ref.Lin, AICSP 1999, 153-162

Willy Sansen 10.05 **1161**

1161

The first type of maximum-current selector and transconductance equalizer circuit are added to the rail-to-rail amplifier, shown first.

Both maximum-current selectors are easily found. Their outputs go to a differential current amplifier with cascodes. Its outputs then drive the Gates of the output transistors. It is still a two-stage amplifier, despite the complexity of the first stage. The equivalent input noise is therefore rather high.

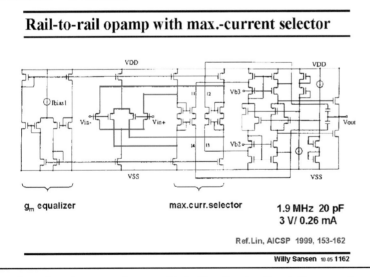

Rail-to-rail opamp with max.-current selector

g_m equalizer   max.curr.selector

1.9 MHz 20 pF
3 V/ 0.26 mA

Ref.Lin, AICSP 1999, 153-162

Willy Sansen 10.05 **1162**

1162

The floating maximum-current selector and transconductance equalizer circuit are added to the same rail-to-rail amplifier, as shown in this slide.

The outputs go to the same second stage as before. It is thus also a second stage amplifier.

However, the input devices have a more symmetrical load. The CMRR is now higher.

Remember that the noise is rather high because of the analog signal processing applied directly to the AC currents, and not to the common-mode circuitry or the biasing.

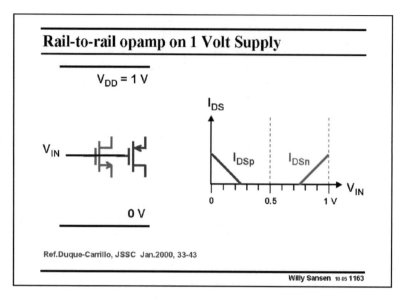

1163

In order to be able to reach a supply voltage, even with V_T's of 0.7 V is not an easy task.

Whatever circuitry is applied to the input devices, an average input voltage of 0.5 V will never allow either the nMOST or the pMOST to conduct. Only the input devices are shown of two folded cascodes.

Only for input voltages larger than about 0.8 V, the nMOST starts conducting. Also, the pMOST can only conduct for average input voltages below about 0.2 V.

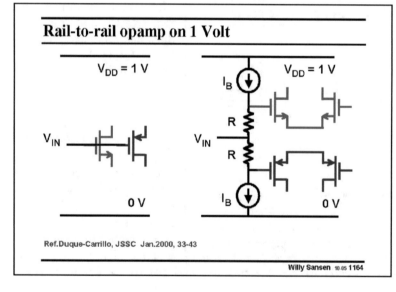

1164

The solution is to insert level shifters.

Indeed, inserting two resistors R between the actual input and the Gates, and two current sources I_B allows the necessary level shifting.

For example, with currents of 10 µA and resistors of 30 kΩ, the level shift is then 0.3 V. For an input voltage of 0.5 V, the nMOST Gate is at 0.8 V and the pMOST Gate at 0.2 V. Both transistors are now operational.

Note that this current source I_B is only needed when the input voltage is 0.5 V. It can disappear for other input voltages. For example, if the input voltage is 0.2 V or lower, current I_B can be zero.

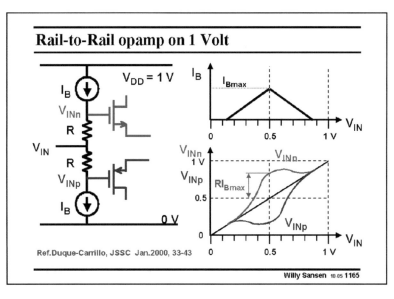

Rail-to-Rail opamp on 1 Volt

Ref.Duque-Carrillo, JSSC Jan.2000, 33-43

Willy Sansen 10.05 1165

1165

In this realization a current is generated which is triangular with its maximum at 0.5 V, and which becomes zero for inputs below 0.2 V or higher than 0.8 V.

The actual input voltage and the Gate voltages are plotted versus average input voltage. It is clearly seen that for input voltages larger than 0.5 V, the nMOST already conducts. Also, the pMOST conducts for input voltages up to 0.5 V. A rail-to-rail input range is now obtained.

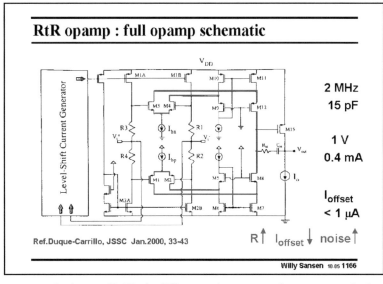

RtR opamp : full opamp schematic

Ref.Duque-Carrillo, JSSC Jan.2000, 33-43

2 MHz
15 pF

1 V
0.4 mA

I_{offset}
< 1 μA

$R\uparrow$ $I_{offset}\downarrow$ noise$\uparrow$

Willy Sansen 10.05 1166

1166

The full schematic is shown in this slide.

Four resistors and current sources are used to level-shift the inputs. The currents I_B are derived in a separate current generator.

The outputs then simply go to two differential current amplifiers. A simple second stage is provided to be able to output the signal.

This rail-to-rail input arrangement also has some drawbacks. First of all, the four current sources I_B, may not match that well. Their difference in current flows out and gives rise to a kind of bias current (as is common in bipolar amplifiers). Moreover, they may not be the same for both inputs. There is thus also an offset current.

Another disadvantage is the presence of noisy resistors in series with all four inputs. The noise performance will now suffer.

It is possible to reduce these resistors but larger currents I_B are required, worsening the input offset currents.

It is the only rail-to-rail input opamp however, which works on 1 V for conventional V_T's.

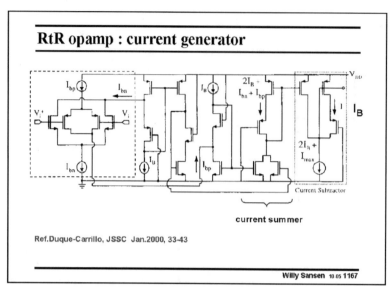

RtR opamp : current generator

Ref.Duque-Carrillo, JSSC Jan.2000, 33-43

Willy Sansen 10-05 **1167**

1167

The level-shift current generator is shown in this slide. It has to generate an output current which first increases with the input voltage and then decreases again.

It is obviously a common-mode block. A replica input stage is used, in which the differential signals are cancelled. The outputs are led to a current summer by means of a number of current sources. Its output thus gives rise to a pseudo-triangular output current I_B.

Comparison rail-to-rail input amplifiers

Type	Ref.	$\Delta g_m/g_m$ %	GBW MHzpF/mW	I_{TOT} µA	V_{DDmin} V
3x Curr.mirr.	JSSC-12-94	15	110	150	3
Electr. Zener	JSSC-7-96	6	70	215	2.7
Curr.switch	AICSP-5-94	8	1.1	500	3.3
Curr.regulat.	CICC 97	4	210	200	1.5
Regulat. VDD	JSSC-10-97	6	43	350	1.3
MOST translin.	AICSP-6-94	8	4.2	800	2.5
Improv.CMRR	JSSC-2-95	9	3	1400	5
Max. current	AICSP-1-99	10	77	260	3
Resistive input	JSSC-1-00	x	75	400	1

Willy Sansen 10-05 **1168**

1168

As a conclusion, a comparative table is given comparing the different specifications of the types discussed.

The first column is the type, followed by the reference.

The error is given in transconductance. Most of them achieve 5–8%. This will give rise to some distortion, which is normally smaller however than the distortion due to the changing offset voltage.

The FOM indicates how much current has been consumed by the addition of circuitry for the rail-to-rail input. Anything that is larger than 30–50 is more acceptable. It is seen that only amplifiers with a large FOM have been discussed in this Chapter. Some other ones are added which are less attractive.

The column with current does not mean much. The effect of the current has been included in the FOM.

The last column lists the minimum supply voltages. For most standard opamps this is about 2.5 V. The two amplifiers with 1.5 and 1.3 V use input devices in weak inversion and are optimized towards low supply voltages.

The only 1 V amplifier is the last one. It compromises however, on specifications which are not listed here, such as input noise and input biasing current.

1169

Table of contents

- **Why rail-to-rail ?**
- **3 x Current mirror rtr amplifiers**
- **Zener diode rtr amplifiers**
- **Current regulator rtr amplifier on 1.5 V**
- **Supply regulating rtr amplifier on 1.3 V**
- **Other rail-to-rail amplifiers**

Willy Sansen 10-05 1169

In this Chapter rail-to-rail input amplifiers have been discussed. Various types of circuit configurations have been analyzed and compared.

Such an input stage will mainly serve as a stage for a class-AB amplifier, to guarantee rail-to-rail performance for both input and output.

1170

References

T. Duisters, etal, "A -90 dB THD rail-to-rail input opamp using a new local charge pump in CMOS", IEEE Journal Solid-State Circuits, Vol. SC-33, pp. 947-955, July 1998.

R. Duque-Carillo, etal, "A 1 V rail-to-rail operational amplifier in standard CMOS technology", IEEE Journal Solid-State Circuits, Vol. SC-35, pp. 33-43, Jan. 2000.

G.Ferri, W.Sansen, "A rail-to-rail constant-gm low-voltage CMOS operational transconductance amplifier", IEEE Journal Solid-State Circuits, Vol. SC-32, pp. 1563-1567, Oct.1997.

R. Hogervorst, etal, "A compact power-efficient 3V CMOS rail-to-rail input/output operational amplifier for VLSI cell libraries", IEEE Journal Solid-State Circuits, Vol. SC-29, pp. 1504-1512, Dec.1994.

R. Hogervorst, etal, "Compact CMOS constant-gm rail-to-rail input stage with gm-control by an electronic Zener diode", IEEE Journal Solid-State Circuits, Vol. SC-31, pp. 1035-1040, July 1996.

R. Lin, etal, "A compact power efficient 3V CMOS rail to rail input/output operational amplifier for VLSI cell libraries", Analog Integrated Circuits and Signal Processing, Kluwer Ac., pp. 153-162, Jan.1999.

Willy Sansen 10-05 1170

1171

References

E. Peeters, etal, "A compact power-efficient 3V CMOS rail-to-rail input/output operational amplifier for VLSI cell libraries", CICC 1997.

W. Wu, etal, "Digital-compatible high-performance operational amplifier with rail-to-rail input and output stages", IEEE Journal Solid-State Circuits, Vol. SC-29, pp. 63-66, Jan 1994.

121

Class AB and driver amplifiers

Willy Sansen

KULeuven, ESAT-MICAS

Leuven, Belgium

willy.sansen@esat.kuleuven.be

Willy Sansen 10-05 **121**

To deliver power to small resistors or large capacitors cannot be achieved with conventional output stages. The output currents are too large. For this purpose we need to bias the output stages in class AB. They have small quiescent currents but can deliver very large currents to the load.

Examples are obviously audio amplifiers for loudspeakers and headphones, but also communications applications such as ADSL and XDSL. All of them require large output currents but very low distortion at the same time.

Audio amplifiers are limited to little over 20 kHz but XDSL amplifiers extend now to 1–3 MHz and even higher in the future. The distortion must be less than -80 dB so as not to mix up channels. This is a very severe specification indeed.

122

Outline

- **Problems of class AB drivers**
- **Cross-coupled quads**
- **Adaptive biasing**
- **I_Q control with translinear circuits, etc.**
- **Current feedback and other principles**
- **Low-Voltage realizations**

Ref.: W. Sansen : Analog Design Essentials, Springer 2006

Willy Sansen 10-05 **122**

Let us first look at the specifications of a class-AB opamp. What are the problems?

A large number of possible solutions are introduced and discussed. A few circuits are added with supply voltages of 1 V or less.

337

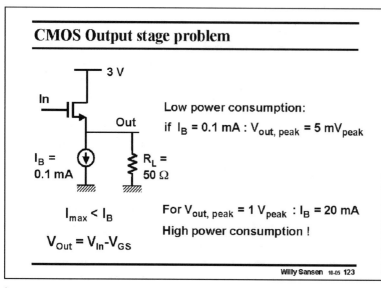

CMOS Output stage problem

3 V

In

Out

Low power consumption:

if $I_B = 0.1$ mA : $V_{out, peak} = 5$ mV$_{peak}$

$I_B =$
0.1 mA

$R_L =$
50 Ω

$I_{max} < I_B$

For $V_{out, peak} = 1$ V$_{peak}$: $I_B = 20$ mA

High power consumption !

$V_{Out} = V_{In} - V_{GS}$

Willy Sansen 10-05 123

123

For a low-resistor load, a low output impedance is required. The source follower is the only simple transistor stage which provides this output resistance. However, its DC current handling is not sufficient.

A source follower is shown, biased at 0.1 mA. A low resistor of 50 Ω is connected to it. It is clear that the maximum output voltage swing can only be 5 mV. For higher output voltages, we would need higher biasing currents as well. This would lead to an excessive power consumption.

We now need a transistor circuit which can deliver large currents only when needed, but with a low quiescent biasing current to lower the power consumption as much as possible.

Note that this transistor stage can deliver (source) a large current but it can only sink the DC biasing current. The positive output swing can therefore be large, but not the negative swing.

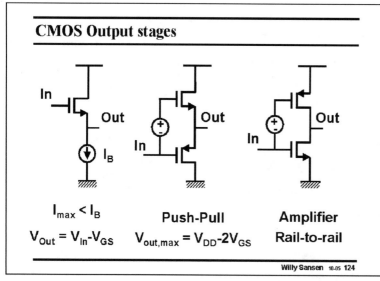

CMOS Output stages

In

Out

I_B

Out

In

In

Out

$I_{max} < I_B$

$V_{Out} = V_{In} - V_{GS}$

Push-Pull

$V_{out,max} = V_{DD} - 2V_{GS}$

Amplifier

Rail-to-rail

Willy Sansen 10-05 124

124

A possible solution is to have two source followers, Source to Source, as shown in the middle. The current out of this stage (source) can again be very large, depending on the transistor size.

The current in this stage (sink) can also be large. The pMOST can now be driven as hard as the nMOST. The main disadvantage of this double source follower is that the output swing can only reach the supply voltage within one V_{GSn}. Also, the output voltage can never be lower than V_{GSp}. For large supply voltages, such as audio amplifiers, this is no problem but for supply voltages of a few Volts, this is not acceptable.

This is why most class-AB output stages for low supply voltages have two output transistors Drain-to-Drain. They constitute an amplifier with al least two stages. Stability will have to be verified. They do guarantee rail-to-rail output swing however, at least for capacitive loads.

In this case, the low output resistance will have to be realized by application of feedback, aggravating the stability issue even more.

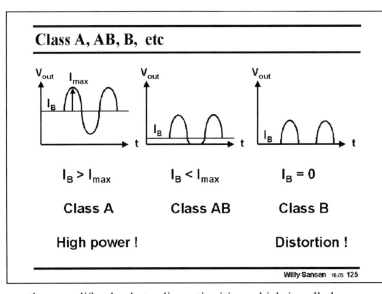

Class A, AB, B, etc

$I_B > I_{max}$ $I_B < I_{max}$ $I_B = 0$

Class A   Class AB   Class B

High power !   Distortion !

Willy Sansen 10-05 125

125

Why do these amplifiers require class AB operation? The answer is that this is the best compromise between current capability and distortion.

A class-A stage is a stage in which the peak current swing never exceeds the DC biasing current. The average current is therefore the DC current.

In a class-B stage the DC biasing current is zero. Connecting these swings to the negative swings from another amplifier leads to discontinuities, which is called crossover distortion.

Class-AB amplifiers are somewhere in between. The DC biasing current (or quiescent current) is small compared to the peak current swings. In this way the connections between the two halves are more smooth. The crossover distortion can be made very small indeed.

One of the specifications will have to do with the predictability and stability of this quiescent current.

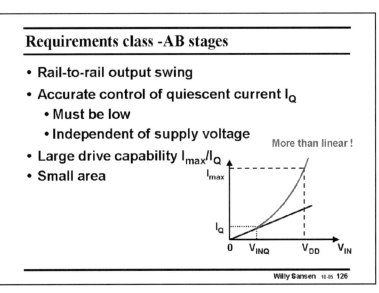

Requirements class -AB stages

- **Rail-to-rail output swing**
- **Accurate control of quiescent current I_Q**
  - **Must be low**
  - **Independent of supply voltage**
- **Large drive capability I_{max}/I_Q**
- **Small area**

More than linear !

Willy Sansen 10-05 126

126

The first requirement is obviously that rail-to-rail swings are possible.

The second one has indeed to do with the quiescent current I_Q.

In addition, large output currents I_{max} must be possible (depending on the application). Their ratio to I_Q is called the drive capability.

The problem is that the transfer curve of such an amplifier is now highly non-linear. For small input voltages, it is perfectly linear, as any class-A amplifier. For higher input voltages, the output current must rise more than linearly with the input voltages. The output current must have an expanding characteristic.

This will generate some distortion as well, which can be reduced by application of feedback. This is why many class-AB stages consist of three stages.

The last specification has to do with complexity. Class-AB amplifiers are the most complicated DC-coupled amplifiers. Some simplicity is still welcome!

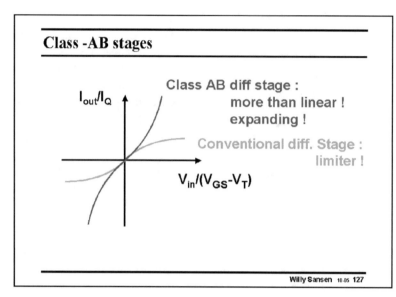

Class -AB stages

Class AB diff stage :
more than linear !
expanding !

Conventional diff. Stage :
limiter !

I_{out}/I_Q

$V_{in}/(V_{GS}-V_T)$

Willy Sansen 10-05 127

127

A conventional differential pair has a limiting characteristic rather than an expanding one. It cannot be used in a class-AB amplifier.

Circuit techniques will have to be devised to convert the limiting characteristic into an expanding one.

The simplest solution is a conventional CMOS inverter amplifier!

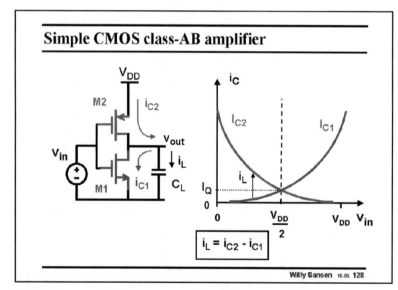

Simple CMOS class-AB amplifier

V_{DD}

M2

i_{C2}

v_{out}

$\downarrow i_L$

V_{in}

M1

i_{C1}

C_L

i_C

I_{C2}

I_{C1}

i_L

I_Q

0

$\dfrac{V_{DD}}{2}$

V_{DD} V_{in}

$i_L = i_{C2} - i_{C1}$

Willy Sansen 10-05 128

128

A simple CMOS inverter is an excellent class-AB amplifier.

Normally it is biased at a small current I_Q. The current through the capacitive load however, can be much larger, because the transistor V_{GS} can be as much as the full supply voltage. The actual load current i_L is the difference between the nMOST current i_{C2} and the pMOST current i_{C1}.

In this circuit the square-law characteristic of a MOST is used. It has an expanding characteristic indeed.

The main disadvantage of this circuit is that its two V_{GS}'s are between supply voltage and ground. As a consequence, the quiescent current depends on the supply voltage. Moreover, all spikes on the supply voltage (from digital blocks) enter this amplifier. Its PSRR is thus zero dB.

Other circuit solutions are now required.

Outline

- **Problems of class AB drivers**
- **Cross-coupled quads**
- **Adaptive biasing**
- **I_Q control with translinear circuits, etc.**
- **Current feedback and other principles**
- **Low-Voltage realizations**

Willy Sansen 10-05 129

129

A better class-AB amplifier is obtained by cross-coupling the input devices. In this way, a complementary differential pair can be constructed with an expanding characteristic.

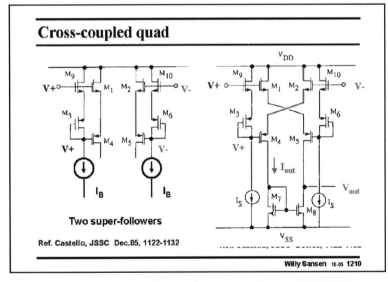

Cross-coupled quad

Two super-followers

Ref. Castello, JSSC Dec.85, 1122-1132

Willy Sansen 10-05 1210

1210

This circuit is shown twice, once without cross-coupling to figure out the biasing, and once with the cross-coupling.

The circuit on the left contains two source followers. Actually they are source followers combined with current mirrors. This is why they are called super source followers.

The nMOSTs are the same and so are the pMOSTs. The current through M1 and M4 will also be I_B. This also applies to the current through M2/M5.

All nodes follow the input voltage. For the positive input voltage V+, the Sources of M1 and M9, but also the Drain of M3 all have obviously the same voltage swing V+.

This also applies to the voltages V− of the other super source follower.

Cross-coupling now the two inner lines, generates a nMOST/pMOST differential pair, with V− and V+ at their Gates, and which exhibit an expanding output current.

This stage has actually four output currents, i.e. the Drain currents of M1 and M5 which increase, and also the Drain currents of M2 and M4 which decrease. They can be combined with current mirrors towards the output. In this example only two output currents are used.

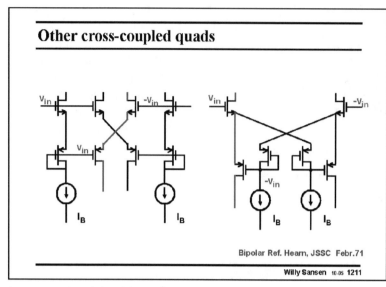

Other cross-coupled quads

Bipolar Ref. Hearn, JSSC Febr.71

Willy Sansen 10-05 **1211**

1211

There are many circuit variations on these cross-coupled quadruples.

In the circuit on the right, which was originally made in bipolar technologies, one transistor is left out on either side. Again all DC currents are set by the current sources I_B. The complementary differential pair is formed again with a nMOST, with

v_{in} at its Gate, and a pMOST, which receives a $-v_{in}$ from the source follower on the right side. An expanding output current is obtained again.

Again, four output currents can be distinguished.

It is less symmetrical than the first cross-coupled quad, which is therefore preferred.

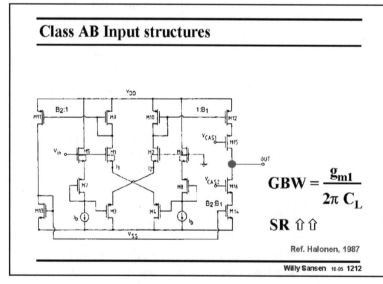

Class AB Input structures

$$GBW = \frac{g_{m1}}{2\pi\, C_L}$$

$$SR \Uparrow \Uparrow$$

Ref. Halonen, 1987

Willy Sansen 10-05 **1212**

1212

An example of such a cross-coupled quad as an input stage is shown in this slide.

After all, this is a symmetrical opamp, with cascodes M15 and M16 for more gain. Normally its maximum output current is limited to $B_1 I_b$. The Slew Rate is quite limited.

However, substitution of the input differential pair by a cross-coupled quad gives an expanding current. The GBW, which is a small-signal specification is the same, but the Slew Rate increases drastically.

Only two output currents are used of the input cross-couple quad.

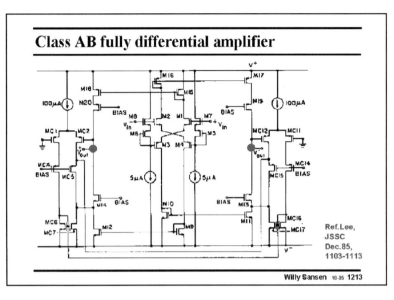

Class AB fully differential amplifier

Ref.Lee, JSSC Dec.85, 1103-1113

Willy Sansen 10-05 **1213**

1213

A realization of all four output currents are used of the input cross-couple quad. It is biased at low currents of 5 μA. The output currents can be much larger however, because of the expanding characteristic.

All four output currents are now used towards the differential outputs. This is also a symmetrical opamp with cross-coupled quads at the inputs rather than conventional differential pairs.

Common-mode feedback is required because the outputs are differential.

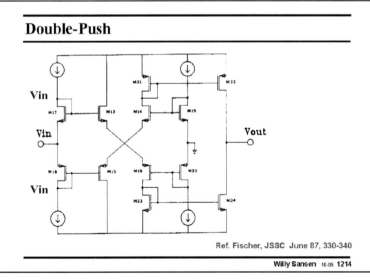

Double-Push

Vin

Vin

Vin

Vout

Ref. Fischer, JSSC June 87, 330-340

Willy Sansen 10-05 **1214**

1214

The same cross-coupling is now used here in the output stage. The eight transistors are M13–20.

The input is taken from the previous stage. It can be either one of the three points indicated. The other input is connected to ground.

Only two of the four output currents are used. They are current-mirrored to the output. Transistors M22 and M24 are quite large, to be able to source and sink large currents.

The quiescent current is well defined as it is directly relayed to the biasing current sources. The total schematic is shown next.

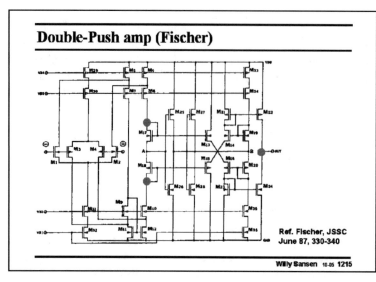

Double-Push amp (Fischer)

Ref. Fischer, JSSC
June 87, 330-340

Willy Sansen 10-05 1215

1215

The output stage is easily recognized because of the cross-coupled quad. It is preceded by a double folded cascode. No g_m equalization is used. The points of high impedance are labeled by big (red) dots. This is clearly a two-stage amplifier.

Point A is one input of the output stage. Point B is now the other one; it is no longer connected to ground. It is derived from point A by an inverter (with transistors M25-M26) which is loaded by the $1/g_m$'s of the two transistors M27 and M28. Its gain is thus about minus unity. The output stage now has a differential drive.

The output stage presents a load to the input stage of about 15 pF (at point A). This capacitor determines the GBW.

Outline

- **Problems of class AB drivers**
- **Cross-coupled quads**
- **Adaptive biasing**
- **I_Q control with translinear circuits, etc.**
- **Current feedback and other principles**
- **Low-Voltage realizations**

Willy Sansen 10-05 1216

1216

Cross-coupled quads are a useful principle for generating expanding differential currents. It is not the only one however. Positive feedback can be used as well. Adaptive biasing is actually used.

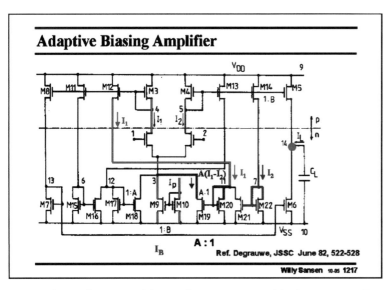

1217

An adaptive biasing amplifier adapts its biasing to be able to provide larger output currents.

The amplifier in this slide is a symmetrical amplifier, which is single ended. Nowadays it would be differential.

Two times two current mirrors are added, i.e. with transistors M11/M12 and M13/M14. Without these transistors the maximum output current would be limited to BI_p.

In order to increase this maximum current, biasing current I_p must be made larger for larger input voltages. This biasing current is adapted to the input signal level. This is why it is in parallel with two more current mirrors through transistors M18 and M19. Let us follow the path to transistor M19.

Transistors M19 forms a current mirror (with current factor A) with M20. This latter transistor take the difference in current $I_1–I_2$, which are proportional to the currents in the input stage. The larger of these two currents wins. If I_1 is larger than I_2 then AI_1 current is added to I_p, increasing the total biasing current of the first stage, and also increasing the maximum output current.

If, however, I_2 is larger than I_1, then it is mirrored by M17/M18, also multiplied by A and also added to I_p.

The adaptive current feedback is a kind of rectifier, as shown next.

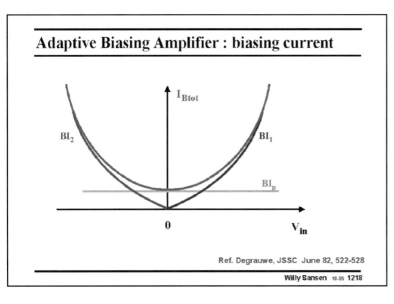

1218

For small input voltages, the biasing current of the first stage I_B is only I_p. The maximum output current is Bi_p. For larger input voltages, for terminal 1 being more positive than terminal 2, current I_1 increases drastically and is fed back to the input biasing current.

With terminal 2 being more positive than terminal 1 however, current I_2 increases by a similar amount and is also fed back to the input biasing current.

This total biasing current increases in either direction. The amount of increase depends on current factor A.

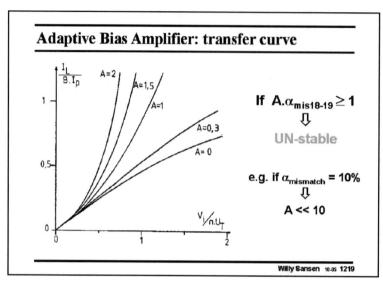

1219
For a current factor A equal to zero, no adaptive biasing takes place. The output current (normalized to Bi_p) is limited for larger input voltages (normalized to nU_T or nkT/q.).

For an increasing factor A however, the expansion of the output current with the input voltage is more and more pronounced. A class-AB behavior is now obtained.

Factor A cannot be increased to very large values, depending on matching. A practical limit is about 10. If cascodes are used however, the matching between all the current sources improves remarkably. Higher factors of A can then be tried.

A disadvantage of this amplifier is that transistors M11–14 are added on the node where the non-dominant pole is formed. They will therefore slow down the amplifier.

Outline

- **Problems of class AB drivers**
- **Cross-coupled quads**
- **Adaptive biasing**
- **I_Q control with translinear circuits, etc.**
- **Current feedback and other principles**
- **Low-Voltage realizations**

Willy Sansen 10-05 1220

1220
A class-AB output stage can also be biased by translinear circuits, as explained previously.

A translinear loop is a circuit which provides a linear relationship by use of nonlinear circuits. The simplest example is a current mirror. Both transistors have a nonlinear current-voltage relationship and yet the current gain is perfectly linear. The voltage between the two transistors is heavily distorted though.

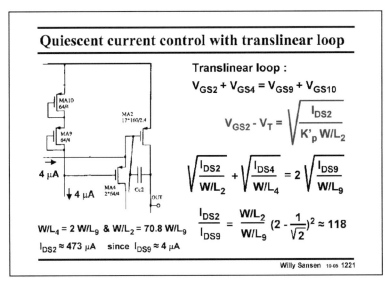

Quiescent current control with translinear loop

Translinear loop :

$$V_{GS2} + V_{GS4} = V_{GS9} + V_{GS10}$$

$$V_{GS2} - V_T = \sqrt{\frac{I_{DS2}}{K'_p \ W/L_2}}$$

$$\sqrt{\frac{I_{DS2}}{W/L_2}} + \sqrt{\frac{I_{DS4}}{W/L_4}} = 2 \sqrt{\frac{I_{DS9}}{W/L_9}}$$

$$\frac{I_{DS2}}{I_{DS9}} = \frac{W/L_2}{W/L_9} (2 - \frac{1}{\sqrt{2}})^2 \approx 118$$

$W/L_4 = 2 \ W/L_9$ & $W/L_2 = 70.8 \ W/L_9$

$I_{DS2} \approx 473 \ \mu A$ since $I_{DS9} \approx 4 \ \mu A$

Willy Sansen 10-05 1221

1221

A translinear loop is formed by transistors MA2/MA4 and MA9/MA10. Their sum of V_{GS}'s is equal.

The currents through MA9–MA10 are set by a DC current source (which is about 4 μA in this example). The current through MA4 is also set by the DC current of the preceding stage (which is also about 4 μA in this example). Only the current through the large output transistor MA2 is not known.

Its current is then defined by the expression in this slide. All transistor sizes W/L's are known. All parameters V_T and K'_p cancel out.

As a result, we obtain an expression linking the currents to the transistor sizes. The current I_{DS2} through transistor MA2 is about 120 times larger than the current through transistor MA9. It is now well defined. It is independent of the supply voltage.

A disadvantage of this loop however, is that I_{DS2} only becomes large when I_{DS4} becomes zero, since I_{DS9} is constant. Transistor MA4 shuts off for large drives and limits the output current.

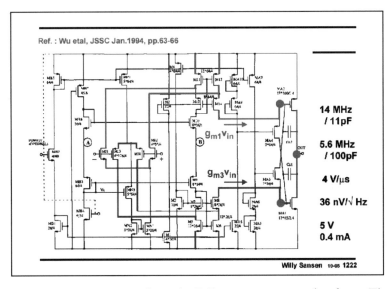

Ref. : Wu etal, JSSC Jan.1994, pp.63-66

$g_{m1}v_{in}$

$g_{m3}v_{in}$

14 MHz / 11pF

5.6 MHz / 100pF

4 V/μs

36 nV/√ Hz

5 V
0.4 mA

Willy Sansen 10-05 1222

1222

A practical example of such a translinear loop is shown in this slide.

The transistors MA2/MA4 and MA9/MA10 have actually been copied on the previous slide. They form a translinear loop indeed.

The same applies to the transistors MA1/MA3 which form a translinear loop with MA5/MA6.

The quiescent current in the output transistors is thus given by the expression on the previous slide.

It is now easy to see where the DC currents are coming from. The current through MA9/MA10 comes form a DC biasing current mirror. The DC current through MA4 comes from the pMOST differential current amplifier M11–14 at the end of the nMOST first stage. This current flows through the nMOST differential current amplifier M5–8 at the end of the pMOST first stage.

Transistors MA3 and MA4 carry this DC current as well but no AC current. They are bootstrapped out for AC behavior. They present an infinite AC impedance to the currents coming from the first stage. This is shown next.

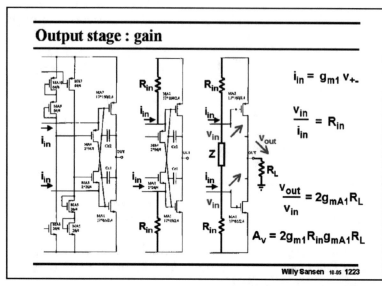

1223

The output stage is shown three times, but more and more simplified.

First, note that the current provided by the current differential amplifiers of the first stage are in phase. The output impedances of the first stage are shown explicitly in the second figure. The transistors MA3 and MA4 are substituted by some impedance Z in the third one.

Since both input currents have the same phase, they both increase the gates of the output transistors by about the same voltage. There is nearly no AC voltage drop across the impedance Z. It is bootstrapped out.

It does not appear in the expression of the gain A_v.

Transistors MA3 and MA4 only play a role in the translinear loop to set the quiescent current trough the output devices. They do not play a role in the gain of GBW.

Transistors MA3 and MA4 carry this DC current as well but no AC current. They are bootstrapped out for AC behavior. They present an infinite AC impedance to the currents coming from the first stage.

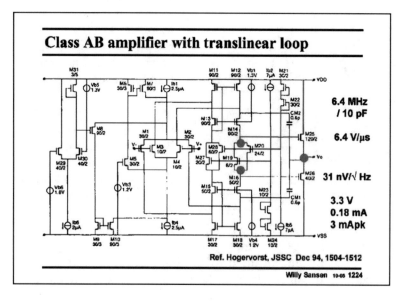

1224

A similar translinear loop to set the quiescent current through the output devices is found in this amplifier. It is a two-stage amplifier. The compensation capacitances are connected to the outputs of the first stage through the cascodes M14 and M16.

The translinear loop with output transistor M25 is highlighted. A similar one is present for output transistor M26.

Again transistors M19/ M20 are bootstrapped out for AC performance. The first stage is a rail-to-rail input stage, which has been discussed in the previous Chapter.

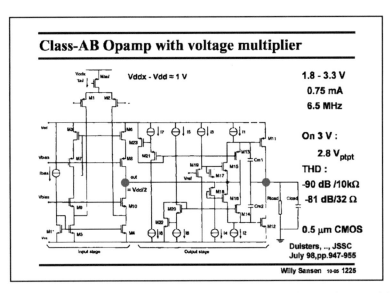

1225

This is again a two-stage amplifier. A single-ended folded cascode is the first stage. The second stage consists of the two output transistors. Transistors M13/M15 and M16/M18 form wideband level shifters between the output of the first stage and the Gates of the output transistors. They are bootstrapped out for AC signals.

The quiescent current in the output transistors is set by two translinear loops. Output transistor M11 with M13 forms a translinear loop with transistors M23 and M21. Transistors M13 and M21 are equal and carry equal currents. The quiescent current in M11 is set by transistor M23.

The same applies to the translinear loop of M12/M14 with M22/M20.

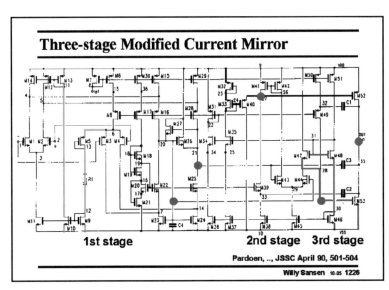

1226

This is a three-stage amplifier with nested Miller compensation. The high impedance nodes are labeled with big (red) dots.

The input stage consists of two folded cascodes. The g_m − equalization is carried out by transistor M5, resistor R1 and the following current mirrors. When the average input voltage increases, the pMOSTs are slowly turned off, but the current through resistor R1 increases, increasing the currents in the input nMOSTs. It is a simple solution. However, the use of a resistor makes this solution depending on the supply voltage.

The second stage is a differential pair, one output of which is directly connected to the gate of the output nMOST M53. The other output has to be inverted first before it can be applied to the output pMOST M52. The output devices of a class-AB stage always have to be driven in phase.

The translinear loops which set the quiescent current are highlighted. They are easily recognized.

This amplifier can drive 4000 pF. It can sink and source about 100 mA. On 2.5 V it takes about 0.6 mA. Its GBW is 1 MHz.

Its main disadvantage is that its Slew Rate is not sufficiently high, causing some cross-over distortion.

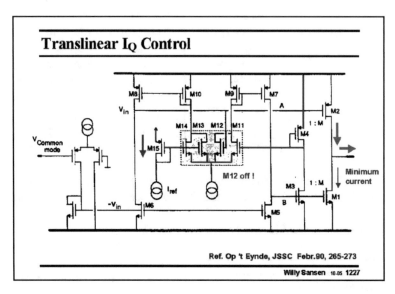

Translinear I$_Q$ Control

M12 off !

Minimum current

Ref. Op 't Eynde, JSSC Febr.90, 265-273

Willy Sansen 10-05 1227

1227

This is the third stage of a three-stage amplifier.

The complementary input voltages v_{in} are shown, as generated by the second stage. The top one is applied directly to the pMOST output transistor M2. The complementary input $-v_{in}$ is inverted and applied to the nMOST output transistor M1. This transistor is copied to M3 but M times smaller. The current through M3 and M4 is therefore a measure of the current through M1.

A three-fold translinear loop is now formed. Two of them include the output transistors. They are M2/M12 and M4/M11. They are added towards current mirror M9/M7. The third one is M15/M13+14. They are added towards current mirror M10/M8.

The purpose of these loops is to prevent one output transistor to go off, when the other one is carrying a large current. Indeed when transistor M2 provides a large output current, M1 may shut off. Even for large output currents, a minimum current through M1 must be guaranteed. This decreases the cross-over distortion and increases the speed.

If the current through M1 is very small, the currents through M3 and M4 are also small. As a result, V_{GS11} becomes larger. For a large current in M2, V_{GS12} becomes smaller. The product of the two currents through M11 and M12 is set by the reference current through M15. They cannot go below a certain value.

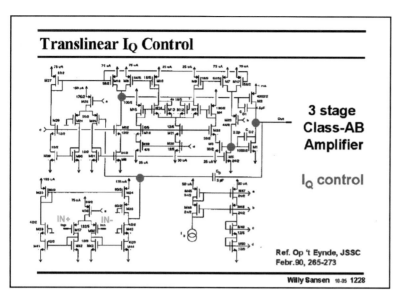

Translinear I_Q Control

3 stage Class-AB Amplifier

I_Q control

Ref. Op 't Eynde, JSSC Febr.90, 265-273

Willy Sansen 10-05 1228

1228

The full amplifier is shown in this slide. The three stages are used to increase the gain, yielding very low distortion.

The first and second stages are symmetrical amplifiers. The third stage consists of two output devices with the quiescent-current control described in the previous slide. Nested Miller compensation is used.

Its GBW is 5 MHz. It gives only -80 dB THD at 10 kHz in 81 Ω in parallel with 15 pF. The quiescent current is 1.4 mA.

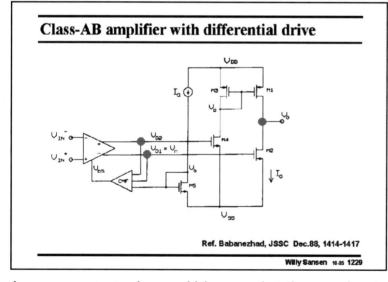

Class-AB amplifier with differential drive

Ref. Babanezhad, JSSC Dec.88, 1414-1417

Willy Sansen 10-05 1229

1229

In this two stage amplifier, a translinear loop is used to control the quiescent current. However, it is used to provide common-feedback at the same time.

The principle is fairly straightforward. A fully differential amplifier is used at the input. One output drives output transistor M2 directly. The other one has to be inverted first.

Since the output has a differential output, CMFB is required. This circuit sets the average output voltages, which are used at the same time to control the V_{GS} values of the output devices, thus also controlling the quiescent current in the output transistors.

The full circuit is given next.

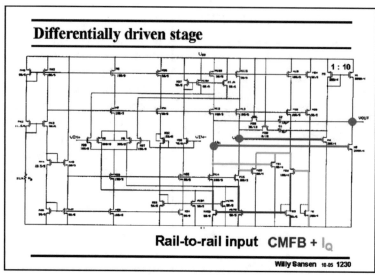

Differentially driven stage

1 : 10

Rail-to-rail input   CMFB + I$_Q$

Willy Sansen 10-05 1230

1230

The input stage is a fully-differential rail-to-rail amplifier. Its outputs are labeled with big (red) dots. One of them goes directly to the nMOST output transistor. The other one is inverted first.

A differential output needs common-mode feedback. This is accomplished by measurement of the two outputs with transistors M20/M21. Their Sources are joined together to cancel the differential signal. This common-mode signal is then fed through a cascode (M22), to a current mirror with transistors M23, M16B and M17B.

The same transistors are part of the translinear loops to set the quiescent current in the output stage. For the nMOST output transistor, the loop consists of transistors M2/M21 with M22/M5. For the pMOST output transistor, M4 is taken instead. The loop then consists of transistors M4/M20, again with M22/M5.

The setting for the average output voltage of the first stage is also used to set the V$_{GS}$'s of the output devices, and hence the quiescent current.

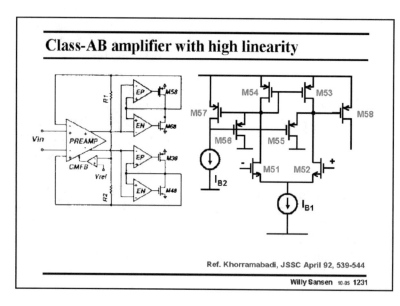

Class-AB amplifier with high linearity

Vin

PREAMP

CMFB

Vref

EP   M58

EN   M58

EP   M38

EN   M48

M54   M53

M57   M58

M56   M55

M51   M52

I$_{B2}$

I$_{B1}$

Ref. Khorramabadi, JSSC April 92, 539-544

Willy Sansen 10-05 1231

1231

This class-AB amplifier uses a separate opamp to drive the Gates of all four output devices. The load is connected between the two output voltages. It is therefore floating.

These opamps are required to provide sufficient gain, even when the output devices enter the linear region. As a result, the distortion is always small.

The amplifier EP which drives output transistor M58 is sketched as well. The feedback loop is not closed.

It is a conventional voltage amplifier with input devices M51/M52. The load current mirror is shunted by two cascodes however M55/M56 to limit the gain (to about 7), again to reduce the distortion.

The quiescent current is set by the translinear loop M58/M55 and M57/M56 such that the current through the output transistor is about I_{B2} times the ratio of M58 to M57.

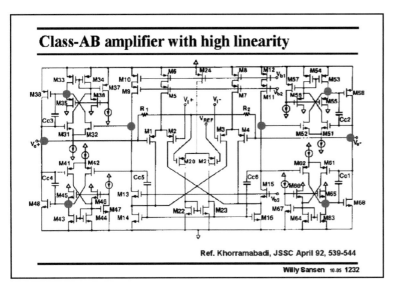

Class-AB amplifier with high linearity

Ref. Khorramabadi, JSSC April 92, 539-544

Willy Sansen 10-05 1232

1232

On the total schematic, the four output transistors with their amplifier, are easily recognized. The amplifier is shown with unity-gain feedback.

Actually, this is a three-stage amplifier. The input stage is a folded cascode. Distributed Miller compensation is applied however, not nested Miller. This method of compensation is much less transparent.

The CMFB amplifier is in the middle, with M20/M21 and R2/R3.

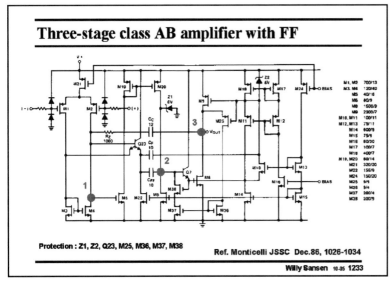

Three-stage class AB amplifier with FF

Protection : Z1, Z2, Q23, M25, M36, M37, M38

Ref. Monticelli JSSC Dec.86, 1026-1034

Willy Sansen 10-05 1233

1233

A three-stage class-AB amplifier is shown in this slide which uses feedforward to boost the high-frequency performance. The first stage is single-ended, which is not good for CMRR. The second stage is a non-inverting amplifier which uses a current mirror. The third stage consists again of pMOST and nMOST devices Drain to Drain. The nMOST is driven by an emitter follower to drive the large C_{GS8} capacitor. The pMOST is driven by a level shifter M10/M11, which is bootstrapped out for AC operation.

The compensation is not a pure case of nested Miller compensation. Compensation capacitor C_c determines the GBW. The other capacitors provide feedforward.

The quiescent current in the output transistors is set by two translinear loops consisting of M9/M11 with M17/M12 for the pMOST output transistor M9, and of M8/M10 with M13/M15 for the nMOST.

The total current consumption (on 5 V) is 0.35 mA. About 22 mA can be delivered to a low-resistance load.

Outline

- **Problems of class AB drivers**
- **Cross-coupled quads**
- **Adaptive biasing**
- **I_Q control with translinear circuits, etc.**
- **Current feedback and other principles**
- **Low-Voltage realizations**

1234

Translinear circuits for the definition of the quiescent currents in the output transistors have the disadvantage that the maximum output current is limited. Several more other principles are therefore investigated. The first one is the application of current feedback to obtain an expanding characteristic.

Current feedback

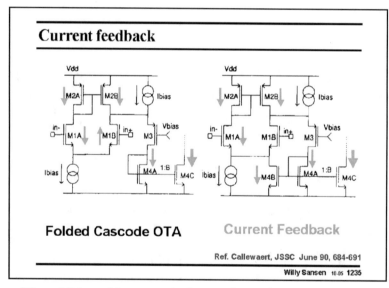

Folded Cascode OTA   Current Feedback

Ref. Callewaert, JSSC June 90, 684-691

1235

A simple but very current-efficient realization is shown in this slide.

On the left is a conventional differential pair loaded with a folded cascode. Its output current is simply B times the circular current of the differential pair. The currents are indicated by arrows.

This output current is limited to the B times I_{bias} however as every differential pair has a limiting characteristic.

The addition of just one single transistor changes the operation drastically and converts this stage into a class-AB amplifier. Transistor M4B is added, which forms a current mirror with M4A as well. It provides current feedback to the differential pair.

Two equal currents now flow from supply to supply. The first one flows through transistors M2A, M1A and M4B. The other one flows through M2B, M3 and M4A, and is multiplied with B towards the output. These currents are not limited by the biasing current I_{bias}. They can be much larger depending on the transistor sizes. They have an expanding or class-AB characteristic.

Clearly, these currents can only increase. Another stage with pMOSTs at the input is now required to have expanding currents in both directions. This is shown next.

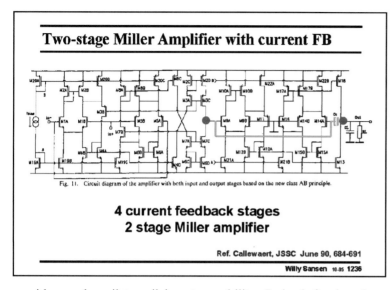

Two-stage Miller Amplifier with current FB

Fig. 11. Circuit diagram of the amplifier with both input and output stages based on the new class AB principle.

4 current feedback stages
2 stage Miller amplifier

Ref. Callewaert, JSSC June 90, 684-691

Willy Sansen 10-05 1236

1236

The first nMOST input current-feedback stage is put in parallel with a pMOST one. Their outputs are current mirrored to a high impedance output node, labeled with a big (red) dot. A second similar nMOST/ pMOST current-feedback stage is then used as an output stage. This is thus a two-stage amplifier. The Miller compensation capacitance is clearly seen.

The parallel nMOST/ pMOST pair at the input provide nearly rail-to-rail input capability. Indeed, for low input common-mode voltages the nMOSTs shut off but the pMOSTs take over and vice versa. The rail itself cannot actually be reached because of a diode-connected transistor M2a (in the first stage). About 0.1 V is lost at both supply lines. No g_m−equalization is provided.

For a 10 kΩ/100 pF load, the GBW is 0.37 MHz and power dissipation 0.25 mW (for ±5 V supply voltages).

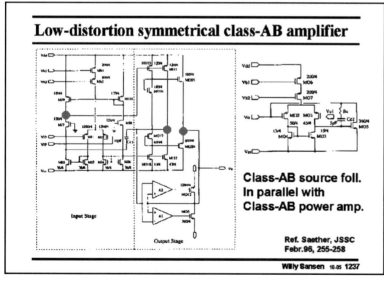

Low-distortion symmetrical class-AB amplifier

Input Stage

Output Stage

Class-AB source foll.
In parallel with
Class-AB power amp.

Ref. Saether, JSSC
Febr.96, 255-258

Willy Sansen 10-05 1237

1237

This amplifier has two output stages in parallel. The top one uses source followers. It can provide moderate voltage swing but with very low distortion.

However, most of the gain and current (power) comes from the bottom amplifier. It consists of two error amplifiers followed by two output transistors, Drain to Drain. Some offset is built in the error amplifiers (shown on the right) so that the output devices are turned off for small output signals. In this way they do not generate cross-over distortion. The source follower amplifier carries out all the tasks.

For large output swing, the source follower amplifier cannot follow any more. The class-AB power amplifier can still provide rail-to-rail output swing, even when the output devices end up in the linear region. The error amplifier still provides sufficient gain.

The quiescent current in the low-distortion source-follower amplifier is set by the translinear loop of transistors MO16/MO19 and MO17/MO20. The DC current through MO17/17 sets the current in the output devices MO19/20.

For a 1 kΩ/150 pF load, the Slew Rate is 7 V/μs. The GBW is 5.5 MHz and power dissipation 6.5 mW ($\pm$ 5 V). The equivalent input noise is 10 nV$_{RMS}$/$\sqrt{Hz}$, which is quite low.

Outline

- **Problems of class AB drivers**
- **Cross-coupled quads**
- **Adaptive biasing**
- **I_Q control with translinear circuits, etc.**
- **Current feedback and other principles**
- **Low-Voltage realizations**

Willy Sansen 10-05 1238

1238

More recent technologies use smaller channel lengths. As a consequence, the supply voltage has become smaller as well. There is a need for class-AB stages for supply voltages of 1.5 V and less.

Translinear circuits cannot be used any more. Several examples are given next.

1.5 V supply voltage class-AB amp.

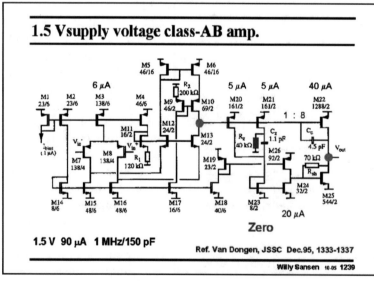

1.5 V 90 μA 1 MHz/150 pF

Zero

Ref. Van Dongen, JSSC Dec.95, 1333-1337

Willy Sansen 10-05 1239

1239

At low supply voltages, the circuitry becomes simpler. In this 1.5 V amplifier, the input stage is a folded cascode. It is followed by an output stage in which the output pMOST is connected directly to the output of the input stage.

The output nMOST drive is very different however. It leads to a current mirror M23/M24 to carry out two inversions. Remember that output transistors have to be driven in phase. This gives rise to extra poles, which have to be compensated for.

Two tricks are used. The first one is local feedback around output transistor M25, with resistor R_{sh}. This shunt-shunt feedback lowers the impedance at input and output indeed.

The second trick is the introduction of a zero with time constant $R_z C_z$. This must be tuned to one of the non-dominant poles, which is not that easy of course.

The quiescent current in the output devices is not that well defined. The variation of this current is decreased by the local feedback of resistor R_{sh}. However, it will never be really independent of the supply or output voltage.

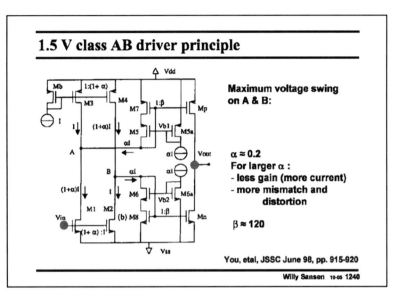

1240

A simpler low-voltage class-AB amplifier is shown in this slide.

Only the output stage is shown. It consists of two current mirrors with current factor β. They are driven by two parallel input devices, with different sizes however. Indeed transistor M1 is $(1+\alpha)$ times larger than transistor M2. The top current mirror with pMOSTs is biased by αI and so is the bottom one with nMOSTs. These are the quiescent currents and are well defined.

The main advantage of this driver is that points A and B can have very large swings. The output stage can sink and source currents which are much larger than the quiescent currents. For a large voltage, on point B for example, V_{GS8} becomes very large but V_{DS8} is limited by the cascode M6. Transistor M8 enters the linear region. This levels off somewhat the increase in output current. This current is still a lot larger than for a translinear loop.

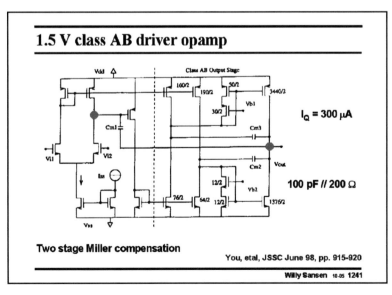

1241

The full amplifier is shown in this slide.

The first stage is a simple differential pair followed by a current mirror to drive the output stage.

Only two high-impedance points can be distinguished. Capacitor C_{m1} sets the GBW, together with g_{m1}. Capacitors C_{m2} and C_{m3} see small resistances only, and are not so effective.

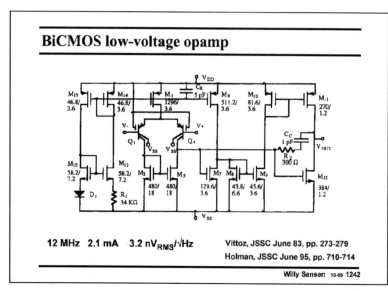

BiCMOS low-voltage opamp

12 MHz   2.1 mA   3.2 nV$_{RMS}$/√Hz

Vittoz, JSSC June 83, pp. 273-279
Holman, JSSC June 95, pp. 710-714

Willy Sansen 10-05 1242

1242

A somewhat similar principle for the quiescent current is used in this amplifier. It is a two-stage amplifier which can operate at low supply voltages.

The input devices are lateral pnp transistors. Actually they are pMOSTs in which the Source-Bulk diode is forward biased (ref. Vittoz). They exhibit very low 1/f noise.

Output transistor M12 is driven directly by the first stage. The other output transistor M11 is driven by two inverters M7-M9 and M10-M11.

The quiescent current in the output transistors is controlled by current source M6. Indeed its currents is split up in two parts. The first part flows through M7, which has the same V_{GS} as output transistor M12, and which has a fixed ratio in current to M12.

The other part flows through M8, which controls the current in output transistor M11 by means of two current mirrors. The currents in the output transistors must be the same. The current through M6 controls this current.

Using the sizes of the transistors in this slide, the quiescent current is about 1.6 times larger than the current in transistor M6.

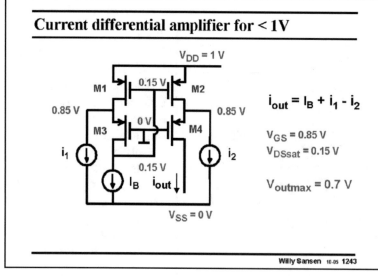

Current differential amplifier for < 1V

V$_{DD}$ = 1 V

M1   0.15 V   M2

0.85 V   0.85 V

0 V

M3   M4

i$_1$ i$_2$

0.15 V

I$_B$ i$_{out}$

V$_{SS}$ = 0 V

$i_{out} = I_B + i_1 - i_2$

$V_{GS} = 0.85$ V
$V_{DSsat} = 0.15$ V

$V_{outmax} = 0.7$ V

Willy Sansen 10-05 1243

1243

A very different principle is based on the current differential amplifier, which has been discussed in Chapter 3. It has three current inputs, actually four, if an extra current input source is applied to the Drain of transistor M4. Moreover, it can operate on very low supply voltages.

If a threshold voltage V_T is taken of 0.7 V, and a $V_{GS} - V_T$ (and V_{DSsat}) of only 0.15 V, then the supply voltage can be as low as 1 Volt.

The maximum output voltage can then be as high as 0.7 V.

If however the V_T is only 0.3 V, and the same $V_{GS} - V_T$ of 0.15 V is taken, then the supply voltage can be a mere 0.6 V!

A similar biasing will be used for the class-AB amplifier discussed next. It can therefore be used at a supply voltage of only 0.6 V!

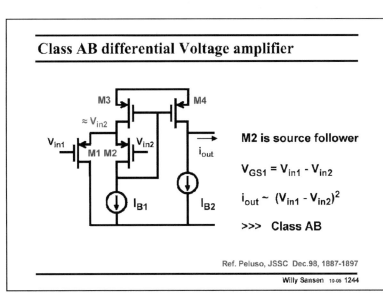

1244

The class-AB amplifier is shown in this slide.

On first sight it seems to consist of a differential pair, the output of which is fed back to the current mirror, biasing this pair. However, feedback from a differential output to a common-mode node is impossible to grasp.

A better way to understand this circuit is to note that transistor M2 has a constant current, i.e. current I_{B1}. If not, the feedback loop to the Gate of M3 will make

sure of that. The only basic single-transistor configuration in which the transistor carries only DC current is the Source follower. Transistor M2 acts as a Source Follower. It passes input voltage V_{in2} unattenuated to the Source of the other input transistor M1.

Input transistor M1 is a differential amplifier by itself. One input voltage V_{in1} is at its Gate and the other, V_{in2} at its Source. It converts this differential input voltage into an AC current which flows from the supply through transistors M3 and M1 to ground. It is mirrored by the current mirror M3/M4 to the output. It could also be mirrored out at the Drain of M1 however, as will be done in the full circuit, shown next.

This AC current is not limited by any DC current. Moreover, it has an expanding characteristic because of the square-law characteristic of a MOST. Transistor M1 acts as a class-AB amplifier.

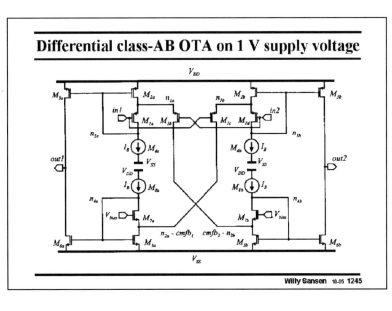

1245

The full schematic is given in this slide. It is fully differential. The common-mode feedback CMFB is not shown.

The class-AB voltage-to-current converting transistors are M1b and M1c.

Transistor M1a passes input voltage in1 to the Source of M1b. An AC current is generated by transistor M1b. This current is mirrored to output out1 by current mirror M2a/M3a and to output out2 by current mirror M5b/M6b.

The same applies to the AC current generated by transistor M1c. It is mirrored to the outputs as well.

The CMFB can be realized by application of a current to the sources of M7a/M7b.

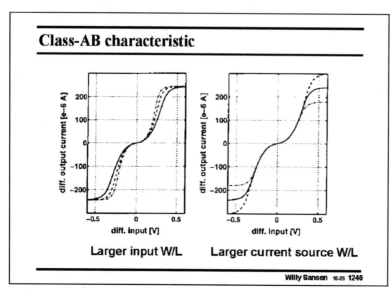

1246

The input transistors M1b and M1c determine the gain from input voltage to output current. Increasing their size increases their gain, as shown left.

For a zero differential input voltage, the differential output current is zero as well. For small input voltages, an expanding characteristic is obtained.

For very large input voltages, the maximum output current saturates, depending on the relative transistor ratios. The quiescent current is set by current sources I_B, which is about 2 μA. The current drive capability is also quite high.

An increase in the DC current sources I_B, evidently increases the maximum output current as well. This is shown on the right.

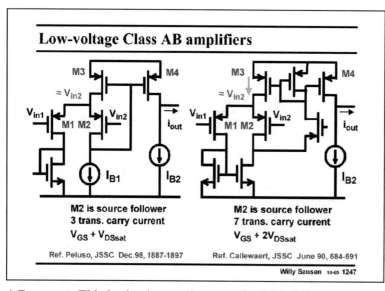

1247

This last class-AB amplifier is actually a simplified version of the class-AB amplifier using current feedback, discussed before. Both are sketched next to each other for sake of comparison. Both use transistor M2 as a Source follower.

Both amplifiers use transistor M2 as a Source follower. In the last amplifier (on the left), only one single transistor provides the voltage-to-current conversion. Only three transistors carry AC current. This is clearly an advantage for high-frequency or for low-power design, or both.

In the right amplifier, seven transistors carry an AC current. In principle the more transistors carry an AC current, the more poles are generated and the slower the circuit will be. The left amplifier is better with this respect.

Moreover, the left amplifier can work at a lower supply voltage. The minimum supply voltage of the right amplifier is $V_{GS} + 2V_{DSsat}$ but only $V_{GS} + V_{DSsat}$ for the left one. For a V_T of only 0.3 V, and a $V_{GS} - V_T$ of 0.2 V is taken, then the minimum supply voltage is 0.7 V for the left amplifier but 0.9 V for the right one.

The left amplifier is clearly superior.

Conclusions

- **Problems of class AB drivers**
- **Cross-coupled quads**
- **Adaptive biasing**
- **I_Q control with translinear circuits, etc.**
- **Current feedback and other principles**
- **Low-Voltage realizations**

Willy Sansen 10-05 1248

1248

A large variety of class-AB amplifiers have been discussed and compared. Many different principles are available depending on the required power levels and output loads. This is only a selection of the amplifiers published. It is a good overview though.

For all amplifiers discussed, the full references are now listed. They allow the reader to study these amplifiers in more detail.

More references are given than have been included in this chapter. This is done to give the reader a more complete list of references to amplifiers, which deserve to be examined.

1249

Reference list - 1

J.Babanezhad, "A rail-to-rail CMOS opamp", IEEE Journal Solid-State Circuits, Vol. SC-23, pp. 1414-1417, Dec.1988.

K. Brehmer, B. Wieser, "Large swing CMOS power amplifier", IEEE Journal Solid-State Circuits, Vol. SC-18, pp.624-629, Dec.1983.

L. Callewaert and W. Sansen, "Class AB CMOS amplifiers with high efficiency", IEEE Journal Solid-State Circuits, Vol. SC-25, pp. 684-691, June 1990.

H.Casier, etal, "A 3.3 V low-distortion ISDN line driver with a novel quiescent current control circuit", IEEE Journal Solid-State Circuits, Vol. SC-33, pp. 1130-1133, July 1998.

R. Castello, P. Gray, "A high-performance micropower switched-capacitor filter", IEEE Journal Solid-State Circuits, Vol. SC-20, pp. 1122-1132, Dec.1985.

M.Degrauwe, J.Rijmenants, E.Vittoz, H.De Man, "Adaptive biasing CMOS amplifiers" IEEE Journal Solid-State Circuits, Vol. SC-17, pp. 522-528, June 1982.

de Langen K., G. Eschauzier, G. van Dijk, J. Huijsing, "A 1-GHz bipolar class-AB operational amplifier with multipath nested Miller compensation for 76-dB gain", IEEE Journal Solid-State Circuits, Vol. SC-32, pp. 488-498, April 1997.

J. Fischer, R. Koch, "A highly linear CMOS buffer amplifier", IEEE Journal Solid-State Circuits, Vol. SC-22, pp. 330-334, June 1987.

Willy Sansen 10-05 1249

1250

Reference list - 2

K.Halonen "Low power high performance switched capacitor circuits for data acquisition systems", PhD KULeuven, October 1987.

R. Hogervorst, etal, "A compact power efficient 3V CMOS rail to rail input/output operational amplifier for VLSI cell libraries", IEEE Journal Solid-State Circuits, Vol. SC-29, pp. 1504-1512, Dec.1994.

W.Holman, A.Connelly, "A compact low noise operational amplifier for a 1.2 micron digital CMOS technology", IEEE Journal Solid-State Circuits, Vol. SC-30, pp. 710-714, June 1995.

J. Huijsing, D. Lineburger, "Low-voltage operational amplifier with rail-to-rail input and output ranges", IEEE Journal Solid-State Circuits, Vol. SC-20, pp. 1144-1150, Dec.1985.

B.Lee, B.Shen, "A high slew-rate CMOS amplifier for analog signal processing", IEEE Journal Solid-State Circuits, Vol. SC-25, pp. 885-889, June 1990.

K.Lee, R. Meyer, "Low-distortion switched-capacitor filter design techniques", IEEE Journal Solid-State Circuits, Vol. SC-20, pp. 1103-1113, Dec.1985.

D. Monticelli, "A quad CMOS single-supply op amp with rail-to-rail output swing", IEEE Journal Solid-State Circuits, Vol. SC-21, pp. 1026-1034, Dec.1986.

F. Op't Eynde, P. Ampe, L. Verdeyen and W. Sansen, "A CMOS large-swing low-distortion three-stage class AB power amplifier", IEEE Journal Solid-State Circuits, Vol. SC-25, pp. 265-273, Febr.1990.

1251

Reference list - 3

M. Pardoen, M.Degrauwe, "A rail-to-rail input/output CMOS power amplifier", IEEE Journal Solid-State Circuits, Vol. SC-25, pp. 501-504, April 1990.

V. Peluso, etal, "A 900 mV low-power Delta-Sigma AD Converter with 77-dB dynamic range" IEEE Journal Solid-State Circuits, Vol. SC-33, pp. 1887-1897, Dec.1998.

T.Saether etal, "High-speed, high-linearity CMOS buffer amplifier", IEEE Journal Solid-State Circuits, Vol. SC-31, pp. 255-258, Febr.1996.

E. Seevinck etal "Low-distortion output stage for power amplifiers", IEEE Journal Solid-State Circuits, Vol. SC-23 pp. 794-801, June 1988.

M.Steyaert, W.Sansen, "A high-dynamic range CMOS opamp with low-distortion output structure", IEEE Journal Solid-State Circuits, Vol. SC-22 pp. 1204-1207, Dec.1987.

R. Van Dongen, V. Rikkink, "A 1.5 V Class AB CMOS buffer amplifier for driving low-resistance loads", IEEE Journal Solid-State Circuits, Vol. SC-30, pp. 1333-1337, Dec.1995.

W.Wu, etal, "Digital-compatible high-performance operational amplifier with rail-to-rail input and output ranges", IEEE Journal Solid-State Circuits, Vol. SC-29, pp. 63-66, Jan.1994

F. You, S.H.K. Embabi and E. Sanchez-Sinencio. "Low-voltage class AB buffers with quiescent current control." IEEE Journal of Solid-State Circuits , Vol SC-33, pp. 915-920, June 1998.

Feedback
Voltage & Transconductance
Amplifiers

Willy Sansen

KULeuven, ESAT-MICAS

Leuven, Belgium

willy.sansen@esat.kuleuven.be

Willy Sansen 10-05 131

131

Feedback is used in almost all analog amplifiers and filters. A thorough understanding is therefore a necessity for whoever wants to build up insight in the art of analog circuit design. This understanding can be gradually developed by reviewing the principles and by applying them to the four basic types of feedback.

Many publications and books are devoted to feedback. Most of them originate from circuit theory, however. They inevitably start from the description of the amplifier and the feedback network by means of matrices. This very formal approach is not always necessary. In most cases, the concepts of open and closed-loop gain and of the loop gain are sufficient to obtain the most important specifications such as the closed-loop gain, the bandwidth and the input-and output impedances. For the impedance at an inner node, the rule of Blackman is necessary. On the other hand, most designers could not care less about the inner nodes. Therefore, Blackman is omitted. The simplest possible approach is envisaged here to learn all about feedback.

In this Chapter, we focus first on voltage and transconductance amplifiers. The next chapter will introduce transimpedance and current amplifiers.

Table of contents

- ◆ **Definitions**
- ◆ **Series-shunt FB for Voltage amplifiers.**
- ◆ **Series-series FB for Transconductance amps.**

Willy Sansen 10-05 132

132

First of all, we want to learn about the definitions. For example, what is the actual loop gain? For large values of the loop gain, it is the ratio of the open-loop gain and the closed-loop gain, or simply the difference if we use dB. For example, an operational amplifier (shortened to opamp) with an open-loop gain of 85 dB, which is used in a negative feedback loop resulting in a gain of 10 or 20 dB, has a loop gain of 65 dB.

This loop gain is the gain by which the characteristics of the amplifier are improved such as the precision of the gain. It also reduces the noise and the distortion, but above all it improves the bandwidth a great deal.

We will look into these phenomena at a later stage. We first want to examine the four cases of feedback. The input can be connected in parallel (shunt) or in series, giving rise to four different cases.

In this Chapter we will focus on voltage and transconductance amplifiers. Both types of feedback circuits take voltages at the input.

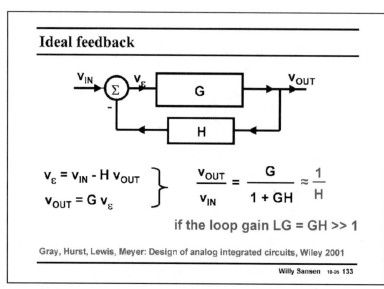

133

An ideal feedback loop consists of a unidirectional amplifier (from left to right) and a unidirectional feedback circuit (from right to left). This amplifier usually consists of a few transistors or even a full operational amplifier. As a result it provides a lot of gain.

The feedback circuit usually consists of a few passive devices. They will set the closed-loop gain as shown next.

Two equations describe the operation of this feedback circuit. The error voltage v_ε is the difference between the actual input voltage v_{in} and the feedback voltage Hv_{out}. It is amplified towards the output itself by G.

The closed-loop gain is then easily extracted from the two equations. Its numerator is simply the gain G itself. The denominator however, is $1 + GH$. The quantity GH is called the loop gain LG. It is the gain, going around in the loop. Since the gain G is always quite large, the loop gain is also quite large. As a result the closed-loop gain can easily be approximated by $1/H$.

This is the reason why H usually consists of passive devices such as resistors or capacitors. Their ratio can be made quite accurate. As a result, the feedback amplifier has a closed-loop gain which is reasonably accurate, whereas the open-loop gain G can vary a lot depending on transistor parameters, temperature, etc. Feedback is thus the most important technique to realize amplifiers with accurate gain.

134

One of the simplest cases of feedback is an operational amplifier with a resistor from the output to the input. Of course, the feedback resistor has to be connected to the negative input. Otherwise the loop gain would build up an ever-increasing output voltage, only to stop at the positive supply voltage. Stable feedback is always negative feedback.

This is a case of shunt (or parallel) feedback at both input and output. Output shunt feedback means that the output terminal is in parallel with the feedback element terminal. This is also the case at the input.

The gain of the amplifier itself is A_0, which is also quite large, between 10.000 and 1.000.000. This is also the loop gain LG, as will be calculated on the next slide.

The output voltage simply equals the input current into the feedback resistor. The closed-loop gain is simply R_F. It is therefore a transresistance amplifier with gain R_F.

The input and output resistances are both affected by the feedback. In the case of shunt feedback, the resistance decreases by an amount equal to the loop gain LG, or actually $1 + LG$.

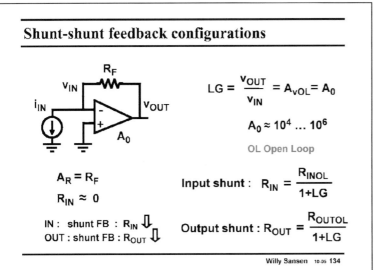

Shunt-shunt feedback configurations

$$LG = \frac{v_{OUT}}{v_{IN}} = A_{vOL} = A_0$$

$$A_0 \approx 10^4 \ldots 10^6$$

OL Open Loop

$$A_R = R_F$$

$$R_{IN} \approx 0$$

$$\text{Input shunt}: \quad R_{IN} = \frac{R_{INOL}}{1+LG}$$

IN : shunt FB : R_{IN} ⇩
OUT : shunt FB : R_{OUT} ⇩

$$\text{Output shunt}: R_{OUT} = \frac{R_{OUTOL}}{1+LG}$$

Willy Sansen 10-95 134

135

It is clear that the loop gain LG is the most important characteristic of a feedback amplifier. Therefore, its value must be calculated first.

The loop gain LG is calculated by breaking the loop and by calculating the gain, going around the loop. The DC conditions must be maintained, only the AC loop is broken.

Ideally it makes no difference where the loop is broken. The loop gain

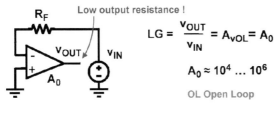

Calculation loop gain or return ratio

Low output resistance !

$$LG = \frac{v_{OUT}}{v_{IN}} = A_{vOL} = A_0$$

$$A_0 \approx 10^4 \ldots 10^6$$

OL Open Loop

Independent sources : voltage source to zero
current source to infinity

Break loop where impedances are very different
Find the loop gain = return ratio

Willy Sansen 10-95 135

should be independent of where the loop is broken. Therefore, we try to find an easy place, a place where the calculations are easy. This is the case for any connection where the difference between the resistance, left and right are the largest.

In the example in this slide, the output resistance of the operational amplifier is quite low, certainly a lot lower than resistor R_F. Therefore, we break in between. We apply a voltage source (as the output resistance of the opamp was low) and we calculate the voltage going around the loop. This gives a value A_0. However, the voltage on both sides of the resistor R_F are the same as there is no current flowing through it.

What happened to the input current source? Since we have applied another input source v_{IN}, we must remove the input current source (called the independent source). For calculating the loop gain, we replace an independent current source by its internal resistance (which is infinity). Independent voltage sources are replaced by their internal resistance as well, which is just about zero or a short-circuit.

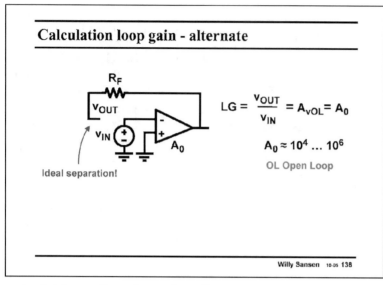

Calculation loop gain - alternate

$$LG = \frac{v_{OUT}}{v_{IN}} = A_{vOL} = A_0$$

$$A_0 \approx 10^4 \dots 10^6$$

OL Open Loop

Ideal separation!

Willy Sansen 10-05 138

136
The gain calculated in this way is therefore the loop gain LG or also called the return ratio.

To illustrate that it does not make any difference where the loop is broken, we calculate the loop gain LG again. This time the loop is broken between the feedback resistor R_F and the input terminal of the opamp. This is an even better place than before as the input resistance of an opamp is just about infinity,

and a lot larger than that resistor R_F.

It is clear that we find the same value of the loop gain LG. It is again equal to the gain of the operational amplifier itself A_0.

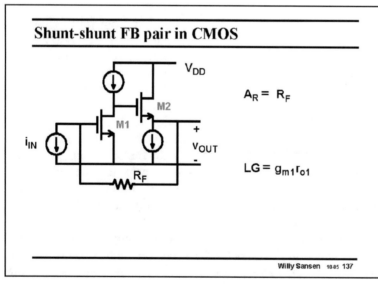

Shunt-shunt FB pair in CMOS

V_{DD}

M2

M1

i_{IN}

$+$

v_{OUT}

$-$

R_F

$$A_R = R_F$$

$$LG = g_{m1} r_{o1}$$

Willy Sansen 10-05 137

137
The amplifier does not need to be a full operational amplifier, with lots of gain. A simple transistor amplifier can do it as well. Here the opamp is replaced by a single-stage amplifier followed by a source follower.

The open-loop gain is simply the gain of that input transistor as a source follower provides a gain of unity only.

This is also the loop gain as easily seen. We can still break the loop where we want. The output resistance is a bit higher now, i.e. $1/g_{m2}$, which is still a lot smaller than R_F.

The closed-loop gain is usually the easiest one to calculate. It is still R_F as for the first feedback amplifier and it is still a transresistance amplifier. In other words, it converts an input current into an output voltage with high accuracy, here with value R_F.

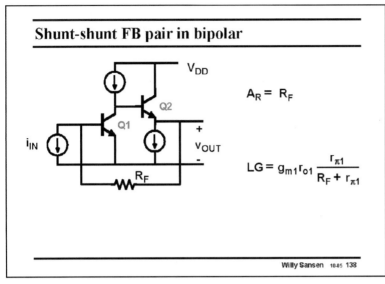

138
Using bipolar transistors instead of MOSTs changes the input resistance considerably. Now that the input resistance of the bipolar transistor is only r_π, instead of infinity for a MOST. As a result, the loop gain will no longer be the same as the feedback resistor may be comparable to that resistance r_π.

The closed-loop gain is still the same however, i.e. R_F.

To calculate the loop gain LG we can break the loop between the emitter of transistor Q2 and resistor R_F as we did before. There is a better place however, as shown next.

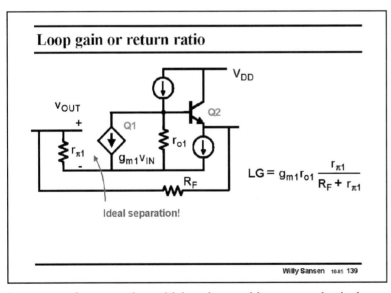

139
To find this better place to break the loop, we have to draw the small-signal schematic. Now it becomes clear that right in between the base terminal of transistor Q1 and its voltage-controlled current source is an excellent place to break the loop. We therefore exploit the fact that the input resistor $r_{\pi 1}$ of transistor Q1 is physically separated from the current source $g_{m1} v_{IN}$ in the small-signal equivalent circuit. They are only linked by means of an equation which only provides a non-physical connection.

The loop-gain is then readily calculated. It shows that at the input, we find a resistive divider of $r_{\pi 1}$ with R_F, which reduces the loop gain somewhat.

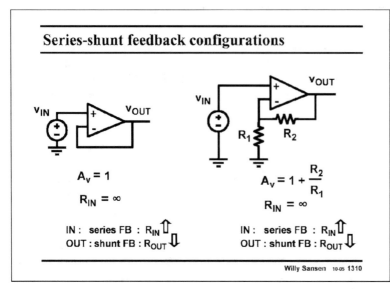

Series-shunt feedback configurations

$A_v = 1$

$R_{IN} = \infty$

IN : series FB : R_{IN} ⇧
OUT : shunt FB : R_{OUT} ⇩

$A_v = 1 + \dfrac{R_2}{R_1}$

$R_{IN} = \infty$

IN : series FB : R_{IN} ⇧
OUT : shunt FB : R_{OUT} ⇩

Willy Sansen 10-05 1310

1310

If we feed back the output voltage in series with the input, we obtain a series-shunt feedback loop. In its simplest case, the output is directly connected to the input, yielding a gain of unity. For an opamp with high gain, the difference between the terminals is approximately zero, whatever the output may be. This circuit is called a buffer amplifier as it can deliver a lot of current without loss in voltage gain.

More often however, a few resistors are used to set the gain at a precise value. The gain is positive as the output is in phase with the input. It is therefore a non-inverting amplifier. Since the input is directly connected to the Gate of a MOST, the input current is zero and the input resistance infinity. No current flows through the input voltage source (or input sensor).

Later we will prove that the input resistance goes up because of the series feedback at the input. Parallel or shunt feedback always causes the resistance to go down. The output resistance therefore goes down. This feedback causes this amplifier to behave as a **voltage-to-voltage amplifier**. Indeed, the voltage is sensed at the input, without drawing current. At the output, the amplifier behaves as a voltage source. Series-shunt feedback turns amplifiers in ideal voltage-to-voltage amplifiers with precise voltage gain!

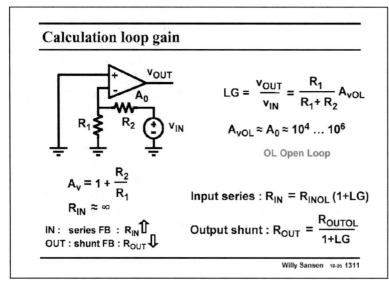

Calculation loop gain

$LG = \dfrac{v_{OUT}}{v_{IN}} = \dfrac{R_1}{R_1 + R_2} A_{vOL}$

$A_{vOL} \approx A_0 \approx 10^4 \ldots 10^6$

OL Open Loop

$A_v = 1 + \dfrac{R_2}{R_1}$

$R_{IN} \approx \infty$

IN : series FB : R_{IN} ⇧
OUT : shunt FB : R_{OUT} ⇩

Input series : $R_{IN} = R_{INOL}(1+LG)$

Output shunt : $R_{OUT} = \dfrac{R_{OUTOL}}{1+LG}$

Willy Sansen 10-05 1311

1311

In order to figure out how much the input and output resistances change, we have to find the loop gain first. For this purpose, we try to find an easy place to break the loop. Outputs of op-amps generally have low output resistances already without feedback. Breaking the loop right after the output is thus a good choice. Of course we could also break the loop right before the minus input of the opamp, as we see an infinite resistance into the amplifier.

The voltage gain around the loop is then easily found. It is attenuated first by the resistor ratio, followed by the total open-loop gain A_0 of the opamp.

As a result, the input resistance, which was already quite high, increases even more. The output resistance decreases by the same amount.

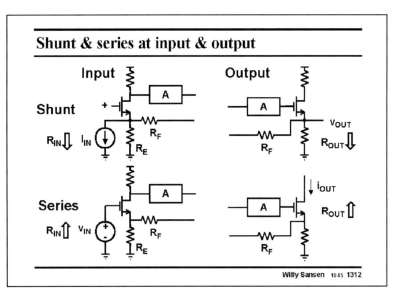

1312

Let us now try to obtain some better insight in why series feedback increases the resistance and shunt feedback decreases it. Also, how can we try to remember this?

All four combinations are shown in this slide. At the input, a transistor is shown to see how exactly the feedback resistors are connected. In all cases some more gain is added in the loop by means of gain block A.

In the first case of input shunt feedback, a cascode transistor is used at the input. The input source is connected at the same node as the feedback resistor R_F. As a result, we will expect the input resistance to go down. The same applies to the shunt output at the right.

For input series feedback, the input signal source is not at the same node as the feedback resistor. Actually, a feedback voltage is added in series with the transistor input voltage. The input resistance is now expected to increase.

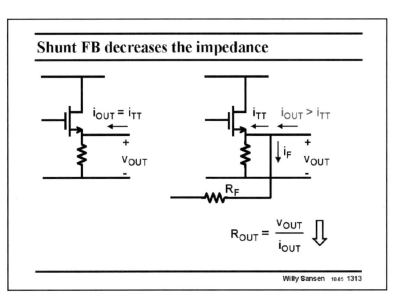

1313

Indeed, shunt feedback, shown here at the output of an amplifier, causes the output current to increase. This output current i_{OUT} is the current flowing through some output load, not shown in this diagram.

This output current i_{OUT} splits up towards the amplifier i_{TT} and towards the feedback resistor i_F. The total output current will be larger than without feedback. As a result, the ratio v_{OUT}/i_{OUT}, which is the output resistance R_{OUT}, will decrease.

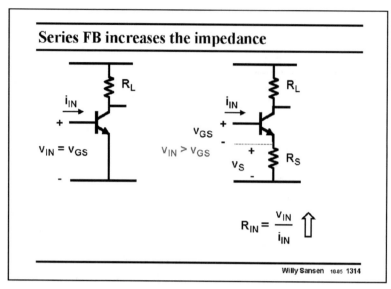

Series FB increases the impedance

$$R_{IN} = \frac{v_{IN}}{i_{IN}} \Uparrow$$

Willy Sansen 10-05 1314

1314

In the same way, series feedback, shown here at the input of an amplifier, causes the input voltage to increase. This input voltage v_{IN} is the total voltage from gate to ground. It is the sum of the actual transistor v_{GS} and the voltage v_S across the feedback resistor R_S.

The total input voltage will therefore be larger than without feedback. As a result, the ratio v_{IN}/i_{IN}, which is the input resistance R_{IN}, will increase by the amount of loop gain realized by the feedback loop.

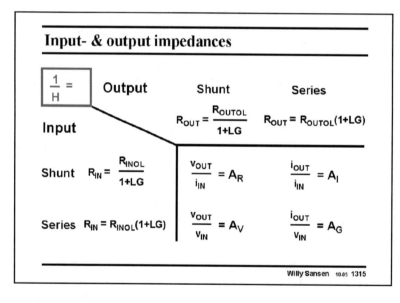

Input- & output impedances

Willy Sansen 10-05 1315

1315

The different kinds of feedback give rise to different kinds of amplifiers. We have already seen that series-shunt feedback provides an amplifier with high precision in voltage gain A_V.

In a similar way, a shunt-shunt feedback amplifier generates an accurate trans-resistance gain A_R. Both input and output resistances will decrease. The input can easily be driven by means of an input current source. The output behaves as a voltage source.

In order to make a good current amplifier, we will have to apply shunt-series feedback. The input resistance is lowered to be able to allow the input current to flow. The output resistance is quite high, as for any current source.

Finally, for a transconductance amplifier, we will need to use series-series feedback.

Why do we need all these different kinds of amplifiers?

Shunt vs series feedback

Shunt feedback
lowers impedance levels : higher bandwidths

Series feedback
increases impedances : lower node poles

Ouput shunt best for interconnect to next stage !

Output series acts as current source !

Willy Sansen 10-05 1316

1316

Shunt feedback will be used when we want to decrease the impedance level of an interconnection between two circuit blocks. Such interconnects can pick up a lot of parasitic capacitance. This causes a severe reduction in bandwidth when the interconnect is at too high an impedance level.

On the contrary, in an operational amplifier we want to create a low-frequency dominant pole by means of one single capacitance. Series feedback is a great help in increasing the node impedance.

Also sometimes a real current source must be built. For example, to carry out an impedance measurement we need to apply a precise current and to measure the voltage generated across it. As a result, we need to generate a circuit with high output impedance and a precise current. Output series feedback is ideal for this kind of application.

Shunt versus series for sensors

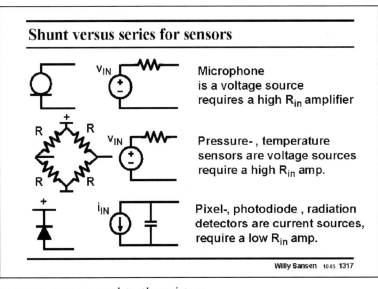

Microphone
is a voltage source
requires a high R_{in} amplifier

Pressure-, temperature
sensors are voltage sources
require a high R_{in} amp.

Pixel-, photodiode , radiation
detectors are current sources,
require a low R_{in} amp.

Willy Sansen 10-05 1317

1317

At the input, it is mainly the kind of sensor which determines whether we need a voltage input or a current sensing input. A dynamic microphone for example, behaves as a voltage source: it has a small internal resistance. The voltage carries the sensor information. We need therefore, to measure the voltage at the input. We also need a high input resistance or series feedback at the input. This also applies to a Wheatstone bridge with

pressure sensors, and to thermisters.

On the other hand, if we have a capacitive pressure sensor or accelerometer, or a photodiode, then we need a current amplifier. They all have a small capacitor as an internal impedance, quite often as low as 10 pF. Its impedance is quite high at low frequencies. It is the current which carries the sensor information. We need a current measurement, or shunt feedback at the input. If we want a voltage output, we will have to take a shunt-shunt feedback amplifier.

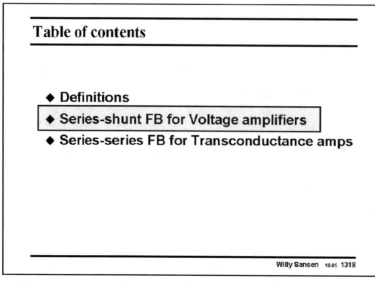

1318

Let us now discuss series-shunt feedback in more detail. It is a kind of amplifier which provides an accurate voltage-to-voltage conversion, as in preamplifiers for pressure sensors.

They are evidently called voltage amplifiers.

This time we will assume a more general case, i.e. the input impedance is not infinite any more but has a limited value. Also, the output resistance is no longer zero.

We want to calculate again the loop gain, the closed and open-loop gains, and finally the input and output resistances.

Afterwards, series-series feedback will be discussed. They will measure a voltage and provide an output current. This is why they are called transconductance amplifiers. They share the high input resistance with voltage amplifiers. Both have series feedback at the input.

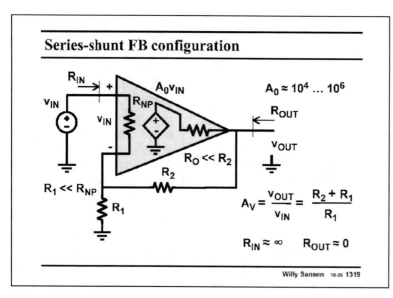

1319

A general-purpose amplifier is shown in this slide, with series-shunt feedback around it. The amplifier has a lot of gain A_0. It is modeled by a voltage controlled voltage source $A_0 v_{IN}$. Its input resistance R_{NP} is high but not infinite. It is certainly larger than resistor R_1.

Its output resistance R_0 is small but not zero.

Feedback resistor R_2 is a lot larger than R_0.

The closed-loop gain is evidently the ratio of the two resistors, as indicated in this slide.

The input resistance will increase as a result of the feedback. It is now quasi infinite.

The output resistance will decrease as a result of the feedback. It is now quasi zero.

What are the actual values?

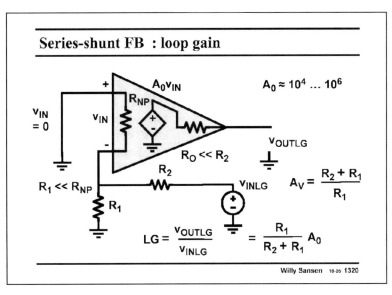

Series-shunt FB : loop gain

$v_{IN} = 0$

v_{IN}

R_{NP}

$R_1 << R_{NP}$

R_1

$A_0 v_{IN}$

$R_O << R_2$

R_2

$A_0 \approx 10^4 ... 10^6$

v_{OUTLG}

v_{INLG}

$A_V = \dfrac{R_2 + R_1}{R_1}$

$$LG = \frac{v_{OUTLG}}{v_{INLG}} = \frac{R_1}{R_2 + R_1} A_0$$

Willy Sansen 10-95 1320

1320

To obtain the actual values we have to find the loop gain LG first.

The loop is broken at the output of the opamp. The input voltage v_{IN} is set to zero.

The loop gain LG is the ratio of the output voltage v_{OUTLG} to the input voltage v_{INLG}. It is easily found to be the open loop gain A_0 divided by the closed-loop gain A_v.

The loop gain LG is now quite large indeed, since A_0 is so large.

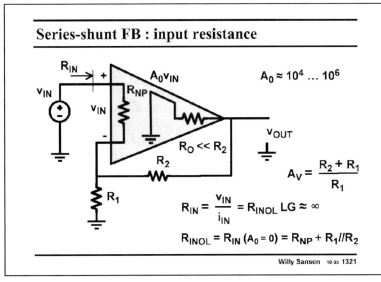

Series-shunt FB : input resistance

R_{IN}

v_{IN}

R_{NP}

v_{IN}

$A_0 v_{IN}$

$R_O << R_2$

R_2

R_1

$A_0 \approx 10^4 ... 10^6$

v_{OUT}

$A_V = \dfrac{R_2 + R_1}{R_1}$

$$R_{IN} = \frac{v_{IN}}{i_{IN}} = R_{INOL} \, LG \approx \infty$$

$$R_{INOL} = R_{IN} (A_0 = 0) = R_{NP} + R_1 // R_2$$

Willy Sansen 10-95 1321

1321

The input resistance R_{IN} is now the open-loop input resistance R_{INOL} multiplied by the loop gain LG.

The open-loop input resistance R_{INOL} is simply the sum of the large input resistor R_{NP} and the feedback resistors in parallel. Output resistance R_O is small and therefore negligible with respect to the others.

This input resistance R_{INOL} is about infinity for a MOST amplifier but not for a bipolar one. Application of series feedback will make the closed-loop input resistance R_{IN} near infinity for a bipolar amplifier.

Note that for the calculation of the open-loop input resistance, the effect of the feedback must be eliminated. This is easily done by setting the gain A_0 to zero.

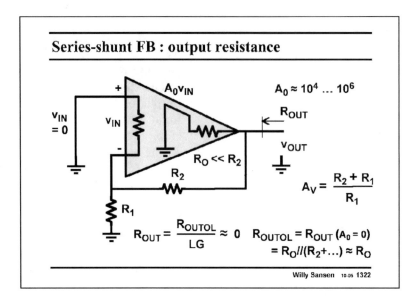

1322
The output resistance R_{OUT} is the open-loop output resistance R_{OUTOL} divided by the loop gain LG.

The open-loop output resistance R_{OUTOL} is the parallel combination of the small output resistor R_O and the feedback resistors in series. Since output resistance R_O is small, the effect of the feedback resistors is negligible.

Application of shunt feedback will make the closed-loop output resistance R_{OUT} nearly zero.

Note that for the calculation of the open-loop output resistance, the effect of the feedback must be eliminated. This is again done by setting the gain A_0 to zero.

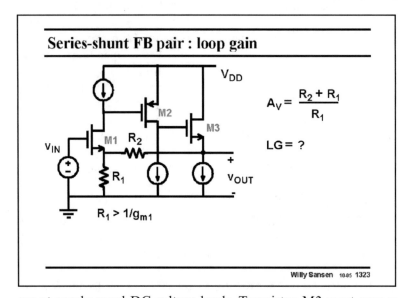

1323
A popular series-shunt feedback amplifier is shown in this slide. It contains only a few transistors. It has two amplifiers M1 and M2 and a source follower. It is called a series-shunt feedback pair. The source follower does not seem to count and yet it provides a lot more loop gain as we will find out next.

Note that a pMOST is used as a second amplifier as it provides easier DC biasing. Indeed the Sources of transistors M1 and M3 are at nearly equal DC voltage levels. Transistor M2 must now provide a lower DC voltage at the output than at the input. This is a lot easier with a pMOST than with an nMOST.

Note also that the feedback resistor is usually larger than $1/g_{m1}$ to make sure that all the feedback current coming from R_2 flows into the transistor, in order to increase the loop gain.

This is not always obvious however, as shown next. We need to know more about the input resistance at the Source of transistor M1. After all, transistor M1 behaves as a cascode transistor for the feedback current. What then is its input resistance?

This is reviewed on the next slide.

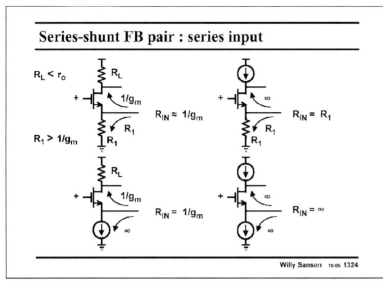

1324

Input transistor M1 behaves as a cascode transistor indeed, for the feedback current. Its input resistance is calculated in all possible cases of Source resistance R_1 (or current source) and all possible cases of load resistor R_L (or current source).

It is clear that the feedback current can only flow into the Source if the cascode input resistance R_{IN} is small (which is $1/g_{m1}$). This is the case when a small load resistor R_L is used.

When a current source is used as a load then input resistance R_{IN} is high. The question is, what would the output voltage of that transistor would be? We need to know, to be able to find the loop gain.

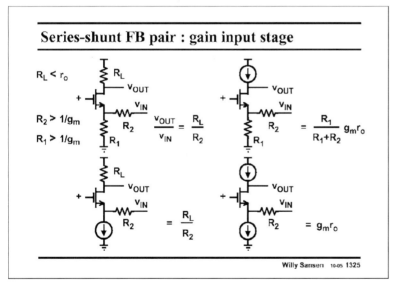

1325

In order to be able to find the loop gain, we must first of all, find the output voltage v_{OUT} at the Drain of input transistor M1, as a result of an input voltage v_{IN} applied to the feedback transistors, as shown in this slide.

Again, four cases can be distinguished.

When a small load resistor R_L is used, the gain is easily found to be the ratio of the two resistors R_L and R_L. It is usually not very large!

This gain is much larger however, when a current source is used as a load. Then the gain of the input transistor M1 comes in.

This is not unexpected. A current source as a load usually provides higher gains!

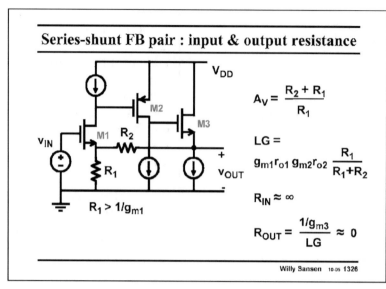

Series-shunt FB pair : input & output resistance

$$A_V = \frac{R_2 + R_1}{R_1}$$

$$LG = g_{m1}r_{o1}\, g_{m2}r_{o2}\, \frac{R_1}{R_1 + R_2}$$

$$R_{IN} \approx \infty$$

$$R_{OUT} = \frac{1/g_{m3}}{LG} \approx 0$$

$$R_1 > 1/g_{m1}$$

Willy Sansen 10-05 1326

1326

Now we are in a situation that we can easily calculate the loop gain.

When a current source is used in the first stage, the loop gain includes the effect of the potentiometric divider with resistors R_1 and R_2, followed by transistor M1, which is $g_{m1}r_{o1}$. Transistor M1 acts as an amplifier for the input signal, but as a cascode for the feedback signal.

This gain is much larger, when a current source is used as a load (right). Then the gains of both transistor M1 and M2 occur in the loop gain LG. It is now much larger indeed!

The input and output resistances are now easily found.

Even without feedback, the input resistance is already infinity. If some Gate current is present, then the input resistance is lower, as for a bipolar transistor amplifier (see later).

The output resistance without feedback is just $1/g_{m3}$. Indeed resistor R_2 is usually a lot larger and can therefore be neglected.

With feedback, this resistor $1/g_{m3}$ must be divided by the loop gain. The closed-loop output resistance R_{OUT} is nearly zero.

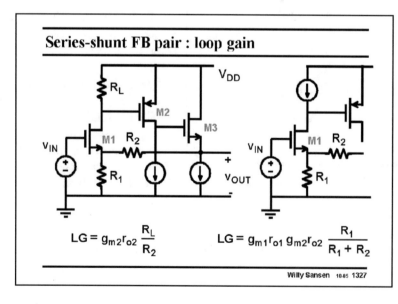

Series-shunt FB pair : loop gain

$$LG = g_{m2}r_{o2}\, \frac{R_L}{R_2}$$

$$LG = g_{m1}r_{o1}\, g_{m2}r_{o2}\, \frac{R_1}{R_1 + R_2}$$

Willy Sansen 10-05 1327

1327

We can again easily calculate the loop gain LG.

When a small load resistor R_L is used (left), the gain is easily found to be the ratio of the two resistors R_L and R_L, multiplied by the gain of transistor M2. For a two-stage amplifier, this is not high at all!

The gain is again large, when a current source is used as a load (right). Then the gains of both transistors M1 and M2 occur in the loop gain LG. It is therefore much larger indeed!

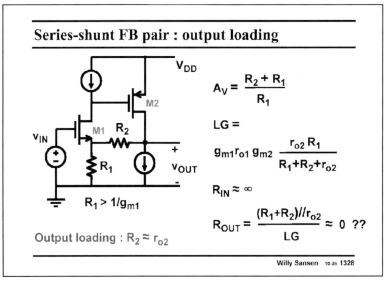

Series-shunt FB pair : output loading

$$A_V = \frac{R_2 + R_1}{R_1}$$

$$LG = g_{m1}r_{o1}\, g_{m2}\, \frac{r_{o2}\, R_1}{R_1 + R_2 + r_{o2}}$$

$$R_{IN} \approx \infty$$

$$R_{OUT} = \frac{(R_1 + R_2)//r_{o2}}{LG} \approx 0 \ ??$$

Output loading : $R_2 \approx r_{o2}$

Willy Sansen 10-05 1328

1328

An important question is, what do we need that Source follower for? After all it takes a lot of current and it only provides a voltage gain of unity!

The difference is that the feedback resistor R_2 is not now any more larger than the output resistance of the output transistor. Resistor R_2 may very well be comparable to output resistance r_{o2}. As a result an additional resistive divider comes in, as clearly shown by the expression of the LG.

The loop gain is smaller than with a Source follower by a ratio of about $(R_1 + R_2 + r_{o2})/(R_1 + R_2)$. It all depends on what the ratio is of R_2 to r_{o2}. For a small R_2, the loss in loop gain is considerable.

In this case, the output is called to be loaded. Feedback resistor R_2 loads the output of the amplifier. It forms a resistive divider with the output resistance r_{o2}. As a result the LG is decreased.

The closed-loop output resistance R_{OUT} is calculated as before. The resulting value is somewhat larger but still close to zero.

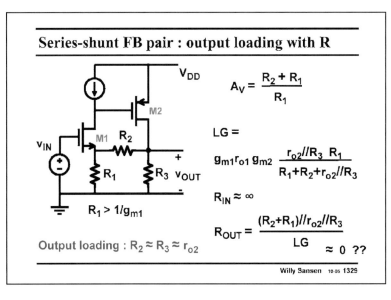

Series-shunt FB pair : output loading with R

$$A_V = \frac{R_2 + R_1}{R_1}$$

$$LG = g_{m1}r_{o1}\, g_{m2}\, \frac{r_{o2}//R_3\ R_1}{R_1 + R_2 + r_{o2}//R_3}$$

$$R_{IN} \approx \infty$$

$$R_{OUT} = \frac{(R_2 + R_1)//r_{o2}//R_3}{LG} \approx 0 \ ??$$

Output loading : $R_2 \approx R_3 \approx r_{o2}$

Willy Sansen 10-05 1329

1329

The loop gain LG is even lower when a resistive load is used for the second amplifier M2, rather than a current source. There is an even stronger resistive division at the output than before.

The result depends on the relative sizes of the feedback resistor R_2 and resistances R_3 and r_{o2}.

The output resistance R_{OUT} is a little bit smaller than before as no current source is used to load transistor M2 but a resistor R_3.

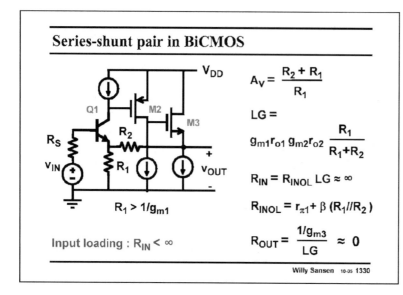

Series-shunt pair in BiCMOS

$A_V = \dfrac{R_2 + R_1}{R_1}$

$LG = g_{m1}r_{o1}\, g_{m2}r_{o2}\, \dfrac{R_1}{R_1 + R_2}$

$R_{IN} = R_{INOL}\, LG \approx \infty$

$R_{INOL} = r_{\pi1} + \beta\,(R_1 /\!/ R_2)$

$R_{OUT} = \dfrac{1/g_{m3}}{LG} \approx 0$

$R_1 > 1/g_{m1}$

Input loading : $R_{IN} < \infty$

Willy Sansen 10-05 1330

1330

When a bipolar transistor is used at the input, then the input resistance is no longer infinity. Because of the series feedback at the input, it is increased by the loop gain LG but not infinity. This is why the input resistance loads the source resistance R_S. In the circuit in this slide a source resistor R_S has been added. This input loading is only present when the input voltage source has an internal source resistance R_S, which is comparable to the input resistance R_{IN}.

In addition, this source resistance R_S forms a low-pass filter with the input capacitance of this amplifier. The input resistance without feedback R_{INOL} is easily found as it is a single-transistor amplifier with emitter degeneration. The emitter resistor is about R_1 in parallel with R_2.

The output resistance is quite small as calculated before. An output capacitive load would cause an output pole at fairly high frequencies.

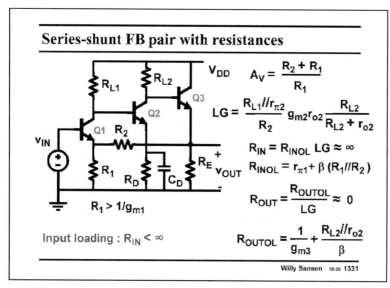

Series-shunt FB pair with resistances

$A_V = \dfrac{R_2 + R_1}{R_1}$

$LG = \dfrac{R_{L1} /\!/ r_{\pi2}}{R_2}\, g_{m2}r_{o2}\, \dfrac{R_{L2}}{R_{L2} + r_{o2}}$

$R_{IN} = R_{INOL}\, LG \approx \infty$

$R_{INOL} = r_{\pi1} + \beta\,(R_1 /\!/ R_2)$

$R_{OUT} = \dfrac{R_{OUTOL}}{LG} \approx 0$

$R_{OUTOL} = \dfrac{1}{g_{m3}} + \dfrac{R_{L2} /\!/ r_{o2}}{\beta}$

$R_1 > 1/g_{m1}$

Input loading : $R_{IN} < \infty$

Willy Sansen 10-05 1331

1331

A practical realization with only bipolar transistors is shown in this slide. All current sources are substituted by resistors. This circuit can also be realized easily with discrete components on a printed circuit board.

Since each stage gives a higher DC voltage at the output, the second stage must be either a PNP transistor or a NPN transistor with an emitter resistor R_D. To avoid the reduction in gain of this resistor, a large capacitance C_D is placed across it. This resistor does not come in for the gain calculations. This is true for all frequencies higher than $g_{m2}/(2\pi C_D)$.

The expressions of the loop gain, input and output resistances are all very much as before. They are all a bit more complicated because a bipolar transistor has a finite input resistance r_π, which shows up in most of the expressions.

Table of contents

- ◆ **Definitions**
- ◆ **Series-shunt FB for Voltage amplifiers.**
- ◆ **Series-series FB for Transconductance amps.**

1332

Let us now discuss series-series feedback in more detail. It is a type of amplifier which provides an accurate voltage-to-current conversion. The main difference with the voltage amplifiers is that they put out a current rather than a voltage. Both input and output resistances are high. They are evidently called transconductance amplifiers.

Such high output impedance is mainly used to drive a variable impedance, for example to carry out an impedance measurement. It also provides current sources to bias any analog circuit by means of current mirrors.

A single-transistor amplifier is also a transconductor. It has a high input resistance and a current output, which is fairly precise. Its transconductance is slightly low, however. An operational amplifier without output stage also behaves as a transconductor. Its transconductance is very high but not very precise. Feedback is used to obtain both, a fairly high and precise transconductance or voltage-to-current converter.

Again we want to calculate the loop gain, the closed and open-loop gains, and finally the input and output resistances.

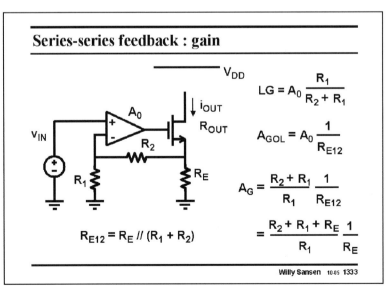

Series-series feedback : gain

$$LG = A_0 \frac{R_1}{R_2 + R_1}$$

$$A_{GOL} = A_0 \frac{1}{R_{E12}}$$

$$A_G = \frac{R_2 + R_1}{R_1} \frac{1}{R_{E12}}$$

$$= \frac{R_2 + R_1 + R_E}{R_1} \frac{1}{R_E}$$

$$R_{E12} = R_E \,/\!/\, (R_1 + R_2)$$

1333

An accurate transconductor is shown in this slide. The output current is provided by a MOST which is driven by an opamp. It is accurately linked to the input voltage v_{IN} by the three resistors used. Indeed, the closed loop gain A_G, which is nothing else than i_{OUT}/v_{IN}, only contains resistor ratios and the absolute value of one resistor R_{E12}!

The loop gain LG itself and the open loop gain A_{GOL} both contain the gain of the opamp A_0. However, this gain does not occur in the closed loop gain A_G!

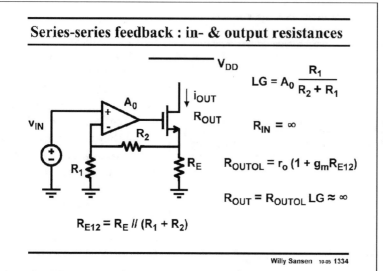

1334
The input resistance of this converter is evidently infinity as no Gate current is flowing.

The output resistance is increased by the loop gain LG. The open-loop output resistance R_{OUTOL} is easily obtained for a single-transistor amplifier with Source resistor R_E. It is therefore already quite high. It is increased even more by the feedback action.

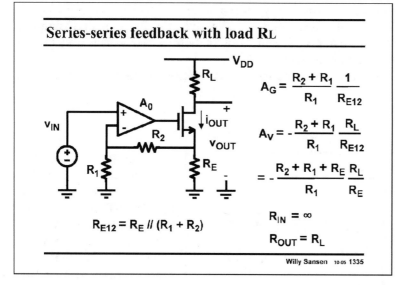

1335
The addition of a load resistor R_L, converts this transconductor into a voltage amplifier again, with very accurate gain A_v.

This gain only depends on resistor ratios, and can be accurately set.

The output resistance towards the Drain of the output transistor is quite high. As a result, the output resistance seen at the output of the whole amplifier is mainly load resistor R_L.

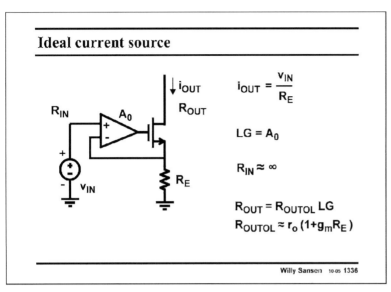

1336

The gain resistors R_1 and R_2 can be left out in the previous transconductor, resulting into the simpler one shown in this slide. It is probably the simplest way to convert a voltage into a current, with high accuracy.

The voltage-to-current conversion only depends on resistor R_E.

The input resistance is again high and so is the output resistance.

This circuit now acts as an ideal current generator.

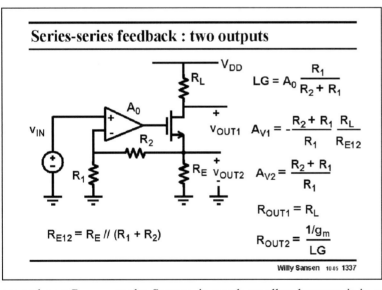

1337

The precise transconductor analyzed before, has two outputs, one at the Drain as before, but also one at the Source of the output transistor. What is the difference?

The first difference is that these two outputs have opposite polarities. Also, the loop gain is the same for both, but not the actual closed-loop voltage gains. The first one A_{v1}, at the Drain, is larger.

The output resistances are also very different. The second one R_{OUT2}, at the Source, is much smaller, because it is only of the order of magnitude of $1/g_m$ and it involves the loop gain LG. Connection of a capacitive load at the second output would give a pole at high frequencies only.

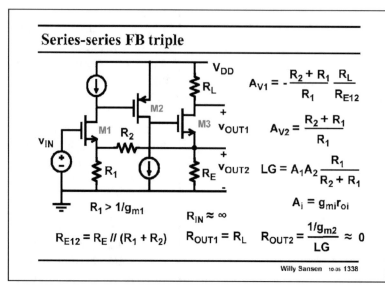

1338

A transistor realization of the last circuit is shown in this slide. Two transistors M1 and M2 are used to realize the operational amplifier. A pMOST is used as a second stage to provide downwards level shifting.

The expressions are exactly as before. The closed-loop gains A_{v1} and A_{v2} are exactly the same as before.

Now the gain A_0 of the opamp has to be substituted by the gains of the two transistors. Each transistor provides a gain $g_m r_o$, as indicated. The loop gain LG depends on both gains.

The input-and output-resistances are as before.

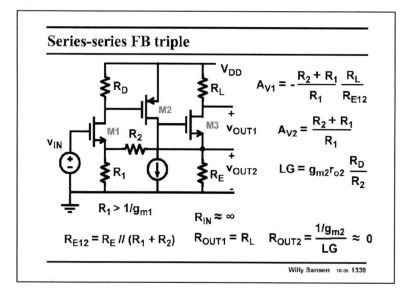

1339

Substitution of the load current source of the input transistor by a load resistor R_L, reduces the loop gain LG somewhat. The gain of the input transistor is not fully exploited any more. However, the closed-loop gains A_{v1} and A_{v2} are exactly the same as before.

Input and output impedances are the same as well.

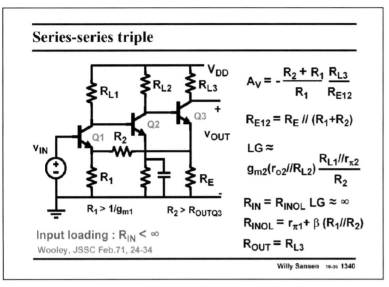

1340

Substitution of all current sources by resistors provides this well-known feedback amplifier with a triple gain stage if the Drain of Q3 is taken as an output.

This is an amplifier which is easy to realize with discrete components, especially with bipolar transistors.

The closed-loop gain A_v is as before.

The loop gain is also similar as before. The resistive division at each node has to be taken into account.

The input resistance R_{IN} is not infinity as a bipolar transistor is used at the input. It is quite high however, because of the feedback action. The output resistance R_{OUT} is mainly load resistor R_{L3}.

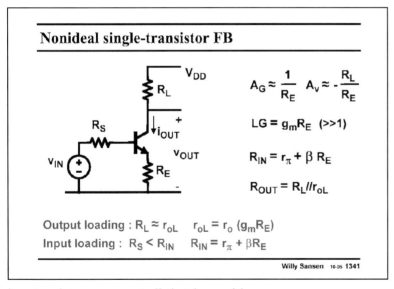

1341

Series-series feedback is also possible on a single-transistor amplifier. However, it is not so easy to distinguish the closed and open-loop gains and the loop gain. This is also called local feedback.

The transconductance and voltage gain are more accurate, the more $g_m R_E$ is larger than unity. This requires a large DC voltage drop across emitter resistor R_E, however!

The loop gain LG and the input resistance are not all that large either.

The output resistance r_{oL} looking into the Collector of the transistor is not all that large either. The output resistance R_{OUT} is thus a parallel combination of both the load resistor R_L and the transistor output resistance r_{oL}.

As a consequence, both input and output loading occur, i.e. the source resistance R_S interacts with the input resistance R_{IN}. Also the load resistor R_L interacts with the transistor output resistance r_{oL}.

This circuit is far from an ideal-feedback circuit. It is better to be analyzed by straight analysis using the two laws of Kirchoff.

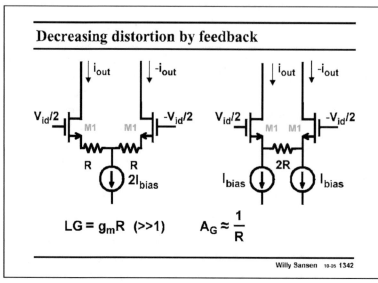

Decreasing distortion by feedback

$$LG = g_mR \quad (>>1) \qquad A_G \approx \frac{1}{R}$$

Willy Sansen 10-05 1342

1342

The previous series-feedback single-transistor amplifier can easily be realized in a differential configuration. Again, the larger the emitter resistors R, the better the feedback works. To avoid the large DC voltage drops across the resistors, the circuit on the right is preferred.

This latter circuit provides the same gains, but no DC current flows through the emitter resistors. It can be used at lower supply voltages. Its noise performance is slightly worse, however (see Chapter 4).

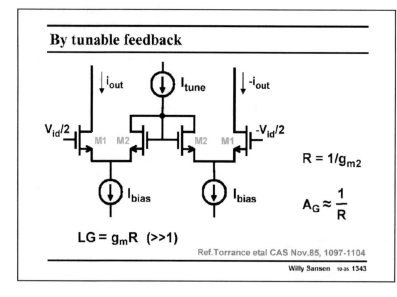

By tunable feedback

$$R = 1/g_{m2}$$

$$A_G \approx \frac{1}{R}$$

$$LG = g_mR \quad (>>1)$$

Ref.Torrance etal CAS Nov.85, 1097-1104

Willy Sansen 10-05 1343

1343

The feedback resistors can be made tunable by replacing them by forward biased diode-connected transistors M2, as shown in this slide.

Each transistor M2 carries a DC current $I_{tune}/2$. It acts as a feedback resistor for transistor M1 with value $1/g_{m2}$.

The transconductance can now be set by setting current I_{tune}. This current must always be larger than $2I_{bias}$ however.

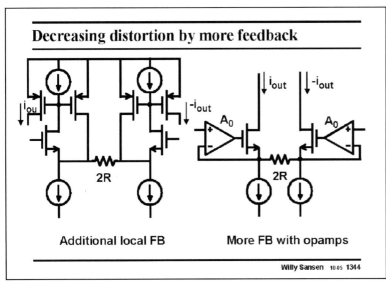

Decreasing distortion by more feedback

Additional local FB

More FB with opamps

Willy Sansen 10-05 1344

1344

More loop gain can be obtained by insertion of more transistors in the feedback loop.

On the left, only one more transistor is included in the feedback loop. Moreover, this configuration allows an elegant way to take the output currents. Indeed, the additional transistors form current mirrors with the output transistors.

Even more loop gain is available if full opamps are inserted in the feedback loops as shown on the right. This is clearly the most accurate way to convert a differential input voltage into a differential output current.

At high frequencies, however, the opamp does not provide all that much gain. At high frequencies the left circuit is now preferred, although it is less accurate.

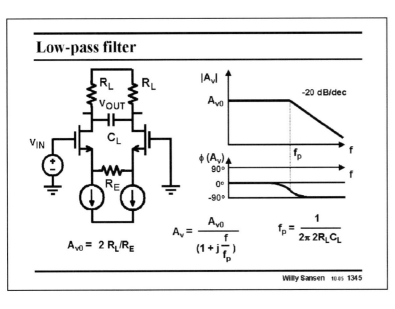

Low-pass filter

$$A_{v0} = 2 R_L/R_E$$

$$A_v = \frac{A_{v0}}{(1 + j\frac{f}{f_p})} \qquad f_p = \frac{1}{2\pi \, 2R_L C_L}$$

Willy Sansen 10-05 1345

1345

Such a single-transistor transconductor is easily modified to make a low-pass filter. A capacitor C_L is connected between the differential outputs.

In this way a first-order pole is obtained with frequency f_p, which creates a -20 dB/decade roll-off starting at the pole frequency.

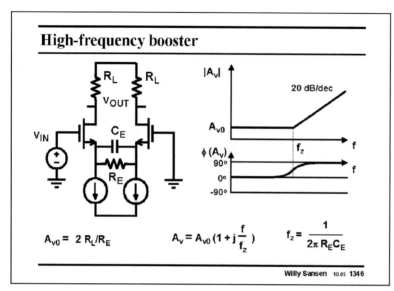

High-frequency booster

$A_{v0} = 2\,R_L/R_E$ $A_v = A_{v0}\left(1 + j\dfrac{f}{f_z}\right)$ $f_z = \dfrac{1}{2\pi\,R_E C_E}$

Willy Sansen 10-05 1346

1346

In a similar way, a first-order zero can be created, by putting a capacitor C_E in parallel with the resistor R_E. The characteristic frequency is now f_z. At this frequency, the gain starts going up with a slope of 20 dB/decade.

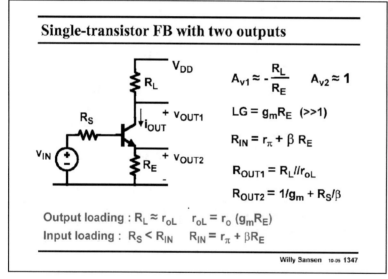

Single-transistor FB with two outputs

$A_{v1} \approx -\dfrac{R_L}{R_E}$ $A_{v2} \approx 1$

$LG = g_m R_E \quad (>>1)$

$R_{IN} = r_\pi + \beta\,R_E$

$R_{OUT1} = R_L // r_{oL}$

$R_{OUT2} = 1/g_m + R_S/\beta$

Output loading : $R_L \approx r_{oL}$ $r_{oL} = r_o\,(g_m R_E)$
Input loading : $R_S < R_{IN}$ $R_{IN} = r_\pi + \beta R_E$

Willy Sansen 10-05 1347

1347

The single-transistor amplifier with local feedback can also provide two outputs as shown in this slide.

The output at the Collector provides non-ideal series-series feedback as discussed in this slide.

The output at the Emitter, however, is the output of an Emitter follower. Its gain A_{v2} is unity and its output resistance R_{OUT2} is small.

Emitter and Source followers do use feedback. They provide an accurate voltage gain, which is closer to unity the larger the loop gain is. They also provide a reduced output resistance.

Clearly, all this applies to a MOST single-transistor amplifier with local feedback. The gain at the Source is unity again, at least if the bulk effect does not come in. The output resistance is the same as for a bipolar transistor with infinite beta. It is simply $1/g_m$.

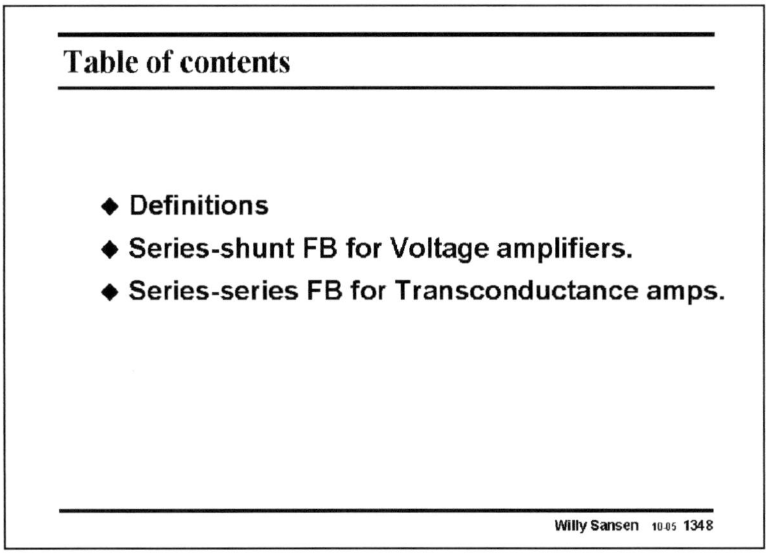

Table of contents

Willy Sansen 10-05 1348

1348

In this Chapter the four types of feedback have been introduced and compared.

Both types of series feedback at the input have been discussed in detail. For a number of circuit realizations the expressions of the loop gain, input and output impedance have been derived.

In the following Chapter, the focus is on circuits with shunt feedback at the input.

141

Feedback Transimpedance & Current Amplifiers

Willy Sansen

KULeuven, ESAT-MICAS

Leuven, Belgium

willy.sansen@esat.kuleuven.be

Willy Sansen 10.05 141

An introduction on feedback has been given in the previous Chapter design. The four basic types of feedback have been identified. Two of them have been discussed in the previous Chapter. They are the types with series feedback at the input as they measure a voltage.

In this Chapter, we will focus on the two other types of feedback, i.e. the ones with shunt feedback at the input. As a consequence, they measure an input current. They are therefore transimpedance amplifiers, if an output voltage is provided, or current amplifiers.

142

Table of contents

Willy Sansen 10.05 142

First of all, we want to review some definitions. In particular we want to review the differences between the four types of feedback amplifiers.

We will then focus on transimpedance amplifiers. They are used mainly for current sensors such as photodiodes, voltammetric sensors, etc.

Current amplifiers are next.

At the end, some different transimpedance amplifiers are added, as they are widely used for photodiode receivers. Attention is paid to low-noise and high-frequency performance.

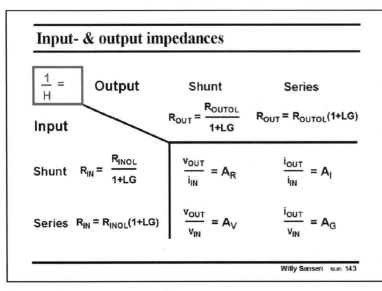

Input- & output impedances

$\frac{1}{H}$ =	Output	Shunt	Series
Input		$R_{OUT} = \dfrac{R_{OUTOL}}{1+LG}$	$R_{OUT} = R_{OUTOL}(1+LG)$
Shunt $R_{IN} = \dfrac{R_{INOL}}{1+LG}$		$\dfrac{v_{OUT}}{i_{IN}} = A_R$	$\dfrac{i_{OUT}}{i_{IN}} = A_I$
Series $R_{IN} = R_{INOL}(1+LG)$		$\dfrac{v_{OUT}}{v_{IN}} = A_V$	$\dfrac{i_{OUT}}{v_{IN}} = A_G$

Willy Sansen 10.05 143

143
The different kinds of feedback give rise to different kinds of amplifiers. Series-shunt feedback provides an amplifier with high precision in voltage gain A_V. It is therefore a voltage amplifier.

In a similar way, a shunt-shunt feedback amplifier generates an accurate trans-resistance gain A_R. Both input and output resistances will decrease. The input can then easily be driven by means of an input current.

The output behaves as a voltage source.

To realize a current amplifier, we have to apply shunt-series feedback. The input resistance is lowered to be able to allow the input current to flow. The output resistance is quite high, as for a current source.

We have seen in the previous Chapter why we need these different kinds of amplifiers. We will now focus on the feedback amplifiers with shunt-feedback at the input, in order to be able to take an input current.

Table of contents

- **Introduction**
- **Shunt-shunt FB for Transimpedance amps.**
- **Shunt-series FB for Current amplifiers**
- **Transimpedance amplifiers for**
  low noise and high frequencies

Willy Sansen 10.05 144

144
Let us now discuss shunt-shunt feedback in more detail. It is a kind of amplifier which provides an accurate current-to-voltage conversion, as in photo-diode detectors or pixel detectors.

They are called transimpedance amplifiers.

First of all, we will assume a more general case, i.e. the input impedance is not infinite but has a limited value. Also, the output resistance is not zero.

We want to calculate again the loop gain, the closed and open-loop gains, and finally the input and output resistances.

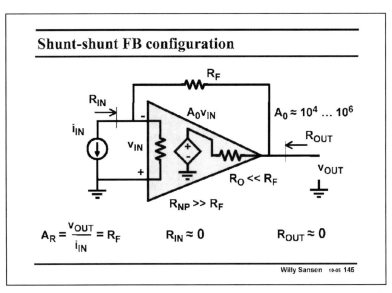

145

This feedback arrangement is very simple indeed. The input signal current i_{IN} will flow through the feedback resistor R_F to create an output voltage $i_{IN}R_F$. The transresistance is simply R_F itself.

This amplifier has a lot of gain (A_0 is between 10^4 and 10^6). As a result, whatever the output voltage is, the differential input voltage v_{IN} will be quite small, comparable to noise. The minus terminal of the amplifier is now at about zero Volt. The current through input resistance R_{NP} is also about zero. This is certainly the case for a MOST for which resistor R_{NP} is infinity. This is also true however, for a bipolar transistor which has a finite input resistance R_{NP}.

All the input current flows through the feedback resistor R_F. This is why the output voltage is quite accurately equal to $i_{IN}R_F$.

The feedback resistor R_F is usually much larger than the output resistance R_O. Later on we will find out what to do if this is not the case. We will call it output loading then.

We want to learn about the loop gain first.

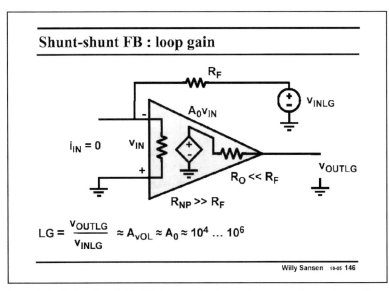

146

There are several places where we can break the loop. We have taken here the output of the amplifier. Now we have to calculate the ratio v_{OUTLG}/v_{INLG}. Remember that we break the loop only for AC performance, whereas the DC conditions are kept the same.

Note that the input current source has been assumed to be ideal. It is left out for this calculation.

No current can flow through R_F; indeed resistor R_{NP} is infinite for MOST input devices. As a result, resistor R_F does not appear in the result. The loop gain is therefore the same as the open-loop gain of the amplifier A_0. Its value is quite high indeed.

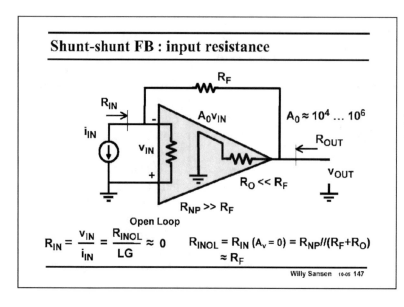

147
Now it is easy to calculate the input resistance R_{IN}, seen by the current input sensor. By definition, the closed-loop input resistance R_{IN} equals the input resistance without feedback, divided by the loop gain. Be aware that the "input resistance without feedback" has to include the components that we use to carry out the feedback. Resistor R_F carries out the feedback and must be included when we calculate the open-loop input resistance R_{INOL}.

This resistance is two resistors in parallel. For a MOST, in which R_{NP} is really high, the open-loop input resistance is mainly R_F itself. For a bipolar transistor, it would be two resistors in parallel

The closed-loop input resistance R_{IN} is R_F divided by the loop gain; its value will be quite small, not to say zero.

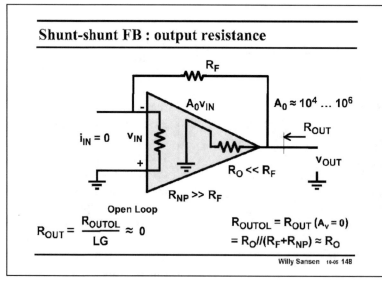

148
The output resistance is readily calculated in a similar way. The closed-loop output resistance R_{OUT} equals the output resistance without feedback, divided by the loop gain. Remember that the "output resistance without feedback" has to include the components that we use to carry out the feedback. Resistor R_F carries out the feedback and must be included again when we calculate the open-loop output resistance R_{OUTOL}. This time however, it does not make much difference as the output resistor R_0 is much smaller than R_F. The open-loop output resistance R_{OUTOL} is mainly R_0 itself.

The closed-loop output resistance is now much smaller as it is the output resistor R_0 divided by the loop gain LG. This amplifier therefore functions as a voltage source.

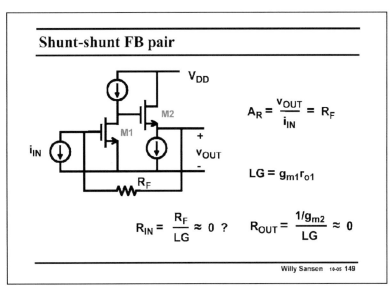

149

A transistor example of such a shunt-shunt feedback amplifier is shown in this slide. The open-loop amplifier consists of a single-transistor amplifier followed by a source-follower, in order to provide a low output resistance. Its value is $1/g_{m2}$.

The gain of the amplifier is readily found to be $g_{m1}r_{o1}$, which is again the loop gain. This loop gain LG is not so large. Its value is barely 100 and yet, all feedback rules still apply.

The closed-loop input resistance is now R_F divided by the loop gain. Output resistor $1/g_{m2}$ is obviously negligible with respect to R_F.

The closed-loop output resistance is then $1/g_{m2}$ divided by the loop gain. It is now very small, or quasi zero.

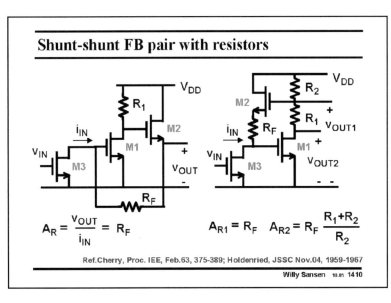

1410

Resistors can be used instead of the current sources as well. Moreover, the input current source can be substituted by a MOST, which is M3 in the examples in this slide. In addition, the DC current source of the source follower M2 can now be left out. The DC current through M2 is the same as the DC current through input transistor M3.

In this case, the transresistance A_R is again R_F. The voltage gain v_{OUT}/v_{IN} is then $g_{m3}R_F$.

In the example on the right, the output v_{OUT1} is now taken at the Gate of the source follower M2, rather than at its Source. The transresistance A_{R1} is again R_F. The loop gain is the same as in the example on the left but the output resistance will be higher.

Another output can be taken, when another resistor R_1 is added, as shown in the example on the right. In this case the gain is increased by a ratio $(R_1 + R_2)/R_2$.

However, both circuits are variations on the theme of shunt-shunt feedback pairs. These variations are mainly used to realize wide-band amplifiers, up to several GHz's.

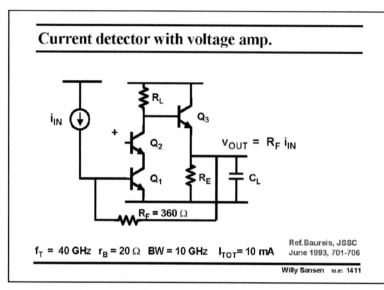

Current detector with voltage amp.

i_{IN}

R_L

Q_3

Q_2

$v_{OUT} = R_F i_{IN}$

Q_1

R_E

C_L

$R_F = 360\,\Omega$

$f_T = 40\,GHz$ $r_B = 20\,\Omega$ $BW = 10\,GHz$ $I_{TOT} = 10\,mA$

Ref.Baureis, JSSC
June 1993, 701-706

Willy Sansen 10.05 1411

1411

A real-life example of such a shunt-shunt feedback amplifier has been published in the *IEEE Journal of Solid-State Circuits*.

Its transresistance is quite small, only 360 Ω. Its bandwidth, however, is impressive, i.e. 10 GHz.

The gain itself is again given by a single-transistor amplifier, with a cascode however, to increase the gain, and to isolate its output better from the input. Again an emitter follower is used to lower the output resistance.

All calculations apply again. The output resistance R_O will again be very small. This is why the bandwidth is so high, even with a fairly large load capacitor C_L. It is simply given by $1/(2\pi R_O C_L)$.

The open loop output resistance is the parallel combination of three resistances, i.e. $R_F + r_{\pi 1}$, R_E and $1/g_{m3} + R_L/\beta_3$. Altogether, this is about $1/g_{m3}$. The closed loop output resistance R_O is then the open loop one divided by the loop gain.

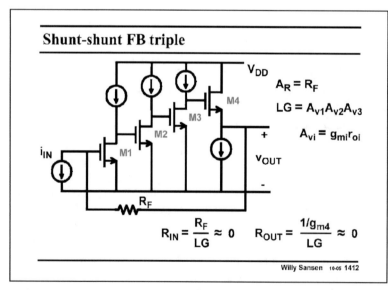

Shunt-shunt FB triple

V_{DD}

M4

M3

M2

M1

i_{IN}

v_{OUT}

R_F

$A_R = R_F$

$LG = A_{v1}A_{v2}A_{v3}$

$A_{vi} = g_{mi}r_{oi}$

$+$

$-$

$R_{IN} = \dfrac{R_F}{LG} \approx 0$ $R_{OUT} = \dfrac{1/g_{m4}}{LG} \approx 0$

Willy Sansen 10-05 1412

1412

In we want to make sure that the feedback is negative, shunt-shunt feedback is only possible either around one single transistor or around three transistors. Indeed each gain transistor inverts the polarity.

A single-gain stage suffers from too small gain, causing too low a loop gain. A triple-gain stage provides a lot more loop gain and is much closer to an ideal feedback amplifier. The gain per stage is simply $g_m r_o$.

As a result, both the input and output resistances will be quite small. The current-to-voltage (closed-loop) gain is again R_F, as expected.

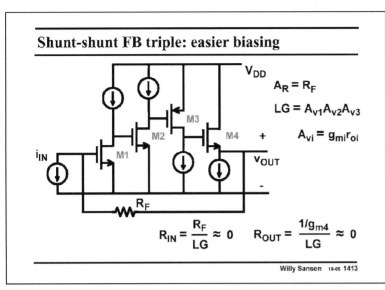

1413

For biasing, or DC setting of the currents, it may be easier to introduce a pMOST as the second or third stage or even as a source follower. In the example in this slide, it is the third stage which is using a pMOST amplifier.

Obviously the expressions for input and output impedance do not change. The biasing is easier however.

Indeed the DC input voltage is about 1 V above ground, and so is the DC output voltage. The other DC voltages are now easily found, depending on the transistor sizes and V_{GS} values chosen.

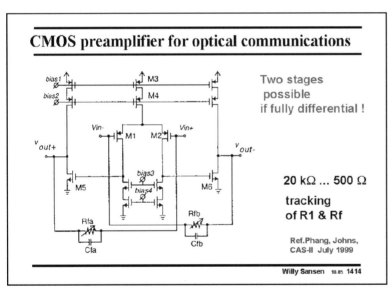

1414

It is possible to apply negative feedback around a two-stage amplifier provided differential pairs are used. Both feedback connections provide negative feedback, yielding a transresistance of R_F, as before.

The advantage of using only two stages is that there are less non-dominant poles so that stability and peaking problems are avoided. Moreover, both inputs and outputs are differential, increasing specifications such as CMRR and PSRR considerably.

The CMRR is the common-mode rejection ratio. It indicates how much less common-mode disturbances in the ground (noise, spikes) are amplified to the output, than the differential input signal. The PSRR then indicates how much less disturbances on the power supply line are amplified to the output, compared to the input signal.

A published realization of this fully-differential two-stage feedback amplifier is shown in this slide. The differential gain is set by R_f. The global bandwidth is then set by capacitance C_f.

Two stages always require an internal compensation capacitance, which here is C_f. It acts as a Miller capacitance. It provides pole splitting to ensure that the second pole is beyond the GBW.

Differential shunt-shunt FB pair

$$A_R = R_F$$

$$LG = A_{v1}A_{v2}$$

$$A_{v1} = g_{m1}r_{o1}$$

$$A_{v2} = g_{m2}r_{o2}$$

$$R_{IN} = \frac{2R_F}{LG} \approx 0 \qquad R_{OUT} = \frac{2/g_{m3}}{LG} \approx 0$$

Willy Sansen 10-05 1415

1415

The problem with such fully-differential amplifiers (i.e. two inputs and two outputs) are the factors of two appearing in the gain. Is the transresistance R_F or $2R_F$? How about the differential input and output impedance?

This amplifier has two input current sources both with value i_{IN}. The differential output voltage will thus be $2R_F i_{IN}$. The transresistance is simply R_F.

The loop gain equals the gains of the two transistor stages. The differential input impedance will now be $2R_F$ divided by the loop gain.

This also applies to the output resistance.

Current detector with voltage amplifier

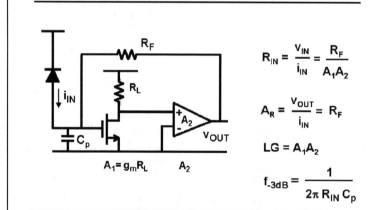

$$R_{IN} = \frac{v_{IN}}{i_{IN}} = \frac{R_F}{A_1 A_2}$$

$$A_R = \frac{v_{OUT}}{i_{IN}} = R_F$$

$$LG = A_1 A_2$$

$$f_{-3dB} = \frac{1}{2\pi R_{IN} C_p}$$

Willy Sansen 10-05 1416

1416

This transimpedance amplifier is usually the first stage of an optical fiber receiver. The photodiode behaves as a current source when it is exposed to light. This current is then multiplied by R_F to generate an output voltage. The first amplifier stage is shown in detail. Its gain is A_1. The subsequent stages, however, are included in the (triangular) black box with gain A_2. The total loop gain LG is simply $A_1 A_2$.

The input resistance R_{IN} is reduced by this loop gain, as expected. As a result, the diode capacitance and the parasitic capacitances C_p at the input of the amplifier, caused by interconnect and the transistor input capacitance, only see a small input resistance $R_F/A_1 A_2$. The bandwidth or f_{-3dB} can now be exceedingly high, to achieve a high bit rate.

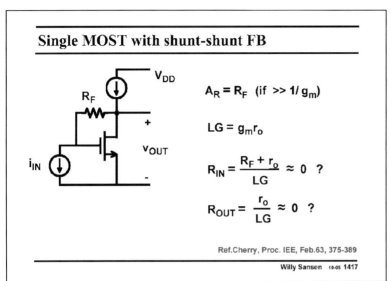

Single MOST with shunt-shunt FB

$A_R = R_F$ (if $\gg 1/g_m$)

$LG = g_m r_o$

$R_{IN} = \dfrac{R_F + r_o}{LG} \approx 0$?

$R_{OUT} = \dfrac{r_o}{LG} \approx 0$?

Ref.Cherry, Proc. IEE, Feb.63, 375-389

Willy Sansen 10-05 1417

1417

Shunt-shunt feedback around a single-transistor amplifier is far from ideal: the loop gain is simply too small.

Note also, that the feedback resistor R_F is comparable in size to the output resistance r_o. This does not decrease the LG because R_F leads to an infinite input resistance at the Gate.

As a result, the simple equations that we have derived before, may not provide accurate results. The closed loop gain is still about right, however, i.e. R_F.

The other quantities, the loop gain and both the input and output impedances can only be approximated in a crude way by the simple expressions given in this slide. To obtain more accurate expressions the method has to be used, which always works but which asks much more analysis. The transistor has to be substituted by its small-signal equivalent circuit (with mainly g_m and r_o at low frequencies) and the equations stating the laws of Kirchoff have to be solved. Needless to say that SPICE or any other similar circuit simulator should provide the same results.

The reference actually describes a combination of a single-transistor amplifier followed by a shunt-shunt single-transistor feedback stage.

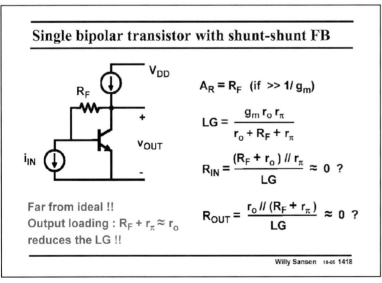

Single bipolar transistor with shunt-shunt FB

$A_R = R_F$ (if $\gg 1/g_m$)

$LG = \dfrac{g_m r_o r_\pi}{r_o + R_F + r_\pi}$

$R_{IN} = \dfrac{(R_F + r_o) \,/\!/\, r_\pi}{LG} \approx 0$?

$R_{OUT} = \dfrac{r_o \,/\!/\, (R_F + r_\pi)}{LG} \approx 0$?

Far from ideal !!
Output loading : $R_F + r_\pi \approx r_o$
reduces the LG !!

Willy Sansen 10-05 1418

1418

Shunt-shunt feedback around a single-transistor amplifier is far from ideal: the loop gain is simply too small. Moreover, the feedback resistor R_F is comparable in size to the output resistance r_o. This causes what is called "output loading". In all previous cases we had a source follower at the output. Now we do not.

As a result, none of the simple equations that we have derived, provide accurate results. The closed loop gain is just about right, i.e. R_F.

The other quantities however, the loop gain and both the input and output impedances can only be approximated in a crude way by the simple expressions given in this slide. This is why question marks are added.

To obtain more accurate expressions the method has to be used, which always works but which asks much more analysis. The transistor has to be substituted by its small-signal equivalent circuit (with mainly g_m and r_o at low frequencies) and the equations stating the laws of Kirchoff have to be solved. Needless to say that SPICE or any other circuit similar circuit simulator should provide the same results.

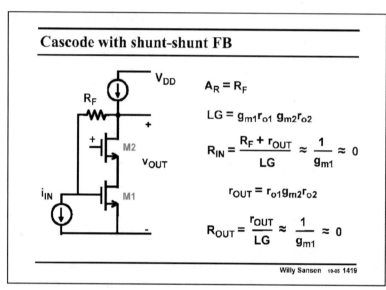

Cascode with shunt-shunt FB

$A_R = R_F$

$$LG = g_{m1} r_{o1} g_{m2} r_{o2}$$

$$R_{IN} = \frac{R_F + r_{OUT}}{LG} \approx \frac{1}{g_{m1}} \approx 0$$

$$r_{OUT} = r_{o1} g_{m2} r_{o2}$$

$$R_{OUT} = \frac{r_{OUT}}{LG} \approx \frac{1}{g_{m1}} \approx 0$$

Willy Sansen 10-05 1419

1419
The loop gain LG can in principle be extended by addition of a cascode transistor M2. Its value will therefore be much larger. As a result, the expression for the closed loop transresistance will be closer to the value of R_F.

Because of the high loop gain LG, fairly accurate values of the input and output impedances can be obtained. The only way to verify them with this circuit is straightforward analysis by use of the two laws of Kirchoff.

Note that both the input and output impedance are the same as for a diode connected transistor, i.e. $1/g_{m1}$. Resistor R_F does not come in as no Gate current is flowing. Also, cascode transistor M2 merely increases the loop gain. Using bipolar transistors rather than MOSTs would cause severe loading at both input and output!

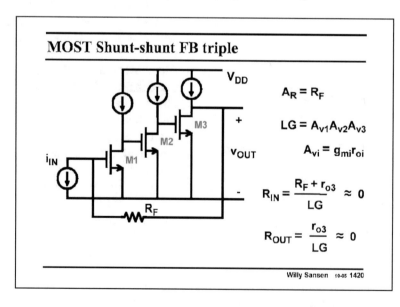

MOST Shunt-shunt FB triple

$A_R = R_F$

$$LG = A_{v1} A_{v2} A_{v3}$$

$$A_{vi} = g_{mi} r_{oi}$$

$$R_{IN} = \frac{R_F + r_{o3}}{LG} \approx 0$$

$$R_{OUT} = \frac{r_{o3}}{LG} \approx 0$$

Willy Sansen 10-05 1420

1420
A shunt-shunt feedback amplifier with three stages is shown in this slide. The closed loop transimpedance will be accurately R_F, because of the high loop gain. All other quantities may have to be found through straightforward analysis.

Since there is no output source follower, the output resistance at the drain of transistor M3 is fairly high. It is reduced by the loop gain LG however, so that the resulting value can still be quite low.

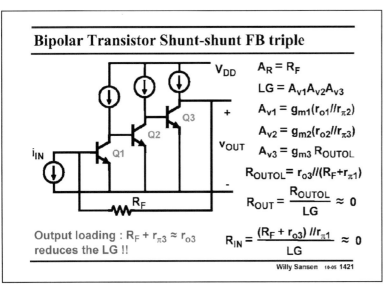

Bipolar Transistor Shunt-shunt FB triple

V_{DD}

$A_R = R_F$

$LG = A_{v1}A_{v2}A_{v3}$

$A_{v1} = g_{m1}(r_{o1}//r_{\pi2})$

$A_{v2} = g_{m2}(r_{o2}//r_{\pi3})$

$A_{v3} = g_{m3}R_{OUTOL}$

$R_{OUTOL} = r_{o3}//(R_F + r_{\pi1})$

$R_{OUT} = \dfrac{R_{OUTOL}}{LG} \approx 0$

$R_{IN} = \dfrac{(R_F + r_{o3})//r_{\pi1}}{LG} \approx 0$

Output loading : $R_F + r_{\pi3} \approx r_{o3}$ reduces the LG !!

Willy Sansen 10-05 1421

1421

Severe output loading is present in this three-stage shunt-shunt feedback amplifier with bipolar transistors. Only the expression of the closed loop transimpedance will be accurately R_F. All other quantities need to be verified through straightforward analysis.

Since there is no output source follower, the output resistance at the drain of transistor M3 is fairly high. Again, it is reduced by the high amount of loop gain.

Resistor R_F comes in now however, as it conducts some base current as well. The open loop output resistance R_{OUTOL} is output resistance r_{o3} in parallel with feedback resistor R_F in series with the input resistance of transistor Q1.

The input resistance will be quite small as well.

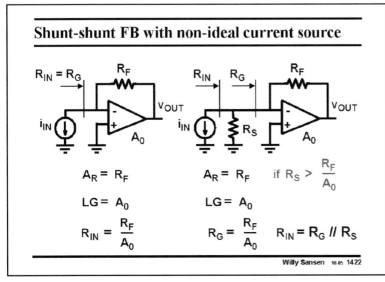

Shunt-shunt FB with non-ideal current source

$R_{IN} = R_G$

R_F

V_{OUT}

i_{IN}

A_0

$A_R = R_F$

$LG = A_0$

$R_{IN} = \dfrac{R_F}{A_0}$

R_{IN} R_G

R_F

V_{OUT}

i_{IN} R_S

A_0

$A_R = R_F$ if $R_S > \dfrac{R_F}{A_0}$

$LG = A_0$

$R_G = \dfrac{R_F}{A_0}$ $R_{IN} = R_G // R_S$

Willy Sansen 10-05 1422

1422

For shunt-shunt feedback with an ideal current source, the calculations are easy. They are given once more in this slide, on the left.

The question now is, what happens if that current source is not that ideal? A source resistance R_S is now in parallel with the current source i_{IN}, as shown on the right. The question is rather how small can we allow that resistor R_S to be such that our simple calculations are still valid.

The answer is obvious as long as the source resistor R_S is larger than the closed-loop input impedance, it will not affect the result. This input impedance is about R_F/A_0. It will be called R_G later on.

We find thus that we can keep the same expressions as before, as long as the source resistor R_S is larger than R_F/A_0 (or R_G).

This means that we can still calculate the loop gain LG as if R_S were not there. This also means that we now have two input resistances, one without R_S, which is R_G, and one with R_S, which is R_{IN}. Evidently, R_{IN} is the parallel combination of R_S and R_G. Since R_S is much larger than R_G however, the input resistances R_{IN} and R_G are about the same.

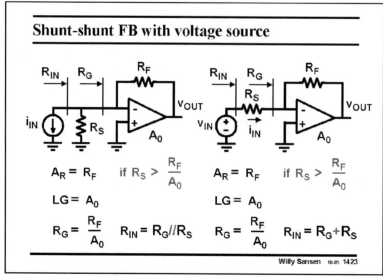

Shunt-shunt FB with voltage source

$A_R = R_F$ if $R_S > \dfrac{R_F}{A_0}$

$LG = A_0$

$R_G = \dfrac{R_F}{A_0}$ $R_{IN} = R_G /\!/ R_S$

$A_R = R_F$ if $R_S > \dfrac{R_F}{A_0}$

$LG = A_0$

$R_G = \dfrac{R_F}{A_0}$ $R_{IN} = R_G + R_S$

Willy Sansen 10.05 1423

1423

This input current source i_{IN} can also be substituted by a voltage source v_{IN} with series resistor R_S provided the input voltage v_{IN} equals $R_S i_{IN}$, as shown on the right.

Again, we have to assume that resistor R_S is much larger than R_G or R_F/A_0.

The loop gain LG and transresistance A_R and resistance at the Gate R_G are now the same as before.

The actual input resistance R_{IN}, seen by the voltage source, is now the sum of resistance R_G and resistor R_S. Since R_S is much larger than R_G, this input resistance R_{IN} will thus be mainly R_S.

All input resistances are now known. The transresistance A_R is also known. Since we use a voltage source however, we would also like to know the voltage gain A_V. This is given next.

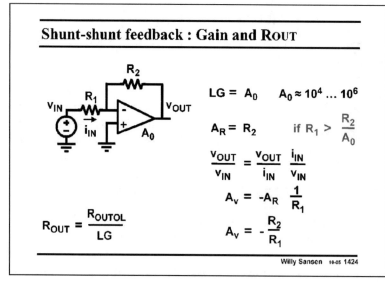

Shunt-shunt feedback : Gain and R$_{OUT}$

$LG = A_0$ $A_0 \approx 10^4 \ldots 10^6$

$A_R = R_2$ if $R_1 > \dfrac{R_2}{A_0}$

$\dfrac{v_{OUT}}{v_{IN}} = \dfrac{v_{OUT}}{i_{IN}} \dfrac{i_{IN}}{v_{IN}}$

$A_V = -A_R \dfrac{1}{R_1}$

$A_V = -\dfrac{R_2}{R_1}$

$R_{OUT} = \dfrac{R_{OUTOL}}{LG}$

Willy Sansen 10-05 1424

1424

The voltage gain A_V then simply becomes the ratio of the two resistors. Indeed it is derived in a simple way from the transresistance A_R and resistor R_1, as shown in this slide.

This voltage gain A_V is quite accurate, and nearly independent of the open loop gain A_0. For this reason this amplifier has become famous. It has become one of the most widely used amplifiers with an operational amplifier.

It is called the inverting amplifier.

The output resistance R_{OUT} will be quite small. Even without feedback, an opamp already provides a low output resistance R_{OUTOL}. This is a result of the use of a source (or emitter) follower at the output, inside the opamp. The application of feedback with a large loop gain LG decreases the output resistance to very small values!

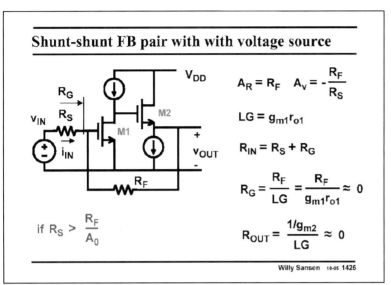

Shunt-shunt FB pair with with voltage source

$$A_R = R_F \quad A_v = -\frac{R_F}{R_S}$$

$$LG = g_{m1}r_{o1}$$

$$R_{IN} = R_S + R_G$$

$$R_G = \frac{R_F}{LG} = \frac{R_F}{g_{m1}r_{o1}} \approx 0$$

if $R_S > \dfrac{R_F}{A_0}$

$$R_{OUT} = \frac{1/g_{m2}}{LG} \approx 0$$

Willy Sansen 10-05 1425

1425
There are many ways to realize that opamp on the previous slide. Normally, an opamp has a differential pair at the input. A differential input is also possible however, by use of one single transistor. The gate is the minus input, whereas the source acts as the positive input. It is connected to ground.

Since the transistor circuit is no more than a transistor realization of the opamp block, the same equations are valid as before. The values of the closed-loop gain and loop gain LG are readily copied. The gain A_0 is now simply $g_{m1}r_{o1}$.

The results will not be as accurate as before however, because the gains are a bit too small with only one single amplifying transistor.

As a result, the input resistance at the minus input R_G will not be that small. It is called R_G, and calculated as before. It will have to be added to the input series resistor R_S to give the input resistance R_{IN} seen by the input voltage source.

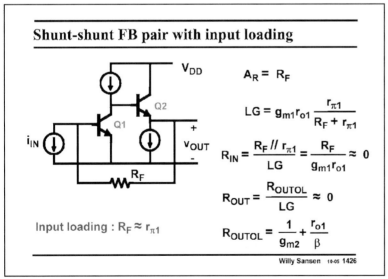

Shunt-shunt FB pair with input loading

$$A_R = R_F$$

$$LG = g_{m1}r_{o1}\frac{r_{\pi1}}{R_F + r_{\pi1}}$$

$$R_{IN} = \frac{R_F // r_{\pi1}}{LG} = \frac{R_F}{g_{m1}r_{o1}} \approx 0$$

$$R_{OUT} = \frac{R_{OUTOL}}{LG} \approx 0$$

Input loading : $R_F \approx r_{\pi1}$

$$R_{OUTOL} = \frac{1}{g_{m2}} + \frac{r_{o1}}{\beta}$$

Willy Sansen 10-05 1426

1426
Even with an ideal input current source, input loading also shows up when the input resistance of the amplifier is not large. In this case, there is an interaction between feedback resistor R_F and input resistance $r_{\pi1}$ – they are in parallel for the input resistance. Moreover, they cause a voltage division in the loop gain LG.

The output resistance will be quite small again, because a Source follower is used at the output. The output resistance without feedback is the output resistance of an emitter follower Q2. It must be divided by the loop gain to obtain the closed-loop output resistance.

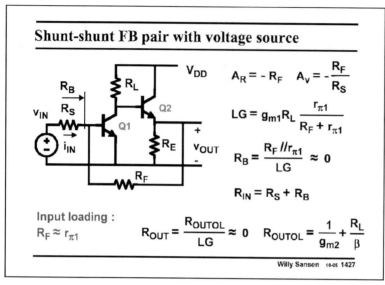

Shunt-shunt FB pair with voltage source

$A_R = -R_F \quad A_v = -\dfrac{R_F}{R_S}$

$LG = g_{m1}R_L \dfrac{r_{\pi1}}{R_F + r_{\pi1}}$

$R_B = \dfrac{R_F /\!/ r_{\pi1}}{LG} \approx 0$

$R_{IN} = R_S + R_B$

Input loading :
$R_F \approx r_{\pi1}$

$R_{OUT} = \dfrac{R_{OUTOL}}{LG} \approx 0 \quad R_{OUTOL} = \dfrac{1}{g_{m2}} + \dfrac{R_L}{\beta}$

Willy Sansen 10-05 1427

1427
Actually, all DC current sources can be substituted by resistances, as was common in discrete amplifiers. As a result, the gains will be somewhat lower. The same equations can still be used but they will be less accurate.

Moreover, the input current source can be substituted by a voltage source with series resistor R_S.

Input loading is present again because bipolar transistors are used. Feedback resistor R_F will interact with the input resistor $r_{\pi1}$ of transistor Q1. The expressions have now become more evident.

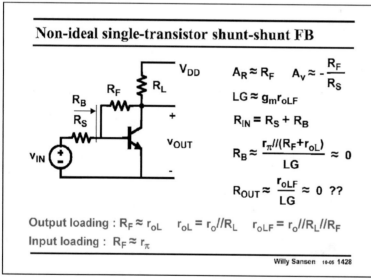

Non-ideal single-transistor shunt-shunt FB

$A_R \approx R_F \quad A_v \approx -\dfrac{R_F}{R_S}$

$LG \approx g_m r_{oLF}$

$R_{IN} = R_S + R_B$

$R_B \approx \dfrac{r_\pi /\!/ (R_F + r_{oL})}{LG} \approx 0$

$R_{OUT} \approx \dfrac{r_{oLF}}{LG} \approx 0 \ ??$

Output loading : $R_F \approx r_{oL} \quad r_{oL} = r_o /\!/ R_L \quad r_{oLF} = r_o /\!/ R_L /\!/ R_F$

Input loading : $R_F \approx r_\pi$

Willy Sansen 10-05 1428

1428
The most non-ideal feedback circuit that can be found, is probably the one using a single transistor only. This is a case of severe input loading AND output loading. Everything will interact with everything.

It is clear that in this case, the feedback equations can hardly be applied. An attempt is made in this slide. Some approximate expressions are given. The only sure way to obtain an accurate result is to substitute the transistor by its equivalent circuit (with parameters g_m and r_o at low frequencies), and solve the Kirchoff equations.

As a result this is one of the most complicated feedback circuits around, despite its simple circuit configuration.

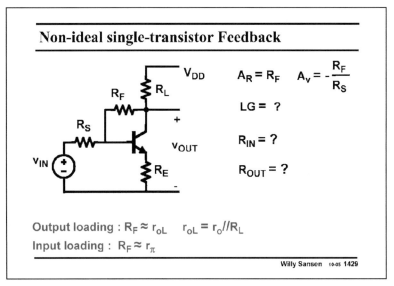

Non-ideal single-transistor Feedback

$A_R = R_F$ $A_v = -\dfrac{R_F}{R_S}$

$LG = ?$

$R_{IN} = ?$

$R_{OUT} = ?$

Output loading : $R_F \approx r_{oL}$ $r_{oL} = r_o // R_L$

Input loading : $R_F \approx r_\pi$

Willy Sansen 10-05 1429

1429

Certainly, the most non-ideal feedback circuit that can be found, is the one using one single transistor and a parallel feedback combined with series feedback. This is a case of severe input loading AND output loading. Everything will interact with everything.

It is clear that in this case the feedback equations cannot be applied. An attempt is not even made.

The only way to obtain an accurate result is to substitute the transistor by its equivalent circuit (with parameters g_m and r_o at low frequencies), and solve the Kirchoff equations.

As a result, this is one of the most complicated feedback circuits, despite its simplicity.

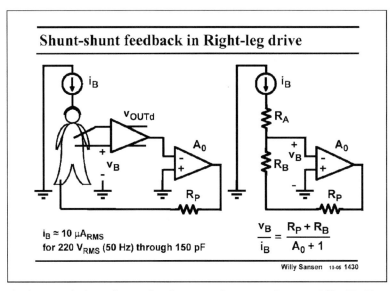

Shunt-shunt feedback in Right-leg drive

$i_B \approx 10 \ \mu A_{RMS}$
for 220 V_{RMS} (50 Hz) through 150 pF

$\dfrac{v_B}{i_B} = \dfrac{R_P + R_B}{A_0 + 1}$

Willy Sansen 10-05 1430

1430

A nice example of shunt-shunt feedback is the right-leg drive used for measurements of small signals such as ECG, EEG, etc. on the human body. Such measurements are carried out by means of a differential amplifier, which provides a differential output voltage v_{OUTd}.

These measurements are disturbed however, by the injection of hum at 50 Hz coming from the mains. Each human body is capacitively coupled to the mains by capacitances of up to 150 pF. For a mains voltage of 220 V_{RMS}, this correspond to an injected current of about 10 μA_{RMS}. In order to suppress the effect of this current, a common-mode shunt-shunt feedback loop is arranged. The voltage v_B on the body, caused by the injected current i_B, is then considerably reduced.

The equivalent circuit is shown on the right. The average output voltage is taken (e.g. by means of two resistors as shown in Chapter 8), and fed to a common-mode amplifier with gain A_0. The common-mode gain of the differential amplifier has been taken to be unity.

The output of amplifier A_0 is applied to the right leg of the body, through a resistor R_P. This resistor is needed for safety, in case the electronics break down. Typical values are 0.5 to 1 MΩ. The body itself is modeled by a few resistances R_A and R_B. Their values are of the order of magnitude of 10 kΩ.

The voltage on the body v_B is reduced by the loop gain!

Table of contents

Willy Sansen 10.05 1431

1431

Now that we know the principles of shunt-shunt feedback, let us learn about the other type of feedback with shunt feedback at the input. It also takes input currents.

It has series feedback at the output. It acts as a current generator. It is therefore a current amplifier with gain A_I. It amplifies the input current to the output with high precision. Its input resistance will be low and its output resistance high.

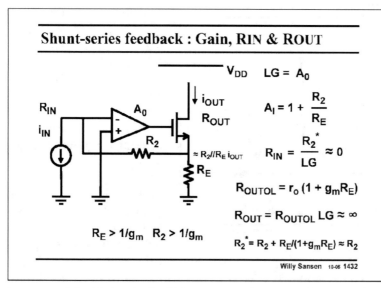

Shunt-series feedback : Gain, RIN & ROUT

$LG = A_0$

$A_I = 1 + \dfrac{R_2}{R_E}$

$R_{IN} = \dfrac{R_2^*}{LG} \approx 0$

$R_{OUTOL} = r_o(1 + g_m R_E)$

$R_{OUT} = R_{OUTOL}\, LG \approx \infty$

$R_E > 1/g_m \quad R_2 > 1/g_m$

$R_2^* = R_2 + R_E/(1+g_m R_E) \approx R_2$

Willy Sansen 10-06 1432

1432

A shunt-series feedback amplifier with an opamp is shown in this slide. Usually, the feedback resistor R_2 is much larger than $1/g_m$. Resistor R_E is also much larger than $1/g_m$.

For the loop gain, the output transistor acts as a Source Follower. This is why the loop gain is the gain of the opamp A_0 itself. The current gain A_I is easily found, once it has become clear that the input voltage of the opamp is about zero, because of the high gain A_0. This current gain is very precise indeed as it only depends on resistor ratios. This is a real current amplifier indeed.

The input resistance is just about R_2 in an open loop. For a closed loop, it must be divided by the loop gain LG. It is therefore small indeed.

The output resistance will be very large, this is large because of the local feedback of resistor R_E. This increases because of the feedback.

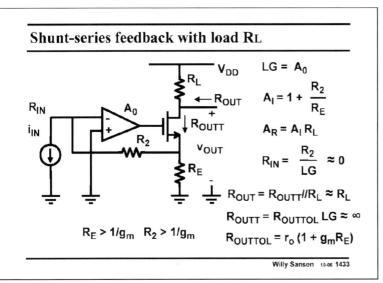

1433

Insertion of a load resistor in series with the output converts this circuit into a transimpedance amplifier with large gain A_R.

The loop gain and current gain are obviously the same as before. The transresistance A_R is simply the current gain A_I times the load resistor R_L. The input resistance is the same as before.

The output resistance R_{OUT} is now the parallel combination of the transistor output resistance R_{OUTT}, which is really large, and the load resistor R_L. It is mainly the load resistor R_L.

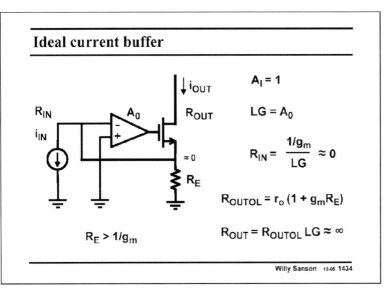

1434

Omission of feedback resistor R_2 gives the circuit in the slide. Its current gain A_I is now unity. Its input resistance is very small and its output resistance very large. It is therefore an ideal current buffer. It can also be called a current mirror, although this latter name is usually reserved for a current buffer with only a few devices, and hence with less ideal specifications.

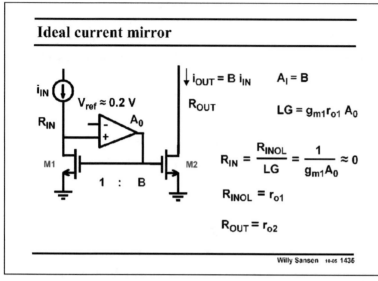

1435

Such a current mirror is shown in this slide. Its current gain A_I is now B, which is the ratio between W_2/L_2 and W_1/L_1.

The simplest current mirror has a short between the Drain and Gate of transistor M1. The specifications can be improved a lot by insertion of an opamp with high gain A_0. This circuit is only one of the many possible configurations of current mirrors, as shown in Chapter 3.

This one has the advantage that it can operate at low supply voltages. Indeed the Drain of transistor M1 is maintained at 0.2 V rather than at 0.9 V for a simple current mirror.

The loop gain LG is very large. The input resistance is therefore very small.

The output resistance, however, is not so large. It is merely the output resistance r_{o2} of transistor M2.

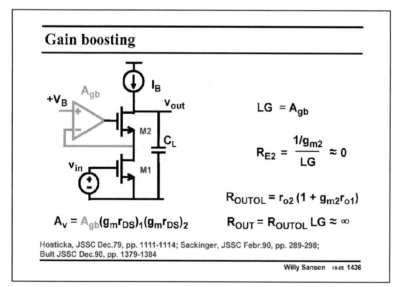

1436

A beautiful example of shunt-series feedback is gain boosting. As explained already in Chapter 2, gain boosting or regulating the cascode means that feedback is applied around the cascode, as shown in this slide.

The loop gain LG is the gain A_{gb} because transistor M2 behaves as a Source follower for the feedback loop.

As a result of the feedback loop, the output resistance R_{OUT} of the amplifier increases by the gain A_{gb} of the gain-boosting amplifier, and so does the gain A_v of the total amplifier.

The resistance R_{E2} between both transistors, at the Source of M2, is then divided by the same loop gain LG or A_{gb}.

Differential current amplifier

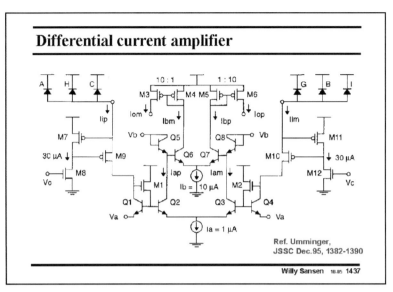

Ref. Umminger,
JSSC Dec.95, 1382-1390

Willy Sansen 10.05 1437

1437
A more complicated current amplifier using mainly current mirrors is shown in this slide. The input currents are provided by photo-diodes. These currents are picked up by regulated cascodes M9 and M10 and are mirrored (and amplified) towards a translinear circuit consisting of transistors Q5–8. Their output currents are mirrored and amplified by a factor of 10.

This is now a current amplifier without feedback.

Linear laser diode driver

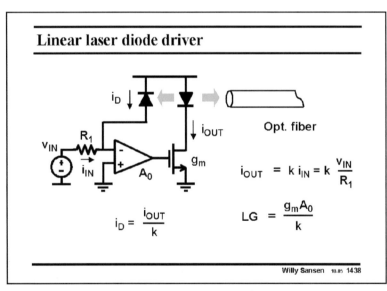

Opt. fiber

$$i_{OUT} = k\, i_{IN} = k\, \frac{v_{IN}}{R_1}$$

$$i_D = \frac{i_{OUT}}{k}$$

$$LG = \frac{g_m A_0}{k}$$

Willy Sansen 10.05 1438

1438
As a final example of shunt-series feedback a linear LED driver is given. A LED (Light Emitting Diode) or laser diode gives light in a very non-linear way depending on the voltage applied. However, the light output is linear versus the current. Also, the MOST driver transistor is nonlinear. This can be solved by creating the shunt-series feedback loop shown in this slide.

The light of the LED is sensed by a photodiode, which acts as a current source. It injects its current at the input of the opamp, where it is added to the current i_{IN} from the input source v_{IN}. At the input, we have the shunt feedback.

The output is the current provided by the output transistor. This is the series feedback. We now have a current amplifier. It is very linear versus the input current, and for the input voltage v_{IN}.

Table of contents

Willy Sansen 10.05 **1439**

1439

We go back to transimpedance amplifiers. The reason is that they are of great importance of all photodiode receivers. Their most important specifications are low noise and high bandwidth. They are now discussed in more detail.

First of all, we want to figure out whether to use a voltage amplifier at the input or a current amplifier.

Optical receiver : Current or voltage amplifier

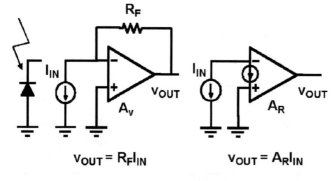

$$v_{OUT} = R_F I_{IN}$$ $$v_{OUT} = A_R I_{IN}$$

Willy Sansen 10.05 **1440**

1440

The photodiode can be modeled by a current source I_{IN}.

It can applied to a transimpedance amplifier, i.e. a voltage amplifier with a feedback resistor R_F or to an amplifier without feedback but with a current input which has a transimpedance A_R. If R_F equals A_R, then both have the same gain.

Which one would have the higher bandwidth BW?

A Figure-of-merit could be the product $A_R BW$, usually expressed in THzΩ. Which one would have the higher $A_R BW$ product?

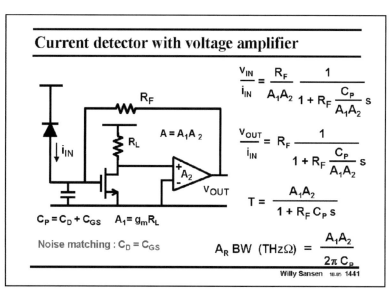

Current detector with voltage amplifier

$$\frac{v_{IN}}{i_{IN}} = \frac{R_F}{A_1 A_2} \frac{1}{1 + R_F \frac{C_P}{A_1 A_2} s}$$

$$\frac{v_{OUT}}{i_{IN}} = R_F \frac{1}{1 + R_F \frac{C_P}{A_1 A_2} s}$$

$$T = \frac{A_1 A_2}{1 + R_F C_P s}$$

$$A_R BW \ (THz\Omega) = \frac{A_1 A_2}{2\pi C_P}$$

$A = A_1 A_2$

$C_P = C_D + C_{GS}$ $A_1 = g_m R_L$

Noise matching : $C_D = C_{GS}$

Willy Sansen 10-05 1441

1441

In order to find the bandwidth, we have to find the node with the largest time constant. This is most likely the input node. Indeed the capacitance C_P at the input is the sum of the diode capacitance C_D and the input capacitance C_{GS} of the input transistor. Since they are about the same, because of noise matching (see Chapter 4), we can as well take $2C_D$ for C_P.

This time constant is then $R_F C_P / A_1 A_2$. It is smaller, or the BW is larger for smaller R_F and larger gains A_1 and A_2.

The $A_R BW$ product however only depends on the diode capacitance C_D and the two gains A_1 and A_2.

In order to increase the gain A_1 we have to increase load resistor R_L. The capacitance at that Drain will cause a second pole however, as shown next.

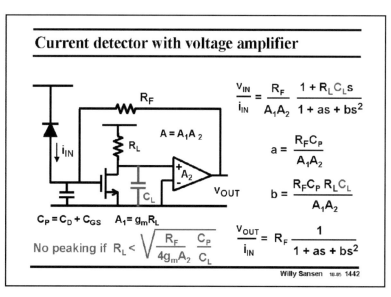

Current detector with voltage amplifier

$$\frac{v_{IN}}{i_{IN}} = \frac{R_F}{A_1 A_2} \frac{1 + R_L C_L s}{1 + as + bs^2}$$

$$a = \frac{R_F C_P}{A_1 A_2}$$

$$b = \frac{R_F C_P R_L C_L}{A_1 A_2}$$

$$\frac{v_{OUT}}{i_{IN}} = R_F \frac{1}{1 + as + bs^2}$$

$A = A_1 A_2$

$C_P = C_D + C_{GS}$ $A_1 = g_m R_L$

No peaking if $R_L < \sqrt{\dfrac{R_F}{4g_m A_2} \dfrac{C_P}{C_L}}$

Willy Sansen 10-05 1442

1442

When we take into account the capacitance C_L at the Drain of the input transistor, we then obtain a second-order expression for the transimpedance. To avoid peaking, we must have two real poles. This condition imposes an upper limit on the value of load resistor R_L. Increasing the gain A_2 too much, would only decrease the R_L and hence gain A_1.

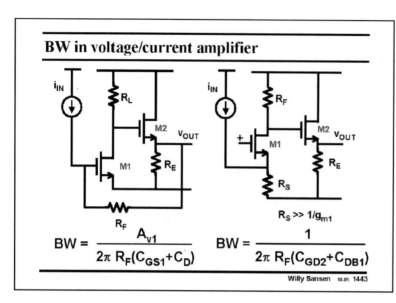

BW in voltage/current amplifier

$$BW = \frac{A_{v1}}{2\pi \, R_F(C_{GS1}+C_D)}$$

$$BW = \frac{1}{2\pi \, R_F(C_{GD2}+C_{DB1})}$$

Willy Sansen   10.05   1443

1443

For sake of comparison let us optimize both the transimpedance amplifier with voltage input (on the left) and the one with current input (on the left). The latter one usually has a cascode at the input or a regulated cascode (see later).

Both amplifiers have the same transresistance R_F.

Both amplifiers are designed for high frequencies. They are simple and carry fairly large currents!

Which one has the larger BW?

The dominant pole in the first amplifier is clearly at the input. The capacitance at the input node is again $C_D + C_{GS}$ or about $2C_D$. Load resistor R_L is sufficiently small such that the second pole does not play.

The dominant pole in the cascode amplifier is not at the input. The input capacitance is the same but the input resistance is only $1/g_m$. This pole is now at about $f_T/2$ of the input transistor. The dominant pole is this time at the Drain.

It is clear, that for about equal input transistors, the BW of the first transistor is higher. The capacitances all have similar values. The gain factor A_{v1} in the first amplifier, or the feedback in the first amplifier, makes the difference!

The main advantage of the second amplifier is that its input impedance is constant (and equal to $1/g_m$) up to high frequencies.

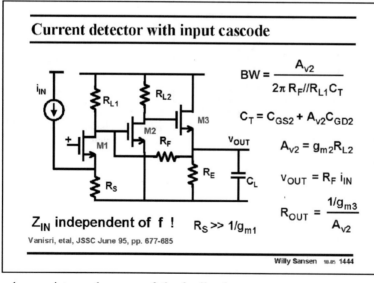

Current detector with input cascode

$$BW = \frac{A_{v2}}{2\pi \, R_F // R_{L1} C_T}$$

$$C_T = C_{GS2} + A_{v2}C_{GD2}$$

$$A_{v2} = g_{m2}R_{L2}$$

$$V_{OUT} = R_F \, i_{IN}$$

$$R_{OUT} = \frac{1/g_{m3}}{A_{v2}}$$

Z_{IN} independent of f ! $R_S \gg 1/g_{m1}$

Vanisri, etal, JSSC June 95, pp. 677-685

Willy Sansen   10.05   1444

1444

This photo current detector is a shunt-shunt feedback pair, preceded by a cascode again. The main reason is to have an input impedance which is independent of frequency, not to interact with the current source. A cascode has an input resistance of $1/g_{m1}$, which goes up to very high frequencies.

The transresistance is simply R_F, as the input current flows through the cascode into resistor R_F. The Gate of transistor M2 is at a low resistance because of the feedback.

The loop gain is A_{v2}. The bandwidth is increased by this loop gain. Also, the output resistance R_{OUT} is decreased by that same loop gain.

The dominant time constant is at the Drain of transistor M1. It has a fairly large capacitance because of the Miller effect across transistor M2. Moreover, resistor R_{L1} is fairly high to have high gain. The time constants at the nodes are smaller. For example, at the input node, the time constant is g_{m1}/C_{GS1}, which is the f_T of the transistor M1. The time constant at the Drain of transistor M2 is too small since the capacitance at this node contains mainly the output capacitance of transistor M2. The input capacitance of transistor M3 is bootstrapped as transistor M3 is a Source follower.

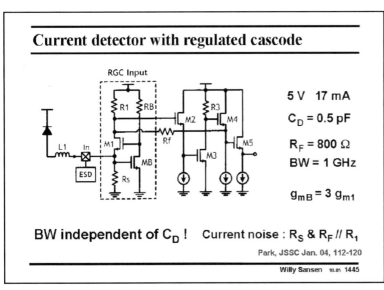

1445

A similar current input is realized here by means of a regulated cascode. Gain boosting is introduced by means of transistor MB. As a consequence, the input resistance will be smaller at low frequencies. At higher frequencies however, the local feedback gain rolls off. Moreover, it can form complex poles at intermediate frequencies.

For very high frequencies, it is better to use single-transistor cascodes. For intermediate or low frequencies a regulated cascode can be advantageous.

Moreover, the more transistors are used, the larger number of noise sources can play a role. Does the current-input amplifier perform as well as the voltage-input amplifier?

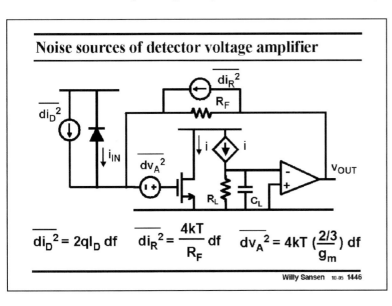

1446

The noise performance is now compared of both types of transimpedance amplifiers. The one with a voltage input comes first.

All the relevant noise sources are shown. The first one is the diode current shot noise, which is in parallel with the photo diode.

The second one is the noise generated by the feedback resistor R_F. It is taken as a current as we have a current input!

The third one is the equivalent input noise voltage of the amplifier. We assume that it is mainly the input transistor which is responsible for the amplifier noise.

Only thermal noise is taken into account.

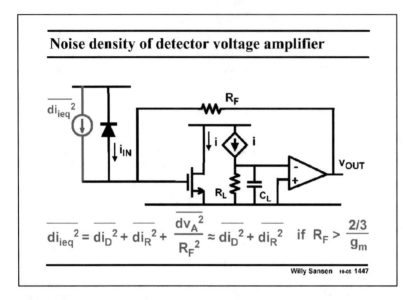

1447
The total equivalent input noise current is now calculated. For this purpose, the gain from each noise source to the output is calculated, added in power, and then divided by the total transconductance, which is R_F.

The expression shows that the noise of the input transistor is actually divided by R_F squared towards the input. Making R_F larger will make the noise of the input transistor negligible. A fairly small value of R_F is already sufficient!

The noise of the feedback resistor R_F is now the dominant noise source, obviously, in addition to the noise of the diode itself. Since the resistor noise is a current, the larger the resistor, the lower its noise!

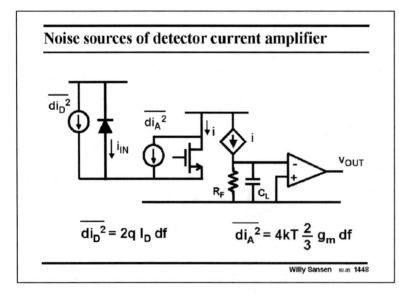

1448
If a cascode is used at the input then the noise current of that cascode transistor comes in. It is the only noise source around, in addition to the noise of the diode itself.

Note that the load resistor R_L is the same as R_F for the voltage-input amplifier so that the transimpedances are the same.

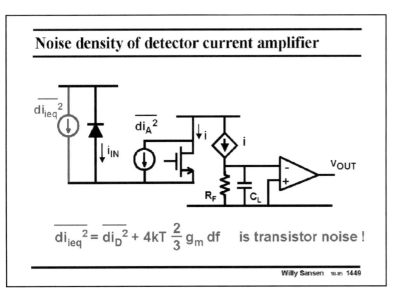

Noise density of detector current amplifier

$$\overline{di_{ieq}^2} = \overline{di_D^2} + 4kT\,\frac{2}{3}\,g_m\,df \quad \text{is transistor noise !}$$

Willy Sansen 10.05 1449

1449

Whether the noise of the cascode plays a role really depends on the load seen by that cascode. In this example the cascode sees the input of a current mirror, which is a low resistor (usually $1/g_m$). In this case, the current noise of the cascode can flow from the supply through the diode to ground. It is now added to the output current and to the noise current of the diode.

If the cascode were loaded with a very high impedance, in order to create a lot of gain, as is common in opamps, then the noise current of the cascode would be negligible. It is difficult to create a high impedance at the output of the cascode however in this kind of circuit, as it is to work at real high frequencies.

The noise current of the cascode will usually be the dominant noise source!

Comparison of noise densities

Voltage amp.: $\overline{i_{IN}^2} = \overline{di_R^2} = \dfrac{4kT}{R_F}\,df$

Current amp.: $\overline{i_{IN}^2} = \overline{di_A^2} = 4kT\,\dfrac{2}{3}\,g_m\,df$

➡ Voltage amplifier better when $R_F > \dfrac{3}{2}\dfrac{1}{g_m}$

Willy Sansen 10.05 1450

1450

A comparison between a voltage-input transimpedance amplifier and a current-input one is now easily made. The expressions of the equivalent input noise current densities are repeated in this slide.

It is clear that the voltage-input amplifier is better, provided its R_F is sufficiently large, i.e. larger than $1.5/g_m$.

This is normally the case!

Comparison of integrated noise

Large I_D : $\quad \overline{i_{IN}^2} = \overline{di_D^2}\,(BW\,\frac{\pi}{2})$

Small I_D :

Voltage amp.: $\quad \overline{i_{IN}^2} = \overline{di_R^2}\,(BW\,\frac{\pi}{2}) = \dfrac{kT}{R_L}\left(\dfrac{R_L}{R_F}\right)^2 \dfrac{g_m}{C_P}$

Current amp.: $\quad \overline{i_{IN}^2} = \overline{di_A^2}\,(BW\,\frac{\pi}{2}) = \dfrac{2}{3}\dfrac{kT}{R_F}\dfrac{g_m}{C_L}$

➡ Voltage amplifier better when $R_F > \dfrac{3}{4}\,R_L$

Willy Sansen 10.05 1451

1451

A similar comparison can be carried out for the total integrated noise.

For this purpose, the noise densities have to be multiplied with the noise bandwidths or the bandwidths themselves, multiplied by $\pi/2$ (see Chapter 4).

For the voltage-input amplifier, the BW is determined at the input node of the input transistor. The total input capacitance C_P is again about twice the diode capacitance. Remember that resistor R_L is the load resistor of the input transistor.

For the current-input amplifier the bandwidth is at the output node of the input transistor. The capacitance at this node is C_L. In the simple example of slide 41, this capacitance is the output capacitance C_{DB} of the input transistor. It has about the same size as its C_{GS} capacitance. This load capacitance C_L is about half of C_P.

The comparison shows that the voltage-input amplifier is better if feedback capacitor R_F is larger than the load capacitor R_L. This is usually the case.

Moreover, care has to be taken to make the resistor R_S form the cascode Source to ground (see slide 41) sufficiently large to avoid its current noise.

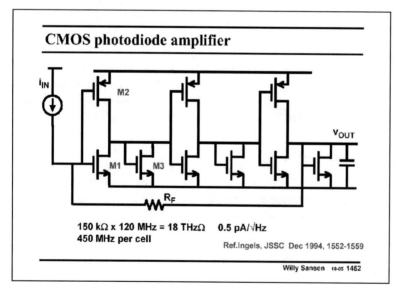

CMOS photodiode amplifier

i_{IN}

M2

M1 M3

v_{OUT}

R_F

150 kΩ x 120 MHz = 18 THzΩ   0.5 pA/√Hz
450 MHz per cell

Ref.Ingels, JSSC Dec 1994, 1552-1559

Willy Sansen 10-05 1452

1452

A good example of a CMOS voltage-input transimpedance amplifier is shown in this slide. It consists of three wide-band CMOS amplifiers with R_F as a feedback resistor. The bandwidth of 120 MHz may not be all that high but the transimpedance of 150 kΩ is fairly high such that the BW.R_F product is quite high, i.e. 18 THzΩ.

The equivalent input noise current is mainly the current noise of feedback resistor R_F.

Each amplifier consists of a CMOS inverter amplifier loaded by the $1/g_m$ of a diode connected nMOST. Input and output have the same DC voltage such that they are easily cascaded.

Moreover, all nodes are at low impedance ($1/g_m$ level) such that the bandwidth can be fairly high, depending on the DC biasing currents flowing.

Such an amplifier is easily optimized as shown next.

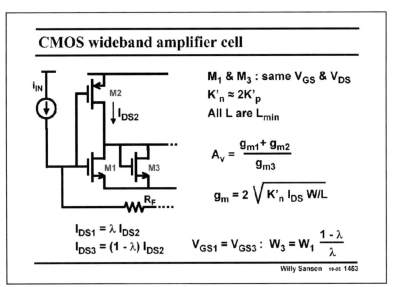

Willy Sansen 10-05 1453

1453
One of the three cells of the previous amplifier is shown in this slide.

The voltage gain A_v only depends on the transconductances. Depending on the DC biasing currents and the widths, a small gain of 5–8 is easily achieved. The total gain for three similar stages is therefore, easily over 100.

For the calculation of this gain, the current I_{DS2} through transistor M2 is assumed to be constant. It is divided over the transistors M1 and M2 by factor λ. Since all lengths are taken the same, this factor λ also determines the ratio of the widths W_3 to W_1.

The gain A_v is now readily calculated. It is shown next.

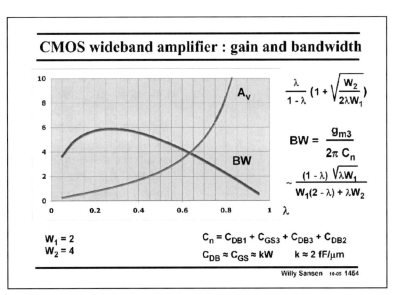

Willy Sansen 10-05 1454

1454
The gain A_v is calculated for $W_1 = 2$ and $W_2 = 4$. These are arbitrary units, for example micrometers or ratio's to the channel length.

The expression of the gain A_v is given in this slide. It increases if the current through M3 is smaller. In this case the output resistance increases and so does the gain. A gain of 5 is reached for a λ of about 0.7. In this case W_3 is about 0.86.

The bandwidth BW is limited by the capacitance at the output node C_n, which is given in this slide. It is also easily calculated, noting that the BW is also proportional to the square root of the total current I_{DS2} and some technological parameters. It reaches a maximum at a low value of the gain however, at about $\lambda = 0.3$.

A compromise needs to be taken. For example for $\lambda = 0.7$, the BW is only about half of this maximum.

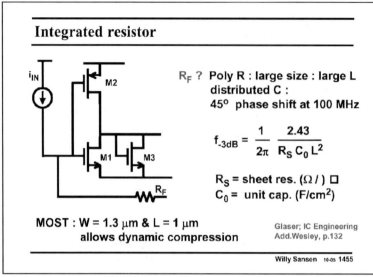

1455

One of the biggest problems with high-speed transimpedance amplifiers is the feedback resistor R_F. High values are difficult to obtain at high frequencies.

A conventional polysilicon resistor with length L for example (the length is the distance between the two contacts), has a certain sheet resistance R_S but also a parallel distributed capacitance C_0 to ground. It thus acts as a kind of transmission line. Its -3 dB frequency heavily depends on the resistor length L. It is easy to calculate that poly resistors are difficult to make beyond about 100 MHz.

A much better solution is to use a MOST in the linear region. Their areas $W \times L$ are quite small and so are their parallel capacitances. Their -3 dB frequency can therefore be much higher.

In this example a nMOST is taken of merely 1.3×1 μm. This is the Gate voltage used which corresponds to about 150 kΩ.

Indeed a MOST resistor can be made larger or smaller depending on the Gate voltage This allows dynamic compression, as shown next.

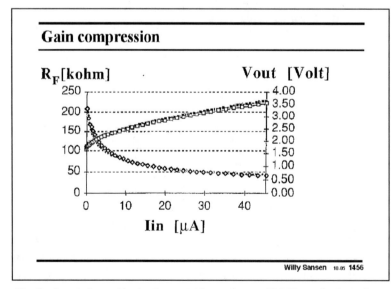

1456

Such transimpedance amplifiers must generate an output signal amplitude which is better by being constant. For a small diode current, a lot of gain is now required, or a large value of feedback resistor R_F. For a large input current, a small value of feedback resistor R_F is preferred. This gain compression is easily realized with a MOST, rather than with a constant resistor.

This is illustrated in this slide. For input currents of 40 μA the resistor R_F is decreased to about 40 kΩ such that the output signal is about 1.6 V. A post amplifier is used with a gain of two, to boost this to 3.2 V. For small input currents, the resistor is over 200 kΩ and the output signal about 1.5 V.

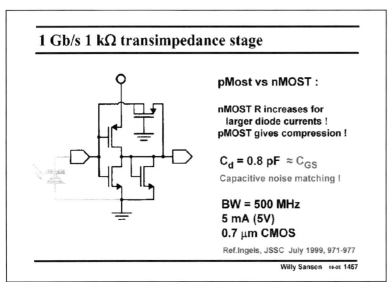

1 Gb/s 1 kΩ transimpedance stage

pMost vs nMOST :

nMOST R increases for
larger diode currents !
pMOST gives compression !

C_d = 0.8 pF ≈ C_{GS}

Capacitive noise matching !

BW = 500 MHz
5 mA (5V)
0.7 µm CMOS

Ref.Ingels, JSSC July 1999, 971-977

Willy Sansen 10-05 1457

1457

Compression can also be realized on each gain cell separately. The question arises of whether to realize the feedback resistor with a nMOST or a pMOST transistor?

It is shown in this slide that a pMOST should be used as it provides compression for larger input currents.

The optimization of such a 4-transistor circuit is a beautiful design project, within a certain CMOS technology. The result given in this slide, show that even in a modest 0.7 µm CMOS technology, 500 MHz can be achieved. Obviously this result depends on the diode capacitance (0.8 pF here).

The current is a result of the capacitive noise matching as discussed in Chapter 4.

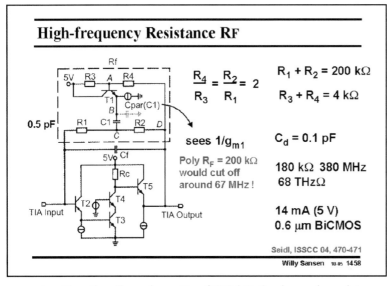

High-frequency Resistance RF

$$\frac{R_4}{R_3} = \frac{R_2}{R_1} = 2$$

$R_1 + R_2$ = 200 kΩ

$R_3 + R_4$ = 4 kΩ

sees $1/g_{m1}$

Poly R_F = 200 kΩ
would cut off
around 67 MHz !

C_d = 0.1 pF

180 kΩ 380 MHz
68 THzΩ

14 mA (5 V)
0.6 µm BiCMOS

Seidl, ISSCC 04, 470-471

Willy Sansen 10-05 1458

1458

Another way to realize a feedback resistor R_F at high frequencies is shown in this slide. It uses the bipolar transistors of a Becomes technology.

The transimpedance amplifier itself has an Emitter follower at the input, followed by a cascode amplifier and another Emitter follower, at the output.

The feedback resistor R_F consists of two resistors R1 and R2 in series, at low frequencies. Together they give a R_F of 200 kΩ. Such a poly resistor would cause a −3 dB frequency of no more than 67 MHz!

At high frequencies, capacitor C1 acts as a short circuit. The result is that resistors R3 and R4 take over the role of resistors R1 and R2. They are much smaller in absolute value however, such that they can provide a similar transimpedance up to much higher frequencies.

The parasitic capacitance at node B only sees a small $1/g_{m1}$ resistance.

For a diode capacitance of 0.1 pF, the bandwidth is now 380 MHz. With a transimpedance of 180 kΩ, this gives an impressive BW.R_F product of 68 TzΩ!

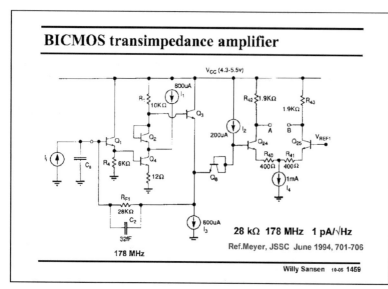

BICMOS transimpedance amplifier

28 kΩ 178 MHz 1 pA/√Hz

Ref.Meyer, JSSC June 1994, 701-706

Willy Sansen 10-05 1459

1459

Another shunt-shunt feedback amplifier for optical fiber receivers is shown in this slide. It uses mainly bipolar transistors again.

The amplifier itself is preceded by an emitter follower Q1. The amplifier itself consists of transistors Q4 and Q2 with resistors R_1 of 10 kΩ and 12 Ω. Again an emitter follower is used with Q3 to lower the output resistance.

The load resistor R1 is shunted by the input resistance of the emitter follower. As a result, its effective value is only 5 kΩ. The transconductance of transistor Q4 is about 1/28 Ω. As a result, the gain of this stage is $A_v = 5000/40 \approx 125$. The bandwidth is set by the parallel $R_{F1}C_2$, which is 178 MHz.

More details are given on the next slide.

The input resistance will now be $R_{F1}/A_v \approx 240\,\Omega$. This is made low to annihilate the effect of the sensor capacitance C_s.

The input noise is equally caused by the feedback resistor R_{F1} and the input base current noise.

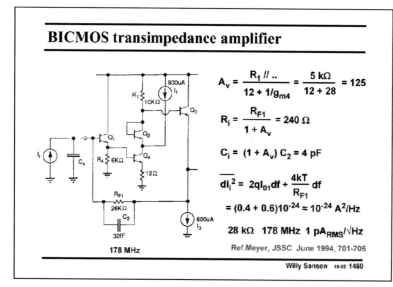

BICMOS transimpedance amplifier

$$A_v = \frac{R_1 \,//\, ..}{12 + 1/g_{m4}} = \frac{5\,k\Omega}{12 + 28} = 125$$

$$R_i = \frac{R_{F1}}{1 + A_v} = 240\ \Omega$$

$$C_i = (1 + A_v)\, C_2 = 4\ pF$$

$$\overline{di_i^2} = 2qI_{B1}df + \frac{4kT}{R_{F1}}\,df$$

$$= (0.4 + 0.6)10^{-24} \approx 10^{-24}\ A^2/Hz$$

28 kΩ 178 MHz 1 pA$_{RMS}$/√Hz

Ref.Meyer, JSSC June 1994, 701-706

Willy Sansen 10-05 1460

1460

The voltage gain of the input stage is $A_v \approx 125$. The bandwidth is limited by the product $R_{F1}C_2$ and is 178 MHz. The input resistance will thus be $R_{F1}/A_v \approx 240\,\Omega$. This is made low to reduce the effect of the sensor capacitance C_s, and to increase the bandwidth to frequencies beyond 178 MHz.

The input capacitance is increased by the same amount. It is now 4 pF.

The equivalent input noise current is mainly caused by two components. The first one is the base current shot noise of the input bipolar transistor Q1. It is slightly smaller than the current noise of the feedback resistor R_{F1}. Together it is about 1 pA$_{RMS}$/√Hz.

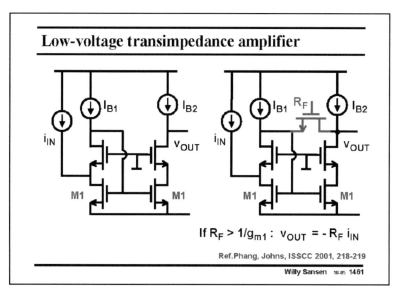

1461

A transimpedance amplifier for low supply voltages is derived from the differential current amplifier, shown on the left. Only one single transistor has been added to create a transimpedance amplifier, shown on the right. This transistor operates in the linear region, with resistance R_F. This resistor provides shunt-shunt feedback to the middle node of the current mirror. The input is current source i_{IN} however, and the output is a voltage. It is therefore a transimpedance amplifier! Its transresistance is R_F itself.

Because of the different feedback arrangement, the input resistance is different from the one without R_F; it is $R_{IN} = 1/2g_{m1}$, which is still quite low. Since the differential current amplifier can operate at supply voltages below 1 V (see Chapter 3), this transimpedance amplifier can do this as well!

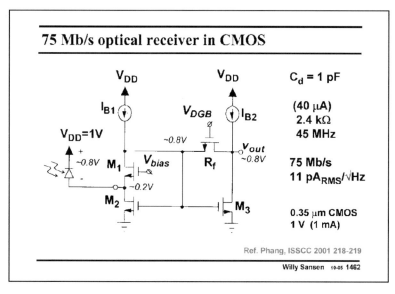

1462

The actual specifications of the realization are shown in this slide.

A photodiode is used with a capacitance of 1 pF. Its maximum input current is about 40 μA. The transimpedance R_F is 2.4 kΩ. The output voltage would then be about 100 mV.

The speed is limited by the input node capacitance.

The noise is determined by the input devices M_1 and M_2, and the feedback resistor R_F, as explained before.

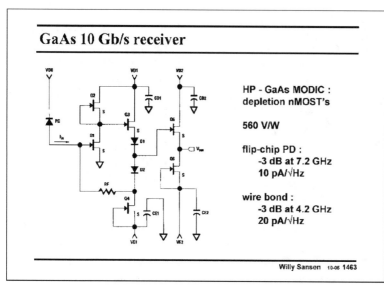

GaAs 10 Gb/s receiver

HP - GaAs MODIC :
depletion nMOST's

560 V/W

flip-chip PD :
-3 dB at 7.2 GHz
10 pA/√Hz

wire bond :
-3 dB at 4.2 GHz
20 pA/√Hz

Willy Sansen   10-05 **1463**

1463

An example of a transimpedance amplifier for real-high speed is shown in this slide. It is realized in GaAs technology to achieve the highest possible speed.

GaAs FET transistors are depletion devices. They conduct for zero Volt V_{GS}. Transistor Q2 acts as an active load (DC current source) for amplifying transistor Q1. Transistor Q3 is just a Source follower. Two diodes are used for level shifting to be able to close the feedback loop by means of resistor R_F. The output is taken through another Source follower.

Resistor R_F is fairly small such that the bandwidth is fairly high. However, this bandwidth depends on the packaging. Bond wires reduce the bandwidth more than flip-chip packaging!

Table of contents

Willy Sansen   10-05 **1464**

1464

In this Chapter we have learned how to link closed-loop gain to open-loop gain and loop gain. Moreover, considerable attention has been paid to the effect of the loop gain on input-and output resistances. In this way we can easily find the pole frequencies caused by capacitances at input and output.

This has been done for all four types of feedback amplifiers, in this Chapter and the previous one.

In addition, noise and high-frequency considerations have been spelled out on a number of published transimpedance amplifiers. Voltage-input and current-input transimpedance amplifiers have been compared as well. For not too small values of feedback resistor RF, the voltage-input transimpedance amplifier is found to be better for both bandwidth and noise.

151

Offset and CMRR : Random and systematic

Willy Sansen

KULeuven, ESAT-MICAS
Leuven, Belgium

willy.sansen@esat.kuleuven.be

Willy Sansen 10-05 151

The main limitation to precision in analog integrated circuits is established by noise and mismatch. Actually, the smaller the channel lengths become, the more severe is the mismatch and this becomes the dominant limitation in precision.

Mismatch is the main reason for high offset and low CMRR (Common-mode rejection ratio). It is also the main reason for low PSRR (Power-supply rejection ratio).

In this Chapter we will investigate the relationship between mismatch and these specifications. Moreover, we will try to find out how to modify the layout of a circuit to improve mismatch and hence to lower the offset. This will be done for both CMOS and bipolar technologies.

152

Table of contents

- **Random offset and CMRR$_r$**
- **Systematic offset and CMRR$_s$**
- **CMRR versus frequency**
- **Design rules**
- **Comparison MOST and bipolar transistors**

Ref: Pelgrom, JSSC Oct.1989, 1433-1439
Croon, JSSC Aug.02, 1056-1064
Croon, Springer, 2005

Willy Sansen 10-05 152

First of all, we start by a few definitions. What is actually offset and what is CMRR?

They can be caused by random effects but also by systematic errors in design.

We now focus on how these phenomena behave at higher frequencies.

Probably the most important part of this Chapter is the list of design rules for good design.

Finally, the differences are highlighted between CMOS and bipolar design.

153

When a single-ended opamp is taken with zero differential input voltage, its output voltage should be zero, whatever the gain is. In practice, this is not the case. The output voltage is not zero.

The offset voltage v_{os} is now by definition the differential input voltage that is required to make the output voltage zero. Since it is a differential input voltage it can be inserted in either

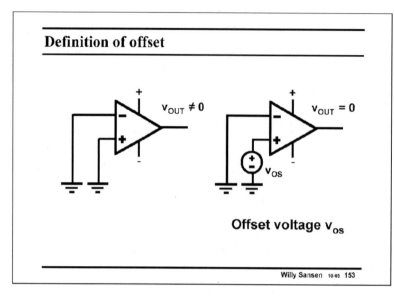

Definition of offset

$v_{OUT} \neq 0$

$v_{OUT} = 0$

v_{os}

Offset voltage v_{os}

Willy Sansen 10-05 153

one of the two input terminals. In this example, the offset voltage is inserted in the plus terminal. When shifted to the minus terminal it has the same value but opposite sign.

It is usually a few mV's for a bipolar amplifier. For a CMOS amplifier it can be up to ten times larger!

What causes this offset voltage? Random effects can play a role, but also systematic ones.

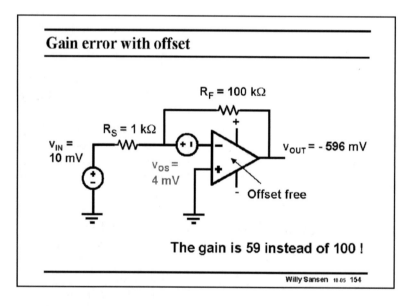

Gain error with offset

$R_F = 100 \text{ k}\Omega$

$R_S = 1 \text{ k}\Omega$

$V_{IN} = 10 \text{ mV}$

$V_{os} = 4 \text{ mV}$

$V_{OUT} = -596 \text{ mV}$

Offset free

The gain is 59 instead of 100 !

Willy Sansen 10-05 154

154

The offset can cause large errors in high-gain opamp configurations.

In the example in this slide, a small DC voltage is amplified coming from a thermocouple. A gain of 1000 is expected. For an input voltage this would give an output voltage of -1 V.

The output voltage is only -596 mV however. An offset voltage of 4 mV leaves only 6 mV as a voltage across the resistor R_S. The voltage across resistor R_F is now 100 times higher, which is 600 mV leading to the output voltage shown.

This offset causes a large error in gain!

155

This offset causes an error in this ADC (analog-to-digital converter) as well. In this flash converter, the input voltage V_{in} is compared with a voltage which is divided from a reference voltage V_{ref}. The comparators indicate at which tap of the reference voltage the input voltage is located.

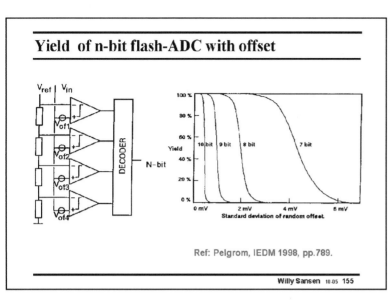

Yield of n-bit flash-ADC with offset

Ref: Pelgrom, IEDM 1998, pp.789.

Willy Sansen 10-05 155

Obviously, when these comparators have an offset voltage they may give an erroneous result. The yield of such an ADC will depend on the offsets present. The graph on the right shows that an 8-bit ADC can be expected to provide a yield of only 60%, if the offset is about 2 mV.

As a consequence, the offset severely limits the resolution of the ADC's if a high yield is required, which is usually the case!

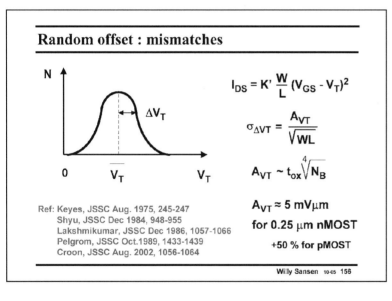

Random offset : mismatches

$$I_{DS} = K' \frac{W}{L} (V_{GS} - V_T)^2$$

$$\sigma_{\Delta VT} = \frac{A_{VT}}{\sqrt{WL}}$$

$$A_{VT} \sim t_{ox} \sqrt[4]{N_B}$$

$$A_{VT} \approx 5 \text{ mV}\mu\text{m}$$

for 0.25 μm nMOST

+50 % for pMOST

Ref: Keyes, JSSC Aug. 1975, 245-247
Shyu, JSSC Dec 1984, 948-955
Lakshmikumar, JSSC Dec 1986, 1057-1066
Pelgrom, JSSC Oct.1989, 1433-1439
Croon, JSSC Aug. 2002, 1056-1064

Willy Sansen 10-05 156

156

This offset is caused by mismatches between transistors which have been laid out equal.

When a large number such as 10,000 equal transistors are evaluated, their threshold voltage V_T are measured, and their K' values, etc. When the number of transistors are plotted versus the actual V_T values, a diagram is obtained as shown in this slide.

Normally, it shows a Gaussian distribution with an average and a spreading or sigma. For a Gaussian distribution, only about 0.5% of the transistors have a V_T more than three sigma's away from the average.

Several models have shown that this sigma is inversely proportional to the square root of the area WL of the transistor. The proportionality constant A_{VT} on itself, depends on the technology used. For smaller channel lengths L, the oxide thickness t_{ox} ($\approx$L/50) decreases but the doping levels increase. Parameter N_B is the doping level of the substrate underneath the transistor. For a nMOST in 0.5 μm CMOS, A_{VT} is about 10 mVμm. For a MOST of 20×0.13 μm the sigma would be about 6.2 mV.

For a pMOST, the A_{VT} is about 50% higher, mainly because of the higher substrate doping level in a n-well CMOS technology.

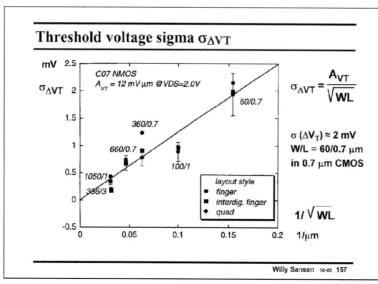

157

In order to illustrate this dependency on size, a measurement curve is shown in this slide. The smaller sizes are on the right of the horizontal axis.

It is clear that for smaller sizes the spreading is larger as well.

The slope on the other hand, corresponds to a A_{VT} factor of about 12 mVμm. This is for a 0.7 micrometer CMOS technology, which has an oxide thickness of about 700/50 or 14 nm. As a rule of thumb, the value of the A_{VT} factor in mVμm can be taken to be about the same as the oxide thickness in nm. This is shown on the next slide.

Note also that the layout style does not make all that much difference. Whether a finger layout is used or an interdigitated one seems to be unimportant. Only the total gate area WL is important.

It is also clear that this curve goes through zero. This would mean that for even very small channel lengths, the oxide thickness would decrease accordingly and therefore the factor A_{VT}. Whether nanometer CMOS processing develops in this way still has to be seen, however.

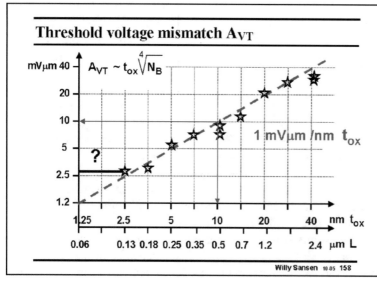

158

The value of the A_{VT} factor in mVμm can (as shown here) be taken to be the same as the oxide thickness in nm. Each star corresponds to a different technology.

Both axes are provided, one with the oxide thickness and one with the channel length. A ratio of 50 is taken between them.

This curve allows prediction of the A_{VT} for future CMOS technologies in a fairly obvious way. Care has to be exerted however, when extrapolating to deep submicron CMOS technologies.

For example, it has already been found (Tuinhout, IEDM 1997, 631–634 and more recently by Croon, Springer 2004), that below 130 nm channel lengths, several phenomena show up such that the A_{VT} does not decrease any more but becomes more or less constant at a value of about 3 mVμm.

In conventional CMOS the main contribution to A_{VT} is the effect of fluctuations in channel doping. Fluctuations in Gate doping and surface-roughness scattering are dominant in A_{WL}. For nanometer CMOS, poly-silicon gate depletion comes in heavily. Metal Gates are expected to remedy this.

More experimental evidence is required, however.

Random offset : mismatches

$$\frac{\overline{\Delta K'}}{K'} = \frac{A_{K'}}{\sqrt{WL}} \qquad A_{K'} \approx 0.0056 \; \mu m \; \text{+50 \% for pMOST}$$

$$\frac{\overline{\Delta W/L}}{W/L} = A_{WL} \sqrt{\frac{1}{W^2} + \frac{1}{L^2}} \qquad A_{WL} \approx 0.02 \; \mu m \; \text{+50 \% for pMOST}$$

$$\frac{\overline{\Delta \gamma}}{\gamma} = \frac{A_{\gamma}}{\sqrt{WL}} \qquad A_{\gamma} \approx 0.016 \; \mu m \; \text{-25 \% for pMOST}$$

Negligible if B = S

Ref.: Pelgrom : JSSC Oct.1989, pp.1430-1440

Willy Sansen 10-05 159

159

The other transistor parameters are subject to similar spreadings.

An expression can be established for the K' parameter in a similar way as for the threshold voltage. However, its parameter $A_{K'}$ is quite small.

The same applies to the dimensions W and L. Photolithography and mask making play a role in this expression. It is clear that the smaller dimension W or L plays the dominant role.

The A_{WL} parameter is larger than A_{VT}. It is also larger for a pMOST than for a nMOST. Its value does not seem to change all that much with technology. It is still around 2%μm for minimum feature sizes.

Finally, the substrate effect parameter g has a similar expression as well. When the Bulk is shorted to the Source, the effect of this parameter spreading can be neglected. We will do so whenever possible. This is why most input stages of operational amplifiers use pMOSTs. They can be put in the same n-well. Their matching can now be expected to improve.

Mismatch coefficients for nMOST

Techno L (μm)	2.5	1.2	0.7	0.5	0.35	0.25
t_{ox} (nm)	50	25	15	11	8	6
A_{VT} (mV μm)	30	21	13	7.1	6	⇒ 0
A_{WL} (% μm)	2.5	1.8	2.5	1.3	2	⇒ 1.8
S_{VT} (mV/mm)	0.3	0.3	0.4	0.2		
S_{WL} (%/mm)	0.3		0.2	0.2		

Willy Sansen 10-05 1510

1510

Some more data is given next.

It is clearly seen that the A_{VT} continues to decrease for smaller channel lengths but not A_{WL}. The latter parameter seems to stabilize around 2%μm. This means that if the spreading on the threshold is the main contributor to mismatch, then spreading in sizing may become the dominant one in the future.

Two more parameters are introduced below. The first

one is S_{VT}. It indicates the spreading of the threshold voltage V_T for two transistors separated by a distance of 1 mm. Values are given for some older technologies but not for more recent ones. The reason is that CMOS processing is now carried out on large wafers (of 12 inch and more) such that homogeneity has been strongly improved. Spreading on such short distances of 1 mm has therefore become negligible. The same applies to the other factor S_{WL}.

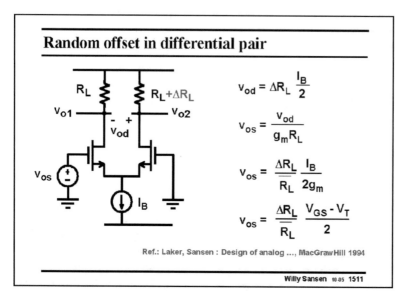

Random offset in differential pair

$$v_{od} = \Delta R_L \frac{I_B}{2}$$

$$v_{os} = \frac{v_{od}}{g_m R_L}$$

$$v_{os} = \frac{\Delta R_L}{R_L} \frac{I_B}{2g_m}$$

$$v_{os} = \frac{\Delta R_L}{R_L} \frac{V_{GS} - V_T}{2}$$

Ref.: Laker, Sansen : Design of analog ..., MacGrawHill 1994

Willy Sansen 10 05 1511

1511
Note that several sources of parameter spreading have been identified, we can try to establish their relationship with the offset.

A simple differential pair is taken first, in which the only source of asymmetry is the spreading in load resistor RL. This will result in a differential output voltage v_{od} and therefore into an offset voltage. It is calculated in this slide.

The differential output voltage v_{od} is readily calculated, as both transistors carry an equal current $I_B/2$. This v_{od} divided by the small-signal gain $g_m R_L$ gives the differential input voltage required to make the differential output voltage zero, which is by definition the offset voltage v_{os}.

The final result for the offset voltage v_{os} shows that the input transistors must be designed for high gain, which means they must be designed for small $V_{GS} - V_T$.

Pushing them into a weak inversion would make the offset voltage even smaller! Indeed for a weak inversion factor $(V_{GS} - V_T)/2$ it can be substituted by nkT/q, which is always smaller.

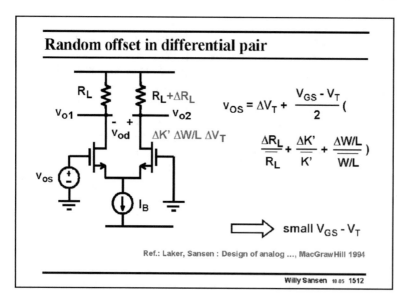

Random offset in differential pair

$\Delta K' \ \Delta W/L \ \Delta V_T$

$$v_{os} = \Delta V_T + \frac{V_{GS} - V_T}{2} ($$

$$\frac{\Delta R_L}{R_L} + \frac{\Delta K'}{K'} + \frac{\Delta W/L}{W/L})$$

⟹ small $V_{GS} - V_T$

Ref.: Laker, Sansen : Design of analog ..., MacGrawHill 1994

Willy Sansen 10 05 1512

1512
A similar calculation can be carried out for the other delta's.

The easiest one to understand is the one for the spreading in V_T. This one simply appears at the input of the differential pair. This is why it can simply be added to the offset voltage v_{os}.

The resulting expression contains four terms. Since all of them can have both positive and negative values, they never all add up. Such a worst case never occurs in practice. They never cancel out either.

Three of them are scaled by $(V_{GS} - V_T)/2$. The offset can now be reduced by designing the transistors with small values of $V_{GS} - V_T$, or by pushing them into weak inversion.

Note that trimming the resistors allows it to compensate for all other terms. It is clear however, that this cancellation point depends on the stability of the biasing point (through $V_{GS} - V_T$) with respect to other biasing and supply voltages and with respect to temperature. This is exceedingly difficult to realize in practice. Trimming of the offset voltage for MOSTs is therefore a real problem.

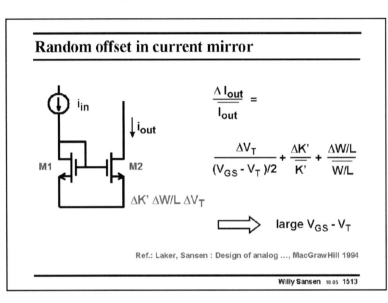

1513

Two matched transistors are also used in current mirrors. The difference with a differential pair however, is that we now have to concentrate on the output currents, rather than on the differential input voltage.

The relative spreading on the output current depends again on the transistor parameter spreadings.

It is easy to understand that when one transistor is 1% larger than the other one, the currents differ accordingly.

This time the spreading in threshold voltage must be scaled by $(V_{GS} - V_T)/2$.

This leads to the conclusion that current source transistors are well matched if they have been designed for large value of $V_{GS} - V_T$. In this case, the A_{WL} is the dominant source of mismatch. Remember that this one does not scale with technology.

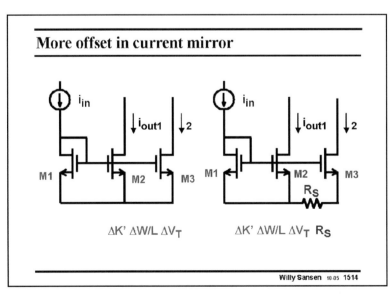

1514

Note however, that current mirrors, which are used for biasing, may have errors in the output currents, caused by resistances in the supply lines. On the left such a current mirror is shown. It is duplicated on the right with some series resistance R_S in the supply line.

The output current I_{out2} will be decreased because a voltage drop $R_S I_{out2}$ has to be subtracted from its V_{GS2}. This voltage drop must always be subtracted. It is now a systematic error, not a random one. More systematic spreadings will be discussed later.

Mismatch in drain current

$$I_{DS} = \frac{\beta}{2}(V_{GS} - V_T)^2 \qquad \beta = \frac{K'}{n}\frac{W}{L}$$

$$\frac{\Delta I_{DS}}{I_{DS}} = \frac{\Delta\beta}{\beta} - \Delta V_T \frac{2}{V_{GS} - V_T}$$

$$\sigma^2\left(\frac{\Delta I_{DS}}{I_{DS}}\right) = \sigma^2\left(\frac{\Delta\beta}{\beta}\right) + \sigma^2(\Delta V_T) \quad \frac{4}{(V_{GS} - V_T)^2}$$

$$\frac{1}{(nkT/q)^2} \quad \text{in wi}$$

$$\left(\frac{g_m}{I_{DS}}\right)^2 \quad \text{in general}$$

Willy Sansen 10-05 1515

1515

The spreading on the total drain current contains both the spreading on the beta and the spreading on the threshold voltage V_T, but in a different way. The spreadings on K' and on W/L are taken up by the spreading on beta for simplicity.

Taking the derivatives allows us to calculate the total effective spreading on the drain current I_{DS}. It is clear that for large values of $V_{GS} - V_T$, the spreading in beta is dominant. This is the case for a current mirror.

For small values of $V_{GS} - V_T$ however, the spreading in threshold voltage may be dominant depending on the actual values, which depend on the actual sizes used.

This also applies to transistors biased in weak inversion. For weak inversion the term $(V_{GS} - V_T)/2$ has to be substituted by nkT/q. Remember that these terms are nothing more than the g_m/I_{DS} values of the transistor, independent of weak or strong inversion.

A plot version weak inversion coefficient is shown next.

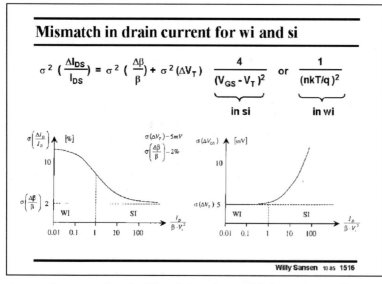

Mismatch in drain current for wi and si

$$\sigma^2\left(\frac{\Delta I_{DS}}{I_{DS}}\right) = \sigma^2\left(\frac{\Delta\beta}{\beta}\right) + \sigma^2(\Delta V_T) \quad \frac{4}{(V_{GS} - V_T)^2} \quad \text{or} \quad \frac{1}{(nkT/q)^2}$$

$$\text{in si} \qquad \text{in wi}$$

Willy Sansen 10-05 1516

1516

A plot of the mismatch of the drain current version weak inversion coefficient is shown below. This is the ratio of the drain current to the si/wi transition current (see Chapter 1). In terms of beta, this transition current is about $2n\beta(kT/q)^2$ or 0.002 β. Note that $V_t = kT/q$.

It is clear that for weak inversion the spreading in threshold voltage is dominant. For strong inversion, the spreading in beta is dominant, most probably

due to the spreading in W and L values, whichever is smaller.

The crossover region is quite large however, as can be expected near the cross-over between weak and strong inversion (explained in Chapter 1).

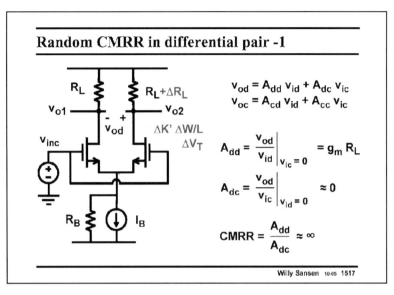

Random CMRR in differential pair -1

$$v_{od} = A_{dd} \, v_{id} + A_{dc} \, v_{ic}$$
$$v_{oc} = A_{cd} \, v_{id} + A_{cc} \, v_{ic}$$

$$A_{dd} = \left. \frac{v_{od}}{v_{id}} \right|_{v_{ic}=0} = g_m \, R_L$$

$$A_{dc} = \left. \frac{v_{od}}{v_{ic}} \right|_{v_{id}=0} \approx 0$$

$$CMRR = \frac{A_{dd}}{A_{dc}} \approx \infty$$

Willy Sansen 10-05 1517

1517

Besides the offset voltage, a differential pair has another specification, which also reflects the influences of the spreadings. It is the CMRR or Common-mode Rejection Ratio.

Remember (from Chapter 3) that a differential pair has two inputs, which are better converted into a differential input v_{id} and a common-mode (or average) input v_{ic}. The same applies to the outputs. In this way insight can be built up about the actual operation.

As a result, we find four different gains. Up till now we have concentrated on the differential-to-differential gain A_{dd}. This gain is easy to calculate as it is the differential output voltage v_{od} obtained for a differential input voltage v_{id}, when the common-mode input voltage v_{ic} is zero. In this case the differential-to-common-mode gain A_{dc} does not play a role.

If the differential pair is driven with a common-mode voltage however, the gain A_{dc} may come in. It is defined as the differential output voltage v_{od} obtained for a common-mode input voltage v_{ic}, when the differential input voltage v_{id} is zero. This situation is sketched in this slide. A common-mode input voltage v_{inc} is applied and the differential output voltage v_{od} is measured.

The ratio of the two gains is the CMRR. It is infinity if gain A_{dc} (and not A_{cc}!) is zero. Note also that the CMRR does not play a role for a purely differential drive ($v_{ic}=0$).

1518

In order to calculate the CMRR, we need to calculate the gain A_{dc}. A common-mode input voltage v_{inc} is then applied and the differential output voltage v_{od} is measured.

It is clear that no differential output voltage v_{od} can be detected if no delta's occur. Both input Gates and the common-Source point carry the same signal. The currents in both transistors are now the same and for equal load resistors R_L the output voltages are also the same. The differential output voltage is then zero.

Let us assume that a difference in load resistor is now present. Both transistors are still equal. In this, case the input voltage v_{inc} causes a small current i_c to flow through the output resistance R_B of the current source. This current is divided equally through both transistors and reaches

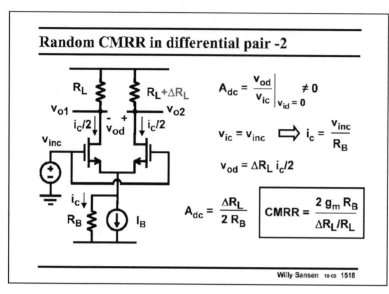

Random CMRR in differential pair -2

$$A_{dc} = \frac{v_{od}}{v_{ic}}\bigg|_{v_{id}=0} \neq 0$$

$$v_{ic} = v_{inc} \implies i_c = \frac{v_{inc}}{R_B}$$

$$v_{od} = \Delta R_L\, i_c/2$$

$$A_{dc} = \frac{\Delta R_L}{2\,R_B} \qquad \boxed{CMRR = \frac{2\,g_m\,R_B}{\Delta R_L/R_L}}$$

Willy Sansen 10-05 1518

the output resistors. A differential output voltage v_{od} thus develops as given in this slide. Gain A_{dc} is thus readily obtained.

Division by the differential gain $g_m R_L$, yields the CMRR.

It is clear that the CMRR depends on the output resistance of the current source. It can be made large by use of cascodes.

As an example, for a $g_m R_B$ of 30 and a $\Delta R_L/R_L$ of 1%, the CMRR is about 6000 or 75 dB.

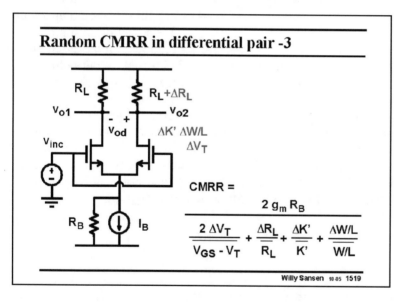

Random CMRR in differential pair -3

$$\Delta K' \; \Delta W/L$$
$$\Delta V_T$$

$$CMRR = \frac{2\,g_m\,R_B}{\dfrac{2\,\Delta V_T}{V_{GS}-V_T} + \dfrac{\Delta R_L}{R_L} + \dfrac{\Delta K'}{K'} + \dfrac{\Delta W/L}{W/L}}$$

Willy Sansen 10-05 1519

1519

If all other parameter spreadings are included, an expression is found for the CMRR as shown in this slide. Note again that we find four terms, which add up algebraically. The sum is never reached and neither is an average of zero. The same combination of terms is found as for the offset.

There must therefore be a simple relationship between offset and CMRR, as shown next.

Relation random offset and CMRR

$$v_{OSr} = \Delta V_T + \frac{V_{GS} - V_T}{2} \left(\frac{\Delta R_L}{R_L} + \frac{\Delta K'}{K'} + \frac{\Delta W/L}{W/L} \right)$$

$$CMRR_r = \frac{2\, g_m\, R_B}{\dfrac{2\, \Delta V_T}{V_{GS} - V_T} + \dfrac{\Delta R_L}{R_L} + \dfrac{\Delta K'}{K'} + \dfrac{\Delta W/L}{W/L}}$$

$$v_{OSr}\, CMRR_r = \frac{V_{GS} - V_T}{2}\, 2\, g_m\, R_B = I_B\, R_B = V_E L_B = 5 \ldots 15\ V$$

$$v_{OSr}\, CMRR_r = \qquad\qquad\qquad\qquad 10\ V$$

Willy Sansen 10-05 1520

1520

Copying the expressions of the random offset and the CMRR shows that their product cancels out all the delta terms. Moreover, the product can be simplified as the input devices are involved in an obvious way, and also the current source transistor.

The resulting parameter is merely the Early Voltage $V_E L_B$ of the current source. Its value depends on the channel length L_B chosen. If an average value for $V_E L_B$ is taken of about 10 V, then some important conclusions can be drawn.

Relation random offset and CMRR

$$v_{OSr}\, CMRR_r \approx V_E L_B \approx 10\ V \quad (\sim L_B)$$

10 mV	60 dB	≈ 10 V	as for MOSTs
1 mV	80 dB	≈ 10 V	as for Bipolar transistors
10 μV	120 dB	≈ 10 V	with trimming : with laser
			with Zener zap
			with fusible links

Low offset = High CMRR

Willy Sansen 10-05 1521

1521

For an average value for $V_E L_B$ of 10 V, it becomes clear that decreasing the offset or increasing the CMRR is the same design task.

If an offset can be expected of about 10 mV, as for many MOST differential pairs and opamps, then a CMRR of approximately 60 dB can be expected.

If, on the other hand, the offset is 10 times smaller, as for bipolar opamps, then the CMRR is 20 dB higher as well.

If the offset is trimmed down to the μV level, then the CMRR increases accordingly. Note however, that a CMRR of 120 dB can only be reached provided the offset is trimmed down to 10 μV. This is not an easy task whatever technique is used.

Table of contents

Willy Sansen 10 05 1522

1522

Up till now only random effects have been examined. They lead to a random offset and random CMRR, which are linked by each other.

Systematic errors can also occur. In general, they are the result of systematic asymmetries. In principle, they can be avoided by proper design, as they are systematic. They give rise to a systematic offset and systematic CMRR which are also linked.

Some examples are given next.

Systematic offset in current mirror

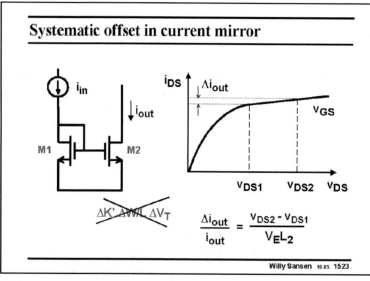

$$\frac{\Delta i_{out}}{i_{out}} = \frac{v_{DS2} - v_{DS1}}{V_E L_2}$$

Willy Sansen 10 05 1523

1523

A first source of systematic error is caused by the systematic asymmetry of a current mirror. Even without random errors, a difference in Drain-Source voltage v_{DS}, causes a small difference in output current Δi_{out}, which always has the same sign.

It can easily be calculated, leading to the expression in this slide.

The larger the channel length, the more horizontal the curves are and the smaller the output current difference. It can never be made zero however, as it is just about impossible to make v_{DS1} exactly equal to v_{DS2}.

1524

Another source of systematic error is a result of the common-mode drive v_{inc} of a differential pair, as shown in this slide. The systematic asymmetry of the current mirror gives a differential output current i_{OUT}. This current can then be compensated at the input by a differential input voltage or offset voltage v_{osc}.

In order to calculate the offset voltage required to compensate for the output current caused by the common-mode input voltage v_{inc}, a small signal equivalent circuit is sketched.

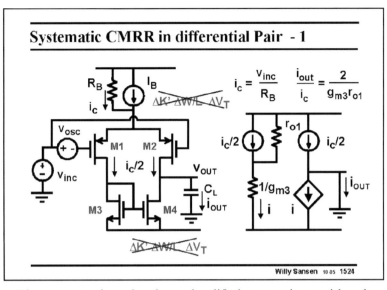

The common-mode input voltage v_{inc} causes a current i_c to flow through the output resistance R_B of the current source. Half of this current flows through both input devices. They can now be represented by two current sources with value $i_c/2$.

The load can be taken to be a short to ground, as intermediate frequencies are considered. The current through it is the differential output current i_{OUT}. As a result, only r_{o1} has to be included.

The current mirror has been simplified to a resistor with value $1/g_{m3}$. The current through it is mirrored to the output.

From this simplified equivalent circuit, the output current is easily calculated. It evidently depends on g_{m3} and r_{o1}.

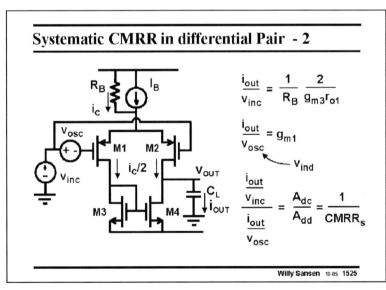

1525

This output current i_{OUT}, can now be referred back to the input by simple division by g_{m1}. The offset voltage v_{osc} is merely this output current divided by g_{m1}.

Note however, that the ratio of the output current caused by the common-mode voltage to the output current caused by the differential input voltage or offset voltage, is nothing else than the inverse of the $CMRR_s$. This is the link between systematic offset voltage and systematic $CMRR_s$.

The actual expression is given next.

1526

Substitution of several gains by means of the transistor parameters yields the $CMRR_s$. Its value is quite large as two $g_m r_o$ products are multiplied.

Again, the product of the CMRR and the offset amounts to some constant value, which is the

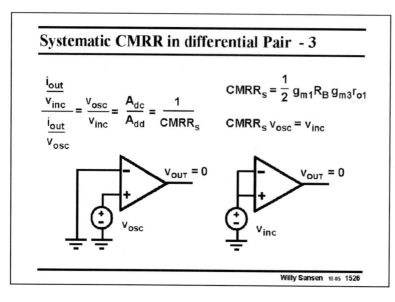

common-mode input voltage in this case. Its value is quite limited. As a result, high values of common-mode rejection can only be reached provided the offset is quite small.

It is now also clear that the output voltage of an opamp can be made zero either by application of a differential input voltage, which is by definition the offset voltage v_{osc}, or by application of a common-mode input voltage v_{inc}. Their ratio is then the CMRR.

An easy measurement technique for the CMRR consists of the application of a certain differential input voltage v_{osc}, and by measurement of the common-mode input voltage v_{inc} required to return the output voltage to the original value. Their ratio is again the CMRR.

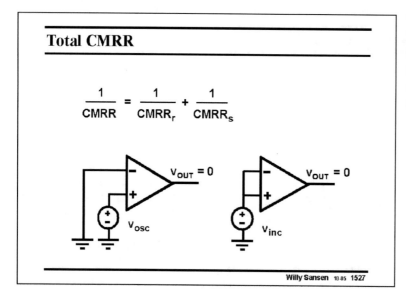

1527

However, in this way both the random $CMRR_r$ and the systematic $CMRR_s$ are measured. The smaller one always dominates.

1528

A CMOS Miller OTA is shown in this slide. The offset of such an amplifier is examined next.

The input devices are normally pMOSTs so that they can share the same n-well bulk. In this way mismatch in substrate parameter γ does not come in. This parameter does come in however, for the nMOSTs. This is why the term ΔV_{T3} has an asterisk. It is larger without the effect of the substrate parameter.

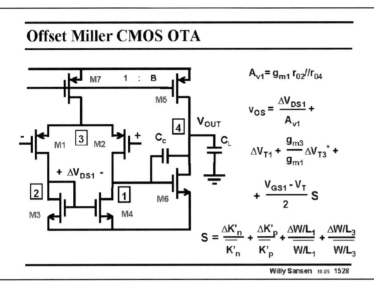

The offset voltage is given in this slide. First of all, it includes any difference between nodes 1 and 2, which is called ΔV_{DS1}. This difference can be caused by a difference between V_{GS6} and V_{GS3}. It also includes the large AC voltage swing at node 1, which is much smaller on node 2.

The second and third term are caused by the mismatches between the V_T's. The last term includes the mismatches between the sizes and the K' factors.

They are obviously scaled by $V_{GS1} - V_T$.

For a small value of $V_{GS1} - V_T$ and for a small g_{m3}/g_{m1}, the term ΔV_{T1} is probably dominant if ΔV_{DS1} can be kept small. If not, its contribution to the offset is $\Delta V_{DS1}/A_{v1}$ or $V_{OUT}/A_{v1}A_{v2}$.

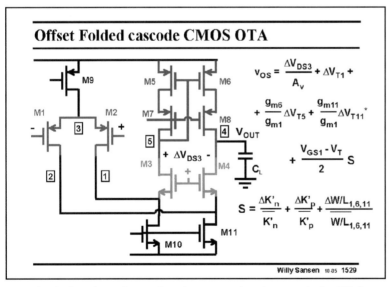

1529

A folded cascode OTA is shown in this slide. Its offset voltage is examined next.

The offset voltage v_{os} is given in this slide. It firstly includes any difference between nodes 4 and 5, which is called ΔV_{DS3}. It is mainly the large AC voltage swing at node 4, which is much smaller on node 5.

The next three terms are caused by the mismatches between the V_T's. They are equally important depending on the g_m's. The last term includes the mismatches between the sizes and the K' factors. They are obviously scaled by $V_{GS1} - V_T$. The cascodes do not come in.

This clearly shows that the offset of a folded cascode can be quite large, as it is made up by the spreading of three differential pairs.

Similar conclusions can be drawn as for the CMOS Miller opamp.

Table of contents

Ref.: Laker, Sansen : Design of analog ..., MacGrawHill 1994

Willy Sansen 10-05 1530

1530

The expression of the CMRR may have led to the conclusion that the output resistance of the current source, feeding the differential pair, is of utmost importance. This is true, at least at low frequencies.

At intermediate and high frequencies, the output capacitance of this current source is of even more importance, as shown next.

CMRR vs frequency

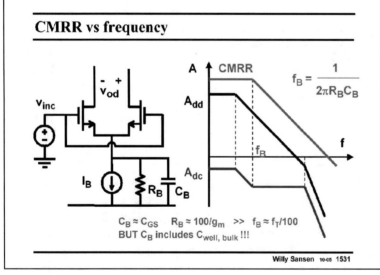

$f_B = \dfrac{1}{2\pi R_B C_B}$

$C_B \approx C_{GS}$ $R_B \approx 100/g_m$ >> $f_B \approx f_T/100$
BUT C_B includes $C_{well, bulk}$!!!

Willy Sansen 10-05 1531

1531

The current source of a differential pair has both an output resistance R_B and capacitance C_B. Its size depends mainly on the Drain-Bulk capacitance C_{DB} of the current source transistor. Its value is close to that of the C_{GS} of that transistor (as explained in Chapter 1). This capacitance C_B also includes the capacitance $C_{well,bulk}$ between the well, in which both input transistors are imbedded, and the substrate. It can therefore be a lot larger than the C_{GS} of the current source transistor.

As a result, a new break frequency f_B shows up. It occurs as a zero in the characteristic of the gain A_{dc}, but as a pole in one of the CMRR. Calculation of this frequency depends on the values of these different capacitances. It will be somewhere between the dominant pole of the amplifier and a fraction of the f_T frequency of the current source transistor.

The best way to design a differential pair with high CMRR at high frequencies is to provide it with a current source with the minimum size of drain area. A small square device is optimal, most probably requiring a high value of $V_{GS} - V_T$, which is good for its f_T as well!

Table of contents

- Random offset and $CMRR_r$
- Systematic offset and $CMRR_s$
- CMRR versus frequency
- Design rules
- Comparison MOST and bipolar transistors

Willy Sansen 10-05 1532

1532

Mismatch is determined mainly by size and a number of other design rules, which are now discussed.

It is already known that matching improves with the square root of the size. This is the main criterion. It always works. There are many other layout conditions however, which help in the reduction of mismatch.

They all work to some extent, even when proof is not easy to find in the literature. They are summarized in 10 rules, as shown next.

Layout rules for low offset

1. Equal nature
2. Same temperature
3. Increase size
4. Minimum distance
5. Same orientation
6. Same area/perimeter ratio
7. Round shape
8. Centroide layout
9. End dummies
10. Bipolar always better !

Hastings,
"The Art of Analog Layout"
Prentice Hall 2001
R. Soin, .."A-D Asics, .. "
Peregrinus, 1991

Willy Sansen 10-05 1533

1533

The first rule for good matching is that the components must be of equal nature. It is, for example, impossible to match a resistor to a $1/g_m$ value.

Another example is that matching between a MOST capacitance and a junction capacitance, will not work either.

Many other examples can be found.

Layout rules for low offset

1. Equal nature
2. Same temperature
3. Increase size
4. Minimum distance
5. Same orientation
6. Same area/perimeter ratio
7. Round shape
8. Centroide layout
9. End dummies
10. Bipolar always better !

Willy Sansen 10.05 1534

1534

A second condition for good matching is that both devices must be on the same isotherm.

Most large chips operate at high temperatures. Silicon is a good thermal conductor and yet temperature differences are easily generated, for example, power devices operate on one end of the chip and the input transistors of an opamp are on the other side.

Such a situation is sketched next.

On same isotherm

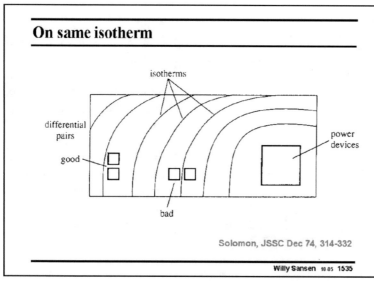

Solomon, JSSC Dec 74, 314-332

Willy Sansen 10.05 1535

1535

Power devices are located on the right. They heat up the chip on the right side, generating isotherms on the other side of the chip.

The input transistors of an opamp or any other differential circuit is better laid out on this isotherm. If not, an internal thermal resistance is active, which limits the open-loop gain in exactly the same way as a feedback resistor would (see reference).

This applies not only to static isotherms but also dynamic ones, for example when power switches (or relays) are integrated on the chip.

Temperature gradients can also induce stress in the chip. The same applies to the package. Matched components have to be laid out on iso-stress lines, which may be hard to be identified!

Layout rules for low offset

1. Equal nature
2. Same temperature
3. Increase size
4. Minimum distance
5. Same orientation
6. Same area/perimeter ratio
7. Round shape
8. Centroide layout
9. End dummies
10. Bipolar always better !

Willy Sansen 10-05 1536

1536

The most important rule for good matching is to increase the size.

Let us first have a look at how size plays for resistors and capacitors. For MOSTs we know already that spreading on V_T (and the other parameters) is inversely proportional to the square root of WL. In other words, if both the channel width W and the channel length L are multiplied by 2, then the spreading (delta, sigma) is divided by 2.

We will see that this also applies to resistors but not to capacitors.

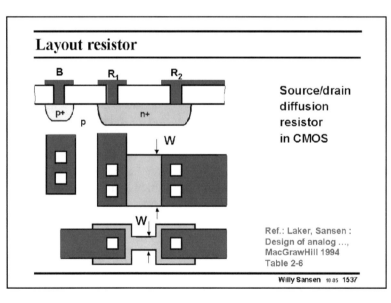

Layout resistor

Source/drain diffusion resistor in CMOS

Ref.: Laker, Sansen : Design of analog ..., MacGrawHill 1994 Table 2-6

Willy Sansen 10-05 1537

1537

Resistors are little more than a conductive island contacted at both extremities. Many different resistors can be realized as many different diffused or ion implanted islands can be realized.

Some examples are given in this slide. Normally the smaller dimension, which is W in the example at the bottom, determines the matching which can be achieved.

In bipolar technologies, several diffusion layers can be identified, such as the ones for the base, for the emitter, for the collector, etc. In CMOS technologies however, the source and drain diffusion is always available. Also the well-diffusion can be used for high-valued resistances. They are all listed next.

1538

This table lists the most common resistances in both bipolar and CMOS technologies.

The sheet resistivities are given, followed by the absolute accuracies and some more specifications.

It is clear that in bipolar technologies a wide variety of resistivities is available. The ones with

Table resistors

Process	Type	$\rho\square$ $\Omega/\square$	absolute accuracy percent	temperature coefficient percent/°C	voltage coefficient percent/V	breakdown voltage V
Bipolar	base diffusion	150	10	0.12	2	50
	emitter diffusion	10	20	0.02	0.5	7
	pinch resistance	5 k	40	0.33	5	7
	epi layer	1 k	10	0.3	1	60
	aluminum	50 m	20	0.01	0.02	90
	ion-implantation	2 k	1	0.02	0.2	20
	ion-implantation	200	0.3	0.02	0.05	20
CMOS	S/D diffusion	20-50	20	0.2	0.5	20
	well	2.5 k	10	0.3	1	20
	poly gate	50	20	0.2	0.05	40
	poly resistance	1.5 k	1	0.05	0.02	20
	aluminum	50 m	20	0.01	0.02	90
Thin film	NiCr(Ta)	200	1	0.005	0.005	90
	aluminum	50 m	20	0.01	0.02	90

Willy Sansen 10-05 1538

the highest accuracy however, are the ion-implanted ones. They are an addition to a standard process however, and are not always available.

In CMOS technologies, the source/drain diffusion yields a resistance of low value which is very imprecise. The well is only a little better. The only precise one is the poly-silicon resistor. Its sigma has an A_R of about 0.04 μm. Again, it is an addition to a standard CMOS process and is therefore expensive. It is not available in a digital CMOS process.

Thin film resistances can also be added on top of the silicon structure. They are normally realized with Tantalum or Nichrome. They are very precise but mainly used for trimming. Aluminum metallization is less precise by lack of control of its thickness. Copper is now also used.

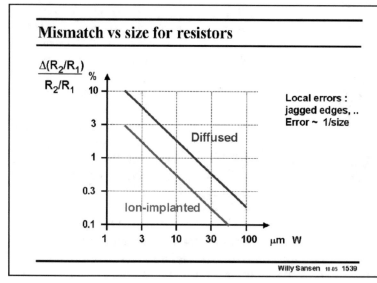

Willy Sansen 10-05 1539

1539

For resistors, the absolute and relative accuracy decrease with size. Above the relative accuracy is sketched versus linear dimension, for a fixed W/L ratio, in which W is the smaller dimension.

This accuracy is about inversely proportional to size, as for MOST devices. This is not surprising at all as MOST devices are actually resistances.

The reason for this dependency is that local errors dominate. They have jagged and rounded edges, and many more local deficiencies in the definition of the layout.

Also, note that ion implanted resistances are a lot better than diffused ones, because of the higher reproducibility involved.

Finally, note that resistances only provide limited accuracy for average sizes. If a minimum dimension is taken to be about 10 μm, then about 0.5% error can be expected. This corresponds to a signal-to-error or signal-to-distortion ratio of about 200 or 46 dB. Divided by 6 this 46 dB yields little over 7 bit. This means that resistive ladders are easy to lay out as more than 7–8 bits accuracy is not required. Many 8 bits ADC's are still realized in this way (see Chapter 20).

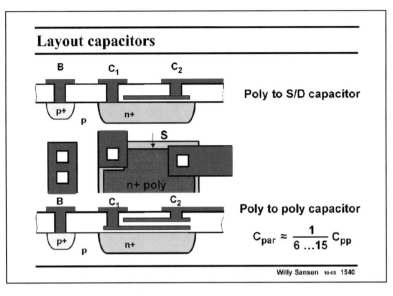

1540
Some layouts of capacitances are shown in this slide.

The top one is a (n+ doped) poly-to-diffusion capacitance C_{pp}. It uses the Gate oxide as a dielectric. It is not all that attractive as it has a large series resistance in the diffusion layer. Moreover, it has a large parasitic capacitance C_{par} between the diffusion layer and the substrate.

A double poly-layer has much less series resistance but still a fair amount of parasitic capacitances. Moreover a double poly-layer is an addition to standard CMOS processing, and is therefore expensive.

Nowadays, CMOS processes offer many metal layers on top. Each pair of metal layers can be used as a capacitance. The main criteria are then the reproducibility of the dielectric thickness and the parasitic capacitance to all other layers. A wide choice is now available.

Table capacitors

Process	Type	C nF/cm^2	absolute accuracy percent	temperature coefficient percent/°C	voltage coefficient percent V	breakdown voltage V
Bipolar	C_{CB}	16	10	0.02	2	50
	C_{EB}	50	10	0.02	1	7
	C_{CS}	8	20	0.01	0.5	60
CMOS	C_{ox}(50 nm)	70	5	0.002	0.005	40
	$C_{m,poly}$	12	10	0.002	0.005	40
	$C_{poly,poly}$	56	2	0.002	0.005	40
	$C_{poly,substrate}$	6.5	10	0.01	0.05	20
	$C_{m,substrate}$	5.2	10	0.01	0.05	20
	$C_{poly,substrate}$	6.5	10	0.01	0.05	20

Ref.: Laker, Sansen : Design of analog ..., MacGrawHill 1994 Table 2-7

Willy Sansen 10-05 1541

1541
Some typical values are collected in the Table in this slide.

In bipolar technologies all capacitances are junction capacitances. They are heavily voltage dependent.

Only CMOS technologies offer good capacitances. The best one is still the Gate oxide capacitance. For a 50 nm oxide thickness (which corresponds to a 2.4 μm CMOS process), the capacitance is quite high. It is even ten times higher for a 0.25 μm CMOS process.

Several other capacitances are gives. Do not forget that many more can be added, depending on the number of metal layers available.

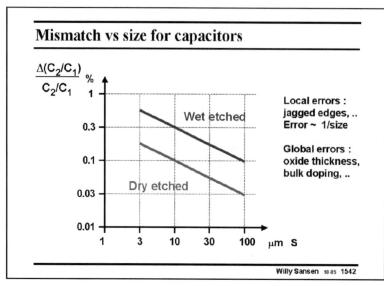

Mismatch vs size for capacitors

$$\frac{\Delta(C_2/C_1)}{C_2/C_1} \%$$

Local errors :
jagged edges, ..
Error ~ 1/size

Global errors :
oxide thickness,
bulk doping, ..

Willy Sansen 10.05 1542

1542

The curve of the relative accuracy versus size is not as steep as for resistors but gives smaller values. Parameter S stands for the side of a square capacitor.

The slope is about half of the one for resistors. The reason is that now a combination is found of local and global errors. Global errors are related to slowly changing oxide thicknesses from one side of a wafer to the other side, doping levels, under-etching, etc. This combination leads to a lower slope.

Note that dry etching defines the capacitors in a better way than wet etching as used before.

Note also that capacitors provide higher accuracy than resistors. If square capacitors are taken with a side of 10 µm, then about 0.1% error can be expected. This corresponds to a signal-to-distortion ratio of about 1000 or 60 dB. Divided by 6 this 60 dB yields about 10 bit.

This means that capacitive ladders can be laid out with 10 bit accuracy. Many 10–12 bits ADC's are realized in this way (see Chapter 20). For higher values, a very large number of capacitances have to be used, but 14 bit has been reached!

Layout rules for low offset

1. Equal nature
2. Same temperature
3. Increase size
4. Minimum distance
5. Same orientation
6. Same area/perimeter ratio
7. Round shape
8. Centroide layout
9. End dummies
10. Bipolar always better !

Willy Sansen 10.05 1543

1543

In addition to size, distance between two devices plays a role, albeit less than in earlier technologies.

From the Table with S_{VT} and S_{WL}, it has become clear that the requirement of short distance has been relaxed. Indeed CMOS processing has been carried out on ever increasing wafer sizes, reaching about 12 inch at the moment, but 15 inch wafers are surely being planned.

As a result, the processing technologies have become more homogeneous over these large wafers. This is why distance on a short range, does not play the same role as before. Only when large fractions of a chip are being covered, then distance will play a role as explained under point 8 centroide layout. Examples are large capacitance banks aiming at 12–14 bits accuracy.

Layout rules for low offset

1. Equal nature
2. Same temperature
3. Increase size
4. Minimum distance
5. Same orientation
6. Same area/perimeter ratio
7. Round shape
8. Centroide layout
9. End dummies
10. Bipolar always better !

Willy Sansen 10-05 1544

1544

A more important point for good matching has to do with crystal orientation. A crystal never has exactly the same crystal structure (density, default density, ...) in different directions. As a result, the mobility and thus K′ parameter will not be quite the same in two different directions. This is illustrated next.

Matching of transistor pairs

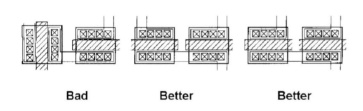

| Bad | Better | Better |

Willy Sansen 10-05 1545

1545

In the first example, the direction of the current in the left transistor is perpendicular to the direction of the current in the second transistor. This is indeed bad for matching.

The other two examples are better ones. The actual positioning of the connections may cause some minor difference in Source contact resistance, but this is not that evident.

Layout rules for low offset

1. Equal nature
2. Same temperature
3. Increase size
4. Minimum distance
5. Same orientation
6. Same area/perimeter ratio
7. Round shape
8. Centroide layout
9. End dummies
10. Bipolar always better !

Willy Sansen 10-05 **1546**

1546

An important improvement in layout style towards good matching is achieved by use of the same area/perimeter ratio, or simply by use of the same shapes.

In this way, the relative amount of jagging of the edges, and of rounding is always the same, as shown next.

Matching of current mirrors

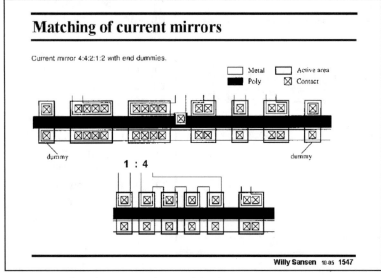

Current mirror 4:4:2:1:2 with end dummies.

Metal / Poly / Active area / Contact

dummy   dummy

1 : 4

Willy Sansen 10-05 **1547**

1547

A number of matched transistors is shown for a current mirror with multiple outputs. The relative transistor sizes are 4: 4: 2: 1: 2, when the first and last dummy transistors are excluded. The second transistor with relative size 4 is connected as a diode indeed.

These ratios won't be very accurate however. The rounding of the corners for a transistor with size 4, is relatively speaking, much less important than the same rounding for a transistor with size 1.

A better solution is to use the same shapes for all transistors and to connect them in parallel, as shown below. This 1:4 ratio will be much more accurate as local errors all have the same relative influence: their area to perimeter ratio is always the same.

This is a typical bipolar transistor layout style. More space is evidently required but we know already that more space leads to better matching.

Even better would be to lay out the single transistor in the middle of the four transistors, so that they have the same point of gravity, as explained in point 8 on the centroide layout.

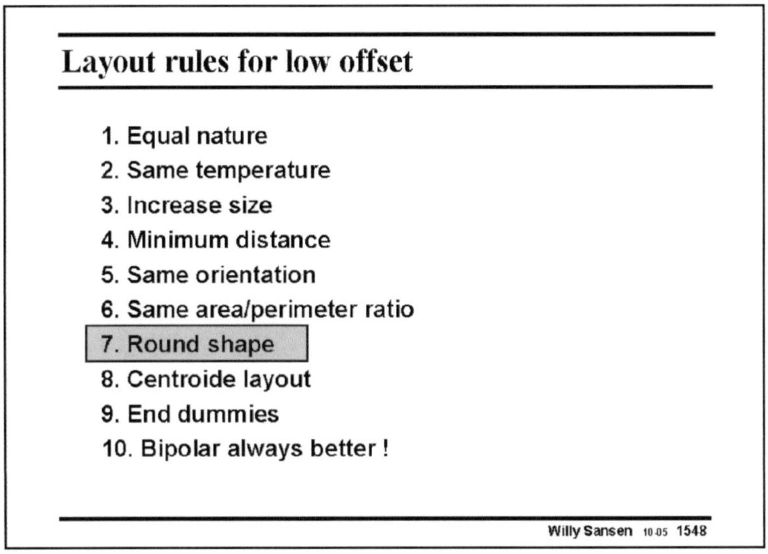

Layout rules for low offset

1. Equal nature
2. Same temperature
3. Increase size
4. Minimum distance
5. Same orientation
6. Same area/perimeter ratio
7. Round shape
8. Centroide layout
9. End dummies
10. Bipolar always better !

Willy Sansen 10-05 1548

1548

A nice way to avoid rounding of corners is to make round transistor shapes. Surrounding the Drain of a MOST with its Gate, and adding a Source all around has been known to yield excellent matching.

Unfortunately, not all layout systems allow such round shapes. An orthogonal or hexagonal shape is not that good!

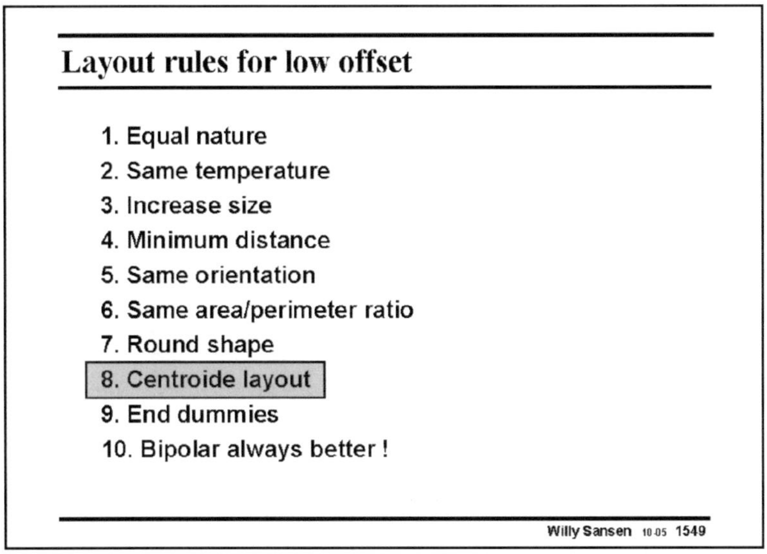

Layout rules for low offset

1. Equal nature
2. Same temperature
3. Increase size
4. Minimum distance
5. Same orientation
6. Same area/perimeter ratio
7. Round shape
8. Centroide layout
9. End dummies
10. Bipolar always better !

Willy Sansen 10-05 1549

1549

A very important layout style for low mismatch is centroide layout.

This means that all matched structures must have the same point of gravity. In this way the effects of global changes (in oxide thickness, ...) are averaged out.

This is illustrated next.

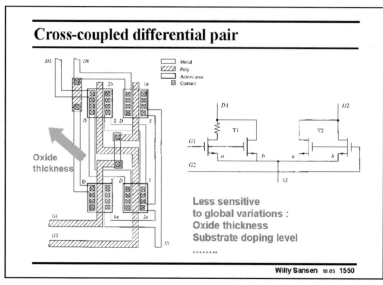

1550
The layout is shown of a differential pair, each transistor of which consists of two equal transistors, connected in parallel but laid out in opposite corners. As a result, the effects of global changes, for example, in oxide thickness, are averaged out. For example, MOST 2b has the lowest K' but MOST 2a the highest one. Putting them in parallel gives an average value for K' which is about the same as for MOSTs 1a and 1b.

Care must be take however, not to insert additional resistances in the Source connections, to connect the transistor terminals. Judicious use of the two or more layers of interconnect is thus mandatory.

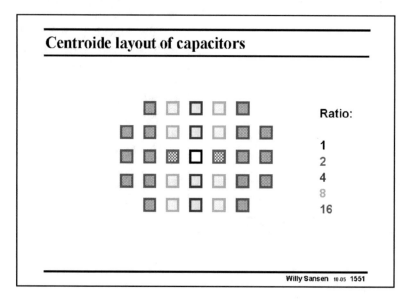

1551
Another example of centroide layout with capacitors is shown in this slide.

A capacitance bank is made with ratios 1:2:4:8:16, etc. The unit capacitor is in the middle. The two capacitors are on both sides.

They are connected in parallel.

The four capacitors are on either end as well. The eight capacitors have as a point of gravity the unit capacitor in the middle, etc. In this way global errors are averaged out.

For a large capacitance bank, a large number of unit-capacitors have to be put in parallel, on either side of the unit capacitor in the middle. At this point it is better to lay out all the unit-capacitors randomly distributed over the whole area. In this way 14 accuracy has been obtained (Van der Plas, JSSC Dec. 99, 1708–1718).

Layout rules for low offset

1. Equal nature
2. Same temperature
3. Increase size
4. Minimum distance
5. Same orientation
6. Same area/perimeter ratio
7. Round shape
8. Centroide layout
9. End dummies
10. Bipolar always better !

Willy Sansen 10-05 1552

1552

In a series of equal structures, the first and last ones never quite match. The reason is that the first and last ones see different neighbors. The effects of processing steps such as under-etching, contact holes, etc. will be different. This is why dummy transistors (or capacitors) have to be added at the beginning and at the end. They are not used. They are dummies.

This is illustrated next.

Matching of current mirrors

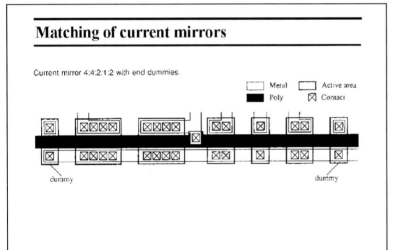

Current mirror 4:4:2:1:2 with end dummies.

Metal · Active area
Poly · ⊠ Contact

dummy   dummy

Willy Sansen 10-05 1553

1553

In this series of transistors, which form a multiple current mirror, the first and last transistors are not connected. They are dummies. They are only present to make sure that the processing is more homogeneous for all the other transistors in between.

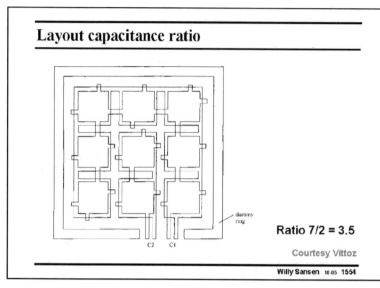

Layout capacitance ratio

Ratio 7/2 = 3.5

Courtesy Vittoz

Willy Sansen 10.05 1554

1554

A good example of a layout in which all rules are properly applied is shown in this slide. It consists of nine equal capacitors, to which tabs have been added. These tabs are used to connect some of the capacitors. In this way a ratio of 7/2 is realized.

All tabs are always present even if they are not used. In this way the parasitic capacitances associated with the four tabs are always the same.

All unit capacitances have the same shape: they have the same area/perimeter ratio.

Moreover, they are laid out in centroide form. The two capacitors are in the middle, surrounded by the seven other ones. They have about the same point of gravity.

Finally, the whole structure is surrounded by a dummy ring to make sure that all capacitors see the same neighbors.

Layout rules for low offset

1. Equal nature
2. Same temperature
3. Increase size
4. Minimum distance
5. Same orientation
6. Same area/perimeter ratio
7. Round shape
8. Centroide layout
9. End dummies
10. Bipolar always better !

Willy Sansen 10.05 1555

1555

All rules in this slide lead to better matching, which means lower offset and higher CMRR. Whatever is being tried in CMOS technology, bipolar is always doing better for the same area. There are several reasons for this.

Table of contents

- **Random offset and CMRR$_r$**
- **Systematic offset and CMRR$_s$**
- **CMRR versus frequency**
- **Design rules**
- **Comparison MOST and bipolar transistors**

Willy Sansen 10 05 1556

1556

Comparison of the models of MOST and bipolar technologies show that MOSTs are actually modulated resistances, whereas bipolar transistors employ a forward biased diode as an input. A bipolar transistor does not have a threshold voltage V_T and always has an exponential I_{CE}/V_{BE} relationship. Its current is a small current I_S multiplied by the voltage V_{BE}, scaled by kT/q.

This is worked out further next.

Offset of MOST and bipolar transistors

MOST :
$$v_{OS} = \Delta V_T + \frac{V_{GS} - V_T}{2} \left(\frac{\Delta R_L}{R_L} + \frac{\Delta K'}{K'} + \frac{\Delta W/L}{W/L} \right)$$

Bipolar :
$$v_{OS} = \frac{kT}{q} \left(\frac{\Delta R_L}{R_L} + \frac{\Delta I_S}{I_S} \right) \text{ is much smaller !!}$$

1) no V_T

2) $kT/q \ll (V_{GS}-V_T)/2$

3) Drift decreases with v_{OS} : $\dfrac{\Delta v_{OS}}{\Delta T} = \dfrac{v_{OS}}{T}$

Bipolar : Base current !

Willy Sansen 10 05 1557

1557

Indeed, the expression of the offset of a bipolar transistor does not include the effect of the spreading of threshold voltage V_T. Moreover the scaling factor by which parameters such as $\Delta R_L/R_L$, etc. come in is only kT/q, whereas it is $(V_{GS} - V_T)/2$ for a MOST.

These are two important reasons why the offset voltage v_{os} for a bipolar is so much smaller.

Moreover, the drift of the offset voltage v_{os} with temperature is well controlled. The derivative of v_{os} to T is the absolute value of v_{os} divided by T. Trimming the offset voltage v_{os} to a small value at the same time reduces the drift with temperature. This is not at all true for MOSTs, in which there is no relationship between offset and offset drift.

As a result, a bipolar transistor is an excellent choice for high-temperature applications, where a low offset is required. If MOSTs have to be used, they will have to be supplemented by offset cancellation circuitry, carrying out chopping or auto-zeroing (Ref. Enz, Temes, Proc. IEEE, Nov. 96, 1584–1614).

The main problem of bipolars, on the other hand, is that they have base currents. This is discussed next.

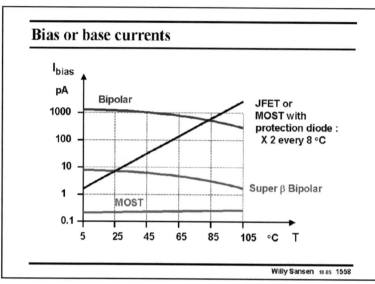

The base currents in bipolar transistors can be quite high. Shown in this slide are the input currents for a number of opamps in which the input devices are realized in different technologies.

The currents in conventional bipolar opamps are of the μA level. Their base currents are therefore of the nA level. Since the beta increases with temperature, the base currents decrease with temperature, which is clearly an advantage in power applications. These base currents can be decreased even further by use of super-beta devices. They have beta values above 3000. As a result, the base currents are smaller. On the other hand, they cannot take collector voltages above a few Volts.

MOSTs have the lowest input currents, at least if they do not have a protection device. Such device includes some diodes, which have a leakage current which increases drastically with temperature (×2 every 8 degrees). This is similar to a Junction-FET.

A conventional bipolar transistor has the highest base current. Circuit techniques can be devised to compensate these currents. Some of them are discussed next.

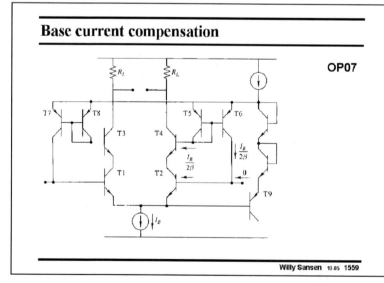

An older solution to provide input current compensation is shown in this slide.

We rely on the matching between transistors T1 and T3. In this case their base currents can be expected to be about the same. A current mirror senses the current into the base of transistor T3 and injects the same current into the base of transistor T1.

A voltage clamp is present between the emitters of transistors T5–8 and the emitters of the input transistors T1,2. As a result the collector-emitter voltage across the input transistors T1,2 never exceeds about 0.7 V. Super-beta devices can now be used for T1,2.

The base currents are compensated by this additional circuitry. No current is required externally. In practice, there is always some input current flowing, depending on mismatching between T1 and T2 and between T7 an T8.

The main disadvantage of this circuit however, is that two different base current cancellation circuits are used on either side of the differential pair. The noise of these additional circuits is now injected at the input terminals. The noise performance is poor.

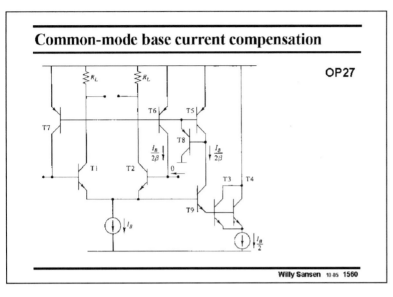

Common-mode base current compensation

Willy Sansen 10-05 1560

1560

A better solution is to construct one common current generator and apply it to both input terminals. The noise generated by this block then cancels out at the differential output.

This is illustrated in this slide.

We rely on the matching between transistors T1,2 and T3,4. The base current is sensed of the latter ones by cascode transistor T9, mirrored by transistors T5–8 and injected into the input bases.

The noise is now generated by all transistors T3–5 and T8–9 and is cancelled out at the differential output. Only the noise by transistors T6 and T7 is still differential and thus injected in the signal path.

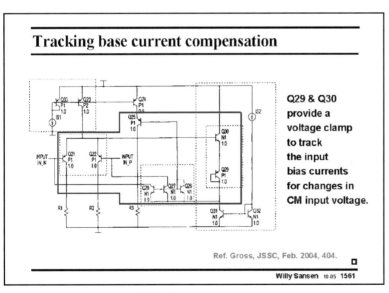

Tracking base current compensation

Q29 & Q30 provide a voltage clamp to track the input bias currents for changes in CM input voltage.

Ref. Gross, JSSC, Feb. 2004, 404.

Willy Sansen 10-05 1561

1561

In order to realize a more precise compensation of the base currents, a more accurate current mirror is required. This means that the transistors which carry out the current mirroring must have the same currents, the same beta's and the same collector-emitter voltages v_{CE}. This latter requirement is fulfilled in the circuit in this slide.

The actual current compensation is realized if the input transistors Q21,22 are matched to transistor Q25. Its base current is mirrored by current mirror Q26–28 and fed to the input transistor bases.

In order to ensure that v_{CE25} equals the input $v_{CE21,22}$, a voltage clamp of about 1.4 V (or 2 V_{BEon}) is introduced with transistors Q29,30. This clamp (or bootstrap circuit) senses the common-mode input voltage at the Sources of the input transistors Q21,22 and keeps the Sources of the current mirror transistors Q26–28 at a constant 0.7 V (or V_{BEon}) below the common-mode input voltage. As result, all the transistors within this bootstrap loop (blue frame) follow the common-mode input voltage.

The base of Q25 is therefore always the same voltage as the average (or common-mode) input voltage. Its collector current is also the same as for transistors Q21,22 and also its v_{CE}, because of equal resistors R1–3.

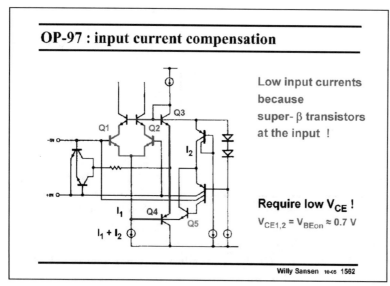

1562
A similar voltage clamp is used in the circuit in this slide. Transistors Q3 and Q4 maintain a 0.7 V voltage drop across the input transistors Q1,2. These devices are normally super-beta transistors. Their v_{CE}'s must therefore be small indeed.

Their base currents are drawn from a current mirror, which derives its input current from the base of transistor Q5. This transistor must therefore be well matched to the input devices. Indeed, it is the same as for the input devices. Its current is also the same.

The actual currents are shown on the next slide.

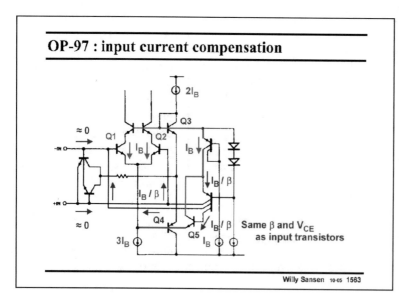

1563
All relevant currents are indicated. The DC current in one single input transistor Q1 (or Q2) is denoted by I_B. As a result, the DC currents in all transistors are I_B as well. The base current of transistor Q5 is now I_B/β, and so are the currents sent to the bases of the input transistors.

The external input currents will therefore be close to zero.

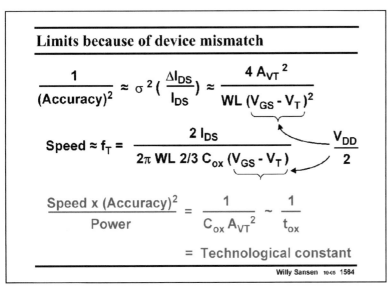

Limits because of device mismatch

$$\frac{1}{(\text{Accuracy})^2} \approx \sigma^2 \left(\frac{\Delta I_{DS}}{I_{DS}} \right) \approx \frac{4 A_{VT}^2}{WL (V_{GS} - V_T)^2}$$

$$\text{Speed} \approx f_T = \frac{2 I_{DS}}{2\pi WL \; 2/3 \; C_{ox} (V_{GS} - V_T)} \qquad \frac{V_{DD}}{2}$$

$$\frac{\text{Speed} \times (\text{Accuracy})^2}{\text{Power}} = \frac{1}{C_{ox} A_{VT}^2} \sim \frac{1}{t_{ox}}$$

$$= \text{Technological constant}$$

Willy Sansen 10-05 1564

1564

It is clear from this slide that mismatch has become the main limitation to the Dynamic Range that can be obtained at a specific frequency and a specific power level.

If the accuracy is limited by the spreading on the threshold voltage only, then it can be described approximately by the top expression.

Moreover, the speed is related to the f_T of the transistor, which has been derived in Chapter 1.

As a result, the Speed Accuracy product for a specific amount of power consumption is determined by factors which are constant within a specific CMOS technology. Within the same technology it is not possible to go beyond a certain signal-to-distortion ratio for a certain speed and power consumption. It is assumed that this distortion is set by mismatch.

Moreover, as A_{VT} is proportional to the oxide thickness, this product improves for deeper submicron or nanometer CMOS. Obviously, this is somehow to be expected!

This means that nanometer CMOS is able to provide lower power levels with similar signal-to-distortion ratios and given specific frequencies.

Limits because of device noise

$$S/N = \frac{V_{pp}^2/2}{4kT \, R \, BW} \qquad\qquad S/N = \frac{V_{pp}^2/8}{kT/C}$$

$$P_{min} = \frac{V_{pp}^2}{R} \qquad\qquad P_{min} = V_{DD} \, BW \, V_{pp} \, C$$

$$P_{min} \approx 8kT \, BW \, S/N$$

Willy Sansen 10-05 1565

1565

It is clear that noise also establishes a fundamental limitation to the Dynamic Range that can be obtained at a specific frequency and a specific power level.

The limit is calculated for both the thermal noise generated by a resistor R and for the noise bandwidth limited by a capacitance C.

In both cases, V_{pp} is the maximum peak-to-peak voltage that can be obtained. It is taken to be the same as the supply voltage V_{DD}. It cancels out of the expression of the minimum power consumption. Only the Signal-to-noise ratio S/N and the bandwidth are left in this expression.

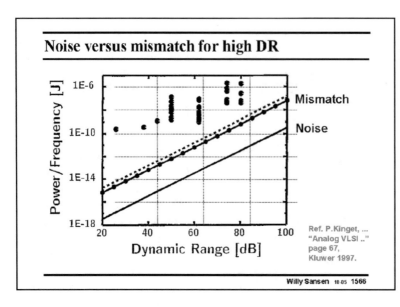

1566

Both the Dynamic Range limited by distortion (mismatch) and by noise are given in the graph in this slide. The power consumption required to carry out a certain function (filter, ADC, ...) at a certain frequency is plotted versus the required dynamic range. The curves are calculated with the expressions given before.

A number of experimental points are given as well. They are taken from papers in the *IEEE Journal of Solid-State Circuits*. They reflect data from ADC's, from continuous-time and switched-capacitor filters (without the power consumption due to the clocks).

It is clear that mismatch data presents a more realistic estimate of the power/frequency ratio for a certain dynamic range than the noise does. Mismatch is now a more severe limit to the accuracy of analog integrated circuits than noise.

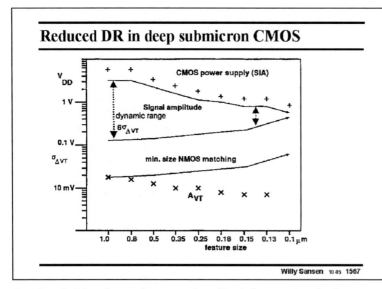

1567

As a conclusion to this Chapter, another point of concern is given about the maximum dynamic range for deep submicron CMOS.

For ever smaller channel lengths, the supply voltage is shrinking, as predicted by the SIA roadmap (see Chapter 1). The maximum signal amplitude is a constant fraction of the supply voltage, determined by the distortion allowed.

The parameter A_{VT} describing the spreading on the threshold voltage decreases but if minimum-size devices are taken, the spreading on the V_T increases. If six times this spreading is taken, then only a small voltage dynamic range is left. It seems to go to zero for CMOS technologies beyond 90 nm.

A few obvious applications can live with such small dynamic ranges. Some biomedical applications are happy with 20 dB, but most communication applications require more than 70 dB! To reach such values, the supply voltage cannot be allowed to decrease, the distortion must be reduced (see Chapter 18) and larger than minimum-size transistors will have to be used.

The analog parts of a mixed-signal chip will consume more and more of the total area, as has already been seen.

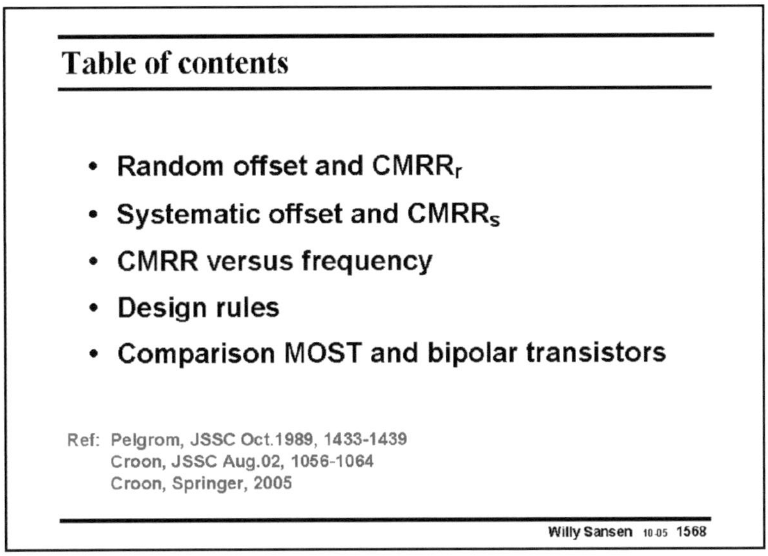

Table of contents

* **Random offset and CMRR$_r$**
* **Systematic offset and CMRR$_s$**
* **CMRR versus frequency**
* **Design rules**
* **Comparison MOST and bipolar transistors**

Ref: Pelgrom, JSSC Oct.1989, 1433-1439
Croon, JSSC Aug.02, 1056-1064
Croon, Springer, 2005

Willy Sansen 10-05 1568

1568

In this Chapter an overview of the mechanisms are given which play a role in mismatch. Both random and systematic mismatch have been discussed.

It is shown that mismatch leads to specification such as offset and CMRR, which are thus related to each other.

Layout techniques are listed which can be applied to improve matching. Inevitably they lead to larger chip sizes.

Finally a short comparison is provided with bipolar transistor matching. Also, an attempt is made to establish the limits in power for dynamic range and frequency in both cases, when mismatch is a limit and when noise is a limit. It is shown that mismatch is the most severe.

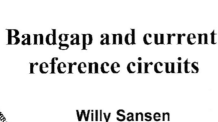

Bandgap and current reference circuits

Willy Sansen

KULeuven, ESAT-MICAS

Leuven, Belgium

willy.sansen@esat.kuleuven.be

Willy Sansen 10-05 161

161

Bandgap references are directly derived from the bandgap of silicon. This is why they provide the only real voltage reference, which is available. It is about 1.2 V.

A current reference does not actually exist. They are derived from the bandgap voltage references and one or two resistors.

In this Chapter, we will see how a voltage reference can be realized, the absolute value of which is highly accurate. Moreover, its temperature coefficient can be reduced to ppm's per degree Celsius. As a consequence, they can be used over very large temperature ranges.

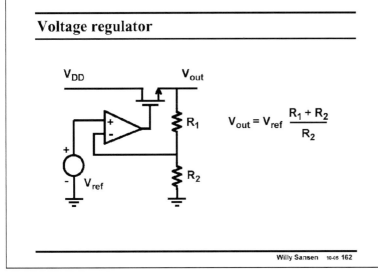

Voltage regulator

$$V_{out} = V_{ref} \frac{R_1 + R_2}{R_2}$$

Willy Sansen 10-05 162

162

First of all, we have to look at what voltage references are actually used for.

They are used in Analog-to-digital converters. They can also be used in both voltage and current regulators. Both schematics are given.

A voltage regulator locks the output voltage to the reference voltage by use of a resistor ratio. Actually it is a two-stage feedback amplifier, with the reference voltage V_{ref} as an input. The first stage is the opamp, whereas the second stage is a source follower. This follower can deliver a large current to the load, depending on its W/L ratio.

The load is not shown. It is usually a combination of resistances and capacitances, which can vary over a very wide range, depending on the current drawn from the regulator.

The supply voltage V_{DD} usually contains a ripple, which will be suppressed by the regulator. The accuracy of the output voltage depends on the accuracy of the resistor ratio and the absolute accuracy of the reference voltage. This resistor ratio can have a smaller error than 0.1% (see Chapter 15), if they are large in area. The final accuracy will therefore depend on the absolute error of the reference voltage. We will see that this can also be 0.1%!

457

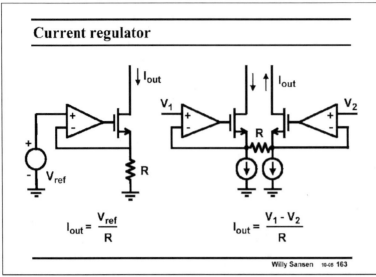

Current regulator

$$I_{out} = \frac{V_{ref}}{R}$$

$$I_{out} = \frac{V_1 - V_2}{R}$$

Willy Sansen 10-05 163

163
A current regulator is easily derived from a voltage reference.

A single-ended version is shown on the left; a differential one on the right.

In both cases, the reference voltage is converted into a current by use of an opamp and a resistor. The absolute accuracy of the output current will depend on both the absolute accuracies of the voltage reference and of the resistor. The latter one is by far the worst.

In the section on current references, several examples are given of which resistances can be used. However, the absolute accuracy will always be a problem.

Table of contents

Ref.: B.Gilbert, G.Meijer, ACD , Kluwer 1995

Willy Sansen 10-05 164

164
The physics of a bipolar transistor is reviewed to examine where the absolute accuracy and temperature coefficient are coming from. A correction circuit is added to equalize the output voltage over temperature.

Several realizations are now discussed in both bipolar and CMOS technologies.

For supply voltages below 1 V, it is still possible to use the same principle, but some more resistors must be added.

Finally, a current reference is derived from the bandgap reference, by proper use of a few resistances.

165

A bandgap reference voltage uses a bipolar transistor, connected as a diode. Its current-voltage expression is then quite accurately given by the exponential. A real pn-junction may have a coefficient of 1.05–1.1 in front of the kT/q; a bipolar-transistor connected as a diode does not.

For a constant-current drive, this diode exhibits a large dependence on temperature. It is about -2 mV/°C. We intend to reduce this value to less than 1/1000th!

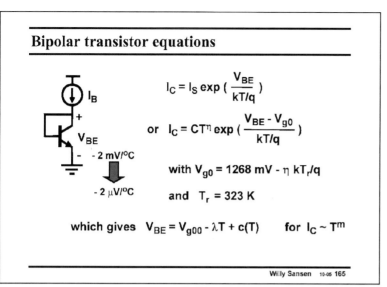

Bipolar transistor equations

$$I_C = I_S \exp \left(\frac{V_{BE}}{kT/q} \right)$$

$$\text{or} \quad I_C = CT^\eta \exp \left(\frac{V_{BE} - V_{g0}}{kT/q} \right)$$

$$\text{with } V_{g0} = 1268 \text{ mV} - \eta \, kT_r/q$$

$$\text{and} \quad T_r = 323 \text{ K}$$

$$\text{which gives} \quad V_{BE} = V_{g00} - \lambda T + c(T) \qquad \text{for } I_C \sim T^m$$

Willy Sansen   10-05 165

In order to do so we need an expression of the current in which the temperature is shown explicitly. The voltage V_{g0} is the diode voltage at zero absolute temperature (Kelvin). It depends on temperature by itself. The values are given for a reference temperature T_r of 323 K or 50°C. This has been chosen to be the middle of the temperature range of interest. This is from 0 to 100°C.

For this range, the values given in this slide are good empirical approximations. Parameter η is about 4. Then the actual value of V_{g0} is about 1.156 V (kT_r/q is about 28 mV).

If we want the current to be dependent on the temperature by exponent m, then the base-emitter voltage V_{BE} can be written as shown in this slide. A linear dependence on the temperature emerges with slope λ, and a correction factor $c(T)$, called the curvature.

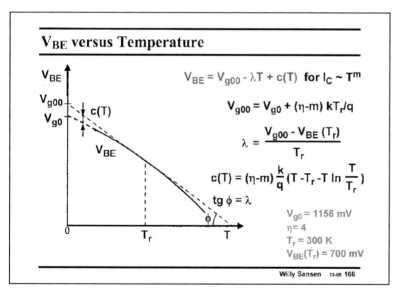

V_{BE} versus Temperature

$$V_{BE} = V_{g00} - \lambda T + c(T) \quad \text{for } I_C \sim T^m$$

$$V_{g00} = V_{g0} + (\eta - m) \, kT_r/q$$

$$\lambda = \frac{V_{g00} - V_{BE}(T_r)}{T_r}$$

$$c(T) = (\eta - m) \frac{k}{q} \left(T - T_r - T \ln \frac{T}{T_r} \right)$$

$$\tan \phi = \lambda$$

$V_{g0} = 1156$ mV
$\eta = 4$
$T_r = 300$ K
$V_{BE}(T_r) = 700$ mV

Willy Sansen   10-05 166

166

A new zero-temperature V_{g00} emerges, which is the extrapolated value, as shown in this slide. It is always larger than V_{g0}. For a constant current (m=0), its value is about 1.268 V.

It is also clear that the curvature is a complicated function of temperature but very much resembles a second-order function, a parabola. Its largest value is a zero Kelvin, and is 112 mV if m=0. It is even smaller if we allow the current to have a larger value of m.

This curvature is sketched separately on the next slide.

167

From this sketch it is clear that this curvature is not always important.

For a current, which is independent of temperature (m=0), the curvature correction is only

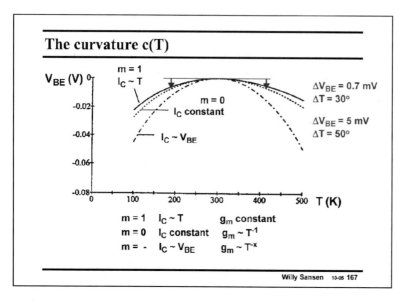

The curvature c(T)

m = 1 $I_C \sim T$
m = 0 I_C constant
m = - $I_C \sim V_{BE}$

$\Delta V_{BE} = 0.7$ mV $\Delta T = 30°$
$\Delta V_{BE} = 5$ mV $\Delta T = 50°$

m = 1	$I_C \sim T$	g_m constant
m = 0	I_C constant	$g_m \sim T^{-1}$
m = -	$I_C \sim V_{BE}$	$g_m \sim T^{-x}$

Willy Sansen   10-05 167

about 5 mV, on 50°C distance from the reference temperature T_r (which is 300 K in this example). This is of the same order of magnitude of mismatch. It is therefore negligible if the temperature range is not all that large.

If we allow the current to have a positive temperature coefficient (m = 1), then this curvature correction is even smaller. This is the case if we need the (bipolar transistor) transconductance g_m to be independent of temperature, for example.

Since, for many applications, the curvature term is negligible, let us concentrate on the linear term with coefficient λ.

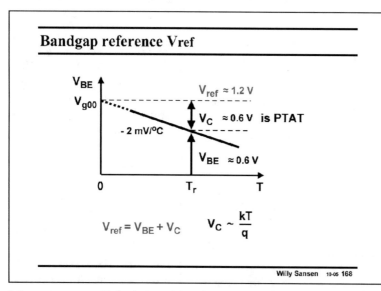

Bandgap reference Vref

V_{BE}
V_{g00}
$V_{ref} \approx 1.2$ V
$V_C \approx 0.6$ V is PTAT
- 2 mV/°C
$V_{BE} \approx 0.6$ V

$V_{ref} = V_{BE} + V_C$ $V_C \sim \dfrac{kT}{q}$

Willy Sansen   10-05 168

168

For a constant current, the voltage V_{BE} across the diode-connected transistor decreases with temperature in a linear way with slope coefficient λ. Its value is about -2 mV/°C.

If we now find a way to add a voltage V_C to this diode voltage V_{BE}, which is Proportional To the Absolute Temperature (PTAT) with the same slope λ, then we obtain a reference voltage V_{ref} which is independent of temperature.

Moreover, we will find that this reference voltage is the bandgap voltage itself. It provides a high absolute accuracy!

Since both V_{BE} and V_C are of similar size, the reference voltage will be around 1.2 V. Since it is difficult to predict the actual voltage V_{BE}, we will need to trim the added voltage V_C such that the reference voltage is constant around our reference temperature T_r. This is the same as saying that we will need to trim the added voltage V_C such that the curvature is symmetrical with respect to the reference temperature T_r.

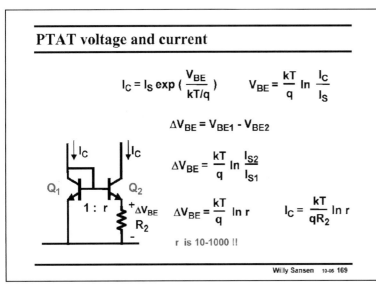

PTAT voltage and current

$$I_C = I_S \exp\left(\frac{V_{BE}}{kT/q}\right) \qquad V_{BE} = \frac{kT}{q} \ln \frac{I_C}{I_S}$$

$$\Delta V_{BE} = V_{BE1} - V_{BE2}$$

$$\Delta V_{BE} = \frac{kT}{q} \ln \frac{I_{S2}}{I_{S1}}$$

$$\Delta V_{BE} = \frac{kT}{q} \ln r \qquad I_C = \frac{kT}{qR_2} \ln r$$

r is 10-1000 !!

Willy Sansen   10-05 169

169

Now let us build a voltage reference with this diode.

We need to generate a circuit which provides a voltage V_C which is PTAT.

Such a circuit is shown in this slide. It consists of a bipolar transistor current mirror. Transistor Q2 is much larger (by a factor r) than transistor Q1 such that its V_{BE} is smaller. This difference in voltage ΔV_{BE} is taken up by a resistor R_2.

The equations show that the voltage across R_2 is PTAT. The current through it is also PTAT. Obviously, this is only true if transistor R_2 has a negligible temperature coefficient.

Note also that both transistor currents I_C are made equal.

For example, if r = 10, then ΔV_{BE} is about 60 mV. For a resistor of $R_2 = 2 \text{ k}\Omega$, the currents are 30 μA.

Clearly this current must fall in the region where the exponential current voltage relationship is precisely exponential, for both transistors. The current density of the smaller transistor Q1 is much higher, which may cause some mismatch problems.

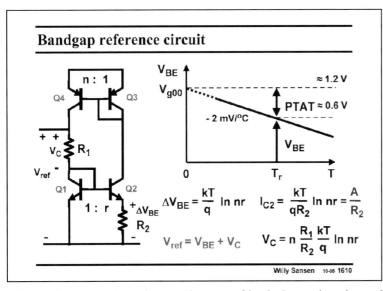

Bandgap reference circuit

$$\Delta V_{BE} = \frac{kT}{q} \ln nr \qquad I_{C2} = \frac{kT}{qR_2} \ln nr = \frac{A}{R_2}$$

$$V_{ref} = V_{BE} + V_C \qquad V_C = n \frac{R_1}{R_2} \frac{kT}{q} \ln nr$$

Willy Sansen   10-05 1610

1610

Remember that both transistor currents I_C must be made equal. This can be achieved by another current mirror on top. Moreover, this pnp current mirror can be used to add another current ratio n.

The result is that now both ΔV_{BE} and the currents are PTAT, with a factor nr. The added voltage V_C is then easily found to be PTAT and proportional to a resistor ratio, which can now be accurately attained.

The reference voltage is now the sum of both. It can be trimmed by adjusting resistor R_1. The resulting value will be around 1.2 V.

Two more specifications are important in such a bandgap reference. The first one is the output impedance. It indicates if current can be pulled out from the reference. For this purpose an emitter follower is usually added, or an additional current mirror, as illustrated in the realizations later on.

The other characteristic is the output noise. Since this reference voltage is probably used to bias a number of circuits, its output noise risks entering the circuitry. This has to be avoided.

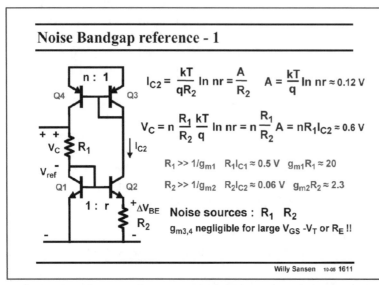

Noise Bandgap reference - 1

$$I_{C2} = \frac{kT}{qR_2} \ln nr = \frac{A}{R_2} \qquad A = \frac{kT}{q} \ln nr \approx 0.12 \text{ V}$$

$$V_C = n\frac{R_1}{R_2}\frac{kT}{q} \ln nr = n\frac{R_1}{R_2}A = nR_1I_{C2} \approx 0.6 \text{ V}$$

$R_1 \gg 1/g_{m1}$ $R_1I_{C1} \approx 0.5$ V $g_{m1}R_1 \approx 20$

$R_2 \gg 1/g_{m2}$ $R_2I_{C2} \approx 0.06$ V $g_{m2}R_2 \approx 2.3$

Noise sources : R_1 R_2

$g_{m3,4}$ **negligible for large $V_{GS} - V_T$ or R_E !!**

Willy Sansen   10-05 1611

1611

The main question with respect to noise, is whether to use large currents with small resistors or vice-versa.

The expressions for the current and the added voltage V_C are repeated in this slide. Parameter A is introduced to shorten the writing. It is about 0.12 V, corresponding to a nr product of 100.

The voltage V_C is still about 0.6 V.

The main noise sources are the two npn transistors Q1 and Q2, and the two resistors R_1 and R_2. The noise contributions of the resistors dominate because they are larger than the $1/g_m$ values of the transistors.

The noise of the top transistors Q4 and Q5 can be neglected by proper choice of their $V_{GS} - V_T$ or by addition of series resistors.

It is clear that by choosing larger currents, the resistors must be smaller. Is this advantageous for the output noise voltage?

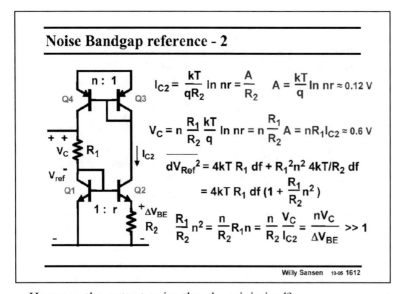

Noise Bandgap reference - 2

$$I_{C2} = \frac{kT}{qR_2} \ln nr = \frac{A}{R_2} \qquad A = \frac{kT}{q} \ln nr \approx 0.12 \text{ V}$$

$$V_C = n\frac{R_1}{R_2}\frac{kT}{q} \ln nr = n\frac{R_1}{R_2}A = nR_1I_{C2} \approx 0.6 \text{ V}$$

$$dV_{Ref}^2 = 4kT\,R_1\,df + R_1^2n^2\,4kT/R_2\,df$$

$$= 4kT\,R_1\,df\,(1 + \frac{R_1}{R_2}n^2)$$

$$\frac{R_1}{R_2}n^2 = \frac{n}{R_2}R_1n = \frac{n}{R_2}\frac{V_C}{I_{C2}} = \frac{nV_C}{\Delta V_{BE}} \gg 1$$

Willy Sansen   10-05 1612

1612

The equivalent input noise voltage due to the two resistors is given in this slide.

The voltage noise of resistor R_1 appears directly in series with the output. The noise of the other resistor R_2 is transformed towards the output by the resistor ratio and factor n. Its contribution to the output is a lot larger than that of resistor R_1.

Actually, the noise of R_2 is amplified to the output whereas the noise of R_1 is not.

How can the output noise then be minimized?

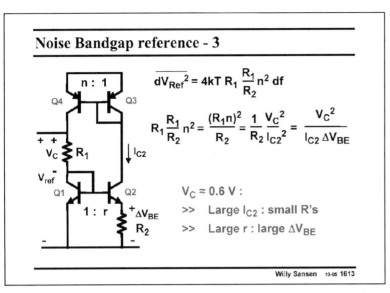

Noise Bandgap reference - 3

$$\overline{dV_{Ref}^2} = 4kT\, R_1\, \frac{R_1}{R_2}\, n^2\, df$$

$$R_1 \frac{R_1}{R_2} n^2 = \frac{(R_1 n)^2}{R_2} = \frac{1}{R_2}\frac{V_C^2}{I_{C2}^2} = \frac{V_C^2}{I_{C2}\,\Delta V_{BE}}$$

$V_C \approx 0.6\ \text{V}$:

$\gg$ Large I_{C2} : small R's

$\gg$ Large r : large ΔV_{BE}

Willy Sansen   10-05 **1613**

1613

If we neglect the noise of R_1 and only focus on the noise generated by R_2, we find that only two parameters play a role. We find that V_C is always about 0.6 V.

The first one is I_{C2}. The larger we make the current I_{C2}, the smaller resistor R_2 becomes, and the smaller its noise at the output.

The other parameter is ΔV_{BE}. A large voltage difference ΔV_{BE} will require a large value of r, but will reduce the output noise.

With this respect, it is better to use a large value of r, than a large value of n.

Table of contents

- **Principles**
- **Bipolar bandgap references**
- **CMOS bandgap references**
- **Bandgap references < 1 V**
- **Current references**
- **LDO Regulators**

Ref.: B. Gilbert, G.Meijer, ACD , Kluwer 1995

Willy Sansen   10-05 **1614**

1614

Now that the principles are known, let us have a look at how they are implemented in bipolar technologies.

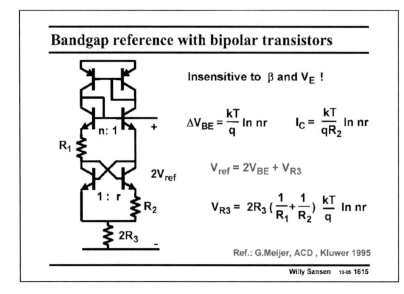

Bandgap reference with bipolar transistors

Insensitive to β and V_E !

$$\Delta V_{BE} = \frac{kT}{q} \ln nr \qquad I_C = \frac{kT}{qR_2} \ln nr$$

R_1

n : 1   +

2V_{ref}

$$V_{ref} = 2V_{BE} + V_{R3}$$

1 : r R_2

$$V_{R3} = 2R_3 \left(\frac{1}{R_1} + \frac{1}{R_2} \right) \frac{kT}{q} \ln nr$$

2R_3

Ref.: G.Meijer, ACD , Kluwer 1995

Willy Sansen   10-05 1615

1615

One of the problems of bipolar current mirrors is that some of the precision is lost because of base currents and output resistances. Moreover, mismatch between the bipolar device sizes leads to offset, and causes error voltages. In this realization, these errors are avoided. The pnp current mirrors on top are is series with npn devices. Also, a double reference voltage is taken to reduce the effect of mismatch. Its output voltage is therefore about 2.4 V.

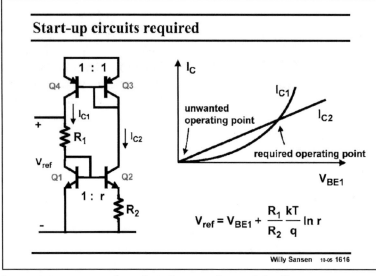

Start-up circuits required

1 : 1

Q4   Q3

I_{C1}

+

R_1 I_{C2}

V_{ref}

Q1   Q2

1 : r

R_2

I_C

I_{C1}

unwanted operating point

I_{C2}

required operating point

V_{BE1}

$$V_{ref} = V_{BE1} + \frac{R_1}{R_2} \frac{kT}{q} \ln r$$

Willy Sansen   10-05 1616

1616

Such a bandgap reference based on two current mirrors has two operating points. A zero current is also perfectly possible. This circuit does not start by itself.

The two operating points are found by plotting the two currents versus the common voltage V_{BE1}. The current of transistor Q2 is more linear as it has a feedback resistor R_2.

The top current mirror has unity gain. The operat-

ing points are now found at the crossings of the lines.

Zero current works as well as the required operating point.

Startup circuits are therefore required.

1617

A few simple startup circuits are shown in this slide.

A capacitance at the base of the pnp current mirror draws a current when the supply voltage is switched on. This current flows through the pnp transistors and starts injecting a current in the bottom npn transistors as well, biasing up the bandgap reference circuit.

However, if for some other reason the current drops to zero, then the supply voltage has to be switched on and off again.

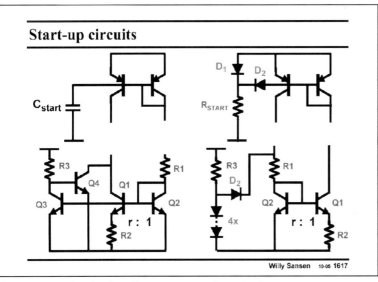

Start-up circuits

Willy Sansen 10-05 1617

The other circuit is better with this respect. When the supply voltage is switched on, diode D_2 is forward biased, drawing current through the pnp transistors, and biasing up the total circuit. However, the current also starts flowing through resistor R_{start}. The voltage across this resistor increases until about 0.7 V below the supply voltage. Diode D_2 is then reverse biased, and disconnected from the actual bandgap circuit. In this way the currents in the bandgap are not disturbed.

A similar arrangement with diodes is shown below.

The startup circuit below left is different. When the supply voltage is turned on, transistor Q4 starts drawing current, biasing up the bandgap reference. This circuit on its turn drives Q3, which switches Q4 off again. As a result transistor Q4 does not influence the current balance in the bandgap circuit.

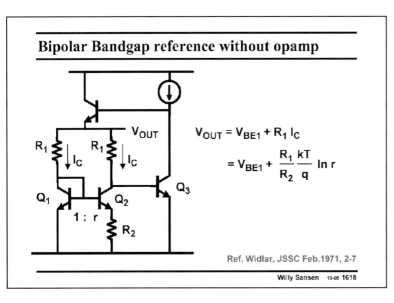

Bipolar Bandgap reference without opamp

$$V_{OUT} = V_{BE1} + R_1 I_C$$

$$= V_{BE1} + \frac{R_1}{R_2} \frac{kT}{q} \ln r$$

Ref. Widlar, JSSC Feb.1971, 2-7

Willy Sansen 10-05 1618

1618

One of the first bandgap references is shown in this slide.

The PTAT current generator Q1-Q2-R2 provides a PTAT voltage across resistor R1. This voltage is added to the V_{BE1} of transistor Q1, towards a bandgap reference voltage indeed.

Again, the output impedance is low, because an emitter follower is used to close the feedback loop at the output.

1619

An all-npn realization of a bandgap circuit is given in this slide. It is less simple but highly symmetrical. Many error terms as a result of too low beta's, output resistances, etc. are cancelled.

A ratio of 100 is used between the sizes of Q3 and Q4, giving a fairly large ΔV_{BE} (of about

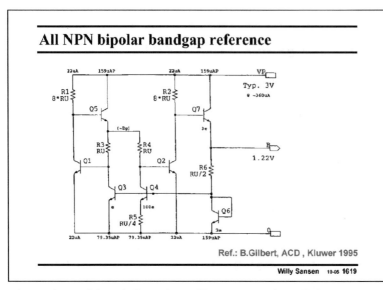

All NPN bipolar bandgap reference

Ref.: B.Gilbert, ACD , Kluwer 1995

Willy Sansen 10-05 **1619**

120 mV). This is good for low sensitivity to offset and noise. The currents are indicated for RU = 6 kΩ.

The currents in both transistors are kept the same because of the equal feedback networks Q1-R1-Q5 and Q2-R2-Q7. The emitters of Q5 and Q7 are therefore at the same voltage, which is the bandgap voltage. Indeed, resistor R6 can here be tuned to provide the exact bandgap voltage of 1.22 V. What is also obviously is the sum of the V_{BE} of transistor Q6 and the voltage V_C across resistor R6.

A low output impedance is obtained because of the use of emitter followers and feedback.

The curvature can be compensated by putting a resistor RU across Q6 and addition of another emitter follower at the output, which lowers the output voltage to about 0.4 V (not shown).

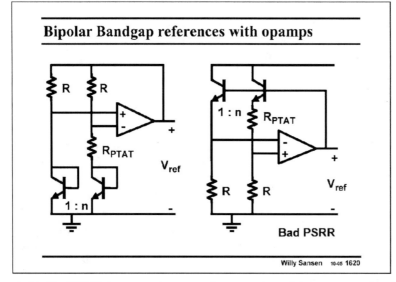

Bipolar Bandgap references with opamps

Bad PSRR

Willy Sansen 10-05 **1620**

1620

Operational amplifiers can provide much more loop gain than a transistor pair.

In this bandgap reference, the opamp finds an output voltage such that the differential input zero becomes zero. If the resistors R are chosen such that they take up about 0.6 V, then the output voltage is about 1.2 V. The output impedance is obviously very low, at least at low frequencies.

The same applies to the bandgap reference on the right. Its PSRR is worse however because transistors are used, with their collectors connected to the power supply.

1621

What is the absolute tolerance that can be obtained with such a bandgap reference?

For the circuit in this slide, the current and reference voltage are copied from before.

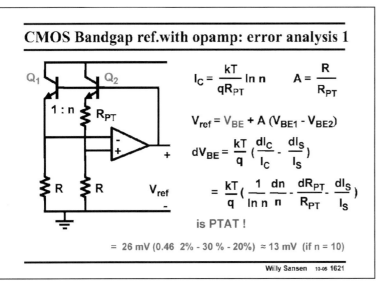

CMOS Bandgap ref.with opamp: error analysis 1

$$I_C = \frac{kT}{qR_{PT}} \ln n \qquad A = \frac{R}{R_{PT}}$$

$$V_{ref} = V_{BE} + A(V_{BE1} - V_{BE2})$$

$$dV_{BE} = \frac{kT}{q}\left(\frac{dI_C}{I_C} - \frac{dI_S}{I_S}\right)$$

$$= \frac{kT}{q}\left(\frac{1}{\ln n}\frac{dn}{n} - \frac{dR_{PT}}{R_{PT}} - \frac{dI_S}{I_S}\right)$$

is PTAT !

$$= 26 \text{ mV } (0.46 \ 2\% - 30 \% - 20\%) \approx 13 \text{ mV (if n = 10)}$$

Willy Sansen 10-05 **1621**

Two terms can be distinguished. The first one is V_{BE}. The second term is $A\Delta V_{BE}$. This takes the first term on this slide.

Taking the total derivative yields three terms, the first one of which is the smallest. The other two are comparable. When we add them we find about 13 mV error, for the numbers given.

This error is PTAT however, and can be trimmed away.

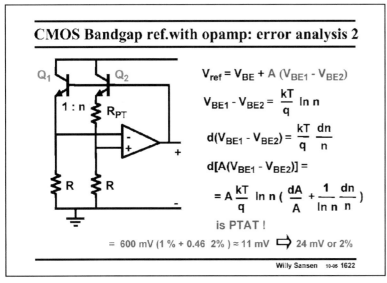

CMOS Bandgap ref.with opamp: error analysis 2

$$V_{ref} = V_{BE} + A(V_{BE1} - V_{BE2})$$

$$V_{BE1} - V_{BE2} = \frac{kT}{q} \ln n$$

$$d(V_{BE1} - V_{BE2}) = \frac{kT}{q}\frac{dn}{n}$$

$$d[A(V_{BE1} - V_{BE2})] =$$

$$= A\frac{kT}{q} \ln n \left(\frac{dA}{A} + \frac{1}{\ln n}\frac{dn}{n}\right)$$

is PTAT !

$$= 600 \text{ mV } (1 \% + 0.46 \ 2\%) \approx 11 \text{ mV } \Rightarrow 24 \text{ mV or 2\%}$$

Willy Sansen 10-05 **1622**

1622

Now we take the total derivative of the second term.

The percentages are now smaller but the scaling factor is V_{BE} rather than kT/q. The absolute value is therefore of the same order of magnitude. It is 11 mV for the numbers given.

When we add both values we obtain 24 mV, which is about 2% of the bandgap reference voltage. It is PTAT gain and can therefore be trimmed by adjusting A for example.

From this analysis, it is clear that the curvature is smaller than the error. Only if trimming is applied, must the curvature be compensated.

Also, the offset of the opamp has not been taken into account. It has the same effect as a difference in V_{BE} of the two transistors. However, this offset can be avoided by using chopper amplifiers.

1623

Operational amplifiers can provide much more loop gain than a transistor pair.

In this bandgap reference, the opamp finds an output voltage such that the differential input zero becomes zero. If the resistors R are chosen such that they take up about 0.6 V, then the

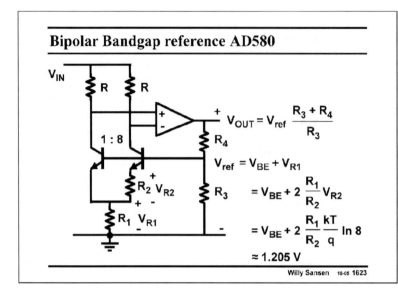

Bipolar Bandgap reference AD580

V_{IN}

R   R

1 : 8

$R_2 V_{R2}$

$R_1 V_{R1}$

R_4

R_3

$$V_{OUT} = V_{ref} \frac{R_3 + R_4}{R_3}$$

$V_{ref} = V_{BE} + V_{R1}$

$$= V_{BE} + 2 \frac{R_1}{R_2} V_{R2}$$

$$= V_{BE} + 2 \frac{R_1}{R_2} \frac{kT}{q} \ln 8$$

$$\approx 1.205 \text{ V}$$

Willy Sansen   10-05 1623

output voltage is about 1.2 V. The output impedance is obviously very low, at least at low frequencies.

The same applies to the bandgap reference on the right. Its PSRR is worse however, because transistors are used, with their collectors connected to the power supply.

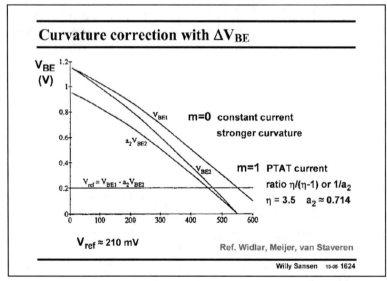

Curvature correction with ΔV_{BE}

V_{BE} (V)

V_{BE1} **m=0 constant current**
stronger curvature

$a_2 V_{BE2}$

V_{BE2} **m=1 PTAT current**
ratio $\eta/(\eta\text{-}1)$ or $1/a_2$

$V_{ref} = V_{BE1} - a_2 V_{BE2}$ $\eta = 3.5$ $a_2 \approx 0.714$

$V_{ref} \approx 210 \text{ mV}$

Ref. Widlar, Meijer, van Staveren

Willy Sansen   10-05 1624

1624

Curvature correction is always possible, if needed. It is always based on taking two references with a different curvature. We have seen before that the curvature for a current which is independent of temperature (m=0) is larger than for a PTAT current (m=1). The ratio of the corresponding nonlinearity is about $\eta/(\eta-1)$. This ratio is actually independent of the current levels used.

A weighted addition of the output voltages then allows perfect compensation of the curvature. Curvature compensation is also possible in another way, as shown next.

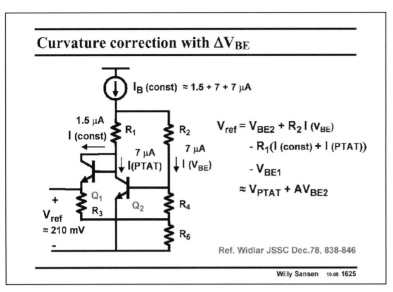

Curvature correction with ΔV_{BE}

I_B (const) $\approx 1.5 + 7 + 7$ μA

1.5 μA
I (const) R_1 R_2 $V_{ref} = V_{BE2} + R_2 I (V_{BE})$

7 μA   7 μA $\quad - R_1(I \text{ (const)} + I \text{ (PTAT)})$
↓I(PTAT) ↓I (VBE)

$\quad - V_{BE1}$

$+$ Q_1 $\quad \approx V_{PTAT} + AV_{BE2}$
V_{ref} R_3 Q_2 R_4
≈ 210 mV

R_6

Ref. Widlar JSSC Dec.78, 838-846

Willy Sansen 10-05 1625

1625

Another way to provide curvature compensation is shown in this slide.

Currents with three different temperature coefficients are available. The total current I_B is independent on temperature.

The expression shows that the reference voltage can also be made independent of temperature. Because it is a combination of currents with different m's and different curvature, it can be made independent of curvature as well.

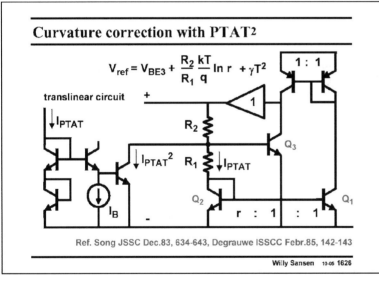

Curvature correction with PTAT²

$$V_{ref} = V_{BE3} + \frac{R_2}{R_1} \frac{kT}{q} \ln r + \gamma T^2$$

1 : 1

translinear circuit $+$

↓I_{PTAT} R_2 Q_3

↓I_{PTAT}^2 R_1 ↓I_{PTAT}

I_B Q_2 Q_1
$-$ r : 1 : 1

Ref. Song JSSC Dec.83, 634-643, Degrauwe ISSCC Febr.85, 142-143

Willy Sansen 10-05 1626

1626

Another way to apply curvature correction is to inject a parabolic current. It has become clear from the earlier slides that the curvature error has a very strong parabolic shape.

Injection of a current with an inverse parabolic shape must be able to correct the curvature.

The derivation of a parabolic or second-order current is carried out by means of the translinear circuit on the left. All transistors work in the weak-inversion region where they have exponential current-voltage relationships. The sum of V_{GS}'s therefore translates in a product of currents. This technique is also used to bias the output transistors of many class-AB stages (see Chapter 12).

A small fraction of the second-order PTAT current is then added to the bandgap voltage. This will now require a second trimming.

1627

The realization of bandgap reference voltages are not so obvious in CMOS technology. A bandgap voltage is essentially connected to a pn diode or rather to a diode connected transistor. This device must have a exponential and reproducible current-voltage characteristic.

Such a device is not easily found in a CMOS technology.

1628

In a n-well technology, both lateral npn and pnp transistors are easily found. They have very low beta's, however. Moreover, the exponential current voltage characteristic is limited to a narrow range of currents.

On the other hand, the vertical or substrate pnp transistor can also be used . Its base is the n-well and its collector is the common p-substrate. As a consequence, all their collectors are connected together.

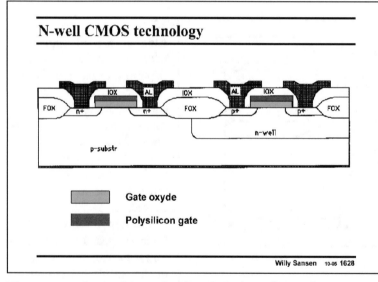

N-well CMOS technology

Gate oxyde

Polysilicon gate

Willy Sansen   10-05 **1628**

They cannot be used in a circuit unless the collector is connected to ground.

Their beta's are reasonable and their output resistances are very high.

1629

Two realizations are shown in this slide.

The collectors are connected to the substrate.

It is obvious that both circuits are only possible in a n-well CMOS technology. This is no problem, as most CMOS technologies are n-well indeed.

The same advantages apply as for the bipolar equivalents.

There is one important difference however. CMOS opamps have a larger offset. This means that they are only self-starting if the offset is right. Moreover, the offset gives a much larger error in the current equalization and therefore in the output voltage.

The circuit on the right has again a worse PSRR.

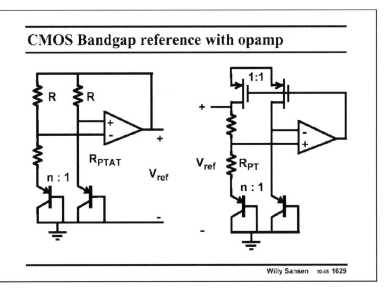

1630
A practical example of a bandgap reference is shown in this slide. The output voltage is trimmed to 1.2 V, with a variation of maximum 20 mV. The temperature coefficient is only 4000 ppm or 0.4% over a very wide temperature range, or 20 ppm/°C.

The PTAT current is generated by transistors Q1/Q2 and resistor R_{PTAT}. Resistors R_1 and R_2 are equal and generate the output voltage, together with R_{TRIM}, which is trimmed.

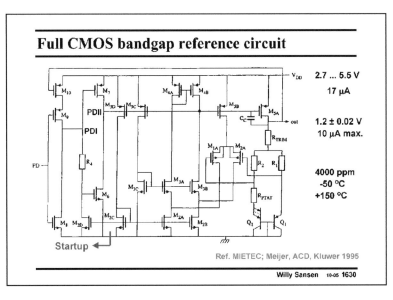

The operational amplifier is a folded cascode, with M1/M2 as input devices and the Drains of M3B/M4B (point PDII) as output. They operate in weak inversion and are now quite large to suppress offset and 1/f noise. It is a two-stage amplifier. A compensation capacitance is therefore required.

The startup circuit is on the left. It includes a Power-Down function. The Drain of transistor M7 (point PDII) biases the amplifier and the bandgap reference. There are two inverters between this point and PD, by means of transistors M8/M9 and M6/M7. When PD is high. The output of the first inverter is low (point PDI) and the output of the next inverter is high (point PDII). All pMOSTs connected at this point are thus off and so is the opamp and bandgap. The output is low.

When PD is low, point PDI is high and PDII is low, biasing all the pMOSTs. The output is regulated by the opamp. Current mirror M2C/M2D and resistor R4 clamp this situation even when PD disappears.

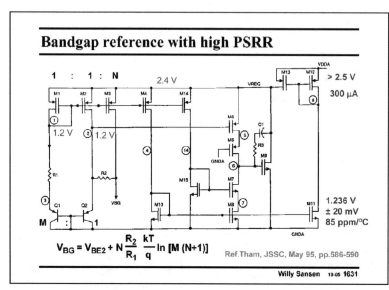

1631

In order to improve the Power-Supply-Rejection Ratio, an internal voltage regulator can be used, as shown in this slide.

The bandgap reference itself is easily recognized on the left.

The goal of the regulator is to ensure that nodes 1 and 2 are at exactly the same voltage. In this way, the PSRR is greatly improved.

For this purpose, a two-stage opamp is used with M5 as an input transistor and M9 as a second stage. As a result, the VREG is adjusted for maximum equality of the voltages at nodes 1 and 2. Clearly, M5 must be matched to M1.

The PSRR is then -95 dB at 1 kHz and still -40 dB at 1 MHz.

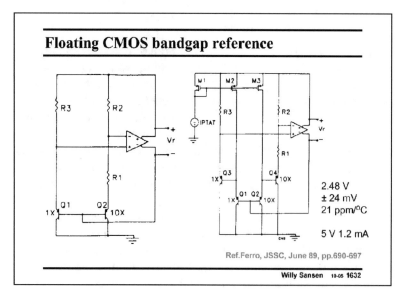

1632

Sometimes a voltage reference is required which is not referred to ground or supply voltage. It is floating. This necessitates a full-differential opamp, as shown in this slide.

The principle is shown on the left. Two pairs of substrate pnp's are used to generate a double bandgap voltage of about 2.4 V on the right. For this purpose, transistors Q1 and Q2 must be biased by a PTAT current, which is not shown here.

1633

It is possible to realize a bandgap reference without resistors. Actually, different sizes of MOSTs are used to amplify differences between diode voltages.

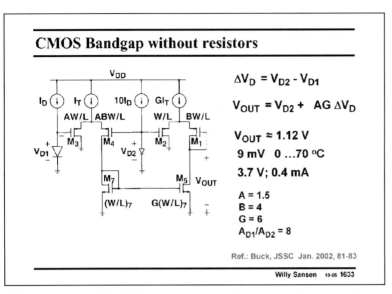

CMOS Bandgap without resistors

$\Delta V_D = V_{D2} - V_{D1}$

$V_{OUT} = V_{D2} + AG\, \Delta V_D$

$V_{OUT} \approx 1.12\ V$

9 mV   0 ...70 °C

3.7 V; 0.4 mA

A = 1.5
B = 4
G = 6
$A_{D1}/A_{D2} = 8$

Ref.: Buck, JSSC Jan. 2002, 81-83

Willy Sansen   10-05 1633

Two diodes are present. A ten times larger current is pushed through the eight times smaller diode D_2 to develop a voltage difference ΔV_D, which is PTAT. This difference is then amplified by a differential pair M3/M4 and mirrored to the output by M7/M5. Another differential pair M1/M2 then converts this current into a voltage again.

The output voltage is a result of many scaling factors A, B and G such that an appropriate PTAT voltage is added to the voltage across diode D_2.

The output voltage is little less than 1.12 V with a variation of only 9 mV over a 70°C temperature range.

1634

Mismatch between the two transistors of the PTAT cell is still a problem. This is why it is better to use the same transistor, provided it can be switched in and out.

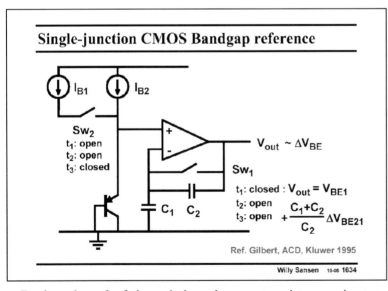

Single-junction CMOS Bandgap reference

I_{B1} I_{B2}

Sw_2
t_1: open
t_2: open
t_3: closed

$V_{out} \sim \Delta V_{BE}$

Sw_1

t_1: closed : $V_{out} = V_{BE1}$
t_2: open
t_3: open $+ \dfrac{C_1 + C_2}{C_2} \Delta V_{BE21}$

C_1 C_2

Ref. Gilbert, ACD, Kluwer 1995

Willy Sansen   10-05 1634

During phase 1 of the switches, the pnp transistors carries current I_{B2} only. The amplifier is in unity gain. The output voltage is therefore V_{BE} only.

During phase 2 of the switches, all switches are open. All voltages are now on hold.

During phase 3 of the switches, the pnp transistor carries a current $I_{B1} + I_{B2}$. The transistor now increases its V_{BE} by ΔV_{BE}. Moreover, the opamp has now a gain, set by the two capacitors. This ΔV_{BE} is therefore amplified and added to the output voltage held before.

As a result, the total output voltage is the bandgap reference voltage.

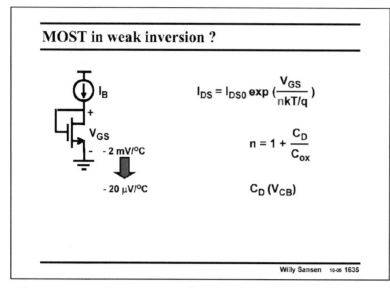

1635

It is clear that a MOST in weak inversion has an exponential current-voltage relation, which is nearly as good as that of a bipolar transistor. In this way the bipolar bandgap references can probably be duplicated in CMOS.

There are some important differences, however. First of all, for weak inversion the currents are small and the resistors are large, which is bad for noise.

Also, MOST have larger offset voltages. The errors are therefore larger as well.

Also, the coefficient I_{DS0} does not have the same temperature coefficient as in bipolar. It is less reproducible.

Finally, the exponential of this MOST in weak inversion contains a factor n, which contains a depletion capacitance, which is voltage dependent. Its value is therefore not very reproducible either.

It is now clear that with MOSTs in weak inversion, the same precision can be obtained as with bipolar transistors.

Table of contents

- **Principles**
- **Bipolar bandgap references**
- **CMOS bandgap references**
- **Bandgap references < 1 V**
- **Current references**
- **LDO Regulators**

Ref.: B. Gilbert, G.Meijer, ACD , Kluwer 1995

Willy Sansen   10-05 1636

1636

As bandgap reference voltages are always 1.2 V, we can wonder how to achieve a reference voltage with value below 1 V.

Indeed, the only physical constant available is this bandgap voltage. How can the properties of this bandgap voltage be exploited at lower supply voltages?

The answer will lie in the conversion of this bandgap reference voltage into currents, which are then summed up to lower output reference voltages. The requirement on the supply voltage will be that we need at least one single V_{BE} and some mV's. We can bias the bipolar transistor at very low currents, which may yield V_{BE} values down to 0.5 V. In this case, a minimum supply voltage could be reached of the same order of magnitude.

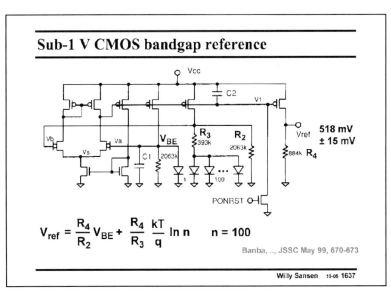

Sub-1 V CMOS bandgap reference

$$V_{ref} = \frac{R_4}{R_2} V_{BE} + \frac{R_4}{R_3} \frac{kT}{q} \ln n \qquad n = 100$$

Banba, .., JSSC May 99, 670-673

Willy Sansen 10-05 1637

1637
The principle of a sub-1 V bandgap reference is given in this slide. The reference voltage is about 0.5 V.

As a pure CMOS technology is used, only vertical pnp's can be used. They are all represented by diodes.

The operational amplifier has sufficient gain to equalize voltages Va and Vb. It is a two-stage amplifier with C2 as a compensation capacitance. Note that this is not a Miller capacitance. Also, all pMOSTs carry equal currents.

Since the voltages Va and Vb are the same, the currents from these nodes to ground must be the same as well. The current through R_3 is therefore PTAT, whereas the current through R_2 is simply V_{BE}/R_2. The sum of these two currents is also flowing through the output pMOST. The value of R_4 then sets the output voltage.

The input differential pair of the opamp does not allow really low supply voltages, however. This limits the supply voltage to $2V_{GS} + V_{DSsat}$. For a V_T of about 0.5 V this is about 1.6 V! The opamp is now the limiting factor.

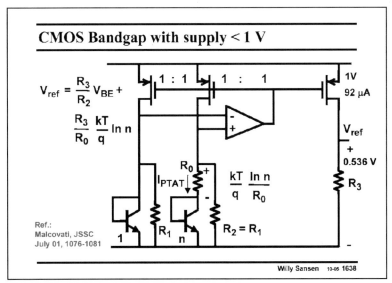

CMOS Bandgap with supply < 1 V

$$V_{ref} = \frac{R_3}{R_2} V_{BE} +$$
$$\frac{R_3}{R_0} \frac{kT}{q} \ln n$$

Ref.:
Malcovati, JSSC
July 01, 1076-1081

Willy Sansen 10-05 1638

1638
In this realization, an opamp is used which operates on supply voltages below 1 V. As a result, a real sub-1V bandgap reference emerges. BiCMOS is used, however, rather than standard CMOS.

The principle is similar to the one previously mentioned. The opamp equalizes its input voltages by use of feedback. This voltage is simply V_{BE}. All pMOSTs have again equal currents.

The current through resistor R_0 is thus PTAT. It is added to a current V_{BE}/R_2. This sum also flows through the output transistor. Resistor R_3 then sets the output reference voltage, which is here about 0.54 V.

Recently, a full CMOS version has been added by the same authors (Cabrini, ESSCIRC 2005) with 7 ppm/°C over −50°C to 160°C consuming only 26 μW at 1 V supply voltage. A folded cascode is then used as an opamp.

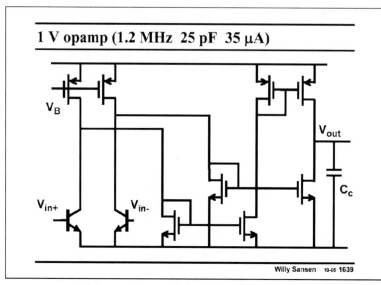

1 V opamp (1.2 MHz 25 pF 35 μA)

Willy Sansen 10-05 **1639**

1639

The BiCMOS opamp which operates at a supply voltage below 1 V is shown in this slide.

It uses a pseudo-differential input stage. This is easy, as the input voltages are connected to the V_{BE}'s of the bandgap. Actually, the input transistors form current mirrors with the diode connected npn transistors of the bandgap reference. The currents are now well defined!

The outputs of the input transistors lead to the output through one or two current mirrors. It is therefore a single-stage opamp. The gain is moderate, but the dominant pole is high, as indicated by the GBW.

The minimum supply voltage is $V_{GS} + V_{DSsat}$. For a V_T of about 0.5 V this is only about 0.9 V.

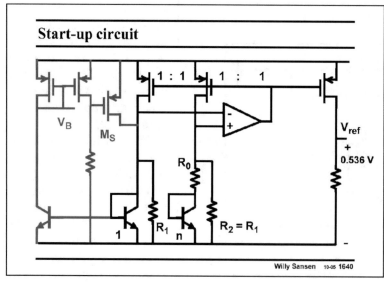

Start-up circuit

Willy Sansen 10-05 **1640**

1640

The startup circuit is added in this slide.

When the supply voltage is turned up, the resistor does not yet carry any current, and turns on transistor M_S. This also turns on the bandgap reference.

The npn current mirror then starts conducting and turns on the pMOST current mirror, the Gates of which are always set at a biasing voltage V_B. The resistor now carries a large current, such that the Gate of M_S is close to the supply voltage. Transistor M_S is turned off again. It does not influence the current balance in the bandgap reference.

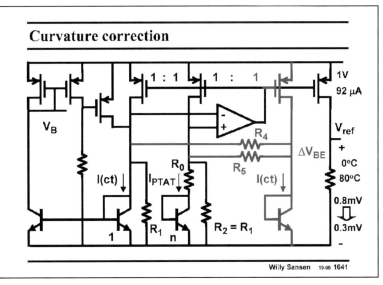

Curvature correction

Willy Sansen 10-05 1641

1641

Curvature compensation is now possible.

For this purpose, a diode is added and two resistors R_4 and R_5. They generate a term which subtracts the V_{BE} of a junction with constant current from the V_{BE} of a junction with the PTAT current. This term is the curvature correction. Its value is set by the two resistors.

In this way, the curvature error is reduced from 0.8 mV to 0.3 mV, over 80°C.

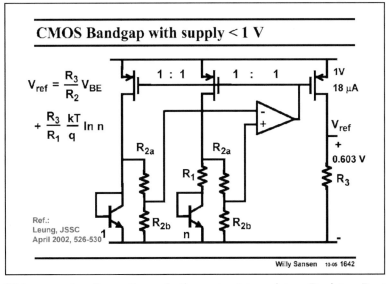

CMOS Bandgap with supply < 1 V

$$V_{ref} = \frac{R_3}{R_2} V_{BE}$$

$$+ \frac{R_3}{R_1} \frac{kT}{q} \ln n$$

Ref.:
Leung, JSSC
April 2002, 526-530

Willy Sansen 10-05 1642

1642

In this realization another opamp is used which operates on supply voltages below 1 V. A real sub-1 V bandgap reference again emerges. Standard CMOS is used this time.

The principle is similar as before. The opamp equalizes its input voltages by use of feedback. This voltage is simply V_{BE}. All pMOSTs have again equal currents.

The current through resistor R_1 is PTAT. It is added to a current $V_{BE}/(R_{2a}+R_{2b})$.

This sum also flows through the output transistor. Resistor R_3 then sets the output reference voltage, which here is about 0.6 V.

As an opamp, a folded cascode is used. Indeed, its input voltage range includes the ground. Also, a symmetrical OTA can be used provided low voltage current mirrors are used.

Several solutions are now possible, provided an opamp can be designed, operating at the right input voltage range.

1643

To derive a current reference from a bandgap voltage reference, requires a resistor.

Most of the uncertainty on the current will depend on this resistor. Also, the temperature coefficient of the resistor will play a role.

Table of contents

Ref.: B. Gilbert, G.Meijer, ACD , Kluwer 1995

Willy Sansen   10-05 1643

1644

The most accurate way to convert a voltage into a current is the use of an opamp.

The reference voltage is imposed across the resistor. The output current is exactly the same as the current through this resistor.

The main error is caused by the offset of the opamp.

The temperature coefficient is determined by the temperature coefficients of the bandgap reference and that of the resistor. One could be made to compensate the other, depending on the application.

Voltage-current converter

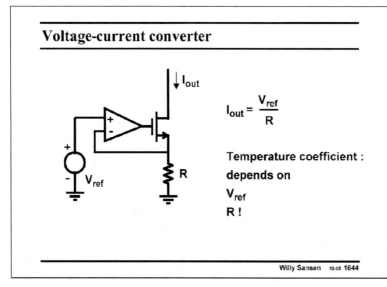

$$I_{out} = \frac{V_{ref}}{R}$$

Temperature coefficient :

depends on

V_{ref}

R !

Willy Sansen   10-05 1644

1645

The temperature coefficient of a resistor strongly depends on the doping level used. Only integrated resistors in silicon are considered.

Highly-doped resistors, with small sheet resistivities, have a small temperature dependence.

This is shown in this slide for n-regions. Resistivity strongly depends on the mobility of the carriers. Examples are emitter regions, but also drain- and source regions in CMOS technologies.

Lowly-doped resistors depend very strongly on temperature. For example the n-well resistor

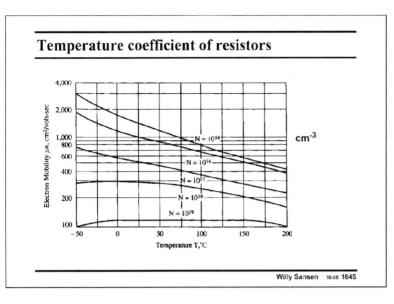

Temperature coefficient of resistors

N = 10¹⁴ cm⁻³
N = 10¹⁶
N = 10¹⁷
N = 10¹⁸
N = 10¹⁹

Willy Sansen   10-05 **1645**

has a strong temperature coefficient. Also, some ion-implanted resistors with high resistivity $(1–2\,\text{k}\Omega/$ square$)$ have high temperature coefficients, depending a bit on the annealing used.

Base resistances are somewhere in between, depending on the technology used.

Also the K′ factor in the MOST current-voltage expression contains the mobility. It decreases to some extent with temperature. A typical value is $K' \sim T^{-1.5}$.

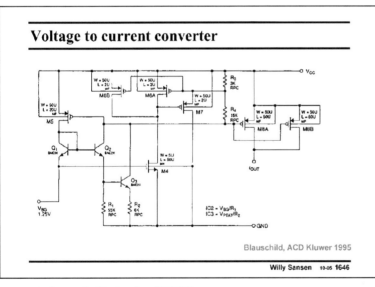

Voltage to current converter

Blauschild, ACD Kluwer 1995

Willy Sansen   10-05 **1646**

1646

For precise conversion of a bandgap reference voltage to a current, the circuit in this slide can be used. A bandgap voltage of 1.25 V is applied to the input.

The same bandgap voltage appears at the emitter of Q_2 and across resistor R_1. The current through Q_2 is therefore independent of temperature.

The voltage at the emitter of Q_3 and across resistor R_2 is PTAT, as it is a bandgap voltage minus a V_{BE}. The current through Q_3 is also PTAT.

Both currents are added and generate a voltage across R_4, which drives the output transistors. Current feedback is applied through M5.

Transistor M6 cancels the threshold voltage of the output devices. In this way, resistor R_3 does not play a role for the precision. Transistor M6 is driven by M4 and M7. Its Gate acts as a virtual ground for the voltage-to-current conversion.

Transistors M6 and M8 are shown double to indicate that they are large and well matched by means of centroide layout, etc. (see Chapter 15).

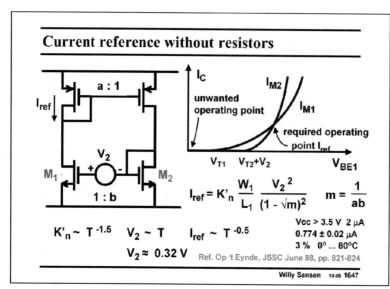

Current reference without resistors

I_{ref} = K'_n $\frac{W_1}{L_1}$ $\frac{V_2^2}{(1 - \sqrt{m})^2}$ m = $\frac{1}{ab}$

K'_n ~ T^{-1.5}   V_2 ~ T   I_{ref} ~ T^{-0.5}

V_2 ≈ 0.32 V   Ref. Op 't Eynde, JSSC June 88, pp. 821-824

Vcc > 3.5 V 2 μA
0.774 ± 0.02 μA
3 %   0° ... 80°C

Willy Sansen 10-05 1647

1647

A current reference can also be realized without resistors. Actually, the resistive channel of a MOST will be used instead.

It consists of current mirrors top-to-top. With size ratios a and b. The nMOST current mirror has an offset voltage V_2. In this way, the only biasing points possible are at zero and at reference current I_{ref}. Note that the bulk effect of the nMOSTs does not come in!

The actual expression is given. It depends on this offset voltage V_2, on the ratios a and b, on the size W_1/L_1 and especially K' which contains the mobility. Needless to say that the latter one is the worst one for high precision.

Moreover, the K' factor has a negative temperature exponent of about -1.5. If we can make V_2 PTAT; then about -0.5 is left as the temperature exponent for the current. This is acceptable indeed!

How can a PTAT offset voltage V_2 be realized by means of MOSTs? A value of about 0.32 V can be realized as follows.

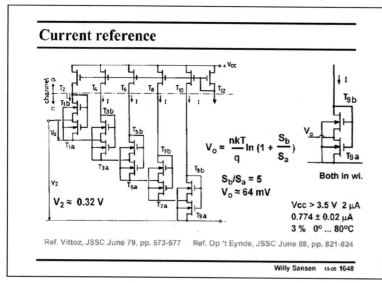

Current reference

$V_o = \frac{nkT}{q} \ln(1 + \frac{S_b}{S_a})$

$S_b/S_a = 5$
$V_o ≈ 64$ mV

Both in wi.

$V_2 ≈ 0.32$ V

Vcc > 3.5 V 2 μA
0.774 ± 0.02 μA
3 %   0° ... 80°C

Ref. Vittoz, JSSC June 79, pp. 573-577   Ref. Op 't Eynde, JSSC June 88, pp. 821-824

Willy Sansen 10-05 1648

1648

A fairly easy way to realize the offset voltage V2 of the previous slide is to use MOSTs in weak inversion.

In the same way, as for a bipolar PTAT cell, two MOSTs are taken in weak inversion with common Gate (on the right). The lower one T_{9a} serves as a resistor for the upper one, however T_{9b}. This is only possible if the upper one is made larger, such that its V_{GS} is smaller. The voltage difference in V_{GS}, or the voltage across T_{9a}, denoted by V_o, is then again PTAT. Parameter S is W/L. Voltage V_o is about 64 mV for the values given. It only depends on W/L ratio's and n.

Five such cells in series then yield a reference voltage of 320 mV. This is large enough to suppress the effect of mismatches. A spreading on the current can be as low as 5%!

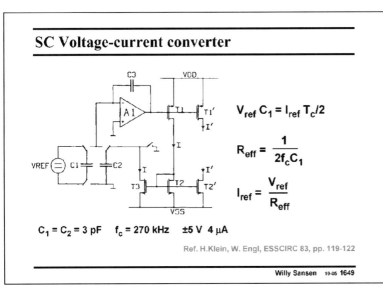

SC Voltage-current converter

$$V_{ref} C_1 = I_{ref} T_c/2$$

$$R_{eff} = \frac{1}{2f_c C_1}$$

$$I_{ref} = \frac{V_{ref}}{R_{eff}}$$

$C_1 = C_2 = 3$ pF $\quad$ $f_c = 270$ kHz $\quad$ ±5 V $\quad$ 4 µA

Ref. H.Klein, W. Engl, ESSCIRC 83, pp. 119-122

Willy Sansen 10-05 **1649**

1649

An accurate current reference can be realized without resistor if switched capacitors are used. We now know that a precise resistor can be realized by means of a switched-capacitor equivalent. This is used in the circuit in this slide.

When the switches are closed as shown, the reference voltage is stored on capacitor C_1. Its charge is then $V_{ref}C_1$.

During this same time, which lasts half a clock cycle $T_c/2$, the current through T3 is discharging C_2, which has the same size as C_1. This current through T3 is equal to the reference current I_{ref}.

Note that both a positive and negative supply is used!

When the switches are closed in the other way, the charges on C_1 and C_2 are made equal by the integrator A1 with capacitor C_3. If not, the integrator adjusts the current through T1, T2 and T3, which also equals the I_{ref}.

The charges on C_1 and C_2 are equal in steady-state. The effective resistor is exactly as expected for a switched-capacitor equivalent.

The current is now very precise, as it only depends on a crystal oscillator clock and the absolute value of a capacitor. This is a lot more precise than the absolute value of a resistor!

Table of contents

- **Principles**
- **Bipolar bandgap references**
- **CMOS bandgap references**
- **Bandgap references < 1 V**
- **Current references**
- **LDO Regulators**

Ref.: B. Gilbert, G.Meijer, ACD , Kluwer 1995

Willy Sansen 10-05 **1650**

1650

Both voltage and current references have been discussed now. Both can be used in voltage and current regulators. Only one more kind of regulator will be discussed here. It is a feedback circuit with variable load. As a consequence the design is not straightforward.

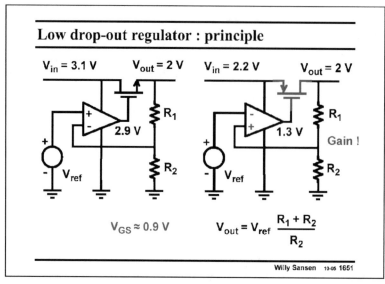

Low drop-out regulator : principle

$V_{in} = 3.1\,V$ $V_{out} = 2\,V$ $V_{in} = 2.2\,V$ $V_{out} = 2\,V$

2.9 V R_1 1.3 V R_1

V_{ref} R_2 Gain ! V_{ref} R_2

$V_{GS} \approx 0.9\,V$ $V_{out} = V_{ref}\dfrac{R_1 + R_2}{R_2}$

Willy Sansen   10-05 1651

1651
Both kind of regulators are shown in this slide.

The left one uses the output transistor in source-follower configuration, the right one as an amplifier. As a consequence the right one has one more stage in the feedback loop. It is therefore much more prone to instability.

It is preferred however, because the voltage drop across the output device can be smaller. Hence, the power dissipation is smaller.

The stability problem is even worse noticing that the load impedance can vary greatly. Output currents can vary over three or more orders of magnitude. The equivalent load resistors vary as much. The transconductance of the output transistor varies a lot as well. Remember that the non-dominant pole is determined by this transconductance.

The compensation devices will have to cover a wide range of output loads. The only way to avoid very large compensation capacitances is to try compensation schemas which track the load.

Table of contents

- Principles
- Bipolar bandgap references
- CMOS bandgap references
- Bandgap references < 1 V
- Current references
- LDO Regulators

Ref.: B. Gilbert, G.Meijer, ACD , Kluwer 1995

Willy Sansen   10-05 1652

1652
In this Chapter, all important aspects of bandgap and current references have been covered. Bandgap references have been discussed in great detail, in both bipolar and CMOS technologies.

They can be made to operate below 1 V supply voltage.

Current references are actually non-existing. Either discrete or MOST channel resistances are required for the voltage-current conversion.

Finally, we have to remember that these references are required in ADCs but also in voltage and current regulators. The stability of these feedback loops is not always as obvious.

References

P. Brokaw, "A simple three-terminal IC bandgap reference"
JSSC Dec.74, pp.388-393

M. Degrauwe, etal, "CMOS voltage references using lateral bipolar transistors",
JSSC Dec.85, pp.1151-1157

K. Kuijk, "A presicion reference voltage source" JSSC June 1973, pp.222-226

G. Meijer etal "An integrated bandgap reference", JSSC June '76, pp.403-406

G. Meijer etal "A new curvature-corrected bandgap reference"
JSSC Dec.82, pp.1139-143

G. Meijer, "Bandgap references", ACD Kluwer, 1995

B. Song, P.Gray, "A precision curvature-compensated CMOS bandgap
reference" JSSC Dec.83, pp.634-643

A. van Staveren etal "An integratible second-order compensated
bandgap reference for 1 V supply", ACD Kluwer 1995.

R. Widlar, "New developments in IC Voltage Regulators", JSSC Febr.71, pp. 2-7.

R. Widlar, "Low-voltage techniques", JSSC Dec.78, pp.838-846.

Willy Sansen   10-05 1653

1653

Only the most important references are collected here. They are in alphabetical order.

Historically, however, Widlar came first followed by Kuijk and Brokaw. Gilbert and Meijer soon followed.

Many more references can be found in the *IEEE Journal of Solid-State Circuits*. They are left to be explored by the reader.

171

Switched-capacitor filters

Willy Sansen

KULeuven, ESAT-MICAS

Leuven, Belgium

willy.sansen@esat.kuleuven.be

Willy Sansen 10 05 N171

Filters at low frequencies, such as for speech and bio-medical signals, require large time constants. They can be realized either with large capacitors or with large resistors. None of them can easily be integrated, however.

Switched capacitors behave like resistors with large values. As a result, they allow integration of low-frequency filters without external components. They have created a revolution in the realization of integrated low-pass filters for these applications.

For this purpose, the capacitors have to be switched in and out at a clock frequency, which is much higher than the filter frequencies.

Many questions arise, such as how the switching affects the filter characteristic, how much higher the clock frequency must be compared to the filter frequency, and finally, what kind of dynamic range can be achieved with such filter.

These questions are answered in this Chapter.

Switched-Capacitor Filters

* **Introduction : principle**
* **Technology:**
  * MOS capacitors
  * MOST switches
* **SC Integrator**
  * SC integrator : Exact transfer function
  * Stray insensitive integrator
  * Basic SC-integrator building blocks
* **SC Filters : LC ladder / bi-quadratic section**
* **Opamp requirements**
  * Charge transfer accuracy
  * Noise
* **Switched-current filters** McCreary, JSSC Dec 75, 371-379
  Gregorian, IEEE Proc. Aug 83, 941-986

Willy Sansen 10 05 N172

172

First of all, we will have a look at the principle of a switched capacitor. The main ingredients of such a circuit block are capacitors, switches and opamps.

Integrator and full first- and second-order filters are looked at next.

Finally, some opamp specifications are discussed which are typical for switched-capacitor circuits.

To conclude, a comparison is made with switched-current filters. The equivalence of a switched-capacitor with a resistor is explained first.

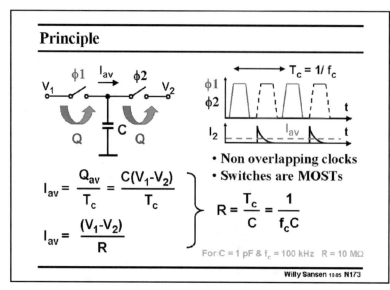

173

Switching a capacitor in and out at the rate of a high-frequency clock passes a charge, which is peaked.

Indeed, a clock is taken with frequency f_c, which has two non-overlapping phases $\Phi1$ and $\Phi2$, which are both somewhat smaller than half the period T_c.

Charging capacitor C to voltage V_1 during phase $\Phi1$ and discharging capacitor C to voltage V_2 during phase $\Phi2$, passes a charge $C(V_1 - V_2)$ from the input to the output terminal during period T_c.

The current which flows out of the output terminal is peaked, as it only flows at the beginning of phase $\Phi2$. Its average however, I_{av}, can be regarded as an average current flowing from the input to the output terminal, as a result of the voltage difference $V_1 - V_2$. It can be regarded as a current flowing between a voltage $V_1 - V_2$ because of a resistor R.

A switched capacitor now behaves as a resistor, provided averages are taken. This is true for low frequencies which are very low compared to the clock frequency.

The equivalent resistance R is $1/f_c C$. It can be increased in size for small values of clock frequency and capacitor. For 100 kHz and 1 pF we already find a resistance of 10 MΩ, a value which is impossible to integrate otherwise.

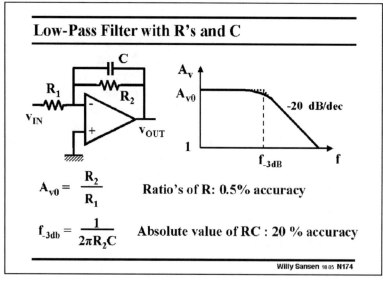

174

Take as an example, this low-pass filter. The low-frequency gain A_{v0} is a ratio of resistors, which can be made quite accurately.

The cut-off frequency f_{-3dB} however depends on a product of a resistor R_2 and the capacitor C. There is no way that this product R_2C can be realized in an accurate way. Errors of more than 20% cannot be avoided.

Substitution of this resistor R_2 by its switched-capacitor equivalent, will involve ratios only, which are easily realized with great accuracy.

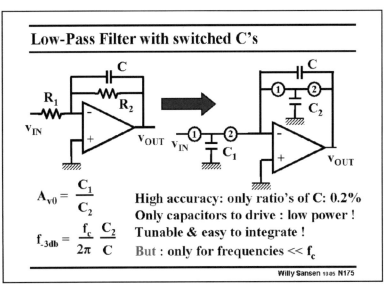

Low-Pass Filter with switched C's

$$A_{v0} = \frac{C_1}{C_2}$$

$$f_{-3db} = \frac{f_c}{2\pi} \frac{C_2}{C}$$

High accuracy: only ratio's of C: 0.2%
Only capacitors to drive : low power !
Tunable & easy to integrate !
But : only for frequencies $\ll f_c$

Willy Sansen 1005 N175

175

Substitution of all resistors by switched capacitors yields a circuit with only capacitors and switches, and an opamp.

Substitution of all R's in the expressions gives only C's. The low-frequency gain A_{v0} is now a ratio of capacitors, which can be made even more accurate, than a ratio of resistors.

The cut-off frequency f_{-3dB} now depends on a ratio of capacitors as well, and on the absolute value of clock frequency f_c. This latter frequency is normally derived from a crystal oscillator and is very accurate indeed (see Chapter 22). The ratio of capacitors can be made quite accurate as well. The larger we make the capacitors in area the better the matching will be (see Chapter 15). Values of less than 0.2% can be reached.

As a result, a fully integratable low-pass filter can be realized at low frequencies. There are only two drawbacks. It can only function at signal frequencies much lower than the clock frequency. Moreover the ratio of the signal frequency to the clock frequency is determined by a capacitor ratio. Large values of this ratio are not easy to realize. The signal frequency cannot thus allowed to be too large but not too small either!

Finally, note how the charges flow in this circuit. On phase 1 the input voltage is stored on C_1. On phase 2 the node goes back to zero as it is connected to the input of the opamp. All the charge of C_1 is now transferred to capacitor C_2, which changes the output voltage accordingly. The charge is conserved and hence $C_1 V_{IN} = C_2 V_{OUT}$.

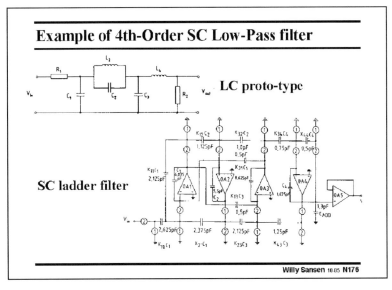

Example of 4th-Order SC Low-Pass filter

LC proto-type

SC ladder filter

Willy Sansen 1005 N176

176

As an example, a 4th order LC ladder filter is shown with its switched-capacitor equivalent. There is an opamp with switched capacitors all around. Opamp OA5 is just an output buffer.

Note that this filter involves only capacitor ratios. The smallest one has to be chosen. In this example, it is 0.5 pF. The smaller this minimum capacitor or unit capacitor is chosen, the more parasitic capacitances will cause errors in the capacitor ratios. Nowadays, unit capacitors of 0.2–0.25 pF are common for errors of the order of 0.05%.

Also, the opamps only drive small capacitors, which are different however in phase 1 compared with phase 2. The largerst has to be taken into account when designing such an opamp.

The smaller the capacitors however, the lower the power consumption.

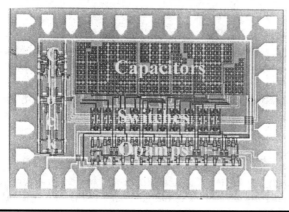

4th-Order SC Low-Pass filter

Capacitors

Switches

Opamps

Willy Sansen 10.05 N177

177

Moreover, the layout of such a switched-capacitor filter can be made quite regular. All opamps are on one side, the capacitors are on the other side and the switches in the middle.

Quite often, all opamps are made equal, which may not be the best solution for minimum power consumption, but which saves design time.

The capacitors are always made up of capacitor banks, using an integer number of equal unit capacitors. These have to be laid out for optimum matching.

Let us have a closer look now at these capacitors and switches.

Switched-Capacitor Filters

- **Introduction : principle**
- **Technology:**
  - MOS capacitors
  - MOST switches
- **SC Integrator**
  - SC integrator : Exact transfer function
  - Stray insensitive integrator
  - Basic SC-integrator building blocks
- **SC Filters : LC ladder / bi-quadratic section**
- **Opamp requirements**
  - Charge transfer accuracy
  - Noise
- **Switched-current filters**   McCreary, JSSC Dec 75, 371-379
                                      Gregorian, IEEE Proc. Aug 83, 941-986

Willy Sansen 10.05 N178

178

In present day technology, many types of capacitors are available. Switched-capacitor filters started off with MOS capacitances. This is why they are discussed first.

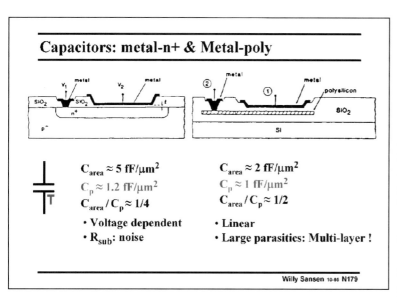

Capacitors: metal-n+ & Metal-poly

$C_{area} \approx 5$ fF/μm²
$C_p \approx 1.2$ fF/μm²
$C_{area}/C_p \approx 1/4$
- Voltage dependent
- R_{sub}: noise

$C_{area} \approx 2$ fF/μm²
$C_p \approx 1$ fF/μm²
$C_{area}/C_p \approx 1/2$
- Linear
- Large parasitics: Multi-layer !

Willy Sansen 10-05 N179

179

A MOS capacitor is formed between the top metal plate (or gate poly) and the source/drain diffusion. The thin gate oxide serves as a dielectricum. Values are given in this slide for a 0.35 micron CMOS technology. Its gate oxide thickness is about 1/50 or 7 nm. This gives a C_{ox} of 5×10^{-7} F/cm² or 5 fF/μm² (see Chapter 1).

The n+ plate has a lot of resistance however, which gives a lot of noise and even some voltage dependence.

It is much better to substitute the bottom layer by a highly doped poly layer. This capacitor is more linear.

Nowadays, many metal layers are available on top of the silicon structure. Any pair can be selected to be used as capacitors. Two criteria have to be fulfilled, however. The dielectricum must be of high quality and its thickness must be reproducible. This is why the technology file usually suggest which pair of metal layers are best used for capacitors and what is the capacitor per unit square.

For each integrated capacitor, the bottom plate has a parasitic capacitance C_p to the underlying layer. For the capacitor on the left, this is the junction capacitance to the substrate. For the one on the right, this parasitic capacitance is between the poly layer and the substrate. This parasitic capacitance is relatively large. It must be taken into account in the design of such a filter.

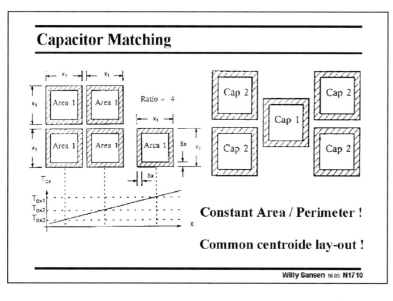

Capacitor Matching

Ratio = 4

Constant Area / Perimeter !

Common centroide lay-out !

Willy Sansen 10-05 N1710

1710

Matching these capacitors is of the utmost importance to achieve accurate filter frequencies. Matching capacitances has been discussed in great detail in Chapter 15. The most important conclusions are repeated here.

The only way to achieve good matching between two or more capacitors is to use integer ratios of unit capacitors. They must all have the same shape, or area to perimeter ratio. They must be laid out in centroide form.

For example, a ratio of 4 is obtained by putting Cap1 in the middle of 4 capacitors Cap2. They now all have the same value.

Moreover, the larger the unit capacitor is, the better the matching is, as shown next.

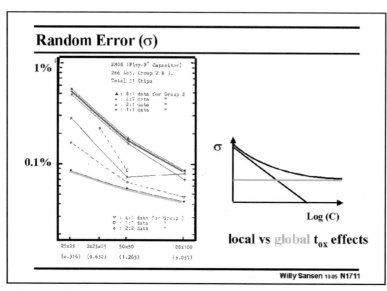

1711
The random sigma or spreading decreases for increasing capacitor size, as shown in this slide. The actual values have to be checked for each different CMOS technology however. It is clear that 0.2% is not difficult to achieve. For areas larger than 50×50 micrometer, 0.05% is feasible as well.

The slope of this curve depends on the ratio of local versus global errors. This is why it is hard to predict. It is preferable to be measured on an extensive number of samples.

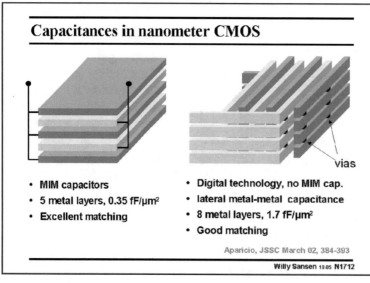

1712
Nowadays, many metal layers have become available. They can be used as horizontal capacitances or as vertical capacitances.

In horizontal capacitances (left), the odd-order plates are connected in parallel to one terminal, and the even order ones to the other terminal. Typical values of the capacitance value obtained are much smaller than what can be obtained with the Gate-oxide capacitance. Their breakdown voltage is higher, however. Matching is also good, depending on size.

Vertical capacitances have also become available (right). The lateral-capacitance value is even larger than on the left. Moreover they are available in purely digital CMOS technologies.

However, the matching is not as good as for MIM horizontal capacitances (0.5% versus 0.2%, respectively).

Switched-Capacitor Filters

- **Introduction : principle**
- **Technology:**
  - **MOS capacitors**
  - MOST switches
- **SC Integrator**
  - SC integrator : Exact transfer function
  - Stray insensitive integrator
  - Basic SC-integrator building blocks
- **SC Filters : LC ladder / bi-quadratic section**
- **Opamp requirements**
  - Charge transfer accuracy
  - Noise
- **Switched-current filters**   McCreary, JSSC Dec 75, 371-379
                                  Gregorian, IEEE Proc. Aug 83, 941-986

Willy Sansen 1095 N1713

1713
Now that we know what capacitors are available and what matching can be expected, let us have a look at what switches can be used. Clearly, a MOST acts as an ideal switch, as shown next.

A MOST as a switch

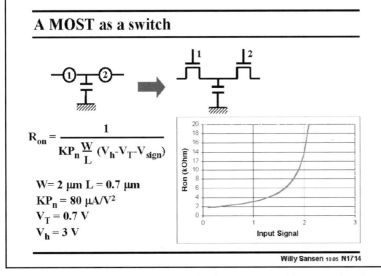

$$R_{on} = \frac{1}{KP_n \dfrac{W}{L} (V_h - V_T - V_{sign})}$$

W = 2 μm   L = 0.7 μm
KP_n = 80 μA/V²
V_T = 0.7 V
V_h = 3 V

Willy Sansen 1095 N1714

1714
A switch, which is closed on clock phase Φ1, is a nMOST which conducts on this phase. This means that its Gate is at a high voltage V_h with respect to ground. As a result, its V_{GS} is high and its V_{DS} is fairly small. This MOST is now in the linear region and behaves as a resistor with value R_{on}. Its expression is taken from Chapter 1 and repeated in this slide.

For a zero input signal voltage V_{sign}, the Source on the right hand side of the MOST is also zero. In this case, the $V_{GS} - V_T$ of the nMOST is $V_h - V_T$. This is the largest possible drive voltage. Its R_{on} has therefore the smallest possible value.

For larger input signals, the $V_{GS} - V_T$ values decrease and the R_{on} increases. This is illustrated for a small switch of 2/0.7 micrometer. For an input signal of about 2.3 V, the $V_{GS} - V_T$ becomes zero and the R_{on} becomes very large. The MOST does not act any more as a switch! The drive voltage is insufficient. The clock voltage V_h is not sufficiently large to turn on the switch! For such input voltages the clock voltage V_h must be larger than 3 V!

The maximum signal voltage that can be switched is therefore $V_h - V_T$.

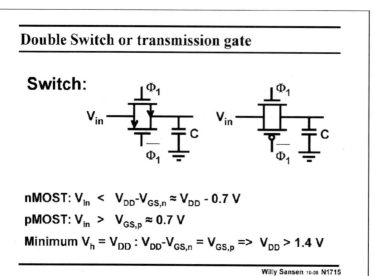

Double Switch or transmission gate

Switch:

nMOST: $V_{in} < V_{DD}-V_{GS,n} \approx V_{DD} - 0.7$ V

pMOST: $V_{in} > V_{GS,p} \approx 0.7$ V

Minimum $V_h = V_{DD}$: $V_{DD}-V_{GS,n} = V_{GS,p} \Rightarrow V_{DD} > 1.4$ V

Willy Sansen 10-06 N1715

1715

In order to be able to switch in a larger input voltage, a second transistor must be added, which is a PMOST. It is driven by the opposite clock phase. This is the lowest voltage available, which is usually ground.

The nMOST will conduct for lower input voltages, whereas the pMOST will conduct for larger input voltages. This parallel connection will thus always have a low R_{on} over the whole range of input voltages, from zero to supply voltage V_{DD}.

This double switch, also called transmission gate, is a good solution for large input voltages. However, it does not work for small supply voltages.

For small supply voltages, none of the two MOSTs can conduct. For example, if the minimum voltage V_{GS} is taken to be the same as the threshold voltage V_T, or 0.7 V, then the minimum supply voltage is about 1.4 V.

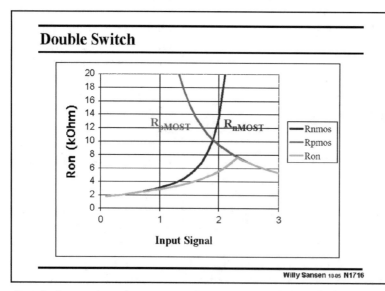

Double Switch

Willy Sansen 10-05 N1716

1716

For a large input voltage, the total on-resistance is the parallel combination of both. These values of R_{on} have been calculated for the same data as before.

It is clear that the maximum value of R_{on} is about 8 kΩ, whatever the input signal is, between zero and 3 V. For small input voltages, the R_{on} is even less than 2 kΩ.

In most cases, the designer knows what the input range will be. It covers the whole supply voltage range. In most cases, it is sufficient to take one switch only, a nMOST one for low input voltages or a pMOST one for high input voltages.

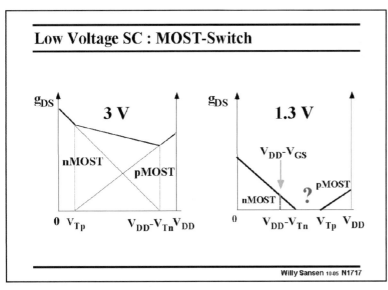

1717

Plotting the conductance of the switches shows clearly where the switches conduct and where they do not.

For low input voltages, only the nMOST conducts. For an input signal larger than V_{Tp}, the pMOST starts conducting. For a high supply voltages, there is a large region in the middle where both conduct. For an input signal larger than $V_{DD} - V_{Tn}$, the nMOST stops conducting.

For small supply voltages, there is obviously a region in the middle where none of the MOSTs conduct. The minimum supply voltage would thus be the sum of V_{Tn} and V_{Tp}.

How can switches be made to work in the middle if the supply voltage is really small?

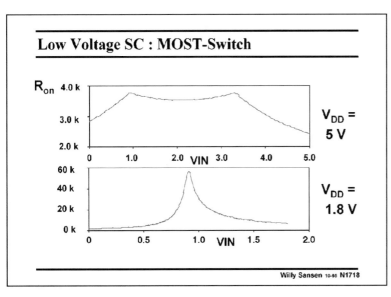

1718

The same message is given when the total on-resistance is plotted rather than the conductance. In this example, both V_T's are close to 0.9 V. Their sum is about 1.8 V.

For a large supply voltage, the on-resistance is always small. For a supply voltage which is a little larger than the sum of the two threshold voltages, the switches do not conduct any more in the middle. The on resistance is too high to be of use for a switch.

For large on-resistances, the time constant becomes too long. It takes too much time to fully charge the capacitances, as shown next.

1719

In such a switch-capacitor filter, we always assume that the charge is fully transferred from one capacitor to the other. Otherwise, the gain accuracy would be lost.

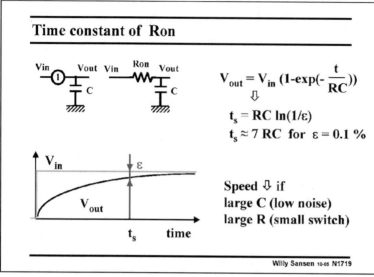

Time constant of Ron

$$V_{out} = V_{in} (1-\exp(-\frac{t}{RC}))$$

$$\Downarrow$$

$$t_s = RC \ln(1/\varepsilon)$$

$$t_s \approx 7\ RC \ \text{ for } \ \varepsilon = 0.1 \ \%$$

Speed ⇩ if
large C (low noise)
large R (small switch)

Willy Sansen 10-05 N1719

It takes time however, to fully charge a capacitor. In theory it takes an infinite time. In practice however a limited number of time constants is sufficient.

Charging a capacitor C by means of a constant resistor gives an exponential time response as shown in this slide. The time required to reach the final value within an error of 0.1% is called the 0.1% settling time t_s. For 0.1% it is the time constant times ln(1000) or 6.9 or about 7. It takes about 7 times the time constant before the final value is reached within 0.1%. This is a considerable amount of time.

For low kT/C noise, larger capacitors will be used and this time will be even longer. For small switches (W = 2L) the on resistances are of the order of 10 kΩ, this time will be longer as well. Half the clock period must now be at least 7 times this time constant. There is therefore a minimum length of clock period and thus a maximum value of clock frequency, as discussed next.

Finally, note that the R_{on} increases because the V_{GS} of the MOST decreases when the voltage comes up. The actual time constant will be even larger. Only a circuit simulator can give accurate values.

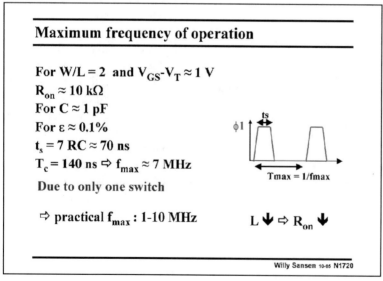

Maximum frequency of operation

For W/L = 2 and V_{GS}-$V_T \approx 1$ V
$R_{on} \approx 10$ kΩ
For C ≈ 1 pF
For ε ≈ 0.1%
$t_s = 7\ RC \approx 70$ ns
$T_c = 140$ ns ⇨ $f_{max} \approx 7$ MHz
Due to only one switch

⇨ practical f_{max} : 1-10 MHz

L ⬇ ⇨ R_{on} ⬇

Willy Sansen 10-05 N1720

1720
The maximum clock frequency also sets the maximum signal frequency, as the clock frequency must always be much larger than the signal frequency. The error which occurs when the clock frequency is not sufficiently large, will be calculated later.

We now need to know on what maximum clock frequencies can be achieved.

A small switch is taken with a R_{on} of 10 kΩ. The capacitor is 1 pF. For an error of 0.1% the settling time is 70 ns and minimum period 140 ns. This corresponds to a clock frequency f_{max} of 7 MHz.

As a consequence, if we need higher clocks than about 10 MHz, the switches must have to be made larger (larger W/L) or the capacitor smaller. Minimum unit capacitors are about 0.2 pF. However, if some gain is required, the other capacitor will be larger.

Larger switches store more charge and will give rise to other side effects, which will be discussed later.

It can be concluded that switched-capacitor filters do not easily work which clocks beyond a few tens of MHz.

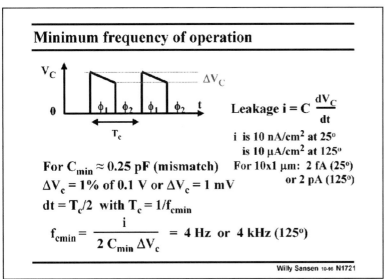

Minimum frequency of operation

Leakage $i = C \dfrac{dV_C}{dt}$

i is 10 nA/cm^2 at 25^0

is 10 μA/cm^2 at 125^0

For $C_{min} \approx 0.25$ pF (mismatch) For 10x1 μm: 2 fA (25^0)

$\Delta V_c = 1\%$ of 0.1 V or $\Delta V_c = 1$ mV   or 2 pA (125^0)

$dt = T_c/2$ with $T_c = 1/f_{cmin}$

$f_{cmin} = \dfrac{i}{2\,C_{min}\,\Delta V_c} = 4$ Hz or 4 kHz (125^0)

Willy Sansen 10-05 N1721

1721

There is also a minimum frequency of operation.

MOST Sources and Drains form junctions with respect to the substrate (or well). They leak. At room temperature these leakage currents are small but they increase drastically at higher temperatures. Remember that nowadays the Gate leaks as well, but this is left beyond consideration!

As a result, the charge stored on a capacitor slowly disappears. The voltage slowly decreases. It "droops". The droop rate dV_C/dt is given in this slide.

If a 100 mV signal amplitude is taken, which we can allow to droop by 1% or 1 mV. Then the maximum half period is about 2 dt. At room temperature the minimum clock frequency f_{cmin} is then about 4 Hz. This increases to 4 kHz at 125°C.

It will be difficult to realize switched-capacitor filters at very low frequencies, unless leakage can be better controlled or the temperature lowered!

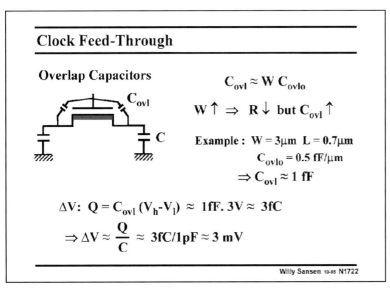

Clock Feed-Through

Overlap Capacitors

$C_{ovl} \approx W\,C_{ovlo}$

$W \uparrow \Rightarrow R \downarrow$ but $C_{ovl} \uparrow$

Example : W = 3μm L = 0.7μm

$C_{ovlo} = 0.5$ fF/μm

$\Rightarrow C_{ovl} \approx 1$ fF

$\Delta V: Q = C_{ovl}(V_h\text{-}V_l) \approx 1\text{fF}.\,3V \approx 3\text{fC}$

$\Rightarrow \Delta V \approx \dfrac{Q}{C} \approx 3\text{fC}/1\text{pF} \approx 3$ mV

Willy Sansen 10-05 N1722

1722

Another problem with MOST switches is that the terminals are connected by parasitic capacitances. In a MOST they are the overlap capacitances. The larger the widths (for smaller R_{on}), the larger the overlap capacitors.

The clock pulses are partially injected in the signal path. Indeed, the overlap capacitor C_{ovl} forms a capacitive divider with the storage capacitor

C. if C_{ovl} is about 1 fF and

C is 1 pF, then about 0.1% of the clock pulse is injected in the signal path. For a clock of 3 V, this is a 3 mV error signal in the signal path. This gives a contribution at the clock frequency and its harmonics! It does not affect the lower frequency bands, however.

The charge itself, transferred by C_{ovl}, is of the order of fC. It is quite small but not negligible.

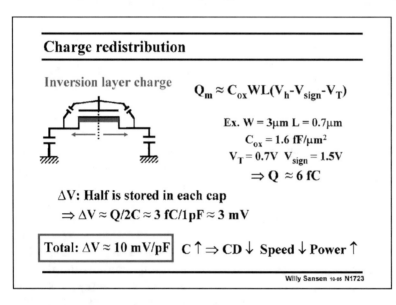

Charge redistribution

Inversion layer charge

$$Q_m \approx C_{ox}WL(V_h\text{-}V_{sign}\text{-}V_T)$$

Ex. W = 3µm L = 0.7µm
C_{ox} = 1.6 fF/µm²
V_T = 0.7V V_{sign} = 1.5V
$\Rightarrow$ Q $\approx$ 6 fC

ΔV: Half is stored in each cap
$$\Rightarrow \Delta V \approx Q/2C \approx 3 \text{ fC}/1\text{pF} \approx 3 \text{ mV}$$

Total: ΔV ≈ 10 mV/pF   C ↑ ⇒ CD ↓ Speed ↓ Power ↑

Willy Sansen 10-05 N1723

1723
Moreover, a MOST, which is switched on, contains a mobile charge Q_m in the channel (inversion layer), which disappears when the MOST is switched off. The charge is redistributed towards both ends.

This charge disappears towards the Source side and towards the Drain side depending on the relative impedances seen. If the capacitances on both sides are the same, then half the charge goes left and half right.

A first-order calculation shows that this charge is also of the order of magnitude of fC. For a 1 pF storage capacitor, it also causes mV's error.

This error depends on the signal and causes distortion.

A rule of thumb says that per pF storage capacitor, about 10 mV error can be expected as a result of clock injection and charge redistribution.

This is a lot. Let us see what circuit techniques can be used to reduce these errors. Making everything fully differential is certainly one way to reduce these errors.

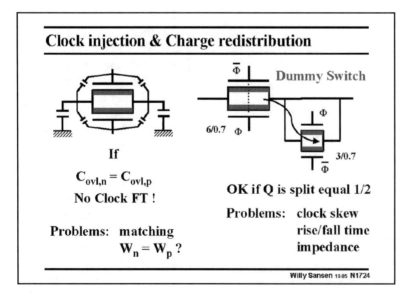

Clock injection & Charge redistribution

Φ̄   **Dummy Switch**

6/0.7 Φ

Φ

3/0.7
Φ̄

If
$$C_{ovl,n} = C_{ovl,p}$$
No Clock FT !

OK if Q is split equal 1/2

Problems: clock skew
             rise/fall time
             impedance

Problems: matching
$$W_n = W_p ?$$

Willy Sansen 10-05 N1724

1724
Using a double switch is a possible remedy for clock injection (see left). When the nMOST receives a positively going clock pulse at its Gate, the pMOST receives a negatively going clock pulse at its Gate. The effects can cancel, provided the overlap capacitances match. Charge redistribution is reduced as well as the electrons of the nMOS recombining with the holes of the pMOST.

The addition of a dummy

switch with specific dimensions (W/L) can help as well (see right). When the nMOST in the signal patch is switched out, its charge is taken up by the dummy nMOST switch, which is switched in. It is only half the size because we assume that the capacitances on both ends are the same. The same applies to the charge of the pMOSTs.

Clock skew (delay in time) and different rise and fall times may render the charge compensation incomplete but these are only second-order effects. The main difficulty with the dummy switch is that the relative impedances (capacitors) must be known on both ends. If not, addition of a dummy switch may make things worse!

Let us have a closer look at this compensation technique.

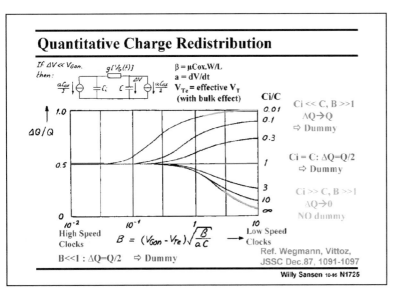

1725

This graph shows what fraction of the charge $\Delta Q/Q$ charge goes left and right, when a MOST switch is switched out. The capacitance at the input is C_i and at the output C.

On the horizontal axis, a parameter is used which includes the steepness of the clock pulses or simply the clock speed. It is normalized by means of some transistor parameters. High speed clocks are on the left, whereas slow clocks are on the right.

It shows that for high-speed clocks (small B), the charge redistribution is always the same at both ends. Half of the charge goes left, and half right, irrespective of what the input and output capacitors are. Dummies are therefore better used.

The picture is very different if the clock has slow edges (B large). In this case, the charge has time to find out what the capacitances (impedances) are on both sides and obviously flows towards the highest capacitance (lowest impedance).

For a large C_i/C ratio, all the charge flows to the output and dummies are required with equally sized transistors. When C_i equals C, then half the charge flows to the output and a dummy MOST is required of half the size. For a small C_i/C ratio, no charge flows to the output and a dummy is better not to be used.

In practice, however, the situation is not as clear cut. It is difficult then to find out the right size of the dummy.

1726

It has become clear from the above discussion that overlap capacitances must be as small as possible. On top of that, some more parasitic capacitances have to be added.

For example, in the layout on the left, the poly Gate lines cross the Source and Drain lines. The crossing areas (black) give coupling capacitances to be added to the overlap capacitances.

Layout considerations

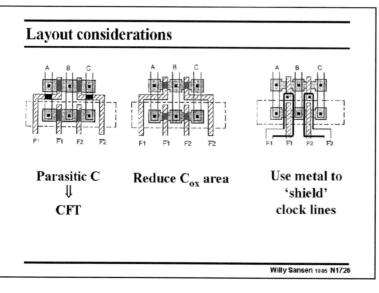

| Parasitic C ⇓ CFT | Reduce C_{ox} area | Use metal to 'shield' clock lines |

Willy Sansen 10-05 N1726

The clock feedthrough will now be increased. Such layout is thus better avoided.

The best that can be achieved is to use as small a MOST switch as possible, as shown in the middle. The overlap capacitances are also as small as possible as they scale with the widths of the MOSTs.

On the right, an example is given on the use of metal shields between the clock lines and the actual MOSTs, reducing the coupling capacitances.

Switched-Capacitor Filters

- **Introduction : principle**
- **Technology:**
  - MOS capacitors
  - MOST switches
- SC Integrator
  - SC integrator : Exact transfer function
  - Stray insensitive integrator
  - Basic SC-integrator building blocks
- **SC Filters : LC ladder / bi-quadratic section**
- **Opamp requirements**
  - Charge transfer accuracy
  - Noise
- **Switched-current filters**

McCreary, JSSC Dec 75, 371-379
Gregorian, IEEE Proc. Aug 83, 941-986

Willy Sansen 10-05 N1727

1727

Now that we know what capacitors can be used and what switches, let us find out what performance can be expected from the filter blocks themselves. Let us find out what happens when the signal frequencies are not all that much smaller than the clock frequencies.

A simple inverter is taken first.

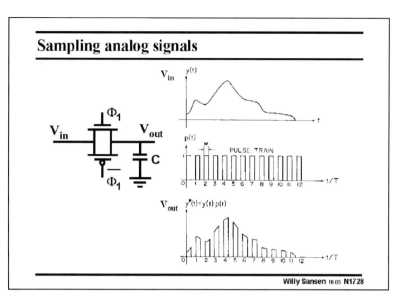

Sampling analog signals

Willy Sansen 10-05 N1728

1728
An analog signal is continuous in amplitude and in time (on top). When it is switched in and out at the rate of a clock frequency f_c (represented by a pulse train in the middle), it becomes discontinuous in time (at the bottom). It is only available when the clock is high. It is a sampled analog signal.

These signals are shown versus time. As Fourier has explained, they have an equivalent representation versus frequency, as shown next.

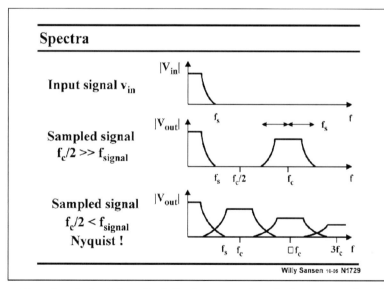

Spectra

Input signal v_{in}

Sampled signal $f_c/2 \gg f_{signal}$

Sampled signal $f_c/2 < f_{signal}$ Nyquist !

Willy Sansen 10-05 N1729

1729
The bandwidth of the analog signal is limited. Its spectral content is limited to f_s.

When this signal is sampled by clock frequency f_c, it is actually multiplied by this clock frequency. Its spectrum appears as two sidebands on both sides of the clock frequency, as shown in the middle.

Note that the signal bands appear on all the harmonics of the clock frequency.

Care has to be taken that the frequency bands do not overlap. This is called aliasing. To avoid overlap, the signal frequency f_s must be smaller than half the clock frequency f_c. This is called the Nyquist criterion.

When the signal bandwidth f_s is too large, aliasing occurs (at the bottom), the information in the overlap frequency band does not know to which band it belongs. It is lost. Aliasing must be avoided at all cost. To achieve this, a low-pass filter is applied before sampling. This filter is normally a passive filter. It is called an anti-aliasing filter.

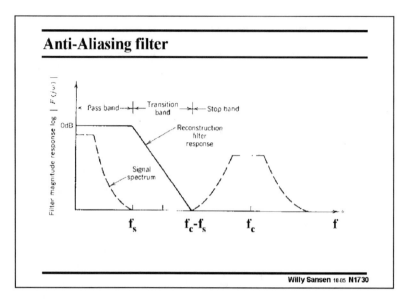

Anti-Aliasing filter

Willy Sansen 10.05 N1730

1730

To avoid aliasing, an anti-aliasing filter is required with a low-pass characteristic to make sure that no high-frequency content is present in the input signal beyond $f_c/2$.

Some distance must be kept between f_s and $f_c - f_s$ so that a first or second-order filter is sufficient. Steep filters require many components and are difficult to match. Remember, that passive filters are usually used to avoid distortion. Higher-order anti-aliasing filters are to be avoided.

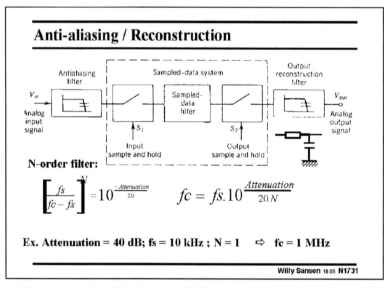

Anti-aliasing / Reconstruction

N-order filter:

$$\left[\frac{fs}{fc - fs}\right]^N = 10^{\frac{-Attenuation}{20}} \qquad fc = fs.10^{\frac{Attenuation}{20.N}}$$

Ex. Attenuation = 40 dB; fs = 10 kHz ; N = 1   ⇨   fc = 1 MHz

Willy Sansen 10.05 N1731

1731

The order of the filter and the amount of attenuation are related as given in this slide. Normally, a first-order filter ($N = 1$) is preferred.

Such a sampled filter has therefore the following building blocks. The analog signal is applied to a anti-aliasing filter. It is sampled by switches. A sampled-data filter is applied, which has the advantage that no external components are required. A clock is necessary, however.

The output signal is then applied to a sample-and-hold circuit, to make it continuous in time. Another low-pass filter is then applied to filter out the clock frequency. It is called a reconstruction filter. A pure analog signal results. The same expressions are valid.

Nowadays, the input signal is filtered and kept in sampled form to be applied to a Analog-to-digital converter, and then eventually to a DSP block.

After all this signal processing, it is applied to a Digital-to-analog converter, the last block of which is a reconstruction low-pass filter.

Sampled Data Basics : z-transform

Analog System: s = jω

Z-TRANSFORM	SEQUENCE
$a\,X(z) + b\,V(z)$	$ax(n) + bv(n)$
$z^{-n_1}\,Y(z)$	$y(n - n_1)$
$Y(z/b)$	$b^n\,y(n)$
$-z\,\dfrac{dY(z)}{dz}$	$n\,y(n)$
$Y(z^{-1})$	$y(-n)$
$X(z)\,V(z)$	$x(n) * v(n)$

$$\frac{V_{out}}{V_{in}} = \frac{1}{1 + sRC}$$

Sampled data: z-transforms

1 delay is z^{-1}

$$z = e^{j\omega T_c} = e^{j\frac{2\pi f}{f_c}}$$

$$e^{j\omega T_c} = 1 + j\omega T_c + \frac{(j\omega T_c)^2}{2} + \ldots\ldots \quad \text{if } \omega T_c \ll 1$$

Willy Sansen 10 05 N1732

1732
In analog systems, the signals are best represented by their Laplace transforms with variable s which is the complex frequency jω. The best way to describe transfer functions is the use of the Laplace transform. An example is given of a first-order low-pass filter with time constant RC. It is easy to transform this expression back to frequency by simple substitution of s by jω.

In sampled data systems, the signals are best represented by their z-transforms. Indeed the only accurate way to describe transfer functions is the use of z-transforms. A delay of one single clock pulse then corresponds to a multiplication by z^{-1}. A few elementary characteristics of z-transforms are given in this slide.

It is easy to transform an expression in z back to frequency by substitution of z by $\exp(j\omega T_c)$ or $\exp(2\pi jf/f_c)$. Since the signal frequency f is much smaller than the clock frequency f_c, this exponential can be developed into a power series, as shown in this slide. It is sufficient to keep the first few terms.

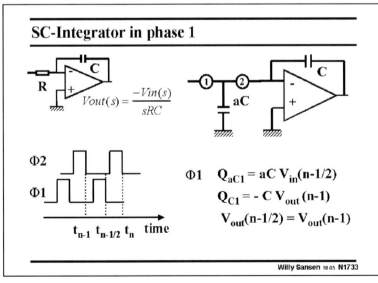

SC-Integrator in phase 1

$$Vout(s) = \frac{-Vin(s)}{sRC}$$

$\Phi 1\quad Q_{aC1} = aC\,V_{in}(n\text{-}1/2)$

$Q_{C1} = - C\,V_{out}(n\text{-}1)$

$V_{out}(n\text{-}1/2) = V_{out}(n\text{-}1)$

Willy Sansen 10 05 N1733

1733
In order to find the transfer characteristic of a sampled-data filter in z, charge conservation is used. There are other more formal techniques (see Laker-Sansen, McGrawHill 1994), but charge conservation is the easiest one, although it may not always work.

An example is given of a simple integrator.

An analog integrator has a transfer characteristic, which is well known, as shown on the top left. What is now the transfer characteristic of the sampled-data inverter shown on top right? The resistor has been substituted by its sampled-data equivalent and called aC.

In order to find the transfer characteristic in z, charge conservation is applied. This means that we add up the charges on the capacitors in phase 1, and equate it to the charges in phase 2. Indeed, charges cannot disappear as currents cannot disappear (laws of Kirchoff).

For this purpose, a clock pulse of phase 2 is considered at time t_n. Note also, this time actually corresponds to the end of the clock pulse, when all charge has been fully transferred. One period earlier, the clock pulse of phase 2 occurs at t_{n-1}. The other phase (phase 1) then occurs at $t_{n-1/2}$ at the end.

The charge on capacitance aC during phase 1 is denoted by Q_{aC1} and is given in this slide. It is available at time $t_{n-1/2}$. The charge on capacitor C is given by Q_{C1}. As switch 2 is open during phase 1, we can as well take this charge at time $t_{n-1/2}$ as at time t_{n-1}.

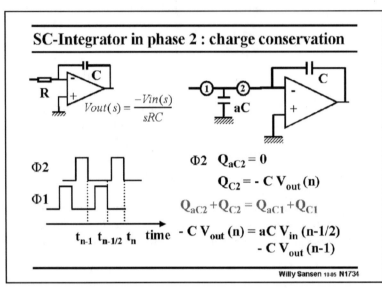

1734

When switch 2 is closed (on phase 2), capacitor aC is fully discharged as the minus input of the opamp goes to zero because of the feedback. The gain is assumed to be sufficiently high to reduce the differential input zero, whatever appears at the output.

In this phase 2, the charge on capacitor ac C is zero and the charge on capacitor C changes to Q_{C2}.

Noting that the sum of the charges in phase 1, equals the sum of the charges in phase 2, gives the charge conservation equation. Note that this equation links the output voltage to the input voltage. The voltages appear at different times, however. In this form, this equation cannot be solved.

1735

The charge conservation equation is copied on top. All voltages are now converted into z-transforms. One delay corresponds to z^{-1}.

The gain V_{out}/V_{in} is now readily written. It shows that the gain equals a, the ratio of the two capacitors, multiplied by a factor in z.

To see what this factor means in the frequency domain, z is substituted by $\exp(j\omega T_c)$. For frequencies much lower than the clock frequency f_c, this exponential can be developed into a power series, which can be cut off after $j\omega T_c$.

The resulting gain is exactly the same as for a purely analog integrator, provided the time constant T_c/a.

This is approximation however, for low frequencies. For higher frequencies, an error is made, discussed next.

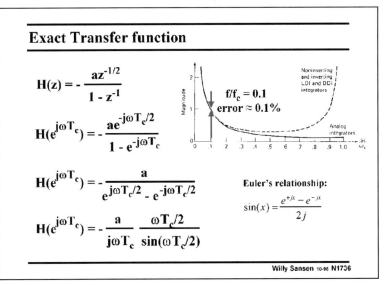

Exact Transfer function

$$H(z) = -\frac{az^{-1/2}}{1 - z^{-1}}$$

$$H(e^{j\omega T_c}) = -\frac{ae^{-j\omega T_c/2}}{1 - e^{-j\omega T_c}}$$

$$H(e^{j\omega T_c}) = -\frac{a}{e^{j\omega T_c/2} - e^{-j\omega T_c/2}}$$

$$H(e^{j\omega T_c}) = -\frac{a}{j\omega T_c} \frac{\omega T_c/2}{\sin(\omega T_c/2)}$$

$f/f_c = 0.1$
error $\approx 0.1\%$

Euler's relationship:

$$\sin(x) = \frac{e^{+jx} - e^{-jx}}{2j}$$

Willy Sansen 10-05 N1736

1736

To find out what the transfer characteristic is for all frequencies, the original expression H(z) in z is taken again.

The exponentials can be rewritten in terms of $\omega T_c/2$. They can be combined in a sine function.

The same expression of an integrator is obtained, but multiplied by a $\sin(x)/x$ function. This function is the error functional which is typical for all sampling of analog signals.

This function is calculated on the next slide. A rule of thumb is that for a frequency one tenth of the clock frequency, the error is about one tenth of a percent. This error grows with the square of the frequency, as shown on the next slide.

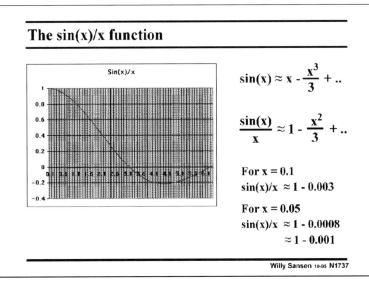

The sin(x)/x function

$$\sin(x) \approx x - \frac{x^3}{3} + ..$$

$$\frac{\sin(x)}{x} \approx 1 - \frac{x^2}{3} + ..$$

For x = 0.1
$\sin(x)/x \approx 1 - 0.003$

For x = 0.05
$\sin(x)/x \approx 1 - 0.0008$
$\approx 1 - 0.001$

Willy Sansen 10-05 N1737

1737

The $\sin(x)/x$ function is easily calculated, as shown in this slide. It starts at 1 for small x and goes through zero at $x = \pi = 3.14$.

For small values of x, the sine function can be represented by a power series and cut off after the first two terms.

For x = 0.1, the function drops to 0.3% below unity. For x = 0.05 as on the previous slide, the function drops to 0.08% below unity. This is taken to be about 0.1%.

Note also that this function changes with x^2. The errors decrease rapidly for smaller values of x. For example, for an error of 0.05% the value is about 0.04. An error of 0.05% is a typical value used in the design of SC filters. It leads to dynamic ranges of about 70 dB, as will be shown later.

Switched-Capacitor Filters

- **Introduction : principle**
- **Technology:**
  - MOS capacitors
  - MOST switches
- **SC Integrator**
  - SC integrator : Exact transfer function
  - Stray insensitive integrator
  - Basic SC-integrator building blocks
- **SC Filters : LC ladder / bi-quadratic section**
- **Opamp requirements**
  - Charge transfer accuracy
  - Noise
- **Switched-current filters**   McCreary, JSSC Dec 75, 371-379
                                        Gregorian, IEEE Proc. Aug 83, 941-986

Willy Sansen 10-05 **N1738**

1738

In this switched-capacitor realization of a resistor, parasitic capacitances play an important role. These capacitors cannot therefore be made small, which leads to excessive power consumption.

This is why a stray-insensitive equivalent is better used.

This is discussed next.

Stray Capacitances

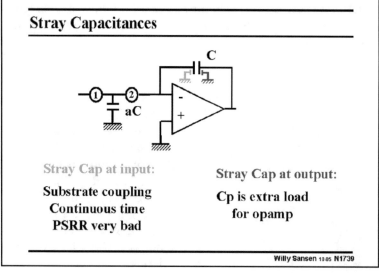

Stray Cap at input:

**Substrate coupling
Continuous time
PSRR very bad**

Stray Cap at output:

**Cp is extra load
for opamp**

Willy Sansen 10-05 **N1739**

1739

Each capacitor has a parasitic capacitor from the bottom plate to the underlying conductor (substrate, ...). For capacitor aC, this bottom plate is obviously connected to ground, where it is shorted out.

For capacitor C however, it is not so clear. If the bottom plate is connected to the minus input of the opamp (green), then the minus node picks up substrate noise more easily, which is bad for the PSRR.

If on the other hand, the bottom plate is connected to the output of the opamp (blue), then the load capacitance increases by this amount, increasing the power consumption. Yet, the latter solution is usually preferred.

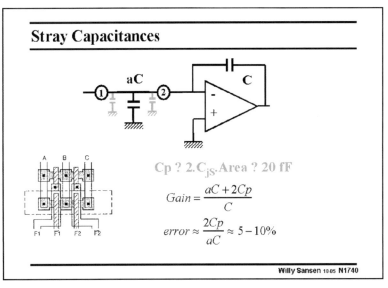

Stray Capacitances

$Cp \approx 2.C_{jS}.Area \approx 20\ fF$

$$Gain = \frac{aC + 2Cp}{C}$$

$$error \approx \frac{2Cp}{aC} \approx 5 - 10\%$$

Willy Sansen 10-05 **N1740**

1740

Moreover, the Source and Drain junction capacitances of the switches have to be added to capacitor aC.

The Source capacitor of switch 1 and the Drain capacitor of switch 2 are shown in this slide (green). Together, they can easily give an error on aC of 5 to 10%.

Remember that a Source junction capacitance is of the same order of magnitude as the C_{GS}, which is about kW (with $k \approx 2\ fF/\mu m$) for minimum-L transistors. For a $W \approx 5\ \mu m$, such a Source junction capacitor is about 10 fF.

A better alternative is to make capacitor aC floating, as shown next.

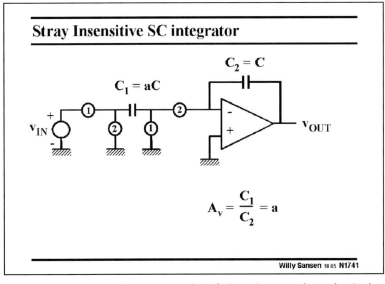

Stray Insensitive SC integrator

$C_2 = C$

$C_1 = aC$

v_{IN} v_{OUT}

$$A_v = \frac{C_1}{C_2} = a$$

Willy Sansen 10-05 **N1741**

1741

Capacitor aC is now floating. Two more switches are necessary. However, the ratio a will be insensitive to the parasitic junction capacitances.

Similar functions are carried out during the two phases. During phase 1, capacitor aC is charged to the input voltage. During phase 2, this capacitor aC is discharged to zero, forcing its charge towards C.

Let us now see how the parasitic junction capacitances of all four switches come in. If they do not, the gain A_v is accurately equal to a.

The operation of this SC integrator is illustrated even better by showing the circuit first with the clocks closed on phase 1, and then on 2.

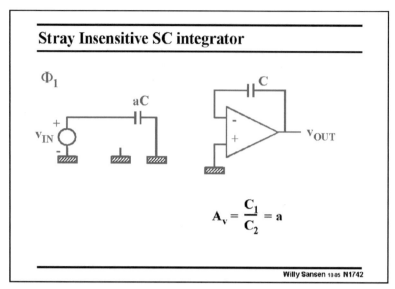

1742

Note that in phase 1, the left side of capacitor aC is charged positively. Its right side is negative. During phase 2 it will be applied to an inverting amplifier configuration. We will therefore have a non-inverting integrator.

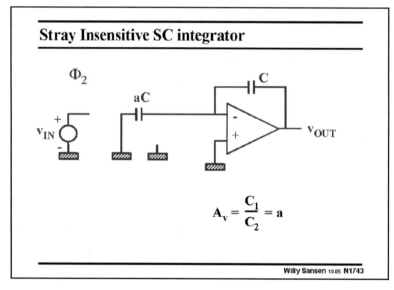

1743

In phase 2, the voltage across aC is applied to the inverting amplifier, generating the voltage gain of a.

Capacitor aC is now fully discharged provided the gain of the opamp is sufficiently high and no offset is present.

1744

Let us now see how the parasitic junction capacitances of all four switches come in.

The full integrator is shown on top, with the parasitic capacitances C_p.

The situation is depicted during clock phase 1. The opamp is not connected and is now left out.

The parasitic capacitor C_p on the left of aC is driven by the output of the previous stage. It is a low-impedance point which easily charge this capacitor without affecting the voltage across aC. As a result, it does not affect charge Q_{aC}.

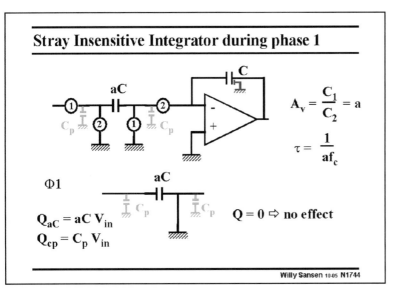

The parasitic capacitor C_p on the right of aC is shunted to ground. As a result, it does not affect charge Q_{aC} either.

The parasitic capacitances have no influence on the charge on capacitor aC. This latter capacitance aC can therefore be smaller without loosing accuracy. Typical values are 0.2 to 0.25 pF.

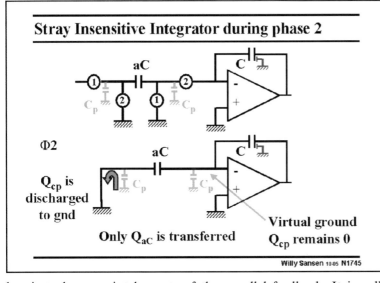

1745

The full integrator is again shown on top.

The situation is now depicted during clock phase 2.

The parasitic capacitor C_p on the left of aC is shunted to ground. Its charge obtained during phase 1, disappears to ground. As a result, it does not affect charge on any capacitor.

The parasitic capacitor C_p on the right of aC is now connected to the minus input of the opamp. It is a low-impedance point because of the parallel feedback. It is called the "virtual ground". As a result, it does not affect charges an any capacitor either. Of course, this only holds if the gain of the opamp is sufficiently high.

The parasitic capacitances have therefore no influence on the charge on capacitor aC or C. Only charge Q_{aC} has been transferred. The capacitors can now be chosen smaller without loosing accuracy. In this way, power can be saved.

Switched-Capacitor Filters

- **Introduction : principle**
- **Technology:**
  - **MOS capacitors**
  - **MOST switches**
- **SC Integrator**
  - **SC integrator : Exact transfer function**
  - **Stray insensitive integrator**
  - **Basic SC-integrator building blocks**
- **SC Filters : LC ladder / bi-quadratic section**
- **Opamp requirements**
  - **Charge transfer accuracy**
  - **Noise**
- **Switched-current filters**

McCreary, JSSC Dec 75, 371-379
Gregorian, IEEE Proc. Aug 83, 941-986

Willy Sansen 10-05 **N1746**

1746
Now that we know how to construct a switched-capacitor integrator, let us investigate at some more complicated filter structures.

Simple low-pass filters come first.

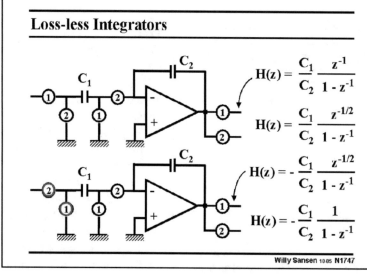

Loss-less Integrators

$$H(z) = \frac{C_1}{C_2} \frac{z^{-1}}{1 - z^{-1}}$$

$$H(z) = \frac{C_1}{C_2} \frac{z^{-1/2}}{1 - z^{-1}}$$

$$H(z) = -\frac{C_1}{C_2} \frac{z^{-1/2}}{1 - z^{-1}}$$

$$H(z) = -\frac{C_1}{C_2} \frac{1}{1 - z^{-1}}$$

Willy Sansen 10-05 **N1747**

1747
Let us first investigate some variations on the integrators.

Switches are now also connected to the output. They are actually the input switches of the next stage. They determine on which switch the output becomes available. The delay through the inverter is obviously affected by them.

All the integrators in this slide have only a single capacitance in the feedback loop. This is why they are called "loss-less". They do not have a resister in the feedback loop which would "dampen" the integration.

The top integrator with output on phase 1, has been discussed before. It has a non-inverting gain C_1/C_2, and half a clock delay.

However, when the output is sampled on phase 2, another half cock delay is added. The gain is again C_1/C_2, but the delay is now a full clock period.

In the bottom integrator, the two switches at the input have been interchanged. In clock phase 2, the well-known inverting amplifier configuration appears. It has gain C_1/C_2 and no delay. Taking the output on phase 1, adds half a clock period delay.

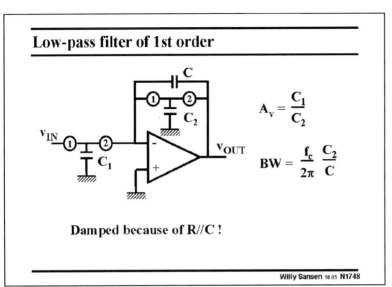

Low-pass filter of 1st order

$$A_v = \frac{C_1}{C_2}$$

$$BW = \frac{f_c}{2\pi} \frac{C_2}{C}$$

Damped because of R//C !

Willy Sansen 10 05 **N1748**

1748

Addition of a switched resistor, with capacitor C2, across integration capacitor C, dampens the integration. It forms a first-order low-pass filter.

The gain at low frequencies is non-inverting and equals C_1/C_2. The pole of this filter (or the bandwidth) is a fraction of the clock frequency f_c, depending on the ratio C_1/C_2.

Many more filters can be constructed in this way.

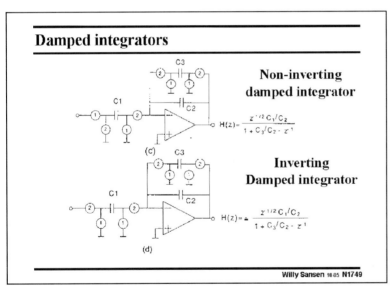

Damped integrators

Non-inverting damped integrator

$$H(z) = \frac{z^{-1/2} C_1/C_2}{1 + C_3/C_2 \cdot z^{-1}}$$

(c)

Inverting Damped integrator

$$H(z) = \frac{z^{-1/2} C_1/C_2}{1 + C_3/C_2 - z^{-1}}$$

(d)

Willy Sansen 10 05 **N1749**

1749

Two more examples are given in this slide.

Both are first-order low-pass filters. Both have the same gain and cut-off frequency. The top one is non-inverting however, because of the different switch

arrangement. Some more complicated filter configurations will be discussed later.

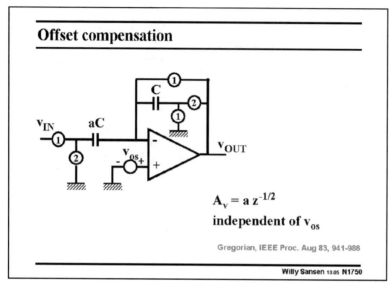

Offset compensation

$$A_v = a\,z^{-1/2}$$

independent of v_{os}

Gregorian, IEEE Proc. Aug 83, 941-986

1750

As the input signal is stored and amplified in the next clock phase, the offset voltage can also be stored and subtracted in the next phase. In this way, the offset of the opamp can be cancelled out.

An example of such a cancellation circuit is shown in this slide.

If the offset v_{os} were zero then it is clear that the gain is accurately zero. In the presence of an offset voltage v_{os}, we want to discover how much of this offset is measured at the output.

For this purpose, charge conservation is applied.

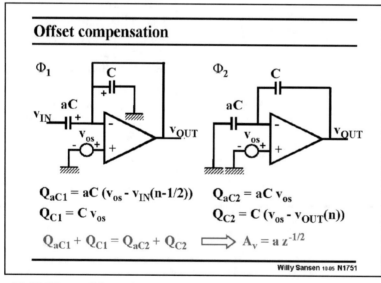

Offset compensation

$$Q_{aC1} = aC\,(v_{os} - v_{IN}(n-1/2))$$

$$Q_{C1} = C\,v_{os}$$

$$Q_{aC2} = aC\,v_{os}$$

$$Q_{C2} = C\,(v_{os} - v_{OUT}(n))$$

$$Q_{aC1} + Q_{C1} = Q_{aC2} + Q_{C2} \implies A_v = a\,z^{-1/2}$$

1751

The same circuit is shown twice, once during clock phase 1 (on the left) and once during clock phase 2 (on the right).

In both clock phases, the charges are written on both capacitors aC and C. The equation of the sum of the charges, shows that the offset voltage v_{os} cancels altogether. Indeed, it appears in all terms. As a result it cancels out.

This technique is also used to cancel the 1/f noise of MOST amplifiers. At very low frequencies, 1/f noise resembles offset. It is now cancelled out at the cost of a small increase of the thermal noise.

Switched-Capacitor Filters

* **Introduction : principle**
* **Technology:**
  * **MOS capacitors**
  * **MOST switches**
* **SC Integrator**
  * **SC integrator : Exact transfer function**
  * **Stray insensitive integrator**
  * **Basic SC-integrator building blocks**
* **SC Filters : LC ladder / bi-quadratic section**
* **Opamp requirements**
  * **Charge transfer accuracy**
  * **Noise**
* **Switched-current filters**

Gregorian, Temes, Analog MOS Integrated
  Circuits for Signal Processing, Wiley, 1986
Laker, Sansen, Design of Analog Integrated
  Circuits and Systems, McGrawHill, 1994
Johns, Martin, Analog Integrated Circuit
  Design, Wiley 1997

Willy Sansen 10 05 **N1752**

1752

More complicated filters can also be constructed.

They can basically be divided into ladder filters and biquads.

Only a few examples are given here. For a more detailed treatment the reader is referred to the references in this slide, the last one of which is the most up-to-date.

4th Order SC low-pass ladder filter

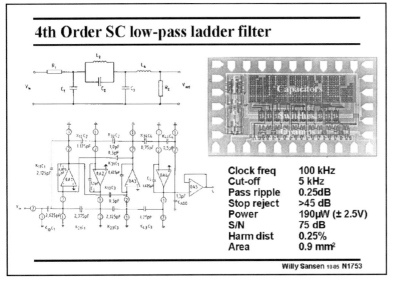

Clock freq	100 kHz
Cut-off	5 kHz
Pass ripple	0.25dB
Stop reject	>45 dB
Power	190µW (± 2.5V)
S/N	75 dB
Harm dist	0.25%
Area	0.9 mm²

Willy Sansen 10 05 **N1753**

1753

One example of a ladder filter is shown in this slide. It has been shown at the beginning of this Chapter.

Ladder filters have the advantage that they are relatively insensitive to errors in the coefficients, or the actual component values. They are normally of odd order although even-order ones are possible as well.

This is a single-ended one. Nowadays, preference is given to fully-differential ones to reject substrate noise.

The power consumption is about 50 µW per pole. Values down to 25 µW have also been achieved. This is only possible however, if class AB opamps are used in both the input AND output stages.

Biquadratic filter

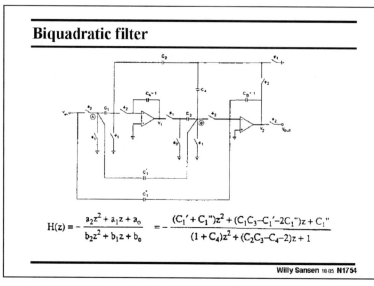

$$H(z) = -\frac{a_2z^2 + a_1z + a_0}{b_2z^2 + b_1z + b_0} = -\frac{(C_1' + C_1'')z^2 + (C_1C_3 - C_1' - 2C_1'')z + C_1''}{(1 + C_4)z^2 + (C_2C_3 - C_4 - 2)z + 1}$$

Willy Sansen 10-05 **N1754**

1754

An alternative circuit configuration is a biquadratic filter. It is a second-order filter configuration with local and overall feedback. It contains two operational amplifiers, as shown in this slide.

The transfer characteristic is of second-order in both the numerator and denominator. It has two poles and zeros. Filter design consists of positioning the poles and zeros accordingly. The reader is referred to the general references on the introductory slide.

For higher-order filters, several biquads can be cascaded.

Finally, note that the C's in the expression refer to capacitor ratios. The unit capacitance has to be selected. It is usually around 0.25 pF.

Nowadays, such filters are realized in a fulyl-differential version to reject substrate noise.

Switched-Capacitor Filters

* **Introduction : principle**
* **Technology:**
    * MOS capacitors
    * MOST switches
* **SC Integrator**
    * SC integrator : Exact transfer function
    * Stray insensitive integrator
    * Basic SC-integrator building blocks
* **SC Filters : LC ladder / bi-quadratic section**
* **Opamp requirements**
    * Charge transfer accuracy
    * Noise
* **Switched-current filters**   McCreary, JSSC Dec 75, 371-379
                                 Gregorian, IEEE Proc. Aug 83, 941-986

Willy Sansen 10-05 **N1755**

1755

Finally, we have to have a look at the specifications of the opamps used in such switched-capacitor filters. After all, they take most of the power consumption.

They are necessary to guarantee full charge transfer from the input (sampling) capacitor to the output (integration) capacitor. They need therefore high gain and low offset. Moreover, they have to settle to within 0.05% in as short a time as possible, to allow use of high-frequency clocks.

This is now discussed in more detail.

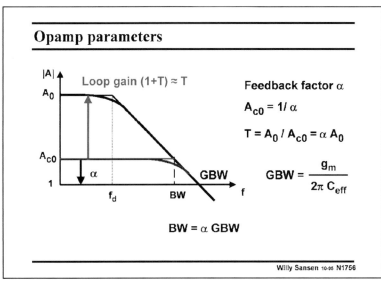

Opamp parameters

Loop gain $(1+T) \approx T$

Feedback factor α

$A_{c0} = 1/\alpha$

$T = A_0 / A_{c0} = \alpha A_0$

$GBW = \dfrac{g_m}{2\pi C_{eff}}$

$BW = \alpha\, GBW$

Willy Sansen 10-95 N1756

1756

When feedback is used around an operational amplifier with gain-band-width product GBW, the closed-loop gain A_{c0} is the inverse of the feedback factor α. This is the ratio between the bandwidth BW and the GBW.

The loop gain T is then the difference (in dB) between the open-loop gain A_0 and the closed-loop gain A_{c0}. It is also given by αA_0.

The loop gain T will determine the accuracy at low frequency or the static accuracy.

The bandwidth will determine the settling time.

For example, take a closed-loop gain of 5, which corresponds to $\alpha = 0.2$. For $A_0 = 10^4$, the loop gain is 2000 or 66 dB. If the GBW = 1 MHz, then the BW is 0.2 MHz. Also, the dominant pole f_d is at 100 Hz. At this frequency, the loop gain starts decreasing, to become unity at 0.2 MHz.

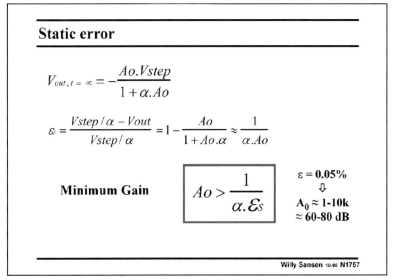

Static error

$$V_{out, t=\infty} = -\frac{Ao.Vstep}{1 + \alpha.Ao}$$

$$\varepsilon = \frac{Vstep/\alpha - Vout}{Vstep/\alpha} = 1 - \frac{Ao}{1 + Ao.\alpha} \approx \frac{1}{\alpha.Ao}$$

Minimum Gain $\boxed{Ao > \dfrac{1}{\alpha.\varepsilon_s}}$

$\varepsilon = 0.05\%$
⇩
$A_0 \approx 1\text{-}10\text{k}$
$\approx 60\text{-}80$ dB

Willy Sansen 10-95 N1757

1757

The static accuracy ε_S for a step input with amplitude V_{step}, is now given by the inverse of the loop gain or by the inverse of αA_0.

For a minimum static accuracy ε_S, there is a minimum amount of gain A_0 required. For example, for $\varepsilon_S = 0.05\%$ we need a loop gain of the inverse or 2000. For a closed-loop gain of 5 $(\alpha = 0.2)$ the open-loop gain A_0 must be 10^4.

Since open-loop gains are not accurately known, a safety factor has to be included of 3–5. The open-loop gain must therefore be 3×10^4 or 90 dB.

Dynamic error

$$\mathcal{E}_D = EXP(-\frac{\alpha.gm.ts}{C_{i,ef}}) \qquad GBW = \frac{gm}{2\pi C_{L,ef}}$$

$$\mathcal{E}_D = EXP(-\alpha.2\pi.GBW.ts) \qquad ts = \frac{1}{2f_c}$$

$$GBW = \frac{1}{\alpha.2\pi.ts}\ln(\frac{1}{\mathcal{E}_D}) = \frac{2f_c}{2\pi.\alpha}\ln(\frac{1}{\mathcal{E}_D})$$

Minimum GBW: $$GBW > \frac{f_c}{\pi.\alpha}\ln(\frac{1}{\mathcal{E}_D})$$

$\varepsilon = 0.05\%$
⇓
$GBW \approx 2\text{-}3*f_c$

Willy Sansen 10-95 N1758

1758

It takes a time t_s, which includes a number of time constants, before the exponential of the voltage across the output capacitor reaches its final value. For a deviation of 0.1%, approximately 7 time constants are required. This deviation is called the dynamic error ε_D. A typical value is again 0.05%.

The time constant itself is $1/(2\pi BW)$ in which the BW equals αGBW. For a single-stage opamp the GBW is determined by the load capacitances, as given in this slide.

The maximum value of settling time t_s is half the clock period, which is the inverse of the clock frequency f_c.

A minimum value is now required for the GBW. The corresponding time constant must be sufficiently small to be able to reach settling with sufficient dynamic accuracy within half a clock period.

The expression is given in this slide.

For example, the term $\ln(1/\varepsilon_D)$ is about 7 for 0.1% but 7.6 for 0.05%. For $\alpha = 0.2$, the GBW must be about 12 times f_c. If α were unity, then 2.4 times f_c would be sufficient. This is where the rule of thumb is coming from that the GBW must be 2–3 times the clock frequency f_c.

Switched-Capacitor Filters

- **Introduction : principle**
- **Technology:**
  - **MOS capacitors**
  - **MOST switches**
- **SC Integrator**
  - **SC integrator : Exact transfer function**
  - **Stray insensitive integrator**
  - **Basic SC-integrator building blocks**
- **SC Filters : LC ladder / bi-quadratic section**
- **Opamp requirements**
  - **Charge transfer accuracy**
  - Noise
- **Switched-current filters**

McCreary, JSSC Dec 75, 371-379
Gregorian, IEEE Proc. Aug 83, 941-986

Willy Sansen 10-05 N1759

1759

Static and dynamic accuracy is one concern, noise is another.

Because of the switching, noise is increased in the low-frequency band, as shown next.

kT/C versus kTR noise

Narrow-band noise >> noise density : $\overline{dv_{ni}^2} = 4kT\,R\,df$

Wide-band noise >> integrated noise : $\overline{v_{ni}^2} = \dfrac{kT}{C}$

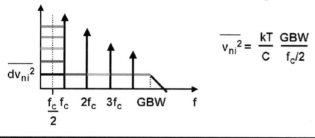

$$\overline{v_{ni}^2} = \frac{kT}{C}\,\frac{GBW}{f_c/2}$$

Willy Sansen 10-05 **N1760**

1760

Because the GBW is always larger than the clock frequency f_c, the noise is folded back towards the lowest frequency band. Actually, this is a heavy case of aliasing.

The total integrated noise of an opamp with load capacitance C (or compensation capacitance for a two-stage opamp) is close to kT/C (see Chapter 4).

The integrated input noise voltage power has to be multiplied by the ratio of the GBW to the clock frequency as shown in this slide. For a GBW which is about 3 times f_c, this gives a multiplication factor of 6 for the noise power or about 2.5 for the noise voltage.

For less noise, larger capacitors must be used. Also, minimum values of GBW must be used.

Switched-Capacitor Filters

- **Introduction : principle**
- **Technology:**
  - **MOS capacitors**
  - **MOST switches**
- **SC Integrator**
  - **SC integrator : Exact transfer function**
  - **Stray insensitive integrator**
  - **Basic SC-integrator building blocks**
- **SC Filters : LC ladder / bi-quadratic section**
- **Opamp requirements**
  - **Charge transfer accuracy**
  - **Noise**
- **Switched-current filters** McCreary, JSSC Dec 75, 371-379
  Gregorian, IEEE Proc. Aug 83, 941-986

Willy Sansen 10-05 **N1761**

1761

Now that switched-capacitor techniques have been discussed, switched-current circuits are added for sake of comparison. They use currents rather than voltages.

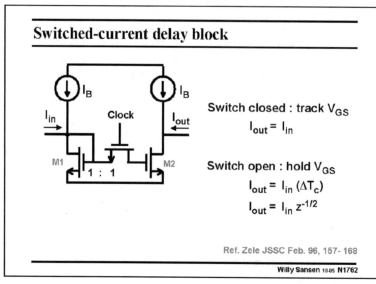

Indeed, when a switch is added to a current mirror as shown in this slide, the operation is exactly as in a current mirror when the switch is closed. For equal transistor sizes, the output current I_{out} equals the input current I_{in}.

When the switch opens however, capacitance C_{GS2} holds the voltage at the Gate of M2. As a result, the output current I_{out} continues to flow, independent of what happens at the input.

This stage has memory. It generates a delay of half a clock period. It can thus be used as a filter in a similar way as a switched-capacitor block acts as a memory and generates delay.

Such a filter is shown next.

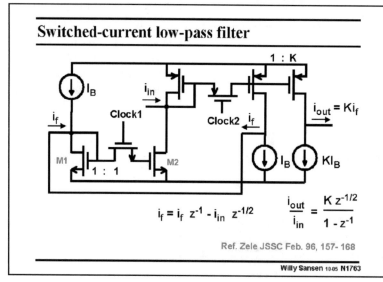

Two switched-current mirrors with feedback yield a low-pass filter, as shown in this slide.

The output current i_{out} equals K times the feedback current i_f.

On clock phase 2, the feedback current i_f is the sum of the input current, applied half a clock period earlier and the feedback current, applied a full clock period earlier.

The current gain is more easily found. It has the same expression in z as a switched-capacitor low-pass filter. The gain is K and the delay is half a clock period.

The difference with a switched-capacitor filter is that currents are used, rather than voltages. The advantage of this filter is that no capacitors are added. The transistor capacitances are used. This type of filter may be more compatible with a digital CMOS process.

Another example is given next.

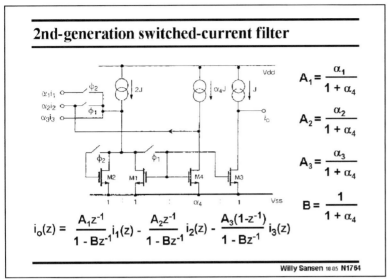

2nd-generation switched-current filter

$$A_1 = \frac{\alpha_1}{1 + \alpha_4}$$

$$A_2 = \frac{\alpha_2}{1 + \alpha_4}$$

$$A_3 = \frac{\alpha_3}{1 + \alpha_4}$$

$$B = \frac{1}{1 + \alpha_4}$$

$$i_o(z) = \frac{A_1 z^{-1}}{1 - B z^{-1}} i_1(z) - \frac{A_2 z^{-1}}{1 - B z^{-1}} i_2(z) - \frac{A_3(1 - z^{-1})}{1 - B z^{-1}} i_3(z)$$

Willy Sansen 10-05 N1764

1764

In this filter more is done with less transistors. A number of second-order effects can be avoided.

Judicious playing with transistor size ratios allows construction of fairly sophisticated filter characteristics.

The actual derivation is left as an exercise for the reader.

Comparison SC - SI

	SC	SI
Signal :	Voltage	Current
	Charge on linear C	Charge on MOST C_{GS}
	$Q = C\,V$	$Q = I\,t$
Accuracy :	Capacitor ratio	MOST area ratio
	0.2 %	2 %
Amps :	Opamps	Current mirrors
S/N+D	70 dB	50 dB

Willy Sansen 10-05 N1765

1765

A comparison between switched-capacitor and switched-current filters is now imperative.

In a switched-capacitor filter a charge is stored on a capacitor. The signal is therefore a voltage. The accuracy depends on the full charge transfer from one capacitor to another. This also depends on the matching of these two capacitors.

Mismatch, clock injection and charge distribution limit the dynamic range to about 70 dB without excessive precautions.

In a switched-current filter a charge is a result of a current flowing during a certain amount of time, determined by the clock period. The signal is a current. The accuracy depends on the matching of transistor sizes.

Mismatch, clock injection and charge distribution limit the dynamic range to about 50 dB without excessive precautions.

The main difference is that in a switched-current filter the charge transfer is less accurate because of the worse mismatch between transistors than between capacitors. The main advantage of switched-current filters is that they reach higher frequencies because they do not use opamps, just current-mirrors.

It is clear that such a comparison can only be of first-order. A full comparison would take a full workshop.

Another comparison between all important filter types will be given at the end of Chapter 19.

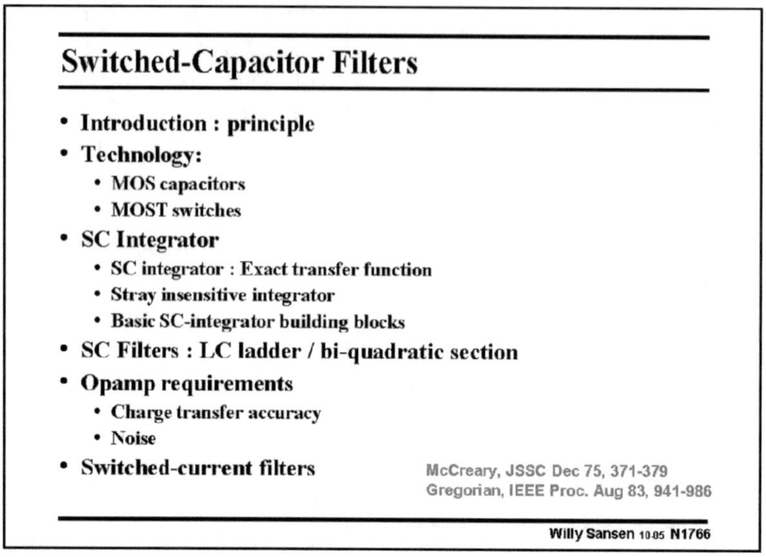

Switched-Capacitor Filters

- **Introduction : principle**
- **Technology:**
  - MOS capacitors
  - MOST switches
- **SC Integrator**
  - SC integrator : Exact transfer function
  - Stray insensitive integrator
  - Basic SC-integrator building blocks
- **SC Filters : LC ladder / bi-quadratic section**
- **Opamp requirements**
  - Charge transfer accuracy
  - Noise
- **Switched-current filters**   McCreary, JSSC Dec 75, 371-379
                                        Gregorian, IEEE Proc. Aug 83, 941-986

                                        Willy Sansen 10 05 **N1766**

1766

In this Chapter switched-capacitor techniques have been introduced. Considerable attention is paid to the capacitors, switches and opamps required to build such filters.

Also the limitations are discussed such as mismatch, clock injection, charge redistribution and noise.

Finally a short comparison with switched-current filters is given.

Several more types of filters are available on silicon. Continuous-time filters are next.

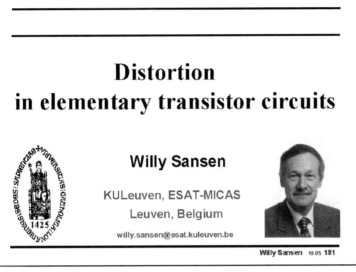

Distortion in elementary transistor circuits

Willy Sansen

KULeuven, ESAT-MICAS

Leuven, Belgium

willy.sansen@esat.kuleuven.be

Willy Sansen 10.05 **181**

181

Distortion has become an important topic. The larger the signal levels, the larger the distortion becomes.

Indeed, the supply voltage decreases with smaller channel lengths. The signal levels have to be made as large as possible to obtain a Signal-to-Noise-and-Distortion ratio, which is as high as possible.

We now need to know which distortion levels correspond with which signal levels.

Unfortunately, distortion is a kind of garbage can for everything that impairs the spectral purity of a signal. Every month new kinds of distortion are discovered. We will concentrate, however, on the main sources. The distortion levels will be described in terms of signal amplitudes.

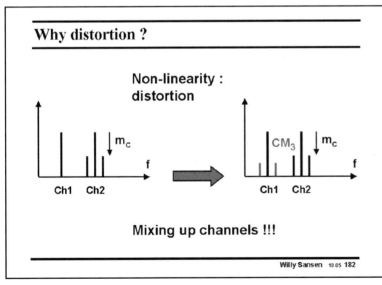

Why distortion ?

Non-linearity : distortion

m_c

CM_3 m_c

Ch1   Ch2   Ch1   Ch2

Mixing up channels !!!

Willy Sansen 10.05 **182**

182

Distortion is most important in communication systems, in which many frequency channels have to be processed.

For example, two channels are shown at two adjacent frequencies. The second of which contains modulation information. Its modulation index is m_c.

Distortion will cause the modulation to be transferred from Channel 2 to Channel 1. This is called cross-modulation. We want

to calculate how the cross-modulation is for a non-linear system.

Table of contents

Willy Sansen 10 05 183

183

Before we engage in calculations, we want to review the definitions commonly used for distortion.

After that we will calculate the distortion generated by a MOST, both as a single-ended amplifier as in a differential amplifier. This is repeated for a bipolar transistor.

The main technique used to reduce distortion is feedback. Calculations are given to predict the distortion when feedback is present.

Finally, some examples are given on how to calculate distortion in more complicated circuits, including operational amplifiers. Both two-stage and three-stage operational amplifiers will be discussed.

Finally some other sources of distortion are mentioned but not discussed in detail.

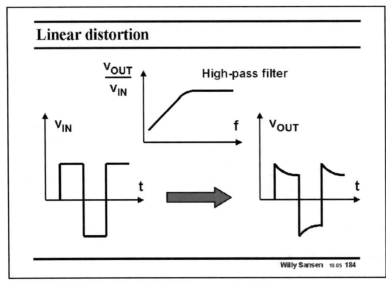

Willy Sansen 10 05 184

184

We will not discuss the effects of linear distortion in this Chapter but of nonlinear distortion.

Linear distortion is a result of filter action. A perfectly linear system which shows some kind of filtering action, will distort the picture of the signal versus time. For example, a high-pass filter will emphasize the steep edges of a square waveform, as they represent the higher frequencies. The output waveform has therefore peaked. It is different from the input waveform, and is distorted. This is linear distortion. It occurs for all filters.

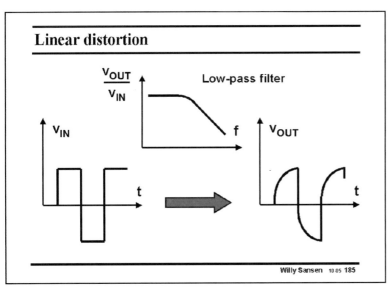

185
A similar effect is seen for low-pass filters. Any low-pass filter will eliminate the steep edges of a square waveform, as they represent the higher frequencies. The output waveform is now rounded. It is again different from the input waveform, and is distorted. This kind of linear distortion occurs for all filters.

It will not be discussed any further.

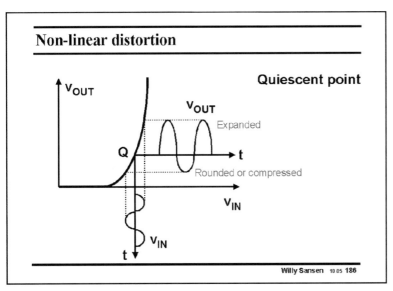

186
Nonlinear distortion on the other hand is generated by a nonlinear transfer curve, as shown in this slide. The output voltage is nonlinear in terms of the input voltage for this amplifier.

This amplifier is biased at a specific point Q, the quiescent point. A sine waveform will generate an output waveform which is distorted. The bottom half is compressed, whereas the top half is expanded.

The larger the amplitude of the input voltage, the more distorted the output voltage will be, as a larger fraction of the transfer curve is covered around point Q.

Distortion levels will always increase for increasing input amplitudes.

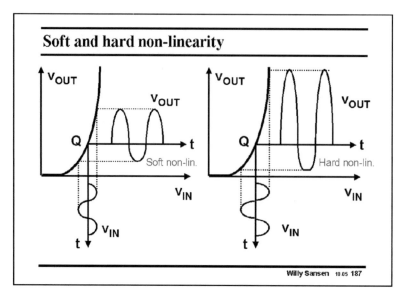

187

When the input voltage becomes too large, clipping occurs of the output voltage. This is called a hard nonlinearity.

At this point, the distortion is very large, more than 10% for example. Few applications can tolerate such large values. We will limit ourselves to low levels of distortion, of the level of 0.1% (−60 dB) down to 0.001% (−100 dB) depending on the application.

We are therefore only interested in soft nonlinearities, which give small amounts of distortion only.

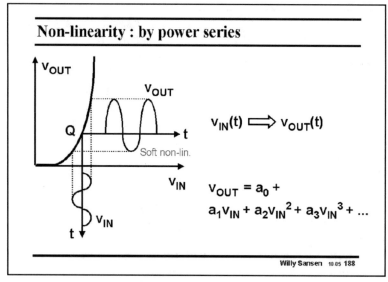

$$v_{IN}(t) \Longrightarrow v_{OUT}(t)$$

$$v_{OUT} = a_0 + a_1 v_{IN} + a_2 v_{IN}^2 + a_3 v_{IN}^3 + ...$$

188

Such soft nonlinearity can be described by means of a power series, as shown in this slide.

The coefficient a_0 gives the DC output voltage in quiescent point Q. Coefficient a_1 gives the small-signal gain. It is the slope of the transfer curve in point Q. Coefficient a_2 gives the second-order nonlinearity. Actually, it represents the even-order nonlinearities, as coefficients a_4, a_4, etc. become gradually smaller.

In the same way, third-order coefficient a_3 represents the higher-order ones a_5, a_7, etc.

How to find a_0, a_1, a_2, a_3, ...

$$y = a_0 + a_1 u + a_2 u^2 + a_3 u^3 + ...$$

$$a_0 = y\big|_{u=0} \qquad a_1 = \frac{dy}{du}\bigg|_{u=0}$$

$$a_2 = \frac{1}{2}\frac{d^2 y}{du^2}\bigg|_{u=0} \qquad a_3 = \frac{1}{6}\frac{d^3 y}{du^3}\bigg|_{u=0}$$

Willy Sansen 10.05 189

189

For any nonlinear transfer curve, the coefficient a_0, a_1, a_2, a_3, ... can easily be found by taking derivatives.

Coefficient a_0 is simply the DC value, which is reached for zero small signal input level u.

Coefficient a_1 is clearly the first derivative of the output signal y with respect to the input signal u.

The other coefficients are obtained as indicated. Corrections have to be made for the coefficients appearing during the derivation. It is clear that for coefficient a_3 for example, the transfer curve must be sufficiently smooth. Several MOST models were not that smooth at the cross-over points, prohibiting the calculation of coefficient a_3.

Definition of harmonic distortion HD

$$y = a_0 + a_1 u + a_2 u^2 + a_3 u^3 + ...$$

With u = U cos ωt $\qquad \cos^2 x = 1/2\,(1 + \cos 2x)$
$\qquad\qquad\qquad\qquad\qquad \cos^3 x = 1/4\,(3\cos x + \cos 3x)$

$$y = a_0 + a_1 u + a_2 u^2 + a_3 u^3 + ... \;\; = a_0 +$$

$$\left(a_1 + \frac{3}{4} a_3 U^2\right) U \cos ωt + \frac{a_2}{2} U^2 \cos 2ωt + \frac{a_3}{4} U^3 \cos 3ωt$$

$$\boxed{HD_2 = \frac{1}{2}\frac{a_2}{a_1} U} \qquad \boxed{HD_3 = \frac{1}{4}\frac{a_3}{a_1} U^2}$$

Willy Sansen 10.05 1810

1810

Once the nonlinearity has been described by a power series, the harmonic distortion can easily be calculated.

In an input signal, u is applied with amplitude U and frequency ω, then a little trigonometry helps us to find the contributions at 2ω and 3ω.

The ratio of the component at 2ω to the fundamental at ω is then by definition the second harmonic distortion. Since coefficient a_3 is usually quite small compared to a_1, the component at the fundamental at ω is just about $a_1 U$.

Note that the second harmonic distortion HD_2 is proportional to the amplitude U of the fundamental. Doubling the input voltage will therefore double the second harmonic distortion.

In the same way, the third harmonic distortion is defined.

The ratio of the component at 3ω to the fundamental at ω is thus by definition the third harmonic distortion. Note that the third harmonic distortion HD_3 is proportional to the square of the amplitude U of the fundamental. Doubling the input voltage will multiply the third harmonic distortion by four.

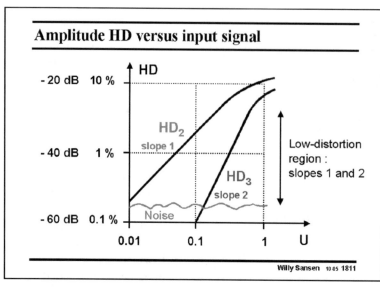

1811

These relationships are easily plotted on log scales as shown in this slide. They give straight lines with slopes of 1 dB/dB for HD_2 and 2 dB/dB for HD_3.

Clearly, this is only valid for low levels of distortion. For high input signals the curves flatten because of the high levels of distortion. We never want to increase the input voltage that much!

For low input signals the distortion is drowned in noise. Clearly, it is impossible to recognize the slopes in that region.

It is clear from the above, that the only way to measure distortion is to find the region where it has a linear dependence (on a log-log scale) on the input signal amplitude and to check whether the slopes are right. The second-order distortion must have a slope of 1 dB/dB and the third-order one 2 dB/dB.

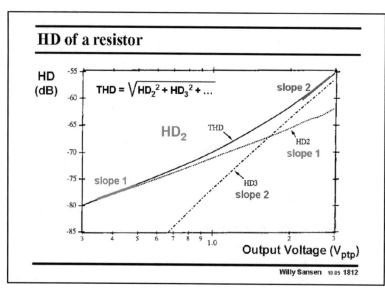

1812

Some measurement equipment only measures total harmonic distortion THD, which is a kind of RMS value (or effective value) of all distortion ratios. In this way it is impossible of course to verify the slopes.

However, quite often it is possible to identify the slopes in different regions of the input voltage.

For example the THD was measured of a diffused resistor. For the low voltage across it, the HD_2 is dominant, whereas for high voltage, the HD_3 is dominant. Indeed, they are clearly identified by their slopes.

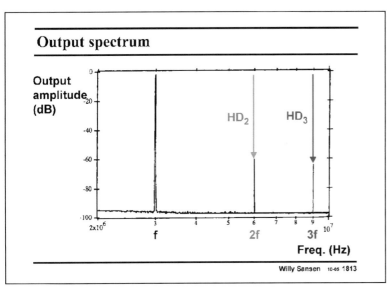

Output spectrum

Output
amplitude
(dB)

HD$_2$ HD$_3$

f   2f   3f

Freq. (Hz)

Willy Sansen 10-05 1813

1813

The distortion is also measured by means of a spectrum analyzer. An example is shown in this slide for a fundamental frequency of 30 MHz. The second harmonic is set at 60 MHz and the third one at 90 MHz.

Note however, that the components are now given, not the ratios of HD$_2$ and HD$_3$. Increasing the fundamental component by 1 dB will raise the second-order by 2 dB and the third-order by 3 dB.

Definition of intermodulation distortion IM

$$y = a_0 + a_1 u + a_2 u^2 + a_3 u^3 + \ldots$$

with $u = U (\cos \omega_1 t + \cos \omega_2 t)$

$$y = a_0 + \ldots$$

IM$_2$ at $\omega_1 \pm \omega_2$

IM$_3$ at $2\omega_1 \pm \omega_2$ and $\omega_1 \pm 2\omega_2$

$$IM_2 = 2\,HD_2 = \frac{a_2}{a_1} U$$

$$IM_3 = 3\,HD_3 = \frac{3}{4} \frac{a_3}{a_1} U^2$$

Willy Sansen 10.05 1814

1814

Another way to characterize distortion is to use intermodulation distortion.

For this purpose, two sine waves have to be applied. In this case, they have equal amplitudes U and frequencies ω_1 and ω_2. This is more common in communication systems where two adjacent channel frequencies are taken. In HiFi systems on the other hand, frequencies are taken of 50 Hz and 4 kHz with widely different amplitudes.

These two fundamental frequencies will generate all intermodulation products, if applied to a nonlinear system, described by a power series.

Substitution of input signal u generates second-order intermodulation products with coefficient a_2 and third-order intermodulation products with coefficient a_3.

IM$_2$ is now the ratio of the two components at $\omega_1 \pm \omega_2$ to the fundamental. In a similar way IM$_3$ is now the ratio of the four components at $2\omega_1 \pm \omega_2$ and $\omega_1 \pm \omega_2$ to the fundamental.

In order to learn where all these components occur on the frequency axis, a picture is given next.

Note, however, that there is a very simple relationship between IM and HD. For example, IM$_3$ is about 10 dB higher than HD$_3$.

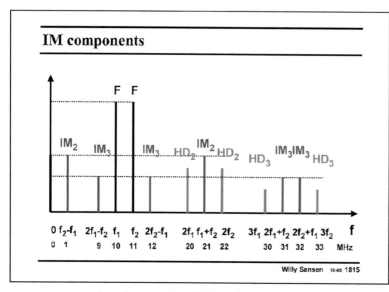

1815

In order to show where all these intermodulation components appear, a simple example is given in which the fundamental frequencies are at 10 and 11 MHz.

Both have their second harmonics at 20 and 22 MHz and their third harmonics at 30 and 33 MHz. They are all green.

The second-order intermodulation components are at 1 and 21 MHz. It is clear that second-order distortion generates intermodulation components at low frequencies. This is important for HiFi amplifiers for example!

Third-order intermodulation shows up in four places, i.e. at 9, 12, 31 and 32 MHz. The ones at 9 and 12 MHz are especially important, as they appear in the adjacent channels. They are also easier to measure with a spectrum analyzer as they appear in the same narrow frequency region as the fundamentals.

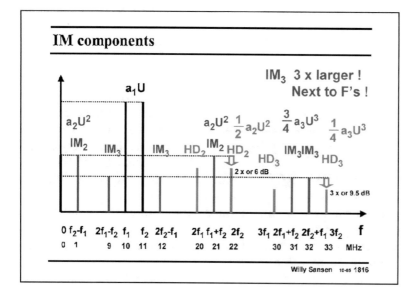

1816

There are thus several reasons why we will focus on IM_3.

First of all, IM_3 is 10 dB larger than HD_3. It gives components next to the fundamentals which are to measure, and finally, they are the only important ones in differential systems, in which second-order distortion is cancelled out, as we will see later.

1817

An example of such a spectrum is shown in this slide.

Two frequencies are applied to this IF filter, of 10.695 and 10.705 MHz. The IM_3 components

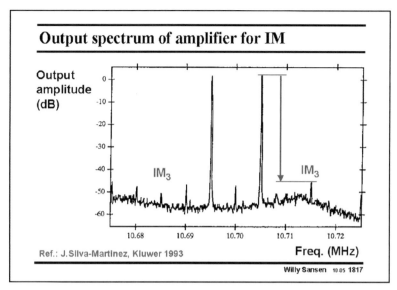

Output spectrum of amplifier for IM

Ref.: J.Silva-Martinez, Kluwer 1993

Willy Sansen 10 05 1817

can therefore be found at equal distances on both sides, i.e. at 10.685 and 10.715 MHz. The other distortion components have nothing to do with this. They are generated by some other mechanisms.

Also, note that both IM_3 components are supposed to be equal. However, small differences are always possible!

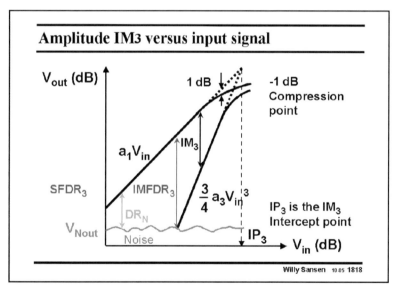

Amplitude IM3 versus input signal

Willy Sansen 10 05 1818

1818
Several more measures exist for IM_3. A few of them are shown in this slide. A differential amplifier is taken with input voltage V_{in}, which gives a nonlinear output voltage V_{out}, described by a power series with coefficients a_1 and a_3. Because of the differential nature $a_2 = 0$.

The components themselves are plotted on a log scale versus (log) input voltage amplitude. The fundamental itself has a slope of 1 dB/dB, whereas the third-order IM_3 components have a slope of 3 dB/dB. The ratio between both is the IM_3. Its maximum value is attained when the IM3 level equals the noise level. This is the largest possible dynamic range that can be reached. It is the third-order Intermodulation Free Dynamic Range $IMFDR_3$. It obviously occurs at only one specific value of the input voltage.

Any other Noise Dynamic Range is smaller than the $IMFDR_3$.

The input voltage, for which their extrapolated values meet is called the IM_3 intercept point IP_3. The input voltage, for which the output is compressed by 1 dB is called the -1 dB compression point. Obviously, both of them must be related to the expression of the IM_3, as shown next.

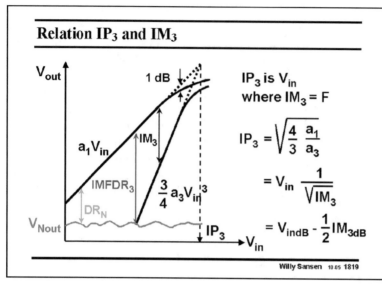

1819

The IP_3 is reached at the input voltage for which the amplitude of the fundamental $a_1 V_{in}$ equals IM_3. Its value is easily calculated from the power series. It is obviously determined by the same coefficients a_3 and a_1 as IM_3 itself.

The relation with IM_3 is now easily found, both in absolute value, as in dB.

For example, for coefficients of $a_3 = 0.01$ and $a_1 = 0.5$, $IM_3 = 3.4 \times 10^{-4}$ or -69.4 dB for 0.15 V RMS

input voltage (-16.5 dB); then $IP_3 = 8.16$ V or 18.2 dB. Note that, 1 V is taken as a reference and hence all dB are actually dBV. In high-frequency systems such as communication systems, 1 V is usually not seen as a reference. It is the Voltage corresponding to 1 mW in 50 Ω. This gives as a reference Voltage, the square root of 0.05 V^2 or 0.2236 V_{RMS}. In this case 13 dB (which is $20 \times \log(0.2236)$) has to be added to the dBV values in this slide, which gives an input voltage of -3.5 dBm, an IM_3 of -56.4 dBm and an IP_3 of 31.2 dBm.

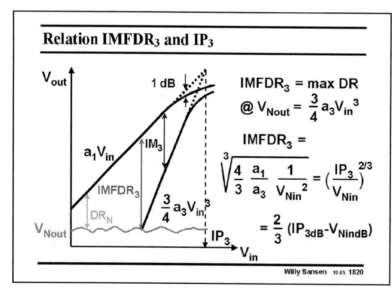

1820

The maximum dynamic range $IMFDR_3$ is also easily found. It is the IM_3 ratio at the input voltage where the IM_3 component at the output equals the output noise level V_{nout}.

Since it also depends on coefficient a_3, it can be described in terms of the IP_3.

For example, for the same values as on the previous slide, the $IMFDR_3 = 48.8$ dB (when the noise level is at -42 dBm).

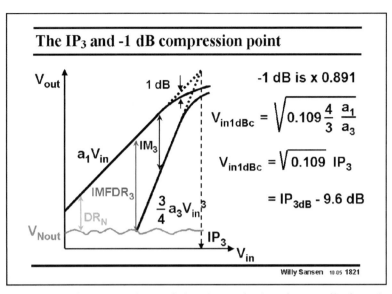

1821

Another way to specify the IM_3 distortion is the -1 dB compression point V_{in1dBc}. It is the input voltage for which the output amplitude is reduced by 1 dB, as a result of the subtraction of the IM_3 component. This is easy to measure, but is not so accurate, as 1 dB is not that easy to distinguish.

It is also given by the a_1/a_3 ratio as both IM_3 and the IP_3.

A reduction by 1 dB corresponds to a coefficient of about 0.9. The expressions of the V_{in1dBc} is therefore easily found.

Note that the V_{in1dBc} is always approximately 10 dB smaller than the IP_3 in dB.

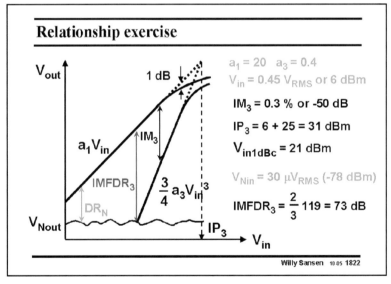

1822

As another exercise, a differential amplifier is taken with a large amount of gain. Its value is 20 or 26 dB. For the third-order coefficient a_3 of 0.4 and input voltage of 0.45 V_{RMS}, all the values are calculated again, which specify the third-order distortion. They are the IM_3, the IP_3 and the V_{in1dBc}.

If the input noise level is given (which is the output noise divided by the gain), then the $IMFDR_3$ can be calculated as well.

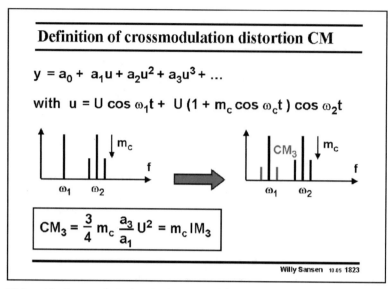

Definition of crossmodulation distortion CM

$$y = a_0 + a_1 u + a_2 u^2 + a_3 u^3 + \ldots$$

with $u = U \cos \omega_1 t + U (1 + m_c \cos \omega_c t) \cos \omega_2 t$

$$CM_3 = \frac{3}{4} m_c \frac{a_3}{a_1} U^2 = m_c IM_3$$

Willy Sansen   10 05   1823

1823

Another advantage of dealing with IM_3 is that it also gives the third-order crossmodulation CM_3. Crossmodulation is what we are really after, as it describes how much modulation information is transferred from one carrier to its adjacent carrier.

Application of an amplitude modulated waveform with modulation index m_c to the same power series, shows that the CM_3 is simply the product of the IM_3 with the modulation index m_c. This is also true to some extent for other modulation systems.

There are plenty of reasons to concentrate on the IM_3 as the most important specification of third-order distortion.

Table of contents

Willy Sansen   10 05   1824

1824

Let us find out what third-order distortion is given by a MOST and then by a bipolar transistor. As a bipolar transistor has the steepest transfer curve, and thus the highest g_m/I_{CE} ratio, it will also give the highest distortion.

A MOST on the other hand, only has a square-law characteristic. It gives much less distortion. This is why it is used at the input of most receivers, as well as in other exotic technologies such as GaAs.

Distortion in a single-MOST amplifier

$$i_{DS} = K (v_{GS} - V_T)^2 \qquad K = K' \frac{W}{L}$$

$$I_{DS} + i_{ds} = K (V_{GS} + v_{gs} - V_T)^2$$

I_{DS} is the DC component

i_{DS} is the DC + ac component

i_{ds} is the ac component

I_{ds} is the amplitude of the ac component

Willy Sansen 10 05 1825

1825

The current of a MOST can still be described by a square-law relationship in its simplest form.

Application of a small-signal v_{gs}, with amplitude V_{gs}, superimposed on a DC biasing voltage V_{GS}, yields a small-signal current i_{ds} superimposed on the DC current I_{DS}.

Note that we have to write voltages and currents in the correct way so as to clarify which components are actually needed.

A graphical explanation is given next.

DC and ac components

I_{DS} : DC component

i_{DS} : DC + ac component

i_{ds} : ac component

I_{ds} : amplitude of the ac component

Willy Sansen 10 05 1826

1826

DC components are always written with capitals: I_{DS} is therefore the DC current through the transistor.

The instantaneous value of the small-signal or AC component is indicated by i_{ds}, which has an amplitude I_{ds}.

Finally, the total DC and AC signal is denoted by i_{DS}. This notation is used to give general model and network expressions.

Distortion in a single-MOST amplifier

$$I_{DS} = K (V_{GS} - V_T)^2 \qquad K = K' \frac{W}{L}$$

$$I_{DS} + i_{ds} = K (V_{GS} + v_{gs} - V_T)^2$$

$$i_{ds} = K (V_{GS} + v_{gs} - V_T)^2 - K (V_{GS} - V_T)^2$$

$$i_{ds} = 2K (V_{GS} - V_T) v_{gs} + K v_{gs}^2$$

Willy Sansen 10.05 **1827**

1827
Subtraction of the DC component I_{DS} from the total current expression i_{DS} or $I_{DS} + i_{ds}$, gives the AC component i_{ds}. It is easily converted in a power series of i_{ds} versus input voltage v_{gs}.

It is not surprising to see that no third-order coefficient is present. Indeed, a third-order component cannot be obtained from a square-law relationship! There is no IM_3, only IM_2!

Coefficients a₁, a₂, a₃ by comparison

$$i_{ds} = 2K (V_{GS} - V_T) v_{gs} + K v_{gs}^2$$

$$\text{or} \quad i_{ds} = g_1 v_{gs} + g_2 v_{gs}^2 + g_3 v_{gs}^3 + \dots$$

$$g_1 = 2K (V_{GS} - V_T)$$

$$g_2 = K \qquad\qquad\qquad K = K' \frac{W}{L}$$

$$g_3 = 0$$

$$IM_2 = \frac{g_2}{g_1} V_{gs} = \frac{V_{gs}}{2(V_{GS} - V_T)} \qquad \& \quad IM_3 = 0$$

Willy Sansen 10.05 **1828**

1828
The coefficients g of the power series are now easily identified.

The first is g_1, obviously the transconductance g_m of the MOST.

The second is g_2, this represents the second-order distortion, described by IM_2. Its value is proportional to the peak amplitude of the input voltage V_{gs}, scaled by the value of $V_{GS} - V_T$ used.

For low distortion, large values must be used of $V_{GS} - V_T$.

No IM_3 is present, which is an enormous advantage in a single-ended MOST amplifier. Of course, this is only true in so far as the current of a MOST can be accurately described by the simple square-law relationship. This is nowadays only true for values of $V_{GS} - V_T$, slightly over 0.2 V.

Normalized current swing

$$i_{ds} = 2K (V_{GS} - V_T) v_{gs} + K v_{gs}{}^2 \qquad i_{DS} = K (v_{GS} - V_T)^2$$

$$\text{or} \quad y = a_1 u + a_2 u^2 + a_3 u^3 + ..$$

$$y = \frac{i_{ds}}{I_{DS}} = \frac{2 v_{gs}}{V_{GS} - V_T} + \frac{1}{4} \left(\frac{2 v_{gs}}{V_{GS} - V_T} \right)^2$$

$$y = \frac{i_{ds}}{I_{DS}} = u + \frac{1}{4} u^2 \qquad\qquad U = \frac{V_{gs}}{(V_{GS} - V_T)/2}$$

y is the relative current swing !

Willy Sansen 10-05 1829

1829

A more general way of describing distortion is to use the relative current swing U rather than the current I_{ds} or the input voltage V_{gs}. It is defined as the ratio of the peak AC current I_{ds} to the DC current I_{DS}.

We will see that distortion can easily be calculated once we know what the relative current swing is. Moreover, we will find that any technique which reduces the relative current swing, such as feedback, can be used to reduce distortion.

The power series for the relative current swing is denoted by y. Its first-order component is denoted by u, the peak value is denoted by U. Its value is obviously given by the same ratio V_{gs} to $V_{GS} - V_T$.

Choosing a large value of $V_{GS} - V_T$ will therefore reduce the relative current swing, and hence the IM_2 distortion.

Numerical example

The peak value of V_{gs} is V_{gsp} = 100 mV

(then V_{gsRMS} = 100 /√2 = 71 mV$_{RMS}$)

if V_{GS}-V_T = 0.5 V then V_{gsp}/[2(V_{GS}-V_T] = 0.1

gives IM_2 = 10 % (HD_2 = 5 %) & IM_3 = 0

The relative current swing U = 0.1/0.25 = 0.4 !

Willy Sansen 10-05 1830

1830

An example is given in this slide.

For a peak relative current swing of 40%, the IM_2 is quite high, i.e. 10%. This is reached for an input voltage of 71 mV$_{RMS}$ if $V_{GS} - V_T = 0.5$ V.

More coefficients a_1, a_2, a_3 ...

In general

$$i_{ds} = g_m v_{gs} + K_{2gm} v_{gs}^2 + K_{3gm} v_{gs}^3 +$$
$$g_o v_{ds} + K_{2go} v_{ds}^2 + K_{3go} v_{ds}^3 +$$
$$g_{mb} v_{bs} + K_{2gmb} v_{bs}^2 + K_{3gmb} v_{bs}^3 +$$
$$K_{2gm\&gmb} v_{gs} v_{bs} + K_{3,2gm\&gmb} v_{gs}^2 v_{bs}$$
$$+ K_{3,gm\&2gmb} v_{gs} v_{bs}^2 +$$
$$............ +$$
$$K_{3gm\&gmb\&go} v_{gs} v_{ds} v_{bs}$$

Willy Sansen   10 05 1831

1831
A MOST has many more distortion components. Only K_{2gm} has been discussed hitherto. In general, K_{3gm} is not zero, leading to IM_3. Finding AC voltages at the Drain or at the Bulk, will also generate distortion, and lots of intermodulation products, most of which are of less importance. However, especially as K_{2go} is K_{3g0} have become more important as the output conductance of deep submicron transistor have become quite small.

Maybe all distortion components of a MOST have never been fully extracted, certainly not in all three regions of operation such as strong inversion, weak inversion and velocity saturation.

We will concentrate mainly on coefficient K_{2gm} which corresponded with K or K'W/L in the original (or simplest) power series.

Distortion of a MOST diode

$$i_{DS} = K (v_{DS} - V_T)^2$$

$$y = \frac{i_{ds}}{I_{DS}} = \frac{2 v_{ds}}{V_{DS} - V_T} + \frac{1}{4} \left(\frac{2 v_{ds}}{V_{DS} - V_T} \right)^2$$

$$y = \frac{I_{ds}}{I_{DS}} = u + \frac{1}{4} u^2 \qquad U = \frac{V_{ds}}{(V_{DS} - V_T)/2}$$

Same as for a MOST transistor amplifier !

Willy Sansen   10 05 1832

1832
A MOST with its Drain connected to the Gate is called a diode-connected MOST. It is as non-linear as a MOST. This is clearly seen when v_{GS} is substituted by v_{DS} in the square-law expression of the MOST.

The relative current swing and the IM_2 are the same as for the MOST.

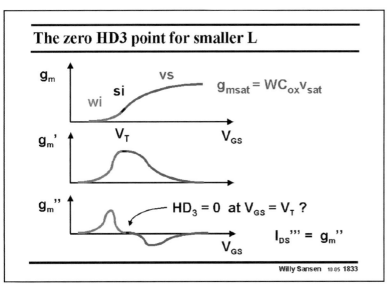

The zero HD3 point for smaller L

g_m

wi si vs

$g_{msat} = WC_{ox}v_{sat}$

V_{GS}

g_m'

V_T

V_{GS}

g_m''

$HD_3 = 0$ at $V_{GS} = V_T$?

$I_{DS}''' = g_m''$

V_{GS}

Willy Sansen 10 05 **1833**

1833

For deep-submicron or manometer CMOS, the strong-inversion has become quite small. It is the region where the curvature is inverted. The exponential of the weak-inversion region curves upwards whereas the flattening in velocity saturation gives a downwards curvature.

As a result, the second derivative of the transconductance goes through zero. This is a point of zero HD_3 and IM_3. It occurs for values just above V_T. Some parameter extraction routines take this point as V_T !

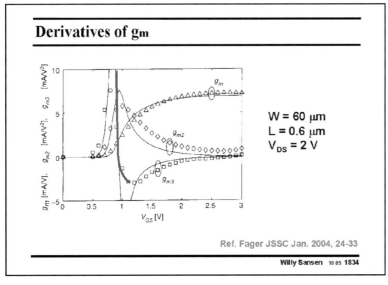

Derivatives of g_m

g_m

g_{m2}

g_{m3}

W = 60 μm
L = 0.6 μm
V_{DS} = 2 V

Ref. Fager JSSC Jan. 2004, 24-33

Willy Sansen 10 05 **1834**

1834

In practice, however, this point of zero HD_3 is quite sharp. Any variation in $V_{GS} - V_T$ or in transistor size gives a large increase in HD_3.

Some experimental curves are shown in this slide. The crossover from weak inversion (red) to velocity saturation (blue) is barely visible. It is exceedingly hard to maintain the biasing ($V_{GS} - V_T$) at this point.

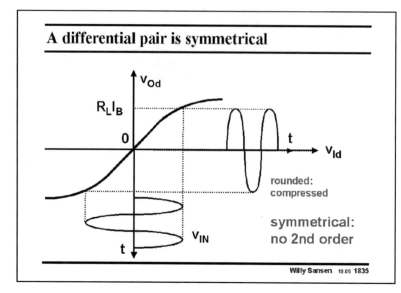

A differential pair is symmetrical

rounded:
compressed

symmetrical:
no 2nd order

Willy Sansen 10.05 **1835**

1835

When two MOSTs are connected as a differential pair, then the limiting transfer characteristic emerges as described in Chapter 3. The input sine wave is therefore rounded or compressed.

If no offset is present, then the rounding is the same on both halves of the sine wave. The DC output voltage will not change and the IM_2 is zero. As long as the rounding on both halves is the same, symmetry is maintained and no IM_2 is found.

However, third-order distortion is now generated as a result of the rounding.

Distortion in MOST differential pair

$$y = \frac{i_{Od}}{I_B} = \frac{v_{Id}}{V_{GS}-V_T} \sqrt{1 - \frac{1}{4}\left(\frac{v_{Id}}{V_{GS}-V_T}\right)^2}$$

v_{Id} is the differential input voltage

i_{Od} is the differential output current ($g_m v_{Id}$) or
twice the circular current $g_m v_{Id}/2$

I_B is the total DC current in the pair

Note that $g_m = \dfrac{I_B}{V_{GS}-V_T} = K' \dfrac{W}{L}(V_{GS}-V_T)$

Willy Sansen 10.05 **1836**

1836

The relative current swing in a differential pair with MOSTs is derived in Chapter 3 and repeated here.

Care must be taken to have the factors of 2 right!

For a small input voltage the output current is linear. The square root with the squared relative input voltage becomes important and causes the flattening of the transfer curve in both directions.

Distortion in MOST differential amplifier

$$y = \frac{i_{Od}}{I_B} = \frac{v_{Id}}{V_{GS}-V_T} \sqrt{1 - \frac{1}{4}\left(\frac{v_{Id}}{V_{GS}-V_T}\right)^2}$$

$$\sqrt{1-x} \approx 1 - \frac{x}{2}$$

$$y = \frac{i_{Od}}{I_B} = U\sqrt{1 - \frac{1}{4}U^2} \approx U - \frac{1}{8}U^3$$

$$IM_2 = 0 \qquad \boxed{IM_3 = \frac{3}{32}U^2} \qquad U = \frac{v_{Id}}{V_{GS}-V_T}$$

U is the relative current swing

$$IP_3 = 4\sqrt{\frac{2}{3}}(V_{GS}-V_T) \approx 3.3(V_{GS}-V_T)$$

Willy Sansen 10-05 1837

1837

For a small relative input voltage, the square root can be developed into a power series, only the first term of which is retained.

The power series of the relative current swing y has become very simple. Coefficient $a_1 = 1$ and $a_3 = 1/8$. Obviously, IM_2 is zero and the IM_3 is simply about one tenth of the peak relative current swing U squared.

Again, for a large value of $V_{GS}-V_T$, the relative current swing is small and so is the IM_3.

This is also obvious from the IP_3, which is proportional to $V_{GS}-V_T$. For a $V_{GS}-V_T$ of 0.5 V, we obtain an IP_3 of 1.65 V or 17.4 dBm.

Distortion in linear region

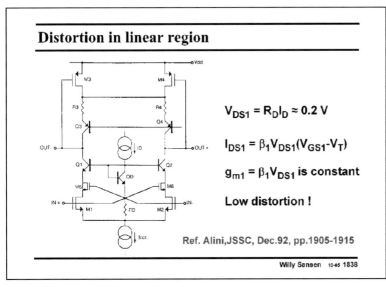

$$V_{DS1} = R_D I_D \approx 0.2 \text{ V}$$

$$I_{DS1} = \beta_1 V_{DS1}(V_{GS1}-V_T)$$

$$g_{m1} = \beta_1 V_{DS1} \text{ is constant}$$

Low distortion !

Ref. Alini,JSSC, Dec.92, pp.1905-1915

Willy Sansen 10-05 1838

1838

The third-order distortion in a differential pair can be reduced by biasing the MOSTs in the linear region.

For this purpose, the values of V_{DS} of the input MOSTs must be kept constant at a value between 100 and 200 mV. This is achieved by means of a constant-current source I_D. If $V_{GS1} = V_{GS2} = V_{GSD}$ then the input transistors M1 and M2 have equal V_{DS}. The resulting values of g_{m1} and g_{m2} are less in absolute value than in the saturation region, but more constant.

Moreover, changing the value of the current source I_D allows a change in the input transconductance for use in a $g_m - C$ filter (see next Chapter).

The transistors M5 and M6 are shorted. They are used to compensate the poles associated with the input capacitances C_{GS1} and C_{GS2}.

Table of contents

- Definitions : HD, IM, intercept point, ..
- Distortion in a MOST
  - Single-ended amplifier
  - Differential amplifier
- **Distortion in a bipolar transistor**
- Reduction of distortion by feedback
- Distortion in an opamp
- Other cases of distortion and guide lines

Willy Sansen 10-05 1839

1839

The distortion in a bipolar transistor is higher because the current-voltage characteristic is steeper.

Several distortion characteristics are now calculated, for a single ended bipolar transistor amplifier and then for a differential pair.

Distortion in a bipolar transistor amplifier

$$I_{CE} = I_S \exp\left(\frac{V_{BE}}{kT_e/q}\right)$$

$$I_{CE} + i_{ce} = I_S \exp\left(\frac{V_{BE} + v_{be}}{kT_e/q}\right)$$

$$1 + y = \exp\left(\frac{v_{be}}{kT_e/q}\right)$$

$$\approx \exp(u) = 1 + u + \frac{u^2}{2} + \frac{u^3}{6} + \ldots \quad \text{if } u << 1$$

I_{CE} DC component
i_{CE} DC + ac component
i_{ce} ac component
I_{ce} amplitude of the ac component

Willy Sansen 10-05 1840

1840

A bipolar transistor is exponential. Application of a DC Base-Emitter voltage V_{BE} with an ac voltage v_{be} superimposed, gives a DC collector current I_{CE} with a AC signal i_{ce} on it. The relative collector current swing is denoted by y. It is obtained by a simple division of both currents.

This exponential can now be developed into a power series, provided the input signal amplitude V_{be} is small compared to kT_e/q.

Distortion in a bipolar transistor amplifier

$$y \approx u + \frac{u^2}{2} + \frac{u^3}{6} + \dots \qquad U = \frac{V_{be}}{kT_e/q}$$

is the non-linear equation

y is the relative current swing !

$a_1 = 1$

$a_2 = 1/2$ $\qquad IM_2 = \dfrac{a_2}{a_1} U = \dfrac{1}{2} \dfrac{V_{be}}{kT_e/q}$

$a_3 = 1/6$

$$IM_3 = \frac{3}{4} \frac{a_3}{a_1} U^2 = \frac{1}{8} \left(\frac{V_{be}}{kT_e/q} \right)^2$$

Willy Sansen 10-05 1841

1841

The coefficients of the power series are now given in this slide.

The distortion components IM_2 and IM_3 are easily derived from them.

It is clear that a bipolar transistor gives both even and odd-order distortion. Also, their values are quite large.

Numerical example

1. Relative current swing is 10 %

$y_p = 0.1$ **gives** $IM_2 = 5 \%$ $(HD_2 = 2.5\%)$

$\qquad\qquad\qquad IM_3 = 0.125 \%$ $(HD_3 = 0.04 \%)$

As a result $V_{bep} = y_p(kT_e/q) = 2.6$ mV$_p$ **(1.8 mV$_{RMS}$)**

$IP_3 = \sqrt{8}\ (kT_e/q) = 74$ mV$_p$ **or 50 mV$_{RMS}$ or -13 dBm**

2. $V_{bep} = 100$ mV

then $y_p = 0.1/0.026 \approx 4$ **(must be << 1 !!)**

gives $IM_2 =$ **?? Too high distortion !!**

Willy Sansen 10-05 1842

1842

For example, if a relative current swing is used with a maximum of 10%, then the IM_3 is 0.125%. Only 1.8 mV$_{RMS}$ input signal will now be needed.

The IP_3 is quite small. It is only 50 mV$_{RMS}$ or -13 dBm.

For an input voltage of 100 mV peak. The bipolar transistor is overdriven. Power series cannot be used any more, but Bessel functions. We do not develop this topic any more.

Distortion in a diode

$$i_D = I_S \exp\left(\frac{v_D}{kT_e/q} \right) \qquad y \approx u + \frac{u^2}{2} + \frac{u^3}{6} + \ldots$$

$$y = \frac{I_d}{I_D} = u + \frac{u^2}{2} + \frac{u^3}{6} \qquad\qquad U = \frac{V_d}{kT_e/q}$$

Same as for a Bipolar transistor amplifier !

Willy Sansen   10-05 1843

1843

If a bipolar transistor is connected as a diode, which is actually the Base-Emitter diode, then the same amount of distortion can be expected.

This diode has been used, for example, at the input of a bipolar current mirror. The distortion at the middle point of a current mirror is therefore quite large.

Distortion in bipolar differential amplifier

$$y = \frac{i_{Od}}{I_B} = \tanh \frac{V_{ld}}{2kT_e/q} \qquad \tanh x = \frac{e^x - e^{-x}}{e^x + e^{-x}}$$

$$\approx x - \frac{1}{3} x^3$$

$$y = \frac{i_{Od}}{I_B} \approx U - \frac{1}{3} U^3 \qquad U = \frac{V_{ld}}{2kT_e/q}$$

U is the relative current swing

$$IM_2 = 0 \qquad \boxed{IM_3 = \frac{1}{4} U^2} \qquad IP_3 = 4 \, kT_e/q$$

Willy Sansen   10-05 1844

1844

When two bipolar transistors are connected as a differential pair, the transfer characteristic now consists of exponentials as well. Together they yield a tanh function, as explained in Chapter 3.

If the applied differential input voltage v_{ld} is small compared to kT_e/q, then these functions can be developed into power series.

Again, the second-order component is zero if there is no mismatch. The third-order component gives rise to the IM_3 shown in this slide. The corresponding IP_3 is also given.

Remember that a MOST differential pair had about $1/10$ of U^2 as IM_3, now it is $1/4$ U^2. A MOST differential pair is 2.5 times better for IM_3 than a bipolar differential pair, for the same relative current swing. Moreover, the input voltage corresponding with it is also larger, as in a MOST differential pair the input voltage is scaled to $V_{GS} - V_T$.

Distortion in a resistor or capacitor

$$R = R_0 (1 + a_1 V + a_2 V^2 + ...)$$ [≈ JFET with large V_P]

For diffused resistors : $a_1 \approx 5 \text{ ppm/V}$
$$a_2 \approx 1 \text{ ppm/V}^2$$

$$C = C_0 (1 + a_1 V + a_2 V^2 + ...)$$

For poly-poly caps : $a_1 \approx 20 \text{ ppm/V}$
$$a_2 \approx 2 \text{ ppm/V}^2$$

Willy Sansen   10-05 1845

1845

Before we end this section on distortion of bipolar transistors, this is probably a good place to comment about distortion in resistors and capacitors.

Diffused resistors are fairly nonlinear as they are separated from the substrate by a depletion layer. The voltage across the resistor modifies the thickness of the depletion layer and the conductivity of the resistor. A resistor behaves as a Junction FET with a high pinch-off (or threshold) voltage. Its effective $V_{GS} - V_T$ is very high and its distortion is relatively low. It is not negligible, however.

Poly resistors and metal resistors do not have this depletion layer and are much more linear. The same applies to capacitances.

Diffusion capacitances are depletion layer capacitances and are very non-linear. Metal-to-metal capacitances, on the other hand, are very linear. This also applies to poly-to-poly capacitances, some values are given in this slide.

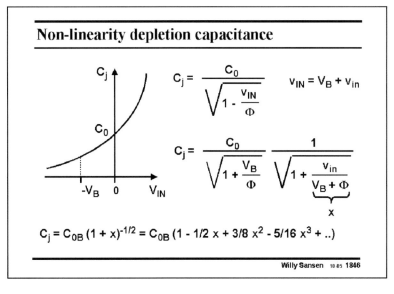

Willy Sansen   10-05 1846

1846

The non-linear characteristic of a diffusion capacitance is illustrated in this slide.

The junction capacitance C_j is C_0 at zero Volt across it. It is modeled by a square-root characteristic. For a bias Voltage $-V_B$, the junction capacitance can be described by a DC factor followed by an AC one. This latter factor can now be developed into a power series. Its coefficients can be used to calculate the values of IM_2 and IM_3.

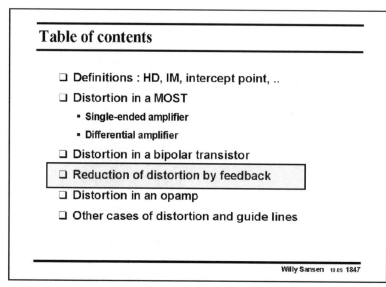

1847

The most common technique to reduce the distortion is by application of feedback.

Feedback reduces distortion in two ways. First of all, it reduces the relative current swing. Secondly, it reduces the distortion by the feedback factor (or loop gain).

Let us investigate how the coefficients of the power series are affected by the feedback. We will now calculate the actual distortion components IM_{2f} and IM_{3f}. Index f is used to indicate that they are valid for a system with feedback.

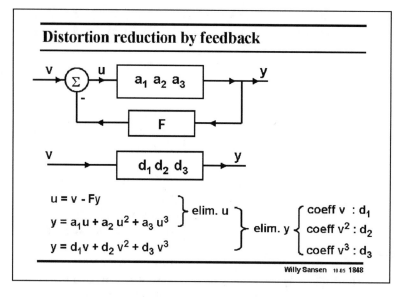

Distortion reduction by feedback

$u = v - Fy$
$y = a_1 u + a_2 u^2 + a_3 u^3$
$y = d_1 v + d_2 v^2 + d_3 v^3$

$\}$ elim. u $\}$ elim. y $\{$ coeff v : d_1
coeff v^2 : d_2
coeff v^3 : d_3

Willy Sansen 10.05 1848

1848

Feedback with factor F is applied around a non-linear amplifier, characterized by coefficients $a_1, a_2, a_3, ...$ The same system can also be described by coefficients $d_1, d_2, d_3, ...$ The question is, what is the relationship between the coefficients d and a?

This is found by writing the network equation and elimination of u. Subsequent elimination of y between this result and the power series with coefficients d yields these d coefficients.

They are given next.

1849

The open-loop gain a_1 is divided by the loop gain $1 + T$, as expected ($1 + LG$ in Chapter 13). For large T, the closed-loop gain d_1 is just about $1/F$, as explained by first-order feedback theory (see Chapter 13).

The second-order coefficient d_2 of the power series with feedback, is given to a_2 but divided by the third power of the loop gain T.

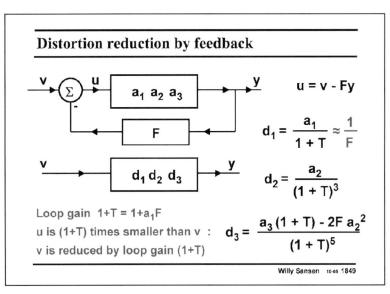

Distortion reduction by feedback

$u = v - Fy$

$$d_1 = \frac{a_1}{1 + T} \approx \frac{1}{F}$$

$$d_2 = \frac{a_2}{(1 + T)^3}$$

Loop gain $1 + T = 1 + a_1 F$

u is $(1 + T)$ times smaller than v :

v is reduced by loop gain $(1 + T)$

$$d_3 = \frac{a_3(1 + T) - 2F\,a_2^{\,2}}{(1 + T)^5}$$

Willy Sansen 10-05 1849

The third-order coefficient d_3 of the power series with feedback, is given to a_3 but is now divided by the fifth power of the loop gain T. Actually, a_2 gives a contribution. The second-order distortion of the nonlinear amplifier can take one more turn through the loop to generate third-order distortion. However, it has the opposite sign of the contribution from a_3. Coefficient a_3 gives compression distortion, whereas a_2 gives expansion of the sine wave.

They can cancel out at one particular value of the coefficients a.
    The distortion components are now easily calculated.

Distortion components with feedback

$$IM_{2f} = \frac{d_2}{d_1}V = \frac{a_2}{a_1}\frac{V}{(1 + T)^2} = \frac{a_2}{a_1}\frac{1}{(1 + T)}\frac{V}{(1 + T)}$$

reduction by loop gain   reduction in current swing

$$IM_{3f} = \frac{3}{4}\frac{d_3}{d_1}V^2 = \frac{3}{4}\left[\frac{a_3}{a_1}\frac{1}{(1 + T)} - \left(\frac{a_2}{a_1}\right)^2\frac{2T}{(1 + T)^2}\right]\frac{V^2}{(1 + T)^2}$$

compression   expansion

reduction in current swing

Willy Sansen 10 05 1850

1850
By use of the coefficients d, the distortion components IM_{2f} and IM_{3f} with feedback (index f) are easily calculated.

    The IM_{2f} contains T^2 in the denominator. If we associate one T to the reduction of the gain, and as a consequence, the reduction of the relative current swing, then we see that the IM_{2f} is reduced by T itself. A rule of thumb is that we calculate the distortion for the relative current swing, taking into account the effect of the feedback on this current swing. The distortion for this stage is now easily calculated. It is divided by the loop gain to take into account the reduction by the loop gain T.

    In the same way, IM_{3f} is calculated. Again, the two terms appear related to a_3 and a_2. If a_3 is dominant, the same rule of thumb applies. An easy way to calculate the distortion is to find the distortion for a relative current swing, taking into account the effect of feedback. The distortion is then obtained by division by the loop gain T itself, and not by T^2.

    The same is true if a_2 is dominant. Indeed the T^2 in the denominator is reduced to T since the numerator also contains T. Clearly, we have assumed all along that T is always larger than unity and that the loop gain can be represented by T, rather than $1 + T$.

Distortion components with feedback : examples

$$IM_{3f} = \frac{3}{4} \frac{d_3}{d_1} V^2 = \frac{3}{4} \left[\underbrace{\frac{a_3}{a_1} \frac{1}{(1+T)} - \left(\frac{a_2}{a_1}\right)^2 \frac{2T}{(1+T)^2}} \right] \frac{V^2}{(1+T)^2}$$

$$\text{For large T :} \quad \frac{a_3 a_1 - 2 a_2^2}{a_1^2} \frac{1}{T} = \frac{a_3}{a_1}\left(1 - \frac{2 a_2^2}{a_1 a_3}\right) \frac{1}{T}$$

MOST : $a_3 = 0$: a_2 dominant

Bipolar : $a_1 = 1$ $a_2 = 1/2$ $a_3 = 1/6$: a_2 dominant

Diff. pair : $a_2 = 0$: a_3 dominant

Willy Sansen   10 05   1851

1851

The factor between square brackets can be rewritten provided T is large, as shown in this slide.

Which term is dominant, depends on the actual transistor configuration.

Three examples are given.

A MOST in the strong-inversion region has a zero a_3. It is clear that in this case, all third-order distortion is due to a_2. Application of feedback around a single-transistor MOST amplifier generates third-order distortion, which is not present without feedback.

For a single bipolar-transistor amplifier the coefficients are given in this slide. This yields as a second term with a_2 a value of 3. Clearly, this latter term is now dominant.

A differential pair, on the other hand, does not have second-order distortion; its a_2 is zero. As a result, its IM_{2f} is zero but not its IM_{3f}. It is obviously caused by the third-order distortion a_3 of the open-loop amplifier.

Emitter resistor to reduce distortion IM_{2f}

$$T = g_m R_E = \frac{V_{RE}}{kT_e/q} \qquad\qquad \frac{a_2}{a_1} = \frac{1}{2}$$

$$IM_{2f} = \frac{1}{2} \frac{1}{(1+T)^2} \frac{V_{in}}{kT_e/q} = \frac{1}{(1+T)} \frac{U}{2}$$

$$U = \frac{1}{(1+T)} \frac{V_{in}}{kT_e/q} \quad \text{is the relative current swing}$$

IM_{2f} decreases linearly with T for constant U !

Willy Sansen   10 05   1852

1852

One of the simplest amplifiers with feedback is a single bipolar-transistor amplifier with a series resistor R_E in the Emitter. The loop gain is now simply $g_m R_E$, which is assumed to be larger than unity. It is also the DC voltage V_{RE} across the resistor R_E, scaled by kT_e/q.

Substitution of the ratio a_2/a_1 into IM_{2f} yields the expression given in this slide. As predicted, the IM_{2f} is linearly proportional to the relative current swing U, and has to be divided by the loop gain $1+T$, or by T itself for large T.

Emitter resistor to reduce distortion IM$_{3f}$

$$IM_{3f} = \frac{1 - 2T}{(1 + T)^2} \frac{U^2}{8} \qquad \frac{a_2}{a_1} = \frac{1}{2} \qquad \frac{a_3}{a_1} = \frac{1}{6}$$

$$U = \frac{1}{(1 + T)} \frac{V_{in}}{kT_e/q} \quad \text{is the relative current swing}$$

Null for T = 0.5

IM$_{3f}$ also decreases with T for constant U
for large T !!

Willy Sansen 10 05 **1853**

1853

Substitution of the ratio a_2/a_1 and of a_3/a_1 into IM$_{3f}$ yields the expression given in this slide.

As predicted, the IM$_{3f}$ is proportional to the relative current swing U squared, and has to be divided by the loop gain $1+T$, or by T itself for large T, not by T^2!

Note that there is a null for T = 0.5. It is a very sharp null however, which cannot be easily used. This is shown in the next slide.

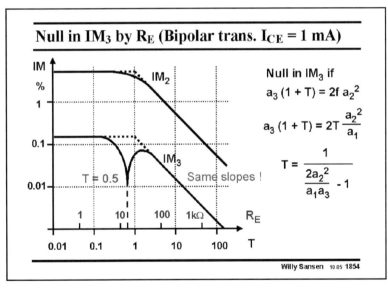

Null in IM$_3$ by R$_E$ (Bipolar trans. I$_{CE}$ = 1 mA)

Null in IM$_3$ if

$$a_3 (1 + T) = 2f\, a_2^2$$

$$a_3 (1 + T) = 2T \frac{a_2^2}{a_1}$$

$$T = \frac{1}{\dfrac{2a_2^2}{a_1 a_3} - 1}$$

Same slopes !

T = 0.5

Willy Sansen 10 05 **1854**

1854

The results are plotted for a bipolar-transistor amplifier with an emitter resistor R$_E$ of increasing value.

The DC current I$_{CE}$ is constant and so is the relative current swing. This means that the input voltage is constant up to T = 1 and then increases with T.

Note that for values of T larger than unity, both IM$_{2f}$ and IM$_{3f}$ decrease with the same slope of -20 dB/ decade. Note also that the IM$_{3f}$ shows a null at T = 0.5, which is quite sharp indeed.

1855

For large loop gain T, simplified expressions can be obtained.

Substitution of T by $g_m R_E$ yields the second set of expressions. Substitution of g_m by the DC current ICE divided by kT_e/q yields the last set of expressions.

They indicate what the relationships are between the IM$_{2fT}$ and IM$_{3fT}$ and the input voltage amplitude V$_{in}$, the DC current I$_{CE}$ and the resistor R$_E$. For constant I$_{CE}$ and input voltage V$_{in}$,

Emitter resistor R_E reduces distortion for large T

$$U = \frac{1}{T} \frac{V_{in}}{kT_e/q} = \frac{V_{in}}{R_E I_{CE}}$$

$$IM_{2fT} = \frac{U}{2T} = \frac{V_{in}}{kT_e/q} \frac{1}{2\,T^2} = \frac{V_{in}\,kT_e/q}{2\,(R_E I_{CE})^2}$$

$$IM_{3fT} = \frac{U^2}{4T} = \left(\frac{V_{in}}{kT_e/q}\right)^2 \frac{1}{4\,T^3} = \frac{V_{in}^2\,kT_e/q}{4\,(R_E I_{CE})^3}$$

Willy Sansen 10 05 1855

and increasing R_E, the relative current swing decreases, decreasing the distortion components considerably.

This is for a bipolar transistor. Let us have a look now at a MOST with a series resistor R_S in the Source.

Source resistor R_S to reduce distortion

$$T = g_m R_S = \frac{V_{RS}}{(V_{GS}-V_T)/2} \qquad\qquad \frac{a_2}{a_1} = \frac{1}{4} \qquad a_3 = 0$$

$$U = \frac{1}{(1 + T)} \frac{V_{in}}{(V_{GS}-V_T)/2} \qquad \text{is the relative current swing}$$

$$IM_{2f} = \frac{1}{(1 + T)} \frac{U}{4} \approx \frac{V_{in}}{(V_{GS}-V_T)/2} \frac{1}{4\,T^2} = \frac{V_{in}\,(V_{GS}-V_T)/2}{4\,(R_S I_{DS})^2}$$

$$IM_{3f} = \frac{T}{(1 + T)^2} \frac{3U^2}{32} \approx \frac{V_{in}^2}{(V_{GS}-V_T)^2/4} \frac{3}{32\,T^3} = \frac{3V_{in}^2\,(V_{GS}-V_T)/2}{32\,(R_S I_{DS})^3}$$

Willy Sansen 10 05 1856

1856

For a MOST, the third order coefficient a_3 is zero. The third-order distortion with feedback, will therefore be caused by the second-order coefficient a_2.

Another big difference between a bipolar transistor and a MOST is that the kT_e/q must be substituted by $(V_{GS} - V_T)/2$, which can be chosen, but which is always larger than kT_e/q. For low distortion, a small value of $V_{GS} - V_T$ must also be selected, because the effect of the increase in g_m, and in loop gain T is more important than the decrease of the $V_{GS} - V_T$.

For large T, similar results are obtained as for a bipolar transistor. Substitution of T by $g_m R_S$ also yields expressions which are very similar to the ones for a bipolar transistor.

For example, for IM_{3fT} the coefficient is about 1/10, whereas 1/4 for a bipolar transistor. It is therefore 2.5 times smaller than for a bipolar transistor. However, if $V_{GS} - V_T = 0.2$ V is chosen, then $(V_{GS} - V_T)/2 = 0.1$ V which is four times larger than $kT_e/q \times 26$ mV. For the same DC current and resistor, the MOST amplifier gives 4/2.5 or about 1.6 times worse IM_3 than a bipolar amplifier, for the same input voltage.

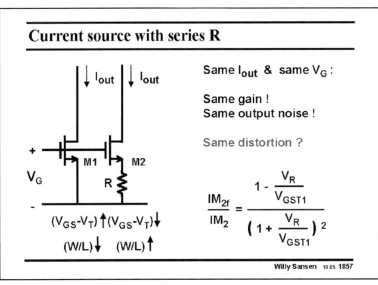

1857

The question arises whether it is better for low distortion, to take a MOST with large $V_{GS} - V_T$ (as for MOST M1), or to take MOST with small $V_{GS} - V_T$ and a series resistor R (as for MOST 2). The difference between the V_{GS1} and the V_{GS2} is the voltage across the resistor V_R.

The Gates are at the same voltage V_G. We also take equal DC currents.

It is clear that for third-order distortion IM_3, the configuration with M1 is the best as it does not give any IM_3. The one with feedback resistor R does give IM_{3f}!

For second-order distortion IM_2 (for M1) and IM_{2f} (for M2), we have to calculate the ratio IM_{2f}/IM_2. The result is given in this slide.

It shows that the voltage across the resistor V_R must be larger than $V_{GS1} - V_T$ (or V_{GST1}) to make a difference. If this is the case, the distortion is inversely proportional to the voltage V_R.

For second-order distortion, it is better to use as large a resistor as possible, but not for third-order distortion!

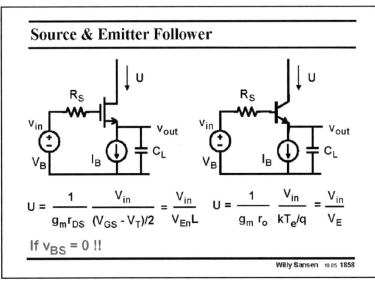

1858

When the series resistor in the Source is greatly increased, i.e. when an ideal current source I_B is used with infinite output resistance, then the distortion due to the non-linearity of the $I_{DS} - V_{GS}$ characteristic is zero.

In this case, the non-linearity of the output conductance is dominant. It can again be described in terms of relative current swing U, as given in this slide. Voltage $V_E L$ is the Early voltage of the MOST. The larger the Early voltage, the smaller the relative current swing and the smaller the distortion is.

The same applies to an emitter follower.

We have silently assumed that the Bulk-Source voltage of the source follower is zero. This is only possible when the Bulk can be connected to the source or when the nMOST is in a p-well. However, CMOS processes have normally a n-well. A nMOST source follower has normally its Bulk connected to ground. This causes a lot of distortion, as shown next.

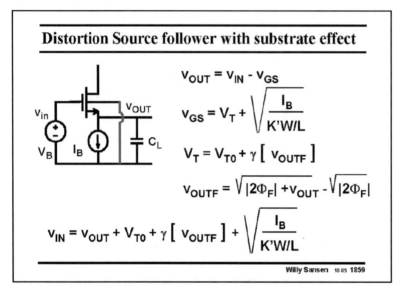

Distortion Source follower with substrate effect

$$V_{OUT} = V_{IN} - V_{GS}$$

$$V_{GS} = V_T + \sqrt{\frac{I_B}{K'W/L}}$$

$$V_T = V_{T0} + \gamma \left[V_{OUTF} \right]$$

$$V_{OUTF} = \sqrt{|2\Phi_F| + V_{OUT}} - \sqrt{|2\Phi_F|}$$

$$V_{IN} = V_{OUT} + V_{T0} + \gamma \left[V_{OUTF} \right] + \sqrt{\frac{I_B}{K'W/L}}$$

Willy Sansen 10 05 1859

1859
Indeed, when the Bulk is connected to ground, the parasitic JFET becomes active (see Chapter 1), and the gain becomes smaller than unity.

In this case, the threshold voltage V_T depends on the output voltage through parameter γ, which represents the body effect as explained in Chapter 1. The relation between input and output voltage is now easily derived. It is clearly a non-linear relationship.

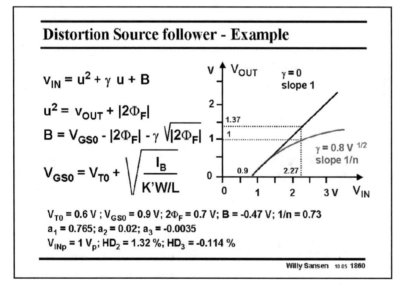

Distortion Source follower - Example

$$V_{IN} = u^2 + \gamma\, u + B$$

$$u^2 = V_{OUT} + |2\Phi_F|$$

$$B = V_{GS0} - |2\Phi_F| - \gamma \sqrt{|2\Phi_F|}$$

$$V_{GS0} = V_{T0} + \sqrt{\frac{I_B}{K'W/L}}$$

$V_{T0} = 0.6$ V ; $V_{GS0} = 0.9$ V; $2\Phi_F = 0.7$ V; $B = -0.47$ V; $1/n = 0.73$
$a_1 = 0.765$; $a_2 = 0.02$; $a_3 = -0.0035$
$V_{INp} = 1$ V_p; $HD_2 = 1.32$ %; $HD_3 = -0.114$ %

Willy Sansen 10 05 1860

1860
This nonlinear relationship is plotted in this slide. The slope of this curve decreases for larger input and output voltages. It is actually $1/n$ in which n contains g (see Chapter 1).

This nonlinear relationship gives rise to a power series. The coefficients are given for specific values of the transistor parameters.

It is obvious that a source follower, which cannot be embedded in a separate well, cannot be used, if distortion is of any concern!

1861
The inclusion of series resistors in the source is used very often to enlarge the input range of a differential pair. In principle this can be done by taking large values of $V_{GS} - V_T$. If this is not sufficient, series resistors R must be inserted.

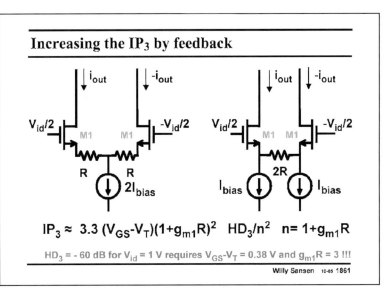

The IM_3 is now reduced or the IP_3 is increased.

Actually, there are two possible realizations of the same principle. For small-signal performance they are almost the same. Actually, the circuit with a single resistor 2R (on the right) has the advantage that we do not have to try to match two separate resistors. On the other hand, the output capacitances of the bias current sources I_{bias} will limit the high-frequency performance, which is not the case for the left circuit.

The main difference however, is that for the left circuit, the DC bias current I_{bias} flows through the resistor R, which is not the case on the right. For large resistors R and low supply voltages, the circuit on the right is the better one.

Differential amplifiers with a large input range, and hence low distortion, are called transconductors. They are used intensively for continuous-time filters, discussed in the next Chapter.

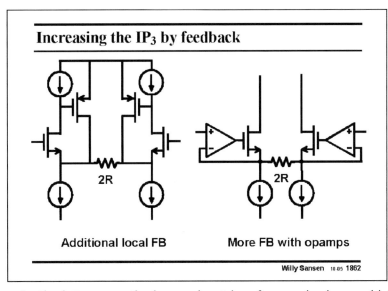

1862

Because of the limited supply voltage, very large value of resistor R cannot be used. Additional feedback is now required as shown in this slide.

The purpose of all such circuits is to increase the loop gain to reduce the distortion. This can be done by addition of just one or two transistors in the feedback loop, as shown on the left, or by an inclusion of a full operational amplifier, as shown on the right.

In the latter case, the loop gain at low frequencies is very high. As a result the differential input voltage is enforced upon the resistor 2R, giving rise to a very linear voltage to current conversion.

For a high-frequency transconductor the circuit on the left may be preferred, as its loop gain may be higher at high frequencies.

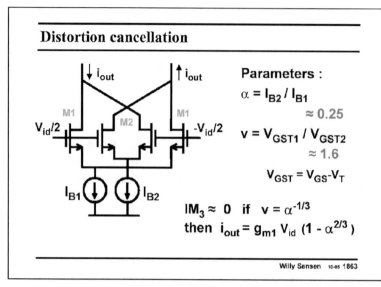

Distortion cancellation

Parameters :

$\alpha = I_{B2} / I_{B1}$

≈ 0.25

$v = V_{GST1} / V_{GST2}$

≈ 1.6

$V_{GST} = V_{GS} - V_T$

$IM_3 \approx 0$ if $v = \alpha^{-1/3}$

then $i_{out} = g_{m1} V_{id} (1 - \alpha^{2/3})$

Willy Sansen   10-05 1863

1863

Transconductors can also suppress the distortion by cancellation, rather than by local feedback. An often used example is given in this slide. It is actually a MOST version of the well known bipolar Gilbert multiplier (JSSC Dec. 68, 365–373). It is asymmetrical, however.

It consists of two differential pairs with transistors M1 and M2, the second of which has smaller g_m's and is cross-coupled.

This cross-coupling allows the reduction of the IM3 to zero but also reduces the signal amplitude itself.

There are actually two design parameters, i.e. the ratio α of the two biasing currents I_B, and the ratio v of the two values of $V_{GS} - V_T$. They are linked by a simple expression for zero IM_3.

For example, for $\alpha = 0.25$, the ratio v must be 1.6 (for example, $V_{GS} - V_T = 0.2$ and 0.32 V).

Obviously, the cancellation will never be perfect. Mismatch will play a role.

Distortion cancellation

$i_{out} = 2 (i_{DS1} - i_{DS2})$

$$\frac{i_{DS}}{I_B} = U - \frac{1}{8} U^3 \qquad U = \frac{V_{id}}{V_{GS} - V_T} \qquad IM_3 = \frac{3}{32} U^2$$

$$IM_3 \approx \frac{3}{32} \left(\frac{V_{id}}{V_{GS1} - V_T}\right)^2 \frac{1 - \alpha v^3}{1 - \alpha v}$$

$IM_3 \approx 0$ if $v_{00} = \alpha^{-1/3}$

at which point $i_{out} = g_{m1} V_{id} (1 - \alpha^{2/3})$

Willy Sansen   10-05 1864

1864

The expressions for the differential output current and the IM_3 are given in this slide.

It is clear that IM_3 depends on the relative current swing U as usual, but also on the two design parameters α and v. This additional fraction with α and v is plotted on the next slide. It is obvious, however, that it becomes zero if v equals $\alpha^{-1/3}$. It gives rise to a reduction of the signal amplitude itself. A compromise is therefore to be taken.

For example, for $\alpha = 0.25$, the ratio v must be 1.6 (for example $V_{GS} - V_T = 0.2$ and 0.32 V). In this case, the gain is reduced by $\alpha^{2/3}$ or by 12%, which is quite reasonable.

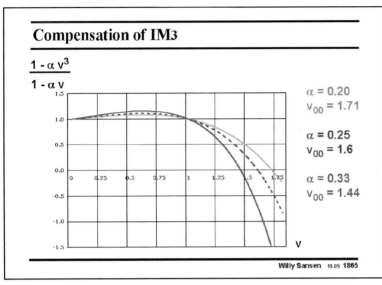

Willy Sansen 10-05 1865

1865

This fraction is plotted for different combinations of parameters α and v.

It can be noted that the crossing point for zero is less sharp if smaller values are taken for α and therefore larger values of v. The range of $V_{GS} - V_T$ values over which the same model holds is limited, however. This limits the value of α as well.

This is why a good compromise for $\alpha = 0.25$, for which the ratio v must be 1.6. This gives very reasonable values for $V_{GS} - V_T = 0.2$ V and 0.32 V.

In this case the gain is reduced by $\alpha^{2/3}$ or by 40%, as shown next.

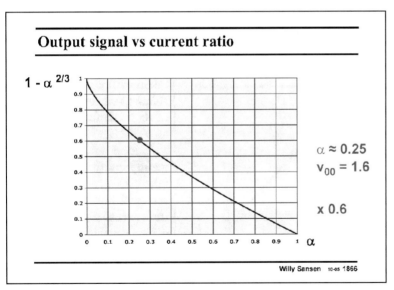

Willy Sansen 10-05 1866

1866

The ratio by which the signal amplitude is reduced as a result of the cross-coupling is shown in this slide.

The larger α is, the less signal amplitude remains. This is another reason not to choose too small a value of α.

For a value of $\alpha = 0.25$, the output signal amplitude is reduced by $\alpha^{2/3}$ or by 40%, as shown in this slide.

Table of contents

Willy Sansen 10 05 **1867**

1867
Now that we know how much the distortion can be reduced by means of feedback, let us apply this to the amplifier, which is always used with feedback, i.e. an operational amplifier. For a large output swing, a two-stage Miller operation is normally used.

First of all, we have to figure out what the signal voltage amplitudes are for both stages. This is achieved first at low frequencies, and then for all frequencies up to the GBW.

It is clear that at high frequencies, the voltage gains are small, such that the input voltages increase. This will have two effects. The distortion per stage increases and secondly, the loop gain, which decreases the distortion, decreases. The distortion will now increase considerably at higher frequencies.

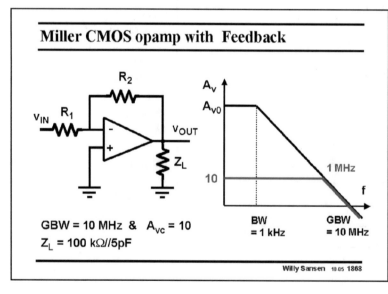

Miller CMOS opamp with Feedback

GBW = 10 MHz & A_{vc} = 10
Z_L = 100 kΩ//5pF

Willy Sansen 10 05 **1868**

1868
As an example, an opamp is taken with a 10 MHz GBW. It is set at a closed-loop gain of 10. Its bandwidth is therefore 1 MHz.

A load is added, which mainly consists of a capacitance. As a result, some current will flow in the output transistors. For higher frequencies, this current will increase, leading to even more distortion.

Let us now discover how distortion is affected by the frequency dependence of the gain blocks in the opamp.

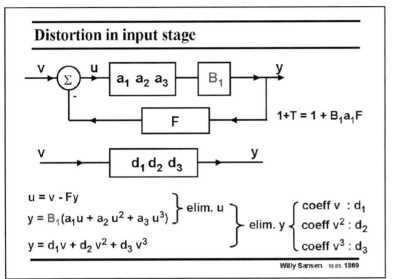

Distortion in input stage

$$u = v - Fy$$
$$y = B_1(a_1 u + a_2 u^2 + a_3 u^3)$$ elim. u
$$y = d_1 v + d_2 v^2 + d_3 v^3$$ elim. y

coeff v : d_1
coeff v^2 : d_2
coeff v^3 : d_3

$1+T = 1 + B_1 a_1 F$

Willy Sansen 10.05 1869

1869

An opamp is taken with two stages. We assume that only the first stage is nonlinear whereas the second stage has a fixed gain factor B1.

By use of the same technique as before, the coefficients d of the resulting power series are easily obtained.

Distortion in input stage

$$IM_{2f} = \frac{d_2}{d_1}V = \frac{a_2}{a_1} \frac{V}{(1+T)^2} = \frac{a_2}{a_1} \frac{1}{(1+T)} \frac{V}{(1+T)}$$

$$IM_{3f} = \frac{3}{4} \frac{d_3}{d_1}V^2 = \frac{3}{4}\left[\frac{a_3}{a_1} \frac{1}{(1+T)} - \left(\frac{a_2}{a_1}\right)^2 \frac{2T}{(1+T)^2}\right] \frac{V^2}{(1+T)^2}$$

Same as before but with different Loop gain :

$$1+T = 1 + B_1 a_1 F$$

Willy Sansen 10.05 1870

1870

The distortion components IM_2 and IM_3 are now easily obtained. We find that they are exactly the same as for a single stage amplifier.

The only difference is that the loop gain is $1+T$ larger. The gains of both stages are obviously present in the loop gain.

The resulting distortion will be smaller because of the increased loop gain.

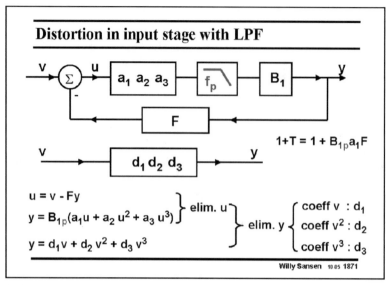

Distortion in input stage with LPF

$1 + T = 1 + B_{1p}a_1F$

$u = v - Fy$

$y = B_{1p}(a_1u + a_2u^2 + a_3u^3)$ } elim. u

$y = d_1v + d_2v^2 + d_3v^3$ } elim. y {
coeff v : d_1
coeff v^2 : d_2
coeff v^3 : d_3

Willy Sansen 10 05 1871

before again yields the coefficients d of the power series.

1871

A different result is obtained when a low-pass filter is inserted between both stages. This better resembles a two-stage amplifier. The compensation capacitance exerts a low-pass filter characteristic to the first stage.

The low-pass filter has a pole at frequency f_p. It is combined with the gain block B_1, to be denoted by gain B_{1p}, which is nothing else than a filter block with low-frequency gain B_1.

A similar analysis as

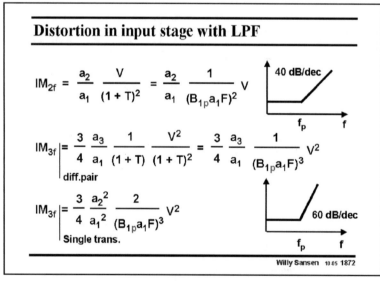

Distortion in input stage with LPF

$$IM_{2f} = \frac{a_2}{a_1} \frac{V}{(1+T)^2} = \frac{a_2}{a_1} \frac{1}{(B_{1p}a_1F)^2} V$$

40 dB/dec

f_p f

$$IM_{3f}\Big|_{\text{diff.pair}} = \frac{3}{4} \frac{a_3}{a_1} \frac{1}{(1+T)} \frac{V^2}{(1+T)^2} = \frac{3}{4} \frac{a_3}{a_1} \frac{1}{(B_{1p}a_1F)^3} V^2$$

$$IM_{3f}\Big|_{\text{Single trans.}} = \frac{3}{4} \frac{a_2^2}{a_1^2} \frac{2}{(B_{1p}a_1F)^3} V^2$$

60 dB/dec

f_p f

Willy Sansen 10 05 1872

1872

Both distortion components IM_2 and IM_3 now contain this low-pass filter characteristic, but inverted. Indeed a low-pass filter characteristic in the feedback loop yields a high pass filter characteristic. This is used to carry out noise shaping in all Sigma-delta modulators (see Chapter 21).

In the IM_3 characteristic, the slope beyond pole frequency f_p, is 60 dB/decade, which is quite steep indeed. Note that the rise starts at frequency f_p itself, without being affected by the loop gain.

Depending on the generator of distortion in the first stage, a different coefficient a emerges. For a differential pair it is evident that the third-order distortion of the input stage dominates.

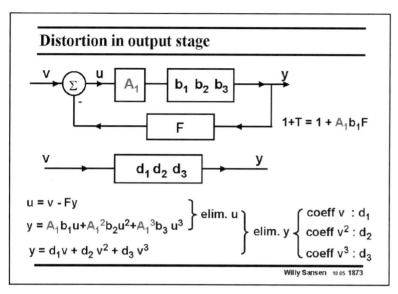

Distortion in output stage

$u = v - Fy$

$y = A_1 b_1 u + A_1^2 b_2 u^2 + A_1^3 b_3 u^3$ } elim. u

$y = d_1 v + d_2 v^2 + d_3 v^3$

} elim. y {
coeff v : d_1
coeff v^2 : d_2
coeff v^3 : d_3

$1+T = 1 + A_1 b_1 F$

Willy Sansen 10 05 1873

1873

Let us now keep the input stage linear. It has a fixed gain A_1. The second stage is now non-linear, with coefficients b.

A similar analysis as before again yields the coefficients d. Note however that the non-linear amplifier has a fairly large input voltage, this is due to the presence of gain block A_1.

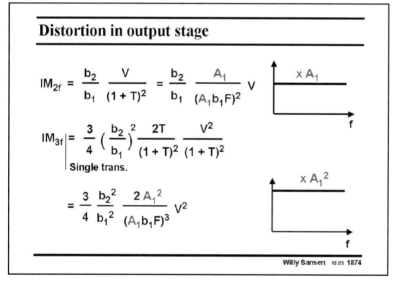

Distortion in output stage

$$IM_{2f} = \frac{b_2}{b_1} \frac{V}{(1 + T)^2} = \frac{b_2}{b_1} \frac{A_1}{(A_1 b_1 F)^2} V$$

$$IM_{3f} \bigg| = \frac{3}{4} \left(\frac{b_2}{b_1}\right)^2 \frac{2T}{(1 + T)^2} \frac{V^2}{(1 + T)^2}$$

Single trans.

$$= \frac{3}{4} \frac{b_2^2}{b_1^2} \frac{2 A_1^2}{(A_1 b_1 F)^3} V^2$$

x A_1

x A_1^2

f

Willy Sansen 10 05 1874

1874

The distortion components IM_2 and IM_3 are easily found.

The big difference with the previous results is that the first-stage gain A_1 also appears in the numerator, even squared for IM_3. This is not unexpected, as the input signal is applied by gain A_1 and then applied to the non-linear amplifier with coefficients b.

1875

If both amplifiers are nonlinear, with a low-pass filter in between, we find the distortion components as given in this slide.

The contributions of the first stage carry coefficients a (black) whereas the second stage coefficients b (red).

It is clear that at low frequencies the distortion of the output stage dominates. The reason is that the input voltage of the output stage is fairly high, because of gain a_1.

At higher frequencies however, the distortion of the input stage dominates, because it has started increasing at frequency f_p, which is the dominant pole of the amplifier.

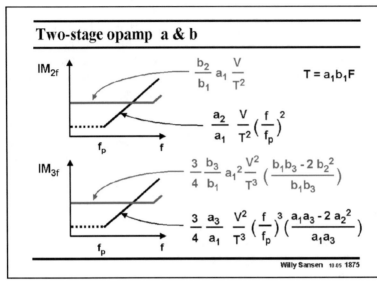

Two-stage opamp a & b

$$\frac{b_2}{b_1} a_1 \frac{V}{T^2} \qquad T = a_1 b_1 F$$

$$\frac{a_2}{a_1} \frac{V}{T^2} \left(\frac{f}{f_p}\right)^2$$

$$\frac{3}{4} \frac{b_3}{b_1} a_1^2 \frac{V^2}{T^3} \left(\frac{b_1 b_3 - 2 b_2^2}{b_1 b_3}\right)$$

$$\frac{3}{4} \frac{a_3}{a_1} \frac{V^2}{T^3} \left(\frac{f}{f_p}\right)^3 \left(\frac{a_1 a_3 - 2 a_2^2}{a_1 a_3}\right)$$

Willy Sansen 10 05 1875

If we take now a Miller opamp with a differential input stage ($a_2 = 0$) and a single-ended output stage, then all IM_2 is due to the non-linearity of the output transistor, at least at low frequencies. This is also true at high frequencies as a differential pair does not generate second-order distortion. In practice, some other sources of distortion take over. The output conductance of the output transistor then starts coming in.

The IM_3 is also due to the non-linearity of the output stage at low frequencies. At high frequencies the non-linearity of the input stage takes over.

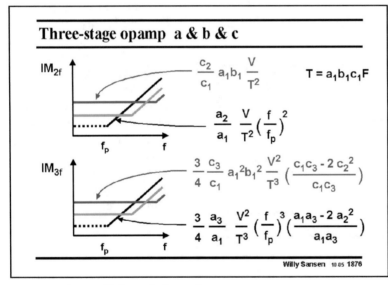

Three-stage opamp a & b & c

$$\frac{c_2}{c_1} a_1 b_1 \frac{V}{T^2} \qquad T = a_1 b_1 c_1 F$$

$$\frac{a_2}{a_1} \frac{V}{T^2} \left(\frac{f}{f_p}\right)^2$$

$$\frac{3}{4} \frac{c_3}{c_1} a_1^2 b_1^2 \frac{V^2}{T^3} \left(\frac{c_1 c_3 - 2 c_2^2}{c_1 c_3}\right)$$

$$\frac{3}{4} \frac{a_3}{a_1} \frac{V^2}{T^3} \left(\frac{f}{f_p}\right)^3 \left(\frac{a_1 a_3 - 2 a_2^2}{a_1 a_3}\right)$$

Willy Sansen 10 05 1876

1876

Similar results can be obtained for three-stage amplifiers ("Distortion in Single-, Two-and Three-stage amplifiers", Hernes, etal, TCAS-I, May 2005, 846–856, "Distortion analysis of Miller-compensated three-stage amplifiers, Cannizzaro, etal, TCAS-1, 2005).

The contributions of the first stage carry coefficients a (black) whereas the output stage coefficients c (red). The intermediate second stage carries coefficients b (green).

It is clear that at low frequencies the distortion of the output stage again dominates. The reason is that the input voltage of the output stage is fairly high, because of gains a_1 and b_1.

At high frequencies, the distortion of the input stage takes over. The contributions of the second stage in the middle are always negligible. This is a result of the compensation scheme in which the lowest non-dominant pole is at a lower frequency than the one of the second stage.

It is clear that a number of less important non-linearities have been neglected. The most important ones are the output conductances of the output transistors.

Note, finally, that a simple rule can be deduced from these expressions. The distortion can always be calculated provided the relative current swing can be calculated. For this purpose, we need to know the input voltage for each stage. The relative current swing then gives the distortion component. This value must now be divided by the value of the loop gain at that frequency. An example will clarify this.

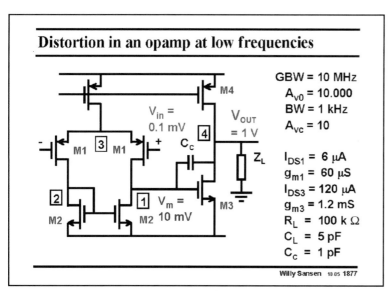

1877
A two stage Miller opamp is taken with a 10 MHz GBW, set at a closed-loop gain of 10. Its bandwidth is therefore 1 MHz. All values of the gains and transistor parameters are given in this slide. Each stage has a low-frequency gain of 100. For a signal output voltage of 1 V, the voltages at the input and in between the two stages, are easily calculated.

What is the main source of distortion?

This is clearly the output stage as it is driven by 10 mV, whereas the input stage only receives 0.1 mV input voltage. This is clear from the calculations.

At higher frequencies however, the input signal required to deliver 1 V will increase, as the open-loop voltage decreases. It is not so obvious however, to determine what will be the signal voltage between the two stages. This is shown in the next slide.

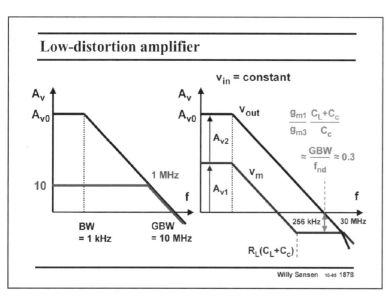

1878
On the left, the Bode diagram is given of the open- and closed-loop gain. It shows what the output voltage is for a small input voltage, which increases in frequency.

On the right, the voltage is added at the intermediate point. It shows that at low frequencies the gain of the first stage decreases. It is followed by a low-pass filter, with a pole equal to the dominant pole of the opamp. The gain across the output stage is constant.

This is true until the opamp reaches the frequency with time-constant $R_L(C_L + C_c)$. From here on the gain of the second stage decreases as well. As a result, the distortion increases. Moreover, the loop gain has become quite small! The distortion will now increase quite steeply.

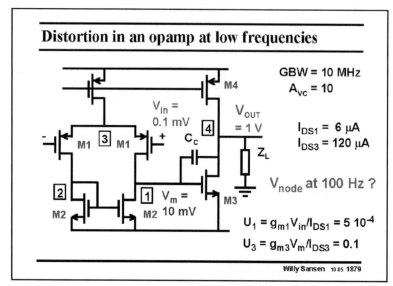

Distortion in an opamp at low frequencies

GBW = 10 MHz
A_{vc} = 10

V_{in} = 0.1 mV

V_{OUT} = 1 V

I_{DS1} = 6 μA
I_{DS3} = 120 μA

V_{node} at 100 Hz ?

$U_1 = g_{m1}V_{in}/I_{DS1} = 5 \ 10^{-4}$

$U_3 = g_{m3}V_m/I_{DS3} = 0.1$

V_m = 10 mV

Willy Sansen 10 05 1879

1879

At low frequencies, for example at 100 Hz. The input signal levels are easily calculated for both the input stage and the output stage.

The relative current swing for the input stage is now only 0.05% but it is 10% for the output stage. It is clear that the output transistor will generate a lot of second-order distortion. It is given next.

Distortion in an opamp at low frequencies

Distortion generation by nonlinear output stage :

$U_3 = g_{m3}V_m/I_{DS3} = 0.1$

$IM_2 = U_3/4 = 0.25 \ 0.1 = 2.5 \ \%$

Distortion reduction by feedback :

T = 1000 IM_{2f} = 2.5 %/1000 = 0.0025 % Negligible !

Willy Sansen 10 05 1880

1880

For a 10% relative current swing, the IM_2 distortion is easily found to be 1.5%. This distortion must now be divided by the loop gain at low frequencies. From the Bode diagram we find that this is 1000.

The resulting IM_{2f} is 0.0025%, which is negligible indeed.

First of all, little distortion is generated. Secondly, the loop gain at this frequency is quite large.

Distortion in an opamp at high frequencies

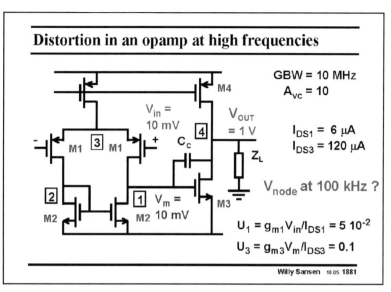

$V_{in} =$ 10 mV

$V_{OUT} = 1$ V

GBW = 10 MHz

$A_{vc} = 10$

$I_{DS1} = 6\ \mu A$

$I_{DS3} = 120\ \mu A$

V_{node} at 100 kHz ?

$U_1 = g_{m1}V_{in}/I_{DS1} = 5\ 10^{-2}$

$U_3 = g_{m3}V_m/I_{DS3} = 0.1$

$V_m = 10$ mV

Willy Sansen 10 05 1881

1881

At a frequency of 100 kHz, on the other hand, which is not such a high frequency for this amplifier, the gain across the input stage decreases to about unity. The gain of the output stage is still approximately 100.

The signal levels at the input and at the intermediate point are shown in this slide.

It is clear that the distortion in the input stage will now be much larger, whereas the distortion on the output stage is the same as at 100 Hz.

The relative current swing in the input stage has increased by a factor of 100. It is now 0.05. This is still too small to be able to play a role for distortion.

The calculations show this.

Distortion in an opamp at high frequencies

Distortion generation by nonlinear output stage :

$U_3 = g_{m3}V_m/I_{DS3} = 0.1$

$IM_2 = U_3/4 = 0.25\ 0.1 = 2.5\ \%$

Distortion generation by nonlinear input stage :

$U_1 = g_{m1}V_m/I_{DS1} = 0.05$

$IM_3 = U_1^2/10 = 0.0025/10 = 0.025\ \%$ Negligible !

Distortion reduction by feedback :

$T = 10$ $IM_{2f} = 2.5\ \%/100 = 0.25\ \%$

Willy Sansen 10 05 1882

1882

Indeed, the IM2 distortion of the second stage is the same as mentioned previously.

The third-order distortion of the differential input stage is still negligible. Higher frequencies would be needed to make the distortion of the first stage dominant.

However, the loop gain at this frequency is only 10. As a result, the IM_{2f} is a lot larger, i.e. 0.25% rather than 0.0025% at low frequencies. This is only a result of the reduced loop gain.

At even higher frequencies, the loop gain becomes even smaller. In addition the input signals of both stages increase further, giving rise to a steep increase of the distortion.

This is shown experimentally next.

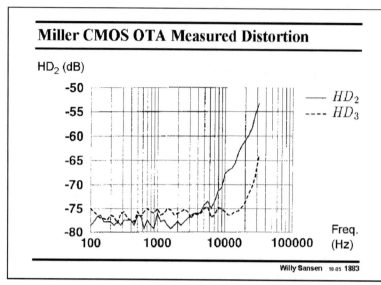

1883

This is an experimental result for a two-stage opamp, with a GBW of 1 MHz.

It is clear that second-order distortion is dominant. It is generated by the output stage. Its slope is close to 40 dB/decade as expected.

The third-order distortion is visible at higher frequencies but never dominant. Its slope is even higher.

For frequencies lower than about 7 kHz, the distortions drown in the noise. No reliable measurements of distortion are possible here.

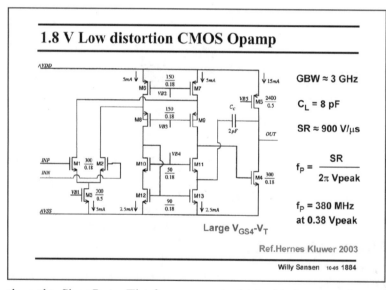

1884

An example of a two-stage opamp with low distortion is shown in this slide.

The first stage is a conventional folded cascode. It is followed by a second-stage without cascodes for large output swing.

For low distortion at intermediate frequencies, the GBW must be as high as possible.

To avoid Slew-Rate induced distortion, the slope of the sine wave going through zero must be less than the Slew Rate. The frequency at which this happens is reached at 380 MHz for a 0.38 V peak output voltage.

It is clear that such an amplifier consumes a lot of power, i.e. 2.5 mA in each input transistor and 15 mA in the output transistors, which is 25 mA altogether on a 1.8 V supply voltage. This gives a FOM of only 1 MHzpF/mA.

The distortion is very low however, as shown next.

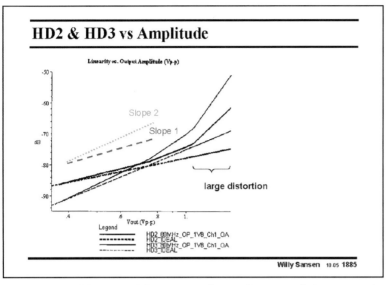

1885

The distortion components HD_2 and HD_3 are plotted on a log-log-scale to verify the slopes. The slopes themselves are also added.

It is clear that the slopes are correct over the whole range of input amplitudes, at least for HD_2.

For HD_3 the slope is correct up to about $0.8\ V_{ptp}$, beyond which there is a slight increase. Beyond $1.2\ V_{ptp}$ there is a sharp increase. We enter the region of hard non-linearity. This is probably caused by a cascode coming out of the strong-inversion region.

Curves at a higher frequency (80 MHz) are given as well. Better curves versus frequency are shown next.

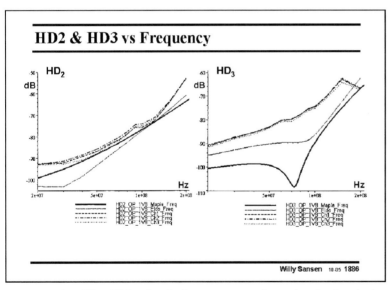

1886

The experimental results versus frequency are shown in this slide. They are all taken for an output swing of $0.75\ Volt_{ptp}$. The labels Ch1, Ch2 and Ch3 refer to different samples.

The simulation results are carried out by means of Maple and by Eldo simulations are shown as well for comparison. The Maple simulations are actually symbolic simulations. They take into account only the first three terms of the power series. The Eldo simulations are transient circuit simulations to which a Fourier analysis is applied. The HD_2 is dominated by the distortion in the output stage. At higher frequencies, it increases because of the reduced loop gain. At lower frequencies it is underestimated because of various other sources of distortion.

The HD_3 shows a larger gap between simulated and measured values. At low frequencies the HD_3 is generated by the output stage whereas at high frequencies it is taken over by distortion of the input stage. They have a different polarity. The cancellation point is clearly visible in Maple but much less in Eldo and in the experimental data.

Also, several more sources of distortion are present, among which the output conductance distortion of the output transistors are the major ones.

Table of contents

Willy Sansen 10 05 1887

1887

As distortion refers to any deviation from the ideal sine wave, for a sinusoidal input waveform, many other sources of distortion can occur.

A few of them are listed here. This list will never be exhaustive however. Each new application may give rise to a new kind of distortion.

Other cases of distortion and guide lines

▪ Distortion caused by limited SR
▪ Distortion of a switch
▪ Distortion at high frequencies :
     Volterra series instead of power series
▪ Distortion in continuous-time filters
▪ Guide lines

Willy Sansen 10 05 1888

1888

Too small a Slew-Rate can prevent to pass the high frequency slope of sine wave, which is reached when the sine wave goes through zero. When the Slew-Rate is much too low, a triangular waveform then results, giving excessive HD_3 (more than 10%). This is clearly to be avoided.

A MOST as a switch will give distortion as well as its resistance depends on the V_{GS}, which is the difference between the Gate drive voltage and the signal output voltage at the Source. This effect is especially detrimental at low supply voltages (< 1.8 V). This will be discussed in more detail in Chapter 21.

At high frequencies, it is actually no longer possible to use power series and phasors to correct the response for frequency dependence. The power series must be substituted by Volterra series. As they can only be applied to simple circuits, they are seldom used.

In continuous-time filters, discussed in the next Chapter, some more techniques are used to cancel the distortion. This is possible provided sufficient matching can be reached. Some examples will be given in the next Chapter.

Finally, some simple guidelines are added to allow reduction of distortion.

Guide lines for low distortion

- Scaling such that voltage amplitudes are limited
- Scaling such that relative current swings are limited
- Feedback
- All fully differential

1889

Since distortion is proportional to both voltage and current swing, the first guidelines simply state that low-distortion performance can be reached when both swings are small.

Distortion can always be reduced by application of feedback. This feedback causes a reduction in distortion, first by reducing the signal swing and on top of that, by dividing the distortion by the loop gain.

Finally, second-order distortion can always be avoided by making all circuitry fully-differential. Obviously, the power consumption also increases and so does the input noise. However, the Signal-to-noise-and-distortion will increase. Mismatch will give some second-order distortion but it will usually be smaller than the third-order one.

Distortion components

Distortion comp.	IM_2 $\times U_p$	IM_3 $\times U_p^2$	$U_p = \dfrac{V_{ip}}{V_{ref}}$ $V_{ref} =$
Bipolar	1/2	1/8	kT_e/q
MOST	1/4	0	$(V_{GS}-V_T)/2$
Bip. diff.pair	0	1/4	$2kT_e/q$
MOST diff.pair	0	3/32	$(V_{GS}-V_T)$

1890

As a final overview, the distortion components are listed for all elementary circuits, first of all, without, and then with feedback. They all have been previously derived.

For example, take the second-order distortion of a MOST. Its IM_2 is 1/4 of U_p, in which U_p is the ratio of the peak input voltage V_{ip} divided by $(V_{GS} - V_T)/2$. Also, its IM_3 is about 1/10 of U_p^2, in which U_p is the ratio of the peak input voltage V_{ip} divided by $(V_{GS} - V_T)$.

Distortion components with Feedback (T > 5)

Distortion comp.	IM_2 $\times U_p$	$-IM_3$ $\times U_p^2$	$U_p = \dfrac{V_{ip}}{V_{ref}}$ $V_{ref} =$
Bipolar	1/2T	1/4T	$kT_e/q \times T$
MOST	1/4T	3/32T	$(V_{GS}\text{-}V_T)/2 \times T$
Bip. diff.pair	0	1/4T	$2kT_e/q \times T$
MOST diff.pair	0	3/32T	$(V_{GS}\text{-}V_T) \times T$

Willy Sansen 10.05 **1891**

1891

A similar Table is easily made for the distortion components with feedback. The loop gain is T.

For example, the IM_3 of a single-MOST amplifier is about $1/10$ of U_p^2, in which U_p is the ratio of the peak input voltage V_{ip} divided by $(V_{GS} - V_T)/2$, and divided by T^3. Two T's comes from the reduction in current swing (through U_p) and one of them from the reduction in loop gain.

References

P.Wambacq, W.Sansen : Distortion analysis of analog Integrated
Circuits, Kluwer Ac. Publ. 1998

W.Sansen : "Distortion in elementary transistor circuits"
IEEE Trans. CAS II Vol 46, No 3, March 1999, pp.315-324

J. Silva-Martinez, etal : High-performance CMOS continuous-time
filters, Kluwer Ac. Publ. 1993

B. Hernes, T. Saether : Design criteria for low-distortion in
feedback opamp circuits, Kluwer Ac. Publ. 2003

G. Palumbo, S. Pennisi : Feedback amplifiers, Kluwer Ac. Publ. 2002

Willy Sansen 10.05 **1892**

1892

A few general references are listed on distortion. They speak for themselves.

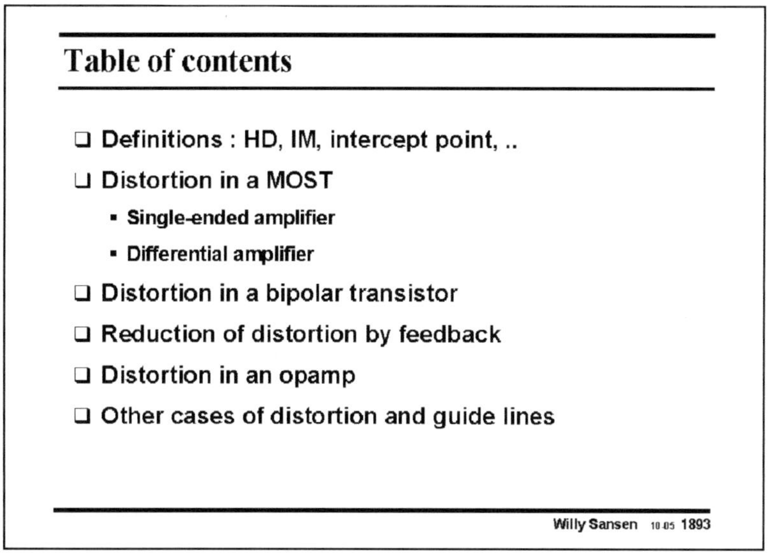

Table of contents

Willy Sansen 10-05 1893

1893

In this Chapter, an introduction is given on distortion caused by the weak nonlinearities of the transistors. The definitions of distortion have been given first. They are followed by distortion calculations of MOST and bipolar devices in both single and differential configurations.

Considerable attention is paid to the reduction of distortion by feedback. Finally some guide lines and overview Tables are added to allow easy prediction of distortion levels in any kind of amplifier.

191

Continuous-time filters

Willy Sansen

KULeuven, ESAT-MICAS

Leuven, Belgium

willy.sansen@esat.kuleuven.be

Willy Sansen 10.05 191

An important class of filters are continuous-time filters. They are filters which don't use switches. They are used mainly at high frequencies, but also for a variety of other applications.

Their biggest problem is distortion. This is why this Chapter follows the Chapter on distortion.

The several filter types are discussed first. Considerable attention is paid to transconductors with low distortion.

Before sampling, anti-aliasing filters are required to limit the signal bandwidth. A prefilter is required, which has to be a continuous-time filter.

High-frequency filters are also usually continuous-time filters. In sampled-data filters, the clock frequencies must be higher than the highest signal frequencies. Exceedingly high clock frequencies would therefore be needed, consuming too much power.

In communications systems (CDMA, UWB, etc.), channel bandwidths reach values of up to 5, 7 and even 10 MHz. Separation of these channels require thus many high-frequency filters.

Finally, high-frequency filters turn into low-frequency filters when the power consumption is reduced drastically. In portable applications, continuous-time filters also play an important role, even at low frequencies (e.g. sensor interfaces, ...).

Applications and problems

- **Applications**
  - **Anti-aliasing filters**
  - **Video and HF filters : hard-disk drives**
  - **Channel select filters**
  - **Low-power filters**
- **Problems:**
  - **Tuning for high precision: mismatch < 5 %**
  - **Distortion : THD < -60 dB**
  - **Low power supply voltages**
  - **High quality factors : Q > 50 ?**

Willy Sansen 10.05 192

192

However, continuous-time filters have many problems. The characteristic frequencies are not very accurate. It is difficult to realize higher-order filters which require accurate matching of the characteristic frequencies. Tuning circuits will thus have to be added, which increase the power consumption.

Linearity is also a problem. Filters are normally specified for fairly large input voltages. To reduce the distortion to below −60 dB, both feedback and cancellation techniques will be required.

This is especially true at low supply voltages. The signal swings cannot be allowed to be smaller. Distortion reduction is now even more important!

Putting al these problems together, especially mismatch and distortion, means that it is difficult to reach filters with high quality factors.

Table of contents

- ◆ **Active RC filters**
- ◆ **MOSFET-C filters**
- ◆ **GmC filters**
- ◆ **Comparison**

Ref.: Tsividis, Voorman, Integrated Cont.-time filters, IEEE Press 1993
J. Silva-Martinez, Kluwer 1993,
W. Dehaene, JSSC, July 1997, 977-988

Willy Sansen 10-05 193

193

Possible solutions are provided by use of RC or MOSFET-C filters. Most attention will be paid to Gm-C filters as they reach the highest frequencies.

At the end of this Chapter, a comparison is given of the dynamic ranges that can be expected for different frequency regions.

Active RC filters

Opamps and passive components (R, C)

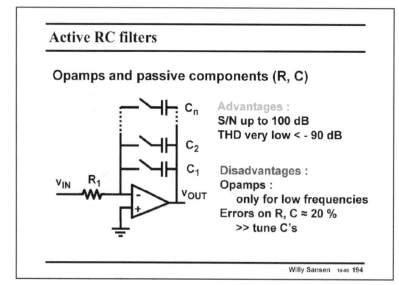

Advantages :
S/N up to 100 dB
THD very low < - 90 dB

Disadvantages :
Opamps :
    only for low frequencies
Errors on R, C ≈ 20 %
    >> tune C's

Willy Sansen 10-05 194

194

Active RC-filters are operational amplifiers with resistors and capacitors in the feedback loop.

If a low-noise opamp is used, large Signal-to-Noise ratios (SNR) can also be achieved, large input signals can be applied with very little distortion. The Signal-to-Noise-and Distortion ratio (SNDR) can also be fairly high. Actually, these filters are capable of the highest SNDR values possible, depending on the power consumption.

This is obviously only correct for the frequencies where the loop gain is high. These OTA-RC filters are not suitable for high frequencies.

Another disadvantage is that passive components are used, which usually have a low absolute accuracies. The absolute errors can be as high as 15 ... 20%. We will therefore need to tune the filter sections.

However, in such an OTA-RC filter no component can be tuned. The only way out is to use resistor or capacitor banks, as shown in this slide. For an 8-bit binary bank of capacitances, the absolute error can be reduced to 0.4%.

Table of contents

- ◆ **Active RC filters**
- ◆ **MOSFET-C filters**
- ◆ **GmC filters**
- ◆ **Comparison**

Ref.: Silva-Martinez, Dehaene, ..

Willy Sansen 10 05 195

195
Tuning is possible when the resistors are substituted by MOST devices. They are used in the linear region (for small V_{DS}) and can be represented by their on-resistance (see Chapter 1).

Changing the Gate voltage changes their on-resistance, allowing tuning of the filter RC time constants.

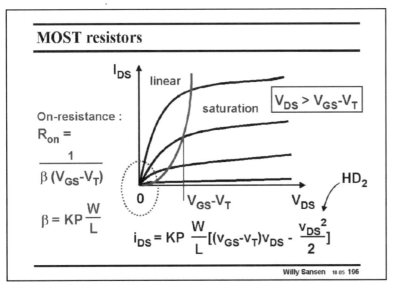

MOST resistors

On-resistance :
$$R_{on} = \frac{1}{\beta(V_{GS}-V_T)}$$

$$\beta = KP\frac{W}{L}$$

$$i_{DS} = KP\frac{W}{L}[(v_{GS}-v_T)v_{DS} - \frac{v_{DS}^2}{2}]$$

$V_{DS} > V_{GS}-V_T$

Willy Sansen 10 05 196

196
Such an on-resistance is inversely proportional to the Gate voltage.

It is very linear provided the V_{DS} is very small. For larger signal-swings however, this is no longer possible and distortion appears.

For larger values of V_{DS}, the term in V_{DS2} has to be included in the $I_{DS}-V_{GS}$ characteristic. This leads to second-order distortion.

The easiest way to avoid this distortion is to adopt differential structures.

Many configurations are possible.

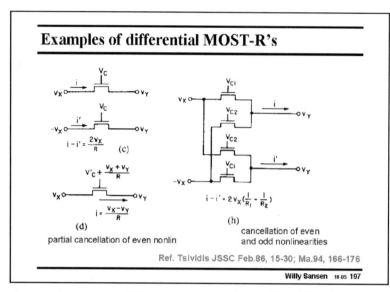

Examples of differential MOST-R's

$$i - i' = \frac{2v_X}{R}$$ (c)

$$v'_C + \frac{v_X + v_Y}{R}$$

$$i = \frac{v_X - v_Y}{R}$$

(d)
partial cancellation of even nonlin

$$i - i' = 2v_X(\frac{1}{R_1} - \frac{1}{R_2})$$

(h)
cancellation of even
and odd nonlinearities

Ref. Tsividis JSSC Feb.86, 15-30; Ma.94, 166-176

Willy Sansen 10.05 197

197
Several examples of MOST resistors are shown in this slide.

The differential one (c) allows the cancellation of the even-order distortion.

Care must be taken to connect the Bulks to the Sources, which requires extra wells. If this is not possible, additional distortion will be generated.

Cancellation of odd-nonlinearities is also possible by addition of cancellation transistors (h), with different sizes and driven by different control voltages V_{c2}. This is the configuration most often used.

It gives higher node capacitances, however. For high frequencies, the two transistor structure (c) may be preferred.

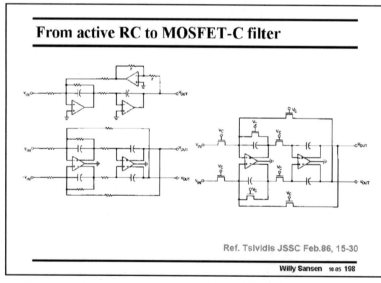

From active RC to MOSFET-C filter

Ref. Tsividis JSSC Feb.86, 15-30

Willy Sansen 10.05 198

198
An example of a second-order filter is shown in the left top-corner. It contains two capacitors indeed and three opamps. It is single-ended, however.

A differential version of the same filter is shown below. One opamp can be left out as it only provides a signal inversion. In a differential circuit both phases are always available. It is sufficient to use two opamps for a second-order filter.

However, the opamps are now fully differential. They require common-mode feedback (see Chapter 8), which takes more power.

A MOSFET-C realization of the same filter is now shown on the right. All resistors have been replaced by MOSFETs. All their Gates are connected together towards control voltage V_c. Tuning of the resistors is thus possible and hence tuning of the filter frequency. A large tuning range also requires a large Voltage range, which may not be easy at low supply voltages.

Moreover, MOSFETs have a limited frequency range of operation, as explained next.

Large R_{ON} values at high frequencies

For low-frequency low-pass filter with f-3dB

$$f_{-3dB} = \frac{1}{2\pi R_{on}C} \approx \frac{KP \; W/L \; (V_{GS}-V_T)}{2\pi \; C}$$

For f_{-3dB} = 4 kHz; KP= 60 μA/V^2; V_{GS}-V_T = 1 V; W = 2 μm; C = 10 pF
R_{on}= 4 MΩ. For matching W = 2 μm: L ≈ 500 μm ! The area is 10^{-5} cm^2

For C_{ox} = 5.10^{-7} F/cm^2 (0.35 μm); C_{GS} = 5 pF;
High-frequency limit at ≈ 8 kHz or f_T ≈ 8 kHz !!!!!!

Willy Sansen 10-05 199

199

Large values of the on-resistance R_{on} are required to realize filters at low frequencies. For example, for a filter at 4 kHz, an R_{on} is required in the MΩYs.

For matching the smallest dimension, which is W in this case, cannot be much smaller than a few micrometers. As a result, the channel length becomes extremely large, reducing the f_T to very low values. The MOST resistor turns into a transmission line with an unpredictable high-frequency characteristics.

LC ladder filter

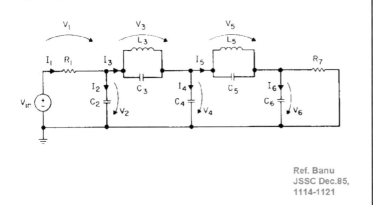

Ref. Banu
JSSC Dec.85,
1114-1121

Willy Sansen 10-05 1910

1910

For example, take this LC ladder filter. It is known to be fairly insensitive to errors on the values of L and C. Except for the source resistor R_1 and the load resistor R_7, five branches can be distinguished.

Each node is connected to ground through a capacitor. It is therefore a low pass filter.

Each branch will be represented by an opamp with feedback elements, as shown next. The circuit must be converted into a differential structure first!

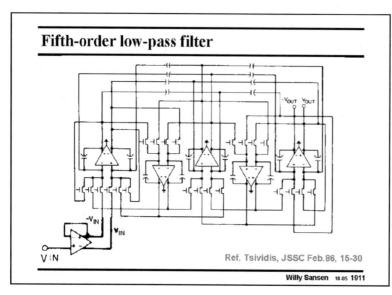

Fifth-order low-pass filter

Ref. Tsividis, JSSC Feb.86, 15-30

Willy Sansen 10.05 1911

1911
A similar fifth-order ladder filter is shown in this slide.

It is a fully differential filter indeed to cancel out even-order distortion. All capacitors are fixed but the MOST resistors can be tuned by means of the control voltage at all the Gates.

A sixth opamp is used to convert the single-ended input into a differential one.

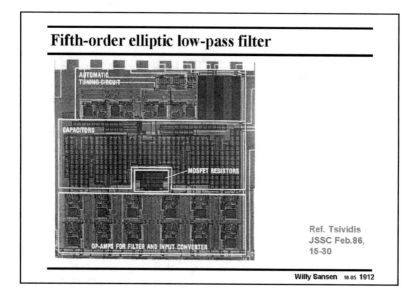

Fifth-order elliptic low-pass filter

Ref. Tsividis
JSSC Feb.86,
15-30

Willy Sansen 10.05 1912

1912
These six opamps are easily seen at the bottom of the layout.

The bank of capacitors takes the larger area, because of better matching. The area required for the MOST resistors is quite small.

The tuning circuitry takes a rather large area. It is discussed next.

1913
The tuning circuit is a Phase-locked loop. It generates the filter control voltage V_c, as a result of a feedback loop.

This loop consists of a Voltage-controlled oscillator (VCO), which generates a frequency (phase), which is compared to the reference frequency (phase) of the clock in the phase detector. The output of this phase detector, which is actually a multiplier, contains the sum and the difference of the input frequencies. The sum frequency is a high frequency and is filtered out by the low-pass loop filter. The difference frequency is a slowly varying signal, which is fed back to the VCO. This is control frequency V_c indeed.

The VCO uses the same opamps as in the filter. Also, the capacitors and MOST resistors are

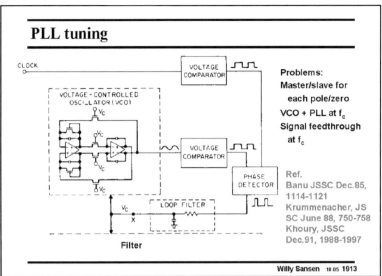

Problems:

Master/slave for
each pole/zero

VCO + PLL at f_c

Signal feedthrough
at f_c

Ref.
Banu JSSC Dec.85,
1114-1121
Krummenacher, JS
SC June 88, 750-758
Khoury, JSSC
Dec.91, 1988-1997

Willy Sansen   10 05 1913

of the same order of magnitude as the ones in the filter, for good matching. A control Voltage V_c is obtained which generates a VCO frequency f_c, which is locked to the external clock frequency. As a result, the filter frequencies are also locked to the external clock frequency. This clock is most often a crystal oscillator with high accuracy (see Chapter 22). The filter frequencies can be expected to have similar high accuracies.

The main disadvantage of such a tuning circuit is that a lot of additional circuitry is required and that the VCO frequency f_c is in the same region as the filter frequency. Leakage from the VCO to the signal path is thus hard to avoid. Other tuning schemes will be discussed later.

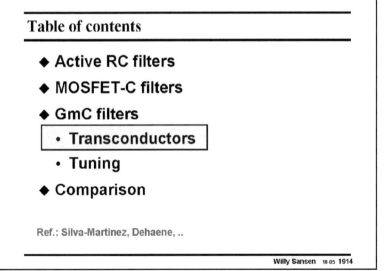

Willy Sansen   10 05 1914

1914

Both active RC and MOSFET-C filters employ operational amplifiers. This means that only at low frequencies, the loop gain is sufficiently large to guarantee accurate performance.

For high frequency filters, the operational amplifiers are reduced to their simplest possible configurations. They only consist of differential pairs, to which some linearization techniques have been applied. They are called transconductors or Gm blocks. Such filters are called GmC filters. They can provide the highest possible filter frequencies with reasonable quality factors.

First of all, some transconductor configurations are discussed, followed by some other tuning circuits.

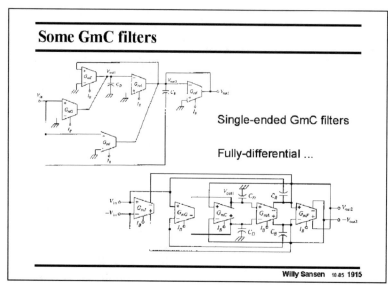

Some GmC filters

Single-ended GmC filters

Fully-differential ...

Willy Sansen 10-05 **1915**

1915

GmC filters consist of Gm blocks, which are not much more than linearized differential pairs and capacitors. A second-order filter is shown on the left. It is single-ended.

To cancel out the even-order harmonics, a differential version is preferred. It is shown on the right. They take more power as the output circuitry is doubled. Moreover, common-mode feedback is required to set the common-mode or average output (and input) voltage levels. Even more power is therefore consumed.

They are now studied in more detail.

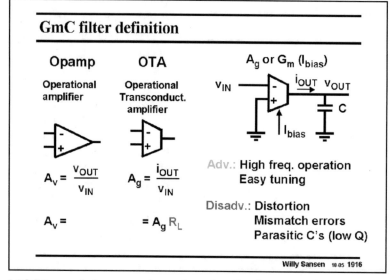

GmC filter definition

Opamp	OTA	A_g or G_m (I_{bias})

Opamp
Operational amplifier

$A_v = \dfrac{v_{OUT}}{v_{IN}}$

$A_v =$

OTA
Operational Transconduct. amplifier

$A_g = \dfrac{i_{OUT}}{v_{IN}}$

$= A_g R_L$

A_g or G_m (I_{bias})

v_{IN} — i_{OUT} v_{OUT} — C — I_{bias}

Adv.: **High freq. operation**
Easy tuning

Disadv.: **Distortion**
Mismatch errors
Parasitic C's (low Q)

Willy Sansen 10-05 **1916**

1916

Such a Gm block differs from an opamp in the sense that it does not include an output stage with low output resistance. A Gm block generates an output current proportional to the input voltage. Comparison with an opamp is only possible provided an output resistor R_L is added.

A considerable advantage of such a Gm block is that its transconductance Gm directly depends on the biasing current I_{bias}. If the input MOSTs work in strong inversion, the Gm is proportional to the square root of the current. If MOSTs are used in weak inversion however, or when bipolar transistors are used, then the Gm is directly proportional to the current.

The tuning is relatively easy. Because of their simplicity, these circuits are able to operate up to high frequencies.

The drawbacks are still the same, distortion and mismatch. Moreover, each node has a parasitic capacitance to ground, limiting the high frequency performance.

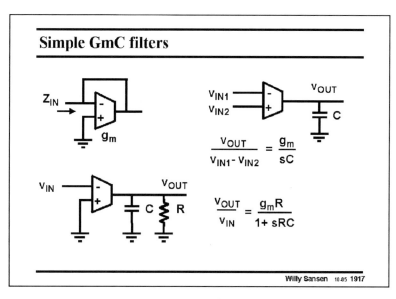

Simple GmC filters

$$\frac{V_{OUT}}{V_{IN1} - V_{IN2}} = \frac{g_m}{sC}$$

$$\frac{V_{OUT}}{V_{IN}} = \frac{g_m R}{1 + sRC}$$

Willy Sansen   10.05   1917

1917

Some simple filter structures with Gm blocks are shown in this slide.

Feedback from the output to the minus input turns the Gm block into a resistance with value $1/g_m$.

Using a Gm block open loop provides a pole determined by the output load capacitance. This is actually an integrator.

If a parallel RC circuit is used as a load, then a voltage amplifier is obtained with gain $g_m R$ and a pole with time constant RC.

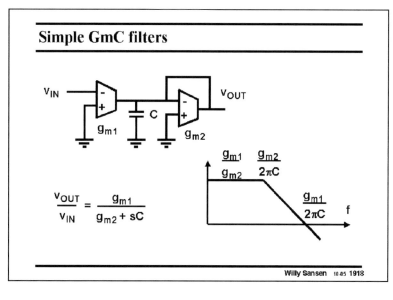

Simple GmC filters

$$\frac{V_{OUT}}{V_{IN}} = \frac{g_{m1}}{g_{m2} + sC}$$

Willy Sansen   10.05   1918

1918

As a load a Gm block or transconductor can be used as well. The voltage gain is now the ratio of the two transconductances. The pole is determined by g_{m2} and the capacitance C at the intermediate node.

The GBW is obviously determined by g_{m1} and the capacitance C at that intermediate node.

This same structure can better be realized in a differential configuration, as shown next.

1919

A differential configuration of a voltage amplifier is shown in this slide. The gain is the ratio of the two transconductances. The pole is determined by g_{m2} and the capacitance C_L at the intermediate node.

In order to make sure this pole frequency is accurate, the parasitic capacitances C_p must be negligible with respect to the load capacitance C_L. This limits the minimum value of C_L and the upper value of the pole frequency.

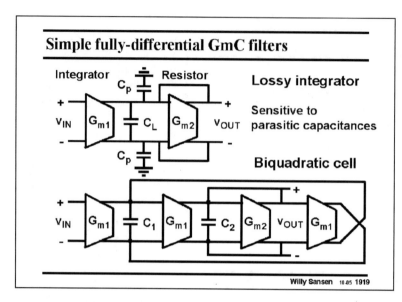

Simple fully-differential GmC filters

Integrator — Resistor — **Lossy integrator**

Sensitive to parasitic capacitances

Biquadratic cell

Willy Sansen 10.05 1919

A more complicated filter structure is shown below. It is a second-order filter which is biquadratic. This means that both the numerator and the denominator of the transfer function are of second-order. For this purpose, two capacitances are used and four transconductors.

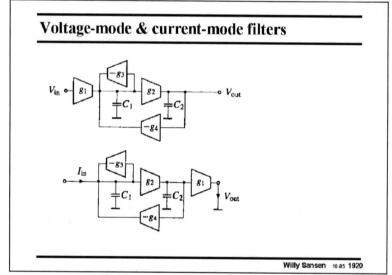

Voltage-mode & current-mode filters

Willy Sansen 10.05 1920

1920

Such filters are usually driven by an input voltage. This is not necessary, however.

As an example, let us take the second-order GmC filter shown in this slide. Moving the input transconductor g_1 towards the output, gives the same transfer function but for the current ratio I_{out}/I_{in}, rather than for the voltage ratio V_{out}/V_{in}.

Voltage GmC filters can easily be converted into current GmC filters. However, we will see later that in general, current-mode filters have a lower SNR (by about 20 dB) than voltage-mode filters. This is why voltage mode filters are normally preferred.

Let us now investigate some important transconductor circuits.

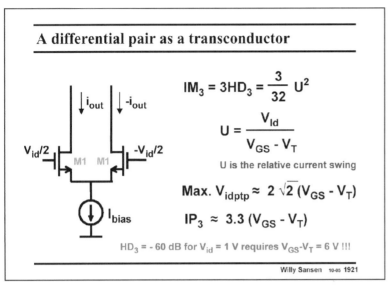

A differential pair as a transconductor

$$IM_3 = 3HD_3 = \frac{3}{32} U^2$$

$$U = \frac{V_{Id}}{V_{GS} - V_T}$$

U is the relative current swing

$$\text{Max. } V_{idptp} \approx 2\sqrt{2}\,(V_{GS} - V_T)$$

$$IP_3 \approx 3.3\,(V_{GS} - V_T)$$

$HD_3 = -60$ dB for $V_{id} = 1$ V requires V_{GS}-$V_T = 6$ V !!!

Willy Sansen 10-05 1921

1921

The simplest transconductor is evidently a differential pair. Its Gm is simply the transconductance of the MOSTs.

Its distortion is quite high, however. Its IM_3 is proportional to the relative current swing squared. It can be reduced for large values of $V_{GS} - V_T$. Exceedingly high values are required, however, to achieve low distortion for a 1 V input signal amplitude.

Additional feedback or other tricks will therefore be required to remedy this.

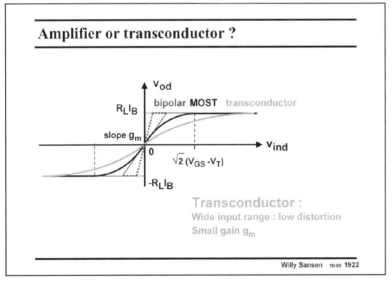

Amplifier or transconductor ?

Transconductor :
Wide input range : low distortion
Small gain g_m

Willy Sansen 10-05 1922

1922

To illustrate once more what the difference is between an amplifier and a transconductor, the transfer curve is illustrated for several of them.

An amplifier has high gain and a small input range. A bipolar-transistor amplifier is a good example.

A transconductor has low gain but a large input range. A MOST amplifier with large $V_{GS} - V_T$ values and a bipolar-transistor amplifier with emitter resistors, are good examples.

Real transconductors have an even larger input range. Their range is extended by means of linearization techniques such as local feedback and cross-coupling. They are now studied.

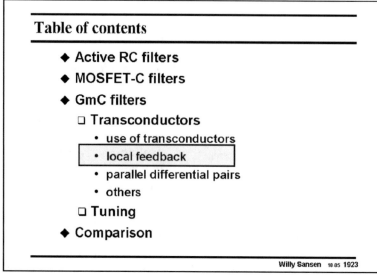

1923
Both active RC and MOSFET-C filters employ operational amplifiers. This means that only at low frequencies, the loop gain is sufficiently large to guarantee accurate performance.

For high frequency filters, the operational amplifiers are reduced to their simplest possible configurations. They only consist of differential pairs, to which some linearization techniques have been applied. They are called transconductors or Gm blocks. Such filters are called GmC filters. They can provide the highest possible filter frequencies with reasonable quality factors.

Firstly some transconductor configurations are discussed, followed by some other tuning circuits.

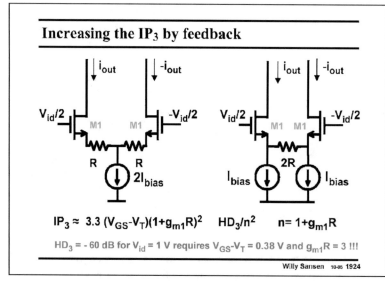

1924
Local feedback is an obvious technique to increase the input range or reduce the distortion (see previous Chapter).

Two practical circuits are shown in this slide. They are equivalent for AC performance but not for DC. In the left case the DC currents also flow through the series resistors R, but not in the right case. More DC voltage drop is thus required for the left case. It is to be avoided at low supply voltages.

Another difference is that, on the left, the two resistors R must be well matched. On the right the two current sources must also be well matched. Which one is easier depends on the area used.

Another difference is that the output capacitance of the current source is connected at a common-mode point on the left. On the right, these output capacitances will limit the high-frequency performance. A small capacitance across 2R may now be needed for compensation.

Both circuits have the same reduction in distortion. If more reduction is required, more loop gain is required, as shown next.

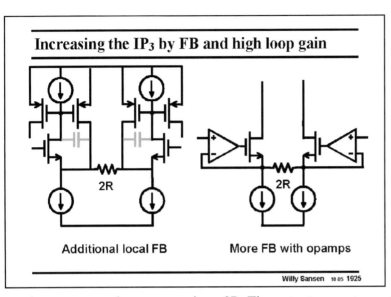

Increasing the IP$_3$ by FB and high loop gain

2R

2R

Additional local FB

More FB with opamps

Willy Sansen 10 05 1925

1925

For larger loop gain in the feedback loop, the distortion is lower and the input range larger.

Local feedback is possible by addition of only a few transistors as shown on the left. The input transistors cannot have any AC current as they have DC current sources in both Drain and Source. Only the pMOST devices thus carry AC current. This current is determined by the differential input voltage which appears nearly unattenuated across transistor 2R. The output currents are then mirrored out.

This voltage-to-current conversion is even more accurate if a full operational amplifier is inserted in the feedback loop, instead of just one single transistor. The loop gain now includes the open-loop gain of the operational amplifier. The distortion will now be very small.

This is only true at low frequencies, however, as the open-loop gain of an opamp drops quite rapidly versus frequency. For high frequencies the circuit on the left may be preferable.

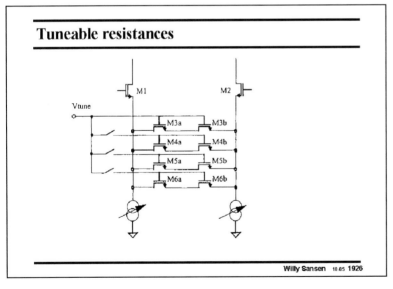

Tuneable resistances

M1   M2

Vtune

M3a   M3b

M4a   M4b

M5a   M5b

M6a   M6b

Willy Sansen 10 05 1926

1926

The previous transconductors all have the disadvantage that they cannot be tuned. Their voltage-to-current conversion is determined by the resistor.

The tuneability can be greatly enhanced by insertion of MOSTs as resistors or even better by a bank of MOSTs, which can be switched in and out for coarse tuning. Changing the voltage V_{tune}, then changes the on-resistance of the MOST, for fine tuning.

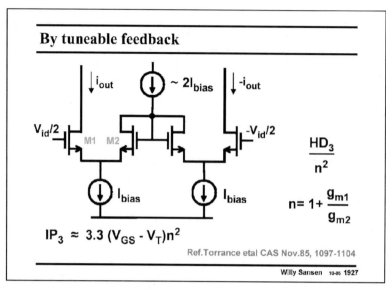

By tuneable feedback

i_{out} ~ $2I_{bias}$ $-i_{out}$

$V_{id}/2$ M1   M2 $-V_{id}/2$

$\dfrac{HD_3}{n^2}$

I_{bias} I_{bias} $n = 1 + \dfrac{g_{m1}}{g_{m2}}$

$IP_3 \approx 3.3\,(V_{GS} - V_T)n^2$

Ref.Torrance etal CAS Nov.85, 1097-1104

Willy Sansen 10-05 **1927**

1927

Rather than use MOSTs as resistors, MOSTs can also be used as diodes, as shown in this slide.

The total resistance 2R between the Sources of the input transistors M1 is now $2/g_{m2}$ for small signals. This value can be tuned by changing the DC current 2 I_{bias} through the diode-connected MOSTs M2.

Evidently, for small I_{bias}, the resistances are larger and the distortion is reduced. The gain is now smaller as well. This is the situation for large input signal amplitudes.

The reduction factor n simply depends on the g_m ratio and hence on the current ratio.

The IP_3 increases accordingly.

Actually, this is a circuit which has been around since the sixties with bipolar transistors. It was used for Automatic Gain Control in some receiver circuits.

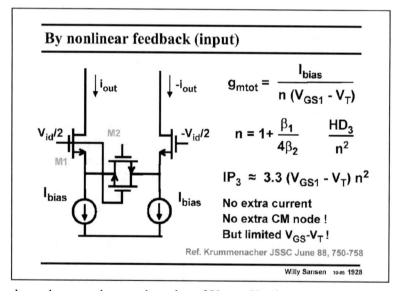

By nonlinear feedback (input)

i_{out} $-i_{out}$ $g_{mtot} = \dfrac{I_{bias}}{n\,(V_{GS1} - V_T)}$

$V_{id}/2$ M2 $-V_{id}/2$ $n = 1 + \dfrac{\beta_1}{4\beta_2}$ $\dfrac{HD_3}{n^2}$

M1

I_{bias} I_{bias} $IP_3 \approx 3.3\,(V_{GS1} - V_T)\,n^2$

No extra current
No extra CM node !
But limited $V_{GS}-V_T$!

Ref. Krummenacher JSSC June 88, 750-758

Willy Sansen 10-05 **1928**

1928

A better realization of a transconductor which uses MOSTs as resistances is shown in this slide. It has the advantage that no extra current is consumed. Also, no extra common-mode node is added.

The reduction factor n is determined by the ratios of the sizes β_1/β_2 of the nMOST transistors. It appears in the total transconductance g_{mtot} but also in the IP_3.

The optimum ratio β_1/β_2 depends somewhat on the value of $V_{GS1} - V_T$ chosen. For example, for a $V_{GS1} - V_T$ of 0.27 V, the optimum value of the ratio is about 6. The transconductance g_{mtot} is then constant (within 1%) for output currents within 80% of the biasing current I_{bias}.

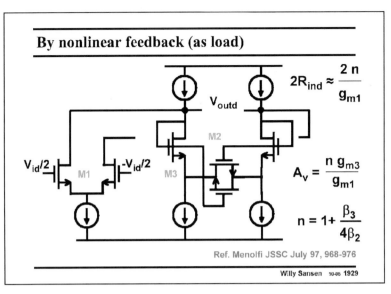

By nonlinear feedback (as load)

V_{outd}

M2

$V_{id}/2$ M1 $-V_{id}/2$ M3

$$2R_{ind} \approx \frac{2n}{g_{m1}}$$

$$A_v = \frac{n\, g_{m3}}{g_{m1}}$$

$$n = 1 + \frac{\beta_3}{4\beta_2}$$

Ref. Menolfi JSSC July 97, 968-976

Willy Sansen 10-05 1929

1929

The same linearization technique can be applied to diode-connected transistors, to be used as loads of a differential pair or transconductor. Transistors M3 are linearized indeed by application of linear transistors M2.

Again, the ratio of the transistor sizes determines factor n, which at the same time reduces the distortion and increases the differential resistor $2R_{ind}$ between the Sources of the diode connected devices M3.

The total voltage gain A_v from the differential input v_{id} to the differential output v_{outd} also increases with factor n.

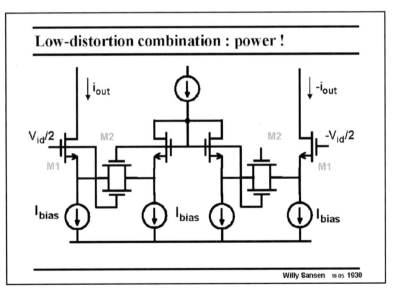

Low-distortion combination : power !

i_{out}

$-i_{out}$

$V_{id}/2$ M2   M2 $-V_{id}/2$

M1   M1

I_{bias} I_{bias} I_{bias}

Willy Sansen 10-05 1930

1930

A combination of both previous techniques leads to the circuit shown in this slide. It combines the advantages of both of them and also the disadvantages. The power consumption is greatly increased and so is the input range.

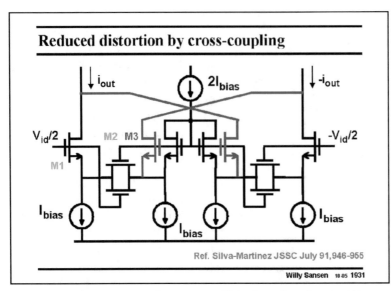

1931

On top of the two previous techniques, cross-coupling can also be added. As will be shown later in this Chapter, cross-coupling has become an often used technique to reduce the distortion, at the cost of a reduction in output signal amplitude.

The combination of these three techniques probably gives the largest possible input voltage range, without using an operational amplifier.

The high-frequency performance will be less than for the Krummenacher circuit, but the input range will be larger, for a certain amount of distortion. This is shown next.

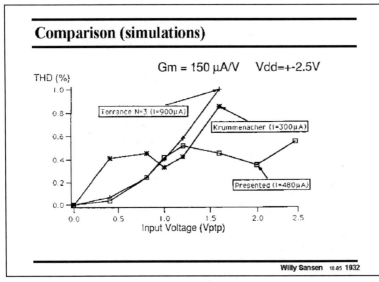

1932

This difference in input range is illustrated in this slide.

The curve of Torrance is flattened by the feedback resistances made up by the diode-connected transistors. The one of Krummenacher realizes distortion cancellation around 1 V_{ptp} and then increases greatly as the Torrance curve.

The one of Silva however, takes the best of both. For low input voltages, it follows the Torrance curve, whereas for high input voltages, it exploits distortion cancellation by both the Krummenacher technique and by cancellation. An input range of 2.5 V_{ptp} is quite large indeed.

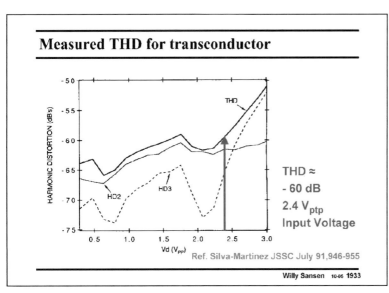

Measured THD for transconductor

THD $\approx$
- 60 dB
2.4 V_{ptp}
Input Voltage

Ref. Silva-Martinez JSSC July 91,946-955

Willy Sansen 10-05 **1933**

1933

It is clear that from a measurement about 0.1% HD_3 can be achieved for as much as 2.4 V_{ptp} input voltage.

For larger input voltages, the distortion increases rapidly, indicating that the region of high distortion is reached. Power series can no longer be used to predict the distortion levels.

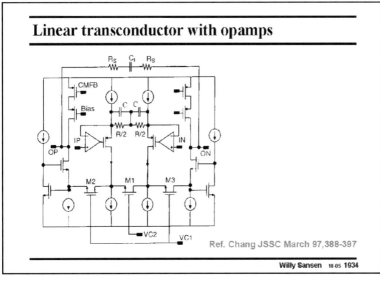

Linear transconductor with opamps

Ref. Chang JSSC March 97,388-397

Willy Sansen 10-05 **1934**

1934

The lowest distortion can be achieved by insertion of a fully operational amplifier in the feedback loop. Fixed resistors R/2 are used to carry out the voltage-to-current conversion. Small capacitances are added in parallel to boost this conversion at higher frequencies. This is necessary as the operational amplifiers have reduced gain at high frequencies.

Tuning of the transconductance is possible by changing the control voltages VC1 and VC2, which control the load resistances M1 and M2.

This transconductor has actually the structure of a folded cascode. The outputs are taken at the Drains of the regulated cascode stages.

With this transconductor a 7th order fully-differential Chebyshev filter has been constructed. It is a bandpass filter which can be tuned from 165 to 505 kHz. The IM_3 is less than -72 dB (at 300 kHz) increasing to less than 61 dB (at 600 kHz). The maximum dynamic range is 75 dB for 0.1% IM_3 at 4 V_{ptp} input voltage. Such large input voltage is only possible with operational amplifiers in the feedback loop.

It was realized in 0.7 micron CMOS.

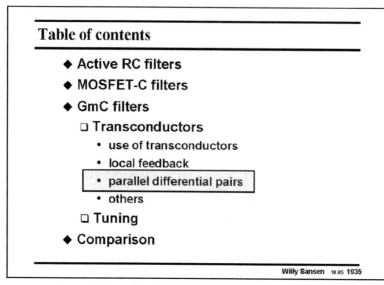

Willy Sansen 10 05 1935

1935
Both active RC and MOSFET-C filters employ operational amplifiers. This means that only at low frequencies, the loop gain is sufficiently large to guarantee an accurate performance.

For high frequency filters, the operational amplifiers are reduced to their simplest possible configurations. They only consist of differential pairs, to which some linearization techniques have been applied.

Now some transconductor configurations are discussed, with parallel differential pairs.

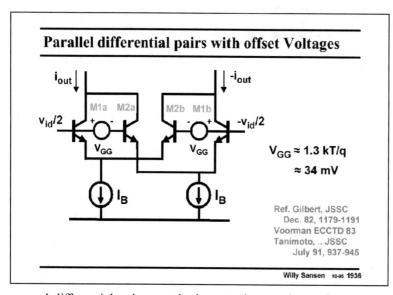

Parallel differential pairs with offset Voltages

$V_{GG} \approx 1.3 \, kT/q$

$\approx 34 \, mV$

Ref. Gilbert, JSSC
Dec. 82, 1179-1191
Voorman ECCTD 83
Tanimoto, .. JSSC
July 91, 937-945

Willy Sansen 10 05 1936

1936
Rather than use feedback by means of local series resistors, differential pairs can be put in parallel, as shown in this slide. The goal is always the same, i.e. to reduce the distortion over a wider input range.

An early example with bipolar transistors is given in this slide. Two differential pairs are put in parallel, with equal biasing currents I_B. An offset voltage V_{GG} is introduced to move the transfer characteristic of the second differential pair over the input voltage axis, as shown next. Note that the output current i_{out} is the sum of the currents of transistors M1a and M2a. The current of this latter transistor M2a is much smaller than the current through M1a, because of the offset voltage V_{GG}. The sum has a smaller slope over a wider input voltage range. The best result is obtained for an offset voltage of about 34 mV (Ref. Tanimoto).

This is explained in more detail next.

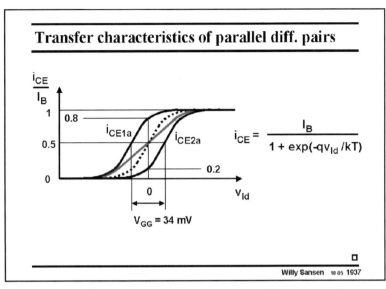

1937

The transfer characteristic of a differential pair without offset is given by the expression in this slide and is represented by the dotted line.

For zero input voltage v_{Id} the output collector current is half the biasing current I_B. For a larger input voltage, the collector current increases until it equals all the biasing current I_B.

This curve was derived in Chapter 3.

If an offset voltage is introduced V_{GG}, then the whole curve shifts over the input voltage axis by exactly this amount V_{GG}. As a result, the current through transistor M1a has increased from 50% to about 80% and has decreased through transistor M2a from 50% to about 20%. The average is again 50%. The transfer characteristic is spread out over a larger range of input voltages however, as shown by the red line.

The transconductance is the slope of the current versus input voltage. The transconductance around zero input voltage has decreased. It is almost 36% smaller than if both differential pairs are put in parallel. The input range however, has increased from about 26 mV$_{ptp}$ for one single differential pair to about 78 mV$_{ptp}$ for the same distortion. This is a factor of three better!

These offset voltages are not easily introduced. They can be realized by means of resistors and DC current sources (Ref. Gilbert). An easier way is to use different transistor sizes, as shown next.

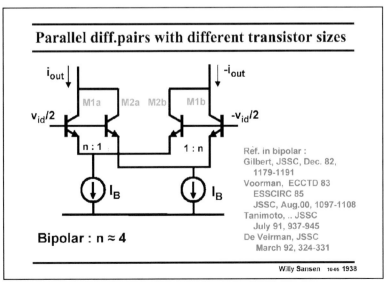

1938

The goal is now to realize an offset voltage of about 34 mV between two bipolar transistors, carrying equal currents.

Use of the exponential expression of the current I_{CE} versus V_{BE} shows that a ratio in I_S is required, or in size, of about 3.7. For simplicity, a factor of four is usually taken. The circuit is shown in this slide.

This circuit has been used by many authors. Also a MOST version has been around for quite some time. This will be shown after a few more bipolar realizations.

Different values of the biasing currents I_B can also be used, leading to different values of n. This will be discussed in more detail for the MOST equivalent.

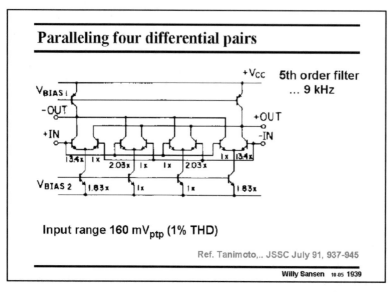

Paralleling four differential pairs

5th order filter
... 9 kHz

Input range 160 mV$_{ptp}$ (1% THD)

Ref. Tanimoto,.. JSSC July 91, 937-945

Willy Sansen 10.05 1939

1939

The same principle can be applied to more than two differential pairs. Four of them are placed in parallel in order to increase the input range. In this example, an input range is obtained of approximately 160 mV$_{ptp}$, which is about ten times larger than for a single differential pair for the same distortion (0.8% variation in transconductance).

The transconductance is reduced to about 35% of the differential pair with the same total current.

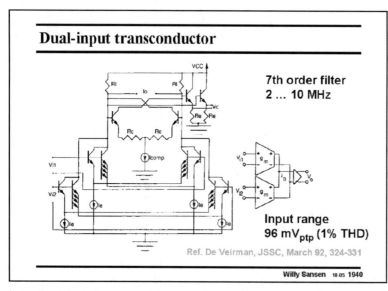

Dual-input transconductor

7th order filter
2 ... 10 MHz

Input range
96 mV$_{ptp}$ (1% THD)

Ref. De Veirman, JSSC, March 92, 324-331

Willy Sansen 10.05 1940

1940

The same input arrangements are used for the two parallel transconductors illustrated in this slide. Again, a factor of n=4 is taken for an equal biasing currents I_e.

The resulting transconductance is now 36% less than the transconductance of a differential pair with the same total current. The input range is about three times higher.

In order to increase the gain, a negative resistance is introduced at the output to cancel out the output resistances. The negative resistance is about $-2R_c$ (see Chapter 3).

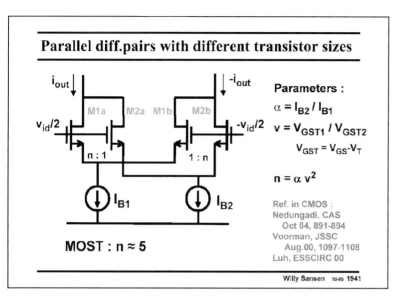

Parallel diff.pairs with different transistor sizes

i_{out} ↓   ↓ $-i_{out}$

M1a M2a M1b M2b

$v_{id}/2$ $-v_{id}/2$

n : 1   1 : n

I_{B1} I_{B2}

MOST : n ≈ 5

Parameters :

$\alpha = I_{B2} / I_{B1}$

$v = V_{GST1} / V_{GST2}$

$V_{GST} = V_{GS} - V_T$

$n = \alpha v^2$

Ref. in CMOS :
Nedungadi, CAS
  Oct 84, 891-894
Voorman, JSSC
  Aug.00, 1097-1108
Luh, ESSCIRC 00

Willy Sansen   10-05 **1941**

1941
The same circuit is easily realized by means of MOSTs as shown in this slide. The optimum value of n however, depends on the square-law characteristic of the MOST. Also, different biasing currents I_B are normally used. Their ratio is α.

Factor n is usually about 5 for MOST. The values of the biasing currents and the $V_{GS} - V_T$ now have to be chosen accordingly, as given in this slide. For equal biasing currents ($\alpha = 1$) the ratio in the $V_{GS} - V_T$ value is about 2.24. Values of $V_{GS} - V_T$ can now be 0.2 V for transistors M1 and 0.48 V for the inner devices M2.

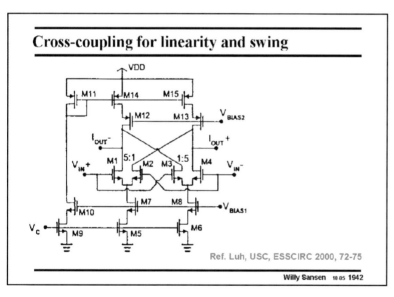

Cross-coupling for linearity and swing

VDD

M11   M14   M15

M12   M13 V_{BIAS2}

i_{OUT}^- i_{OUT}^+

V_{IN}^+ M1 5:1 M2 M3 1:5 M4 V_{IN}^-

M7   M8 V_{BIAS1}

V_c M10   M9   M5   M6

Ref. Luh, USC, ESSCIRC 2000, 72-75

Willy Sansen   10-05 **1942**

1942
An example of such a realization is shown in this slide. The factor n is 5 indeed. Equal biasing currents are used.

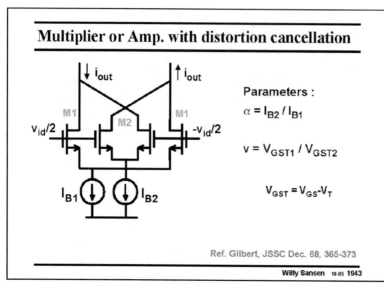

Multiplier or Amp. with distortion cancellation

Parameters :

$\alpha = I_{B2} / I_{B1}$

$v = V_{GST1} / V_{GST2}$

$V_{GST} = V_{GS} - V_T$

Ref. Gilbert, JSSC Dec. 68, 365-373

Willy Sansen 10.05 1943

1943
Note that all previous transconductors have their outputs connected in a different way compared to multipliers.

In multipliers which have been common in bipolar technologies since the sixties, the biasing currents I_{B1} and I_{B2} are equal and all transistors have the same size (or $V_{GS} - V_T$). As a result, the differential output current is always zero, even for a non-zero input voltage v_{id}.

Only a difference in biasing current AND differential input voltage v_{id} can give a contribution to the output.

A similar configuration is used to cancel out the third harmonic distortion, as shown in the previous Chapter. For a certain relationship, the two parameters α and v, the IM$_3$ cancels out.

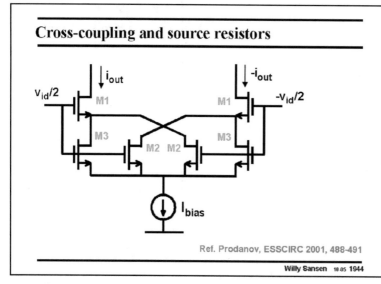

Cross-coupling and source resistors

Ref. Prodanov, ESSCIRC 2001, 488-491

Willy Sansen 10.05 1944

1944
All previous transconductors use two (or more) differential pairs to increase the input voltage range. Cross-coupling of the output Drains allows linearization of the total output current, which leads to a constant transconductance over a wider input range.

Cross-coupling can also be used in single differential pairs to improve the input range.

The first example uses MOSTs M3 as series resistances for the input transistors M1. These MOSTs M3 drive a cross-coupled pair which increases the input voltage range.

Another example is given next.

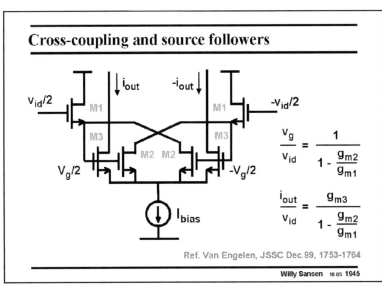

Cross-coupling and source followers

$$\frac{v_g}{v_{id}} = \frac{1}{1 - \dfrac{g_{m2}}{g_{m1}}}$$

$$\frac{i_{out}}{v_{id}} = \frac{g_{m3}}{1 - \dfrac{g_{m2}}{g_{m1}}}$$

Ref. Van Engelen, JSSC Dec.99, 1753-1764

Willy Sansen 10 05 **1945**

1945

In this transconductor, cross-coupling is used but also source-followers at the input. The negative resistance $-1/g_{m2}$ created by the transistors M2, is subtracted from the positive output resistance $1/g_{m1}$ of the input source followers M1, such that gain is achieved at the Gates of the transistors M2/M3. This gain is given by ratio v_g/v_{id}.

The ratio of the transconductances g_{m3}/g_{m2} is the same as the ratio of their W/L ratios as their V_{GS} values are the same. The signal output current is therefore determined by the g_{m3}, multiplied by that same gain factor.

This transconductor has high gain and is at the same time very linear, because two V_{GS} are put in series. Also, it has excellent high-frequency performance as only one node is added, i.e. at the output of the input Source followers. Remember, that output resistances of Source followers are usually $1/g_m$ and therefore low.

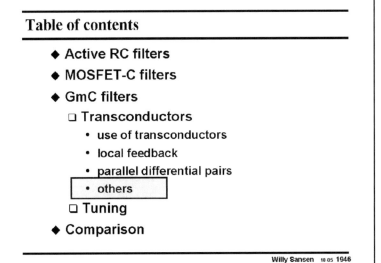

Table of contents

Willy Sansen 10 05 **1946**

1946

Not all transconductors use either feedback or cross-coupling. Some others exploit the inherent linear relationship of a CMOS inverter amplifier. The advantage is that they do not add any more nodes. They are capable of very-high-frequency performance.

A few examples are given next.

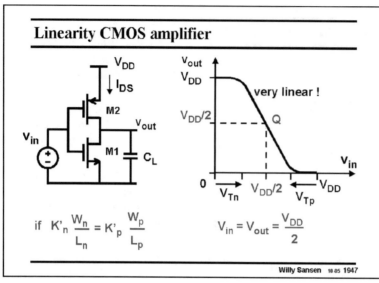

Recall from Chapter 3 that a CMOS inverter amplifier is very linear around the quiescent operating point Q.

From this region both transistors operate in saturation and exhibit the same square-law relationship. This non-linearity cancels out.

The input range is now the supply voltage V_{DD} minus the two threshold voltages V_{Tn} and V_{Tp}. It can be quite large. The linear output voltage range is the supply voltage V_{DD} minus the two threshold voltages V_{DSsatn} (which is close to $V_{GS} - V_T$) and V_{DSsatp}.

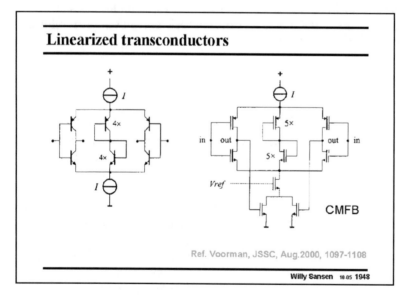

1948

Two such inverter amplifiers are used for differential operation as shown in this slide. Both a bipolar and a CMOS version are shown. The tuning is carried out by current I.

In the bipolar one (on the left) the common-mode current through the diodes in the middle is four times larger than any other inverter branch. It is a differential circuit, as each change of current in one inverter causes the opposite current in the other inverter.

The CMOS realization (on the right), is very similar. CMFB is used to avoid common-mode offset caused by the mismatch between the top and bottom current sources I.

In both realizations, the equivalent input noise is quite low as only the four input transistors contribute to it.

1949

Similar CMOS inverters are used in the transconductor shown in this slide. It is a pseudo-differential realization as a change in current in one inverter does not cause an opposite change of current in the other inverter. They have to be driven in a differential way.

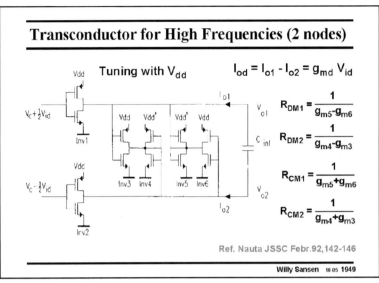

Transconductor for High Frequencies (2 nodes)

Tuning with V_{dd}

$I_{od} = I_{o1} - I_{o2} = g_{md} V_{id}$

$$R_{DM1} = \frac{1}{g_{m5} - g_{m6}}$$

$$R_{DM2} = \frac{1}{g_{m4} - g_{m3}}$$

$$R_{CM1} = \frac{1}{g_{m5} + g_{m6}}$$

$$R_{CM2} = \frac{1}{g_{m4} + g_{m3}}$$

Ref. Nauta JSSC Febr.92,142-146

Willy Sansen 10 05 1949

The differential resistive load R_{DM1} for the top inverter Inv1 are the two diode-connected MOSTs of Inv5 and a contribution from the output of the other inverter Inv2 through inverter Inv6. If all g_m's are the same, this differential resistive load is very large, providing a very large differential gain. The same is true for the resistive load R_{DM2} of the bottom inverter Inv2.

The average or common-mode output resistance R_{CM1} with respect to ground is low on the other hand. Indeed, R_{CM1} equals the sum of the transconductances of Inv5 and Inv6. This also applies to R_{CM2}.

This transconductor has the advantage that the differential gain is high. It can be tuned by changing the supply voltages V_{dd}. The common-mode gain is low however, because the common-mode output resistances are low. Parasitic capacitances at the output nodes are of little importance. Moreover, this circuit has only an input and output node. It is capable of a very-high-frequency performance.

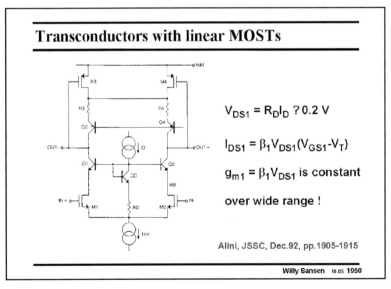

Transconductors with linear MOSTs

$V_{DS1} = R_D I_D ? 0.2$ V

$I_{DS1} = \beta_1 V_{DS1}(V_{GS1} - V_T)$

$g_{m1} = \beta_1 V_{DS1}$ is constant

over wide range !

Alini, JSSC, Dec.92, pp.1905-1915

Willy Sansen 10 05 1950

1950

A different way of reducing the distortion of a differential pair and increasing the input range is to bias the input transistors themselves in the linear region, as shown in this slide.

For this purpose, the voltage V_{DS1} across the input MOSTs must be kept constant. A typical value is 0.2 V.

The transconductance g_m is then constant as well. It can be tuned by changing the value of V_{DS1} or current I_D.

The circuit to keep V_{DS1} constant is quite simple indeed. Biasing (or tuning) current I_D creates a constant voltage $R_D I_D$ across resistor R_D. This voltage imposes the same voltage across the input MOSTs, as all three bipolar transistors Q1, Q2 and QD have similar V_{BE} values.

The main disadvantage is that MOSTs in the linear region exhibit lower values of transconductance. Local feedback by means of series resistance in the Source also reduces the transconductance. It is not that obvious which way is more efficient to reduce distortion or to increase the input voltage range.

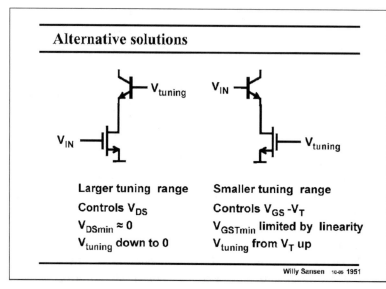

1951

Indeed, a comparison of both alternatives leads to some more insight.

On the left, one side of the input stage is depicted of the previous transconductor. A MOST is used as an input device in the linear region, to avoid distortion, at the cost of reduced transconductance. Tuning is carried out by changing the VDS across the input MOST.

On the right side, the input device is a (bipolar) amplifier with a series resis-

tor in its Emitter to reduce the distortion and to increase the input voltage range. Tuning is now carried out by changing the value of the MOST Emitter resistance.

Which is better in the reduction of the distortion or in the extension of the input voltage range is not obvious.

The one on the left has certainly a larger tuning range.

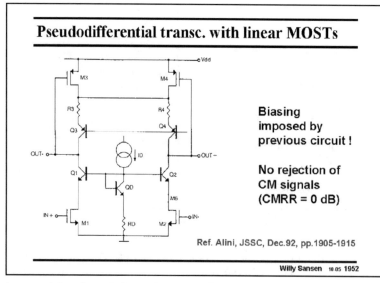

1952

A pseudo-differential realization of the same transconductor is shown in this slide. The current source is left out. As a result, the minimum supply voltage can be smaller by one V_{DSsat} or about 0.2 V.

The disadvantage however, is that this circuit must be driven in a differential way. Moreover the average (or common-mode) input voltage determines the DC currents in this circuit. The common-mode biasing is

imposed by the previous circuit, which is not shown here!

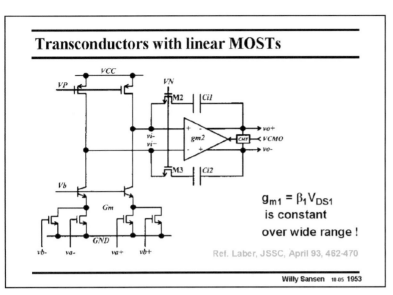

Transconductors with linear MOSTs

$$g_{m1} = \beta_1 V_{DS1}$$
is constant
over wide range !

Ref. Laber, JSSC, April 93, 462-470

Willy Sansen 10 05 **1953**

1953

Another example of a transconductor which uses MOSTs in the linear region at the input is shown in this slide. It also has a pseudo-differential input.

The cascode biasing voltage V_b is sufficiently low to keep all input devices in the linear region.

MOSTs in the linear region are also used as feedback resistors across the differential opamp. Tuning is therefore carried out by means of tuning voltage VN.

Table of contents

◆ **Active RC filters**

◆ **MOSFET-C filters**

◆ **GmC filters**

  • **Transconductors**

  • **Tuning**

◆ **Comparison**

Ref.: Silva-Martinez, Dehaene, ..

Willy Sansen 10 05 **1954**

1954

All previous transconductors can be tuned by changing a biasing current or voltage. This allows tuning of the transconductance such that the time constant of the transconductor is accurately determined. This is the only way to be able to realize higher-order filters. Such circuits are discussed next.

1955

Filters are constructed by putting a number of transconductors in series or in a feedback loop. At the beginning of this Section on Transconductors a good example is given of a biquad consisting of four transconductors.

The characteristic frequencies of such a filter are determined by the Gm/C ratios. In order to be able to match these frequencies, accurate values of Gm must be achieved. Indeed, accurate ratios of capacitances are already available (see Chapter 15). The ratios have to be realized between the load capacitances C_L of several Gm blocks. The parasitic capacitances C_p will render these ratios less accurate. This is compensated however, by adjusting the Gm values.

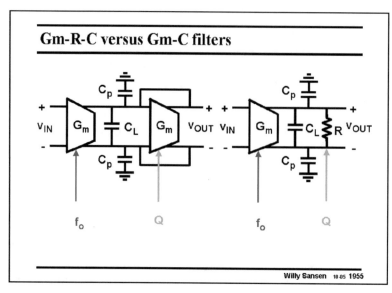

Willy Sansen 10-05 1955

This accuracy can be achieved by adding another Gm block, which is matched to the ones used in the filter, this is tuned to a reference by a tuning circuit. Such circuits will be described next.

It is clear however, that besides the characteristic frequency f_o, sometimes tuning is required of a quality factor Q as well. This can be done in two ways, either by adjusting the Gm of the diode-connected block (shown left) or by insertion of a tunable damping resistor R (shown right).

The latter one is called the Gm-RC filter. The resistor R must have high values and be tunable over a wide range.

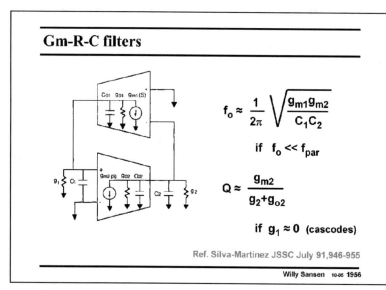

$$f_o \approx \frac{1}{2\pi}\sqrt{\frac{g_{m1}g_{m2}}{C_1 C_2}}$$

if $f_o \ll f_{par}$

$$Q \approx \frac{g_{m2}}{g_2 + g_{o2}}$$

if $g_1 \approx 0$ (cascodes)

Ref. Silva-Martinez JSSC July 91,946-955

Willy Sansen 10-05 1956

1956

For sake of illustration, a filter is shown consisting of only two Gm blocks in a feedback loop.

It is clear that the resonant frequency f_o depends on the g_m's and the capacitances, at least if the parasitic capacitances do not come in yet.

Also, the Q factor depends on the ratio of conductances, as shown.

For a given set of small capacitances C_1 and C_2, the frequency f_o can be tuned by adjusting g_{m1} and g_{m2}. Quality factor Q can then be tuned by adjusting the output conductances g_2 and g_{o2}. Especially the former one g_2 can be tuned as it is realized by a separate tunable resistor R, as shown next.

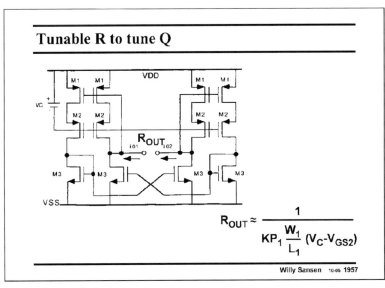

1957

One example of such a tunable resistor R is shown in this slide.

The output resistance R_{OUT} is a floating (or differential) resistance. It is fairly high because is sees a cascode of transistors M1 and M2, with feedback to the gates of M1. The resistance upwards is about $1/g_{m1}$. However, transistors M1 are in the linear region. Their g_{m1} is therefore $KP_1 W1/L_1 V_{DSsat1}$.

It is tunable by means of control voltage V_C. Indeed, V_{DSsat1} is simply $V_C - V_{GS2}$. The smaller V_C, the higher the output resistance R_{OUT}. Really high values of R_{OUT} are obtained provided transistors M1 and M2 enter the weak inversion region.

Downwards it sees the output of a current mirror, with small output conductance.

Let us now concentrate on the tuning circuits themselves.

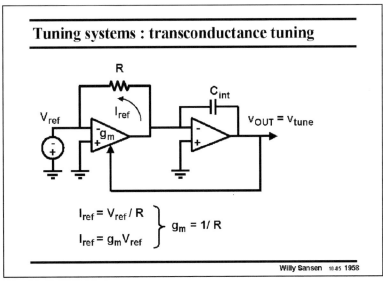

1958

In order to tune a transconductance g_m to an external resistor R, the circuit shown in this slide can be used.

The transconductor uses the resistor R in the feedback loop, this is driven by a reference or biasing voltage V_{ref}. It is followed by an integrator, the output of which is the tuning voltage, which adjusts the transconductance g_m of this transconductor and of all other transconductor blocks connected to it.

The integrator has sufficient gain to determine that its input voltage is always zero. As a result, the transconductance g_m is adjusted so that it equals $1/R$.

Resistors are not readily available, however. Switched capacitors can also be used, as shown next.

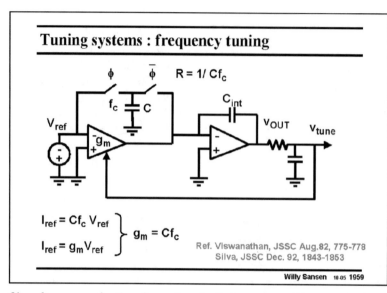

Tuning systems : frequency tuning

$$R = 1/ Cf_c$$

$$I_{ref} = Cf_c V_{ref}$$
$$I_{ref} = g_m V_{ref}$$
$$\left.\right\} \quad g_m = Cf_c$$

Ref. Viswanathan, JSSC Aug.82, 775-778
Silva, JSSC Dec. 92, 1843-1853

Willy Sansen 10 05 **1959**

1959

A switched-capacitor resistor R is given by $1/Cf_c$, in which f_c is the clock frequency. As a result, the transconductance g_m is adjusted such that it equals Cf_c. In other words, the time constant g_m/C is accurately locked to the clock frequency, which can be made very accurate indeed (for example with a crystal oscillator).

Obviously, some scaling will have to be introduced to be able to separate the filter frequency further from the clock frequency. The required accuracy is achieved, however.

Note also, that a low-pass filter is added at the output to suppress the ripple from the clock generator.

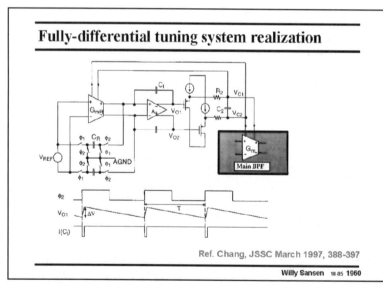

Fully-differential tuning system realization

Ref. Chang, JSSC March 1997, 388-397

Willy Sansen 10 05 **1960**

1960

An example of such a tuning system is shown in this slide. The Gm block to be tuned is G_{mR}. It is matched to all other Gm blocks in the Bandpass Filter.

The switched-capacitor resistor with capacitor C_R is a better version than the one on the previous slide, as it is less sensitive to parasitic capacitances.

Everything is fully differential to be insensitive to disturbances on the ground and supply lines.

This system used a clock of 1.5 MHz to tune filter frequencies between 150 and 800 kHz. These frequencies are still quite close, however. Indeed the capacitance C_R must be close to the capacitances used in the BPF for good matching.

A technique to separate the clock frequency from the filter frequencies is shown next.

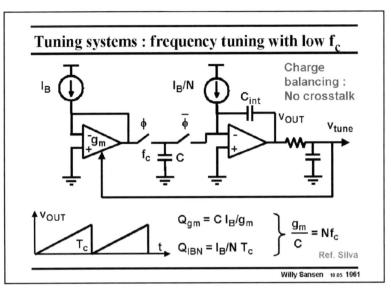

1961

In order to lower the clock frequency, compared to the filter frequency, the circuit in the slide can be used.

Two DC current sources are used with ratio N. The first one generates a voltage I_B/gm across the capacitor C. In the next phase current I_B/N discharges the capacitor C to zero. In the steady state condition, the charge introduced in the first phase must equal the charge taken away in the second phase. Charge balancing is achieved by the integrator.

As a result, the time constant g_m/C is accurately locked to the clock frequency f_c, but different by a factor N.

The main advantage of such a system is that the clock oscillator frequency can be positioned far away from the filter frequencies, such that it does not leak. Oscillator leakage is also the problem with PLL tuning. Charge balancing is therefore a better technique for tuning.

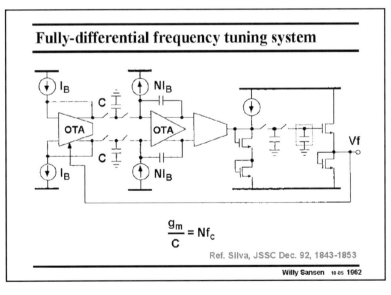

1962

A practical realization of such a tuning system is shown in this slide. It is a differential realization for better rejection of substrate noise. The resistor in the low-pass filter at the output is a switched-capacitor realization as well.

This was used for a band-pass filter around 10.7 MHz with N = 148, such that the clock is only at 450 kHz. This is far removed from the useful filter frequencies indeed.

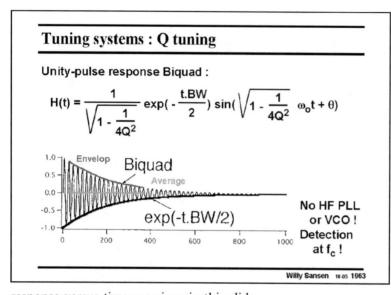

1963
Tuning Q is much more complicated and therefore not used very often.

As explained before it can be achieved by tuning a resistor which is in parallel with the filter load capacitance C_L.

The problem is however, how to measure the Q. For this purpose, an underdamped second-order system is realized by means of two Gm blocks. For low Q its response is oscillatory. Both the expression and the response versus time are given in this slide.

The oscillatory behavior can be detected by taking the difference between the envelop and the average output signal. The measurement has to be carried out with two different time constants, one which provides the average, and one which takes the envelop.

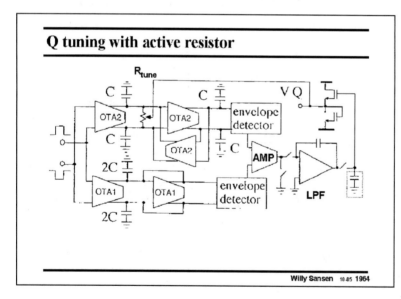

1964
In this realization, two paths lead to the differential amplifier AMP and low-pass filter LPF at the output. The top one has an oscillatory behavior because of low Q. The bottom one has a Q close to unity, which gives a flat response. This difference is amplified and fed back over a low-pass filter to the tunable resistance R_{tune}.

1965
For sake of comparison, three realizations for the same application are listed. They all aim at an IF filter for FM. The center frequency is 10.7 MHz and the passband is about 250 kHz. The first is realized by a switched-capacitor techniques, the second uses RC components around opamps and the third uses Gm blocks such as differential pairs, capacitors and tunable resistors.

It is clear that the first two realizations suffer from insufficient gain at this fairly high frequency.

Comparison of 10.7 MHz filters

	SC	OTA-C	Gm-RC
f_c (BW = 250 kHz)	10.7 MHz	12.5 MHz	10.7 MHz
Order filter	6	4	4
Vin @ IM3= 1%	0.24 V_{RMS}	0.32 V_{RMS}	0.71 V_{RMS}
DR @ IM3= 1%	34 dB	51 dB	68 dB
Power (± V)	500 mW(± 5)	360 mW(± 6)	220 mW(± 2.5)
Chip area	2 mm^2	7.8 mm^2	6 mm^2

Willy Sansen 10 05 1965

As a result, the distortion increases and the dynamic range decreases. This is especially true for the SC realization, where the DR suffers from clock injection and charge redistribution.

Moreover, the power consumption of the Gm-RC realization is better as simple differential pairs provide much better performance at high frequencies than full operational amplifiers.

It can be concluded that from 10 MHz on Gm-C and Gm-RC filters are the best choice.

A more elaborate comparison follows at the end of this Chapter.

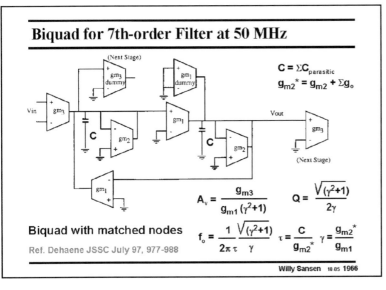

Biquad for 7th-order Filter at 50 MHz

Biquad with matched nodes

Ref. Dehaene JSSC July 97, 977-988

$$C = \Sigma C_{parasitic}$$
$$g_{m2}^* = g_{m2} + \Sigma g_o$$

$$A_v = \frac{g_{m3}}{g_{m1}(\gamma^2+1)} \qquad Q = \frac{\sqrt{(\gamma^2+1)}}{2\gamma}$$

$$f_o = \frac{1}{2\pi\tau} \frac{\sqrt{(\gamma^2+1)}}{\gamma} \qquad \tau = \frac{C}{g_{m2}^*} \qquad \gamma = \frac{g_{m2}^*}{g_{m1}}$$

Willy Sansen 10 05 1966

1966

As a final example of frequency and Q tuning, a 7th order filter is discussed consisting of three biquads and one first-order section. Such a biquad is shown in this slide. Its goal is accurate frequency and phase behavior at high frequencies (here 50 MHz).

In order to increase the parasitic pole frequencies as much as possible, no capacitances have been added. The node capacitance C consists of the sum of all parasitic node capacitances. In order to make sure that capacitance C is the same on both nodes, two dummy g_m blocks are added. Each node now sees three input capacitances (of blocks g_{m1}, g_{m2} and g_{m3}) and three output capacitances (also of blocks g_{m1}, g_{m2} and g_{m3}).

For high frequency performance, the g_m blocks consist of full-differential pairs with local CMFB. They have large $V_{GS} - V_T$ values (0.5 V) for low distortion.

The gain A_v, the characteristic frequency f_o and Q are readily obtained. Note that g_{m2}^* also includes the output conductances.

The Q factor can be tuned by tuning factor γ, which is a ratio of two transconductances. The frequency f_o can be tuned by tuning the time constant τ.

Two parameters γ and constant τ need a tuning system. This is discussed next.

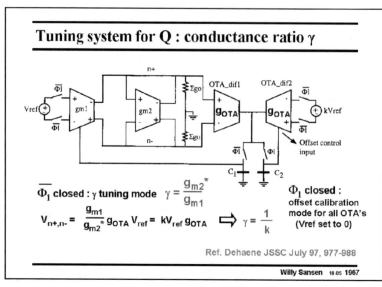

Tuning system for Q : conductance ratio γ

$\overline{\Phi_1}$ closed : γ tuning mode $\quad \gamma = \dfrac{g_{m2}^{*}}{g_{m1}}$

$V_{n+,n-} = \dfrac{g_{m1}}{g_{m2}^{*}} g_{OTA} V_{ref} = k V_{ref} g_{OTA} \implies \gamma = \dfrac{1}{k}$

Φ_1 closed :
offset calibration
mode for all OTA's
(Vref set to 0)

Ref. Dehaene JSSC July 97, 977-988

Willy Sansen 18 05 1967

1967

This tuning system has inputs V_{ref} and kV_{ref}. Ratio k is very accurate as it is set by resistor ratios (not shown). Its output is the current control of transconductor g_{m1}, which is also used for all other filter transconductors.

When all switches Φ_1 dash are closed, the input voltage $V_{n+,n-}$ to the OTA_dif1 is $2V_{ref}g_{m1}/g_{m2}^{*}$, which is actually $2V_{ref}/\gamma$. The input voltage to the OTA_dif2 is $2kV_{ref}$. Both voltages experience the same amplification as amplifiers OTA_dif1 and OTA_dif2 have the same gain g_{OTA}. The difference is amplified and stored on capacitor C_1, which closes the feedback loop and adjusts g_{m1} such that this difference is zero. As a result, parameter γ equals $1/k$. It is set accurately by the value given to k.

For higher accuracy, an offset calibration cycle is introduced. For this purpose, the switches Φ_1 dash are open and switch Φ_1 is closed. The offset error voltage is stored on capacitor C_2 and subtracted in the other phase.

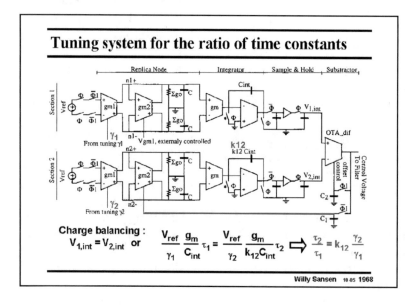

Tuning system for the ratio of time constants

Charge balancing :
$V_{1,int} = V_{2,int}$ or $\quad \dfrac{V_{ref}}{\gamma_1} \dfrac{g_m}{C_{int}} \tau_1 = \dfrac{V_{ref}}{\gamma_2} \dfrac{g_m}{k_{12}C_{int}} \tau_2 \implies \dfrac{\tau_2}{\tau_1} = k_{12} \dfrac{\gamma_2}{\gamma_1}$

Willy Sansen 18 05 1968

1968

The system to tune the time constants is shown in this slide.

Actually, it does not tune the absolute value of one single time constant but rather a ratio of time constants. Indeed, for a higher-order filter, the ratio of characteristic frequencies must be more accurate than the absolute value of one single time constant.

The ratio of time constants τ_1 and τ_2 will be locked to the ratio of γ values, which have been tuned by circuits as on a previous slide, and a constant k12, which is the ratio of the two integrating capacitors.

Again, a charge balancing feedback circuit is used with capacitor C_1 to make the input of the differential amplifier OTA_dif zero. Again, an offset calibration cycle is added, using capacitor C_2.

The two voltages at the input of the OTA_dif are provided by two matched circuits, both are driven by V_{ref} but with different integration capacitors. The one at the top has C_{int} whereas the one at the bottom has $k12C_{int}$.

The voltage at the output of the first g_{m1} block on top is $V_{n1+,n1-}$. It is actually $2V_{ref}/\gamma_1$. This voltage is integrated over time τ_1. As a result, the input voltage $V_{1,int}$ equals the expression given in this slide. In the same way, the voltage $V_{2,int}$ is derived for the block at the bottom. The equation of both, as a result of the feedback loop, shows that indeed the ratio of the time constants is kept constant.

Despite the tuning circuits, the total power consumption of this CMOS realization is quite low. Moreover, no external trimming is required.

Table of contents

♦ **Active RC filters**

♦ **MOSFET-C filters**

♦ **GmC filters**

♦ **Comparison**

Ref.: Tsividis, Voorman, Integrated Cont.-time filters, IEEE Press 1993
　　　 J. Silva-Martinez, Kluwer 1993,
　　　 W. Dehaene, JSSC, July 1997, 977-988

Willy Sansen　10 05　1969

1969

Since several types of filters have been introduced up till now, a comparison in terms of dynamic range and frequency capability is mandatory. As a third axis the power consumption could be added. This is left to the reader however.

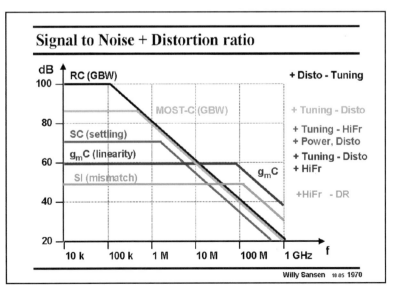

Willy Sansen　10 05　1970

1970

In this graph, a first-order comparison is carried out between the different types of filters in terms of dynamic range DR versus frequency.

For each type, the DR is sketched versus frequency. The main positive and negative points are listed on the right.

At low frequencies, an operational amplifier with RC's in the feedback loop offers the highest dynamic range. For higher power consumption, even values higher than 100 dB can be reached. The distortion is very low because of the high loop gain. At higher frequencies however, the loop gain decreases and the distortion increases. The DR already decreases at rather low frequencies. Moreover tuning is a problem.

Tuning is easy when the resistors are substituted by MOST resistors. The distortion is higher however. The DR can still be as high as 80 dB.

Switched-capacitor filters do not easily offer more than about 70 dB dynamic range, because of clock injection and charge distributions. This is even worse at higher frequencies when settling has to be achieved in shorter time. Tunability is good, however.

Gm-C filters rarely achieve more than 60 dB because of distortion. They reach the highest frequencies however, because they use the simplest circuit configurations. They can be tuned with dedicated circuitry.

Finally, switched-current filters reach nearly the same DR as Gm-C filters and nearly the same high frequencies. Both are lower, however.

1971

In this Chapter, continuous-time filters have been discussed. Considerable attention has gone to Gm-C filters as they are capable of a reasonable performance at the higher frequencies.

At low frequencies however, OTA-RC filters provide the highest dynamic range. The other types of filters provide a performance in between. This is also true for switched-capacitor filters.

Table of contents

- ◆ **Active RC filters**
- ◆ **MOSFET-C filters**
- ◆ **GmC filters**
- ◆ **Comparison**

Ref.: Tsividis, Voorman, Integrated Cont.-time filters, IEEE Press 1993
   J. Silva-Martinez, Kluwer 1993,
   W. Dehaene, JSSC, July 1997, 977-988

Willy Sansen 10.05 **1971**

The main disadvantage of Gm-C filters is that they need tuning circuitry and that they give more distortion. For dynamic ranges up to about 60 dB, however, they provide lower power consumption at higher frequencies.

CMOS ADC & DAC Principles

Willy Sansen

KULeuven, ESAT-MICAS

Leuven, Belgium

willy.sansen@esat.kuleuven.be

Willy Sansen 10-05 201

201

An important class of analog circuits are analog-to-digital and digital-to-analog converters as they provide the conversion from the analog to digital signals and vice-versa.

The number of converter types is legion however, as many different resolutions and speeds are involved.

They can thus be classified in terms of resolution and speed capability.

This chapter is limited to Nyquist converters. The input signal frequency can reach half the clock frequency. Moreover, each sample is quantized to full precision. In oversampling converters the input signal frequency is much lower than the clock frequency. Also, the precision is obtained as a result of averaging in a feedback loop. These will be discussed in the next Chapter.

However, a number of definitions have to be given first.

Table of contents

Willy Sansen 10-05 202

202

After the definitions, we will focus on DACs first. The main limitations in DACs are a result of mismatch, either between transistors or between passives as resistors and capacitors.

The AD Converters will be discussed. The principles of the more important types will be introduced, followed by some realizations. Considerable attention is paid to the limitations in both speed and resolution.

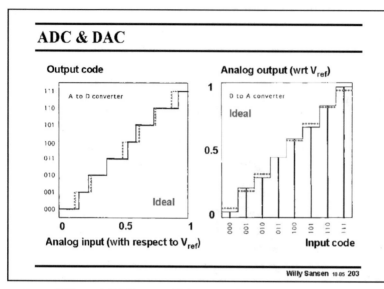

203
An ADC converts a continuous input Voltage into a number of discrete steps, which are labeled by means of a digital code. The input voltage is quantized.

In the example in this slide, any voltage between zero and the reference Voltage V_{ref}, corresponds to a three-bit code. Ideally, all steps are equally wide and high. In practice however, some irregularities occur. These nonidealities will be described by specifications such as INL and DNL, as will be explained later.

A DAC, on the other hand, converts a digital code into a number of voltages. This is shown in this slide for a three-bit input code. The voltages are all fractions of a reference voltage V_{ref}. Again the steps are ideally all equal, but some non-idealities are hard to avoid in practice.

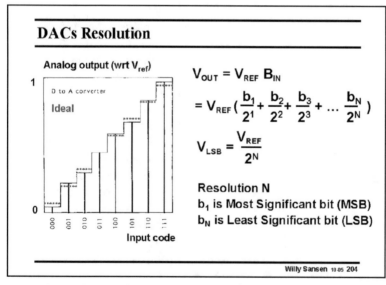

204
The number of steps taken is called the resolution.

The value of the analog output voltage now depends on which bits b are 1 or 0. For example, for a code 110 and a reference voltage of 0.6 V, the output voltage is $0.6 \times (2^{-1} + 2^{-2})$ or 0.45 V. For a resolution N, the smallest step V_{LSB} is the reference Voltage divided by 2^N. For example this is 1/256 or 0.4% for a 8 bit converter, or 2.3 mV for a 0.6 V reference voltage. The error must always be smaller than the resolution.

The coefficient of 1/2 is the most significant bit, whereas the coefficient of the last one $1/2^N$ is the least significant bit.

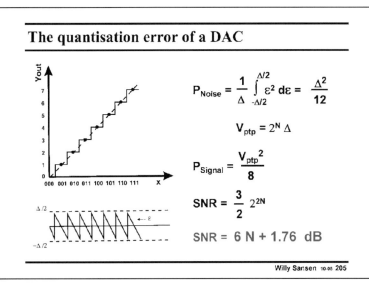

The quantisation error of a DAC

$$P_{Noise} = \frac{1}{\Delta} \int_{-\Delta/2}^{\Delta/2} \varepsilon^2 \, d\varepsilon = \frac{\Delta^2}{12}$$

$$V_{ptp} = 2^N \Delta$$

$$P_{Signal} = \frac{V_{ptp}^2}{8}$$

$$SNR = \frac{3}{2} \, 2^{2N}$$

$$SNR = 6\,N + 1.76 \ \ dB$$

Willy Sansen 10-05 205

205

The difference between the actual input voltage and the quantized voltage is a signal ε, which has the shape of a sawtooth with peak-to-peak amplitude Δ.

Its RMS value can be easily calculated to be $\Delta^2/12$. It is called quantization noise, because it is small.

It is obvious that the quantization noise is smaller if the resolution N is larger.

The Signal-to-noise ratio SNR can then easily be calculated to be $3/2$ times 2^{2N}.

Obviously, the larger the resolution, the larger the SNR.

A rule of thumb is that for a resolution N, the SNR is about $6N+2$. For a resolution of (N=) 8 bit, the SNR is about 50 dB. The SNR increases by 6 dB for each extra bit.

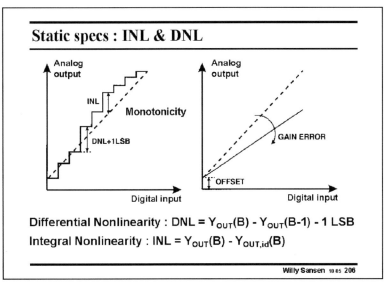

Static specs : INL & DNL

Differential Nonlinearity : DNL = $Y_{OUT}(B) - Y_{OUT}(B-1)$ - 1 LSB

Integral Nonlinearity : INL = $Y_{OUT}(B) - Y_{OUT,id}(B)$

Willy Sansen 10 05 206

206

The non-idealities of the DA conversion give rise to several specifications, such as the DNL and the INL.

The DNL is the largest step ever minus one LSB. It is the largest deviation from a regular step.

The INL is the largest deviation from the average slope.

If this slope is not right, then a gain error is found. This slope must go through zero. If not, an offset occurs.

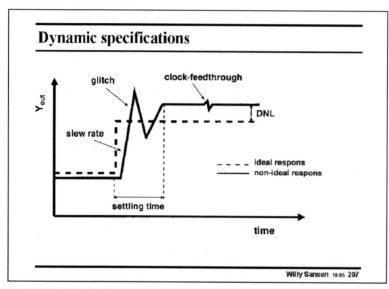

207

When the output voltage changes in time, as a result of change of bit b, then the transition may not be instantaneous.

The change in voltage may experience a limited rise time, or slew rate. Moreover an overshoot may occur, called glitch. The energy in this glitch (which is the integral under this glitch waveform) must be smaller than the energy of a LSB.

The time required to reach the final value (within 0.1% for example) is called the settling time.

As clocks are used, some clock feedthrough may occur as well.

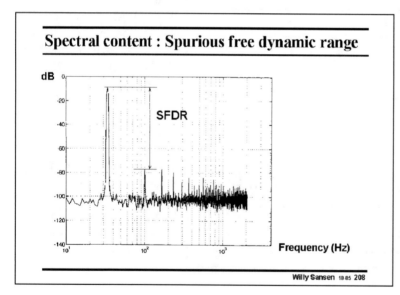

208

In the frequency domain, the analog output voltage of a DAC may contain various harmonics.

The ratio of the output, fundamental to the largest harmonic is called the SFDR. This is usually the second harmonic. For fully-differential systems however, this is most probably the third harmonic.

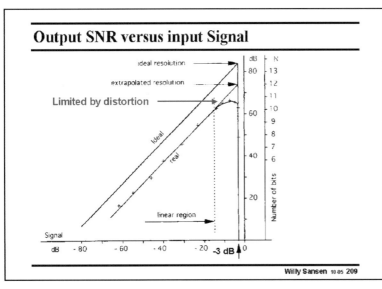

Output SNR versus input Signal

Willy Sansen 10 05 209

209

For a certain resolution, the quantization (and other) noise is about constant. The SNR increases for increasing input voltage. The maximum SNR is reached for the largest possible input signal amplitude.

This latter signal is normally limited by distortion. In the example in this slide the maximum SNR is theoretically 74 dB, corresponding to 12 bit. Because of distortion, only 66 dB is found however, corresponding to about 10.6 bit. The SNDR (Signal-to-noise-and-distortion ratio) is thus only 66 dB.

The ideal values are always somewhat higher. As a rule of thumb about 1 bit is lost from theory to practice (measurement).

Table of contents

- **Definitions**
- **Digital-to-analog converters**
  - **Resistive**
  - **Capacitive**
  - **Current steering**
- **Analog-to-digital converters**
  - **Integrating**
  - **Successive approximation**
  - **Algorithmic**
  - **Flash / Two-step**
  - **Interpolating / Folding**
  - **Pipeline**

Willy Sansen 10 05 2010

2010

Now that all definitions have been given, let us examine which principles of DACs are most used.

There are three of them, the current-steering one being used most often.

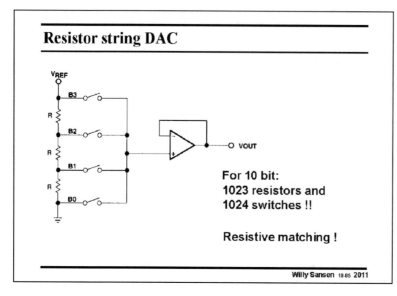

2011

Possibly the simplest DAC is shown in this slide.

It uses a string of equal resistors, which allows all possible voltage fractions of the reference Voltage V_{REF} to be obtained. These fractions are added depending on the switches B and buffered toward the output.

It is clear that many resistors are required for high resolution DAC's of this type. Moreover, matching between the resistors R limits the resolution to 6–8 bit.

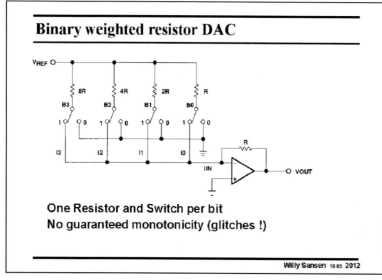

2012

Much fewer resistors are required if they are binary weighted. Currents are also added rather than voltages.

It is clear that matching plays an important role again. This is a problem as the resistors have widely different values: for a 8 bit converter the largest resistor is 256 times larger than the smallest one.

This is remedied by the arrangement on the next slide.

The main disadvantage is probably that this converter is prone to glitches. Consider the transition from (B0B1B2B3 =) 0111 to 1000. In the first case, the current flowing into the opamp is $1/8 + 1/4 + 1/2 = 0.875$ times V_{REF}/R. In the other case the current is V_{REF}/R itself. There is smooth transition expected from 0.875 to 1 indeed.

If however, mismatch is such that 0.875 is higher than 1, then the conversion curve is no longer monotonic. A glitch now occurs!

2013

In this DAC only resistors are used of two sizes r and 2R. Actually, all resistors have equal size but for 2R two resistors are put in series. This greatly facilitates the matching between the resistors (see Chapter 15). A resolution of 10 bit is fairly easily achieved.

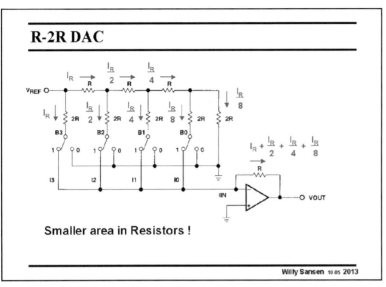

R-2R DAC

Smaller area in Resistors !

Willy Sansen 10 05 2013

Moreover, the total sum of resistor area is much smaller indeed!

To see how this 4-bit converter works, the currents are indicated, through all the branches.

The current through the most left resistor 2R, and switch B3, is current I_R. This is actually $V_{REF}/2R$. This current always flows as it is directed by switch B3, either to the input of the opamp (as shown) or to ground.

The current through the most left (horizontal) resistor R is also current I_R. Indeed, it is followed by another resistor 2R through switch B2 in parallel with another (horizontal) resistor R, which again sees a resistor R. As a result, each (horizontal) resistor R sees a resistance R to the right.

At the end of the resistor string, it is clear that the two most right resistors 2R are in parallel and offer a resistance R to the most right (horizontal) resistance. At all nodes, a resistance R is seen to the right with respect to ground.

When the switches are in position 1111 (as shown), then all currents are directed into the opamp and through the feedback resistor R towards the output. If a switch is in position 0, then its current flows to ground and does not contribute to the output voltage.

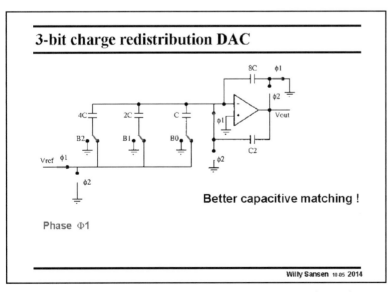

3-bit charge redistribution DAC

Better capacitive matching !

Phase Φ1

Willy Sansen 10 05 2014

2014

Nevertheless better matching can be obtained by means of capacitors rather than with transistors. If a resolution of 10 bit can be easily obtained with a resistor bank, then 12 bit can be obtained with a matched capacitor bank (see Chapter 15).

An example of such 3-bit capacitor bank is shown in this slide. The larger capacitor 4C consists of 4 equal capacitors C laid out properly (see Chapter 15).

By means of charge redistribution, the output voltage is a sum of binary fractions of the reference voltage V_{ref}.

In clock phase Φ1 (as shown) all binary capacitors are charged to V_{ref}.

The output voltage holds the voltage obtained during the previous clock phase.

In clock phase Φ2, the charge of the capacitors which belong to the switches connected to ground, is transferred to the output capacitor 8C.

The output voltage then changes depending on the charge transferred from these capacitors.

Note that this circuit suffers from the same problems as switched capacitors filters. They can easily achieve a 70 dB dynamic range corresponding to about 12 bit, but not much more.

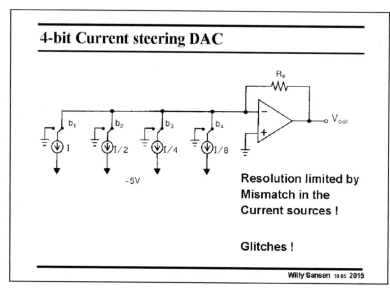

4-bit Current steering DAC

Resolution limited by Mismatch in the Current sources !

Glitches !

Willy Sansen 10-05 2015

2015

Binary currents can also be switched directly as shown in this slide.

The binary code determines which currents are added towards the input of the opamp.

Now, of course, the mismatch between the binary current sources will play a role. In principle, current sources do not match as well as capacitors. Limited resolution is therefore obtained.

Moreover glitches can occur during transitions to higher bits. As this is also a result of mismatch, careful layout will have to be ensured.

Glitches can be avoided however, by using a thermometer code rather than a binary code.

The Binary and thermometer codes

Decimal	Binary			Thermometer Code						
	b_1	b_2	b_3	d_1	d_2	d_3	d_4	d_5	d_6	d_7
0	0	0	0	0	0	0	0	0	0	0
1	0	0	1	0	0	0	0	0	0	1
2	0	1	0	0	0	0	0	0	1	1
3	0	1	1	0	0	0	0	1	1	1
4	1	0	0	0	0	0	1	1	1	1
5	1	0	1	0	0	1	1	1	1	1
6	1	1	0	0	1	1	1	1	1	1
7	1	1	1	1	1	1	1	1	1	1

Monotonicity guaranteed !

Willy Sansen 10-05 2016

2016

This table shows the difference. Each time the binary code changes to the next number, the next digital

value in the thermometer code goes from 0 to 1. Monotonicity is now guaranteed. Glitches can therefore be avoided.

The price to pay however, is the larger number of digital variables. For eight values, only three bits are necessary in the binary code but seven in the thermometer code.

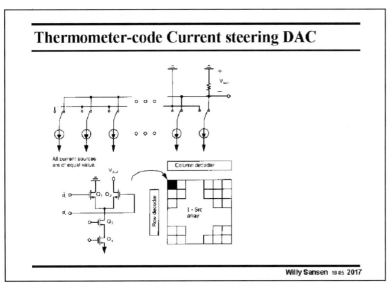

2017
In this current-steering DAC thermometer code is used for both row and columns. This results in good monotonicity.

Current is switched to the output when both the row and column lines for a specific cell are high. These currents add up toward the input of an opamp with a resistor in the feedback loop. Also, simple resistor can be used as shown in this slide. This is especially true for high-frequency realizations where a resistor is 50 Ω.

The switch itself is also shown in this slide.

It consists of a differential pair, the current source of which carries the binary current. How can the matching between these transistors Q_4 be improved, will be explained in the next slide.

The transistors of the differential pair Q_1 and Q_2 are driven by the digital control signals. They are on or off, and conduct the current either to ground or to the output.

Note that a cascode transistor Q_3 is used to better isolate the analog current sources (transistors Q_4) from the digital switches Q_1 and Q_2.

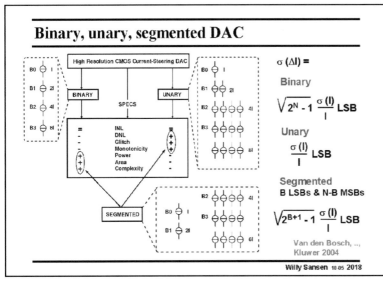

2018
The matching of the current sources is now discussed in more detail.

In the binary implementation, each bit directly steers a current source with a value that is twice as large as that of the next less significant bit, while in the unary implementation each bit steers a number of unit current sources.

Comparing the performance of these two architectures shows that the binary architecture has a larger DNL and glitch energy error, but since this architecture requires no thermometer decoder (as the unary implementation does) it has a lower power and area consumption.

A combination of the advantages of the two architectures is found in the segmented architecture. Here the LSBs are implemented in a binary way while the MSBs are implemented in a unary way.

As a conclusion, one can state that in CMOS the best architecture to achieve a high update rate and a high linearity is the segmented current steering architecture.

In the unary implementation, the DNL error equals about $\sigma(I)/I$, in which $\sigma(I)$ is the standard deviation on a unary current source. In the binary implementation however, at half-scale transition, 2^{N-1} unit current sources are switched off. The DNL error is thererfore much larger. For the segmented architecture, the DNL error is in between.

How much standard deviation $\sigma(I)/I$ is actually required?

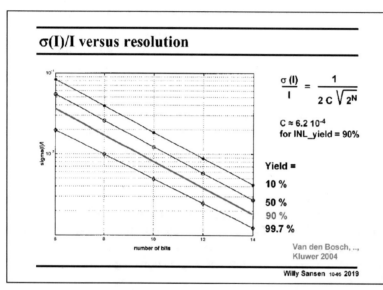

2019

This standard deviation $\sigma(I)/I$ can be predicted in terms of yield, by means of a fairly complicated function, of which a value is given in this slide.

For a 90% yield in INL, the constant C is given in this slide, giving rise to an easy expression, which is plotted in this slide.

It shows that for 10 bit resolution, a standard deviation $\sigma(I)/I$ is required of about 0.8% if a yield of 90% is needed.

Other values are easily found as well.

A standard deviation $\sigma(I)/I$ of 0.8% can easily be achieved in present day CMOS processes. This number will be used to find the sizes of the transistors for the current sources.

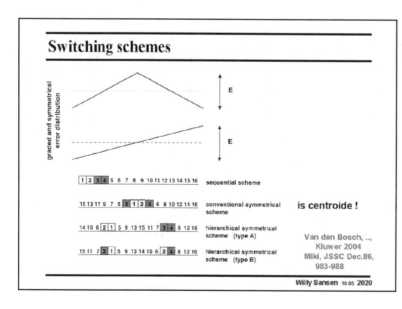

2020

Sizing is only one technique to improve the matching. Other layout techniques have been discussed in Chapter 15. For example, centroide layout is of utmost importance to improve the matching.

This is shown in this slide, for the switching schemes for the current sources.

In general, the error distribution can be modeled by a linear or symmetric gradient E, or a combination of both. A linear gradient is caused by, for example, a variation of oxide thickness or a drop in ground line. A symmetric gradient, on the other hand, can be caused by temperature or packaging stress.

The switching schemes show how the digital codes are transformed into decimal values (or thermometer code).

Several schemes are compared.

In the sequential switching scheme (on top), the current sources in a given row are turned on sequentially from the left to the right. The effect on the INL is given in the next slide. It is clear that this scheme causes large linearity errors due to the accumulation of both graded and symmetrical errors.

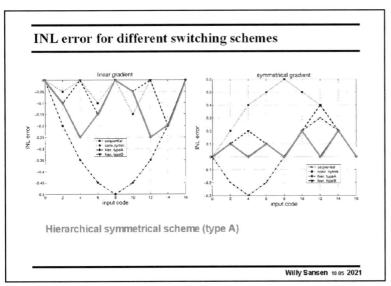

2021

In the conventional symmetrical switching scheme, the current sources are turned on in a symmetrical way around the center of the row. The graded errors are now cancelled at every two increments of the digital input (see left) but the errors generated by a symmetrical gradient will accumulate as indicated on the right.

In the hierarchical symmetrical switching scheme, the current sources are turned on around the first and the third quarter of the current sources row. Two different schemes are now possible. In scheme A, the symmetrical error generated by current source 1 is cancelled by current source 2, while the graded error caused by the current source pair (1,2) is cancelled by the current source pair (3,4).

In switching scheme B, the errors caused by the linear gradient are cancelled out at current source level while the symmetrical errors are cancelled at the current source pair level.

As a result, in type A, the INL error is about the same for both the graded and the symmetrical error distribution whereas in type B, the INL error caused by the symmetrical error distribution is twice as large as the one caused by the graded error distribution.

Therefore, the hierarchical symmetrical scheme A has been preferred.

It is obvious that such heuristics may not yield optimum results. Optimization algorithms such as the Q^2 random walk switching scheme have been used to achieve 14 bit conversion (Ref. Van Der Plas, JSSC Dec. 1999, 1708–1718).

2022

Since two D/A converters that are processed in the same technology do not necessarily have the same specifications due to technological variations, it is of the utmost importance to know the precise relationship that exists between the specifications of the circuit and the matching properties of the used technology.

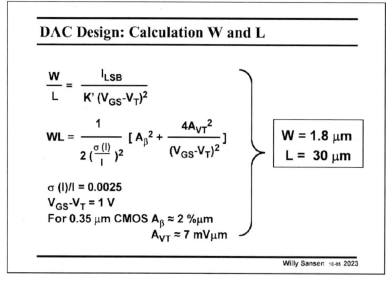

DAC Design: Static Accuracy

INL_yield = percentage of functional D/A converters with an INL specification smaller than half an LSB.

$$INL\_yield = f\ (mismatch) = f\ (\frac{\sigma\ (I)}{I})$$

$$WL = \frac{1}{2\ (\frac{\sigma\ (I)}{I})^2}\ [\ A_\beta^2 + \frac{4A_{VT}^2}{(V_{GS}-V_T)^2}\ ]$$

High yield
⇩
small $\frac{\sigma\ (I)}{I}$
⇩
Large current source area

$$\frac{\sigma\ (I)}{I} \leq \frac{1}{2\ C\ \sqrt{2^N}} \quad \Rightarrow \quad \boxed{\sigma\ (I_{unit})/I_{unit} = 0.25\ \%}$$

Willy Sansen 10-05 2022

For a current-steering D/A converter, the INL is mainly determined by the matching behavior of the current sources. The parameter that is best suited for expressing this technology versus DAC-specification relation is the INL yield. This INL yield is defined as the ratio of the number of D/A converters with an INL smaller than ½ LSB to the total number of tested D/A converters.

The random variations are modeled using a normal distribution with expected value zero and a relative standard deviation $\sigma(I)/I$. The statistical relationship has been investigated analytically, resulting in an accurate formula expressing directly the relationship between the INL yield specification, the resolution and the relative unit current standard deviation for the D/A converter. This was given before.

Note that the gate area of the current sources is inversely proportional to the relative unit current standard deviation. Since a high yield requires a small value for this standard deviation, this directly implies a large current source area.

To obtain a 12 bit accuracy, a relative unit current source standard deviation of 0.25% is necessary.

DAC Design: Calculation W and L

$$\frac{W}{L} = \frac{I_{LSB}}{K'\ (V_{GS}-V_T)^2}$$

$$WL = \frac{1}{2\ (\frac{\sigma\ (I)}{I})^2}\ [\ A_\beta^2 + \frac{4A_{VT}^2}{(V_{GS}-V_T)^2}\ ]$$

$\boxed{\begin{array}{l} W = 1.8\ \mu m \\ L = \ 30\ \mu m \end{array}}$

$\sigma\ (I)/I = 0.0025$
$V_{GS}-V_T = 1\ V$
For 0.35 µm CMOS $A_\beta \approx 2\ \%\mu m$
$A_{VT} \approx 7\ mV\mu m$

Willy Sansen 10-05 2023

2023
Using the following equations, the dimensions of the current source can be easily found. For a 1V peak-to-peak output voltage over a double terminated (coax)cable, a full-scale current I_{FS} of 20 mA is designed. The current equation is the first equation on the right.

The mismatch equation is copied from the previous slide together with the value for the relative unit current standard deviation and the gate overdrive voltage. As a result, the width of the unit current source transistor equals 1.8 µm and its length is 30 µm.

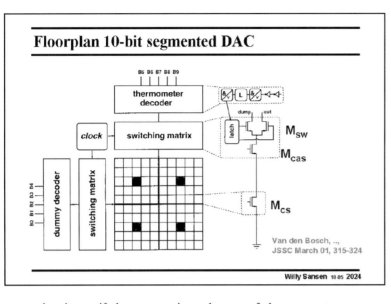

Floorplan 10-bit segmented DAC

B5 B6 B7 B8 B9

thermometer decoder

clock   switching matrix

M_{sw}

M_{cas}

B0 B1 B2 B3 B4

dummy decoder

switching matrix

M_{cs}

Van den Bosch, ..,
JSSC March 01, 315-324

Willy Sansen 10 05 2024

2024

As an example, a realization is discussed of a high-speed 10-bit current steering DAC.

At high frequencies, attention has to be paid to the dynamic performance. This suffers from several effects such as

– timing errors
– capacitive feedthrough from the digital control signals
– lack of stability of the Drain voltage of the current source transistors.

This latter effect is even more dominant if the output impedances of the current sources are too low. Indeed, switching the transistors M_{sw} on and off causes the voltage across the current source transistor M_{cs} to vary. As a result, the effective resolution is decreased.

This is shown in the next slide.

Note that this realization consists of 5 bit binary and 5 bit unary sub-DACs. For the five most significant bits, the input bit streams are converted to a 32 bit thermometer code output. For the 5-bit binary LSB processing, the digital outputs are the same as the input. A dummy decoder has been added however, to minimize latency problems between the signals generated by the MSB decoder and the binary LSB bits.

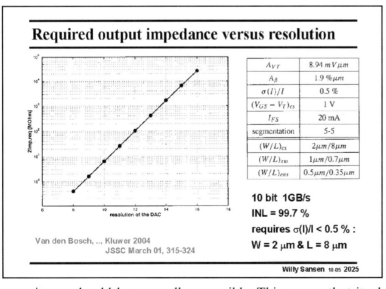

Required output impedance versus resolution

A_{VT}	8.94 $mV\mu m$
A_β	1.9 $\%\mu m$
$\sigma(I)/I$	0.5 %
$(V_{GS} - V_T)_{cs}$	1 V
I_{FS}	20 mA
segmentation	5-5
$(W/L)_{cs}$	$2\mu m/8\mu m$
$(W/L)_{sw}$	$1\mu m/0.7\mu m$
$(W/L)_{cas}$	$0.5\mu m/0.35\mu m$

10 bit 1GB/s

INL = 99.7 %

requires $\sigma(I)/I < 0.5$ % :

W = 2 μm & L = 8 μm

Van den Bosch, .., Kluwer 2004
JSSC March 01, 315-324

Willy Sansen 10 05 2025

2025

The output impedance to achieve a specific resolution is shown for a 25 Ω load. For a resolution of 10 bits, the output impedance must be at least 6.4 MΩ. For a 12 bits resolution, this must be 100 MΩ!

It is clear that the only way to achieve such a high value is to insert a cascode transistor M_{casc} between the switching pair and the current source. The main characteristic of this cascode transistor is that its output capacitance should be as small as possible. This means that its drain area must be as small as possible. As a result its W/L is made small (2–3) and its $V_{GS} - V_T$ is large, as large as 1 V, provided the power supply allows this.

To achieve a 99.7% INL, the current source matching must be at least 0.5%. The dimensions of the current source transistor M_{cs} are then easily calculated. The cascode has been made as small as possible in this 0.35 μm CMOS technology.

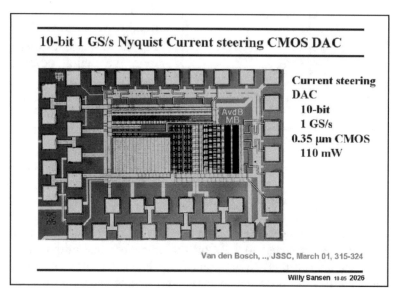

10-bit 1 GS/s Nyquist Current steering CMOS DAC

Current steering DAC
10-bit
1 GS/s
0.35 μm CMOS
110 mW

Van den Bosch, .., JSSC, March 01, 315-324

Willy Sansen 10.05 2026

2026
A layout of this realization is shown in this slide.

Many blocks have been laid out by hand to reduce the dimensions to a minimum. This also applies to the dummy decoder mentioned previousl.

As a result, the dynamic performance is excellent, as shown in the next slide.

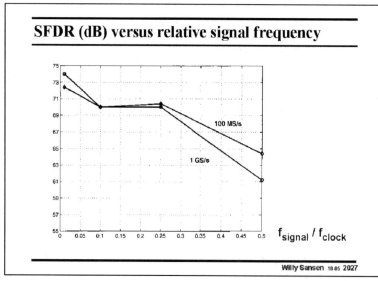

SFDR (dB) versus relative signal frequency

100 MS/s

1 GS/s

f_{signal} / f_{clock}

Willy Sansen 10.05 2027

2027
A single-tone spectrum is shown in this slide.

At low frequencies, an effective resolution is obtained of 74 dB or 11.7 bits.

For a clock frequency of 1 GHz, the maximum frequency on this plot is 500 MHz. A resolution of 70 dB is still achieved up to 250 MHz signal frequency, which is still about 11.3 bits.

At the highest frequency, the SFDR is 61 dB or about 10 bits indeed. At this point the analog part consumes 60 mW and the digital part 62 mW, all from 1.9 V power supply. This is for a 20 mA full-scale current. For a 16 mA full-scale current, the power consumption drops to 110 mW.

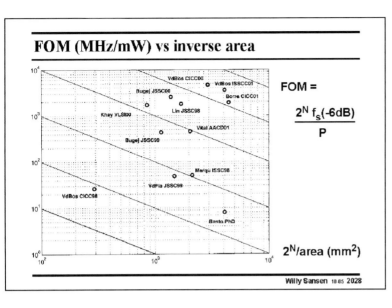

FOM (MHz/mW) vs inverse area

$$FOM = \frac{2^N f_s(-6dB)}{P}$$

2^N/area (mm^2)

Willy Sansen 10 05 2028

2028

For comparison, a Figure of Merit has to be devised.

This FOM includes the resolution, the maximum signal frequency at which this resolution has been obtained and the power consumption.

As it is plotted versus inverse area, it emphasizes the high-frequency performance of this converter.

Table of contents

- **Definitions**
- **Digital-to-analog converters**
  - Resistive
  - Capacitive
  - Current steering
- **Analog-to-digital converters**
  - Integrating
  - Successive approximation
  - Algorithmic
  - Flash / Two-step
  - Interpolating / Folding
  - Pipeline

Willy Sansen 10 05 2029

2029

Now that the most important principles have been discussed of DACs, we look at ADC's.

Again, the most important principles are introduced. They are first of all, compared in terms of resolution and speed.

2030

The conversion speed is certainly the highest for flash ADCs. It is the lowest for integrating ADCs. However, the resolution of the latter ones can be much higher.

Sigma-delta (or oversampling) converters can also reach high resolutions. All the others suffer from matching and are limited to 12–14 bits.

Various Figures of Merit are used for the comparison of the performance. The one shown is mainly used for oversampling converters, sometimes with the peak SNR instead of the DR.

Instead of this FOM the energy per conversion is quite often used or P/f_s. For example, a low-power successive-approximation ADC (Scott, JSSC July 2003, 1123–1129) reaches a power consumption of 3.1 μW for a sampling frequency (or twice the BW) of 100 kHz which yields 31

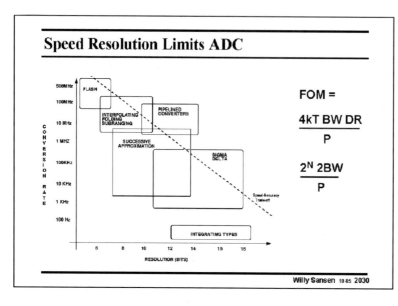

FOM =

$$\frac{4kT\ BW\ DR}{P}$$

$$\frac{2^N\ 2BW}{P}$$

pJoule/Sample. At this point the resolution is only 4.5 bit, however.

A more often used Figure of Merit also includes P/f_s (or $P/2BW$), but supplemented with the dynamic range 2^N. For 4.5 bit, 2^N is 22.6, which yields one over 1.4 pJoule/conversion. Low power ADCs presently reach values below 1 pJ/conversion.

We will start with the integrating ones.

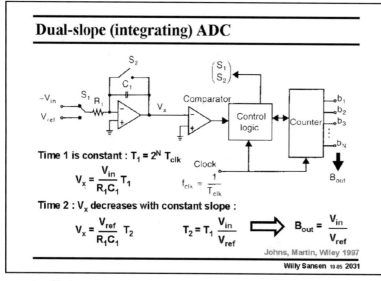

2031

Integrating or dual-slope converters can reach high resolutions because matching does not come in. The same components are used by the same integrator twice.

It uses two time periods T_1 and T_2. The first one is constant whereas the second depends on the input signal.

During the first time T_1, the input Voltage $-V_{in}$ is integrated with time constant R_1C_1 during a constant period T_1, which equals 2^N times the clock period T_{clk}. A voltage V_x is reached, as is shown in the next slide.

During the second period T_2, the input Voltage V_{ref} is integrated with the same time constant R_1C_1 during such a time T_2, that the output voltage is again zero.

A counter is used to measure the times T_1 and T_2. It counts up during time T_1 and down during time T_2.

A simple relationship exists between the two time periods T_1 and T_2, as shown in this slide and in the next slide.

This counter therefore generates the digital equivalent B_{out}.

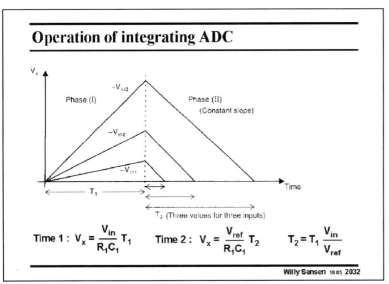

2032

A counter counts up during time T_1 and down during time T_2. In the first time T_1 the slope depends on the input voltage V_{in}, whereas during the second time T_2, the slope is constant, such that this time T_2 depends on the input signal V_{in}.

A simple and very linear relationship exists between the two time periods T_1 and T_2, as shown in this slide.

Integrating ADC

Advantages:
High resolution
High linearity
Low circuit complexity
Mainly for voltmeters, ...
Eliminates mains supply 50 Hz if T1 is n x 20 ms

Disadvantages:
Very slow :
Worst case for $V_{in} = V_{ref}$: 2^{2n+1} clock cycli required !
 Ex. For n = 16 bit (64000) and F_{clock} = 1 MHz :
  7.6 s conversion time
Mainly for voltmeters, ...

Willy Sansen 10-05 2033

2033

Such integrating or dual-slope ADC has many advantages.

The result is independent of the actual values of the resistor R_1 and capacitor C_1. As long as the opamp has sufficient gain, it does not affect the result. The linearity is high and so is the resolution.

The circuitry is actually fairly simple. Only opamps are used, an RC circuit and some switches.

The converter is fairly slow, however. For a large input voltage, such as V_{ref} itself, the counter has to count up over 2^N clock pulses and also down over as many pulses. The conversion time is thus quite long.

For slow digital voltmeters, this is an excellent solution. This is especially true if the hum of the 50 Hz mains (60 Hz in USA) is synchronized by the clock. In this way the effect of the hum is cancelled.

At the input of an oscilloscope, higher speeds are required however, and less resolution.

2034

For such applications a SAR ADC is used.

It consists of a Sample-and-hold at the input, to maintain a constant voltage during the conversion.

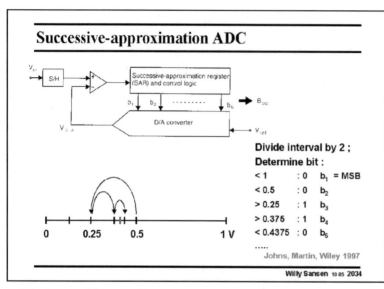

Divide interval by 2 ;
Determine bit :

< 1	: 0	b_1 = MSB
< 0.5	: 0	b_2
> 0.25	: 1	b_3
> 0.375	: 1	b_4
< 0.4375	: 0	b_5

.....

Johns, Martin, Wiley 1997

Willy Sansen 10-05 2034

This is followed by a comparator, which generates the bits by successive approximation. The bits are fed back through a DAC to close the loop. The successive-approximation register compares the incoming voltage to the next binary value as shown in this slide.

Assume that the input voltage is 0.4 V (the reference voltage is 1 V).

In a first comparison 0.4 V is found to be less than 1 V. This yields a digital 0, which is the MSB.

The interval is now divided by 2 and the incoming voltage is compared to 0.5 V. This again yields a digital 0.

The interval is now divided by 2 and the incoming voltage is compared to 0.25 V. This now yields a digital 1. Moreover, this 1 is used as a control signal to indicate that in the next comparison the interval 0.25 to 0.375 has to be selected, not the one between 0.125 and 0.25 V.

This sequence is continued for N comparisons or N bits.

This successive approximation takes only N clock cycles. It is therefore much faster than an integrating ADC. It is sensitive to the offset of the comparator, however.

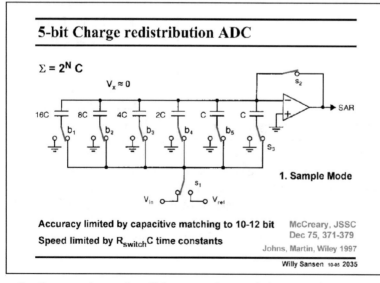

Accuracy limited by capacitive matching to 10-12 bit

Speed limited by $R_{switch}C$ time constants

McCreary, JSSC
Dec 75, 371-379

Johns, Martin, Wiley 1997

Willy Sansen 10-05 2035

2035

As a DAC, a charge redistribution DAC is often used as shown in this slide.

It consists of an opamp and a binary bank of capacitors. In this example a 5-bit capacitor bank is selected for a 5-bit AD Conversion.

It uses 3 phases, i.e. the sample mode, the hold mode and the bit cycling phase.

The picture is repeated three times, each time with adjusted positions of the switches.

In the sample mode, all bottom plates of the capacitors are connected to the input voltage V_{in}. All top plates are connected to the virtual ground of the opamp. All capacitors thus sample the input voltage V_{in}.

The voltage V_x at the input of the opamp is therefore zero and so is the output voltage.

Note that such ADC is quite simple as it consists of one single opamp (comparator) and a capacitor bank and some logic. The total number of unit capacitors is only 2^N. Its power consumption can be quite low, especially if only low-frequencies have to be processed (see Scott, JSSC July 2003, 1123–1129).

Because of the capacitor bank, the accuracy is limited by its matching, which is 10–12 bit, depending on the sizes of the capacitors (see Chapter 15). Its speed is limited by the speed of the opamp and the RC time constants of the switches, as in any switched-capacitor system (see Chapter 17).

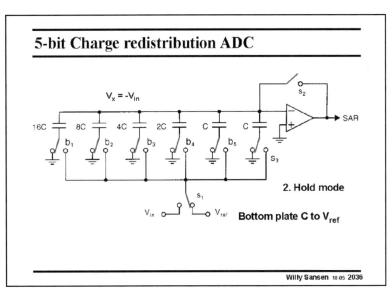

2036
In the hold mode, the feedback loop around the opamp is opened. As a result, the voltage V_x at the input of the opamp can take any value.

The bottom plates of the capacitors are now all connected to ground, pushing V_x to $-V_{in}$.

At the same time, switch S1 switches in V_{ref} rather than V_{in}.

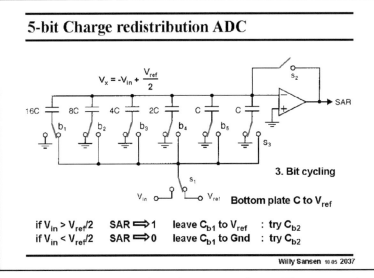

2037
The successive approximation algorithm now takes place.

Each bit is now cycled through a similar comparison sequence, in order to find out the position of the input voltage on the binary scale. We start with the MSB by switching the bottom plate of the largest capacitor 16C to the reference voltage (see slide). As the sum of all other capacitors 8C, 4C, 2C, ... equals 16C, only half of the reference voltage V_{ref} is added to the $-V_{in}$ voltage. The opamp input voltage V_x is illustrated, as shown in this slide.

If the V_{in} is larger than $V_{ref}/2$, then V_x is negative. In this case, the SAR register stores a digital 1; switch b_1 remains as shown.

If, however, V_{in} is smaller than $V_{ref}/2$ then V_x is positive. In this case the SAR register stores a digital 0; switch b_1 is switched back; the bottom plate of capacitor 16C goes back to ground.

The sequence is now repeated with switch b_2: the bottom plate of capacitor 8C is connected to V_{ref}. As a result, an additional $V_{ref}/4$ is added to V_x for comparison.

This sequence is continued until all capacitors have been switched in, until all bits have been cycled through.

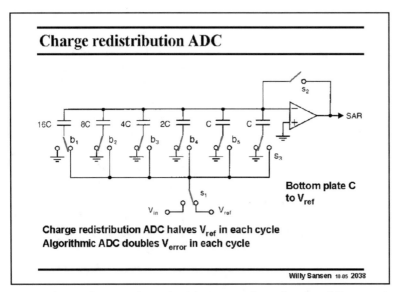

Charge redistribution ADC

Charge redistribution ADC halves V_{ref} in each cycle
Algorithmic ADC doubles V_{error} in each cycle

Willy Sansen 10-05 2038

2038
Note that in such charge redistribution ADC, the input voltage V_{in} is compared with a series of increasingly smaller fractions of V_{ref}.

This algorithm can be continued until the smallest fraction of V_{ref}, which is $V_{ref}/2N$ has become smaller than the offset of the opamp, or smaller than the mismatch error on the smallest capacitors C.

This limits the resolution to 10–12 bit.

An alternative consists of taking at each cycle, the difference between the input voltage and the fraction of the V_{ref} obtained before, and multiplying it by two, to carry out a new comparison. This is called an algorithmic ADC.

It is discussed in more detail next.

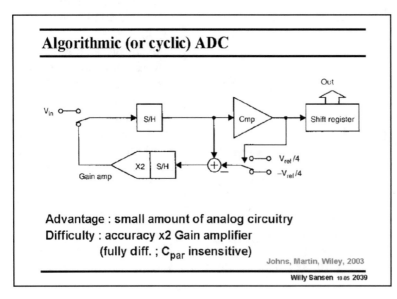

Algorithmic (or cyclic) ADC

Advantage : small amount of analog circuitry
Difficulty : accuracy x2 Gain amplifier
(fully diff. ; C_{par} insensitive)

Johns, Martin, Wiley, 2003

Willy Sansen 10-05 2039

2039
In an algorithmic ADC even less circuitry is required.

It also contains a S/H as well. An amplifier with a precise amplification of two is also included. It is used to amplify the difference between the input voltage V_{in} and the fraction of the reference voltage $V_{ref}/4$.

A register keeps track of the code generated by the subsequent comparisons.

Since in each cycle, this difference is amplified by 2, there is now no need for a binary capacitor bank or a binary set of fractions of the V_{ref}.

We assume however, that the multiplication by two has an error which is smaller than an LSB. Some examples will be given when pipeline ADCs are discussed.

Normally such converters are all realized in a fully-differential way, in order to improve the accuracy.

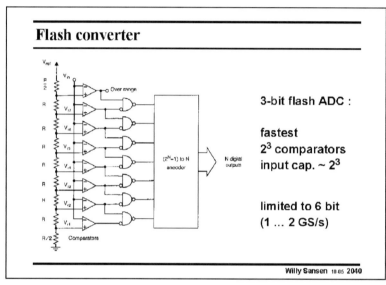

Flash converter

3-bit flash ADC :

fastest
2^3 comparators
input cap. ~ 2^3

limited to 6 bit
(1 ... 2 GS/s)

Willy Sansen 10 05 2040

2040

The fastest ADCs are undoubtedly flash converters, as shown in this slide. The reason is that they process the input voltage in parallel. Only one single clock cycle is required. The previous ones process the bits in series, and are therefore slower.

A large number of input comparators are used to compare the input signal with a (thermometer) fraction of the reference voltage V_{ref}. The outputs of the comparators generate a thermometer code, which is converted into binary code by an encoder.

The comparators are followed by NAND gates. Their outputs are all zero except for the one which sees a difference at its inputs, i.e. where the thermometer code changes from zero to one.

The main disadvantage of such an ADC is the large number of comparators required. For a 6-bit flash converter 2^6 or 64 comparators are required! This is why such converters have very limited resolution.

Moreover, the input voltage sees all comparators in parallel, which represents a very large input capacitance. It will take a lot of power to drive this converter.

2041

To give an idea of what flash ADC's have been realized, a plot is given with the flash ADCs published in the JSSC.

Both the 6-bit and 8-bit examples, show a trend of increasing speed, as a result of the availability of ever deeper submicron CMOS technologies.

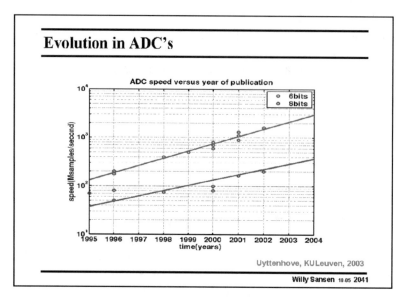

Evolution in ADC's

ADC speed versus year of publication

Uyttenhove, KULeuven, 2003

Willy Sansen 10.05 2041

Extrapolation shows that by 2008, the 8 bit flash ADCs will probably reach 1 GS/s and the 6-bit ones 12 GS/s, which is quite impressive indeed!

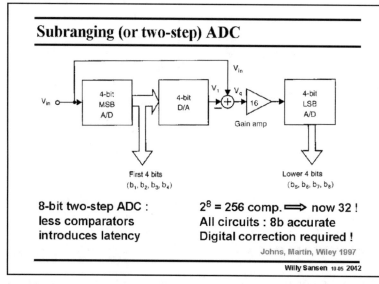

Subranging (or two-step) ADC

8-bit two-step ADC :
less comparators
introduces latency

2^8 = 256 comp. ⟹ now 32 !
All circuits : 8b accurate
Digital correction required !

Johns, Martin, Wiley 1997

Willy Sansen 10.05 2042

2042

In order to reduce the number of comparators in a flash converter, two flash converters can be selected, with lower resolution.

This is called a sub-ranging or two-step ADC and is shown in this slide for an 8-bit converter.

Two 4-bit flash ADC's are used instead of a single 8-bit ADC. The number of comparators is reduced from 256 to two times 16 or 32, plus a DAC. The power consumption will be smaller, but the input capacitances will also be smaller. Indeed only 16 input capacitances are in parallel at the input.

It works as follows. The first 4-bit flash ADC gives the first four MSBs. The resulting quantization error is obtained by taking the difference of the input voltage and the analog value of a 4-bit DAC connected to the first 4-bit ADC. This difference is multiplied by 16 to make it easier for the second 4-bit ADC, which then provides the four LSBs.

Such a two-step ADC takes more clock cycles than a single flash ADC. It has more latency. Moreover, both 4-bit ADCs must have 8-bit accuracy. Digital correction can be used to alleviate this problem.

2043

To further reduce the number of input comparators, interpolation and folding can be used. Interpolation is discussed first.

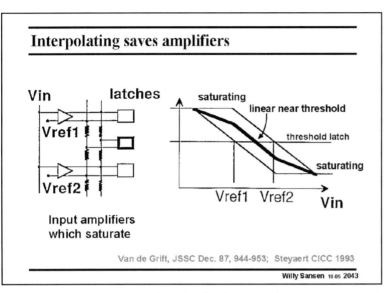

Interpolating saves amplifiers

Input amplifiers which saturate

Van de Grift, JSSC Dec. 87, 944-953; Steyaert CICC 1993

Willy Sansen 10 05 2043

Interpolating converters use analog preprocessing. They have amplifiers at their inputs, which are linear near the threshold of the latches, but which saturate on both ends. In the linear region the gain is typically ten. The actual value is not that important as long as the crossover point is accurate. Low offset is important indeed.

The transfer characteristic of the top amplifier crosses the latch threshold at voltage Vref1 because its second input is connected to Vref1. The latch now changes state at Vref1. This also applies to the bottom amplifier. Its transfer characteristic crosses the latch threshold at voltage Vref2 because its second input is connected to Vref2. The latch now changes state at Vref2.

The outputs of the amplifiers are averaged out by four equal resistors. The transfer characteristic for the latch tin the middle will be the average of the two previous transfer characteristics. It is the bold line in the middle. It crosses the latch threshold at a voltage which is exactly halfway between Vref1 and Vref2. The latch in the middle thus changes state at this voltage halfway between Vref1 and Vref2.

As a result, three latches are required and three levels are detected, but only two input amplifiers are needed. This is the result of interpolation.

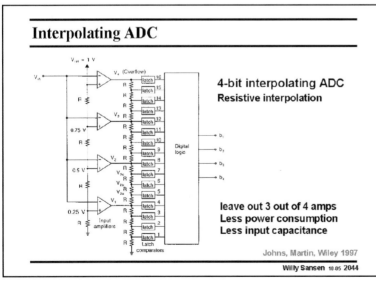

Interpolating ADC

4-bit interpolating ADC
Resistive interpolation

leave out 3 out of 4 amps
Less power consumption
Less input capacitance

Johns, Martin, Wiley 1997

Willy Sansen 10 05 2044

2044

A more elaborate example is shown in this slide.

Again, linear amplifier are used at the inputs, which saturate.

Resistors are connected between the outputs of the amplifiers to create threshold voltages for the latches, which lie in between the reference voltages 0 V, 0.25 V, 0.5 V, 0.75 V and 1 V. Four resistors are inserted. As a result, three out of four amplifiers can be left out.

This results in consider-able savings in power consumption and especially input capacitance.

The actual operation is discussed next.

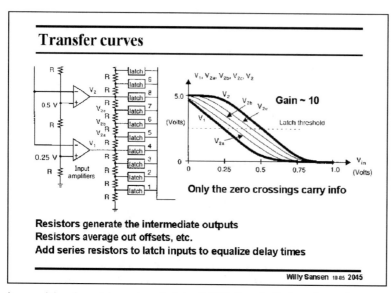

Transfer curves

Resistors generate the intermediate outputs
Resistors average out offsets, etc.
Add series resistors to latch inputs to equalize delay times

Willy Sansen 10·05 2045

2045

The bottom part of the converter of the previous slide is repeated, together with the transfer characteristics.

The transfer characteristic of the bottom amplifier (voltage V_1) crosses the latch threshold at 0.25 V because its positive input is connected to 0.25 V. Latch 4 now changes state at 0.25 V. This also applies to the next amplifier with output V_2. Its transfer characteristic crosses the latch threshold at 0.5 V because its positive input is connected to 0.5 V. Latch 8 therefore changes state at 0.5 V.

The outputs of the amplifiers are averaged out by four equal resistors. The transfer characteristics for the latches 5, 6 and 7 are between $V_1 = 0.25$ V and $V_2 = 0.5$ V cross the latch threshold at equal distances at voltages $V_{2a} = 0.3125$ V, $V_{2b} = 0.375$ V and $V_{2c} = 0.4375$ V. In this way three more reference voltages are generated without the use of preamplifiers.

At high frequencies, the latches which are directly connected to the preamplifiers such as latch 4, 8, ..., are driven at a lower impedance and are therefore faster. The ones in between such as 2, 6, ... have a larger series resistance at their input and are therefore slower. Series resistance has to be added to the inputs of latches 1, 3, 4, 5, 7, 8, to equalize the input delays.

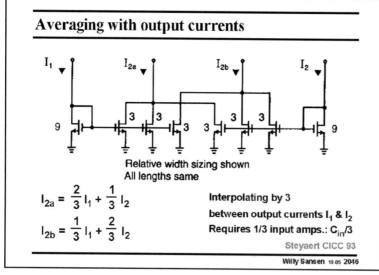

Averaging with output currents

Relative width sizing shown
All lengths same

$$I_{2a} = \frac{2}{3} I_1 + \frac{1}{3} I_2$$

$$I_{2b} = \frac{1}{3} I_1 + \frac{2}{3} I_2$$

Interpolating by 3
between output currents I_1 & I_2
Requires 1/3 input amps.: $C_{in}/3$

Steyaert CICC 93

Willy Sansen 10·05 2046

2046

Averaging the output voltages of the preamplifiers was previously carried out with resistors, but can also be carried out with capacitors or with current sources.

A simple example of averaging with current sources is shown in this slide.

The output currents I_1 and I_2 are interpolated by three. Indeed, when the additional currents I_{2a} and I_{2b} are generated (as given in this slide), perfect interpolation is achieved.

As a result, the number of input amplifiers is divided by three, and so is the input capacitance.

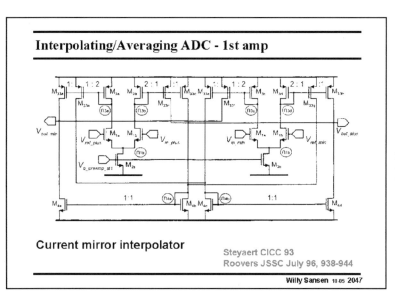

Interpolating/Averaging ADC - 1st amp

Current mirror interpolator

Steyaert CICC 93
Roovers JSSC July 96, 938-944

Willy Sansen 10 05 2047

2047

A more elaborate example of interpolation by means of current sources is given in this slide.

It is clear that the number of transistors connected at the node of each current mirror increases if the interpolation is increased. This reduces the speed at which such ADC can work, as shown next.

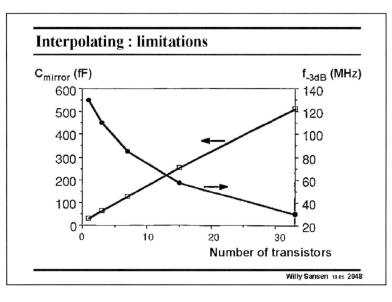

Interpolating : limitations

Willy Sansen 10 05 2048

2048

More interpolation, to save more amplifiers and to further reduce the input capacitance, increases the capacitance C_{mirror} at the inner node of the current mirrors. As a result, the maximum frequency at which the interpolation can be carried out decreases.

An example is given in this slide of the node capacitance and resulting $-3\,dB$ frequency for an interpolator realized in $0.5\,\mu m$ CMOS.

A compromise is to be taken.

2049

Interpolation allows to reduce the number of preamplifiers and the input capacitance, the number of latches being the same, however. Folding can be used to reduce the number of latches. Most often interpolation is used in combination with folding to reduce both preamplifiers and latches, which leads to a more drastic reduction in power consumption.

The principle of folding is illustrated in this slide.

Again, a preamplifier is used, but a different one, however. It folds the input signal in a number of voltage or folding regions. In this example, there are eight folding regions. The folding rate is

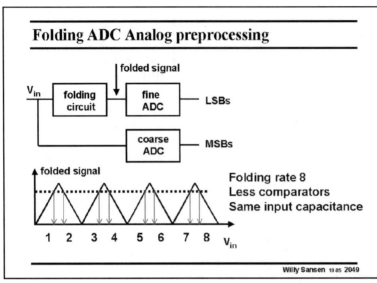

Folding ADC Analog preprocessing

folded signal

V_{in} → folding circuit → fine ADC → LSBs

coarse ADC → MSBs

folded signal

1 2 3 4 5 6 7 8 V_{in}

Folding rate 8
Less comparators
Same input capacitance

Willy Sansen 10 05 2049

therefore eight. The output voltage of the folding circuit is the same for eight different values of the input voltage v_{in}. A separate MSB ADC is required to discover in which of the eight folding regions the input voltage v_{in} is actually present. This is a 3-bit ADC in this example.

The LSBs are the determined by a fine ADC, which is the same for all eight folding regions. The number of comparators is used drastically.

We will see later that the folding circuit is made up of as many differential pairs as the folding rate indicates. The input capacitance is not decreased! This is why interpolation is usually added, to reduce the input capacitance.

A more detailed example of a 4-bit folding ADC is given next.

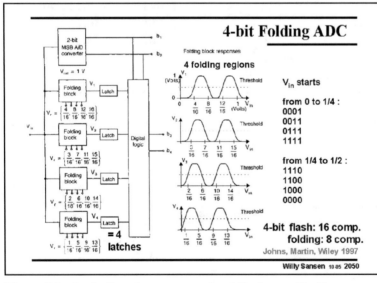

4-bit Folding ADC

2-bit MSB A/D converter → b_1, b_2

$V_{ref} = 1$ V

Folding block → V_1 → Latch

$V_r = \left\{\frac{4}{16}, \frac{8}{16}, \frac{12}{16}, \frac{16}{16}\right\}$

Folding block → V_2 → Latch

$V_r = \left\{\frac{3}{16}, \frac{7}{16}, \frac{11}{16}, \frac{15}{16}\right\}$

Folding block → V_3 → Latch

$V_r = \left\{\frac{2}{16}, \frac{6}{16}, \frac{10}{16}, \frac{14}{16}\right\}$

Folding block → V_4 → Latch

$V_r = \left\{\frac{1}{16}, \frac{5}{16}, \frac{9}{16}, \frac{13}{16}\right\}$

= 4 latches

Digital logic → b_3, b_4

Folding block responses

4 folding regions

V_{in} starts

from 0 to 1/4 :
0001
0011
0111
1111

from 1/4 to 1/2 :
1110
1100
1000
0000

4-bit flash: 16 comp.
folding: 8 comp.

Johns, Martin, Wiley 1997

Willy Sansen 10 05 2050

2050

In this example there are 4 folding regions. A 2-bit MSB ADC is thus required to indicate in which folding region is the input voltage V_{in}.

Four folding circuits (blocks) are required, each followed by a latch. Only four latches are thus required, rather than 16 for a 4-bit flash ADC.

The folding transfer characteristic is given for each folding circuit, giving output voltages V_1, V_2, V_3 and V_3.

They all look similar; they are merely shifted over $V_{in}/4$.

The 2 LSBs are obtained as follows. If the input voltage V_{in} starts increasing from zero, voltage V_4 is the first one to cause a change of state (at 1/16 V), changing the output of its latch to go from 0000 to 0001. At 2/16 V, also V_3 causes a change of stage, changing 0001 into 0011. In this way, a thermometer code is generated until 1111 is obtained, at the end of the first folding region.

When V_{in} increases further beyond 4/16 V, voltage V_4 is the first one to cause a change of state (at 5/16 V), changing the output of its latch to go from 1111 to 1110. At 6/16 V, also V_3 causes a change of stage, changing 1110 into 1100. As a result, a thermometer code is generated, but in reverse order.

When V_{in} increases beyond 9/16 V, the same code is generated as when it crossed 1/16 V, i.e. 0001 and so on. The same LSBs are thus generated.

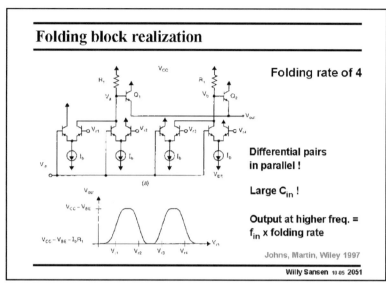

2051
Such a folding block is usually realized by means of a number of differential pairs in parallel, the outputs of some of which are also connected in parallel.

An example of a folding block with folding rate 4 is shown in this slide in bipolar technology.

All four differential pairs are connected in parallel at the input. The appropriate outputs are connected two by two. Four equidistant references are connected to the inputs. They set the inversion points of the transfer characteristics of the differential pairs and hence determine the four folding regions.

Note that the input sees four differential pairs in parallel. This is why folding does not decrease the input capacitance. Each folding block has as many differential pairs in parallel as the folding rate indicates. The input capacitance is the same as for a flash ADC.

Note also that the output frequency of a folding block is at a higher frequency than the input voltage. This establishes the maximum frequency of operation.

Finally, the 2-bit MSB ADC is easily realized with a circuit similar to the one in this slide. As reference voltages 4/16 V, 8/16 V, 12/16 V and 16/16 V have to be taken. The two MSBs are available at the output of the second differential pair and at the V_{out} terminal in this slide.

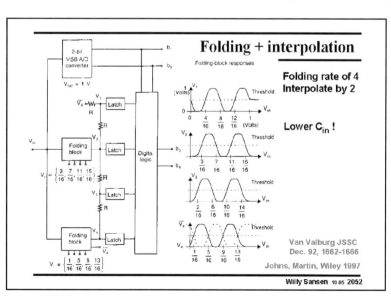

2052
The only way to reduce the input capacitance is to involve interpolation.

Application of interpolation by two to the previous 4-bit folding ADC yields the system shown in this slide.

In this way, an 8-bit ADC has been realized for a 150 MHz input signal consuming only 0.8 W in bipolar technology (Van Valburg). Four folding blocks are used, each with a folding rate of eight. Interpolation is done each time with four resistors. The zero crossings are now at 1/256 V intervals, with a 1 V reference voltage.

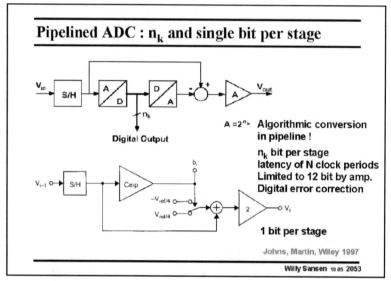

2053
A pipelined converter is a pipeline of ADC stages, each providing n_k bits. One such ADC stage is shown in this slide (on top). It is a similar algorithmic conversion as before.

After a sample-and-hold operation, the input signal is converted into n_k bits. The quantization error (or residue) is obtained after a DAC conversion, by taking the difference with the input signal itself. This difference is then multiplied by a factor of 2^{nk} to be submitted to the next stage.

A single-bit pipelined ADC converter is a converter with as many stages as bits, each stage doing a 1-bit conversion, starting with the MSB. Such a stage is shown in this slide (at the bottom). The ADC is now just a comparator. The DAC is merely a switch from the reference voltage. The amplifier has now a gain of two.

It is clear that the weakness is the amplifier. This is why a single-bit pipelined ADC is preferred. It is fairly easy to realize an amplifier with a precise gain of two, as will be shown later.

Nevertheless, the resolution is normally limited to about 12 bits. Digital correction allows the resolution to be expanded to 15 bits.

In order to speed up the conversion, the first stage starts immediately converting the next sample, after completion of the first one. This applies to the other stages as well, as shown in the next slide.

2054
In this block diagram a converter stage, as shown in the previous slide, is called a Digital approximator DAPRX. It thus carries out the conversion, starting with a S/H and ending up with an amplifier.

Each clock sample a new sample can be introduced to this ADC. It still takes N clock samples however to complete a conversion. The latency is thus N, but the processing rate is one sample per clock cycle.

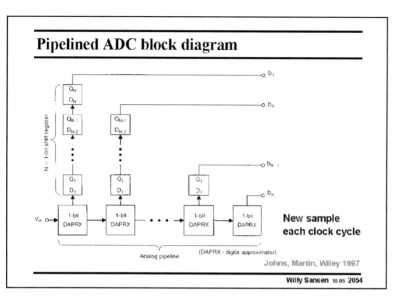

Pipelined ADC block diagram

N - 1-bit shift register

V_{in} ○

1-bit DAPRX ... 1-bit DAPRX 1-bit DAPRX 1-bit DAPRX

○ b_1
○ b_2
○ b_{N-1}
○ b_N

Analog pipeline

(DAPRX - digital approximator)

New sample each clock cycle

Johns, Martin, Wiley 1997

Willy Sansen 10-05 **2054**

They are used especially for low area and low power consumption, with limited resolution. Typical values are 12 bit up to 50 MS/s.

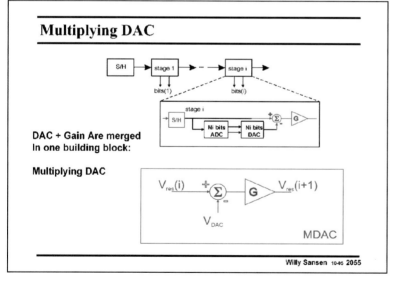

Multiplying DAC

S/H → stage 1 → -- → stage i →

bits(1)   bits(i)

stage i

→ S/H → Ni bits ADC → Ni bits DAC → $+\Sigma$ → G →

DAC + Gain Are merged In one building block:

Multiplying DAC

$V_{res}(i)$ $+\Sigma-$ G $V_{res}(i+1)$

V_{DAC}

MDAC

Willy Sansen 10-05 **2055**

2055

An efficient realization of the ADC/DAC and gain of two (for 1 bit per stage) can be carried out by use of a multiplying DAC.

The output voltage of the DAC must be subtracted from the residue of the previous stage, after multiplication by two. All these functions can be easily integrated with high-precision in a switched-capacitors circuit shown next.

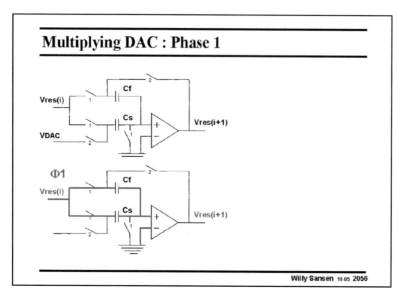

2056

This circuit is shown twice, once for phase 1 (in red) and once for phase 2 (in blue). Two capacitors are used Cf and Cs and several switches, closing on phase 1 or on phase 2.

When the phase 1 switches are closed, both capacitors are in parallel. They sample the input voltages Vres(i) which is the output of the previous stage.

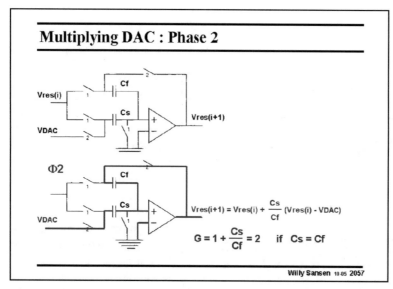

2057

On phase 2, the capacitors are switched in as for an inverting amplifier with gain Cs/Cf. The output voltage Vres(i+1) will be the sum of the voltage initially stored on Cf, which is Vres(i), plus the result of the amplification by Cs/Cf, which is Cs/Cf times the voltage difference Vres(i) − VDAC.

When both capacitors Cf and Cs are made equal, the gain is two. As a result, Vres(I) is multiplied by two and VDAC is subtracted.

The precision can be quite high because matching two equal capacitors can be fairly easily achieved. For better matching they have to be made as large as possible.

Some other factors limit the sizes of the capacitors however, as shown next.

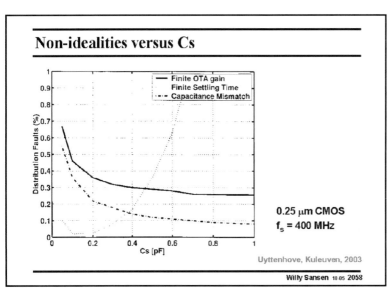

Non-idealities versus Cs

0.25 μm CMOS
f_s = 400 MHz

Uyttenhove, Kuleuven, 2003

Willy Sansen 10.05 2058

2058

The capacitors are better matched when they are made larger. Also the error as a result of finite gain improves somewhat. However, larger capacitors cause the settling time to increase.

A compromise is necessary. This is illustrated for a high-speed pipelined converter in 0.25 μm CMOS technology. The sampling frequency is 400 MHz. The gain block has a transconductance of 20 mS. Capacitors of about 0.4 pF have been selected.

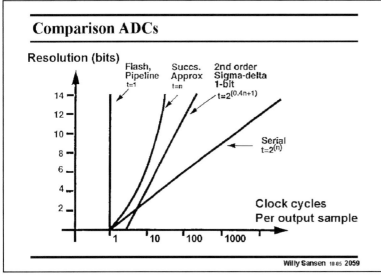

Comparison ADCs

Willy Sansen 10.05 2059

2059

A good way of compareing ADCs is to check how many clock cycles are required to carry out the conversion. Power consumption is not considered here.

It is clear that only two ADC's need one single clock period. They are clearly the fastest ones. They are the flash converter and the pipeline converter. This is not quite true for the latter one as one byte is only available after one cycle in continuous operation.

Moreover, their throughput is independent of the resolution.

This is not true for the other, which all exhibit some relationship with the resolution. The worst one is a serial converter in which each voltage is converted into all bits before the next is started.

The others are in between (i.e. successive approximation and sigma delta ones). Since they can be realized with very little power, they often offer the best compromise between speed and resolution.

2060

Speed and resolution are actually linked to the power consumption. Their relation is even fixed for a certain CMOS technology.

Impact of device mismatch on resolution/power

Two transistors : $\sigma^2(\text{Error}) \sim \dfrac{1}{WL}$ $\boxed{\sigma_{VT} = \dfrac{A_{VT}}{\sqrt{WL}}}$

$(\text{Accuracy})^2 \sim WL$

By design : increasing W increases I_{DS} and Power

decreasing L increases the speed

$\dfrac{\text{Speed x (Accuracy)}}{\text{Power}} = \text{Technol. constant}$

Ref. Kinget, ..."Analog VLSI .."
pp 67, Kluwer 1997.

Willy Sansen 10-05 2060

If resolution is linked to the accuracy, as determined by matching, or the error, then we recall from Chapter 15 on offset, that this error is inversely proportional to the area WL of the MOST.

On the other hand, the transistor width W is proportional to the current or the power consumption. Also the channel length L is inversely proportional to the speed. As a result, for a certain CMOS technology, the speed accuracy over power ratio is about constant.

Before we calculate this constant, let us first of all, try to better understand what this means. If this is true, then the realization of an ADC at higher frequency always requires more power. Also, a higher resolution always requires more power.

Power and mismatch/noise

Accuracy	$1/\sigma^2(V_{os}) \sim \text{Area} / A_{VT}$
Dynamic range	$DR = V_{sRMS} / (3\,\sigma(V_{os}))$
Capacitance	$C \sim C_{ox}\,\text{Area}$
Power	$P = 8\,f\,C\,V_{sRMS}^2$

Mismatch : $P = 24\,C_{ox}\,A_{VT}^2\,f\,DR^2$

Noise : $P = 8\,kT\,f\,DR^2$

Willy Sansen 10-05 2061

2061

Let us try to find the value of this constant so that real numbers can be used to predict this relation between resolution, speed and power.

If we take mismatch in threshold voltage as the main culprit for the lack of accuracy, the A_{VT} gives the relation between error and area WL (see Chapter 15).

As a dynamic range, the ratio is taken of the powers of the signal and this error is due to mismatch.

The power consumption depends on the speed and the capacitances to be charges. This latter capacitance depends on area as well.

Finally, the power consumption can be written in terms of all preceding parameters. It is plotted in the next slide.

Before we investigage, however, this exercise is repeated for noise as an error rather than mismatch.

The resulting expression is given as well.

Both are now plotted.

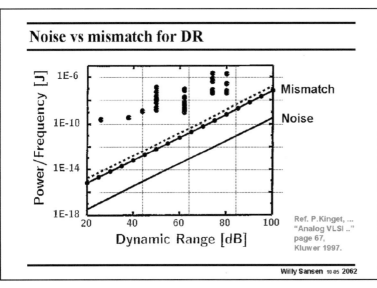

Ref. P.Kinget, ...
"Analog VLSI .."
page 67,
Kluwer 1997.

2062
The power consumption is plotted for an ADC for a certain input signal frequency, versus dynamic range.

The bottom line is the one for noise as a source of error. The other ones are for mismatch, for two different technologies 0.7 and 0.5 μm CMOS.

All the dots on top correspond with ADCs published in the JSSC.

It is clear that these first-order estimates of the power for a certain dynamic range and frequency, are still far away for most realizations.

It is also clear however, that mismatch line is a lot closer to the realizations than the noise curve. This means that mismatch as a source of error is a more realistic approximation of what can be achieved other than the noise.

Also, these curves allow to estimate beforehand what power consumption is to be expected whenever an ADC is to be realized with a certain resolution and speed.

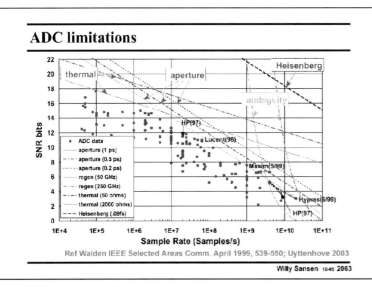

Ref Walden IEEE Selected Areas Comm. April 1999, 539-550; Uyttenhove 2003

2063
Another way to see how physical limitations come in trying to beat the power-resolution-speed barrier, a more general plot is shown in this slide.

It shows that at low frequencies, thermal noise establishes a limit to what SNR can be obtained (red). The lower curve of the two is for a resistance of 2 kΩ.

At higher frequencies however, aperture uncertainty of the sampling comes in (blue). This is mainly jitter of the clocks. The lower curve of the three is for a jitter of 1 ps.

At really high frequencies other ambiguities show up (green). They have to do with the uncertainty of the regenerative switching of the comparator. The left curve is for a 50 GHz clock.

At the highest frequencies Heisenberg comes in (black). This is still far out, however.

All the realizations (red dots) are still fairly far away from these physical limits. Remember however that between the noise curve and the dots, a mismatch curve has to be added.

References

- D. Johns & K. Martin, "Analog Integrated Circuit Design", Wiley 1997
- P. Jespers, "Integrated Converters", Oxford Univ. Press, 2001
- B. Razavi, "Principles of Data Conversion System Design", IEEE Press 1995
- K. Uyttenhove, "High-speed CMOS Analog-to-digital converters", PhD KULeuven, 2003
- A. Van den Bosch, ... "High-resolution high-speed CMOS current-steering Digital-to-Analog Converters, Kluwer Ac. Press 2004.
- R. Van de Plassche, "Integrated Analog-to-digital and Digital-to-Analog converters", Kluwer Ac. Press, 1994

Willy Sansen 10-05 2064

2064

Above are the references given which have been used throughout this Chapter. Many figures have been used from the first reference. All of them have contributed to this Chapter however, in one sense or another.

The most general introduction is given by the last reference.

Table of contents

- Definitions
- Digital-to-analog converters
  - Resistive
  - Capacitive
  - Current steering
- Analog-to-digital converters
  - Integrating
  - Successive approximation
  - Algorithmic
  - Flash / Two-step
  - Interpolating / Folding
  - Pipeline

Willy Sansen 10-05 2065

2065

This Chapter has given an overview of the different types of DAC's and Nyquist ADCs. The operating principles have mainly been discussed. Some of the major compromises have been introduced.

Oversampling ADCs are discussed next. The focus is on low-power realizations.

Low-power Sigma-Delta AD Converters

Willy Sansen

KULeuven, ESAT-MICAS

Leuven, Belgium

willy.sansen@esat.kuleuven.be

Willy Sansen 10-05 211

211

Low power consumption is not always a result of low supply voltages. The additional signal processing required to maintain the same or maybe an even better dynamic range despite the reduction in supply voltage, may cause additional power consumption.

Good examples are sigma-delta (or delta-sigma) analog-to-digital converters. Their lowest supply voltage is 0.6 and even 0.5 V, which is largely sufficient to be embedded in 65 and even 45 nm CMOS. The problem in these converters is that they have to maintain the same dynamic range as their 1.8 V and higher counterparts.

A low supply voltage always assumes that all circuit techniques are also applied to reduce the power consumption. This Chapter explains what circuit techniques are available to realize low power Delta-Sigma converters at very low supply voltages.

Many of these circuit techniques will become important once nanometer CMOS have become mainstream.

Table of contents

- **Delta-sigma modulation**
- **The switch problem**
- **The switched-opamp solution**
- **Other low-power Delta-sigma converters**

Ref. Norsworthy, Delta-Sigma Converters, Wiley 1996
Ref. Op 't Eynde, Peluso, Geerts, Marquez, Geerts, Yao, Kluwer/Springer

Willy Sansen 10-05 212

212

To initiate this Chapter, a few essential aspects of Delta-Sigma converters are reviewed. This is a very short introduction to Delta-Sigma converters. For more thorough understanding, the reader is referred to more detailed references such as Northsworthy (Wiley), Schreier (Wiley) and Johns-Martin (Wiley).

Most low-power Delta-Sigma converters use switched-capacitor filters for the noise shaping. Switches become a problem however, for such low supply voltages. They are discussed first.

Rather than use switches, the opamps can be switched themselves. Today, many very-low-voltage Sigma-Delta converters use this technique. The principle will be explained, followed by a number of design examples.

Finally, some of the most recent low-voltage Delta-sigma converters are discussed. Some of them use series resistors at the input, some others use full feedforward. They are discussed and compared on the basis of a commonly agreed Figure of Merit.

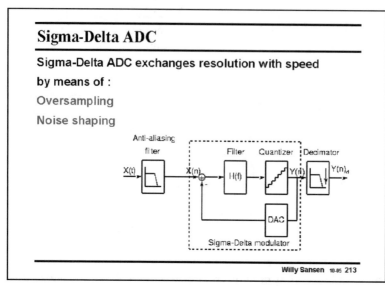

213
A Sigma-Delta Analog-to-digital converter samples the incoming analog signal at a frequency, which is much higher than needed. The minimum sampling frequency according to Nyquist, is twice the maximum frequency of the incoming signal. For low-quality speech, the bandwidth is limited to 3.4 kHz. The minimum sampling frequency would be 6.8 kHz. An ECG (Electro Cardio Gram) is limited to about 150 Hz. Its minimum sampling frequency would be 300 Hz.

In Sigma-Delta ADCs the sampling frequency is much higher, 20 to 1000 times, depending on the application. The ratio of this sampling frequency to the minimum Nyquist sampling frequency is called the oversampling ratio (OSR). It is the first design parameter in a Sigma-Delta converter. The more the incoming signal is over-sampled, the more the signal information is emphasized with respect to the noise. As a result, the SNR (Signal-to-Noise Ratio) will be higher for an higher oversampling ratio (OSR). The resolution of the ADC will be higher for an higher oversampling ratio (OSR).

We can state that the high oversampling leads to higher resolution. A Sigma-Delta ADC exchanges speed for resolution.

To effectively obtain this higher SNR however, the noise must be filtered out. This is achieved by noise shaping. This is realized by a filter H(f).

A Sigma-Delta converter consists of a feedback loop with a filter and a quantizer, which carries out the AD Conversion. The feedback loop is closed over a DAC. A prefilter (anti-aliasing filter) is required to make sure the incoming bandwidth is limited. A post filter (decimator filter) is applied to lower the sampling frequency to what satisfies Nyquist.

214

The largest error in such an ADC converter is the quantization noise. This is actually not noise but the difference between the original analog signal and the analog signal, quantized with a limited number of bits.

An example is shown in this slide. The original analog signal is the dotted sine wave. The sampled and quantized analog signal for 4 bit (or 16 steps) is shown in full line. The difference is in the middle. It is small in amplitude, depending on the number of bits (B) taken. Moreover, it contains many different frequencies. This is why it is called noise. It is the quantization noise. It is the noise which determines the Signal-to-Noise ratio of the converter and hence its resolution.

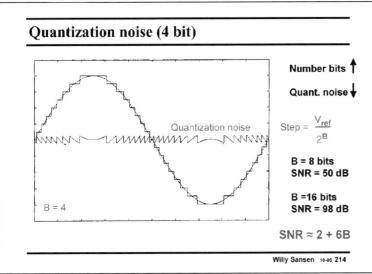

Quantization noise (4 bit)

Quantization noise

B = 4

Number bits ↑
Quant. noise ↓

$\text{Step} = \dfrac{V_{ref}}{2^B}$

B = 8 bits
SNR = 50 dB

B =16 bits
SNR = 98 dB

SNR ≈ 2 + 6B

Willy Sansen 10-05 214

The size of one single step is the reference voltage divided by 2^B in which B is the number of bits; for 8 bits this is 256. For a reference voltage of 1 V, this step is about 4 mV. This is barely larger than mismatch or offset. Additional techniques will be required to enhance the SNR. Noise shaping is such a technique. For this purpose, a noise filter is included in the feedback loop. This is explained in the next slide.

For 8 bits, the number of 256 corresponds to 48 dB. The precise value is 2 dB higher, which is 50 dB. For 16 bits we find, in a similar way 98 dB, which is a very high number indeed, this is required by a limited number of applications such as audio CD's and some others.

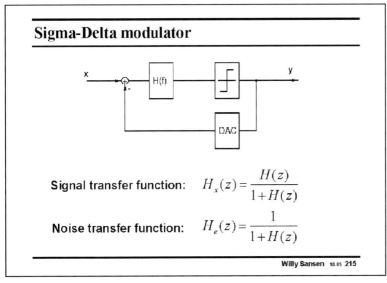

Sigma-Delta modulator

x   H(f)   y

DAC

Signal transfer function: $H_x(z) = \dfrac{H(z)}{1 + H(z)}$

Noise transfer function: $H_e(z) = \dfrac{1}{1 + H(z)}$

Willy Sansen 10-05 215

215

The minimum number of components of the Delta-sigma feedback loop are the noise filter H(f), the quantizer and the DAC.

For a single-bit converter, the quantizer is only a comparator. The DAC then switches between two DC reference voltages.

For multi-bit converters, the quantizer is a real ADC, such a flash ADC or any other type, depending on the speed required. The DAC is then a real multi-bit DAC, in which linearity is one of the main specifications. The higher number of bits, the larger the resolution can be. This is the second design parameter of a Sigma-delta converter, i.e. the number of bits.

Many Sigma-Delta converters stick with one single bit conversion as a comparator and a switch from DC reference voltages are so easy to realize. Matching is hardly an issue for single-bit converters.

It is clear that for the incoming input signal x, the transfer function is as given in this slide. If H(z) has a high magnitude, the gain is about unity.

For noise signals, generated in the feedback loop by the quantizer (comparator) for example, the transfer function is very different. If H(z) has a high magnitude, the quantization noise is reduced considerably. The SNR also increases considerably.

A simple example with a low-pass filter is given next.

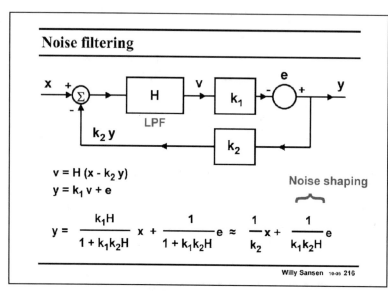

216
The simplest possible feedback loop with such a noise filter H is shown in this slide. Its gain is k_1. The (quantization) noise or error signal in general is represented by e. The feedback loop is closed over a block with gain k_2.

From the network equations it is clear the gains for the incoming signal x and for the error signal e are very different indeed.

For a large loop gain $1 + k_1 k_2 H$, the gain for the incoming signal is $1/k_2$, whereas the gain for the error signal (quantization noise) can be quite small. This latter filter action is the actual noise shaping.

This is evident if a first-order low-pass filter is taken for filter H, as shown next.

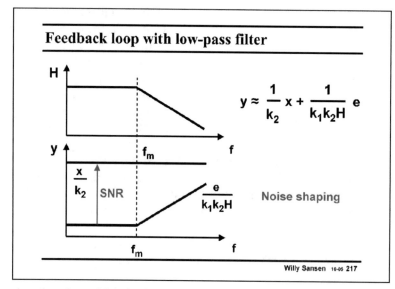

217
When a low-pass filter of first order is taken for H, with cut-off frequency f_m and slope -20 dB/decade, the contribution of the error (noise) signal in the output y shows an inverted characteristic. Its contribution is low at frequencies below f_m and it increases at 20 dB/decade for higher frequencies. The noise is pushed towards higher frequencies.

At low frequencies, the SNR is the ratio of x/k_2 and $e/k_1 k_2 H$. This is proportional to k_1, which is the gain in the forward path of the feedback loop. The higher this gain, the higher the SNR.

Also, the higher the order of the filter, the steeper the slope of the noise beyond frequency f_m.

Higher-order filters allow higher values of k_1. As a result, higher-order filters give rise to higher SNR values.

The order of this filter is therefore the third design parameter of a Sigma-delta converter. Recall that the other two are the oversampling ratio OSR and the number of bits B of the quantizer.

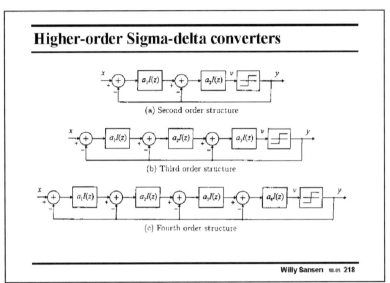

Higher-order Sigma-delta converters

(a) Second order structure

(b) Third order structure

(c) Fourth order structure

Willy Sansen 10.05 218

218
Normally, third-and fourth-order filters are used. A first-order Sigma-Delta is rarely used because it shows patterns in the output spectrum which are hard to remove.

Second-order Sigma-delta converters are easy to realize. Since only a second-order filter is involved, no problems arise with respect to stability. Their noise shaping capability is somewhat limited, however.

Third-order and fourth-order converters are used most (see slide). To go higher is difficult because of stability problems.

Fourth-order filters in a single feedback loop as shown in this slide, are risky but feasible. It is better to organize them in a Mash or cascaded topology as shown next.

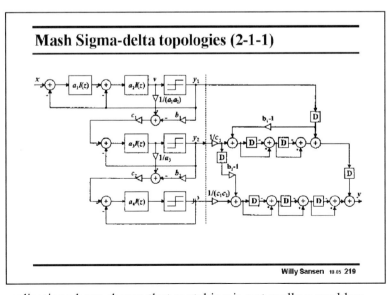

Mash Sigma-delta topologies (2-1-1)

Willy Sansen 10.05 219

219
A 4th-order Mash (or cascaded) topology has four filter sections for 4th order noise shaping. Only two of them are in a single feedback loop. Stability is easily achieved.

It has three outputs however, which are fed to a so-called noise-cancellation circuit. This means that the three outputs must receive the right gain and the right delay so that the noise cancels. Matching will now be an issue. However, many realizations have shown that matching is not really a problem.

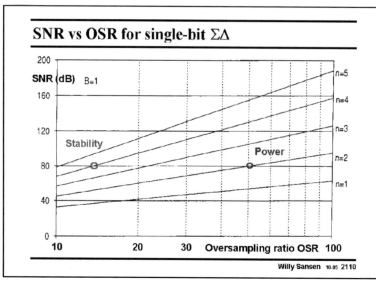

2110

The values of the SNR which can be achieved depending on the values of the three design variables are given in this slide. However, for all of them single-bit quantization is taken.

The SNR is given versus the OSR and for different orders of noise-shaping. The higher the OSR and the noise-shaping order, the higher the SNR.

The same SNR can therefore be achieved for either higher OSR or for higher noise-shaping. For example, 80 dB SNR (corresponding to 13 bit resolution) can be realized by either an OSR of only 14 requiring 4th order noise shaping, or by an OSR of 50 allowing only 2nd order noise shaping. Which one is preferable?

Fourth-order noise shaping may require a Mash topology, which requires tougher matching conditions. High OSR values on the other hand, require higher-speed opamps which require more power consumption.

This is the trade-off. For high-speed Sigma-Delta converters low values of OSR's are preferred, leading to stability and matching considerations. For low-power Sigma-Delta converters, low values of OSR's are also preferred, leading to similar considerations.

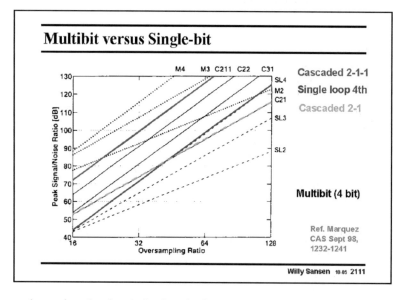

2111

The use of more bits for the quantization also increases the maximum SNR.

In this graph M stands for multibit;

- M4 thus stands for 4 bit quantization,
- C211 stands for 4th-order noise shaping with a 2–1–1 Mash or cascaded topology; note that a 2–2 Mash topology offers less SNR,
- SL4 stand for a 4th-order single-loop topology.

It is clear that a 4-bit 4th-order noise shaping is by far the best.

It is also clear that 3rd-order noise shaping provides lower SNR values. This is true for both a 2–1 Mash topology (C21) and for a 3rd order single-loop topology (SL3).

From this it can be concluded that multibit Sigma-delta realizations offer the highest SNR values for low OSR values. If matching problems have to be avoided however, single-bit solutions are quite attractive provided 4th order noise shaping is used in a 2–1–1 Mash topology.

Table of contents

Willy Sansen 10-05 2112

2112

Now that we know which topology is most attractive, we have to focus on the implementation.

Most noise-shaping filters are realized with switched-capacitor techniques. However, for low supply voltages, the switches are difficult to realize.

This is discussed next.

Low Voltage SC: problem

Switch:

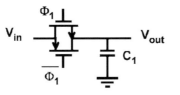

Φ_1

V_{in} —— V_{out}

C_1

Φ_1

nMOS: $V_{in} < V_{DD} - V_{GSn} \approx V_{DD} - 0.8$ V

pMOS: $V_{in} > V_{GSp} \approx 0.8$ V

Limit : $V_{DD} - V_{GSn} = V_{GSp}$

➡ $V_{DDmin} > 1.6$ V

Willy Sansen 10-05 2113

2113

The supply voltage decreases for decreasing channel lengths. For 0.25 μm channel length, the supply voltage was 2.5 V; for 0.18 μm CMOS it was 1.8 V and so on. For 90 nm CMOS the supply voltage is about 1.2 V. Supply voltages will now be used below 1 V.

With a 1 V supply voltage the MOST switch is difficult to turn on.

Assume that a transmission gate is used as a switch, with a nMOST in parallel with a pMOST, as shown in this slide, The nMOST Gate is driven with a positive (supply) voltage to turn it on. The pMOST is at the same time driven with the most negative voltage or ground.

For zero input voltage V_{in}, the Gate drive of the nMOST is the full supply voltage V_{DD}, its V_{GSn} also equals V_{DD}. This is sufficient to provide a low ON-resistance.

Indeed, assume that the V_T is about 0.6 V and the minimum $V_{GS} - V_T$ for conduction is about 0.2 V, the minimum V_{GS} is now about 0.8 V. As a result, the input voltage V_{in} can only increase until it reaches $V_{DD} - 0.8$ V. For higher input voltages, the nMOST can no longer be turned on.

The same is true for the pMOST device. The input voltage V_{in} can only decrease until it reaches 0.8 V. For lower input voltages, the pMOST can no longer be turned on.

This establishes a minimum value for the supply voltage V_{DDmin} which is the sum of both V_{GS} values. In this example this is 1.6 V.

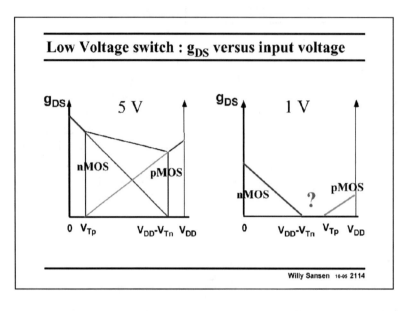

2114
The same story is given in this slide. The output conductance (inverse ON resistance) is given for input voltages from zero to the supply voltage V_{DD}.

In both cases, the nMOST conducts for low input voltages and the pMOST for high input voltages. For a large supply voltage (5 V), there is a large region in the middle where both the nMOST and pMOST conduct, giving rise to a small total ON-resistance.

However, for a small supply voltage (1 V) there is a hole in the middle. None of the MOSTs can be turned on. For such a low supply voltage it is therefore not possible to construct a nMOST/pMOST switch combination which conducts for all input voltages.

How can we solve this problem?

Note that in the plots in this slide, the V_{GS}'s have been substituted by the V_T's themselves. In this way, the lowest possible V_{GS}'s values have been taken into account. There may still be some leakage because of the weak-inversion operation but this has been neglected.

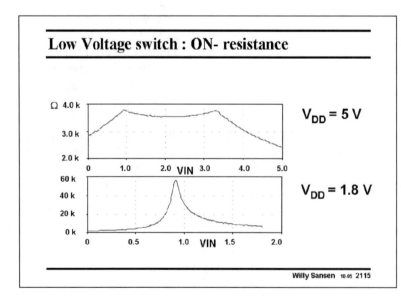

2115
The same story is repeated once more but for the ON-resistances themselves. The numerical values of the ON-resistances are plotted for input voltages going from zero to the supply voltage V_{DD}.

The numerical values correspond to an older CMOS process with rather large V_T values (0.9 V). The W/L ratios are five.

It is clear that for a supply voltage of 5 V, there is a large region in the middle where both transistors conduct. This region disappears when the supply voltage is lowered to a value which is roughly the sum of the two V_T's or 1.8 V in this example.

As a consequence, such a transmission gate with a nMOST/pMOST parallel combination cannot be used any longer for supply voltages below the sum of the V_T's. This is 1.8 V in this example. For present days V_T's of

0.35 V, this is about 0.7 V. This is lower than 1 V, but not significantly!

Low Voltage SC: solutions

◆ **Low V_T techology**
  • special technology : cost
  • switch-off leakage

◆ **On-Chip voltage multipliers**
  • poor power efficiency
  • applicability in submicron technologies ?

◆ **Switched Opamp**   Ref.Crols, ESSCIRC 93, JSSC Aug.94

Willy Sansen 10.05 2116

2116
What solutions are available for using switches at supply voltages below the sum of the threshold voltages?

The first one is to ask for a modification in technology. Normally the V_T's are determined for optimum performance of the digital part. This means that nowadays two different oxide thickness are often available. The thinner ones offer lower V_T's and higher speeds. The thicker ones provide larger values of V_T and lower leakage currents. This latter is the low-power process.

The high-speed process now provides lower V_T values. It is ideally suited to provide switches which can switch all input voltages. The leakage current is higher, however. As a result, the switch may be difficult to switch off!

Another solution is to use voltage multipliers. They are circuits which provide DC output voltages which are higher than the supply voltage V_{DD}. This higher voltage can now be used to drive the Gates of the switches.

The main problems of such multipliers are their low power efficiency. Moreover, they may cause reliability problems for the thin Gate oxides.

Another alternative is not to use these switches any more, but to switch the preceding opamp. This is called the switched-opamp approach.

All three alternatives are now discussed in more detail.

2117

The first alternative is to lower the value of the V_T. Smaller V_{DD} values require smaller V_T values! How far can this go?

A small value of V_T causes the weak-inversion part of the $i_{DS} - v_{GS}$ characteristic to cross the axis of zero V_{GS}. In other words, even for zero V_{GS}, some current flows. This is called the leakage current. This means that all digital gates conduct current even when switched off.

Lowering the V_T causes excessive power consumption. A minimum value seems to be about 0.3 V (Ref. Rabaey).

Most chips operate at higher temperatures, however. Chip temperatures of up to 100°C higher

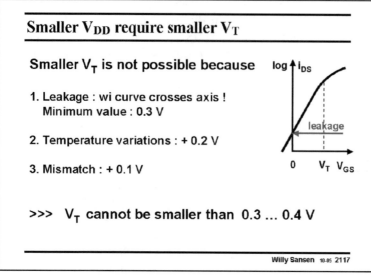

Smaller V_{DD} require smaller V_T

Smaller V_T is not possible because

1. Leakage : wi curve crosses axis !
   Minimum value : 0.3 V

2. Temperature variations : + 0.2 V

3. Mismatch : + 0.1 V

>>> V_T cannot be smaller than 0.3 ... 0.4 V

Willy Sansen 10-05 2117

than room temperature are common. Since the threshold voltage V_T decreases with about $2\,mV/°C$, it can be 0.2 V lower than at room temperature. The leakage current is weak inversion depends on the V_{GS} in an exponential way, increasing by nearly a factor of ten per 100 mV. At high chip temperatures, the leakage current can now be about 2 orders of magnitude larger. The choice of the V_T must compensate this. The V_T cannot be chosen too small!

Values of 0.3 to 0.4 V have become common nowadays.

Finally, mismatch causes a large spreading of the V_T values. The actual V_T value can thus be a lot lower than expected. Again, the leakage current depends on the V_{GS} in an exponential way. The V_T cannot be selected too small!

2118

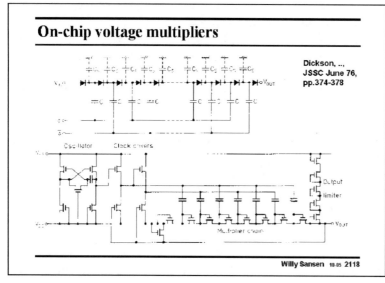

On-chip voltage multipliers

Dickson, ..,
JSSC June 76,
pp.374-378

Willy Sansen 10-05 2118

Voltage multipliers are used to generate DC voltages which are higher than the supply voltage V_{DD}. This high voltage is now used to drive the Gates of the nMOST switches. Little current is required as they only drive Gates. They can therefore be realized with little additional power.

They consist of a series of diodes and capacitors organized in a number of stages. Instead of diodes, diode-connected MOSTs can be used. They are driven by an oscillator.

The larger the number of stages, the higher the output voltage. The larger the capacitors the larger the output power. The most important parameter is doubtless the power efficiency. This has improved many times since the original reference.

This is why more attention is paid to it.

Voltage multipliers : power efficiency

$$P_{loss} \approx R_{eq} I_{out}^2$$

$$P_{VDD} \approx I_{out} V_{DD} \qquad \eta \approx 1 - \frac{R_{eq} I_{out}}{V_{DD}} \approx 50\%$$

$$R_{eq} \approx \frac{n}{fC} \frac{1}{\tan(2f R_{on,sw} C)}$$

Willy Sansen 10-05 2119

2119

The actual power efficiency η can be expressed in terms of an equivalent resistor R_{eq}. It is also clearly a function of the output current.

This loss resistor R_{eq} is proportional to the number of stages n, but inversely proportional to the clock frequency f and the size of the capacitor C used. There is an additional non-linear factor in this expression, which includes the switch resistance. This factor is less important to have an idea about the orders of magnitude.

It can be concluded that a voltage multiplier generates an output voltage with the highest efficiency, provided as few stages are used as possible, but with the highest possible capacitor and clock frequency.

Unfortunately, these are exactly the same conditions for a maximum of injection of spikes in the substrate. Coupling to the sensitive analog parts can now be expected. This will be discussed in more detail in Chapter 24.

Drawbacks of voltage multipliers

♦ **High voltage technology:**
  • In deep submicron : $V_{DD} < 1.8$ V in 0.18 µm CMOS
  • Oxide cannot take more !! 800 V/µm or 0.8 V/nm

♦ **Requires high-speed clock drivers**

♦ **Injection in substrate : coupling to Analog**

♦ **Low power-efficiency**

Willy Sansen 10-05 2120

2120

Voltage multipliers show several more drawbacks.

They generate higher voltage than the supply voltage. Care must be taken not to exceed the maximum voltage on the thin Gate oxides.

At about 0.8 V/nm thickness an oxide breaks down. A CMOS technology of 0.18 µm has an oxide thickness of about 1/50th or 3.6 nm. This would break down under a voltage of about 2.9 V. The standard supply voltage for 0.18 mm CMOS is 1.8 V. This is sufficiently far from the breakdown voltage.

However, if voltage multiplies are added, care has to be take not to end up too close to the breakdown voltage. If not, reliability issues appear.

A CMOS technology of 90 nm has an oxide thickness of about 1.8 nm, corresponding to a breakdown voltage of about 1.5 V. The supply voltage is 1.3 V. Voltage multipliers are therefore excluded!

Another disadvantage of voltage multipliers is that clock drivers are required to drive all the capacitors. They take additional power and cause large spikes into the substrate. Substrate noise and coupling is now hard to avoid.

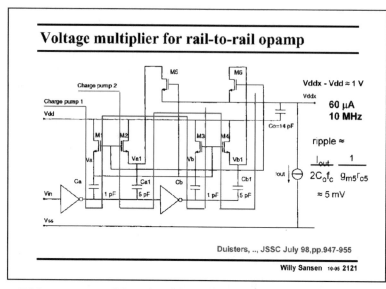

Voltage multiplier for rail-to-rail opamp

Charge pump 2

Charge pump 1

Vdd

$Vddx - Vdd \approx 1\ V$

Vddx

Co=14 pF

60 µA
10 MHz

ripple $\approx$

$$\frac{I_{out}}{2C_o f_c} \cdot \frac{1}{g_{m5} r_{o5}}$$

$\approx 5\ mV$

Duisters, .., JSSC July 98,pp.947-955

Willy Sansen 10-05 2121

2121

An interesting example of a voltage multiplier, which has been designed according to the rules of the previous slides, is given in this slide.

It provides an output voltage which tracks the supply voltage V_{DD} but is always 1 V higher. This difference of 1 V is used to realize a rail-to-rail input amplifier with only one single input pair (see Chapter 11). As a result, the distortion can be made less than -80 dB.

This operates with only two stages but with a fairly high clock frequency of 10 MHz. The capacitors are selected fairly small. The ripple on this DC supply has been reduced by taking a large capacitor C_o. Together with the voltage gain of transistor M5, this capacitor C_o allows a reduction of the ripple to a mere 5 mV, for a 60 µA output current. This was necessary as the voltage multiplier is not used only for Gate drives but to supply DC current to input differential pair.

It works for supply voltages of 1.8 V to 3.3V. It has been realized in 0.5 µm CMOS.

For more details the reader is referred to the paper in reference.

Table of contents

Willy Sansen 10-05 2122

2122

If switches cannot be used any more because of the low supply voltage, the whole opamp can be switched. This is called the switched-opamp approach.

This is discussed next.

The principles are explained first, followed by a few realizations.

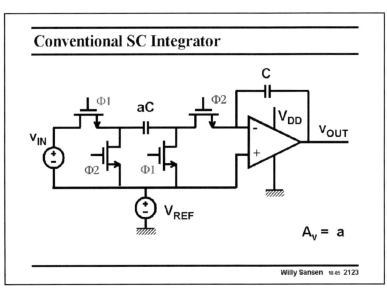

2123

In a conventional switched-capacitor integrator, the input voltage v_{IN} is sampled on capacitor aC, when phase $\Phi 1$ is high. The two MOSTs labeled $\Phi 1$ then conduct whereas the other two are off.

On phase $\Phi 2$, the charge aCv_{IN} is transferred to capacitor C. The resulting output voltage is now a multiplied by the input voltage; the voltage gain is a.

When a single supply voltage V_{DD} is used, a DC reference voltage V_{REF} may have to be added, to make sure that the voltages at both the minus and plus inputs of the opamp never exceed the common-mode input range. A typical value is 0.2 V depending on the type of opamp.

The main problem of this integrator when the supply voltage is small, is the input switch. Indeed, if we take as an input voltage 0.5 V for a supply (and clock) voltage of 1 V ($V_{REF} = 0.2$ V), then all MOST switches receive a V_{GS} of 0.8 V when switched in, except the input transistor. This latter one only receives 0.3 V and is not on at all! This is the problem.

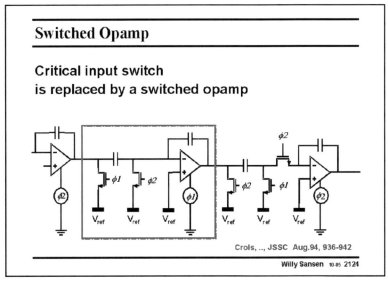

2124

A solution is obtained by switching the preceding opamp. This obviously only works if several switched-capacitor integrators are put in series.

Take the integrator in the frame. Its input switch on phase $\phi 2$ has been shifted to the preceding opamp. The whole opamp is switched in and out at clock phase $\phi 2$.

In a similar way, the opamp within the frame is switched in and out at clock phase $\phi 1$, in order to be able to provide the input voltage to the next integrator.

The last opamp shown is now switched at clock phase $\phi 2$.

It is easier to realize an opamp with a low supply voltage than to realize an input switch with a low supply (or clock) voltage. Moreover, it is fairly straightforward to switch an opamp. This is discussed next.

The only problem left is the very first input switch. At the input, a compromise will have to be taken for the maximum input voltage. Alternative solutions will be discussed later in this Section.

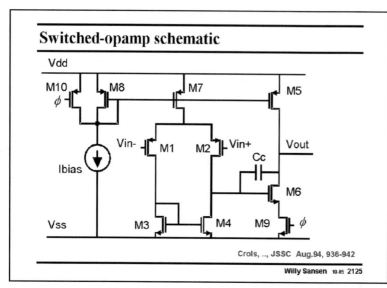

2125

The very first switched opamp is shown in this slide.

It is a two-stage Miller opamp, to which two switches have been added, i.e. nMOST M9 and pMOST M10.

When the voltage of clock phase ϕ is high, M10 is off and M9 is on. The opamp functions as expected.

When the voltage of clock phase ϕ is low however, M10 is on and shorts the Gate voltage of transistors M8, M7 and M5 to the

supply lines. All these transistors are therefore off.

Also, M9 is off. Both stages of the opamp are switched off.

How fast can an opamp be switched on and off depends on the loop gain of the feedback arrangement. If the gain around the opamp is a, then its bandwidth is GBW/a. The corresponding time constant for the output voltage is also the time constant for the switching in and out of the whole opamp.

This is shown next.

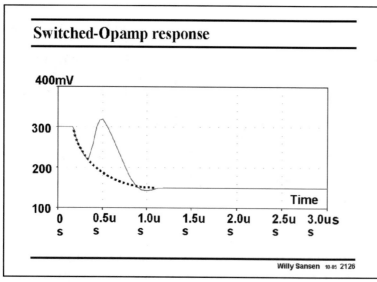

2126

This is the output voltage of the switched opamp of the previous slide, when it is switched out.

The expected exponential is given by the dotted line. The time constant is indeed one over two times π times the bandwidth of the opamp.

A big hump occurs however, after about 0.5 μs. This is a result of the discharge of the compensation capacitance C_c. It is a great deal better to insert a switch in

series with this capacitor, not to allow it to discharge and charge, as will be shown later.

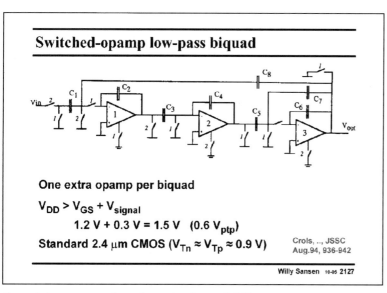

Switched-opamp low-pass biquad

One extra opamp per biquad

$V_{DD} > V_{GS} + V_{signal}$

1.2 V + 0.3 V = 1.5 V (0.6 V_{ptp})

Standard 2.4 μm CMOS ($V_{Tn} \approx V_{Tp} \approx 0.9$ V)

Crols, .., JSSC
Aug.94, 936-942

Willy Sansen 10-05 2127

2127

A low-pass biquadratic filter using this technique, is shown in this slide.

Each opamp is switched according to the switch of the next sampling capacitor. An extra switched opamp is added to provide the feedback on phase 2. The minimum supply voltages for this realization is about 1.5 V. In this older CMOS technology the V_T is about 0.9 V and the V_{GS} is 1.2 V (for $V_{GS} - V_T \approx 0.3$ V). For an input voltage of 0.6 V_{ptp}, the minimum supply voltage is about 1.5 V.

The input switch on phase 2 is the main problem. It limits the input range to V_{DD} minus V_T or about 0.6 V.

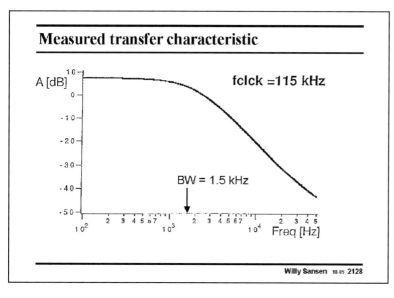

Measured transfer characteristic

fclck =115 kHz

A [dB]

BW = 1.5 kHz

Freq [Hz]

Willy Sansen 10-05 2128

2128

The low-pass filter characteristic is as expected. It shows a −40 dB/decade roll-off for frequencies beyond the bandwidth (which is 1.5 kHz), up to about half the clock frequency of 115 kHz.

At low frequencies the gain is a factor of 2 or 6 dB.

Let us observe the noise and distortion.

2129

The input noise is mainly the kT/C noise, typical for a switched-capacitor filter. It is therefore fairly high.

The distortion comes up rather strongly once the input signal has become so large that the input switch cannot be fully switched on any more. This occurs at about 0.6 V_{ptp} input amplitude. The resulting DR is close to 70 dB which is typical for switched-capacitor filters.

No additional disadvantages are present as a result of the switched opamp approach.

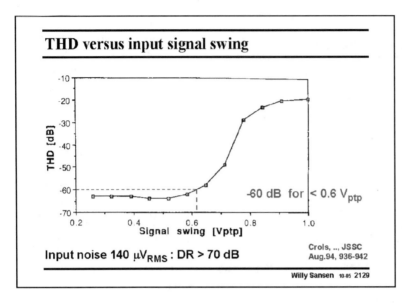

THD versus input signal swing

THD [dB]

-60 dB for < 0.6 V$_{ptp}$

Signal swing [Vptp]

Input noise 140 µV$_{RMS}$: DR > 70 dB

Crols, .., JSSC
Aug.94, 936-942

Willy Sansen 10-05 2129

The total power consumption (at this 2.4 µm CMOS technology) is 110 µW at 1.5 V supply voltage. This is quite low. Indeed, an additional advantage of the switched-opamp approach is that the opamps are off 50% of the time. The power consumption is therefore halved as well, which is a considerable advantage.

Even more power can be consumed when a class-AB stage is used at the output. This will be shown later.

Table of contents

♦ **Delta-sigma modulation**

♦ **The switch problem**

♦ **The switched-opamp solution**

   • Principle : Switched opamp filter

   • Improved switching

   • 0.9 V - 40 µW 12 bit CMOS SO ΣΔ

♦ **Other low-power Delta-sigma converters**

Willy Sansen 10-05 2130

2130
The input switch is still a problem in the switched-opamp approach discussed before. Improved input switching is now required.

Moreover, the switching of the opamp causes transients which are unacceptable. Possible remedies are discussed next.

2131
The first remedy is to make everything fully differential. A fully-differential opamp is now used, as shown in this slide.

It is a two-stage Miller opamp with a folded cascode as a first stage. A second stage is needed to provide rail-to-rail output swing. The gain can be quite high, 75 dB in this example.

For a 1 pF load, the GBW is about 30 MHz, for only 80 µW power consumption. This gives a high FOM indeed!

The opamp can be switched in and out because of the four switches (in blue). Only the output stage is switched in and out, by transistors M11. The input stage remains on, which improves the settling time.

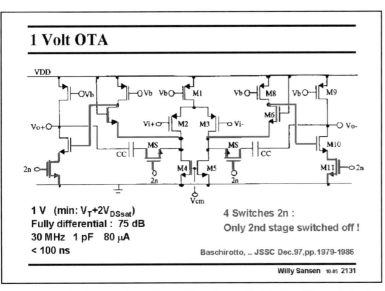

1 Volt OTA

VDD

1 V (min: $V_T + 2V_{DSsat}$)
Fully differential : 75 dB
30 MHz 1 pF 80 µA
< 100 ns

4 Switches 2n :
Only 2nd stage switched off !

Baschirotto, .. JSSC Dec.97,pp.1979-1986

Willy Sansen 10-05 2131

Of more importance, is that the compensation capacitance CC has a series switch MS. The voltage across CC is therefore constant irrespective of the switching. The settling time is greatly improved by this arrangement.

The minimum supply voltage V_{DD} is only $V_{GS} + V_{DSsat}$ or $V_T + 2 V_{DSsat}$. For a V_T of 0.6 V, a V_{DD} of 1 V is easily achieved ($V_{GS} - V_T = 0.2$ V).

Note that the inputs operate close to ground. As a result the whole supply voltage V_{DD} is available for V_{GS} of the input switch V_{GS}.

The average output voltage is about 0.5 V however, for maximum output swing. A level shifter is required between output and input.

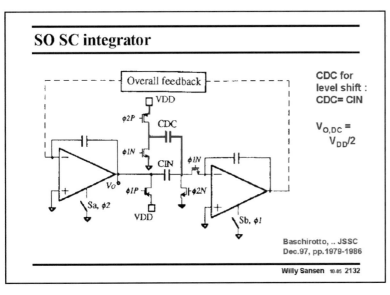

SO SC integrator

Overall feedback

VDD

CDC for level shift :
CDC= CIN

$V_{O,DC} = V_{DD}/2$

Baschirotto, .. JSSC Dec.97, pp.1979-1986

Willy Sansen 10-05 2132

2132

Such a level shifter is implemented by addition of capacitor CDC. This capacitor is taken to be the same as capacitor CIN. The DC output voltage V_O is now found to be half the supply voltage V_{DD}.

This is explained in more detail next. The circuit is repeated twice, once at phase $\Phi 1$ and once at phase $\Phi 2$.

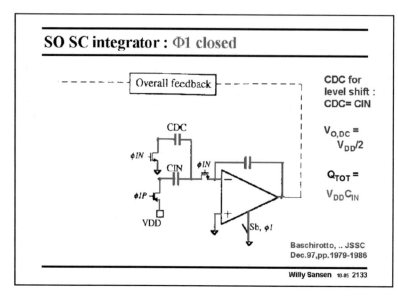

SO SC integrator : Φ1 closed

CDC for
level shift :
CDC= CIN

$V_{O,DC} = V_{DD}/2$

$Q_{TOT} = V_{DD}C_{IN}$

Baschirotto, .. JSSC
Dec.97,pp.1979-1986

Willy Sansen   10-05   2133

2133

Under phase Φ1, the first opamp is switched off and is therefore left out. Capacitor C_{DC} finds itself between ground and the virtual ground, at the minus input of the opamp. It carries no charge.

The total charge at the input is found on capacitor C_{IN}, which is between supply voltage V_{DD} and the virtual ground.

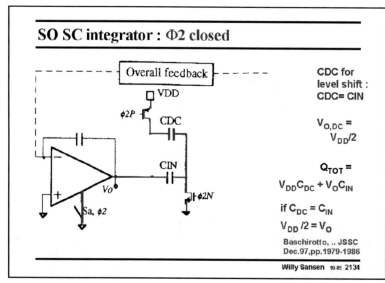

SO SC integrator : Φ2 closed

CDC for
level shift :
CDC= CIN

$V_{O,DC} = V_{DD}/2$

$Q_{TOT} = V_{DD}C_{DC} + V_OC_{IN}$

if $C_{DC} = C_{IN}$

$V_{DD}/2 = V_O$

Baschirotto, .. JSSC
Dec.97,pp.1979-1986

Willy Sansen   10-05   2134

2134

Under phase Φ2, the second opamp is switched off and is therefore left out. Capacitor C_{DC} now finds itself between the supply voltage V_{DD} and ground.

Capacitor C_{IN} now carries charge V_OC_{IN}. The total charge is therefore found, as given in this slide. If capacitors C_{DC} and C_{IN} are made equal, the DC output voltage V_O ends up at half the supply voltage.

This sets the output and input common-mode voltages. All switches now have the full supply-voltage V_{DD} as their V_{GS}. This is also true for the input switches.

2135

Similar level shifting is required in the Common-mode feedback circuit.

Again, a capacitor C_{CM} is added to provide 0.5 V level shifting, between the output of the first opamp and the input of the CMFB opamp. The outputs are sampled and summed to cancel out the differential output signals by capacitors C_P and C_M (see Chapter 8). The CMFB amplifier has a pMOST differential pair at the input such that its input voltage is close to ground as for the other opamps.

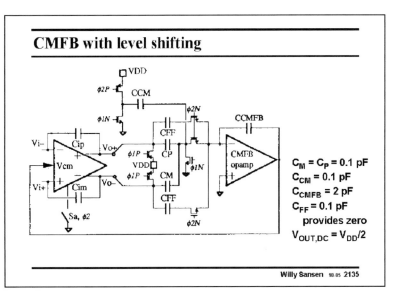

CMFB with level shifting

$C_M = C_P = 0.1$ pF
$C_{CM} = 0.1$ pF
$C_{CMFB} = 2$ pF
$C_{FF} = 0.1$ pF
provides zero
$V_{OUT,DC} = V_{DD}/2$

Willy Sansen 10.05 2135

Capacitors C_{FF} are in parallel with C_P and C_M to provide zeros in the CMFB gain characteristic, to speed up the CMFB performance without adding more power (see Chapter 8).

These circuits have been used to construct a second-order filter with a 450 kHz bandwidth (at 1.8 MHz clock frequency). It consumes only 160 µA at 1 V supply voltage (with V_TYs of 0.65 V in 0.5 µm CMOS). Its peak SNR is about 58 dB. The maximum input voltage is 1.6 V_{ptp}, which is very high indeed.

Table of contents

Willy Sansen 10.05 2136

2136
Even less power consumption has been reached when a class-AB amplifier is used as an output stage.

Moreover, some other design tricks can be applied to minimize the power consumption.

They are collected in this design example.

2137
The switched-opamp integrator is shown in this slide.

It is fully-differential. The common-mode feedback is carried out by means of capacitors $C_{CMS,eq}$. The CMFB circuitry itself is not shown.

The functions of the various capacitors is fairly obvious.

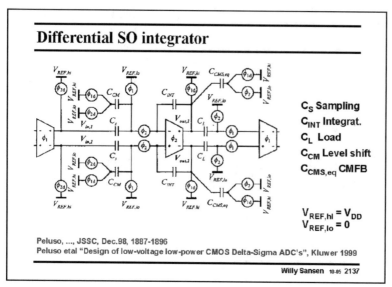

Differential SO integrator

C_S Sampling
C_{INT} Integrat.
C_L Load
C_{CM} Level shift
$C_{CMS,eq}$ CMFB

$V_{REF,hi} = V_{DD}$
$V_{REF,lo} = 0$

Peluso, ..., JSSC, Dec.98, 1887-1896
Peluso etal "Design of low-voltage low-power CMOS Delta-Sigma ADC's", Kluwer 1999

Willy Sansen 10-05 2137

A DC level shift is again required as the average output voltage is at 50% of the supply voltage and the average input voltage close to ground. This is again realized by means of capacitors C_{CM}.

Note that the switches are such that the outputs of the opamp are at V_{DD}, not to forward bias the junctions of the switch transistors.

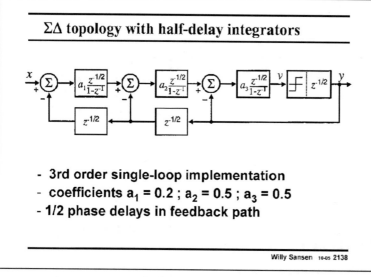

ΣΔ topology with half-delay integrators

- **3rd order single-loop implementation**
- **coefficients $a_1 = 0.2$; $a_2 = 0.5$; $a_3 = 0.5$**
- **1/2 phase delays in feedback path**

Willy Sansen 10-05 2138

2138

The schematic of the 3rd-order Delta-Sigma converter is shown in this slide. A single-loop topology is selected as it consumes somewhat less power than a 2–1 cascaded topology. Moreover, the gain requirements per stage are less severe.

The coefficients have been obtained from Matlab simulations. Care is taken to optimize the output swings for all integrators. For an input swing of 0.2 V, the output swings are respectively 0.36, 0.5 and 0.5 V.

In order to avoid an additional opamp to set the timing of the clocks right, half delays are introduced. They are digital and consume very little power.

2139

The opamp used is a class-AB amplifier as shown in this slide. Class AB is preferred as it lowers the quiescent current.

It contains two input transistors M1 and M2 and a low-voltage current source M2, M3 and M4. Transistors M2 serve as a source follower as its current is constant and equal to I_{B1}. This is ensured by the feedback loop around transistors M2 and M3. As a result V_{GS2} is constant as well. Input voltage V_{in2} is transferred unattenuated to its Source.

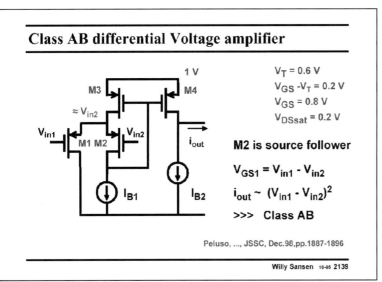

Input transistor M1 receives thus the differential input voltage $V_{in1} - V_{in2}$ as its V_{GS}. This voltage is converted into a signal current by transistor M1 only. In this amplifier, only one single transistor M1 converts the differential input voltage into a current. Moreover this transistor provides a class-AB characteristic.

The signal current flows through transistors M3 and is mirrored out by transistor M4. It can also be taken out at the Drain of transistor M1 as shown in the full schematic, shown next.

Finally, note that this amplifier can operate on less than 1 V supply voltage. It only takes one single V_{GS} and V_{DSsat} to operate properly. If a V_T is taken of 0.6 V and a V_{DSsat} of 0.2 V, then the minimum supply voltage V_{DD} is 1 V. If however, a V_T is available of 0.3 V, then V_{DD} can be as low as 0.7 V. This is quite low indeed!

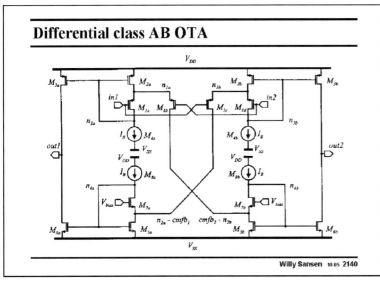

2140

In the full schematic, the differential nature is clearly recognized.

Input pMOST transistors M_{1b} and M_{1c} carry out the input voltage to current conversion. The signal current through transistor M_{1b} is now fed to the output by current mirror M_{2a} and M_{3a}, but also by current mirror M_{5b} and M_{6b}. A differential output current is therefore obtained.

Common-mode feedback is applied as currents injected at the Drains of transistors M_{5a} and M_{5b}.

The opamps can be switched in and out by disconnecting all connections to the power supply lines. This ensures a very fast recovery time.

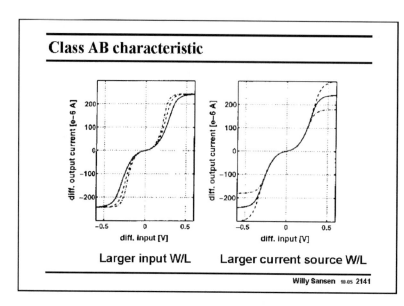

Class AB characteristic

Larger input W/L   Larger current source W/L

Willy Sansen 10.05 2141

2141

The class-AB characteristics are clearly visible in this slide. They give the differential output current versus differential input voltage. They also show a curved class-AB characteristic.

The biasing currents I_B are only of the order of 1 μA. The W/L ratio in the current mirrors is about 120. Relatively large output currents are now available.

In the first plot, several values are taken of the input transistors M_{1b} and M_{1c}.

The larger the input sizes, the steeper the characteristic.

In the other plot the biasing current I_B is varied. As a result, the maximum output current varies with it.

These two parameters allow shaping of the class-AB characteristic.

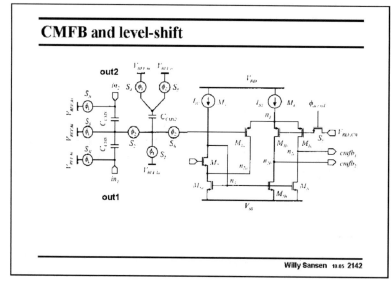

CMFB and level-shift

Willy Sansen 10.05 2142

2142

Since this is a fully-differential amplifier, CMFB is required.

This is affected by two sampling capacitors C_{CMS}, which cancel out the differential signal at their summation point, which is then fed to a differential amplifier. As two different currents are required to close the feedback loop, one side of the differential amplifier is doubled.

In order to realize the half-supply level shift, capacitor C_{CMS2} is added, with equal size as capacitor C_{CMS}.

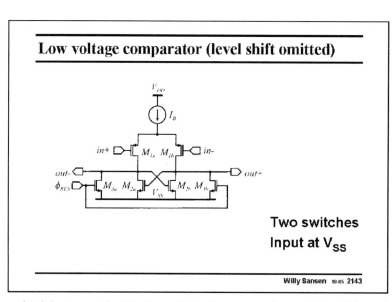

Low voltage comparator (level shift omitted)

Two switches

Input at V$_{SS}$

Willy Sansen 10.05 2143

2143

The comparator for the single-bit conversion is shown in this slide.

The comparator consists of a differential pair with transistors M1, loaded by a negative resistance (because of the positive feedback) with transistors M2. The gain is sufficiently high to cause a regenerative action yielding a logic on one side and a zero on the other. It consumes about 6 μA.

In such a comparator, normally a switch is required between the Drains of the input transistors M1. This switch is closed before the input voltage is applied. As soon as the switch is opened, the regenerative action causes a logic one or zero at the output, depending on the input signal.

Because of the low supply voltage however, such a switch is not possible. This function is taken up by the two switches M3. When they are switched off, regenerative action takes place.

Note that the average input voltage is again close to ground.

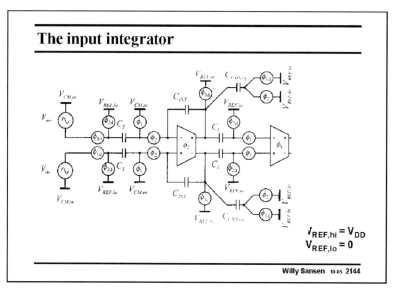

The input integrator

$I_{REF,hi} = V_{DD}$
$V_{REF,lo} = 0$

Willy Sansen 10.05 2144

2144

The input integrator does not have a preceding opamp which can be switched. The input signal will now be limited in amplitude by the input switch.

In this realization, the supply voltage is 0.9 V. The minimum voltage drive for the input switch is about 0.65 V, barely larger than the threshold voltage. As a result, the maximum differential input voltage is about 500 mV$_{ptp}$.

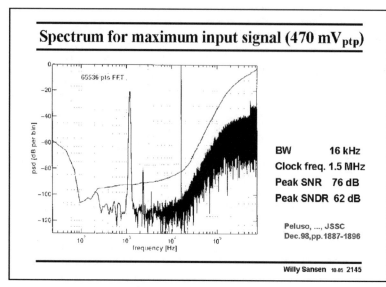

2145

The output spectrum for the maximum input signal is shown in this slide.

The noise floor at low frequencies is quite low. The integrated noise is shown as well. For an input signal of 1 kHz, the SNR is about 76 dB.

Distortion is already clearly visible, however. Second-order distortion is present because of the single-ended input drive used for this measurement. The SNDR (signal to noise and distortion ratio) is then only about 62 dB.

The bandwidth is about 16 kHz, corresponding to an OSR of 48.

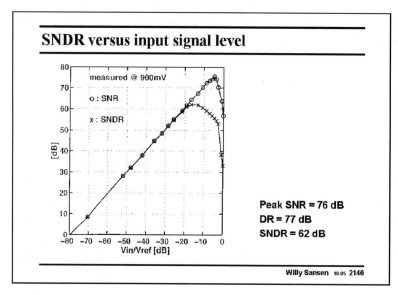

2146

Both the SNR and SNDR are given versus the input signal, for a supply voltage of 0.9 V.

It is clear that distortion shows up at the high end, limiting the SNDR to values about 14 dB below the maximum SNR.

2147

A microphotograph of this realization is shown in this slide. The first integrator uses larger capacitors (4 pF) to reduce the thermal kT/C noise. This means that for the first integrator the thermal noise is about equal to the quantization noise. For the other stages, the sampling capacitor is only 0.6 pF. A unit capacitor of 0.2 pF has been taken.

The power consumption has been minimized by reduction of the currents in the opamps to their minimum. The current consumption of the first stage is larger (33 μA) than for the other stages (6 μA). The current consumption of the CMFB amplifiers is about the same as for the differential ones.

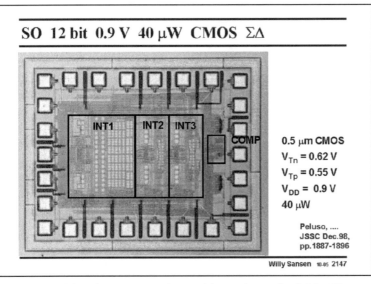

SO 12 bit 0.9 V 40 µW CMOS ΣΔ

INT1 INT2 INT3 COMP

0.5 µm CMOS
$V_{Tn} = 0.62$ V
$V_{Tp} = 0.55$ V
$V_{DD} = 0.9$ V
40 µW

Peluso, ...
JSSC Dec.98,
pp.1887-1896

Willy Sansen 10.05 2147

Their GBW values are only 2–3 times the clock frequency. They are all 4 MHz, to be compared with a clock frequency of 1.5 MHz.

This means that for the first integrator, the thermal noise is about equal to the quantization noise.

Remember that in a switched-opamp approach the current consumption is halved by itself.

As a result, the power consumption for this resolution and this bandwidth of 16 kHz is very low. This will be illustrated by the comparative Table at the end of this Chapter.

Table of contents

Willy Sansen 10.05 2148

2148
Several other low-voltage Sigma-Delta converters exist, which all address the problem of the series switches and in particular of the input switch.

They all operate at supply voltages below 1 V.

The operating principles are discussed next, followed by some discussion on their advantages and disadvantages.

The unity-gain-reset principle is first.

2149
In this principle, the opamp is always reset to ground. In this way, the input switch always has the maximum V_{GS}, even for small supply voltages.

In order to understand the principle, let us neglect the level shifters $V_{dd}/2$. Also, the voltage presented by the previous integrator is called V_{in}.

During phase 1 (red), the charge on capacitor C_1 is Q_1, which is $C_1 V_{in}$. The signal voltage at the output V_{out} is zero as this opamp is connected in unity-gain configuration (except for a DC level shift).

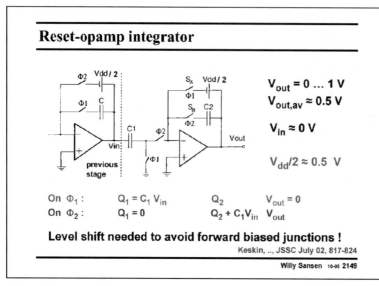

Reset-opamp integrator

$V_{out} = 0 \ldots 1\ V$

$V_{out,av} \approx 0.5\ V$

$V_{in} \approx 0\ V$

$V_{dd}/2 \approx 0.5\ V$

On Φ_1: $Q_1 = C_1 V_{in}$ Q_2 $V_{out} = 0$

On Φ_2: $Q_1 = 0$ $Q_2 + C_1 V_{in}$ V_{out}

Level shift needed to avoid forward biased junctions !

Keskin, .., JSSC July 02, 817-824

Willy Sansen 10-05 2149

During phase 2 (blue), this charge $C_1 V_{in}$ is transferred to capacitor C_2. The charge on this capacitor changes by an amount $C_1 V_{in}$ such that the output voltage V_{out} changes by an amount $V_{in} C_1/C_2$.

During this phase the previous opamp is now reset to zero, as it is connected in unity-gain configuration (except for a DC level shift).

Each opamp has zero at its output during one phase and the output voltage during the other phase.

There is no problem with switches which have to pass a high signal level, and which have too small a drive voltage V_{GS}.

The disadvantage however, is that the output voltages have to swing over a large range for each new clock phase. The Slew Rate must be quite high, leading to more power consumption in the opamps.

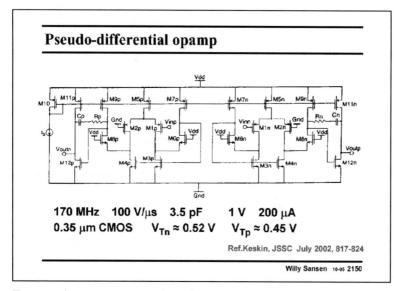

Pseudo-differential opamp

170 MHz   100 V/µs   3.5 pF   1 V   200 µA

0.35 µm CMOS $V_{Tn} \approx 0.52\ V$ $V_{Tp} \approx 0.45\ V$

Ref.Keskin, JSSC July 2002, 817-824

Willy Sansen 10-05 2150

2150

The opamp configuration is shown in this slide.

It is a pseudo-differential amplifier. It consists of two equal but separate amplifiers. Two stages are used for high gain and for large output swing.

The GBW and the SR are quite high in view of the clock frequency of 10 MHz. The power consumption is quite low.

Note that again DC level shifters are required between outputs and inputs.

For maximum output swing the average output voltage is 0.5 V (for a 1 V supply voltage) whereas the average input voltage is close to ground. Such a level shifter can be realized by a few switches and an additional capacitor, as shown before.

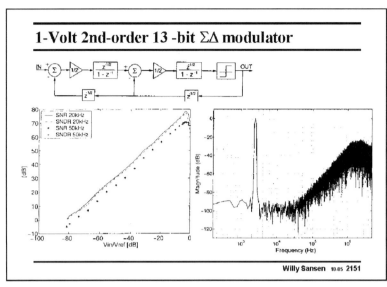

1-Volt 2nd-order 13-bit ΣΔ modulator

Willy Sansen 10-05 2151

2151

This integrator is used in a second-order sigma-delta modulator as shown in this slide. Half-delays are used for correct timing.

The input sampling capacitance is 2 pF. The maximum SNR is 78 dB for 20 kHz signal bandwidth (clock at 10 MHz). The total power consumption is 5.6 mW at 1 V supply voltage.

An 8 dB lower SNR is obtained for a 50 kHz input signal.

Note that the distortion is quite low, at the cost of high current consumption in the amplifiers, however.

Table of contents

- ♦ **Delta-sigma modulation**
- ♦ **The switch problem**
- ♦ **The switched-opamp solution**
- ♦ **Other low-power Delta-sigma converters**
  - • Unity-gain-reset
  - • Optimized input switching
  - • Switched input resistor
  - • Full feedforward

Willy Sansen 10-05 2152

2152

Another low-voltage, low-power Sigma-delta modulator is discussed next.

It has been optimized for low power and has been realized in 90 nm CMOS. A low-power opamp is used with a class-AB output stage.

2153

ΣΔ Modulator on 1 Volt in 90 nm CMOS

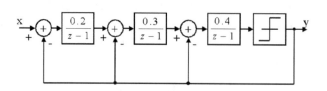

- ♦ **Single-loop third-order single-bit topology**
  - • **Simple and robust**
  - • **Tolerance to building block non-idealities**
- ♦ **Coefficients selected not sensitive to capacitance mismatches**

Yao, ..., JSSC Nov.04, 1809-1818
Yao. etal. "Low-Power Low-Voltage ΣΔ modulators in Nanometer CMOS", Springer '06

Willy Sansen 10-05 **2153**

A single-loop topology is preferred for low-power designs since it is less sensitive to circuit non-idealities such as low opamp gain, switch resistance and capacitor mismatch.

A clock is used of 4 MHz and an OSR of 100 to reach a signal bandwidth of 20 kHz.

The coefficients are such that the same output swing is obtained for all integrators, which is about 80% of the reference voltage (which is 0.6 V) at an input voltage of −3 dB.

The minimum-gain requirement is only about 30 dB. This is easily achieved by a low-voltage opamp as discussed next.

2154

Gain enhancement

$$A = \frac{2}{(1-k)(V_{GS} - V_T)_1 \cdot \lambda_3} = \frac{A_0}{1-k}$$

Willy Sansen 10-05 **2154**

As an opamp a symmetrical OTA is used.

Current starving is used however, to increase the gain (see Chapter 7). At such a low supply voltage, too little room is available for cascodes.

Current starving means that a DC current source takes away most of the DC current from the load transistors M2. As a result, the AC impedance of these transistors increase, giving rise to more gain. A large factor B is used as well.

The first amplifier of this sigma-delta converter uses a k value of 0.8 and a B factor of 10.

Actually, factor k indicates what fraction of the current through the input transistor M1 is taken up by the DC current source. The small-signal resistance $1/g_{m2}$ of transistors M2 increases by a factor $(1-k)$ in weak inversion, and so does the small-signal gain A.

Stability

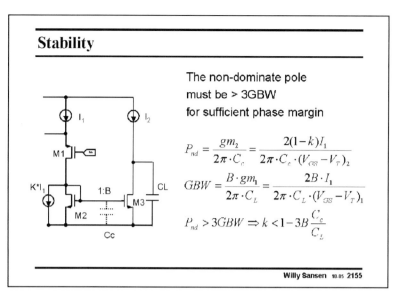

The non-dominate pole must be > 3GBW for sufficient phase margin

$$P_{nd} = \frac{gm_2}{2\pi \cdot C_c} = \frac{2(1-k)I_1}{2\pi \cdot C_c \cdot (V_{GS} - V_T)_2}$$

$$GBW = \frac{B \cdot gm_1}{2\pi \cdot C_L} = \frac{2B \cdot I_1}{2\pi \cdot C_L \cdot (V_{GS} - V_T)_1}$$

$$P_{nd} > 3GBW \Rightarrow k < 1 - 3B\frac{C_c}{C_L}$$

Willy Sansen 10-05 **2155**

2155

Care must be taken however, not to increase the small-signal resistance $1/g_{m2}$ too much. Indeed, a non-dominant pole p_{nd} is formed at this node.

This pole p_{nd} must be kept at 3 GBW to ensure sufficient phase-margin (see Chapter 5). As a result, an upper limit is established on the value of k.

Full OTA circuit

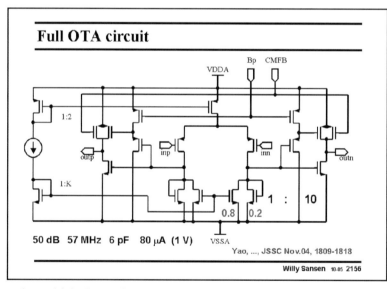

50 dB 57 MHz 6 pF 80 μA (1 V) VSSA

Yao, ..., JSSC Nov.04, 1809-1818

Willy Sansen 10-05 **2156**

2156

The total opamp schematic is shown in this slide.

The current-starving symmetrical OTA serves as a first stage of a two-stage amplifier. An output stage is required to reach rail-to-rail output swing. It is a class-AB amplifier at the same time.

It is a fairly simple class-AB amplifier with a main purpose of increasing the Slew-Rate, without too much distortion.

Node Bp is a fixed biasing point, which determines the class AB operating point. CMFB is the output of a separate CMFB amplifier, realized by switched capacitors.

For load capacitors of 6 pF, as used in the first integrator, the GBW is 57 MHz for only 80 μA current consumption.

The gain is about 50 dB, which is a lot more than 30 dB.

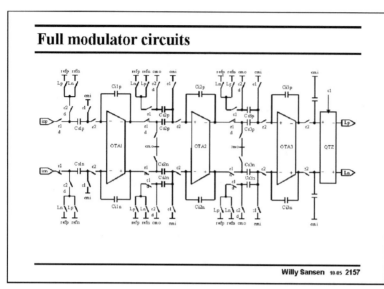

Full modulator circuits

2157

The full schematic of the sigma-delta modulator is shown in this slide. The first integrator takes most of the current as it is designed for 6 pF sampling capacitors. The other two integrators only use 0.4 pF capacitors. Their power consumption is much lower. All common-mode output voltages are 0.5 V, whereas the input ones are 0.2 V.

All switches are implemented as transmission gates. Despite the low supply voltage, sufficient V_{GS} is available for all switches. Indeed, the threshold voltages are about 0.3 V.

Vertical metal-wall capacitors are used with values of 1.7 fF/mm². No horizontal capacitors were available.

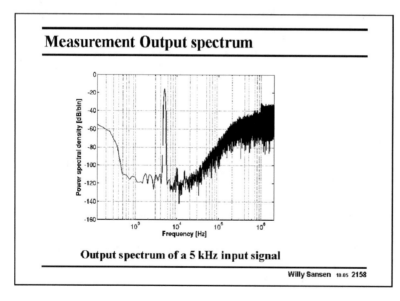

Measurement Output spectrum

Output spectrum of a 5 kHz input signal

2158

The output spectrum for a 5 kHz input signal is shown in this slide.

A peak SNR is reached of 85 dB while the peak SNDR is 81 dB.

The consumption is then 130 µA (at 1 V supply voltage) for the analog part

and 10 µA for the digital part.

Measured SNR and SNDR vs input amplitude

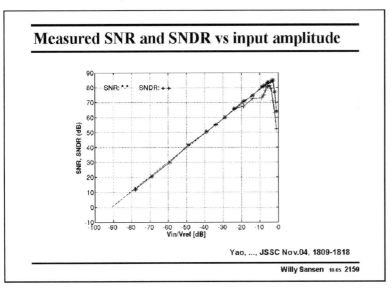

Yao, ..., JSSC Nov.04, 1809-1818

Willy Sansen 10.05 **2159**

2159

Both the peak SNR and SNDR are shown in this slide.

They reach values of 85 and 81 dB respectively.

The reference voltage is 0.6 V.

Table of contents

♦ **Delta-sigma modulation**

♦ **The switch problem**

♦ **The switched-opamp solution**

♦ **Other low-power Delta-sigma converters**

  • Unity-gain-reset

  • Optimized input switching

  • Switched input resistor

  • Full feedforward

Willy Sansen 10.05 **2160**

2160

Other low-power sigma delta converters are discussed. They again address the input switch to improve the SNDR ratio.

In this one, a series resistor is used at the input, which replaces the input switch.

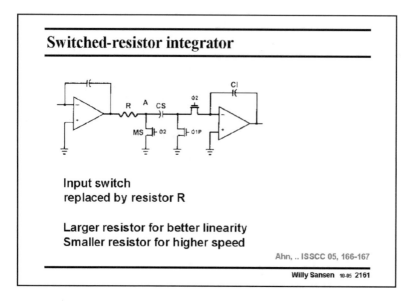

Switched-resistor integrator

Input switch
replaced by resistor R

Larger resistor for better linearity
Smaller resistor for higher speed

Ahn, .. ISSCC 05, 166-167

Willy Sansen 10-05 2161

2161

In the switched-capacitor integrator, shown in this slide, the input switch is replaced by a resistor R. On phase Φ1P the sampling capacitor C_S is charged to the output voltage of the first opamp. On phase Φ2 the charge is then transferred to integration capacitor C_I.

The use of this resistor R has several advantages. First of all, it avoids the need to use a switch. This switch is difficult to drive because of the low supply voltage.

Moreover, the linearity is quite high provided the ON-resistance of switch Φ1P can be made small.

The drawback is that an additional time constant RC_S comes in, which may limit the high-frequency performance. This is a low-frequency solution.

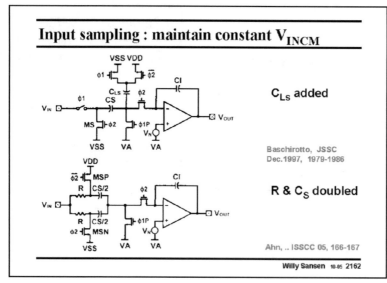

Input sampling : maintain constant V_{INCM}

C_{LS} added

Baschirotto, JSSC
Dec.1997, 1979-1986

R & C_S doubled

Ahn, .. ISSCC 05, 166-167

Willy Sansen 10-05 2162

2162

For low supply voltages, it is difficult to maintain the input common-mode voltage constant during the switching. In order to suppress this change in common-mode input voltage, several solutions exist.

In the top one, a capacitor C_{Ls} is added, connected to the supply lines. It acts as a DC level shifter, as previously described.

In the solution at the bottom, the input resistor R and sampling capacitor C_S are doubled, allowing the cancellation of the input common-mode voltage. This results in less kT/C noise.

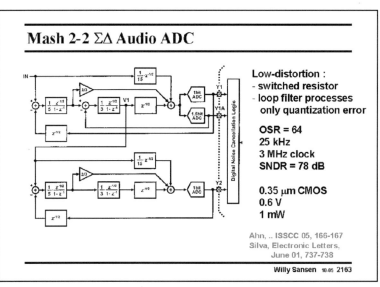

2163

The sigma-delta converter itself is a fourth-order MASH 2–2 converter. In the first second-order loop, the loop filter only processes the quantization error. The signal amplitudes are therefore much smaller and much lower distortion levels are obtained. As a result, the second stage does not contain the signal any more. No additional subtraction is now required in the coupling between both stages.

It is clear from the coefficients used, that the feedforward causes quantization noise only in the feedback loop.

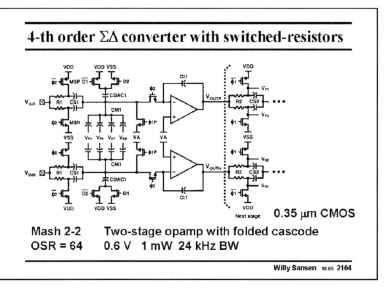

2164

The circuit implementation is shown in this slide.

The switched input resistors are easily found. Everything is fully differential. The CMFB is applied through the capacitors CM1.

On phase $\Phi 1P$ the signal is sampled onto sampling capacitors C_{S1}. On phase $\Phi 2$ the charge is then transferred to the integration capacitors C_{I1} by connecting the bottom plate of each capacitor to V_{DD} or V_{SS} (which is ground).

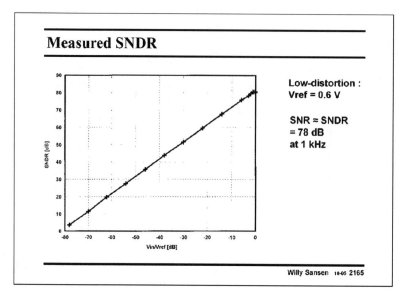

2165
The measured SNDR is shown in this slide. The SNR is the same, since the distortion is very low up to the reference voltage. This means that both the supply voltage and the reference voltage are 0.6 V. For an input voltage of 0.6 V_{ptp} a SNDR is obtained of 78 dB. This is quite impressive indeed.

The maximum signal bandwidth is about 24 kHz. It has been realized in 0.35 μm CMOS and consumes only 1 mW at 0.6 V.

Table of contents

♦ **Delta-sigma modulation**

♦ **The switch problem**

♦ **The switched-opamp solution**

♦ **Other low-power Delta-sigma converters**

　• Unity-gain-reset

　• Optimized input switching

　• Switched input resistor

　• Full feedforward

Willy Sansen 10-05 2166

2166
The main advantage of feedforward is that only quantization noise is processed by the feedback loop. As a result, the input swing can be made larger without distortion. This is also the case in the next and last sigma-delta modulator of this Chapter.

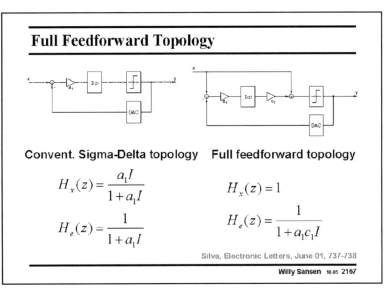

Full Feedforward Topology

Convent. Sigma-Delta topology

$$H_x(z) = \frac{a_1 I}{1 + a_1 I}$$

$$H_e(z) = \frac{1}{1 + a_1 I}$$

Full feedforward topology

$$H_x(z) = 1$$

$$H_e(z) = \frac{1}{1 + a_1 c_1 I}$$

Silva, Electronic Letters, June 01, 737-738

Willy Sansen 10-05 2167

2167

Full feedforward means that the input signal is fed directly to the quantizer. As a result, the noise transfer function $H_e(z)$ is the same as in a conventional topology but the signal transfer function $H_x(z)$ is unity. This suggests that the distortion is a lot less in the full-feedforward topology.

Indeed, the signal goes directly to the quantizer without passing through the loop filter integrators. These filters only process the quantization error, which is much smaller in amplitude than the signal itself. The distortion will be a lot less indeed. The overload level can be higher and so is the dynamic range.

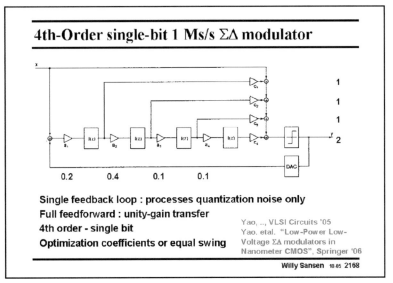

4th-Order single-bit 1 Ms/s ΣΔ modulator

0.2   0.4   0.1   0.1

Single feedback loop : processes quantization noise only
Full feedforward : unity-gain transfer
4th order - single bit
Optimization coefficients or equal swing

Yao, .., VLSI Circuits '05
Yao. etal. "Low-Power Low-
Voltage ΣΔ modulators in
Nanometer CMOS", Springer '06

Willy Sansen 10-05 2168

2168

A 4th-order single-bit sigma-delta converter with full feedforward is shown in this slide. Eight gain coefficients have to be determined, which is only possible by optimization of the SNR and by equalization of the swings at the integrator outputs. This is achieved by means of behavioral simulation. One of the many possible solutions is indicated.

Remember that the signal gain is unity and that the loop filters only process the quantization noise. This reduces the distortion considerably, as shown next.

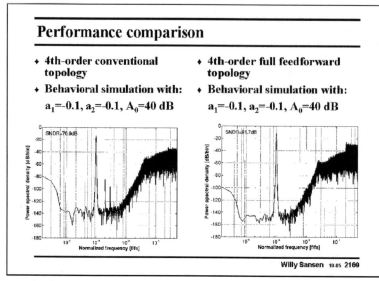

2169

First of all, the gain itself can be relaxed from 60 to 30 dB, which is easily achieved, even in nanometer CMOS at 1 V supply voltage.

Moreover, the effect of the distortion is much smaller.

The gain of each opamp is modeled as given by

$$A = A_0(1 + a_1 v_o + a_2 v_o^2)$$

in which the a_1 represents the second-order non-linearity and a_2 the third-order one.

It is shown in this slide that for equal nonlinearities in the opamps, full feedforward considerably reduces the distortion at the output. Only a modest gain of 40 dB is used. This is not sufficient for a conventional sigma-delta converter, but more than sufficient for the full-feedforward one.

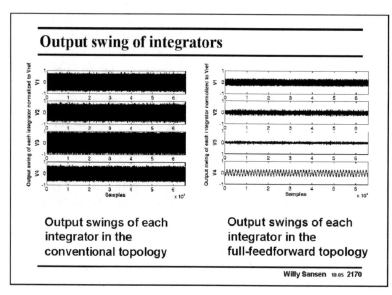

Output swings of each integrator in the conventional topology

Output swings of each integrator in the full-feedforward topology

2170

This is also evidenced by investigating the signal levels at the outputs of the four integrators.

In a conventional topology the signal swings are high at each output node. In a full-feedforward topology the signal swings are much smaller as they mainly contain the quantization errors. The distortion generated by these opamps will thererfore be a great deal smaller.

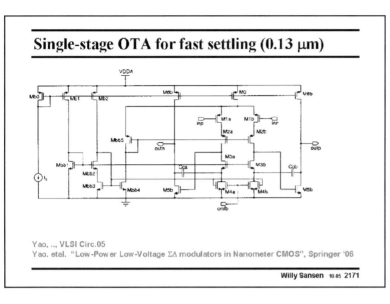

Single-stage OTA for fast settling (0.13 μm)

Yao, .., VLSI Circ.05
Yao. etal. "Low-Power Low-Voltage ΣΔ modulators in Nanometer CMOS", Springer '06

Willy Sansen 10-05 2171

2171

The opamps used in full-feedforward configuration can now have less gain; 30 dB is sufficient.

The opamp schematic is shown in this slide.

It is a conventional two-stage opamp with a telescopic cascode at the input. The settling time has been optimized. A switched-capacitor CMFB is used.

The first one reaches a GBW of about 200 MHz and consumes 4 mA at 1 V. The load capacitor is now about 3 pF.

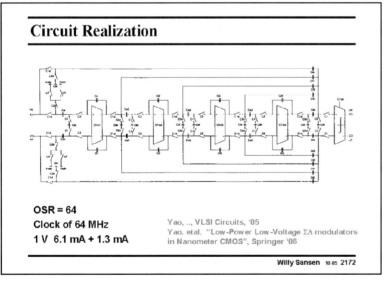

Circuit Realization

OSR = 64
Clock of 64 MHz
1 V   6.1 mA + 1.3 mA

Yao, .., VLSI Circuits, '05
Yao. etal. "Low-Power Low-Voltage ΣΔ modulators in Nanometer CMOS", Springer '06

Willy Sansen 10-05 2172

2172

The circuit schematic of the 4th-order sigma delta converter is shown in this slide.

The four stages are clearly seen. All switches are implemented as transmission gates. Since the threshold voltages are only about 0.35 V (in this 0.13 μm CMOS technology), there is no need for clock boosting circuitry.

The capacitors are realized with sandwich structures using five layers. As a result, the capacitance is $0.35 \, fF/\mu m^2$.

Clocked at 64 MHz, the modulator consumes 6.1 mA in the analog part and 1.3 mA in the digital part including the output buffer. The reference voltage is 0.8 V, with a supply voltage of 1 V. The oversampling ratio is 64, resulting in a 0.5 MHz maximum signal bandwidth.

Measured output spectrum

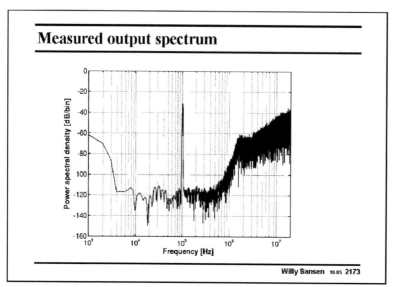

Willy Sansen 10.05 2173

2173

The output spectrum shows the response of a 100 kHz sinusoidal input signal. A peak SNR of 86 dB has been obtained.

The maximum signal bandwidth is 500 kHz.

Measured SNR versus Input voltage

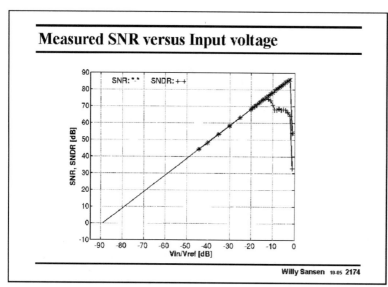

Willy Sansen 10.05 2174

2174

The maximum SNR is 86 dB but the maximum SNDR is reduced to about 75 dB. This is a result of the distortion generated in the feedforward switch, which has to handle the full input signal swing. This must be improved further.

2175

As a conclusion, a table is made up for comparison. A Figure of Merit is used as in Rabii (JSSC June 97, 783–796).

Only low-power sigma-delta converters are considered in the top list. After the name the year of publication (in the JSSC or ISSCC) is given. The type mentions which technique is used to arrive at 1 V supply voltage.

SwR stands for switched-resistor, SO for switched-Opamp, LV for reduced threshold voltage, VM for voltage multiplier.

The supply voltage is listed, followed by the dynamic range, bandwidth and power consumption.

Comparison of Low-power $\Sigma\Delta$ converters						
Ref.	**Type**	**V_{DD}**	**DR**	**BW**	**P**	**FOM**
		V	dB	kHz	µW	x 10^{-6}
Ahn 05	SwR	0.6	78	24	1000	20
Sauerbrey 02	SO,LV	0.7	75	8	80	53
Peluso 98	SO	0.9	77	16	40	330
Dessouky 01	LV	1	88	25	950	275
Keskin 02	ResetOp.	1	74	20	5600	6
Yao 04	LV	1	88	20	140	1490
Rabii 96	SC, VM	1.8	92	25	5400	121
Yin 94	211	5	97	750	180k	346
Geerts 00	211	5	92	1100	200k	144
Vleugels 01	221	2.5	95	2000	150k	700
Gaggl 04	4	1.5	88	300	8k	400
Yao 05	4	1	88	500	7.4k	706
Doerrer 05	Track	1.5	74	2000	3k	280
Hezar 05	5	1.3	86	600	5.4k	737

FOM = 4kT DR BW / P

Willy Sansen 10-05 2175

This shows that for supply voltages of 1 V and below, Yao04, Peluso98 and Dessouky01 are the best.

On the other hand, the lowest supply voltages have been reached by Ahn05 (0.6 V) and Sauerbrey02 (0.7 V). Quite often, this deals with a reduction of the absolute value of the threshold voltage, however.

In the second list, a series of high-frequency sigma-delta converters are added. It is clear that only a few of them operate at 1 V supply voltage or less. Moreover, the FOM's are in general, higher. This illustrates that compromises have to be taken to be able to reach a 1 V supply voltage or less.

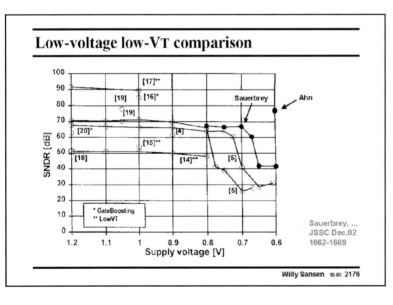

2176

If only the supply voltage itself is taken as a measure for comparison, the graph in this slide results.

Again, the lowest supply voltages have been reached by Ahn05 (0.6 V) and Sauerbrey02 (0.7 V). The two techniques used are also indicated, i.e. gate boosting and reduced threshold voltages.

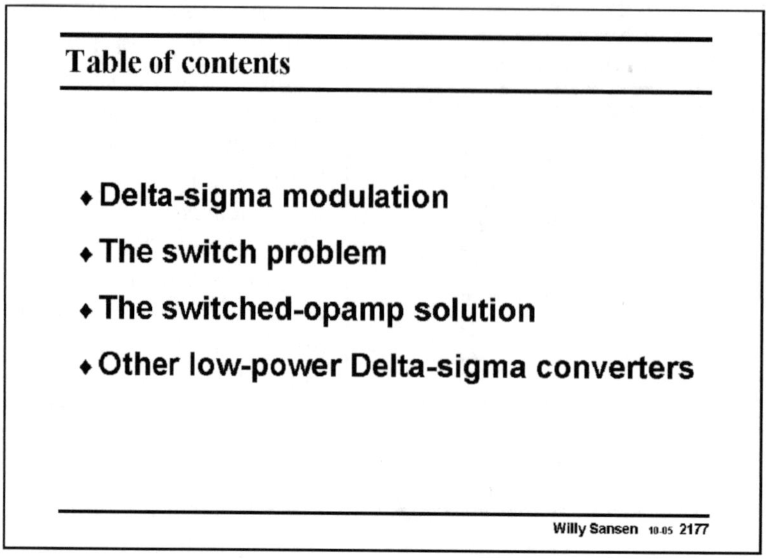

Table of contents

♦ **Delta-sigma modulation**

♦ **The switch problem**

♦ **The switched-opamp solution**

♦ **Other low-power Delta-sigma converters**

Willy Sansen 10.05 **2177**

2177

In this Chapter, an overview is given of the techniques which are available to reduce the supply voltage and the power consumption in sigma-delta converters.

The main problem at low supply voltages is the input sampling switch and the increase of the distortion. This can be remedied by the switched-opamp technique and also by a number of clever circuit techniques.

They have all been discussed and compared for a generally accepted FOM.

221

Design of crystal oscillators

Willy Sansen

KULeuven, ESAT-MICAS
Leuven, Belgium

willy.sansen@esat.kuleuven.be

Willy Sansen 10-05 221

Oscillators are required as references of frequency. All clocks are for (micro)processors but all timing circuits also need a clock.

The highest precision is obtained by using a crystal as a reference. Without a lot of effort, 0.1% precision is easily achieved.

This is why we start this chapter with crystal oscillators. It will be shown that only one single transistor is required to make a crystal oscillator.

222

Table of contents

- ◆ **Oscillation principles**
- ◆ **Crystals**
- ◆ **Single-transistor oscillator**
- ◆ **MOST oscillator circuits**
- ◆ **Bipolar-transistor oscillator circuits**
- ◆ **Other oscillators**

Willy Sansen 10-05 222

Before this oscillator is designed, we have to review the oscillation conditions.

After this we will have a look at what the electrical model is of such a crystal. It can now be used to construct a single-transistor oscillator with minimum power consumption.

Both MOST and bipolar transistors can be used to realize this kind of oscillator.

Finally, the principles of oscillation can be extended to construct other types of oscillators, such as VCO's, etc.

223

An oscillator is a kind of feedback amplifier. The signal that is fed back is exactly what the amplifier requires, to sustain oscillation. Its input is now zero.

This is called the Barkhausen criterion.

The amplifier has a gain $A(j\omega)$ which depends on frequency. Also, the feedback block has an attenuation $F(j\omega)$ which is frequency dependent. The loop gain $F(j\omega)A(j\omega)$ must be large enough so that the signal v_f which is fed back, exactly equals v_ε, which is required.

As a consequence, the loop gain must be slightly larger than unity in amplitude and its phase zero.

The Barkhausen criterion

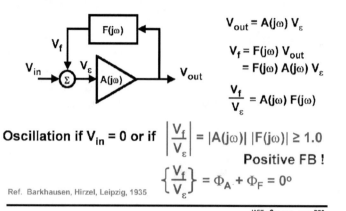

$$V_{out} = A(j\omega)\, V_\varepsilon$$

$$V_f = F(j\omega)\, V_{out}$$
$$= F(j\omega)\, A(j\omega)\, V_\varepsilon$$

$$\frac{V_f}{V_\varepsilon} = A(j\omega)\, F(j\omega)$$

Oscillation if $V_{in} = 0$ or if $\left|\dfrac{V_f}{V_\varepsilon}\right| = |A(j\omega)|\,|F(j\omega)| \geq 1.0$

Positive FB !

$$\left\{\frac{V_f}{V_\varepsilon}\right\} = \Phi_A + \Phi_F = 0°$$

Ref. Barkhausen, Hirzel, Leipzig, 1935

Willy Sansen 10-05 **223**

This means that $A(j\omega)$ must be an amplifier if $F(j\omega)$ is an attenuator. This also means that $F(j\omega)$ must be inductive if $A(j\omega)$ is capacitive. All amplifiers that we have seen contain capacitances. As a consequence, we are looking for an inductor for $F(j\omega)$.

Clearly, these two conditions are a result of the complex nature of both $A(j\omega)$ and $F(j\omega)$. Complex numbers are always pairs of numbers!

Split analysis

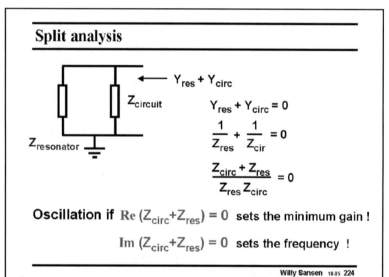

$$Y_{res} + Y_{circ}$$

$$Y_{res} + Y_{circ} = 0$$

$$\frac{1}{Z_{res}} + \frac{1}{Z_{cir}} = 0$$

$$\frac{Z_{circ} + Z_{res}}{Z_{res}\, Z_{circ}} = 0$$

Oscillation if $\mathrm{Re}\,(Z_{circ} + Z_{res}) = 0$ **sets the minimum gain !**

$$\mathrm{Im}\,(Z_{circ} + Z_{res}) = 0 \text{ **sets the frequency !**}$$

Willy Sansen 10-05 **224**

224
Another way to write the Barkhausen criterion is given by the split analysis.

The amplifier is now represented by the impedance $Z_{circuit}$ and the feedback element by the resonator impedance.

As the circuit maintains oscillation by itself, no current is needed from outside. Its total input admittance is zero. The sum of the impedances must also be zero.

The Barkhausen criterion can now be stated as given by the two expressions in this slide. Rather than the amplitude and phase, the Real and Imaginary parts are used now. They are obviously related as reminded in the Appendix.

We will see later, that the first expression determines the minimum gain required, and the other the actual frequency of oscillation.

Willy Sansen 10-05 225

225

Now that the conditions for oscillation are understood, let us find inductors to form a feedback loop with capacitive amplifiers. We know that they can form an oscillator together.

The first inductor we will use is the one embedded in a crystal.

Crystal as resonator

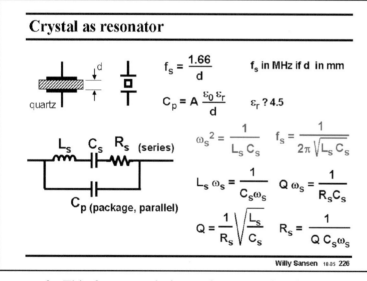

$$f_s = \frac{1.66}{d} \qquad f_s \text{ in MHz if d in mm}$$

$$C_p = A \frac{\varepsilon_0 \varepsilon_r}{d} \qquad \varepsilon_r \approx 4.5$$

quartz

$L_s \quad C_s \quad R_s$ (series)

C_p (package, parallel)

$$\omega_s^2 = \frac{1}{L_s C_s} \qquad f_s = \frac{1}{2\pi \sqrt{L_s C_s}}$$

$$L_s \omega_s = \frac{1}{C_s \omega_s} \qquad Q \omega_s = \frac{1}{R_s C_s}$$

$$Q = \frac{1}{R_s} \sqrt{\frac{L_s}{C_s}} \qquad R_s = \frac{1}{Q C_s \omega_s}$$

Willy Sansen 10-05 226

226

A crystal consists of a plate of piezoresistive material with a certain thickness d. Piezoresistive material allows exchange of mechanical and electrical energy. Examples are quartz and ZnO and some nitrides. Application of a mechanical pressure to it, generates a voltage across it and vice versa.

This exchange of energy is particularly efficient at one particular frequency, called the resonant frequency f_s. This frequency is inversely proportional to the thickness of the quartz. Values of 100 kHz to 40–50 MHz are commonly fabricated. For higher values, the quartz becomes too thin and fragile.

Around this resonant frequency the electrical model of this crystal is a series resonant LRC circuit, the resonant frequency of which is f_s. It is damped by the series resistor R_s, which causes the quality factor Q to be finite. All relevant expressions are given in this slide.

Note that at resonance, the impedance of the inductor equals that of the capacitance. Actually, at resonance the series RLC circuit is just R_s, itself. The inductor L_s and capacitor C_s cancel each other.

In addition to this series RLC circuit, which represents the electro-mechanical operation of the crystal, a plate capacitance C_p has to be added. It is the capacitance between the two plates used to contact the crystal, with the dielectric constant of quartz (4.5 times larger than air). It includes the package and mounting capacitances as well.

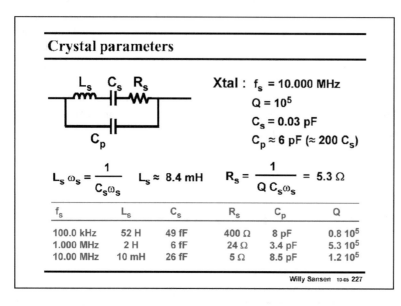

227
For example, a crystal of 10.00 MHz can be modeled by a series LRC with an inductance of about 10 mH in series with 26 fF and a damping resistor of 5 Ω. Note that the inductors are fairly large and the capacitors very small. A rule of thumb says that they are about 1/200 to 1/250 of the package capacitance C_p. Plate or package capacitances are always of the order of magnitude of pF's. The series capacitance C_s is always in the fF range.

The resistors are very small because the quality factors Q are so high, of the order of 10^5!

It is clear that the package capacitance C_p also forms a parallel resonant circuit with L_s. We have both a series and a parallel resonant circuit!

We will try to make an oscillator at the series resonant frequency f_s however, as this is the internal crystal frequency, independent of the package or mounting.

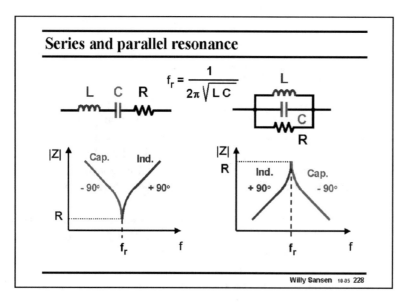

228
To recall what series and parallel resonance means, both are sketched in this slide.

Both have the same expression for the resonant frequency f_r.

The impedances versus frequency look very, different however.

A series resonant circuit has a sharp null at resonance. At resonance the impedance is reduced to the resistor R. The crystal is purely resistive.

For frequencies lower than f_r, the impedance of the capacitor increases and therefore determines the current. The impedance is capacitive, and the phase becomes $-90°$. For higher frequencies, the impedance turns inductive, and its phase 90°.

A parallel resonant circuit has a sharp peak at resonance. At resonance, the impedance is limited by the resistor R. The crystal is again purely resistive.

For frequencies lower than f_r, the impedance of the inductor decreases and therefore determines the current. The impedance is inductive, and the phase becomes 90°. For higher frequencies, the impedance turns capacitive, and its phase $-90°$. This is exactly the opposite compared to a series resonant circuit.

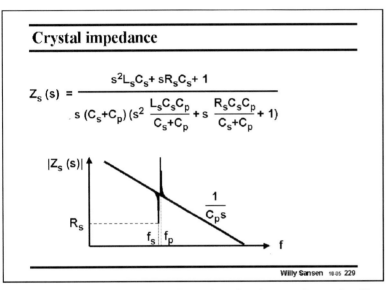

Crystal impedance

$$Z_s(s) = \frac{s^2 L_s C_s + s R_s C_s + 1}{s\,(C_s + C_p)\,\left(s^2\,\dfrac{L_s C_s C_p}{C_s + C_p} + s\,\dfrac{R_s C_s C_p}{C_s + C_p} + 1\right)}$$

Willy Sansen 10-05 229

229

Let us now return to the crystal and plot its impedance versus frequency. It is given in this slide.

It is clearly described by a third-order expression and is plotted below.

In general it shows the impedance of the package capacitor C_p. It decreases with frequency indeed.

Around the resonant frequency, we notice a null and a peak however, very close together. The null comes first and will represent the

series resonance with resonant frequency f_s, whereas the peak will represent the parallel resonance. Let us zoom in, in this region.

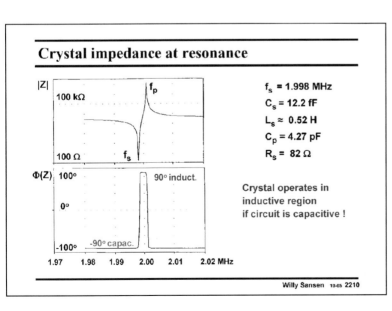

Crystal impedance at resonance

f_s = 1.998 MHz
C_s = 12.2 fF
$L_s \approx$ 0.52 H
C_p = 4.27 pF
R_s = 82 Ω

Crystal operates in inductive region if circuit is capacitive !

Willy Sansen 10-05 2210

2210

Now the resonant frequencies are easily distinguished. The series resonant frequency f_s is the smaller one.

The values have actually been calculated for the crystal given.

The top diagram shows the amplitude, whereas the bottom one shows the phase.

The crystal behaves as a capacitor at the left of the series resonant frequency f_s and at the right of the parallel resonant frequency f_p as explained before.

However, between both resonant frequencies the crystal behaves as an inductor. The transitions are very steep because the quality factor is so high. The crystal now behaves as an inductor from the series resonant frequency f_s to the parallel resonant frequency f_p.

We will use this inductor to make an oscillator together with a capacitive amplifier. We want this oscillator to operate as close as possible to the series resonant frequency f_s as this is the frequency which is the closest to the internal electromechanical operation of the crystal. Moreover, it is the frequency which is the least dependent on the package and mounting capacitances, which are hard to predict.

We will see however, that it is impossible to make an oscillator at the series resonant frequency f_s. It would take infinite current! We will try to be as close as possible however, depending on the current that we are allowed to use. This will be the only free design choice!

2211

The actual expressions of the series resonant frequency f_s and the parallel resonant frequency f_p are given in this slide. The series one is as described in this slide. The parallel resonant frequency f_p on the other hand is determined by both capacitances in series, as clearly found by inspection of the model. This frequency is always somewhat larger than the series one. If we find that C_p is about 200 times larger than C_s, then

Series and parallel resonance

$$Z_s(\omega) = \frac{-j}{\omega C_p} \frac{\omega^2 - \omega_s^2}{\omega^2 - \omega_p^2} \qquad \omega_s^2 = \frac{1}{L_s C_s} \qquad \omega_p^2 = \frac{1}{L_s}\left(\frac{1}{C_p} + \frac{1}{C_s}\right)$$

series   parallel

$$Z_s(\omega) = R_s + j\omega L_s + \frac{1}{j\omega C_s}$$

$$Z_s(\omega) = R_s + \frac{j}{\omega_s C_s}\left(\frac{\omega}{\omega_s} - \frac{\omega_s}{\omega}\right)$$

Frequency pulling factor

$$p = \frac{\omega - \omega_s}{\omega_s}$$

$$Z_s(\omega) \approx R_s + j\frac{2p}{\omega C_s}$$

Ref. Vittoz, JSSC June 88, 774-783

Willy Sansen 10-05 2211

ω_p^2 is approximately 0.5% larger than ω_s^2. Also, ω_p is now about 0.25% larger than ω_s.

The impedance of the series RLC circuit can now be described as given in this slide. We will rewrite this impedance now after introduction of the pulling factor p.

This pulling factor p is the dimensionless parameter which says how far the actual operating frequency is from the series resonance frequency f_s.

Introduction of this factor p and of the frequency f_s, gives another expression of the impedance of the series RLC circuit. It says that this impedance is a resistor R_s in series with an inductor. This inductor is larger the more we deviate from series resonance. We will use this simple model to find the oscillation condition of the oscillator.

2212

The terms series or parallel resonance are often attributed to the circuit configuration. We will see that this is wrong, however!

Series and parallel resonance have only to do with the operating point of the oscillator.

For zero or very small pulling factor p, the operating point is close to f_s. In this case, we clearly have a series oscillator. For a fairly large p, the operating point is close to f_p. We will

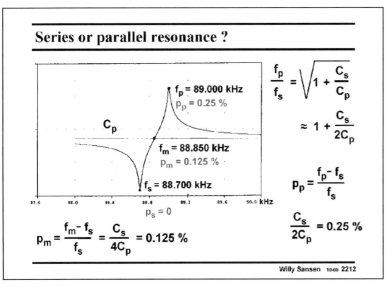

Series or parallel resonance ?

f_p = 89.000 kHz
p_p = 0.25 %

C_p

f_m = 88.850 kHz
p_m = 0.125 %

f_s = 88.700 kHz
p_s = 0

$$\frac{f_p}{f_s} = \sqrt{1 + \frac{C_s}{C_p}}$$

$$\approx 1 + \frac{C_s}{2C_p}$$

$$p_p = \frac{f_p - f_s}{f_s}$$

$$\frac{C_s}{2C_p} = 0.25 \%$$

$$p_m = \frac{f_m - f_s}{f_s} = \frac{C_s}{4C_p} = 0.125 \%$$

Willy Sansen 10-05 2212

call this a parallel oscillator, although C_s is still 200 times more important than C_p for the determination of the oscillation frequency.

The point in the middle is the crossover value f_m, between series and parallel resonance. Since at f_p we had a pulling factor p of about 0.25%, the point in the middle must have a pulling factor of about half or 0.125%.

We can now conclude that the pulling factor p must be less than about 0.1%, to have a oscillator with good stability and predictability!

Table of contents

Willy Sansen 10-05 2213

2213
Let us now apply the oscillation condition to the combination of a single-transistor amplifier, which is capacitive, and a crystal, which is inductive.

The main question is, which circuit components to use to set the pulling factor at sufficiently low values.

Clearly one single transistor can provide sufficient gain to carry out this task. For a differential operation we would need two transistors. Obviously if one transistor can do it, many more transistors can also do it. Many transistor oscillators now exist. There is only one single-transistor oscillator, however.

2214
This circuit is shown on top. The crystal is connected between Drain and Gate to provide gain. The biasing components are omitted.

This basic single-transistor oscillator gives rise to three different oscillator circuits depending on which node is connected to ground. The output terminals obviously also depend on which node is connected to ground.

In the Pierce oscillator, the source is grounded such that the transistor seems to function as

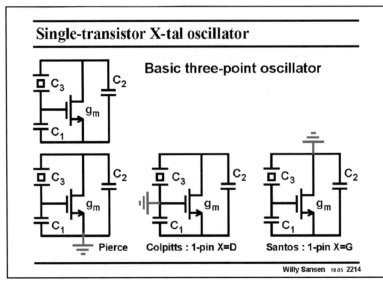

an amplifier. At resonance, the crystal behaves as a small resistor. The voltages at the Drain and at the Gate are almost identical. The output can therefore be taken either at the Gate or at the Drain.

In the Colpitts oscillator, the Gate is grounded. The crystal is also grounded. It is a single-pin oscillator with the crystal connected to the Drain. The transistor looks more like a source follower. The output can only be taken at the Source. Indeed the Drain only carries a very small signal as it is connected to ground by the crystal, which behaves as a small resistor.

In the third oscillator, the Drain is grounded. It is a single-pin oscillator with the crystal connected to the Gate. The transistor looks more like a cascode. The output can only be taken at the Source again. The Gate is connected to ground by the crystal, which behaves as a small resistor. The current output can also be taken by insertion of a current mirror in the Drain to ground connection.

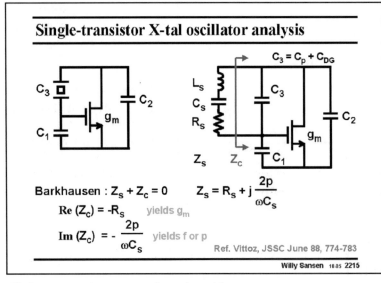

2215

The analysis of the oscillation condition now applies to all three configurations.

For this purpose, the package capacitance C_p is included in capacitance C_3. The transistor capacitance C_{GS} is included in C_1 and the total output capacitance is C_2.

What value of g_m is required for sustained oscillation?

We carry out a split analysis with the crystal on one side and the circuit with all three capacitances on the other side.

Barkhausen requires the sum of both impedances to be zero, which can be written in both the Real and Imaginary part.

Since the Real part of the crystal impedance is only R_s, we find that the circuit must present a negative resistance $Re(Z_c)$ equal to $-R_s$. This will yield a minimum value of transconductance g_m.

Also, the imaginary part of the crystal must equal the negative Imaginary part of the circuit $Im(Z_c)$. Since this is an inductor, the circuit must present a capacitance, as expected. This expression will yield the actual oscillation frequency or the pulling factor p.

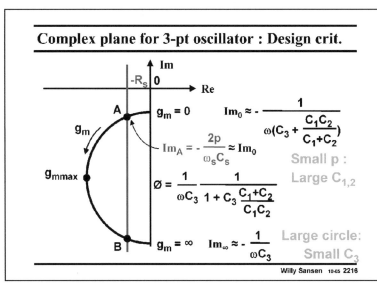

2216

To avoid having to solve complicated expressions, we resort to a graphical method. The impedance Z_c presented by the circuit is plotted in the polar diagram.

It is a half circle (see Appendix Polar diagrams) which starts on the Imaginary axis for zero g_m and ends up on the same axis for infinite g_m. Where this half circle intersects the line of $-R_s$, the first Barkhausen oscillation condition is satisfied. Indeed, in this point A, the circuit presents a negative resistance $-R_s$, which exactly compensates the damping resistor R_s of the crystal. Oscillation is therefore guaranteed.

Moreover, the second Barkhausen condition teaches us that in point A, the Imaginary part gives the pulling factor and therefore the actual frequency. This Imaginary part is very close however, to the Imaginary part at zero g_m. This latter one is easily calculated as it is a combination of capacitances (see Appendix Polar diagrams).

Stable oscillation is achieved for a well defined crossing point A. The circle must be large, which now requires a small capacitor C_3.

For a small pulling factor p, we need to have large capacitors C_1 and C_2. They are usually taken at the same. Large capacitors $C_{1,2}$ will require a large current, however. This is the compromise to be taken!

2217

A numerical example is shown in this slide. Also, the expressions of the pulling factor p_A and the transconductance g_{mA} at point A are added (see Appendix Polar diagrams).

Capacitance C_3 is taken to be as small as possible. It cannot be smaller than the crystal package capacitance, however!

The circle has a diameter of $12\,k\Omega$. For zero g_m, the Imaginary part is $-4\,k\Omega$. Point A is actually much closer to the Imaginary axis than drawn, since R_s is relatively small.

At point A, the transconductance g_{mA} is $11\,\mu S$, and increases along the half circle to infinity. It is obviously proportional to the damping resistor R_s, and also to the square of capacitor C_1 ($=C_2$) and the series resonance frequency. The realization of a GHz oscillator will require a large current!

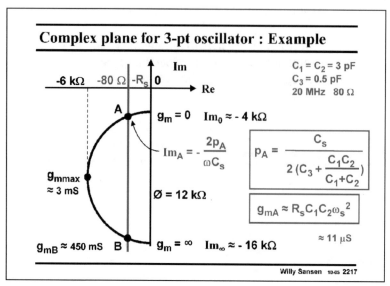

The pulling factor p_A is about C_s/C_1. It can only be made small by increasing C_1, which increases g_{mA} considerably indeed.

The choice of C_1 ($= C_2$) is the only design choice to be made. It sets at the same time the pulling factor and the current consumption.

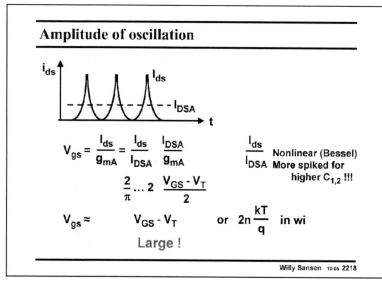

2218
What amplitude can now be expected from such an oscillator? Remember that the amplitude at the Gate and Drain (with the Source as reference) are just about equal, as Gate and Drain are nearly shorted together by a small resistor R_s.

The Gate voltage is related to the Drain current by the transconductance. Separating DC and AC components gives an expression, in which the peak-to-average current ratio I_{ds}/I_{DSA} appears and the $V_{GS} - V_T$.

This ratio depends on how much the transistor is overdriven. For large overdrive or large oscillation amplitude, the current is very nonlinear. This ratio can now be fairly large.

The transistor must certainly be designed for large $V_{GS} - V_T$. This is a problem for bipolar circuits unless some emitter resistors are inserted. Also, weak inversion operation yields only small signal voltages.

Start-up of oscillation

τ_{min} occurs at $g_m \approx g_{mmax}$

$$\tau_{min} = \frac{L_s}{Re\,(Z_s) + R_s}$$

Re (Z_s) is half circle Ø

$$Re\,(Z_s) = \frac{1}{2}\,\frac{1}{\omega_s C_3} \quad \text{if } C_3 << C_1$$

$$R_s << Re\,(Z_s)$$

$$\tau_{min} \approx \frac{2\,C_3}{\omega_s C_s} \approx \frac{400}{\omega_s} \quad \text{since } C_3 \approx 200\,C_s$$

or also $\quad \tau_{min} \approx 2Q\,R_s\,C_3$

Willy Sansen 10-05 2219

2219

The startup time constant is determined by the inductor and by the negative resistance, seen by it. The minimum time constant is obtained for the maximum resistance $Re(Z_c)$ at g_{mmax}. This is the radius of the half circle.

Substitution of the inductance yields an expression which shows that the startup time constant is about 400 periods. We have assumed that the minimum C_3 equals the package capacitance, which is about 200 times C_s.

In order to reach quiescent operation (within 5%), about 3 time constants are required or 1200 periods. A crystal oscillator is therefore a very slow starter!

This is the same as saying that crystal oscillators have very high quality factors!

Power dissipation

In MOST : $\quad g_{mA} \approx \omega_s^2 R_s C_1 C_2 \approx R_s (C_1 \omega_s)^2$

$$I_{DSA} \approx g_{mA}\,\frac{V_{GS} - V_T}{2} \quad \approx 2\,\mu A \implies 6\,\mu W$$

In X-tal : $\quad I_c = \dfrac{V_{gs}}{Z_{C1}} = |V_{gs}|\,C_1 \omega_s \approx |V_{GS} - V_T|\,C_1 \omega_s$

$$P_c = \frac{R_s I_c^{\,2}}{2} = \frac{R_s}{2}\,|V_{GS} - V_T|^2\,(C_1 \omega_s)^2$$

$$= |V_{GS} - V_T|^2\,\frac{g_{mA}}{2} \quad \approx 0.2\,\mu W$$

Willy Sansen 10-05 2220

2220

The power dissipation in the transistor and in the crystal depend to a large extent on the values of the capacitors chosen.

For capacitors C_1 in the order of magnitude of pF's, the currents are quite small. The power dissipation is then quite small and no self heating is expected.

Only when large values of C_1 are selected, to be as close as possible to the series resonance frequency, the currents rise with the square of the capacitance value. The power dissipation then rises quickly as well.

Design procedure for X-tal oscillators - 1

X-tal : f_s f_p R_s C_p (or f_s Q C_s C_p) ($Q = 1/\omega_s C_s R_s$)

1. Take : $C_3 > C_p$ but as small as possible

$$\text{Pulling factor } p = \frac{1}{2} \frac{C_s}{C_3 + \dfrac{C_1 C_2}{C_1 + C_2}} \approx \frac{1}{2} \frac{C_s}{C_L} \qquad C_L = \frac{C_1}{2} = \frac{C_2}{2}$$

If $p < \dfrac{C_s}{4C_p}$ it is a series oscillator (best !)

If $p >$ it is a parallel oscillator (not stable !)

Choose C_L large ($> C_3$), subject to power dissipation !

Willy Sansen 10-05 **2221**

2221
From the discussion, it is easy to derive a design procedure.

A crystal is characterized by its two resonance frequencies, the series resistor and the package capacitance, which are all easily measured. We first choose the capacitor C_3 as small as possible. Obviously, it cannot be smaller than the package capacitance C_p.

We now have to select a value for $2C_L = C_1 = C_2$ as a compromise to reduce the pulling factor and to avoid too much power consumption.

Design procedure for X-tal oscillators - 2

2. Calculate $g_{mA} \approx R_s C_L^2 \omega_s^2$ ($\approx \dfrac{\omega_s}{C_s Q} C_L C_L$)

    and take $g_{mStart} \approx 10\, g_{mA}$

3. Choose $V_{GS} - V_T$, which gives the amplitude V_{gs}

    and current $I_{DS} = \dfrac{g_m(V_{GS} - V_T)}{2}$ and $\dfrac{W}{L}$

    and power $P = (V_{GS} - V_T)^2 \dfrac{g_m}{2}$

4. Verify that biasing $R_B > 1/(R_s C_3^2 \omega_s^2)$

Willy Sansen 10-05 **2222**

2222
The required transconductance is then easily calculated.

As a starting value we take about ten times more transconductance.

The $V_{GS} - V_T$ will determine the current itself and the output amplitude. A large value is selected.

Finally, if we want to connect the Gate of the oscillator to a biasing voltage over a resistor R_B, we have to make sure that the additional damping due to this resistor is negligible. It must be sufficiently large. This expression is taken from Ref. Vittoz (see slide 15).

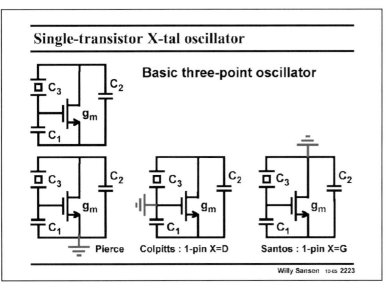

2223
Let us now consider how these three types of grounding can be realized. Attention is paid to biasing as well.

Simple discrete realizations are considered first.

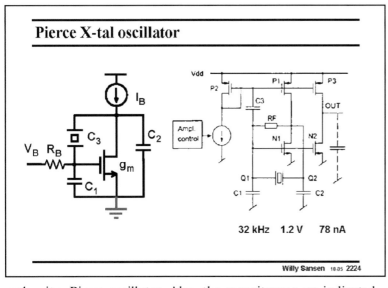

2224
A Pierce oscillator is best biased by a current source I_B and a Gate biasing resistor V_B.

This current source is the best guarantee that the circuit is isolated from the supply line. Otherwise, the oscillator will superimpose spikes at the oscillation frequency on all supply lines connected to it.

A CMOS realization is shown on the right. The crystal is connected between Drain and Gate, which makes it a Pierce oscillator. Also, the capacitances are indicated.

The current source I_B is set by an external current, which is determined by an Automatic-Gain-Control system. Such a system measures the output signal and adjusts the DC current to keep the oscillator at point A.

The output signal is taken by duplicating the amplifier transistor. In this way the load at the output cannot influence the oscillation condition.

One subtlety is capacitor C3. Actually, it is in parallel with capacitor C1 and smoothens somewhat the output current.

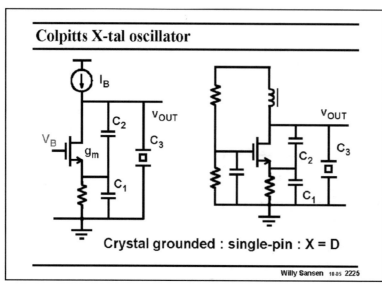

Colpitts X-tal oscillator

Crystal grounded : single-pin : X = D

Willy Sansen 10.05 2225

2225

An example of a Colpitts oscillator is shown in this slide. Indeed it is a single-pin oscillator with the crystal connected to the Drain.

Again, it is biased by a current source I_B to provide adequate isolation from the supply line.

A discrete equivalent is shown on the right. The current source is replaced by a "choke" or large inductor. The capacitors are exactly where they are expected.

The output voltage is taken at the Drain, as shown by most publications. It may not be the best place, however. At resonance, the crystal is a little more than a small resistor. The signal swing at the Drain is therefore quite small.

A better output node is the Source or Emitter, between capacitors C_1 and C_2.

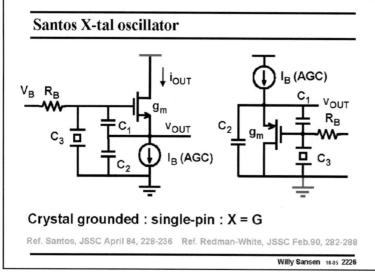

Santos X-tal oscillator

Crystal grounded : single-pin : X = G

Ref. Santos, JSSC April 84, 228-236   Ref. Redman-White, JSSC Feb.90, 282-288

Willy Sansen 10.05 2226

2226

The third type of oscillator is shown in this slide. Some more recent references are added on these integrated oscillators.

It is again a single-pin oscillator but now with the crystal connected to the Gate.

The Source is now the output. The transistor is biased by a current source, which is part of an Automatic Gain Control system, as will be discussed later.

The output can also be taken by a current source inserted in the Drain. Since the Drain is grounded, this is a very elegant way to obtain an output signal without disturbing the oscillator itself.

Both last oscillator types are single-pin oscillators. This means that the crystal is grounded. This may add more parasitic capacitance in parallel to the crystal however than in the Pierce oscillator. This is why Pierce is often preferred despite the fact that it requires two pins for the crystal.

Table of contents

Willy Sansen 10-05 2227

2227

Let us now observe some full oscillator circuits. Circuitry is added to provide AGC but also to buffer the output.

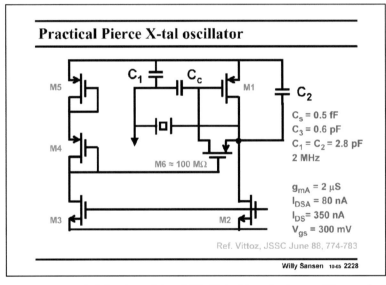

Practical Pierce X-tal oscillator

C_1 C_c C_2

$C_s = 0.5$ fF
$C_3 = 0.6$ pF
$C_1 = C_2 = 2.8$ pF
2 MHz

$g_{mA} = 2$ μS
$I_{DSA} = 80$ nA
$I_{DS} = 350$ nA
$V_{gs} = 300$ mV

M5 M4 M3 M6 ≈ 100 MΩ M2 M1

Ref. Vittoz, JSSC June 88, 774-783

Willy Sansen 10-05 2228

2228

A practical realization of a Pierce oscillator is shown in this slide.

Transistor M_1 is the actual oscillator amplifier. The capacitors C_1 and C_2 are indicated. Capacitor C_c is a coupling capacitor to separate DC and AC. The output is take at the left side of the crystal.

The current is set by transistor M2, which is part of an AGC system shown on the next slide.

The biasing resistor R_B is now represented by transistor M6. It is biased at zero V_{DS} such that its resistance can be quite high although inaccurate. This is achieved by connecting two diodes to the Gate of M6. Both Source and Drain of M6 are at one single V_{GS} below the supply voltage.

The numerical values indicate that all capacitances are quite small in order to limit the power consumption to a minimum. All transistors work in weak inversion!

2229

The AGC circuit is now easily found. It is called an Amplitude Regulator.

The output of the oscillator, at the left of the crystal, is connected to the Q1 input of the AGC circuit, in series with a coupling capacitor C7. It is also connected to the Q1 input of the Output Amplifier, towards a series of inverters.

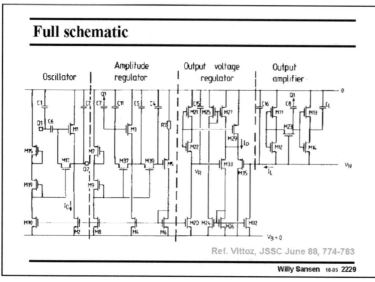

Full schematic

Ref. Vittoz, JSSC June 88, 774-783

Willy Sansen 10-05 2229

In the Amplitude Regulator, the input Q1 drives the Gate of transistor M3, which operates somewhat as a rectifier. When a MOST is overdriven, its average current increases, which is a measure for the input voltage amplitude.

This simple rectifier is followed by a low-pass filter made up by capacitors C_4 and C_5 and transistor M39. This latter transistor operates as a large resistor because its V_{DS} is again zero.

The voltage at transistor M5 is now converted to a current, by use of resistor R7, and fed back to the oscillator transistor M1 by means of current mirror M6–M2.

The AGC loop works as follows. When no signal is present, the Gate voltage of M5 is fairly low and a large current is sent to M1. When the oscillation has come up, more current is flowing through M3 such that the gate of M5 increases. As a result, the current in M3 decreases and so does the current in M1. In equilibrium, the AGC circuit maintains the minimum current for which oscillation is sustained, which corresponds with point A.

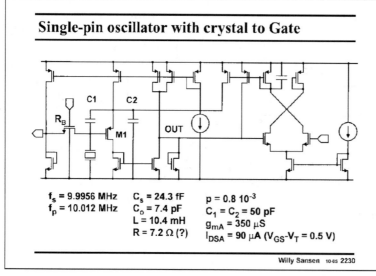

Single-pin oscillator with crystal to Gate

f_s = 9.9956 MHz	C_s = 24.3 fF	$p = 0.8 \; 10^{-3}$
f_p = 10.012 MHz	C_o = 7.4 pF	$C_1 = C_2$ = 50 pF
	L = 10.4 mH	g_{mA} = 350 μS
	R = 7.2 Ω (?)	I_{DSA} = 90 μA (V_{GS}-V_T = 0.5 V)

Willy Sansen 10-05 2230

2230

Another practical oscillator of the third type is shown in this slide. It is a single-pin oscillator, with the crystal connected to the Gate.

The output current is measured at the Drain of transistor M1. The output stage is a wideband amplifier.

The AGC amplifier is shown at the right of the output terminal. A differential pair is used as a rectifier with just one capacitor as a low-pass filter. The output current is now sent to the oscillator transistor M1.

All numerical values are added to allow the reader to fully explore this circuit.

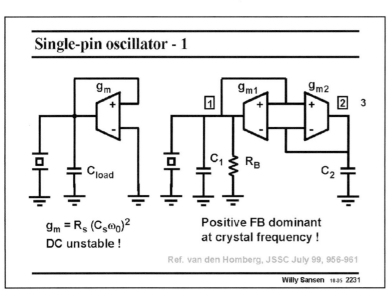

Single-pin oscillator - 1

$g_m = R_s (C_s \omega_0)^2$
DC unstable !

Positive FB dominant
at crystal frequency !

Ref. van den Homberg, JSSC July 99, 956-961

Willy Sansen 10-05 2231

2231

A negative resistance can be generated by a single transistor amplifier but obviously also by a few more transistors. One example is given which uses CMOS as a technology. Many more will be discussed later, which are realized in bipolar technology.

In general, positive feedback always gives rise to negative resistances. In the example in this slide on the left, the negative resistance, because of the positive feedback, must be sufficiently large to compensate for the positive resistor in the crystal, to sustain oscillation. The minimum transconductance is obviously the same as before. However, this circuit is unstable for DC operation. A better realization is shown on the right.

At all frequencies, except around the crystal frequency, the negative feedback with g_{m2} is dominant over the positive feedback with g_{m1}. As a result, the circuit is stable. At the frequency where the crystal offers a large impedance, the positive feedback dominates and the oscillation is sustained.

If the design is such that $g_{m1}/C_1 = g_{m2}/C_2$, and $g_{m1} = g_{m2}$, then the transconductance is again given by the expression in this slide.

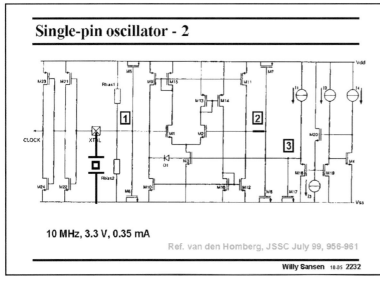

Single-pin oscillator - 2

10 MHz, 3.3 V, 0.35 mA

Ref. van den Homberg, JSSC July 99, 956-961

Willy Sansen 10-05 2232

2232

The circuit realization is shown in this slide. The two g_m blocks are the two input MOSTs of the differential pair M1/M2. The capacitors C_1 and C_2 are realized by means of MOSTs.

The AGC is carried out as follows. If the signal amplitude at node 1 is small, then diode D1 is always reverse biased.

If the amplitude rises, then the diode becomes forward biased for the negative tips of the sine wave. This threshold is set by the reference voltage at node 3.

Because of the forward biasing of diode D1, the voltage at node 3 is pulled down. The current in transistor M3 decreases. The transconductances g_{m1} and g_{m2} also decrease.

The capacitance of transistor M3 reduces the ripple at node 3.
A point of equilibrium is reached in this way, which depends on the voltage at node 3.

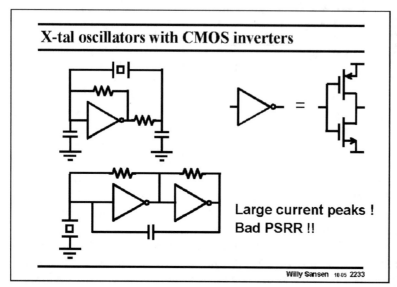

2233
Of course it is also possible to realize crystal oscillators by means of CMOS inverters as amplifiers. A few examples are given in this slide. Since the gain is usually too high, resistors are added to reduce the gain. Moreover, resistors are added in series with the outputs to limit the output currents. If not, large current spikes are drawn from the supply voltage, which causes large voltage spikes on the supply voltages. They propagate to all other circuits connected to the same supplies. As a result, the Power-supply-rejection-ratio is poor.

Moreover the gain setting is not so precise. The transistors are always largely overdriven. The oscillation frequency is thus not precise at all, despite the high quality factor of the crystal used.

Table of contents

- Oscillation principles
- Crystals
- Single-transistor oscillator
- MOST oscillator circuits
- Bipolar-transistor oscillator circuits
- Other oscillators

Willy Sansen 10.05 2234

2234
Crystal oscillators are easy to design with bipolar transistors. Actually, the oldest crystal oscillators have been made with bipolar devices as discrete devices on a printed circuit board. They usually consist of one single transistor with fixed biasing, rather than with an AGC circuit.

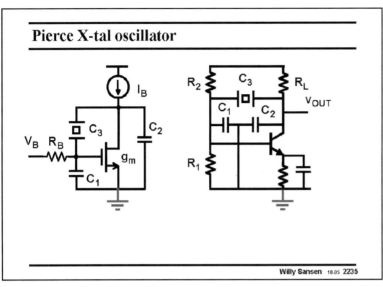

2235

A Pierce oscillator is best biased by a current source I_B and a Gate biasing resistor V_B.

A current source is used to isolate the circuit from the supply line.

A discrete realization is shown on the right. It uses a bipolar transistor. The crystal is connected between collector and base, which makes it a Pierce oscillator. Also the capacitances are indicated.

The current source is replaced by a large resistor R_L. A resistor is used in the emitter for thermal stabilization. The capacitor across it has to alleviate the gain reduction caused by the resistor.

The resistors R_1 and R_2 provide biasing to the base.

It is clear that such a circuit can never be biased exactly at point A of the polar diagram. It has fixed biasing at a transconductance which must be a lot larger than g_{mA}, to always ensure oscillation. As a result, the power consumption is never minimum and the oscillation frequency is not very precise.

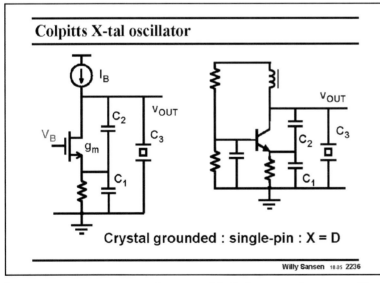

2236

A Colpitts oscillator is again shown in this slide. Indeed it is a single-pin oscillator with the crystal connected to the Collector.

A discrete equivalent with a bipolar transistor is shown on the right. The current source is replaced by a "choke" or large inductor. The capacitors are where they are expected.

The output voltage is taken at the Collector, as shown by most publications. This is not the best place however, as at resonance the crystal is little more than a small resistor. The signal swing at the Collector is therefore quite small.

A better output node is the Emitter, between capacitors C_1 and C_2.

The base is decoupled to ground to make it an AC ground. If not, the effective transconductance would be reduced by the base resistances.

A Heartley oscillator is very similar to this one. Both capacitances C_1 and C_2 are substituted by inductances, and the crystal operates in the capacitive region. Needless to say that the frequency region of operation is now much wider as the crystal has a much wider region where it behaves as a capacitance.

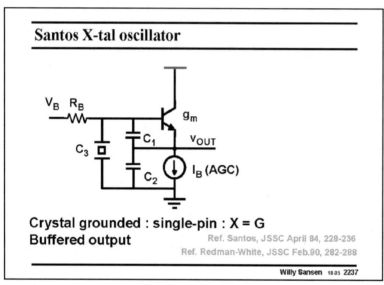

2237
The third type of oscillator is again a single-pin oscillator but now with the crystal connected to the Base.

The Emitter is now the output. The transistor is biased by a current source, which is part of an Automatic Gain Control system.

The output can also be taken by a current source inserted in the Collector. Since the Collector is grounded, this is a very elegant way to obtain an output signal without disturbing the oscillator feedback loop itself.

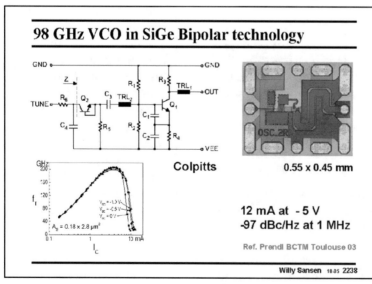

2238
Another example of the same type of oscillator is shown in this slide. It uses high-speed SiCGe bipolar transistors with f_T's up to 200 GHz. As a result, the oscillation frequency is really high, nearly 100 GHz!

Instead of a crystal a series LC circuit is used, the capacitance of which consists of several capacitances in series. One of them in transistor Q2 can be tuned by an external DC voltage.

The output is taken again at the Collector, not to disturb the oscillation at the Base-Emitter nodes.

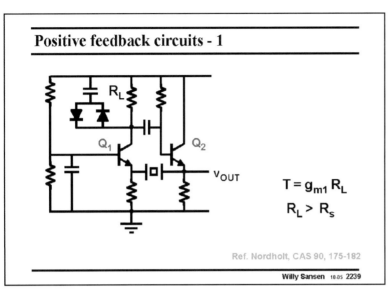

2239

In bipolar technology, many more oscillators have been realized. They all use several transistors and they all aim at the realization of a negative resistance by means of positive feedback.

In the circuit in this slide, transistor Q1 is a cascode, which provides voltage gain at its Collector. Transistor Q2 is just an emitter follower, which provides the current through the small series resistor R_s of the crystal. The loop gain is therefore $g_{m1}R_L$.

As soon as R_L is larger than R_s, the oscillation builds up. This is a good feature of this oscillator. The signal amplitude is limited by the two diodes across the load resistor R_L.

All three capacitors serve as (de)coupling capacitances. They are all large and present negligible impedances at the frequency of oscillation.

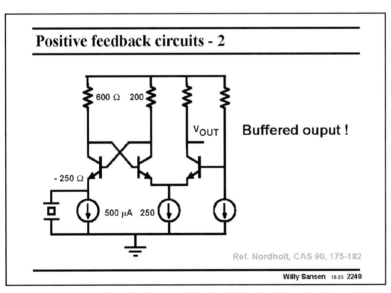

2240

Another bipolar crystal oscillator is shown in this slide. It is again a single-pin oscillator. The positive feedback is easily recognizable. Indeed, a cross-coupled transistor pair presents a negative resistance of $-2/g_m$ (see Chapter 2). This resistance now compensates the series resistance of the crystal.

The output is nicely buffered from the oscillation circuit itself, by means of a cascode.

The biasing is fixed. The amplitude is limited by the Base-Emitter diodes of the cross-coupled transistor pair.

2241

Cross-coupling is used again in this high-frequency oscillator. This is a high frequency as the bipolar transistors have an f_T of only about 400 MHz.

The crystal is replaced by a parallel LC tank. The oscillation frequency is now close to the

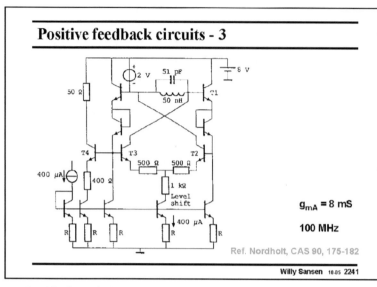

Positive feedback circuits - 3

g_{mA} = 8 mS

100 MHz

Ref. Nordholt, CAS 90, 175-182

Willy Sansen 10.05 2241

parallel resonance frequency of this tank.

Cross-coupling is used to generate a negative impedance to compensate the parallel resistance across the tank. Emitter followers and diodes are used to increase the output swing. This signal amplitude is also increased by insertion of emitter resistances of 500 Ω. Their $g_m R_E$ factor is about 4. This is the answer of a bipolar transistor to the increase of the $V_{GS} - V_T$ in a MOST.

The biasing of the input devices is fixed. A safety margin must now be expected.

Table of contents

Willy Sansen 10.05 2242

2242

Finally, some more oscillators are added.

They are not necessarily crystal oscillators but other types. Moreover, there are quite often many more transistors.

They have been added to show that the principles discussed before, are still very much applicable.

2243

This oscillator again uses an LC tank to set the oscillation frequency. The inductors are planar spiral inductors. The capacitances are diode capacitances. They can therefore be tuned by changing control voltage V_c. This is why they are called VCO's. The tuning range must be large enough to compensate for the variations on the inductor (20%).

For example, an inductor with about three windings, and with a hollow layout, is about 10 mm long. Its inductance is about 10 nH. With a diode capacitance of 1 pF, the oscillation frequency is about 1.6 GHz. Also, the minimum g_{mA} is about 1 mS for a coil resistance R_L of 10 Ω (Q= 10). For a $V_{GS} - V_T$ of 0.5 V, this would require a transistor current of 0.25 mA.

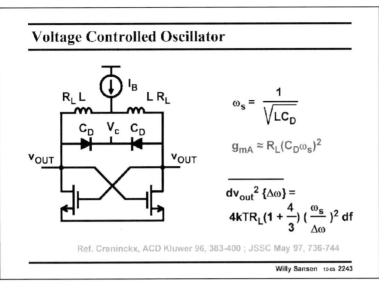

Voltage Controlled Oscillator

$$\omega_s = \frac{1}{\sqrt{LC_D}}$$

$$g_{mA} \approx R_L(C_D\omega_s)^2$$

$$\overline{dv_{out}^2}\{\Delta\omega\} = 4kTR_L(1 + \frac{4}{3})(\frac{\omega_s}{\Delta\omega})^2 df$$

Ref. Craninckx, ACD Kluwer 96, 383-400 ; JSSC May 97, 736-744

Willy Sansen 10-05 2243

It is obvious that the GHz range is easily achieved. However, to go lower or higher in frequency is a problem.

One of the most important specifications of such a VCO is the phase noise. This is mainly the thermal noise of the transistors and coil series resistors R_L, converted to sidebands of the oscillation frequency. Since this resistor R_L is linked to the transconductance g_{mA}, it is the main parameter in the expression of the phase noise. Actually, the term 4/3 is due to the transistor g_m.

A rule of thumb for phase noise is -100 dBc/Hz at 100 kHz distance from the carrier. In this example it is about -120 dBc/Hz at 100 kHz.

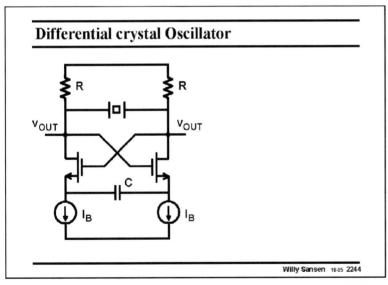

Differential crystal Oscillator

Willy Sansen 10-05 2244

2244

A differential version of the single-transistor Pierce oscillator is shown in this slide.

For low frequencies, the impedance of the capacitor C is too large and no oscillation can build up. At high frequencies, it acts as a short and a negative impedance is presented to the crystal.

The resistors have to be sufficiently high not to dampen the oscillation. They are better when replaced by current sources.

The current setting sources I_B are better driven by an AGC loop, which measures the output voltages. If not, the currents may be set too high, resulting in larger deviations in frequency or distortion.

2245

Leaving out the crystal yields a relaxation oscillator. The frequency is now set by the currents I_B, capacitor C and the limiting voltage, which is here a V_{BE} of a bipolar transistor. Actually, it is $I_B/(4CV_{BE})$. Square waveforms are obtained at the outputs but a triangular one across the time setting capacitor C.

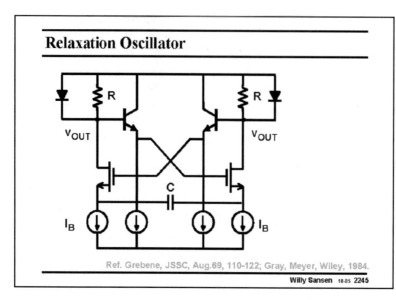

Relaxation Oscillator

V_{OUT} V_{OUT}

I_B I_B

Ref. Grebene, JSSC, Aug.69, 110-122; Gray, Meyer, Wiley, 1984.

Willy Sansen 10-05 2245

Needless to say that this frequency is not all that accurate and very temperature dependent.

Many such relaxation oscillators have been built and published. This is one of the earlier ones.

Its main advantage is its tunability by means of current I_B. This range is limited however by the base currents of the emitter follower transistors. Substituting them by MOST source followers allows a current and hence a frequency range of over 8 decades!

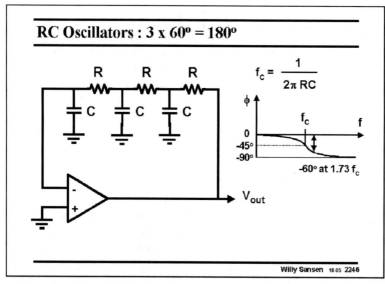

RC Oscillators : 3 x 60° = 180°

R R R

C C C

$f_c = \dfrac{1}{2\pi RC}$

ϕ

f_c f

0
-45°
-90°

-60° at 1.73 f_c

V_{out}

Willy Sansen 10-05 2246

2246

A low-frequency oscillator is easily built by means of an opamp. An example is given in this slide.

Without capacitors we would have negative feedback and no oscillation. However, at the frequency f_c, sufficient phase shift is taken up by the RC circuit to turn the negative feedback into positive feedback. Barkhausen is satisfied provided there is sufficient gain around the loop at this frequency. The oscillation then builds up with a frequency of about 1.7f_c.

No amplitude limitation is shown. Two diodes could be added at the output to limit the output swing.

2247

A well-known oscillator with an opamp is the Wien bridge oscillator, shown in this slide. It consists of an opamp with a series and a parallel RC circuit around the feedback loop. Two more resistors 3R_1 and R_1 ensure a voltage gain of about 3.

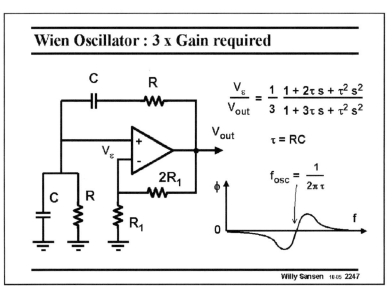

From the expression of the loop gain, it is clear that this circuit will oscillate at the frequency f_{osc}. The attenuation of 1/3 is compensated by the voltage gain of 3. Moreover, the phase shift around the loop is zero at this frequency.

Again no amplitude limitation is added. Two diodes could again be added at the output to limit the output swing.

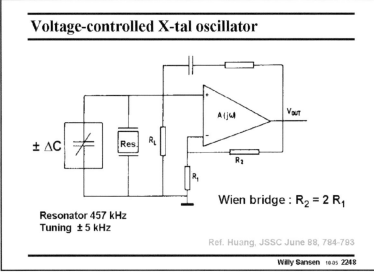

2248

The same Wien bridge oscillator can be used to build a Voltage Controlled Crystal oscillator.

Obviously, this is a contradiction. Normally, a crystal oscillator is used to fix the frequency with high accuracy. A VCO on the other hand is used to vary the frequency over 20–30%.

However, sometimes, especially for timing purposes, we want to set the crystal frequency to an accurate value which is slightly different from the crystal frequency itself.

In this example, a resonator is used, which is similar to a crystal but with a lower Q factor, to set the frequency at exactly 460,00 kHz. The resonator has only 457 kHz. How do we solve this problem?

The resonator is part of the Wien bridge. The solution is therefore to add a parallel capacitance to the resonator to detune it. The oscillation will obviously take place much closer to parallel resonance than to series resonance. Changing the parallel capacitance is now an easy way to slightly change the oscillation frequency.

In order to be able to tune the frequency in both directions, we have to be able to add a capacitance DC, which can be both positive and negative. Moreover, we want to be able to control this capacitance value with a voltage or current.

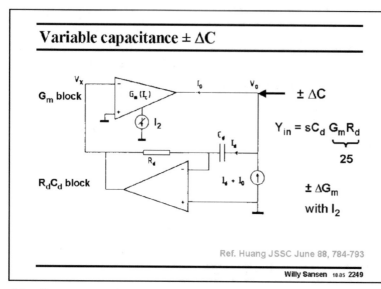

2249

How to realize a capacitance that can be varied by means of a voltage or current?

The Miller effect could be used. An alternative is the circuit in this slide. A differentiator with components R_d and C_d is inserted in a feedback loop with a variable G_m block. The input admittance is easily calculated. It is capacitance C_d, multiplied by $G_m R_d$. This factor is about 25 in this example. With 4 pF for C_d,

this allows a capacitance change ΔC of about 100 pF.

If G_m can now be made positive and negative, then we have a capacitance which is positive and negative.

Current I_2 will be used to control the transconductance G_m from positive, through zero to negative values.

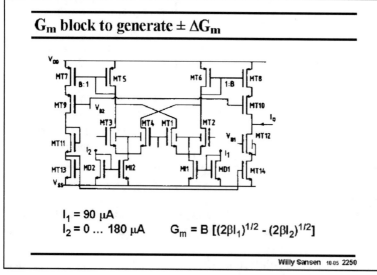

2250

To realize a transconductance G_m from positive to negative values, a symmetrical opamp is used, in which the input stage is doubled.

Moreover, the two input stages are cross-coupled. When both input pairs have the same biasing current, nothing can come out. The output currents cancel perfectly. If one current, for example current I_1 through MI1, is larger than I_2 through MI2, then the differential pair MT1/MT2

provides a larger output current I_o, which flows into the circuit.

If, on the other hand, current I_2 is larger than I_1, than the output current changes polarity. As a consequence we can obtain positive and negative output currents by allowing changes to I_2 with respect to I_1, as indicated in this slide.

R_dC_d block as differentiator

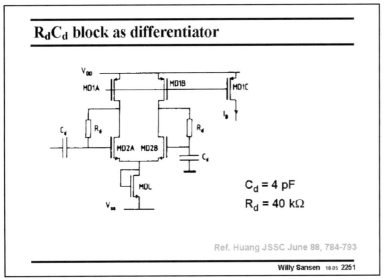

$C_d = 4 \text{ pF}$

$R_d = 40 \text{ k}\Omega$

Ref. Huang JSSC June 88, 784-793

Willy Sansen 10-05 2251

2251

The differentiator is made differential. It is little more than a pseudo-differential pair, to which the resistor R_d and capacitor C_d are applied. With the values of R_d and C_d as given, a time constant is reached of about 160 ns, which is less than the oscillation period of 348 ns. The time constants in the opamp itself are not negligible however, and the picture is slightly more complicated.

Table of contents

♦ **Oscillation principles**

♦ **Crystals**

♦ **Single-transistor oscillator**

♦ **MOST oscillator circuits**

♦ **Bipolar-transistor oscillator circuits**

♦ **Other oscillators**

Willy Sansen 10-05 2252

2252

In this Chapter, the design of oscillators is discussed.

Crystal oscillators have been discussed in great detail, followed by VCO's

and even Wien bridge oscillators. All of them obey the Barkhausen criteria however, which is the main theory behind them.

2253

References X-tal oscillators -1

A.Abidi, "Low-noise oscillators, PLL's and synthesizers", in R. van de Plassche, W.Sansen, H. Huijsing, "Analog Circuit Design", Kluwer Academic Publishers, 1997.

J. Craninckx, M. Steyaert, "Low-phase-noise gigahertz voltage-controlled oscillators in CMOS", in H. Huijsing, R. van de Plassche, W.Sansen, "Analog Circuit Design", Kluwer Academic Publishers, 1996, pp. 383-400.

Q.T. Huang, W. Sansen, M. Steyaert, P.Van Peteghem, "Design and implementation of a CMOS VCXO for FM stereo decoders", IEEE Journal Solid-State Circuits Vol. 23, No.3, June 1988, pp. 784-793.

E. Nordholt, C. Boon, "Single-pin crystal oscillators" IEEE Trans. Circuits. Syst. Vol.37, No.2, Feb.1990, pp.175-182.

D. Pederson, K.Mayaram, "Analog integrated circuits for communications"", Kluwer Academic Publishers, 1991.

2254

References X-tal oscillators - 2

W. Redman-White, R. Dunn, R. Lucas, P. Smithers, "A radiation hard AGC stabilised SOS crystal oscillator", IEEE Journal Solid-State Circuits Vol. 25, No.1, Feb. 1990, pp. 282-288.

J. Santos, R. Meyer, "A one pin crystal oscillator for VLSI circuits", IEEE Journal Solid-State Circuits Vol. 19, No.2, April 1984, pp. 228-236.

M. Soyer, "Design considerations for high-frequency crystal oscillators", IEEE Journal Solid-State Circuits Vol. 26, No.9, June 1991, pp. 889-893.

E. Vittoz, M. Degrauwe, S. Bitz, "High-performance crystal oscillator circuits: Theory and application", IEEE Journal Solid-State Circuits Vol. 23, No.3, June 1988, pp. 774-783.

V. von Kaenel, E. Vittoz, D. Aebischer, " Crystal oscillators", in H. Huijsing, R. van de Plassche, W.Sansen, "Analog Circuit Design", Kluwer Academic Publishers, 1996, pp. 369-382.

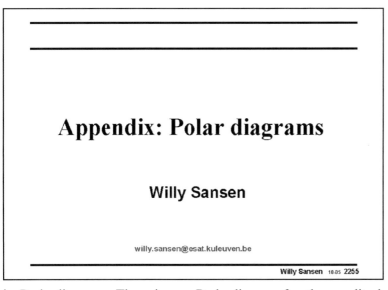

2255

The complex plane is one way to represent complex numbers. After all, complex numbers are numbers which take up a position in a plane rather than on a line. A complex number always contains two numbers. They can be represented by amplitude and phase, or by Real and Imaginary part.

Impedances which contain capacitors or inductors, are also complex numbers. Such impedances can now be plotted versus frequency in Bode diagrams. There is one Bode diagram for the amplitude (or magnitude) and one for the phase.

Such impedance can also be plotted in the complex plane. It is now called a polar plot (in Electronics) or a Cole-Cole plot (in Biochemistry). These curves can then be labeled in terms of frequency, or any other parameter used.

To make the reader familiar with such polar plots, some simple examples are given.

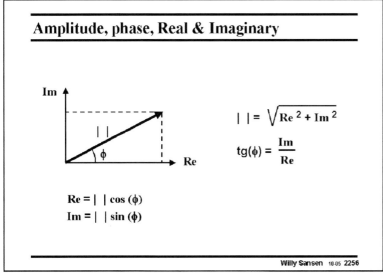

2256

As a reminder to the reader, the relationship is given between amplitude and phase on one hand, and the Real and Imaginary part on the other hand.

It is clear that complex numbers are situated in a plane. We always need two numbers to know where exactly they are.

We can choose however, between amplitude and phase or the Real and Imaginary part!

For example, the complex number (4,3) has a Real part of 4 and an Imaginary part of 3. Its amplitude is 5 and its phase is 0.64 radians or about 37°.

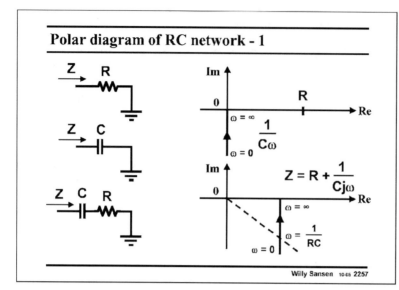

2257

A simple resistor R is just one point on the Real axis.

A simple capacitor C is on the negative part of the Imaginary axis, depending on the actual value of the frequency. For a frequency zero, it is infinitely far away but for a frequency infinity, it is in the zero of the axes.

A series combination of a resistor R and a capacitor C, is also a straight line, but shifted over the Real axis by an amount R. At the circular frequency 1/RC, the point (locus) is at 45° with respect to the two axes.

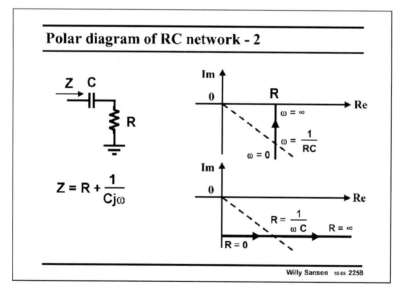

2258

The same impedance can also be plotted with resistor R as a parameter, rather than the frequency. A different locus results.

It shares the same point at frequency 1/RC but is now a horizontal line through this point. For zero resistor, it is on the axis, but obviously goes to infinity for infinite resistor.

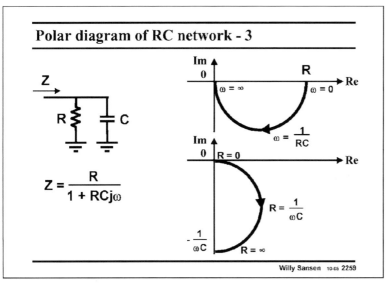

2259

A parallel combination of a resistor and a capacitor gives an impedance which yields a half circle in the complex plane.

In the top one, the frequency is used as a parameter. For zero frequency, the impedance of the capacitor is infinite and thus disappears from the picture. The impedance is now just a resistor R.

For infinite frequency, the impedance is zero and therefore in the zero (or origin) of the axes. At circle frequency 1/RC, the locus is at 45° with respect to the two axes, or at the bottom of the circle.

If the resistor is taken as a parameter, a different half circle is described by the impedance. They go through the same point at RCω = 1. For zero resistor, the locus is at the origin of the axes. For infinite resistor, we just have a capacitor, which is located on the Imaginary axis.

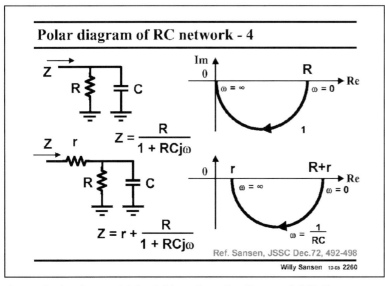

2260

Addition of a resistor r to the parallel combination of R and C, causes the half circle to shift to the right, over a distance r, at least if the frequency is taken as a parameter.

This is a well known polar plot, as it is similar to the input impedance of a bipolar single-transistor amplifier. It is also called the circle diagram.

Measurement of the input impedance versus frequency, allows tracing a circle through the data, which yields values for R, r and 1/RC.

Polar diagram of RC network - 5

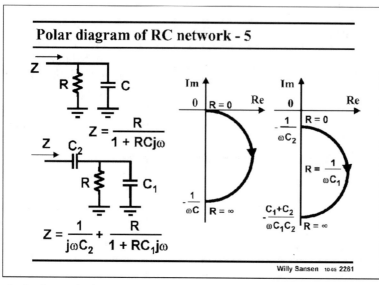

$$Z = \frac{R}{1 + RCj\omega}$$

$$Z = \frac{1}{j\omega C_2} + \frac{R}{1 + RC_1 j\omega}$$

Willy Sansen 10-05 2261

circle through the data, which yields values for R, C_1 and C_2.

2261

In a similar way, addition of a capacitor C_2 to the parallel combination of R and C_1, causes the half circle to shift on the Imaginary axis, over a distance C_2, at least if the resistor is taken as a parameter.

For a zero resistor, we simply have capacitor C_2 left. For an infinite resistor however, both capacitances are in series.

Again measurement of the input impedance versus frequency, allows tracing a

Circuit input impedance Zc

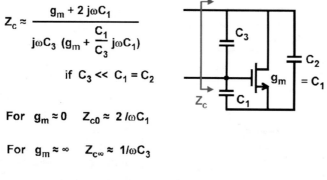

$$Z_c \approx \frac{g_m + 2\,j\omega C_1}{j\omega C_3\left(g_m + \dfrac{C_1}{C_3}\,j\omega C_1\right)}$$

if $C_3 \ll C_1 = C_2$

For $g_m \approx 0$ $Z_{c0} \approx 2\,/\omega C_1$

For $g_m \approx \infty$ $Z_{c\infty} \approx 1/\omega C_3$

Willy Sansen 10-05 2262

2262

The question now is, what is the locus of the input impedance Z_c with g_m as a parameter?

The approximate expression is readily calculated and given in this slide. Also the values at zero and infinite transconductance are easily found.

The corresponding polar diagram is given on the next slide.

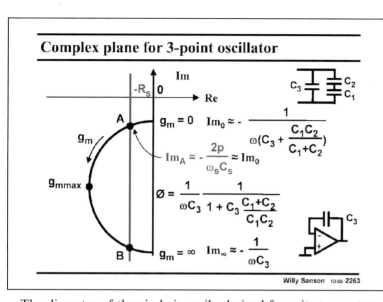

2263

The polar plot of impedance Z_c with g_m as a parameter, is again a half circle. It is on the left side of the Imaginary axis however, because it presents a negative Real part.

For zero g_m, we simply have a combination of the three capacitors. For increasing g_m, the half circle is described until the locus hits the imaginary axis again, for infinite g_m. The capacitor is now only C_3, however.

The diameter of the circle is easily derived from its two extremities.

When used for an oscillator, only one point is of importance, i.e. point A. It is the point where the negative Real part equals resistor R_s. The Imaginary Part Im_A in point A then yields the pulling factor p. It is about the same as the Imaginary part Im_0 for zero g_m. The line of constant negative R_s is actually much closer to the Imaginary axis than sketched. Point A therefore nearly lies on the Imaginary axis!

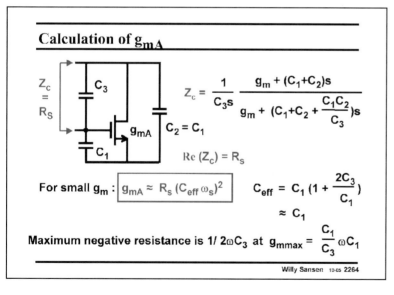

2264

In order to find the actual value of g_{mA} in point A, we have to equate the Real part of impedance Z_c to the resistor R_s itself.

For small values of R_1 (= R_2) the transconductance g_{mA} will be small. It can then be approximated as given in this slide. It is in a frame as it is the most basic expression of the required transconductance of any single-transistor amplifier. It shows why RF oscillators require a large amount of current!

From the same expression of Z_c, we can derive what is the expression of g_{mmax}, which is reached at the utmost left point of the circle. This is the point where the negative Real part is maximum.

231

Low-Noise Amplifiers

Willy Sansen

KULeuven, ESAT-MICAS
Leuven, Belgium

willy.sansen@esat.kuleuven.be

Willy Sansen 10.05 231

Low-noise amplifiers (LNA's) are the first amplifiers in receivers. They must operate at the same high frequencies as the carriers themselves. For GSM, CDMA, etc., these frequencies are beyond 1 GHz, reaching 5 GHz today. In car electronics, frequencies are reached beyond tens of GHz.

Moreover, such an amplifier must be able to handle very small signals close to the noise level and at the same time very high signal levels, close to the transmitter antenna. Noise and distortion are both of importance at the same time.

Finally, the reception antenna is usually connected to the amplifier over a transmission line, usually with 50 Ω characteristic impedance.

Such an amplifier is full of compromises, despite its low count of transistors. Indeed, only a few MOSTs are normally used.

In this Chapter, these compromises will be discussed. Design equations and guide lines will be given and checked on their accuracy.

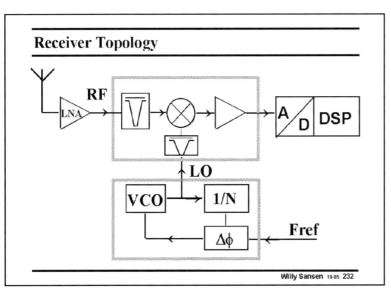

Receiver Topology

Willy Sansen 10.05 232

232

An example of such a receiver is shown in this slide.

The LNA is the first amplifier. It is an RF amplifier followed by a mixer, which translates the modulation content to low frequencies. After some filtering, this signal is then converted in digital form, towards a DSP.

The mixer needs a local oscillator, which is normally derived from a phase-locked loop (PLL). In this feedback loop a VCO generates a frequency, which is locked to an external reference frequency Fref, after a divider by N.

It is clear that the LNA interacts with the antenna. This is why both the antenna output and LNA input must show a characteristic impedance, to avoid reflections. This is discussed next.

Table of contents

Willy Sansen 10.05 233

233

Two of the most important characteristics of a LNA is the noise performance within the constraints of impedance matching.

This is why they are discussed first.

Some attention is now paid to the linearity of a LNA, as large input signals are as common as small ones.

A comparison follows between the most popular configurations. The first has an amplifier (Common-Source Configuration) at the input whereas the second has a cascode (Common-Gate Configuration).

MOST transistors can exhibit non-quasi-static behavior at high frequencies. We have to investigate when this effect is important.

Finally, a large number of configurations are discussed, including components for electrostatic-discharge protection.

Transmission line effects

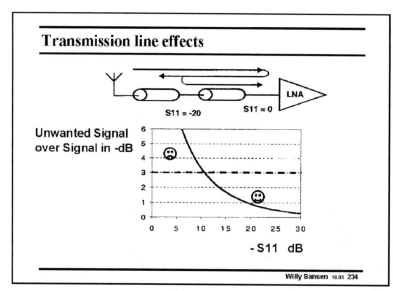

Willy Sansen 10.05 234

234

A voltage wave traveling along a transmission line will be partially reflected at the end of this line if the termination impedance is different from the line impedance. This reflected or unwanted signal is zero if the termination is perfect. Parameter s_{11} gives the amount of reflected signal to the input signal; it is therefore minus infinity.

This is only true if the input impedance of the LNA is exactly the same as the characteristic impedance of the (transmission) line, which is usually 50 Ω.

For slight deviations, reflection occurs and s_{11} is not minus infinity any more. For example, if the reflected signal is −3 dB compared to the sign 1 itself, then s_{11} is only about −10 dB.

Values of s_{11} up to −10 dB are acceptable for LNA's, although higher values are preferred.

This impedance matching obviously applies to the antenna. An example is given in this slide of an LNA with high input impedance (as for MOST) and antenna which exhibits about 50 Ω;

its s_{11} is -20 dB. It is clear that reflections will especially occur at the LNA input and less at the antenna terminal.

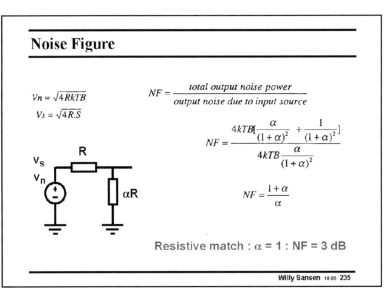

235
If a transmission line with characteristic impedance is terminated in a resistor with the same value, then a division by two occurs.

This is shown in this slide. A voltage source with value v_s has an internal resistance R. It is matched by a transmission line with load αR. It now presents a resistance αR to the voltage source.

For perfect matching the load resistance equals the source resistance and $\alpha = 1$. Usually, the load resistance is somewhat different; α is larger or smaller.

In this case, what is the Noise Figure?

Both the signal power S_{out} and noise power N_{out} are readily calculated at the output.

The Noise Figure, which is defined as the ratio of the total noise at the output, divided by the noise due only to the source resistance (see Chapter 4), is obtained as given in this slide.

For perfect impedance matching, α is unity and the Noise Figure is 2 or 3 dB.

This is a typical value for resistive terminations, as shown next.

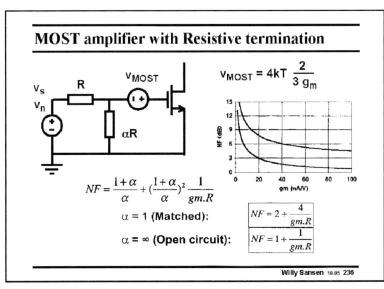

236
If a single transistor is taken as a LNA, to which a resistor αR is added for proper termination, then it is clear that the only design parameter left is the transconductance g_m. Indeed, the Noise Figure is given in two cases, with perfect matching ($\alpha = 1$) and without αR.

Matching ($\alpha = 1$) is required to avoid reflections. The NF is higher, however. If no matching resistance is present ($\alpha = 0$), then the NF can be higher.

Both conditions are plotted versus transconductance g_m.
Deviations from the matching condition can give higher Noise Figures.

Table of contents

Willy Sansen 10.05 *237*

237

Input impedance matching and noise are not the only concerns of a LNA.

For large input signal amplitudes, distortion will also occur. This is why this aspect is discussed in detail.

A compromise will now have to be taken. To have a better feel for these compromises, some attention is paid to all other important specifications.

For this purpose, the schematic of a receiver has to be looked at in detail. A receiver in combination with a transmitter is a transceiver. This is shown next.

Transceiver

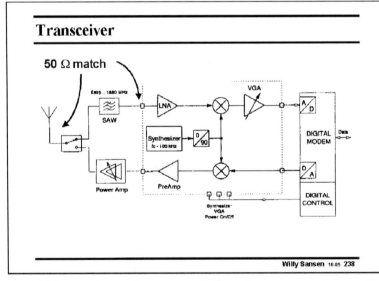

50 Ω match

Willy Sansen 10.05 *238*

238

A transceiver consists of a receiver and a transmitter as shown in this slide.

The input impedance matching is important for the LNA of the receiver for several reasons. First of all, reflections have to be avoided over the transmission line between the channel (SAW) filter and the LNA. Secondly, the LNA must provide the right load impedance to this channel filter.

The input impedance of the LNA must be as close as possible to the source resistance R_S, which is usually 50 Ω. This is called impedance matching.

Moreover, the equivalent input noise must also be as small as possible. After all, the LNA is the first active amplification block. Going for the minimum Noise Figure is called noise matching. This has nothing to do with impedance matching. In practice, noise will be reduced as much as possible within the constraint of impedance matching.

The LNA leads to the mixer. Again a 50 Ω transmission line is used depending on the distance between LNA and mixer. If they are next to each other there is no need for a 50 Ω line. Higher output impedances can now be used in the LNA.

Normally, a zero-IF or low-IF architecture is used. In any case, the LNA must have sufficient gain to avoid the noise of the mixer to appear at the antenna. Gains of 12–16 dB are typical.

Minimum NF and IIP3 for DCS-1800

Sensitivity	-100 dBm
SNR	9 dB
Input noise	-109 dBm
kT =	-174 dBm
Bandwidth (200 kHz)	+ 53 dB
NF : -109 - (-174+53) =	12 dB
Attenuating blocking filter : 3 dB	NF < 9 dB
+ 3 dB Sensitivity	-97 dBm
SNR (-49 dBm sine)	9 dB
IIP3 = -49 + (-49- (-106/2)) =	-20.5 dBm
With attenuating blocking filter : 3 dB	IIP3 < -23.5 dBm

Willy Sansen 10.05 239

239
Let us take as an example, a DCS-1800 receiver. It operates at a carrier frequency of 1.8 GHz.

Several sensitivities and values of Noise Figures are shown in this slide. Remember that 1 dBm is 1 mW in 50 Ω, which corresponds to 224 mV$_{RMS}$. A sensitivity is specified of −100 dBm for an input SNR of 9 dB. The input noise must be at −109 dBm.

The bandwidth is 200 kHz. The Noise Figure is 12 dB. If 3 dB loss is taken into account for the channel filter, then a Noise Figure of 9 dB is required.

Let us now try to find out about the distortion.

Distortion is important to avoid leakage from one channel to the adjacent channels. Especially intermodulation and cross-modulation distortion are to be avoided. Both are described by the 3rd-order intermodulation intercept (IIP3) as shown in Chapter 18.

A minimum input signal is taken 3 dB higher than the sensitivity specifies. For a high input signal at −49 dBm, the SNR must still be 9 dB.

The distortion specification now leads to a IIP3 of −20.5 dBm.

Again, to take into account the attenuation in the preceding channel filter, 3 dB has to be subtracted, leading to an IIP3 of −23.5 dBm.

Normally, MOST amplifiers can easily satisfy such a specification, depending on the choice of the $V_{GS} - V_T$.

Linearity CMOS amplifier

Velocity saturation
$v_{max} \approx 10^7$ cm/s
$\Theta L \approx 0.2$ μm/V

$$I_{ds} = \frac{\mu_0 C_{ox}}{2n} \cdot \frac{W}{L} \cdot \frac{(V_{GS} - V_T)^2}{1 + \Theta \cdot (V_{GS} - V_T)}$$

$$\Theta = \theta + \frac{\mu_0}{L_{eff} \cdot v_{max} \cdot n}$$

$$IM2 = \frac{v}{V_{GS} - V_T} \cdot \frac{1}{(1 + r) \cdot (2 + r)}$$

$$r = \Theta \cdot (V_{GS} - V_T)$$

$$IM3 = \frac{3}{4} \frac{v^2}{(V_{GS} - V_T)} \cdot \frac{\Theta}{(1 + r)^2 \cdot (2 + r)}$$

$$IIP3 \cong 11.25 + 10 \cdot Log_{10}\big((V_{GS} - V_T) \cdot (1 + r)^2 \cdot (2 + r) / \Theta\big)$$

Willy Sansen 10-05 2310

2310

The expression of the current I_{DS} of a MOST has to be taken for high $V_{GS} - V_T$. This means that the fitting parameter Θ must be added to model the linearization of the $I_{DS} - V_{GS}$ characteristic, as a result of mainly velocity saturation.

The second and third derivatives of this expressions give rise to the IM2 and IM3 intermodulation expressions. From the latter one the IIP3 is readily derived.

The IIP3 depends to a large extent on the choice of the $V_{GS} - V_T$, as shown next.

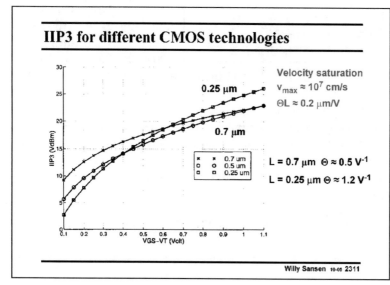

IIP3 for different CMOS technologies

Velocity saturation
$v_{max} \approx 10^7$ cm/s
$\Theta L \approx 0.2$ μm/V

	0.7 um
	0.5 um
	0.25 um

$L = 0.7$ μm $\Theta \approx 0.5$ V^{-1}
$L = 0.25$ μm $\Theta \approx 1.2$ V^{-1}

Willy Sansen 10-05 2311

2311

The IIP3 is plotted versus the value of $V_{GS} - V_T$ for different CMOS technologies.

It is clear that the higher the $V_{GS} - V_T$, the higher the IIP3 is but in a nonlinear way.

It is also clear that for a $V_{GS} - V_T$ of 0.5 V, 15 dBm IIP3 is readily available, independently of the technology.

At $V_{GS} - V_T$ of 0.2 V, the older CMOS technology provides the best IIP3. This is a result of the ever stronger effect of velocity saturation (parameter Θ) in deeper submicron technologies.

Table of contents

Willy Sansen 10.05 2312

2312

Now that the input impedance matching, the Noise Figure and IIP3 have been introduced, they all have to be derived for various circuit configuration.

At these high frequencies, only simple circuits can be used. A single transistor amplifier is used more often, followed by a cascode amplifier. This is why they are discussed first.

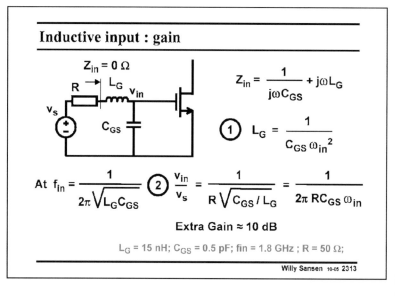

Inductive input : gain

$$Z_{in} = 0 \ \Omega$$

$$Z_{in} = \frac{1}{j\omega C_{GS}} + j\omega L_G$$

$$\text{①} \quad L_G = \frac{1}{C_{GS} \omega_{in}^2}$$

$$\text{At } f_{in} = \frac{1}{2\pi \sqrt{L_G C_{GS}}} \quad \text{②} \quad \frac{v_{in}}{v_s} = \frac{1}{R\sqrt{C_{GS}/L_G}} = \frac{1}{2\pi R C_{GS} \omega_{in}}$$

Extra Gain ≈ 10 dB

$L_G = 15$ nH; $C_{GS} = 0.5$ pF; fin = 1.8 GHz ; R = 50 Ω;

Willy Sansen 10-05 2313

2313

A single-MOST amplifier is shown in this slide. Its equivalent input noise voltage source is added.

A series inductor L_G is added to tune out the effect of the input capacitance C_{GS}. This inductor has been selected so that the operating frequency ω_{in} fulfills the expression labeled 1. This is the first design equation.

At this frequency f_{in}, the ratio of Gate voltage v_{in} to the signal voltage v_s is readily calculated as given in this slide.

It is clear that some gain can easily be realized, provided the input capacitance C_{GS} is sufficiently small (for a given standard R), or the added inductance L_G sufficiently large. This is the second design equation.

Note that, the input capacitance C_{GS} is effectively tuned out such that the input impedance Z_{in} is ideally zero.

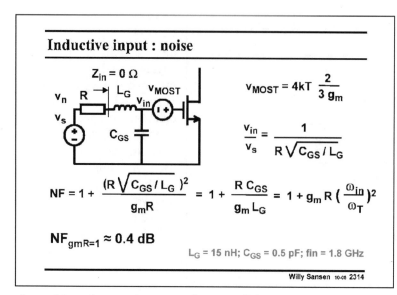

2314
The single-MOST amplifier is shown again in this slide. Its equivalent input noise voltage source v_{MOST} is added.

The Noise Figure is now readily calculated. It can be rewritten in terms of the several parameters involved.

The last expression shows that for low NF the f_T of the MOST must be made high. This means that, first of all, the MOST must be selected and then the corresponding inductor L_G. For high f_T the channel length must be as small as possible and its $V_{GS} - V_T$ high (see Chapter 1).

The Noise Figure is determined by this choice. This is the third design equation. No additional independent design parameters are present.

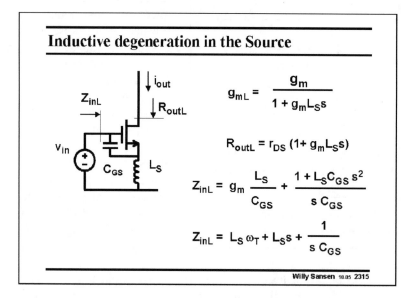

2315
An inductor can also be added in the Source as shown in this slide.

This gives rise to a reduction of the transconductance and an increase in output resistance.

The input impedance however, shows a real part $L_S\omega_T$ and an imaginary part. The real part can be used to tune the input resistance to the source resistance R (of usually 50 Ω) as shown next.

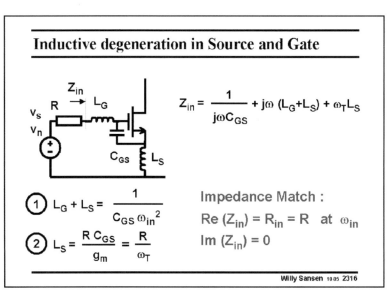

Inductive degeneration in Source and Gate

$$Z_{in} = \frac{1}{j\omega C_{GS}} + j\omega (L_G + L_S) + \omega_T L_S$$

① $L_G + L_S = \dfrac{1}{C_{GS} \, \omega_{in}^2}$

② $L_S = \dfrac{R \, C_{GS}}{g_m} = \dfrac{R}{\omega_T}$

Impedance Match :

$Re \, (Z_{in}) = R_{in} = R \quad$ at ω_{in}

$Im \, (Z_{in}) = 0$

Willy Sansen 10.05 2316

2316

The matching of the input impedance to the source resistor R (usually $50 \, \Omega$) is not carried out with a resistor because it kills the Noise Figure. Normally, two inductances are used, one in the Source L_S and one in the Gate L_G. The latter one L_G is usually the bonding wire.

Both of them are used to tune out the input capacitance C_{GS}, as indicated by the equation number one. This occurs only around the operating frequency f_{in} ($= \omega_{in}/2\pi$).

Under this condition, the input impedance is purely resistive and equal to $\omega_T L_S$ which is made equal to R. This is the design equation number two.

For a given transistor (C_{GS} and ω_T, which is $g_m/2\pi C_{GS}$) and a given frequency f_{in}, both L_S and L_G are now determined.

How much is the gain, and the Noise Figure?

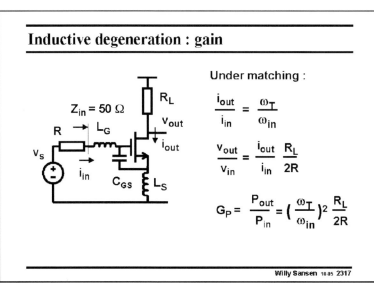

Inductive degeneration : gain

$Z_{in} = 50 \, \Omega$

Under matching :

$\dfrac{i_{out}}{i_{in}} = \dfrac{\omega_T}{\omega_{in}}$

$\dfrac{v_{out}}{v_{in}} = \dfrac{i_{out}}{i_{in}} \dfrac{R_L}{2R}$

$G_P = \dfrac{P_{out}}{P_{in}} = \left(\dfrac{\omega_T}{\omega_{in}}\right)^2 \dfrac{R_L}{2R}$

Willy Sansen 10.05 2317

2317

Several gains are readily calculated. The transistor model only includes the g_m and C_{GS}; the output resistance r_{DS} is usually a lot larger than R_L and can therefore be neglected.

In this case, and remembering the two matching conditions, the current gain is readily calculated. Also, the voltage gain is easily derived.

The power gain is the product of both. It contains the ω_T squared. A transistor must now be taken with high ω_T. This means that its channel length L must be made as small as possible and its $V_{GS} - V_T$ high (see Chapter 1).

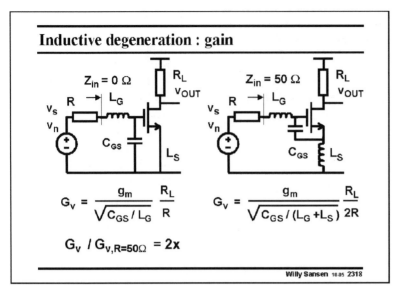

2318

Note that the voltage gain G_v is different whether an inductor L_S is inserted in the Source or not.

Without this inductor (on the left) the input impedance is zero. The input impedance is not matched to the antenna.

With this inductor (on the right), the input impedance is purely resistive and equal to R (or 50 Ω). As a result, there is a division by two at the input such that the voltage gain is a factor of two smaller.

Inductive degeneration : Noise Figure

$$NF = 1 + \frac{dv_{in}^2}{dv_R^2} = 1 + \frac{R\, C_{GS}}{g_m\,(L_G+L_S)}$$

$$dv_{in}^2 = 4kT\, \frac{2/3}{g_m}\, df \approx 4kT\, \frac{1}{g_m}\, df$$

$$dv_R^2 = 4kT\, R\, df$$

$$NF = 1 + g_m\, R\, \left(\frac{\omega_{in}}{\omega_T}\right)^2$$

2319

The Noise Figure NF is defined as in Chapter 4.

The equivalent input noise of the input transistor is as given in this slide. The term 2/3 has been left out to take into account that this MOST is probably operating close to velocity saturation where its thermal noise is probably larger. It may help somewhat to lower the Drain-Source voltage to avoid velocity saturation.

The noise of the load R_L is neglected.

Under the two matching conditions, the NF can then be rewritten as given in this slide.

It is striking that the NF decreaes for larger values of ω_T, but not for larger g_m. Parameter ω_T is therefore clearly the dominant transistor parameter.

Also, note that the NF is quite low at lower frequencies, but increases versus frequency. At the highest frequencies of interest the NF may not be all that attractive.

Inductive degeneration : IIP3 vs IDS

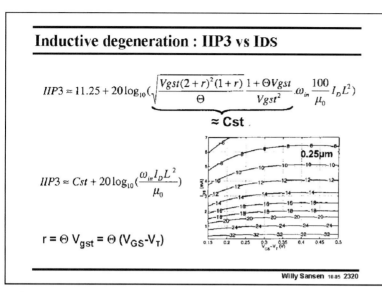

$$IIP3 \approx 11.25 + 20\log_{10}\left(\underbrace{\sqrt{\frac{Vgst(2+r)^2(1+r)}{\Theta}}\frac{1+\Theta Vgst}{Vgst^2}.\omega_{in}\frac{100}{\mu_0}I_D L^2}\right)$$

$$\approx Cst$$

$$IIP3 \approx Cst + 20\log_{10}\left(\frac{\omega_{in}I_D L^2}{\mu_0}\right)$$

$$r = \Theta\, V_{gst} = \Theta\,(V_{GS}-V_T)$$

0.25µm

Willy Sansen 10.05 2320

DC current I_{DS}. The higher the current, the higher the IIP3.

2320

The 3rd-order intermodulation intercept point IIP3 can be calculated using the simple model of Chapter 1, in which fitting parameter θ is used (or Θ).

Because of the two derivatives to be taken, the expression is quite evolved.

A plot is given of the IIP3 versus the transistor current I_{DS} and its $V_{GS}-V_T$. It is striking to see that the value of $V_{GS}-V_T$ is not all that important. The only real important parameter is the

Cascode or current input : Z_{in}

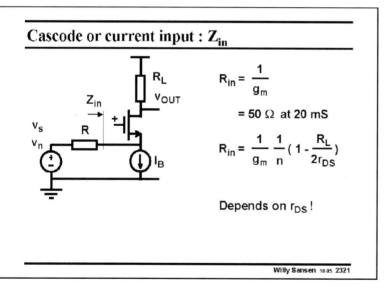

R_L

Z_{in}

V_{OUT}

V_s
V_n

R

I_B

$$R_{in} = \frac{1}{g_m}$$

$$= 50\,\Omega \text{ at } 20\text{ mS}$$

$$R_{in} = \frac{1}{g_m}\frac{1}{n}\left(1 - \frac{R_L}{2r_{DS}}\right)$$

Depends on r_{DS} !

Willy Sansen 10.05 2321

2321

Instead of an amplifier, a cascode can be taken as well. Its advantage is that its input impedance is resistive up to high frequencies and can be set by the current. Indeed, it is the inverse transconductance.

However, a closer look reveals that several other factors come in such as the parameter n ($\approx 1 + g_{mB}/g_m$), and the output resistance r_{DS}. They render the input resistance much less accurate.

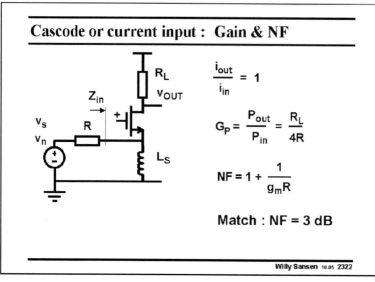

2322

If the LNA is matched, i.e. if $R \approx 1/g_m$, then it is easily found that the current gain is unity and the voltage and power gain are as given in this slide. It shows that the load resistor R_L is much (up to ten times) larger than R.

The Noise Figure is also easily derived. It is given in this slide provided the gain is large enough to neglect the noise from R_L.

Note that this expression yields a NF of 2 or 3 dB under matched conditions.

This value is rather high but is independent of frequency.

For very high frequencies this NF may even be better than for an amplifier configuration, as shown next.

The linearity is always better than for an amplifier configuration because this stage is actually current driven.

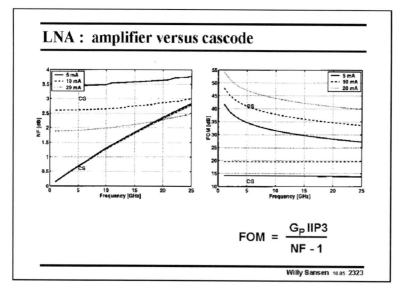

2323

For sake of comparison, the Noise Figure is given for three different currents versus frequency for a standard 0.13 μm CMOS technology.

In addition, a FOM is added as described in this slide.

The amplifier (Common-Gate) configuration shows a NF which deteriorates when the frequency increases, which is not true for the cascode (Common Gate). For a small current (5 mA), the cascode performs better than the amplifier from about 12 GHz. This is about $f_T/2$ however, and may not be so useful.

The FOM of an amplifier is always higher than for a cascode.

Table of contents

Willy Sansen 10-05 2324

2324

As LNA's reach high frequencies, up to $f_T/2$, some other phenomena show up. In addition to distributed capacitors, both to the substrate as between parallel lines, the distributed nature of the channel of a MOST starts playing as well.

This is explained next.

Non-quasi static MOST model

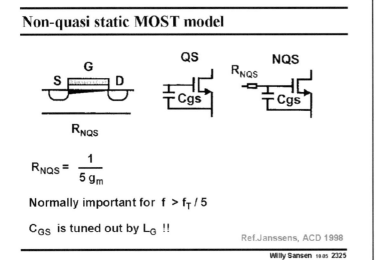

$$R_{NQS} = \frac{1}{5\,g_m}$$

Normally important for f > f_T / 5

C_{GS} is tuned out by L_G !!

Ref.Janssens, ACD 1998

Willy Sansen 10-05 2325

2325

In a classical quasi-static model of a MOST, we assume that any change in Gate voltage is followed by an instantaneous change in channel charge. In practice, there is some delay, however. To change the charge in the channel, or in the inversion layer, carriers must be drawn from Source and Drain, which takes time.

In order to model this time delay to a first degree, a low-pass filter is added at the Gate. It is realized by addition of a resistor R_{NQS}, which forms a low-pass filter with input capacitance C_{GS}.

The value of this resistor R_{NQS} must be about $1/5g_m$ (see Chapter 1, Tsividis 1987).

Clearly, this effect is important for really high frequencies, higher than $f_T/5$. LNA's and VCO's operate up to frequencies of $f_T/3$, however. This effect must now be taken into account if precise predictions are important.

Also, if an inductor is added in series with the Gate, then the effect of this resistor is even more important.

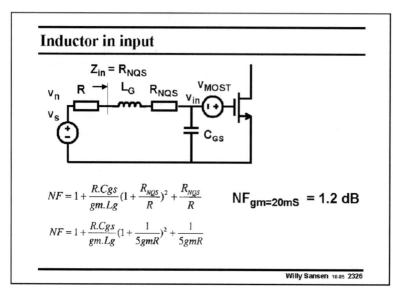

Inductor in input

$Z_{in} = R_{NQS}$

$$NF = 1 + \frac{R.Cgs}{gm.Lg}(1 + \frac{R_{NQS}}{R})^2 + \frac{R_{NQS}}{R}$$

$$NF = 1 + \frac{R.Cgs}{gm.Lg}(1 + \frac{1}{5gmR})^2 + \frac{1}{5gmR}$$

$NF_{gm=20mS} = 1.2$ dB

Willy Sansen 10.05 **2326**

2326

This additional resistance R_{NQS} in series with the Gate deteriorates the NF as shown in this slide. As a resistance $1/5g_m$ is taken.

It is a single-MOST amplifier with only a Gate inductor.

Several terms have to be added to the Noise Figure. The most important one is the term which is squared.

Note that the input impedance is no more zero but equal to this resistance R_{NQS}.

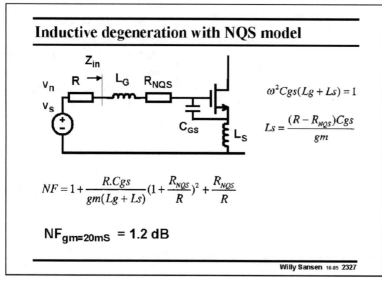

Inductive degeneration with NQS model

Z_{in}

$$\omega^2 Cgs(Lg + Ls) = 1$$

$$Ls = \frac{(R - R_{NQS})Cgs}{gm}$$

$$NF = 1 + \frac{R.Cgs}{gm(Lg + Ls)}(1 + \frac{R_{NQS}}{R})^2 + \frac{R_{NQS}}{R}$$

$NF_{gm=20mS} = 1.2$ dB

Willy Sansen 10.05 **2327**

2327

When two inductors are used, the addition of the NQS resistor influences the matching equations. The first one (on top right) is about the tuning out of capacitor C_{GS}, and is the same as before.

The second one however, includes R_{NQS}, such that a smaller value of L_S is obtained.

The Noise Figure will be worse, as an additional resistor R_{NQS} is present in series with the Gate of the amplifier. In its simplest approximation, the R_{NQS} simply has to be added to source resistor R. The most important addition however, is the factor which is squared.

Noise matching

$$NF = 1 + \frac{R.Cgs}{gm.(Lg + Ls)}(1 + \frac{1}{5gmR})^2 + \frac{1}{5gmR}$$

Optimum $\quad R = \frac{1}{5gm} \cdot \sqrt{1 + \frac{5.(Lg + Ls)gm^2}{Cgs}} = 80\Omega$

NF ≈ 1 dB (g$_m$ = 20 mS)

$L_G + L_S$ = 15 nH; C_{GS} = 0.5 pF; f = 1.8 GHz

Willy Sansen 10-05 2328

2328

This LNA can now be further optimized, taking into account this R_{NQS}. First of all this R_{NQS} is substituted by $1/5g_m$. Then the optimum value of R can be found, as a function of g_m.

The value of R is slightly higher than 50 Ω but the NF is lower than before.

Noise matching (Optimum design)

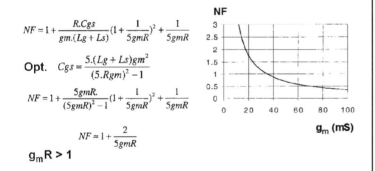

$$NF = 1 + \frac{R.Cgs}{gm.(Lg + Ls)}(1 + \frac{1}{5gmR})^2 + \frac{1}{5gmR}$$

Opt. $\quad Cgs = \frac{5.(Lg + Ls)gm^2}{(5.Rgm)^2 - 1}$

$$NF = 1 + \frac{5gmR}{(5gmR)^2 - 1}(1 + \frac{1}{5gmR})^2 + \frac{1}{5gmR}$$

$$NF \approx 1 + \frac{2}{5gmR}$$

$g_m R > 1$

Lower NF requires more power !

Willy Sansen 10-05 2329

2329

Further optimization shows that now the transistor size has to be adapted as well, giving a different value for C_{GS}.

The expression of the NF has now become very simple now. It is the lowest value possible.

It is obvious that lower values of NF can only be obtained, provided higher g_m values are obtained, and hence more power.

A plot of this expression is also given.

The larger the transconductance, the lower the NF.
For 1 dB Noise Figure, about 33 mS transconductance is required.

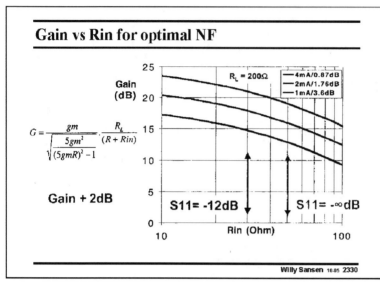

Gain vs Rin for optimal NF

$$G = \frac{gm}{\sqrt{\dfrac{5gm^2}{(5gmR)^2-1}}} \cdot \frac{R_L}{(R+Rin)}$$

Gain + 2dB

S11= -12dB   S11= -∞dB

Willy Sansen 10.05 2330

but gives about 2 dB more gain.

This is clearly advantageous provided some reflection (-12 dB s_{11}) can be allowed.

2330
This optimization illustrates the compromises to be taken.

For a constant load resistor R_L, the gain is given for different currents, causing at the same time, deviations for the ideal input matching conditions.

For a $50\,\Omega$ input resistance, the s_{11} is zero. The gain then increases with the current.

Taking a slightly smaller input resistance however, increases the s_{11} to -12 dB

Table of contents

• **Noise Figure and Impedance Matching**

• **LNA specifications and linearity**

• **Input amplifier or cascode**

• **Non-quasi-static MOST model**

• **More realizations**

• **Inductive ESD protection**

Willy Sansen 10.05 2331

2331
Now that the design equations have been identified for low-noise LNA's, taking into account Non-quasi-static effects, we focus on some realizations.

It is clear that the configurations only differ a small amount. The transistor sizes and currents however, are largely different depending on the frequencies. Also, the sizes of the inductors which are added, can be very different.

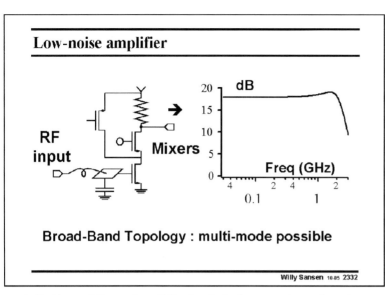

Low-noise amplifier

Mixers

RF input

Broad-Band Topology : multi-mode possible

Willy Sansen 10.05 2332

2332

Our overview starts with a very simple LNA. A cascode is used to improve the isolation between output and input. This is especially required to avoid leakage from the next block, which is usually a mixer, to the input of the LNA.

In order to avoid noise coming from the cascode, its DC current is reduced. A current source with pMOST, provides some of the DC current to the input transistor. In this way, the contribution of the noise of the load devices can be reduced somewhat.

At the input, the bond-wire is shown. It is used to tune out the effect of the input MOST capacitance and of the bonding pad capacitance.

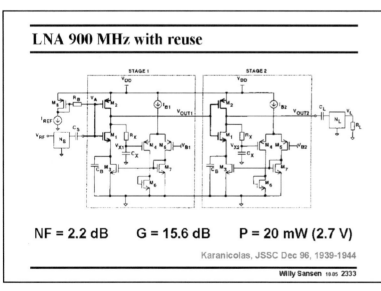

LNA 900 MHz with reuse

NF = 2.2 dB   G = 15.6 dB   P = 20 mW (2.7 V)

Karanicolas, JSSC Dec 96, 1939-1944

Willy Sansen 10.05 2333

2333

This LNA consists of two equal stages.

The input transistors M1 and M2 are connected as a common CMOS inverter amplifier. Both of them contribute to the transconductance. The current in the nMOST is now reused in the pMOST.

However, the biasing current of a CMOS inverter amplifier heavily depends on the supply voltage. This problem is solved here by a biasing block consisting of transistors M3-M7. This block provides DC feedback such that the DC output voltage equals V_{B1} for the first LNA. It is fully decoupled from the AC operation by means of two decoupling capacitors C_B and C_X.

At the input a matching network N_S is provided.

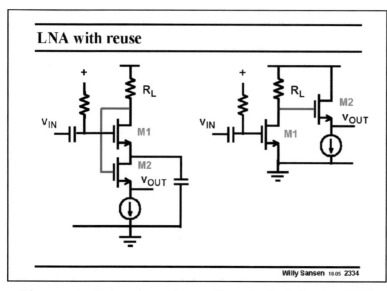

The same technique is used three times in the following LNA.

2334
Current reuse can also be effectuated in another way.

If a two-stage amplifier is required, or an amplifier followed by a source follower, then current is used twice, as shown on the right.

However, a judicious set of capacitors allows limiting the current to just one single path. This is shown on the left.

Transistor M1 sees a signal ground at its Source. Its output is coupled to the Source follower underneath.

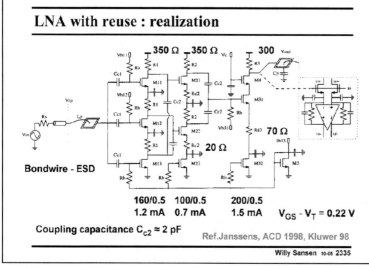

2335
In this LNA, three input nMOSTs M11-M13 are placed in parallel to increase the transconductance. They all share the same DC current, however. The outputs are placed in parallel two by two and applied to a second stage with transistors M21–M22. The same method is then applied in the second stage. Their two outputs are put in parallel again and applied to a third amplifier stage with transistor M31, which provides the output voltage through cascode transistor M4.

The DC biasing is applied by means of current mirrors M5 and M13, M23 and M32.

The transistor sizes, currents and resistors are all given in this slide. The resulting specifications are given next.

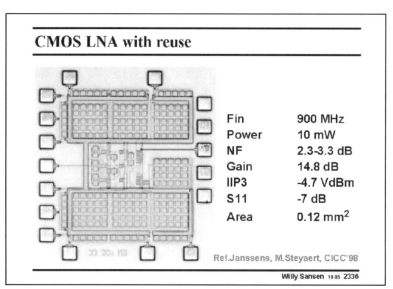

CMOS LNA with reuse

Fin	900 MHz
Power	10 mW
NF	2.3-3.3 dB
Gain	14.8 dB
IIP3	-4.7 VdBm
S11	-7 dB
Area	0.12 mm²

Ref.Janssens, M.Steyaert, CICC'98

Willy Sansen 10-05 2336

2336

The actual gain is 14.8 dB, which is relatively high. The NF is quite reasonable for this power consumption.

The layout shows the large area taken up by the decoupling capacitance. As this LNA has a single-ended input, it easily picks up noise from the ground and supply lines. A decoupling capacitance at the supply line is therefore mandatory.

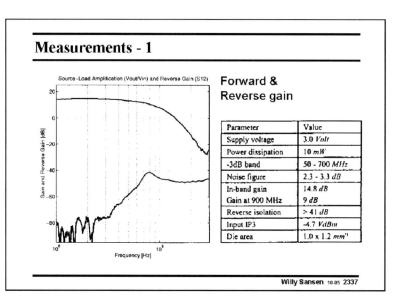

Measurements - 1

Source-Load Amplification (Vout/Vin) and Reverse Gain (S12)

Forward & Reverse gain

Parameter	Value
Supply voltage	3.0 *Volt*
Power dissipation	10 *mW*
-3dB band	50 - 700 *MHz*
Noise figure	2.3 - 3.3 *dB*
In-band gain	14.8 *dB*
Gain at 900 MHz	9 *dB*
Reverse isolation	> 41 *dB*
Input IP3	-4.7 *VdBm*
Die area	1.0 x 1.2 *mm²*

Willy Sansen 10-05 2337

2337

The measurements show that the gain and reverse isolation are quite good indeed.

The bandwidth is less than expected, 700 MHz instead of 900 MHz.

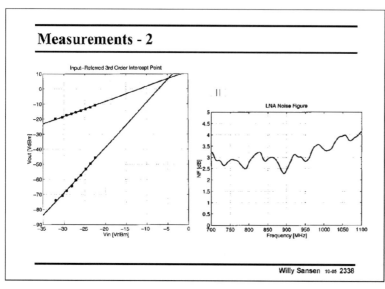

2338

The IIP3 and Noise Figure are shown.

For the IIP3 the gain is measured and the 3rd-order intermodulation distortion. The extrapolated curves meet at an input signal level of about -4.7 dBm. This is a normal value for a MOST input, which has a $V_{GS} - V_T$ of only 0.22 V.

The Noise Figure is not that flat but is never much higher than 3 dB up to 900 MHz.

In this LNA, the ESD protection is not part of the design procedure. A better approach is given next.

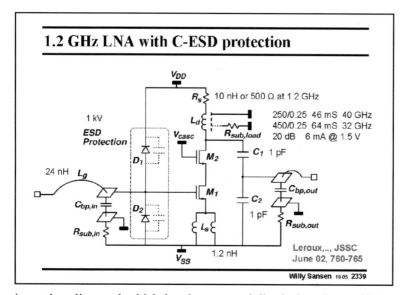

2339

This LNA is matched to 50 Ohms at both input and output. The supply voltage is 1.5 V. At the input, it is an amplifier using inductive source degeneration for input matching. The source inductor L_s is implemented as two parallel bonding wires. It also uses a cascode transistor.

At the input, L_g is the input bonding wire which serves as the inductor for input resonance. To the right of that we have the input bonding pad which has been especially designed, as will be explained further. It basically consists of only the top metal layer. The bottom metal layer serves to shield the pad from the substrate and increase its Quality factor. After that we have the ESD-protection diodes. The upper diode conducts the ESD-charge in case of a positive pulse vs. V_{DD}. The lower diode operates in case of a negative pulse vs. ground.

The output features a load inductor L_d with its series resistance, R_s. This inductor is implemented on-chip and has a patterned ground shield beneath to shield it from the substrate. The 50 Ohms output is ensured through the use of a capacitive divider made up of C_1 and C_2. The output bonding pad also takes part in the matching as it is just in parallel to C_2. Also, this bonding pad has been shielded from the substrate in order to increase coupling through the substrate which would degrade the reverse isolation. It also ensures a fixed and known value for the pad capacitance.

LNA Micrograph

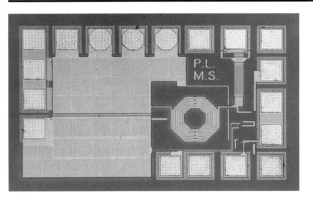

P.L.
M.S.

Leroux,.., JSSC June 02, 760-765

Willy Sansen 10.05 2340

2340

This is a microphotograph of the LNA.

The input and the output are easily recognized.

Also, the protection diodes and the transistors are visible.

Finally, the load inductor and C_1 and C_2, the capacitive divider can be found.

These large objects on the left are the decoupling capacitors.

Noise Figure

- Min. NF = 0.8 dB
- BW (NF<1dB) = 130 MHz

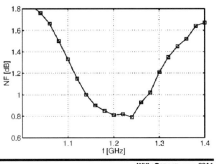

Willy Sansen 10.05 2341

2341

The Noise Figure of the LNA was measured using a noise figure analyzer. This measurement was performed while the LNA was operating in its nominal 9 mW (at 1.5 V supply voltage) regime.

The Noise Figure of the LNA reaches a minimum of 0.79 dB at the frequency of operation. The NF stays below 1 dB for a bandwidth of about 130 MHz and stays below 2 dB for more than 400 MHz around the working frequency.

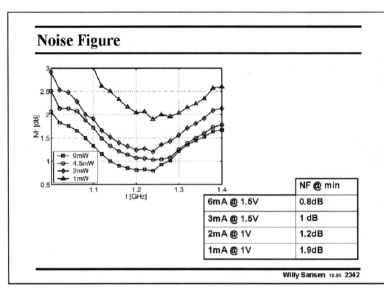

Willy Sansen 10.05 **2342**

2342
Further measurements of the noise figure were performed in different power regimes. The bottom curve is the curve shown on the previous slide. The next curve when going up is measured when reducing the bias current to 3mA. The Noise Figure here reaches a minimum of 1 dB. Above that we have the Noise Figure for the LNA while drawing 2 mA from a 1 V supply. The minimum is now 1.2 dB. The final curve was measured drawing 1 mA from a 1 V supply, i.e. consuming only 1 mW. The minimum noise figure is 1.9 dB.

Note, however, that the input matching conditions are no longer fulfilled in the last three measurements.

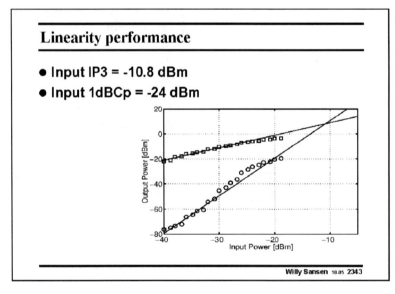

Willy Sansen 10.05 **2343**

2343
Finally, the linearity performance of the LNA was measured. The signal power of the first-order output and the third-order intermodulation products are plotted vs. the input power. Linear regression shows an input referred IP3 of -10.8 dBm. Although linearity is not always that important, this value would be sufficient for the very stringent linearity specification of the GSM system.

The -1 dB compression point is measured to be at -24 dBm.

Performance summary

Parameter	Specification	Measurement
Supply voltage	1.5 *Volt*	1.5 *Volt*
Power dissipation	10 *mW*	9 *mW*
Noise figure	1 *dB*	0.79 *dB*
Power gain @ 1.23 GHz	*Max.*	20 *dB*
S11 at 1.23 GHz	-10 *dB*	-11 *dB*
S22 at 1.23 GHz	-10 *dB*	-11 *dB*
Reverse isolation	30 *dB*	31 *dB*
Input IP3	-20 *dBm*	-10.8 *dBm*
HBM ESD-protection	0.5 *kV*	0.6 *kV* / -1.4 *kV*
Die area	-	0.6 x 1.1 *mm²*

Willy Sansen 10.05 2344

2344

This is a summary of the performance of the LNA compared to the specifications.

The supply voltage is 1.5 V. The power dissipation is 9 mW which is lower than the 10 mW specification. The noise figure was measured 0.79 dB, lower than the 1dB specification. The power gain was maximized and is measured to be 20 dB. Input and output reflection are −11 dB. Reverse isolation is larger than 31 dB over the entire spectrum of the network analyzer. The input IP3 is −10.8 dBm which is more than sufficient.

A final measurement which was done is a HBM test for the level of ESD protection. The specification put forward was 0.5 kV. Measurements have shown that the LNA input is able to withstand pulses from −1.4 to +0.6 kV, exceeding the specification.

Die area is 0.6 by 1.1 mm².

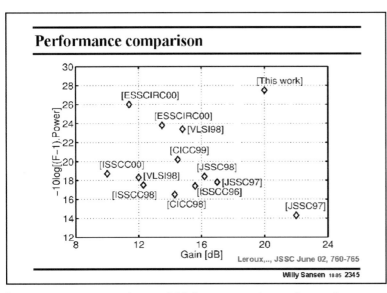

Willy Sansen 10.05 2345

2345

In this figure, a Figure of merit is plotted vs. the power gain for published CMOS LNAs. The Figure of merit is minus ten times the log of F minus one times power, where F is the Noise Figure. A high Figure of merit means that the LNA scores well in the area of low Noise Figure and low power consumption.

As such, the three main performance parameters are shown in this plot. Circuits with a low Noise Figure, a low power consumption and high power gain are located more towards the upper right corner. Most LNAs with very low Noise Figure and/or power consumption also have low power gain and vice versa. This last realization however, succeeds in combining good performance in all three areas.

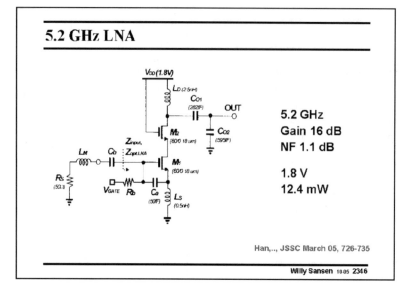

5.2 GHz
Gain 16 dB
NF 1.1 dB

1.8 V
12.4 mW

Han,.., JSSC March 05, 726-735

Willy Sansen 10.05 **2346**

2346

LNA's at higher frequencies have very much the same design flow as their lower frequency counterparts. Only the inductors are smaller to be able to tune out higher frequencies.

An example of an LNA at 5 GHz is given in this slide.

It is a conventional single-ended cascode amplifier. Two inductors are used again, with only 0.5 nH in the Source.

An output matching network is added to obtain 50 Ω output impedance.

The Noise Figure is now quite low.

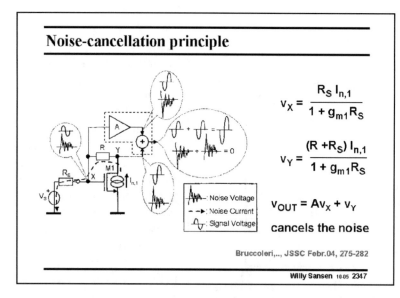

$$v_X = \frac{R_S I_{n,1}}{1 + g_{m1} R_S}$$

$$v_Y = \frac{(R + R_S) I_{n,1}}{1 + g_{m1} R_S}$$

$$v_{OUT} = A v_X + v_Y$$

cancels the noise

Bruccoleri,.., JSSC Febr.04, 275-282

Willy Sansen 10.05 **2347**

2347

An interesting technique to reduce the noise contributed by the input MOST M1, is the circuit schematic in this slide.

In this amplifier, the signal itself is inverted from input to output, as in most amplifiers. The voltage caused by the noise current $I_{n,1}$ of input transistor M1 is not inverted, however. Summation of both signals, with the proper scaling will cancel out the contribution of the noise current $I_{n,1}$ to the output.

The appropriate expressions are given in this slide.
The full circuit is given next.

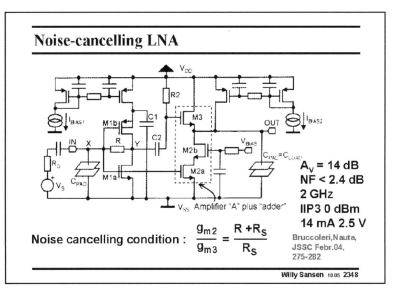

2348

Noise cancellation is subject to the condition given in this slide. The summation or adding amplifier consists of transistors M2a, M2b and M3. The resulting Noise Figure is quite low without a lot of additional power consumption.

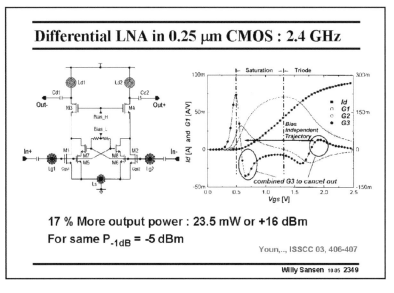

2349

A differential LNA is shown in this slide. Its advantage is that it is less susceptible to spikes and noise on the substrate and the supply lines. However, it takes two times the DC current.

It consists of two cascode amplifiers, to which four transistors M5–M8 have been added to suppress the 3rd-order distortion. As a result the -1 dB compression point is at -5 dBm, meaning that the IIP3 is about at 5 dBm, which is quite large indeed.

The distortion cancellation operates as follows.

For one single MOST the current is given on the right, together with its first three derivatives. The 3rd-order one G3 has a negative peak around a V_{GS} of 0.7 V. It has a positive peak as well at 1.8 V.

Another transistor combination M5–M7 is now added which shows a positive peak at only 0.7 V. Addition of its current to the current of the previous transistor allows (partial) cancellation of the 3rd-order distortion components.

It is clear that cancellation techniques always suffer from mismatch. Complete cancellation is always hard to achieve. However, an increase of the IIP3 with 5 dBm is normally sufficient. This is the case here.

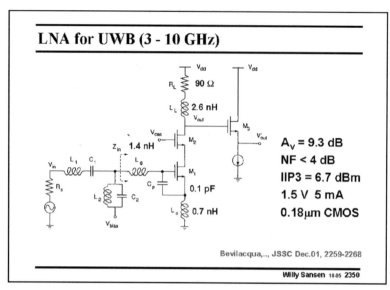

2350

A wideband LNA for an Ultra-Wide Band receiver is shown in this slide.

The frequency extends from 3 to 10 GHz. It has been realized in a fairly conservative 0.18 µm CMOS. The gain is rather low indeed.

The capacitance C_p of 0.1 pF has been added to C_{GS1} to be tuned out more accurately by means of inductors L_g and L_s.

A parallel L_2C_2 network is added to broadband the frequency response without impairing the input impedance matching.

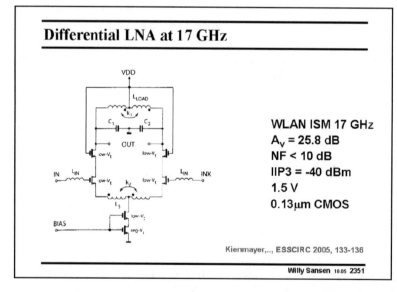

2351

A differential LNA for a WLAN ISM receiver is shown in this slide.

The advantage of a differential configuration is that it rejects substrate noise much better.

The input antenna must then be followed by a balun. This is a toroidal RF transformer which converts the single-ended input into a symmetrical output.

The configuration is a well known cascode stage.

The gain is fairly high. As a result the IIP3 is rather low.

Differential LNA at 5 GHz

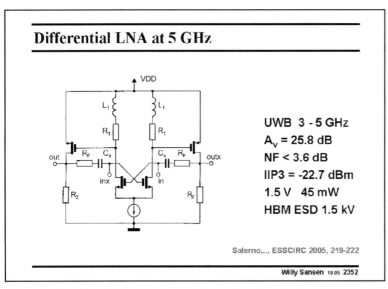

UWB 3 - 5 GHz
$A_v = 25.8$ dB
NF < 3.6 dB
IIP3 = -22.7 dBm
1.5 V 45 mW
HBM ESD 1.5 kV

Salerno,.., ESSCIRC 2005, 219-222

Willy Sansen 10.05 **2352**

2352

Another differential LNA for UWB up to 5 GHz is shown in this slide.

As it is the differential, the sensitivity to substrate coupling is quite small. It consists of two gain stages with resistive feedback. As a result, the linearity is improved over the full bandwidth.

The gain is quite high. As a result, the IIP3 is rather low. A gain reduction mode is provided however, which yields a better IIP3.

Note that the ESD protection is included. The input can withstand 1.5 kV. This is why we focus now on the impact of ESD protection networks on the performance of the LNA's.

Table of contents

- **Noise Figure and Impedance Matching**

- **LNA specifications and linearity**

- **Input amplifier or cascode**

- **Non-quasi-static MOST model**

- **More realizations**

- **Inductive ESD protection**

Willy Sansen 10.05 **2353**

2353

The LNA is normally the first active building block in a receiver.

As it is connected directly to the input pad over a bonding wire, it receives electrostatic discharge voltages from outside.

These voltages can be quite high (kV's) and can easily zap the Gate of the input transistor. A protection device is now required.

Normally, this consists of series resistances to the Gate and parallel diodes to ground. A low-resistance path is now realized for overvoltages. This is discussed in more detail next.

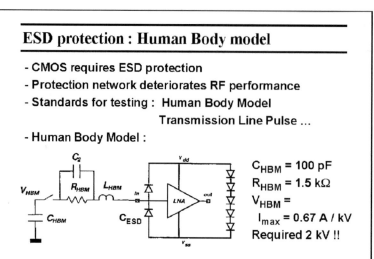

2354

ESD protection networks can be tested in various ways. The simplest model of an ESD source is probably the Human Body Model. It models a human who discharges into a pin. It consists of a capacitor C_{HBM} which is charged to a high voltage (kV's) and discharged over a small resistor R_{HBM} (about 1500 Ω) and bonding wire inductor L_{HBM} to the input pad. A small parasitic capacitor C_2 is present as well.

A protection diode is therefore connected to supply and ground. It must be able to take large currents to avoid large voltages at the Gate. In this example, the diode must be able to conduct 0.67 A/kV overvoltage. It takes a fairly large area, and gives a fairly large capacitor at the input of the LNA.

ESD protection diodes add capacitances to ground however, after the Gate inductance L_{HBM}. The tuning out of the C_{GS} capacitance is disturbed somewhat, leading to smaller values of R and larger transistor sizes.

These capacitances can be tuned out as well. It is always preferable however, to keep the input matching network as simple as possible.

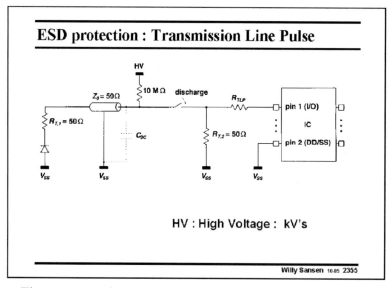

2355

A HBM models a human. Another way to stress the input protection is to charge a transmission line ($Z_o = 50$ Ω) to a high voltage and to discharge it through a large resistor R_{TLP} to the input pin.

Resistors $R_{T,1}$ and $R_{T,2}$ are added to avoid reflections on both ends. The diode on the left end is added to avoid leakage to ground during charging.

The following test procedure is normally adopted. The stress test is executed between two pins (here pin1 and pin 2). First of all, the leakage current is measured between these two pins. Then a high voltage discharge is applied after which the leakage current is measured again. The discharge voltage is stepped up until the leakage current has become excessive. An example of such a TLP test is given at the end of this Chapter.

Simulated deterioration by ESD diodes

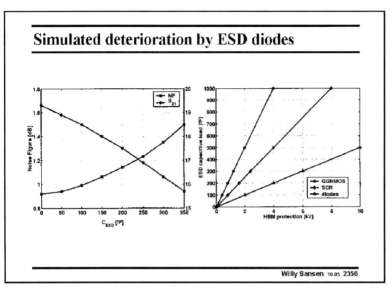

Willy Sansen 10.05 2356

2356

For sake of illustration, the effect is shown of the ESD capacitance on the Noise Figure for a 1.6 GHz LNA (in 0.25 μm CMOS). The current is 6 mA. The input C_{GS} is about 0.15 pF.

For a C_{ESD} increasing to 0.35 pF, the NF increases from 0.9 dB to 1.5 dB. The power gain then decreases from 19.3 to 15.7 dB.

This ESD protection provides better protection, as shown in the second plot.

Π-network with Capacitive ESD protection

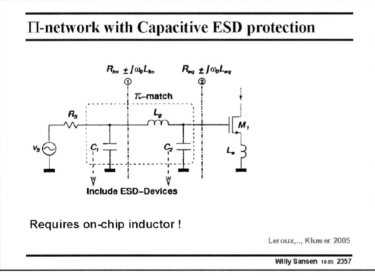

Requires on-chip inductor !

Leroux,.., Kluwer 2005

Willy Sansen 10.05 2357

2357

If capacitive protection devices are used, it is now better to connect them in a π-network, as shown in this slide. The addition of an inductor L_g inbetween increases the protection somewhat. This inductor is actually too small to play a role at the ESD-frequencies. It does play a role for the input impedance matching at the RF frequencies.

Also, any parasitic capacitance at the Gate of the MOST to ground, can be absorbed by the ESD-device capacitance C_2. This latter capacitor must always be included in the design process, for high-frequency LNA's.

The design of such a π-network is not obvious, as even more degrees of freedom are available. However, the main concern is to make C_1 as small as possible.

The Noise Figure will be reduced somewhat as the inductor L_g has some series resistance (typically 2 Ω/nH).

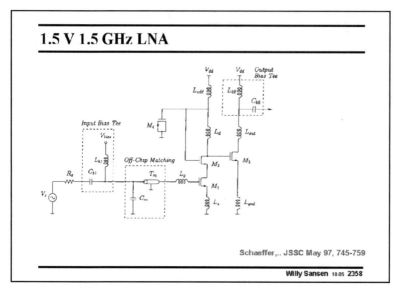

1.5 V 1.5 GHz LNA

Schaeffer,.. JSSC May 97, 745-759

Willy Sansen 10.05 **2358**

2358

An example of such a π-network is shown in this slide.

The capacitance C_m of the ESD protection is also part of the matching network. It forms a π-network with the input capacitance of M1.

The LNA is a single-ended cascode amplifier, with a second amplification stage to provide a 50 Ω output resistance.

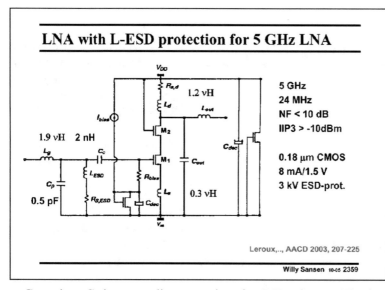

LNA with L-ESD protection for 5 GHz LNA

5 GHz
24 MHz
NF < 10 dB
IIP3 > -10dBm

0.18 μm CMOS
8 mA/1.5 V
3 kV ESD-prot.

Leroux,.., AACD 2003, 207-225

Willy Sansen 10-05 **2359**

2359

As an ESD-protection, an inductor can be used instead of a capacitor as well. This is shown in this slide.

Inductor l_{ESD} is used to short the low-frequency ESD currents to ground while it forms a parallel resonant circuit with parasitic capacitance C_p, such that it is invisible for RF voltages.

Clearly, the ESD inductor has some series resistance as well $R_{S,ESD}$. It is clear that this resistance must be minimized.

Capacitor C_c is a coupling capacitor for RF voltages. The low-frequency ESD signals see a low-pass filter to ground. The high-frequency RF signals see a high-pass filter to the input Gate.

The advantage of this protection is that for a capacitive protection an inductor must be added to tune out the parasitic capacitances of the protection components. For an inductive protection, nothing must be added. The protection inductor can be used to tune out existing parasitic capacitances.

For noise, the inductive protection is better as no resistances are added in series with the input Gate.

Inductor L_{ESD} is 2 nH; it is realized by a string of five diodes, giving 3 Ω series resistance only. It takes an area of about 130 μm². This inductor forms a parallel resistance of 1 kΩ together with C_p, which is sufficiently high not to impair the noise performance.

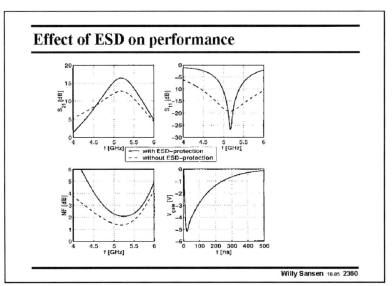

2360

The addition of an ESD protection changes the characteristics considerably as shown in this slide.

The gain (s_{21}) increases but the Noise Figure decreases around the center frequency of 5 GHz. The input impedance (s_{11}) deviates less from 50 Ω except at one particular frequency, evidently depending on the values of the tuning components.

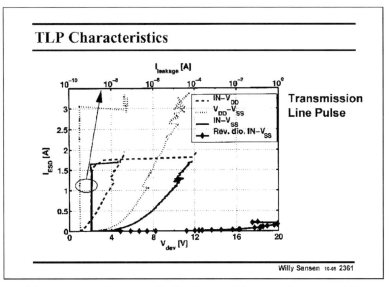

This corresponds to static discharge voltages of up to 3 kV.

2361

Measured results on the protection network as a result of a TLP test are shown in this slide.

The leakage currents before and after the large ESD currents are shown. This is completed for four combinations of pins.

Both the available currents from the input terminal towards the positive supply (V_{DD}) and ground (V_{SS}) are quite large. About 1 A to ground is reached for a 10 V input voltage.

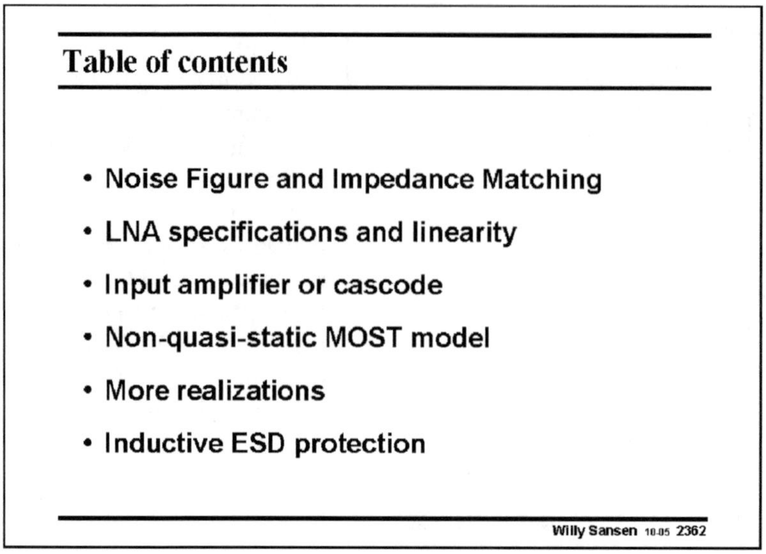

Table of contents

- Noise Figure and Impedance Matching

- LNA specifications and linearity

- Input amplifier or cascode

- Non-quasi-static MOST model

- More realizations

- Inductive ESD protection

Willy Sansen 10.05 2362

2362

In this Chapter, Low-noise amplifiers (LNA's) have been discussed which operate at high (RF) frequencies.

They have been optimized for low noise and distortion within the constraints of input (and output) impedance matching.

The elementary expressions have been derived for impedance matching and noise matching. The two main configurations have been compared. It has been shown that the amplifier input almost always performs better than the cascode input.

At these high frequencies the Non-quasi-static behavior of the MOST has sometimes to be taken into account. This is certainly true if the operating frequency is higher than $f_T/3$ of the MOST.

Many of the realizations given are commented on. They all exhibit similar circuit configurations but take different compromises in speed, noise and distortion.

Finally, considerable attention is paid to the ESD-protection. Indeed, as the LNA is the first active stage in a receiver, it is subject to lightning and other static discharges. Protection components such as diodes and MOSTs must be added, which shift the compromise in the direction of more power consumption and/or noise.

Coupling effects in Mixed analog-digital ICs

Willy Sansen

KULeuven, ESAT-MICAS

Leuven, Belgium

willy.sansen@esat.kuleuven.be

Willy Sansen 10.05 241

Today, most integrated circuit realizations include both digital and analog functions, on the same substrate. Processors include ADCs and DACs. RF functions are added to digital processors. Many more examples can be found.

This means that coupling occurs between both types of circuits. Most often, the digital blocks generate spikes on the ground and supply lines, which are sensed by the analog amplifiers.

To avoid deterioration of the dynamic range of these high-performance analog functions, coupling must be minimized.

This is discussed in this Chapter. First of all, several sources of coupling are categorized. Layout techniques are now discussed to reduce coupling. Attention is also paid to the specification power-supply-rejection-ratio and how it can be predicted for some specific amplifier blocks.

242

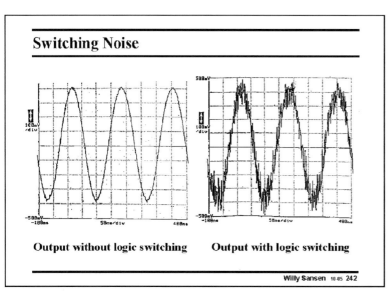

Switching Noise

Output without logic switching   Output with logic switching

Willy Sansen 10.05 242

Coupling is best visible when a sinusoidal waveform is analyzed with high spectral purity. Switching on the logic functions on the same substrate, induces high-frequency noise on the waveform. This is a result of coupling. It deteriorates drastically the SNR. This is what has to be avoided in present-day mixed-signal circuits.

243

In coupling, three phenomena have to be investigated.

The first one is the generator of the noise. The digital circuits usually work with a clock. They draw current spikes through their supply lines, which cause voltage spikes on the supply linesand ground.

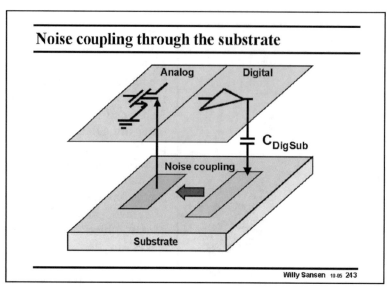

Noise coupling through the substrate

Analog   Digital

C_{DigSub}

Noise coupling

Substrate

Willy Sansen 10-05 243

These spikes are then injected into the substrate, as each output has a small capacitance C_{DigSub} to the substrate. They have a wide frequency spectrum. This is why they are called noise.

The second phenomenon is the transmission of this noise to the other side of the substrate, where the analog circuits are present. This obviously depends on the nature of the substrate, on the presence of an epitaxial layer, etc.

The third phenomenon is the pickup of this noise by the sensitive analog circuits. Each transistor also has the substrate as an input. It converts the substrate noise into a drain current. As a result, the SNR suffers badly from substrate noise. The most important specification with this respect, is the PSSR or Power-supply-rejection ratio.

All analog circuits are better made fully-differential to reject the substrate noise as much as possible. Mismatch limits this rejection however, and limits the SNR, which can be reached in a mixed-signal circuit.

Outline

- **Circuit noise generation**

- **Circuit noise coupling**
  - **Power supply pinning**
  - **Substrate coupling**
  - **Circuit placement**

- **Rejection of circuit noise**
  - **PSRR**

Willy Sansen 10-05 244

244

Let us concentrate first on the generation of the noise in the substrate.

Transmission in the substrate, or coupling to the analog parts through the substrate comes next.

The pickup by the analog circuits is the final Section.

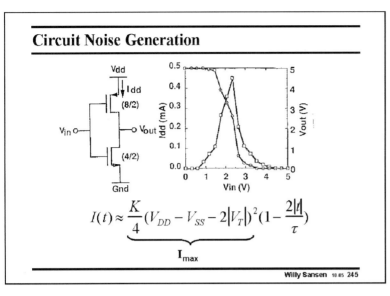

245

Any digital gate causes transitions at its output from a logical zero to one, or vice-versa. This transition causes a current spike from substrate to ground. As an example, a simple CMOS inverter is used. This current spike is then easily described, as shown in this slide.

The size of this spike evidently depends on the sizes of the transistors used. Since thousands of digital gates are put together in a processor, the total current can be quite large and can have a wide frequency spectrum.

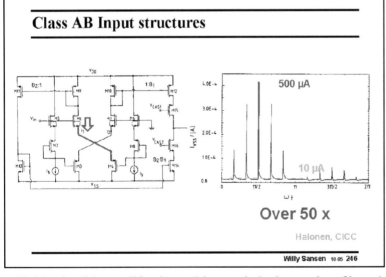

246

However, analog circuits can also cause current spikes. For example, class-AB stages draw currents which are larger than their quiescent current.

As an example, look at the class-AB input stage shown in this slide. For large input voltages, the currents can become quite large. When the current is monitored through the positive supply line. Large current spikes are visible, nearly 50 times the quiescent current.

Evidently, this amplifier is used in a switched-capacitor filter. At each clock cycle, the opamp is driven hard, causing these current spikes to flow.

An important compromise is now to be taken. Class AB circuits allow considerable reduction in quiescent power consumption. On the other hand, they cause current spikes which may impair the SNR.

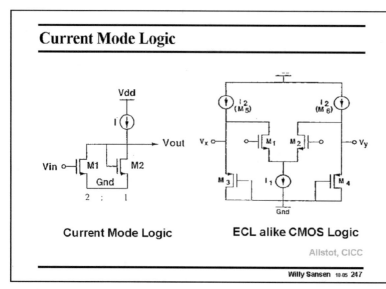

247

Do all digital gates cause current spikes? The answer is negative. Current spikes are typical for CMOS logic. No current is consumed in either the zero output state or the output state. Current is only consumed during the transition. This is the strength of CMOS logic. The static power consumption is very low but increases with the clock frequency.

Several other loc families exist which consume current continuously. ECL in bipolar technology or current-mode logic in CMOS (see slide) are good examples. As current is consumed continuously, the power consumption is high. The advantage however, is that the current spikes are hardly visible.

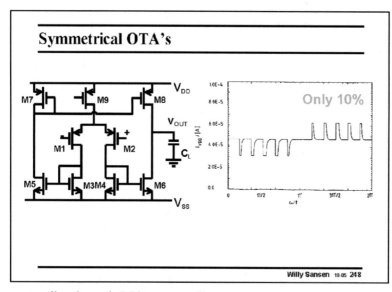

248

Even class-A amplifiers can cause current spikes. An example is shown in this slide of a symmetrical class-A OTA.

When it is used in a switched-capacitor filter, current spikes are found at each clock cycle. Indeed, charging load capacitance C_L fast, generates a current spike from the positive supply through M8 and C_L.

In the same way, will the discharge of this load capacitance C_L cause a current spike through M6 to ground?

These spikes are not as large as in a class-AB circuit, they are not invisible, however.

Outline

- ■ **Circuit noise generation**
- ■ **Circuit noise coupling**
  - ■ **Power supply pinning**
  - ■ **Substrate coupling**
  - ■ **Circuit placement**
- ■ **Rejection of circuit noise**
  - ■ **PSRR**

Willy Sansen 10-05 249

249

Now that it is clear that both digital and analog circuits generate plenty of current spikes, it has to be figured out how they can be prevented from traveling to the other side of the substrate, where the analog input stages are detecting whatever reaches their gates or substrates underneath.

The distribution of the pins plays an important role. It is discussed first.

The longitudinal transmission through the substrate is another important factor.

Finally, layout techniques can be used to reduce coupling.

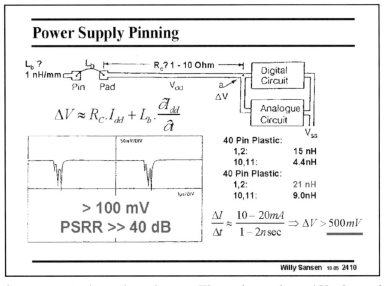

Power Supply Pinning

$$\Delta V \approx R_C . I_{dd} + L_b . \frac{\partial_{dd}}{\partial}$$

> 100 mV

PSRR >> 40 dB

40 Pin Plastic:
| 1,2: | 15 nH |
| 10,11: | 4.4nH |

40 Pin Plastic:
| 1,2: | 21 nH |
| 10,11: | 9.0nH |

$$\frac{\Delta I}{\Delta t} \approx \frac{10 - 20mA}{1 - 2n\sec} \Rightarrow \Delta V > 500mV$$

Willy Sansen 10-05 2410

2410

A pin of a chip package is normally connected over a bonding wire to a bonding pad. This pad then leads to the active circuits by means of metal lines.

Both bonding wires and long metal lines, on chip and in the package, represent some resistance but especially inductance. At high frequencies (or clocks) this inductance can represent impedances of many Ohms. Indeed, fast clocks have short rise times and therefore generate large impedances. The voltage drop ΔV along this pad or line can now be considerable.

A bonding wire has an inductance of about 1 nH/mm. Depending on which pin is contacted, the connection to the active circuit on the chip, can reach values up to 21 nH, for a 40-pin plastic package. A clock current spike of 20 mA with a 2 ns rise time causes a voltage drop of about 210 mV. Depending on the logic, values can be allowed up to 500 mV.

This voltage drop is generated by the current to the digital circuit and appears on the supply line of both, the digital and analog circuit block. As a result the PSRR of the analog block will suffer heavily and will probably be no more than 40 dB.

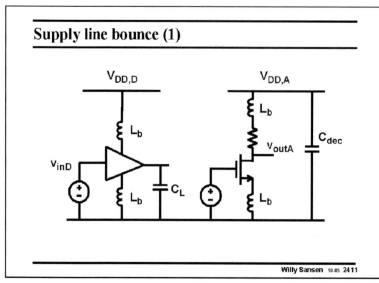

Supply line bounce (1)

Willy Sansen 10-05 2411

2411
The bonding wires also cause ground bounces and supply line bounces. This is illustrated in this slide.

A digital gate with input voltage v_{inD} and supply voltage $V_{DD,D}$ is in the neighborhood of an analog amplifier, this is biased at a certain current and has $V_{DD,A}$ as a supply voltage. The latter supply line has a decoupling capacitance C_{dec}, to suppress noise and ripple.

All connections to the supply lines carry bonding wires. We monitor the output voltage of the analog amplifier v_{outA} as a result of a digital input signal v_{inD}.

The only coupling between both circuits is along the ground line, which is common to both.

Digital transitions at the output of the digital gate, are visible at the output of the analog amplifier, as shown next.

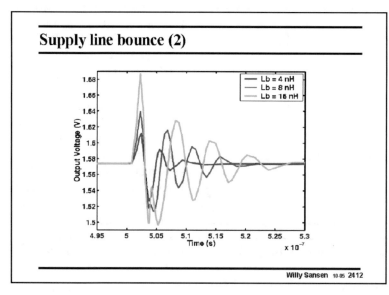

Supply line bounce (2)

Willy Sansen 10-05 2412

2412
Digital transitions of the digital gate, cause ringing indeed at the analog output. The size of it depends heavily on the values of the bond wire inductances. The larger the inductances, the larger the ringing.

They have to be shortened, or replaced by flip-chip bonding, as will be shown later.

A more elaborate analysis of ground bounce is given in Badaroglu, JSSC July 2004, 1119–1130.

2413
The best strategy for pinning is therefore to avoid common supply lines or common ground lines altogether.

The top solution is the best one. It costs one extra pin, however.

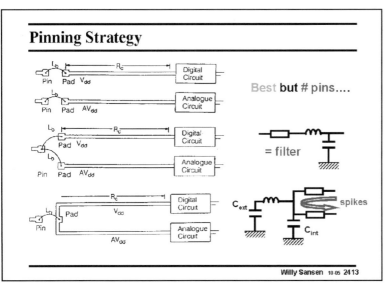

The other two solutions are poor. Each pad has a capacitance to ground and has an inductor in series to model the bond wire.

It is now clear that both solutions form a filter from one supply to the other. The middle one is better than the bottom one.

In the bottom connection, the pin itself and all external decoupling capacitors connected to that pin, are isolated from the supply lines because of the bonding-wire inductance. This is clearly the worst possible solution.

2414

The bonding wires again play an important role in this sensor preamplifier followed by a DSP block. Both the input and output are actually analog. The core is a DSP block preceded by a ADC and followed by a DAC. All this is on-chip, whereas the sensor itself and the output are off-chip.

How to connect the supply lines and ground?

The substrate is separated from the analog ground, which is reserved for the

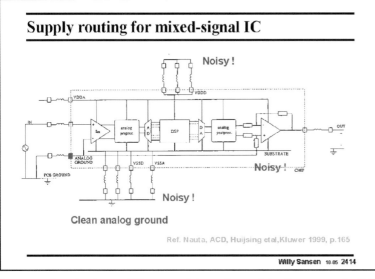

input and output amplifiers on-chip only and to the ground of the sensor off-chip.

The substrate is taken out separately and connected externally to the supply ground line. Many parallel bonding-wires are used in parallel as this ground connection is common to both the analog circuits and digital blocks. Two separate pins are used however, for these ground connections VSSA and VSSD.

The supply voltages also come in over two separate pins VDDA and VDDD. Again, multiple bonding wires are used for the digital supply pin.

All necessary precautions have now been taken to avoid coupling. The most sensitive point for coupling is at the analog ground of the input amplifier. As no differential sensor is used, this point will pick up noise from the substrate and from the external PCB ground. Differential sensors are therefore always preferred.

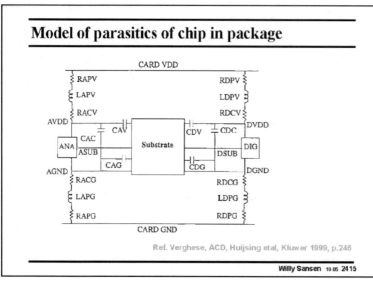

Model of parasitics of chip in package

Ref. Verghese, ACD, Huijsing etal, Kluwer 1999, p.246

Willy Sansen 10-05 2415

2415
It is not always easy to model the coupling between the analog and digital blocks. A network of resistors and capacitors can be used, as shown in this slide. However it is a non-trivial task to obtain realistic values of all these components.

Both the analog and the digital blocks are coupled by means of capacitors to a common substrate, which can be modeled as a grid of resistors and capacitors.

Both the analog and the digital blocks have separate pins for supply lines and ground. They all consist of inductors in series with small resistors.

The external supply line and ground are on top and at the bottom.

This applies to analog and digital blocks which are sufficiently small such that they do not have to be subdivided over more blocks. If this is the case, then the simulation time grows considerably.

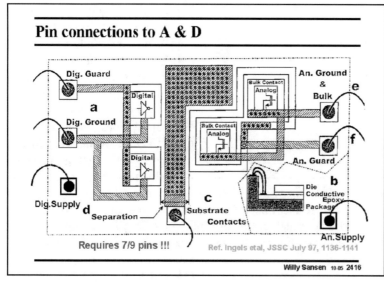

Pin connections to A & D

Requires 7/9 pins !!!   Ref. Ingels etal, JSSC July 97, 1136-1141

Willy Sansen 10-05 2416

2416
In this example, we try to find out what is a reasonable number of pins.

The maximum number of pins is nine. This is evidently rarely acceptable. Some of the pins can be combined to reduce this number, as explained now.

The nine pins can be divided into four for the analog part, four for the digital part and one for the screen in the middle. This screen is a metal plate with a certain width to physically separate the analog blocks from the digital ones as far as necessary. It should also go as deep as possible. It is therefore connected to an underlying deep diffusion.

The four pins of the analog part are the following:

– the positive supply line
– the ground line

– the substrate (bulk)
– the guard rings.

Occasionally, the substrate can be connected to one of the supply lines (positive supply or ground). This is not always advisable however, as will be shown later.

The guard rings are diffusions surrounding the most sensitive analog amplifiers. They are meant to take up noise coming laterally from the digital parts. Of course, they do not pick up a great deal of noise arriving vertically. This will also be discussed later in more detail.

Digital blocks can also have four pins, although in practice only two of them are used, the positive supply line and ground.

Rules for pin connections

- The analog and the digital power supply are separated
- The analog ground and the power supply are connected to the outside world with multiple bondwires
- The respective power supplies' bondpads are placed closely to each other to prevent ground loops
- Integrated decoupling capacitors are provided for both the analog and the digital power supplies
- All biasing voltages are internally decoupled to the correct power supply
- The optical input is differential with a dedicated ground bondwire
- The input bondwire is far from the noisy output and power supplies
- A large substrate contact provides a good connection with the heavily doped bulk
- All analog transistors are closely surrounded by substrate contacts that are biased with the analog ground
- All digital transistors are closely surrounded by a guard-ring that is biased with a dedicated clean voltage
- The analog and the digital circuits are separated by a distance that corresponds to approximately 4 times the epi-layer thickness

- A supplemental guard-ring biased with a dedicated voltage is provided between the analog and the digital subcircuits.

Willy Sansen 10.05 2417

2417
These conclusions are repeated, together with some useful suggestions about the layout.

They speak for themselves.

Noise reduction techniques

- **At noise sources side**
  - **Reduce substrate noise generated by the cells, Switching activity reduction techniques**
  - **Switching activity spreading techniques**
- **At noise receiver part**
  - **Design techniques (fully differential design, etc …)**
  - **Layout techniques (fully differential implem. …)**
  - **Separate, and multiple, supply bonding pads**
  - **Guard ring close to the transistors**
  - **Buried layers under the transistors**
  - **On chip decoupling capacitances**

Willy Sansen 10.05 2418

2418
Another way to summarize the techniques to reduce noise pick up is given in this slide.

Some effort is also being made to reduce the generation of noise. The spreading of the digital activity may help to reduce the spectral content of the noise generated.

On the analog (receiver) side we notice that buried layers are suggested to screen the transistors from the noise coming in vertically. This is a very effective technique, indeed. On the other hand, buried layers are not always available.

Quite often, considerable space is allotted to decoupling capacitances. These are small capacitances connected locally to provide low impedances to ground at high frequencies.

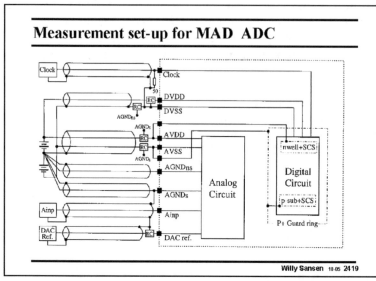

2419

Actually, the same techniques to avoid coupling of noise is applicable for printed-circuit boards (PCB) as for silicon chips.

This is why an example is given of a printed-circuit board with an analog block and a digital block. It is to be connected to several pieces of equipment.

It is clear that all supply lines are separated. Moreover, the shielding of all connecting cables is connected to the external

ground separately. The substrate or the ground plane of the PCB is also connected separately to the external ground.

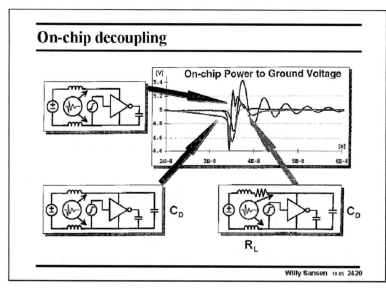

2420

On-chip decoupling may not always work.

An example is given in this slide where decoupling does more harm than good.

A single digital gate is used. It is switched by application of a logic one. It is loaded by a load capacitor. Its supply lines are connected over bonding wire inductors. The supply voltage is measured at the terminals of the gate.

Without decoupling capacitance, high-frequency

ringing is detected on the supply line. The addition of a decoupling capacitance C_D across the gate terminals, causes ringing of this capacitor with the bonding wire inductors. It has a lower frequency than before, but is larger in amplitude.

Addition of series resistors R_L dampens this ringing but also causes a DC voltage drop along the supply lines, which is to be avoided.

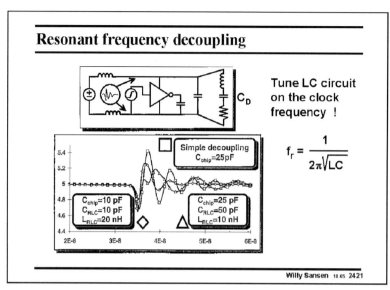

Resonant frequency decoupling

Tune LC circuit on the clock frequency !

$$f_r = \frac{1}{2\pi\sqrt{LC}}$$

Simple decoupling $C_{chip}=25pF$

$C_{chip}=10$ pF
$C_{RLC}=10$ pF
$L_{RLC}=20$ nH

$C_{chip}=25$ pF
$C_{RLC}=50$ pF
$L_{RLC}=10$ nH

Willy Sansen 10-05 2421

2421

A better solution is to add a series resonant RLC circuit across the decoupling capacitor C_D.

A series resonant RLC circuit provides a null in the impedance (see Chapter 22) at its resonant frequency f_r. It is therefore an ideal circuit to dampen oscillation at this frequency.

If its resonant frequency can be tuned at the ringing frequency of the decoupling capacitor with the bonding wire inductors, then the damping will stop the ringing.

If this ringing occurs at the clock frequency, then it is fairly easy to tune this series RLC circuit to this clock frequency. It is not as critical as it sounds. Some typical values are given in this slide.

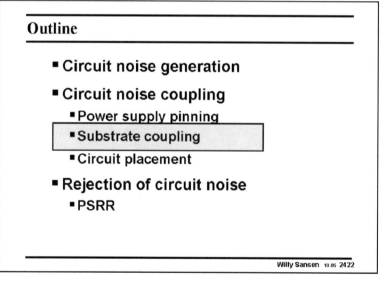

Outline

- **Circuit noise generation**
- **Circuit noise coupling**
  - **Power supply pinning**
  - **Substrate coupling**
  - **Circuit placement**
- **Rejection of circuit noise**
  - **PSRR**

Willy Sansen 10-05 2422

2422

Now that we have a better view on how to cope with the bonding wire inductors, let us see how the substrate itself plays a role.

It is clear that the substrate is most responsible for the transmission of the noise from the digital part to the analog part.

2423

Several possible substrates are shown in this slide.

The top one has an epitaxial layer, whereas the other one has not.

The epitaxial layer is usually doped fairly low (with high resistivity) so as to set the MOST characteristics. The substrate however, is highly doped (with low resistivity) to allow current flow with low voltage drops. As a consequence, the current reaches the p+ diffusion from the Injector to the Receiver in a vertical direction. Also, as the substrate has low resistivity, the

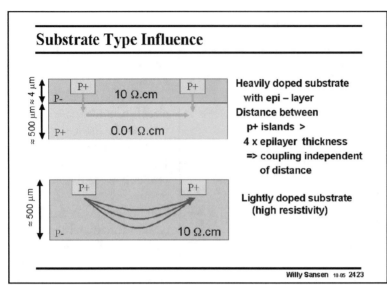

Substrate Type Influence

Heavily doped substrate
with epi – layer
Distance between
  p+ islands >
  4 x epilayer thickness
  => coupling independent
     of distance

Lightly doped substrate
(high resistivity)

Willy Sansen 10-05 **2423**

distance between Injector and Receiver does not matter.

One of the questions is, how close can the Injector be to the Drain before horizontal current flows in the epitaxial layer? The answer is that if the distance between both diffusions is smaller than about 4 times the thickness of the epitaxial layer, then horizontal current can also be expected in the epitaxial layer.

This is the case if no epitaxial layer is present. The whole substrate is now lowly doped. As a consequence, the current from the left p+ island (the Injector) spreads into the substrate to be collected again at the right p+ island, or the Receiver. Both horizontal and vertical currents are then present around the p+ island.

In this case, the distance will play a role: the larger the distance, the larger the resistance between both islands.

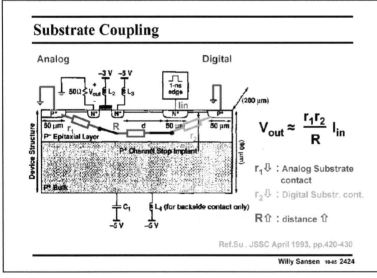

Substrate Coupling

Analog   Digital

$$V_{out} \approx \frac{r_1 r_2}{R} I_{in}$$

$r_1 \Downarrow$: Analog Substrate contact

$r_2 \Downarrow$: Digital Substr. cont.

$R \Uparrow$: distance $\Uparrow$

Ref.Su , JSSC April 1993, pp.420-430

Willy Sansen 10-05 **2424**

2424

In this experiment, the influence of the distance between the Injector (at the right) and the Receiver (at the left) is investigated in more detail.

A short current pulse is injected at a n+ island. This is expected to cause a current to both p+ grounds, the digital ground at the right, and the analog ground at the far left. A nMOST transistor is added at the left. It is biased as a source follower with a 50 Ω Source resistor, at which node the output voltage is measured.

The left p+ island is connected to the analog ground, but is actually the substrate contact for the analog part, which is here the nMOST Source follower. The lateral resistor between this substrate contact and the Source of the nMOST is denoted by r_1. It must be made smaller in size.

In a similar way, the lateral resistor between the digital injector and its substrate contact is denoted by r_2. It must also be made smaller.

The lateral distance between the analog nMOST and the digital injector is denoted by resistor R. The larger this R is the better.

Indeed the Voltage V_{out} sensed at the Source of the nMOST Source follower, as a result of the input current I_{in}, is as given in this slide. This current flows mainly to its digital ground and causes a voltage $r_2 I_{in}$, in the epitaxial layer, underneath the digital p+ island. This voltage is then sensed at the output through a potentiometric divider of R and r_1.

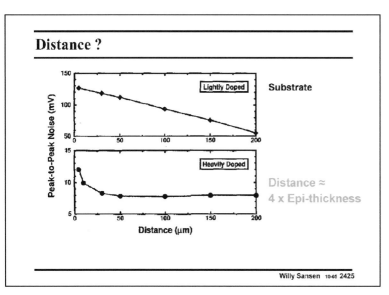

2425

The output voltage V_{out} is given in this slide, versus distance between Injector and Receiver.

On a lightly doped substrate, without epitaxial layer, the currents flows through a large resistance. As a result, the output voltage is large and inversely proportional to the distance.

When a lightly doped epitaxial layer is added on top, the result is the same. Indeed, the current still encounters a large resistance, which is proportional to the distance.

Only when a heavily-doped substrate is used, all the current flows directly to the substrate and along the substrate to the other island. The distance is no longer of any importance. Moreover, the output voltage is small because the substrate resistance is small.

Only when the distance between both islands is smaller than 3–4 times the epitaxial thickness, some current now flows through the epitaxial layer. The resistance is higher and so is the output voltage.

A designer must be aware of the kind of substrate that is being used in this particular CMOS technology.

2426

If no epitaxial layer is present, the extension of the distance is now about the only remedy left to reduce coupling.

This is not quite true. The third dimension can be used as well.

An example is given in this slide. Between the injector (on the right) and the receiver, a deep diffusion is added (such as the n-well), connected to the supply voltage. As a result, a wide depletion region is added.

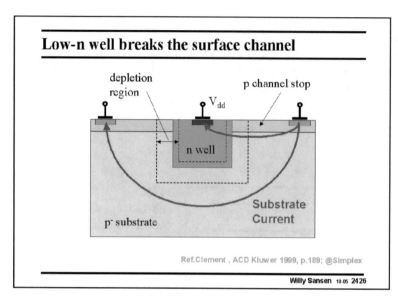

Low-n well breaks the surface channel

depletion region

p channel stop

V_{dd}

n well

Substrate Current

p⁻ substrate

Ref.Clement , ACD Kluwer 1999, p.189; @Simplex

Willy Sansen 10-05 **2426**

This n-well diffusion with its depletion region acts as an isolator between Injector and Receiver.

In this way, the current path from Injector to Receiver is made longer, the resistance larger and the coupling weaker.

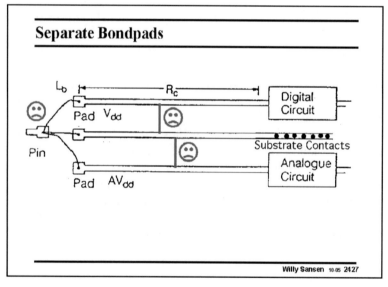

Separate Bondpads

L_b

R_c

Pad V_{dd}

Pin

Pad AV_{dd}

Digital Circuit

Substrate Contacts

Analogue Circuit

Willy Sansen 10-05 **2427**

2427

Remember that all these contacts need separate pins.

As an example, a supply line (or ground line) of a digital block which runs in parallel with a supply line (or ground line) of analog block, and gives capacitive coupling.

A deep diffusion with metal on top between both reduces significantly the coupling. Even better, is to lead the three bonding wires to different pins.

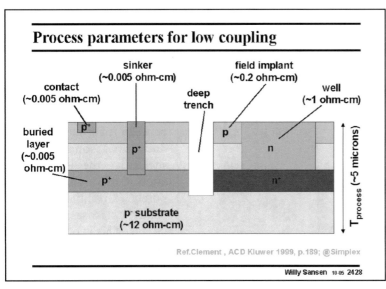

2428

Deeper isolation is obtained by deep trench isolation between Injector and Receiver.

In the example in this slide, a deep trench is etched between the areas on the left and the right. Moreover, both areas have a buried layer, which are ideal to screen away currents coming in vertically.

The right n-well houses pMOST devices, whereas the area on the left houses nMOST devices. The isolation between both is quasi perfect as it cuts through most of the silicon.

Taking two different chips is even more perfect, but requires more expensive packaging.

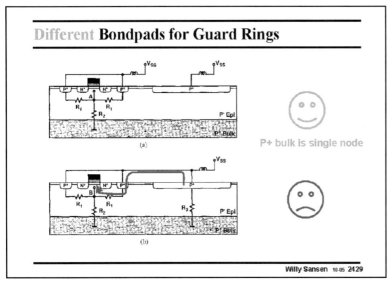

2429

Guard rings are ideal to screen away currents coming in horizontally. This is illustrated in this slide.

Guard rings are highly-doped diffusion rings around a sensitive transistor or amplifier stage, with the same doping as the underlying material. In a p− epitaxial layer, they are shallow p+ diffusions, as shown in this slide.

Their purpose is to pick up the current which comes in horizontally along the epitaxial layer and along the surface and to lead it outside.

The substrate contact (on the right) is normally separate. Moreover it is better connected to a separate pin as well, as shown on top.

If the substrate contact and guard ring contact are connected together, to save a pin, as shown at the bottom, the current from the substrate contact can re-enter the silicon through the guard-ring contact and reach the transistor. This is better to be avoided.

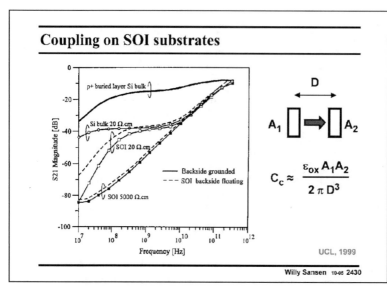

2430

Horizontal coupling between two parallel lines is purely capacitive. However, this is only true for non-conductive substrates.

The coupling is shown between two parallel conductors at some distance, such that the isolation is about 60 dB at 1 GHz. It is purely capacitive. It evidently decreases at 20 dB/decade. This is only true for perfectly isolating SOI substrates.

Actually, this capacitance can be calculated, as shown in this slide for two areas A, a distance apart D, which is much larger than the dimension of area A. It is assumed that the dielectric is pure silicon oxide.

These curves also show that the isolation becomes more resistive the higher the doping level is. For partially doped SOI (around 20 Ωcm), the coupling is resistive between 200 MHz and 10 GHz. The same applies to bulk silicon with the same low resistivity (20 Ωcm). Taking SOI does not pay off in this frequency region!

Bulk silicon is a lot worse. For frequencies higher than 100 MHz, it provides very little isolation. The coupling is resistive. It can be modeled with resistors as done before.

At very low frequencies, the isolation can be very large. It is now mainly limited by the leakage currents of the reverse biased diodes (Drain-bulk, etc. ...).

Outline

- **Circuit noise generation**
- **Circuit noise coupling**
  - **Power supply pinning**
  - **Substrate coupling**
  - **Circuit placement**
- **Rejection of circuit noise**
  - **PSRR**

Willy Sansen 10-05 2431

2431

Now that we know what isolation to expect in the substrate, let us see how transmission in the substrate can be modeled and what layout rules can be extracted. This is mainly based on simulations of larger substrates with noise sources and sensitive preamplifiers combined. Sometimes experimental data is available as well, however.

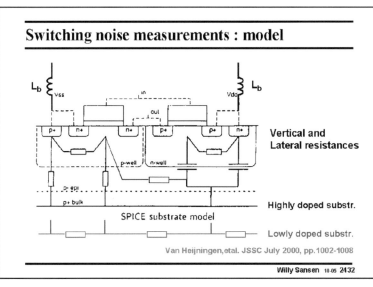

2432
Experimental work to model all substrate resistances and capacitances is a tedious job. It is the only way however, to obtain realistic results with simulations. In this way, the effect of bond wire inductances can be evaluated. An example is given in this slide.

A cross-section is given of a CMOS inverter in n-well CMOS. Actually, this is twin-well CMOS as an additional p-well is provided in the p-substrate for the nMOSTs. Both cases are considered, i.e. a highly doped substrate which can be regarded as a equipotential plane, and a lowly-doped substrate, which has to be modeled by horizontal resistors or impedances.

The nMOST in the p-well has a p+ substrate contact, connected with the Source to ground over a bonding wire L_b. There is a horizontal resistor (or impedance) between the channel area of the nMOST and its substrate contact. There are also two resistors (or impedances) connecting the channel area and the contact area to the substrate.

Similarly, for the pMOST, a horizontal resistor (or impedance) must be included between the channel area and the substrate contact. Moreover, two separate capacitors are required to model the depletion region of the n-well on the p-substrate. One additional horizontal resistor (or impedance) is required to model the area between both transistors.

All these resistors (or impedances) are distributed resistors and are therefore not easy to model.

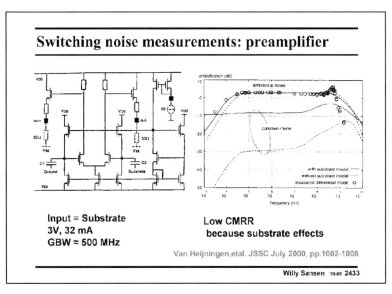

2433
In order to measure the substrate noise, a wide-band amplifier is required, as shown in this slide.

It is a single differential pair with output Source followers. The CMRR is somewhat limited by uneven substrate impedances. However, it does provide a GBW of 500 MHz.

The difference is measured between the external ground and the substrate. Both are capacitively coupled. The output voltage includes all noise picked up from the substrate after redistribution by all horizontal and vertical impedances, as discussed previously.

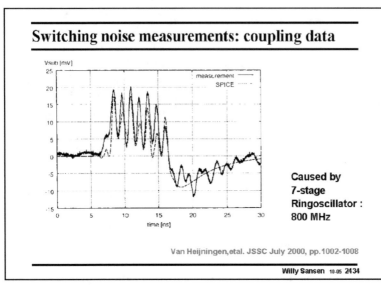

2434
The output voltage is shown in this slide.

The noise is generated by a ring oscillator, which can be switched on and off. Its frequency is about 800 MHz.

It is clear that this frequency is clearly visible at the substrate of the analog MOSTs. It can now be regarded as an input signal at the back gate of the MOSTs.

Moreover, it also shows that the horizontal and vertical resistors (or impedances) can be modeled fairly well. The measurement results do not deviate a significant amount from the SPICE results.

In conclusion, the modeling of the distributed horizontal and vertical impedances is a necessity to be able to predict coupling by means of simulation.

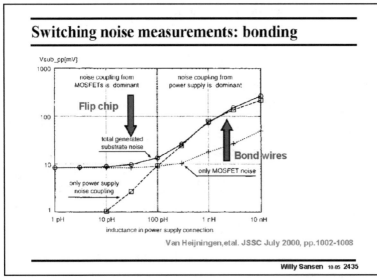

2435
From the same data, the influence of the bonding wires can be extracted in great detail.

The noise is generated in the substrate and reaches the analog CMOS inverter through substrate coupling.

If bonding wires are used with large inductances, then the noise currents cannot escape from the substrate contacts. As a consequence, the noise levels in the substrate underneath the analog circuits are high.

This is illustrated in this slide. It shows the voltage obtained on the substrate, versus bond wire inductance. Very low inductance can only be obtained with flip-chip bonding. In this case the supply line and ground contacts carry zero resistance. What is left is due to direct coupling to the Sources and Drains of the MOSTs.

The cross-over point seems to be around 0.1 nH which corresponds to about 0.1 mm bonding wire. This is very short indeed, and not practical.

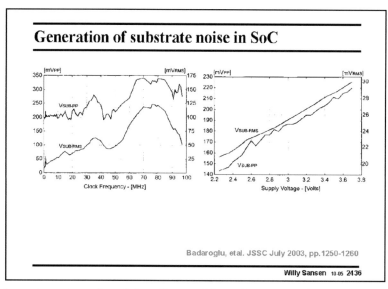

Generation of substrate noise in SoC

Badaroglu, etal. JSSC July 2003, pp.1250-1260

Willy Sansen 10-05 2436

2436

In the previous example, a 7-stage ring oscillator is used as a noise source and only a single-CMOS inverter-amplifier as receiver. This has been extended to a 220-Kgates telecom chip as a noise source and a WLAN modem as receiver. The bonding wires are 6 nH.

Both the peak-to-peak and the RMS noise are shown (on the left). The largest peak around 75 MHz can also be found by simulation. Accurate values have been introduced for all parasitic resistances and capacitances.

There is a fairly linear relation between the noise and the supply voltage. This is a result of the increased current injection in the substrate, which increases linearly with the supply voltage.

It is also found that as a great deal of noise is generated by the Input/Output buffers as by the digital gates themselves!

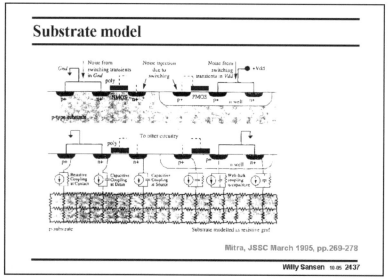

Substrate model

Mitra, JSSC March 1995, pp.269-278

Willy Sansen 10-05 2437

2437

Another example of a substrate model to carry out simulations of the noise coupling in more complex circuitry is shown in this slide.

A digital CMOS inverter is shown together with its substrate contacts. It causes the injection of a number of currents into the substrate. The substrate itself is modeled as a mesh of resistors.

The noise currents have various origins, such as noise from the switching transients on the supply lines and noise from the switching on the output Drains. These noise currents are modeled as separate independent input current sources, in parallel with either resistors (if no junctions are involved) or capacitors (in the case of junctions).

Long simulations are required to discover which are the nodes in the circuit with higher noise coupling and with lower noise coupling.

Examples of the results are shown next.

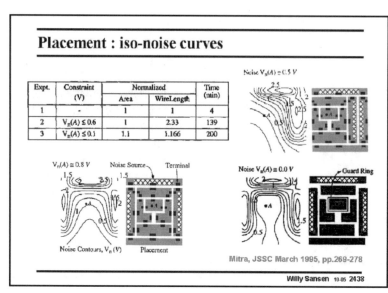

Placement : iso-noise curves

Expt.	Constraint (V)	Normalized		Time (min)
		Area	WireLength	
1	-	1	1	4
2	$V_n(A) \leq 0.6$	1	2.33	139
3	$V_n(A) \leq 0.1$	1.1	1.166	200

Mitra, JSSC March 1995, pp.269-278

Willy Sansen 10-05 2438

2438

It is obvious that the simulation times heavily depend on the resolution of the mesh. The simulation time increases with a higher order of the number of nodes. Only partial circuits can therefore be handled, depending on the computer used.

Examples are given of iso-noise curves (or contours) obtained for an amplifier surrounded by three noise sources (bottom left). Clearly, these noise sources cause a great deal of noise, even in the middle where the amplifier is positioned (point A). When a guard ring is added however (bottom right), the noise contours are compressed closer to the noise sources, leaving the center (point A) nearly noiseless.

If, on the other hand, all noise sources are on top and on the right (top right) the noise contours are clearly asymmetrical.

Such simulations show that it is possible to obtain quantitative data about noise coupling. However, depending on the computer power available, such simulations will always be limited to partial circuits.

Outline

- **Circuit noise generation**
- **Circuit noise coupling**
  - Power supply pinning
  - Substrate coupling
  - Circuit placement
- **Rejection of circuit noise**
  - PSRR

Willy Sansen 10-05 2439

2439

If everything fails, if a lot of noise is generated by the digital or class-AB circuits and if the substrates under digital and analog are not isolated at all, then we can still try to make the analog circuits insensitive to the substrate and power supply noise.

This specification is called the power-supply rejection ratio (PSRR). We will see that the PSRR is never as good for single-ended amplifiers. The conclusion is that all mixed-signal circuits must be made full-differential. As a result, the power consumption increases drastically (more than doubles!).

For full-differential circuitry, the limit is now reached by mismatch. This is also the case for CMRR. Actually, what the PSRR is for the rejection of power-supply-line noise is the CMRR for the rejection of ground noise.

Its definition is therefore easily established.

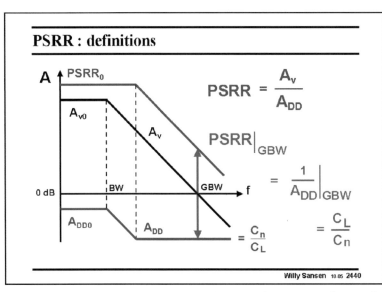

2440

This Bode diagram shows the gain A_v of an amplifier for a differential input. It is high at low frequencies (A_{v0}) but decreases with a slope of $-20\,\mathrm{dB/decade}$ until the GBW is reached.

The gain A_{DD} of this amplifier is also shown with the positive power supply v_{DD} as an input. It is rather low at low frequencies (A_{DD0}) but does not change all that much at high frequencies. The reason is that at high frequencies, capacitive coupling prevents the output voltage to become really small. The output becomes constant as a result of a ratio of some capacitors, usually the load capacitor C_L and some parasitic coupling capacitor C_n. If C_L is 2 pF and C_n is 0.1 pF, the ratio is now 20 or 26 dB.

The PSRR is now defined as the ratio of the gain A_v to the differential input to the gain A_{DD} of the supply voltage. It is high at low frequencies and low at high frequencies. At the GBW however, the gain A_v is unity. At this frequency, the PSRR equals the inverse of the gain A_{DD}. It is an excellent measure of the PSRR at high frequencies.

For the example in this slide, the PSRR is 26 dB at the GBW.

At this frequency, the PSRR is rather small, because it is a ratio of two capacitors, the load capacitor and some parasitic coupling capacitor.

2441

What components play an important role in PSRR?

An example is given in this slide of a current amplifier. The impedance at the Gates of transistors M1 and M2 is low. Does this mean that the PSRR is large at high frequencies?

At low frequencies, all capacitances can be left out. In this case, output resistor r_o determines the current to the current mirror. The output current is therefore small.

At high frequencies, the parasitic coupling capacitance C_p takes over the role of r_o. The current to the current mirror now increases with frequency and so does the output current. The gain of

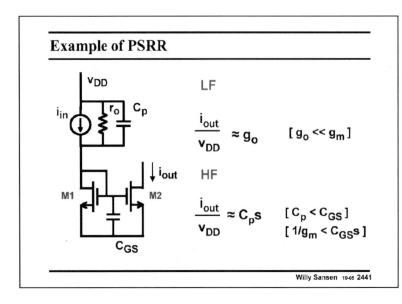

Example of PSRR

V_{DD}

i_{in} r_o C_p

LF

$$\frac{i_{out}}{v_{DD}} \approx g_o \qquad [\, g_o \ll g_m \,]$$

HF

$$\frac{i_{out}}{v_{DD}} \approx C_p s \qquad [\, C_p < C_{GS} \,]$$
$$[\, 1/g_m < C_{GS}s \,]$$

M1   M2 $\downarrow i_{out}$

C_{GS}

Willy Sansen 10-05 2441

the current mirror is constant up to a frequency $g_m/2\pi C_{GS}$ or f_T. This is evidently higher than all frequencies of interest.

Note that capacitance C_p is a coupling capacitor between the positive supply line and the output of the input transistor. It is usually a lot larger than the output capacitor of the transistor only.

2442

Let us now have a look at the $PSRR_{DD}$ of a simple OTA, as shown in this slide.

A detailed analysis has shown that the most important components are r_{o1} and r_{o5}. Indeed, the current caused by a small signal on the positive supply voltage v_{DD} flows through M4, through resistor r_{o1} and through M2 to the output. The current through r_{o5} reaches the output directly.

The $PSRR_{DD}$ of this gain block is now the ratio of

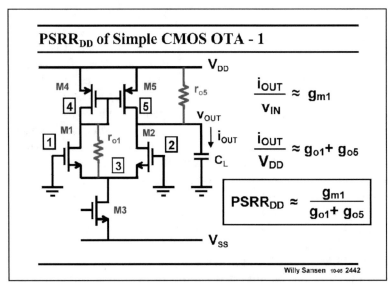

PSRR$_{DD}$ of Simple CMOS OTA - 1

V_{DD}

M4   M5 r_{o5}

$$\frac{i_{OUT}}{v_{IN}} \approx g_{m1}$$

V_{OUT}

M1 r_{o1} M2 $\downarrow i_{OUT}$ C_L

$$\frac{i_{OUT}}{v_{DD}} \approx g_{o1} + g_{o5}$$

M3

$$PSRR_{DD} \approx \frac{g_{m1}}{g_{o1} + g_{o5}}$$

V_{SS}

Willy Sansen 10-05 2442

both gains given in this slide. It has a very similar expression as the small-signal voltage gain. It is also large at low frequencies as it includes the $g_m r_o$ products of the transistors in the signal path.

It will be reduced at higher frequencies.

2443

The $PSRR_{DD}$ at higher frequency is of much higher importance as digital blocks on chips are more likely going to work at high clock frequencies. These are the frequencies which are expected to be rejected by the analog circuits.

At high frequencies, the coupling is mainly carried out by the capacitors C_{n4} and C_{n5DD}. The first one C_{n4} is a capacitance from node 4 to ground. Capacitor C_{n5DD} is the coupling capacitance between node 5 (at the output) and the V_{DD} supply line. Again, this capacitance can be a great

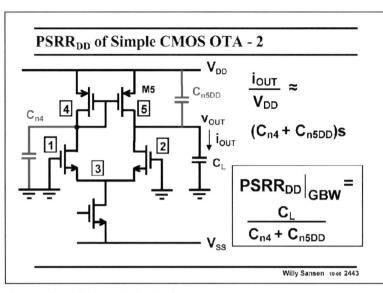

deal larger than the output capacitor C_{DS} of transistor M5. It is made up by the coupling between the V_{DD} supply line and the output line of the amplifier.

For the $PSRR_{DD}$ at high frequencies the g_{m1} comes in of the input devices. At the GBW however, the g_{m1} is substituted by C_L, as shown in this slide.

As expected, this $PSRR_{DD}$ at the GBW is not very large, as it is a ratio of small capacitances. For a frequency at 1/10th of the GBW, the $PSRR_{DD}$ is 20 dB larger, etc.

Note that the $PSRR_{DD}$ can only be high if both the capacitances C_{n4} and C_{n5DD} are small.

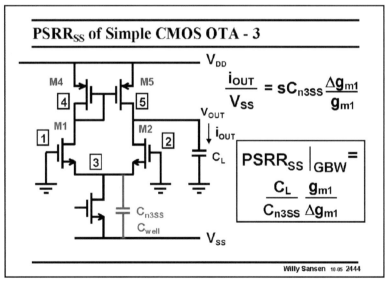

2444

The $PSRR_{SS}$ with respect to the other supply voltage is significantly different.

It is caused by the coupling capacitor C_{n3DD} at the common-Source point. For perfect matching of the transistor pair M1,2 and M4,5 none of this current can flow through the output load C_L.

Mismatch between, for example, transistors M1 and M2, will cause some of this current to flow through the output load, however.

The $PSRR_{SS}$ will be a ratio of capacitors, multiplied by a matching factor. It will be a lot larger than the $PSRR_{DD}$ of the positive supply line.

On the other hand, this coupling capacitance can be quite large as the capacitance C_{well} between the p-well and the substrate has to be added. Both input devices are in this p-well. Its area is therefore quite large and so is the capacitor C_{well}.

Note that nowadays mainly n-well CMOS technologies are used. The structure is then inverted – the input devices are then pMOSTs.

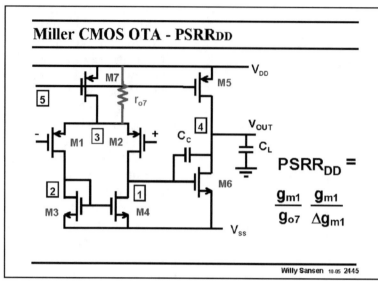

2445

The situation is somewhat more complicated in a two-stage Miller OTA.

After all, when the input signal is applied at the positive supply voltage, it is not clear whether the current through the first stage is added or subtracted from the current through the second stage.

A thorough calculation at low frequencies shows that the current through output resistor r_{o7} of the current source transistor M7 of the first stage is the dominant one. This current can only provide a contribution to the output voltage of the first stage provided there is some mismatch in the first stage, as previously explained.

At low frequencies, the $PSRR_{DD}$ is then readily found. It can be fairly large as it contains two factors.

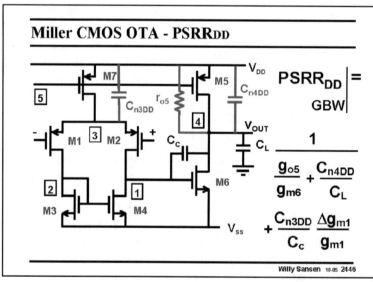

2446

At high frequencies, the situation is very different.

First of all, capacitances take over but also resistor r_{o5} remains in the expressions. Calculations show that the $PSRR_{DD}$ at the GBW now contains three terms.

Two of them are simply related to the two dominant coupling capacitors. For the first C_{n3DD} it is easy to see that this performs exactly the same role as resistor r_{o7}. The mismatch in the first stage comes in. Also, C_{n3DD} can be quite large as it includes the well to substrate capacitance of the input transistors.

Capacitor C_{n4DD} is the direct coupling capacitance between the supply line and the output.

The first term g_{o5}/g_{m6} is somewhat harder to understand. It is actually the resistive divider from the supply line to the output, made up by r_{o5} and the resistance $1/g_{m6}$ offered by M6 at high frequencies.

It is impossible to predict which one of the three terms is dominant. It is probably the first or the second one as they only have one single factor involved. The last one has two factors.

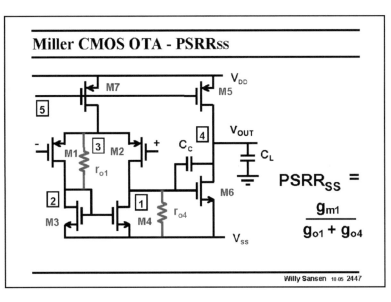

2447
The PSRR_{SS} from the negative supply does not involve matching. Indeed at low frequencies, it is a result of currents through resistors r_{o1} and r_{o4}.

The current through resistor r_{o1} flows through M3 (and the input devices M1 and M2). It is therefore mirrored to M4 and injected as a current on node 1. From here it is amplified to the output in the same way as the signal current from the input transistors M1 and M2.

The current through r_{o4} is also injected as a current on node 1. It is now added to the previous current.

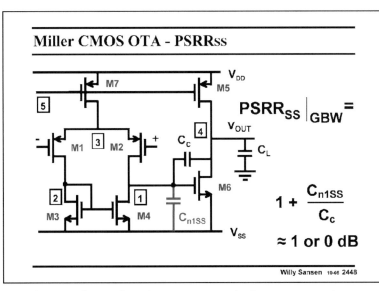

2448
At high frequencies, the situation is again somewhat more complicated.

The dominant capacitances are added in the circuit schematic. They are the coupling capacitors between the supply line and node 1, the output of the input stage.

Calculation of the PSRR_{SS} shows however, that this capacitor is not that important. The PSRR_{SS} is zero dB anyway. This means that any signal at the negative supply reaches the output terminal unattenuated.

This clearly proves that single-ended opamps cannot be used for mixed-signal analog processing.

What is the origin of this 0 dB?

At high frequencies, transistor M6 behaves as a small resistance with value $1/g_{m6}$. As a result the negative line is nearly shorted to the output line. The attenuation from supply line to output is therefore zero.

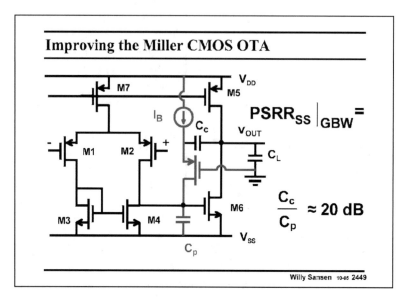

Improving the Miller CMOS OTA

$$PSRR_{SS}\Big|_{GBW} =$$

$$\frac{C_c}{C_p} \approx 20\ dB$$

Willy Sansen 10-05 2449

2449
One way to improve the $PSRR_{SS}$ is to a make compensation capacitor C_c unidirectional. After all, it is this capacitor C_c which turns transistor M6 into a resistor.

The addition of a cascode, used very much to get rid of the positive zero (see Chapter 5), increases the $PSRR_{SS}$ from zero to a capacitance ratio, as for the other amplifiers. Values of the order of 20 dB (at the GBW) can now be expected.

The actual calculation is shown next.

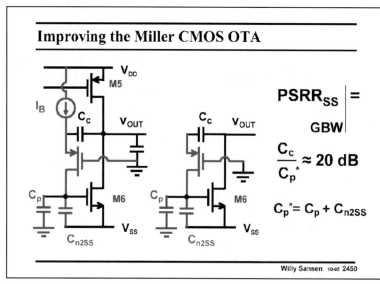

Improving the Miller CMOS OTA

$$PSRR_{SS} =$$

$$GBW\Big|$$

$$\frac{C_c}{C_p^*} \approx 20\ dB$$

$$C_p^* = C_p + C_{n2SS}$$

Willy Sansen 10-05 2450

2450
A simplified schematic is shown in this slide, and an even more simplified one to the right.

Calculation of the gain from the negative supply line V_{SS} to the output shows that the parasitic capacitance at the input node of transistor M6 now plays a dominant role.

Its ratio to the compensation capacitor C_c determines the $PSRR_{SS}$. For this reason, we may want to make C_c larger. Remember, however, that the value of C_c determines the GBW, the stability and the integrated noise. So many compromises have already come together in C_c!

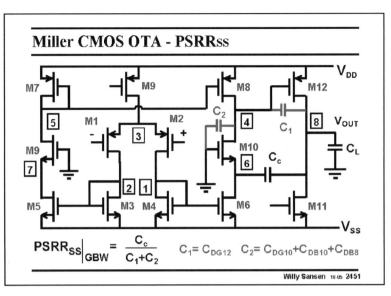

2451

Another two-stage Miller amplifier is shown in this slide. It uses a symmetrical OTA as a first stage. Cascodes are used as well. It is therefore evident that the compensation capacitor is led through a cascode, which is here M10.

As a result the $PSRR_{SS}$ is enhanced as well.

The actual expression is shown in this slide.

It is a ratio of capacitors involving a large number of parasitic capacitances. It will be larger than 0 dB, but not significantly!

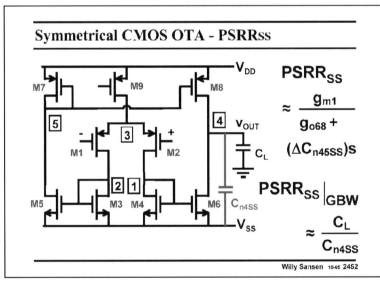

2452

The $PSRR_{SS}$ for a symmetrical OTA by itself is shown in this slide. It is mainly determined by mismatches.

At low frequencies it is determined by the resistance at node 5.

At high frequencies (at GBW) however, it is determined by the difference in coupling capacitance from the supply line to the nodes 4 and 5. Because node 4 is the output node and node 5 is not, the coupling capacitor C_{n4SS} is likely to be much larger than the one on the other side C_{n5SS}.

As a result, the simplest expression of the $PSRR_{SS}$ is the ratio of C_L to this coupling capacitor C_{n4SS}. Again, values of 20 dB can be expected.

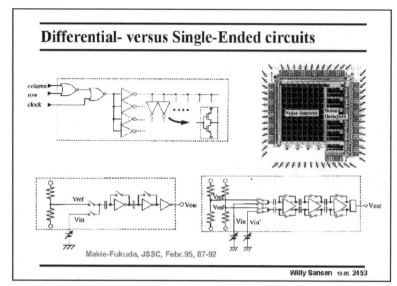

2453

Experimental evidence on PSRR is not easily found. One example is shown in this slide.

The PSRR has been measured for both a single-ended as for a fully-differential switched-capacitor amplifier. They act as a receiver.

The noise sources are a number of logic gates and inverters, as shown on top. A smaller or larger number of logic gates can be switched in.

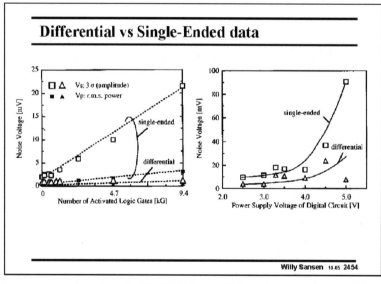

2454

The noise voltage is measured at the output of the analog amplifier structure.

It is clear that the single-ended amplifier is much more sensitive to the digital noise than the fully-differential one.

Increasing the supply voltage increases the currents in the logic gates and also the spike current transfer to the analog circuits. This is clearly visible on the right. Even the differential noise pick up increases rather drastically. This is probably due to mismatch in substrate contacts.

Quite often many sources of mismatch are often overlooked. For example for a fully-differential input stage of an opamp, good matching not only requires perfectly symmetrical layout but also identical surroundings. For example, the substrate contacts of the input transistors must also have the same size and an equal distance to the axis of symmetry. If not, different substrate resistances result, and different body effects.

Conclusions

- **Reduce circuit noise generation**
  - **Use linear circuits**
  - **Current mode logic**
  - **Avoid class AB amplifiers**
- **Reduce substrate coupling**
  - **Use different power supplies for A, D, G and S**
  - **Reduce drain areas**
  - **Guard rings close to A with dedicated pin : high-R substr.**
  - **Buried layers under A : low-R substrate**
  - **Use decoupling capacitances on A**
  - **Create distance : high-R substrate**
- **Improve PSRR by use of differential circuits : matching !!**

Willy Sansen 10-05 2455

2455

As a conclusion, several measures are again repeated, to avoid coupling from the digital noise sources to the analog amplifiers. They are listed in this slide.

Most of these have been discussed previously and therefore speak for themselves.

It is clear that only fully-differential circuits can be used in mixed-signal systems. Moreover, the layout must be carried out with extreme attention towards symmetry.

Outline

- **Circuit noise generation**
- **Circuit noise coupling**
  - **Power supply pinning**
  - **Substrate coupling**
  - **Circuit placement**
- **Rejection of circuit noise**
  - **PSRR**

Willy Sansen 10-05 2456

2456

In this Chapter, an overview has been given on what are the possible mechanisms of coupling between digital and analog blocks. They have all been identified.

Also, a number of design considerations have been added to improve the isolation. Some of these have to do with technology and some others with design.

Again, mismatch seems to have emerged as the main stumbling block.

This is hardly surprising as mismatch has been the stumbling block for many other specifications as well. It is therefore one of the main concerns for any analog designer!

Index of subjects

The numbers refer to the slide numbers, not to page numbers. Each slide has a number in the bottom right corner. For example slide 1523 is the 23rd slide of Chapter 15; slide 024 is the 4th slide of Chapter 2, and slide 113 is the 3rd slide of Chapter 11.

Printed in the United States
By Bookmasters